HOW TO UNDERSTAND, SERVICE AND MODIFY

CORVETTE FUEL INJECTION

& ELECTRONIC ENGINE MANAGEMENT

1982 THROUGH 2001

L83, L98, LT1, LT4, LS1, LS6, ZR-1

BY CHARLES O. PROBST

BentleyPublishers®
.com

BENTLEY PUBLISHERS® | Automotive Reference™

Bentley Publishers, a division of Robert Bentley, Inc.
1734 Massachusetts Avenue
Cambridge, MA 02138 USA
800-423-4595 / 617-547-4170

Information that makes the difference®

BentleyPublishers®.com

WARNING—Important Safety Notice

Do not use this manual unless you are familiar with basic automotive repair procedures and safe workshop practices. This manual illustrates the workshop procedures required for most service work. It is not a substitute for full and up-to-date information from the vehicle manufacturer or for proper training as an automotive technician. Note that it is not possible for us to anticipate all of the ways or conditions under which vehicles may be serviced or to provide cautions as to all of the possible hazards that may result.

The vehicle manufacturer will continue to issue service information updates and parts retrofits after the editorial closing of this manual. Some of these updates and retrofits will apply to procedures and specifications in this manual. We regret that we cannot supply updates to purchasers of this manual.

We have endeavored to ensure the accuracy of the information in this manual. Please note, however, that considering the vast quantity and the complexity of the service information involved, we cannot warrant the accuracy or completeness of the information contained in this manual.

FOR THESE REASONS, NEITHER THE PUBLISHER NOR THE AUTHOR MAKES ANY WARRANTIES, EXPRESS OR IMPLIED, THAT THE INFORMATION IN THIS BOOK IS FREE OF ERRORS OR OMISSIONS, AND WE EXPRESSLY DISCLAIM THE IMPLIED WARRANTIES OF MERCHANTABILITY AND OF FITNESS FOR A PARTICULAR PURPOSE, EVEN IF THE PUBLISHER OR AUTHOR HAVE BEEN ADVISED OF A PARTICULAR PURPOSE, AND EVEN IF A PARTICULAR PURPOSE IS INDICATED IN THE MANUAL. THE PUBLISHER AND AUTHOR ALSO DISCLAIM ALL LIABILITY FOR DIRECT, INDIRECT, INCIDENTAL OR CONSEQUENTIAL DAMAGES THAT RESULT FROM ANY USE OF THE EXAMPLES, INSTRUCTIONS OR OTHER INFORMATION IN THIS BOOK. IN NO EVENT SHALL OUR LIABILITY WHETHER IN TORT, CONTRACT OR OTHERWISE EXCEED THE COST OF THIS MANUAL.

Your common sense and good judgment are crucial to safe and successful service work. Read procedures through before starting them. Think about whether the condition of your car, your level of mechanical skill, or your level of reading comprehension might result in or contribute in some way to an occurrence which might cause you injury, damage your car, or result in an unsafe repair. If you have doubts for these or other reasons about your ability to perform safe repair work on your car, have the work done at an authorized Corvette dealer or other qualified shop.

Before attempting any work on your Corvette, read the warnings and cautions, and any warning or caution that accompanies a procedure in the service manual. Review the warnings and cautions on pages vii and viii each time you prepare to work on your Corvette.

Special tools required to perform certain service operations are identified in the manual and are recommended for use. Use of tools other than those recommended in this service manual may be detrimental to the car's safe operation as well as the safety of the person servicing the car.

Copies of this manual may be purchased from most automotive accessories and parts dealers specializing in Corvette automobiles, from selected booksellers, or directly from the publisher. The publisher encourages comments from the reader of this manual. These communications have been and will be carefully considered in the preparation of this and other manuals. Please write to Bentley Publishers at the address listed at the top of this page or e-mail us through our website.

Corvette Fuel Injection & Electronic Engine Management 1982 Through 2001: How to Understand, Service and Modify by Charles O. Probst

ISBN 978-0-8376-0861-7
Bentley Stock No. GFCV
Mfg. code: GFCV-04-D-1504

Library of Congress Cataloging-in-Publication Data

Probst, Charles O.
Corvette Fuel injection & electronic engine management 1982 through 2001 : how to understand, service and modify / by Charles O. Probst.
p. cm.
Includes index.
ISBN 0-8376-0861-9 (pbk. : alk paper)
1. Corvette automobile--Motors--Fuel injection systems. 2. Corvette automobile--Motors--Computer contol systems. I. Title: Corvette fuel injection and electronic engine management 1982 through 2001. II. Title.

TL215.C6 P76 2001
629.25'3--dc21

2001025370

Cover credits: 5.7 liter LS1 V-8 engine courtesy of Chevrolet Motor Division; yellow '92 ZR1, red '91 ZR1, and blue '98 convertible photos courtesy Ron Sessions

The paper used in this publication is acid free and meets the requirements of the National Standard for Information Sciences-Permanence of Paper for Printed Library Materials. ∞

Manufactured in the United States of America

HOW TO UNDERSTAND, SERVICE AND MODIFY

CORVETTE
FUEL INJECTION
& ELECTRONIC ENGINE MANAGEMENT
L83, L98, LT1, LT4, LS1, LS6, ZR-1

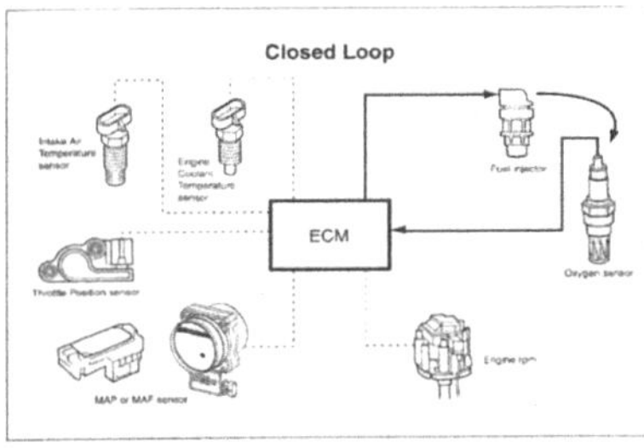

Chapter 2: Closed Loop operation...

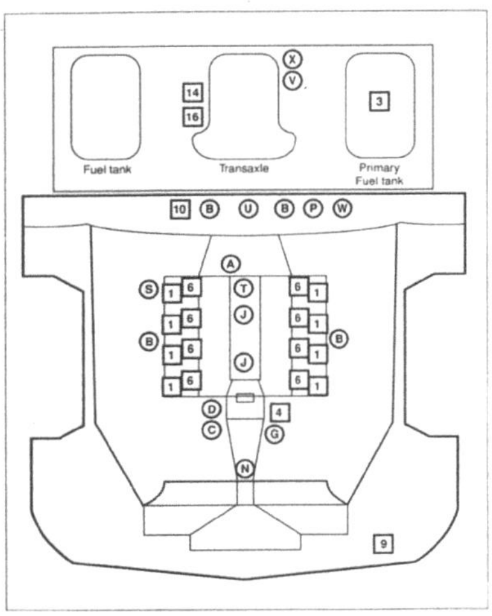

Chapter 4: Sensor locations in a **'97-02 LSI** engine

Chapter 9: **LT1** engine with optional Super Ram intake installed

Table of Contents

Please read these warnings and cautions before proceeding with maintenance and repair work.

WARNINGS and CAUTIONS—

- Some repairs may be beyond your capability. If you lack the skills, tools and equipment, or a suitable workplace for any procedure described in this manual, we suggest you leave such repairs to an authorized Corvette dealer service department or other qualified shop.

- Before starting a job, make certain that you have all the necessary tools and parts on hand. Read all the instructions thoroughly, do not attempt shortcuts. Use tools appropriate to the work and use only replacement parts meeting Porsche specifications. Makeshift tools, parts and procedures will not make good repairs.

- Be mindful of the environment and ecology. Do not pour any leftover automotive liquids onto the ground, down a drain, or into a stream, pond or lake. Dispose of in accordance with Federal, State and Local laws.

- Never work under a lifted car unless it is solidly supported on stands designed for the purpose. Do not support a car on cinder blocks, hollow tiles or other props that may crumble under continuous load. Never work under a car that is supported solely by a jack. The vehicle lifting jack supplied with the vehicle is intended for tire changes only. A heavy duty floor jack should be used to lift vehicle before installing jack stands. Never work under the car while the engine is running.

- If you are going to work under a car on the ground, make sure that the ground is level. Block the wheels to keep the car from rolling. Disconnect the battery negative (–) terminal (ground strap) to prevent others from starting the car while you are under it.

- Never run the engine unless the work area is well ventilated. Carbon monoxide kills.

- Tie long hair behind your head. Do not wear a necktie, a scarf, loose clothing, or jewelry when you work near machine tools or running engines. If anything were to get caught in the machinery, severe injury could result.

- Do not attempt to work on your car if you do not feel well. You increase the danger of injury to yourself and others if you are tired, upset or have taken medication or any other substance that may keep you from being fully alert.

- Illuminate your work area adequately but safely. Use a portable safety light for working inside or under the car. Make sure the bulb is enclosed by a wire cage. The hot filament of an accidentally broken bulb can ignite spilled fuel or oil.

- Always observe good workshop practices. Wear goggles when you operate machine tools or work with battery acid. Gloves or other protective clothing should be worn whenever the job requires working with harmful substances. Read any manufacturer's instructions and warnings carefully. Avoid direct skin contact. And always keep sparks, lighted matches and open flame away from the area.

- Disconnect the battery negative (–) terminal (Ground strap) whenever you work on the fuel system or the electrical system. Do not smoke or work near heaters or other fire hazards. Keep an approved fire extinguisher handy.

- Do not re-use any fasteners that are worn or deformed in normal use. Many fasteners are designed to be used only once and become unreliable and may fail when used a second time. This includes, but is not limited to, nuts, bolts, washers, self-locking nuts or bolts, circlips and cotter pins. Always replace these fasteners with new parts.

- Use pneumatic and electric tools only to loosen threaded parts and fasteners. Never use these tools to tighten fasteners, especially on light alloy parts. Always use a torque wrench to tighten fasteners to the tightening torque specification listed.

- Friction materials (such as brake pads or shoes or clutch discs) contain asbestos fibers or other friction materials. Do not create dust by grinding, sanding, or by cleaning with compressed air. Avoid breathing dust. Breathing any friction material dust can lead to serious diseases and may result in death.

- Connect and disconnect battery cables, jumper cables or a battery charger only with the ignition switched off, to prevent sparks. Always make sure ignition is off before disconnecting battery.

- The air conditioning system is filled with chemical refrigerant, which is hazardous. The A/C system should be serviced only by trained technicians using approved refrigerant recovery/recycling equipment, trained in related safety precautions, and familiar with regulations governing the discharging and disposal of automotive chemical refrigerants.

- The ignition system produces high voltages that can be fatal. Avoid contact with exposed terminals and use extreme care when working on a car with the engine running or the ignition switched on.

- Aerosol cleaners and solvents may contain hazardous or deadly vapors and are highly flammable. Use only in a well ventilated area. Do not use on hot surfaces (engines, brakes, etc.).

- Do not remove coolant reservoir or radiator cap with the engine hot. Danger of burns and engine damage.

- Label battery cables before disconnecting. On some models, battery cables are not color coded.

- Disconnecting the battery may erase fault code(s) stored in control module memory. Using special diagnostic equipment, check for fault codes prior to disconnecting the battery cables.

- If a normal or rapid charger is used to charge battery, the battery must be disconnected and removed from the vehicle in order to avoid damaging paint and upholstery.

- Do not quick-charge the battery (for boost starting) for longer than one minute. Wait at least one minute before boosting the battery a second time.

- Connect and disconnect a battery charger only with the battery charger switched off.

- Sealed or "maintenance free" batteries should be slow-charged only, at an amperage rate that is approximately 10% of the battery's ampere-hour (Ah) rating.

- Do not allow battery charging voltage to exceed 16.5 volts. If the battery begins producing gas or boiling violently, reduce the charging rate. Boosting a sulfated battery at a high charging rate can cause an explosion.

Charles Probst
a memorial

Charles O. Probst
1918-2000

My father died a week after sending the last changes for this book to his editor. Nothing otherworldly about that: He didn't want to have to keep thinking about the book while recovering from heart surgery. He was already planning his next book. But it won't be written. Charles Probst died on September 28, 2000, at the age of 82.

Chuck Probst grew up surrounded by cars. His father, Karl Probst, was a consulting automotive engineer who regularly brought home pre-production cars. He would tell his sons, Chuck and older brother Jack, to take a car out and see how it handled. The boys interpreted that as an invitation to conduct destructive testing on the roughest streets of Toledo, Ohio. A few times they had to abandon a car where it broke and take the trolley home.

Chuck's other boyhood passion was photography, and his first job out of college – making films for automotive dealers – combined these two interests. Then in 1940, he saw World War Two approaching and joined the Army Air Force, producing training films for pilots and bombardiers.

That same year, Chuck's father also contributed to the war effort. Karl Probst was contracted to design a small military car for the Army. That car was the original Jeep.

Chuck loved to tell stories, and one of my favorites was how my grandfather designed the Jeep in his head. Bantam Automotive, nearly bankrupt, bid on the Jeep contract as a save-the-company strategy. The Army expected a modification of an existing car, and required delivery to the testing grounds in two months. Bantam called Karl and asked him to design a car from scratch.

Karl drove from Toledo to Butler, Pennsylvania. His route was not direct; he zig-zagged across the state of Ohio, visiting the suppliers he would use for the vehicle. He picked up transmission blueprints in one city and spring specifications in another. And as he drove on those pre-Interstate roads, the Jeep came together in his mind.

When Karl arrived at Bantam in Butler, he sat at a drafting table for two days without sleep, capturing the plans that were in his head. Bantam's team built the first Jeep in 49 days, just making the Army's two-month deadline. The rest is history.

My father inherited that same ability to take a huge body of technical information, turn it over in his mind, and produce a clean, timeless exposition. Karl Probst did it with automotive blueprints; Chuck Probst did it with films and books.

Chuck continued his military service through the Korean conflict, retiring as a Colonel from the Air Force, having designed the training programs that touched hundreds of thousands of airmen.

Chuck started a small business writing and producing instructional and technical documentary films. Many contracts were with the Air Force and NASA, but his company also made films for dental students, spinning-mill workers, and on how to operate a high-speed pie-making machine. I was lucky enough to work for my father on some of these films, as an assistant cameraman, key grip, or designated pie-filling cleaner.

Chuck was proud of his ability to communicate the essence of a complicated topic.

For one film, a missile guidance engineer sent him a 30-page report – on the problems of targeting an airplane in a squadron – with half of the words underlined, saying the script must include these points. My father read the report, thought a bit, and then had his model builder create a rig of flashing lights, which he photographed by pulling the camera away from the rig, running the film backwards, and changing the camera speed during the shot. It lasted all of 15 seconds.

Chuck loved to tell how the engineer watched the scene and said, "Well, it seems no words were needed after all..."

One of my father's NASA films intersected with Probst family history. The film was about the Lunar Rover, which newspapers called the "moon jeep." NASA had built a cratered lunarscape in a quarry near Detroit, to test the vehicle and train astronauts (when Neil Armstrong returned, he said that quarry was "the best moon on Earth"). Chuck's film showed the moon jeep cresting craters and kicking up dirt – although not quite as impressive as the rooster-tails the real Lunar Rover made on the moon.

As Chuck grew older, the long days of film production became physically demanding, and he began a new career, as an author of books on electronic fuel injection. You are holding one of these in your hands.

Chuck Probst was a wonderful father for David and me, and a great husband of 57 years to my mother Sally. But most people know him as you do, through his ability to explain something you care about in a way that makes it easy to understand.

Richard Probst
November 12, 2000

Foreword

The rapid advances in digital technology over the past two decades have revolutionized the automotive industry. This digital revolution was driven by emission requirements and it started with digital electronic fuel injection (EFI), but it has now spread like wildfire to every conceivable corner of the automobile. Where it will end is anybody's guess.

Metering fuel at a constant air fuel ratio was first accomplished with a carburetor. The carburetor is one of the key devices that made the automobile possible and served it well for over 100 years. Typical of the analog devices that automotive engineers invented to manage relational functions, the carburetor was at best an approximate solution. And, given the limits of manifold distribution, this solution always used more fuel than necessary.

Engineers dreamed of a better way. WWII saw the application of diesel fuel injection technology to the Daimler-Benz engine of the Messerschmitt BF109. With fuel injection, its pilots could pull negative g's, a maneuver unavailable to the carbureted Allied fighter aircraft.

After WWII several attempts were made to introduce fuel injection into racing and then into road cars. Corvette fuel injection, introduced in 1957, is an excellent example of these developments. Though it used different principles to meter fuel proportional to air, it was still an analog device. In the end its complexity trumped its improved performance and it disappeared in favor of increasingly complex carburetors.

The late '70s and early '80s were a sorry tale of sacrificing performance for emission control. Over time, Corvette engines went from an optional 390 hp to a low of 180 hp in California. In the early 80's the limitations of the carburetor had brought Chevrolet to its knees. Even so, it was reluctant to embrace the emerging but more costly electronic fuel injection. In 1982, the Cadillac unit injector solution was adapted for the Corvette in the hope that it would work adequately and be less costly. It lasted only two years, during which time electronic fuel injection (EFI) was finally developed, tooled and certified for the Corvette. With EFI the Corvette recovered to 230 then 245 hp. It also left poor starting and stumbling behind.

Having brought the Corvette back from the brink, the engine designers who developed EFI for the Corvette rested. They thought they had done their job. But those of us guiding the Corvette program had only begun. We had to go outside the formal organization to Lotus Engineering for the next step. Lotus designed a new four-cam EFI engine for the Corvette (LT5, also known as the ZR-1) that produced 375 and then 405 hp. We had our power back but at a prohibitively high price to the consumer.

Interestingly, this challenged the engine designers, who then went back to work on improving the small block V8. The Generation II (340 hp LT4) and then Generation III (345 hp LS1) engines were the result of this effort. And there is no end in sight for the development of this now all aluminum engine.

To put this achievement into better perspective, the 390 hp of 1970 was achieved only as a gross rating with super premium (103 octane) fuel to stave off detonation. This engine will not run on modern, 93 octane, premium fuel. Modified with a lower compression ratio to run on modern fuel, and using today's more conservative test procedure, it might pump out 300 hp.

While the breathing gains of the four valve head of the LT5 (ZR-1) and the roller cam of the small block are important, none of this performance recovery would have been possible without the microprocessor and its complex of fuel and ignition timing maps and sensors that now optimize performance, fuel economy and emissions simultaneously.

These microprocessors are intruding everywhere into our lives as the digital revolution continues. The future advances in automotive technology are now inextricably linked to digital technology. We can allow ourselves to be intimidated by these advances, and choose to remain illiterate about microprocessors, or we can try to understand what they do and the impact they have on the cars we drive. Read Charles Probst's book and you will have taken an important step toward literacy.

Dave McLellan
Corvette Engineering Director 1975-1992
Holly, Michigan
March 30, 2000

Chapter 1

Overview

1

Contents

IN THIS CHAPTER, LOOK FOR:

- how it all began—the starts and stops of Corvette fuel injection
- what's in this book—a preview of each chapter
- the basics of electronic fuel injection
- the differences among Corvette systems
- how engine control includes much more than just fuel injection
- applications by model year—which Corvettes use which system

Some of you will read this book cover to cover, and you'll find a logical development. Others will pick and choose from the chapters. So at the start of each chapter, you'll find a box that tells you what's in it, and at the end of each chapter, what's ahead. You may also pick and choose based on model year. I've organized the information to specify the model-year changes in fuel injection/electronic engine management.

Fig. 1-2. In the Corvette, GM's Ram Jet was part of Ed Cole's drive for technological breakthroughs. The design and manufacture of the mechanical mass air-flow fuel-injection system improved during its Corvette application from 1957 to 1965. The 1963 engine is shown.

1. INTRODUCTION

In July 1956, as Ed Cole became Vice President and General Manager of the Chevrolet Division, he let it be known that he valued fuel injection, especially on the 1957 Corvettes. The cars were due in showrooms in a few months, and the GM fuel injection system had not yet been cleared for production.

Fig. 1-1. Ed Cole was an engineer's engineer who "pushed the envelope." As Chief Engineer of Cadillac, Cole lead the way to the 1949 lighter-weight, high-compression V-8 engines. As Chevrolet Chief Engineer, he introduced the 1955 small-block engine that served over 40 years, including **LT1/LT4**. Moving up to VP/General Manager, he got fuel injection in the 1957 Corvette.

In the time of carburetors, performance demanded four or more barrels. Ed Cole had enough clout to push fuel injection for the new Chevrolet small-block V-8. He wanted to make engineering history. And he did make history in many ways, but specifically with early U.S. production fuel injection.

In January 1957, GM engineers laid it out for the Detroit Section of the Society of Automotive Engineers. They said their fuel injection system could correct carburetor deficiencies, improve fuel economy, and reduce hydrocarbon pollution, both exhaust and evaporation.

Impressive, if you think about it. Here's GM in *1956*, designing a fuel injection system to reduce hydrocarbon pollution. That's eleven years before exhaust pollution problems drove VW to adopt the first production (electronic) fuel injection, in 1967. (See my *Bosch Fuel Injection,* Robert Bentley, Inc., 1988.) For several years, GM's engineering staff had been experimenting in great secrecy with different types of fuel injection. A few hundred of these Corvette "fuelies" reached the streets in late 1956.

1.1 The Beginnings of Corvette Fuel Injection

GM's early, experimental fuel injection followed aircraft-engine practice: it was mechanical, not electronic. Fuel was delivered into the intake manifold ahead of the intake valves—they referred to it as "port injection." Tests showed that constant-flow injectors operated on less fuel compared to timed, or pulsed, injection, with almost as much torque.

The first systems determined the density of intake air, metering the amount of fuel to be injected for the proper air/fuel ratio. Working from the density of the air, the system measured the speed of the engine and the amount of air, known as "speed density." But it looked as if it would be expensive.

John Dolza, of the engineering staff, quickly designed an alternative. It metered fuel according to the intake vacuum readings, which were directly related to mass airflow, although they didn't call it that back then.

Under what must have been intense pressure from the top, GM's engineering staff and GM's Rochester Carburetor Division hand-fabricated the first few production "Ram Jet" systems, barely in time for the new-car announcement of the 1957 Chevrolet Corvettes, in the fall of 1956.

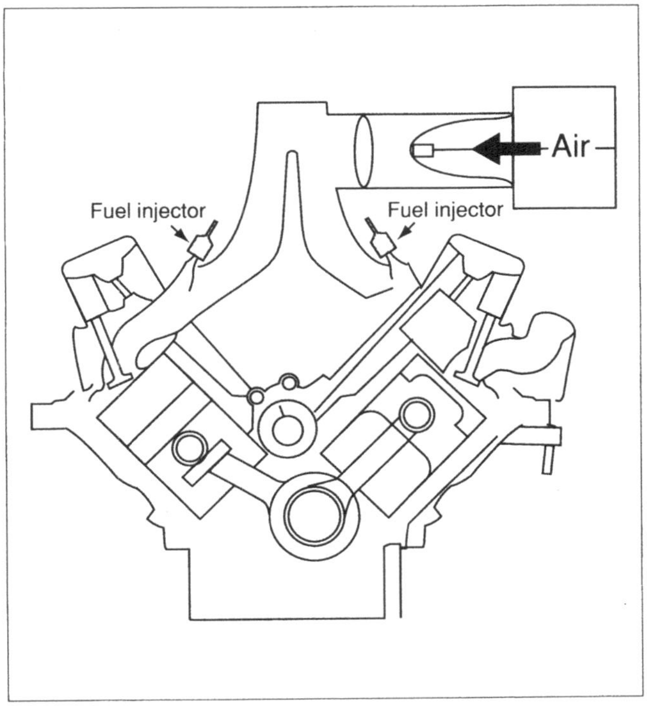

Fig. 1-3. Rochester fuel injection depends on mechanical measurement of the intake air as well as mechanical metering of the fuel injected.

The 1957 engine, the so-called "small-block," was bored out to 283 cubic inches. One fuel-injected version of the engine was rated at 283 horsepower. It made engineering history with the outstanding figure of one horsepower per cubic inch.

Compare the different Corvette 5.7-liter engine ratings, shown in Table 1-1. The 1971 "bare" horsepower rating dropped because of the lower compression ratio (required by reduced-lead, lower-octane fuels). In comparing the outputs, keep in mind that before 1973, horsepower ratings were measured with the bare engine on a dynamometer, there was no drag from a water pump or generator, and it was free from exhaust restrictions. The 1973 "installed" rating, with the accessory loads, was more realistic, considering that most people usually operate their engines *inside* the vehicle. Horsepower also dropped with the addition of emission controls, in 1966 for California and in 1968 for the rest of the country.

Table 1-1. 5.7-liter Corvette engine ratings

Model year	Compression ratio	Horsepower @ rpm	Fuel
1970 (bare engine)	11:1	370 @ 6000	Leaded
1971 (bare engine)	8.5:1	270 @ 4800	Low-lead
1973 (installed)	8.5:1	195 @ 4400	Low-lead

GM's engineering staff had said that their fuel injection was not ready for production in 1956 ('57 models). Some of the problems they needed to solve included:

- rough idling, caused by fuel nozzles absorbing heat from the manifold
- flat spots during acceleration
- street users found it hard to service

Of the 6,339 1957 Corvettes, 1040 of them, about 16%, were sold with the Ram Jet (most of them with serial numbers above 2000). It was a $484 option. These days that would be the equivalent to about $3000 extra, just for fuel injection!

Early tests on the dynamometer, comparing fuel injection to carburetors (at steady engine speeds), showed no increase in

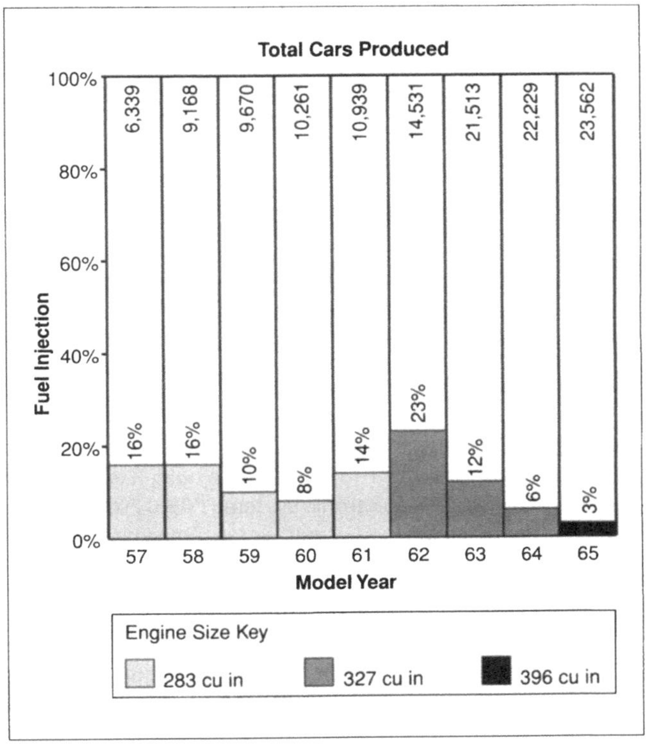

Fig. 1-4. Rochester's mechanical fuel injection continued as a production option from 1957–65—roughly one out of every six Corvettes. It peaked at 23% in 1962, and orders had faded away to about 3% in 1965.

horsepower output. But street acceleration figures measured almost 10% better. How's that? The first fuel-injection payoffs resulted from the fact that, compared to carburetion, fuel injection kept a better air/fuel mixture during transitions—as during shifting and acceleration.

The engineers realized that they could increase power by "tuning" lengths of intake manifolds, carrying only air instead of a carbureted air/fuel mixture. The fuel injection advantage increased with more testing and development time.

The fuel-injected 283 engines, with the original 3-speed manual transmission, turned 0–60 in 6.5 seconds—not too shabby. The 4-speed manual box, new in 1957's mid-year Corvettes,timed 0–60 at 5.7 seconds, with a maximum speed of 132 mph. How about that, in 1957!

Faced with low demand, related to its high cost of $484 (in 1965 again!), Corvette stopped production installations. For only $331, you could buy the more powerful Mark IV four-barrel-carbureted, 427-cubic-inch (7-liter) engine—the "big-block." Gas was cheap, so when in doubt, "add cubic inches". After 1965, some Ram Jet fuel injection continued in Corvette racing engines, but as an RPO (Regular Production Order), fuel injection was dropped.

OF FURTHER INTEREST
Today, the early, mechanically fuel-injected Corvettes have been modified and improved by owners and specialty shops until they are among the most valuable collectibles—they're known as "fuelies." Because they differ individually by aftermarket modification, they're not covered in this book.

1.2 What's in This Book

On the inside front cover you'll find a brief year by year comparison chart of engine codes, horsepowers, fuel injection/electronic engine control, and quantities built.

Chapter 1: Overview

Chapter 1 introduces the idea of fuel injection and engine control. In contrast to the 1957–65 "fuelies," which control only fuel injection, Corvette electronic systems (**'82-01**) control fuel injection, spark timing and dwell, some emission systems, and idle rpm. In olden days, when these systems operated independently of each other, we often limited troubleshooting to one system or the other. But not now!

The **LS1** is a "clean-sheet" engine, changing the bore, stroke, and firing order—just about everything except the designation "5.7-liter." Special attention was paid to the emissions (in anticipation of tighter limits) and smoothness. Each cylinder is matched to assure identical cylinder-head volume and shape, airflow direction, and balanced air/fuel ratios.

Fig. 1-5. L98: Tuned Port Injection is a major advance in electronic fuel injection. In **'85–88**, Bosch parts were used, including the main sensor (the Mass Air Flow sensor) and the injectors. The beautiful plenum chamber feeds the eight tuned intake runners, to deliver air to individual cylinders.

Fig. 1-6. ZR-1: This 4-cam, 32-valve, limited-production engine carried tuned port injection to new levels with additional intake runners for high-power operation. Delivering new heights of acceleration and speed, it inspired nicknames like "Corvette from Hell" and "King of the Hill." It came in the **ZR-1**—a total-performance option which cost over $31,000 *more* than the base Corvette. This engine is so linked with this car that I'm going to refer to the it the way Corvette guys do—even though the engine would otherwise be referred to as the "**LT5**".

Fig. 1-7. **LT1:** This engine is a major redesign of the small block, with shorter intake runners. The electronic ignition controls and distributes sparks, replacing the distributor.

Fig. 1-8. **LS1:** This engine has an aluminum block and an Electronic (distributorless) Ignition, with eight coils. One of the principal differences in engine control is the Throttle Actuator Control (TAC). The **'01 LS6** is similar.

Chapter 2: Fundamentals

Chapter 2 will help you understand the principles behind fuel injection and engine control. Some people regard fundamentals as dry theory. But you'll find this valuable if you're going to do even simple troubleshooting, or talk with your technician about your engine. Engine-control systems are necessarily complex. You cannot hope to understand troubleshooting if you go by the numbers without understanding what they mean.

I'll discuss different engine needs for each driving condition, giving a broad picture of engine control, and including the different strategies of engine control. Different conditions are sensed by the computer, which then determines output to the injectors and other actuators of the powertrain. You'll see the intricate relationships among fuel metering, spark timing, idle-speed control, and emissions.

In this book, you'll find an approach that follows the engine management system, itself. This sets it apart from most other automotive books. I haven't necessarily structured the book with separate chapters for the "ignition system," "fuel system," "emission system", etc., because they don't actually function separately. You'll want to understand the intricate relationships in modern electronic engine control, to modify your Corvette for performance, as in Chapter 9, and/or diagnose symptoms and trouble codes, as you'll see in Chapter 10.

Pressure is one of the most important factors in understanding fuel injection. I'll spend some time on the many different air pressures and fuel pressures in fuel injection and the different units of measure you'll deal with and compare.

Chapter 3: Emissions and Fuels

Government standards for emissions and fuel economy are becoming increasingly important. You'll see why computerized engine control is a key factor in meeting emission limitations, while still providing the "good ole" power and driveability, that Corvette owners have come to expect. Electronic engine control is a useful way to avoid the "Gas-Guzzler" tax. And, as corporate policy, GM has vowed not to incur such taxes.

Fuels are also increasingly considered part of the solution to automobile pollution. Recent changes to pump fuels, the so-called "reformulated gas", include alcohol additives and blends based on seasonal changes. These new fuels can affect engine operation, whether you are driving your trusty old '82 or the very latest models.

Chapters 4–7: The Different Parts of the Systems

In Chapters 4, 5, 6, and 7, I'll show you how each part operates in the system.

Sensors (Chapter 4) are the "nerves" of the system, and are the parts that send important information to the computer. Sensors advise of engine conditions, such as engine coolant temperature.

Control Module (Chapter 5) is the "brain." It receives and analyzes the signals from the sensors, calculates, and sends the necessary commands to the engine. Electronic Control Module (ECM) is the GM term; Powertrain Control Module (PCM) is used when the computer controls both the engine and the transmission.

Actuators (Chapter 6) are the "muscles"—they are the parts which move or react to output signals from the computer, fuel injectors, for example.

Fuel Delivery (Chapter 7) is the "circulatory system"—the different electric fuel pumps, filters, regulators, and controls which get pressurized fuel into the injectors.

Chapter 8: Strategies

Strategies are the details of engine management that are stored in the Control Module. In Chapter 8, I'll show how the engine-control systems combine the components to satisfy the driver's needs. Specifically, I'll describe how each engine condition—cruising, starting, and acceleration—is monitored by the sensors, computed by the control module, and satisfied by the control signals to the injectors and other actuators. This is where it all comes together.

Chapter 9: Tuning for Performance and Economy

In Chapter 9, for those who want more power from their engine-control systems (and what Corvette owner doesn't?), I'll show you some modifications for street-legal use and some for the track. I'll discuss which ones are likely to work—and which are probably not worth your time and money.

Chapter 10: Diagnosis

Chapter 10 covers simple diagnostic routines and testing that can be done by the average owner. The **'94-01** Corvettes provide greatly expanded diagnostic information, using the system that became standard on all manufacturers' passenger cars in 1996: On Board Diagnostic, the second generation. The five-digit trouble codes number over 100, and you will need a scan tool to read them.

In many cases, you can do some diagnosis and testing using simple tools—like a digital volt-ohmmeter (DVOM), and a fuel pressure gauge. Some complain that these modern engines mean the end of the shade-tree mechanic. "Hoo boy, I'll never touch that baby," I've heard owners say, as they peer underhood. But after you've read this book, you won't be saying that.

General Motors chose the '94 Corvette, its most high-tech car, as its bellwether, one of the first GM cars to begin adoption of OBD-II. So the 1994–95 Corvettes are transition cars, or "partial OBD-II."

Chapter 11: Servicing

Chapter 11 shows some of the specific servicing procedures. You can improve your dialogue with service technicians, even if you don't want to do much of the work yourself. Knowing about troubleshooting will help you communicate with your technician who will use special tools to service your car.

Chapter 12: Schematics

In Chapter 12, you'll find technical data to aid you in diagnosis and service or just to add to your know-how about why engine things happen. You'll find schematics with engine-control wiring diagrams, components, connectors, leads, and control modules.

1.3 Terminology

Car makers and the Society of Automotive Engineers (SAE) have heard the complaints from service technicians: the use of different names for the same part, in different cars, is a servicing headache.

The push for common terminology came for emission-control reasons, spearheaded by both the U.S. Environmental Protection Agency (EPA) and the California Air Resources Board (CARB). Technicians had a hard time applying the training they received, in diagnosing emission-control problems, because until 1993 the same part could be called by different names and acronyms (initials), and the same acronym could refer to two completely different kinds of parts!

Beginning in 1993, GM and most U.S. and foreign car companies cooperated with SAE in adopting a set of part names and acronyms that can become common to all cars (*SAE Recommended Practice J1930*). Here are some examples of names and acronyms from the J1930 paper, compared to the automakers own terms:

Throttle Body Fuel Injection

- J1930 term: TBI
- Corvette term: Cross-Fire Injection

Multiport Fuel Injection

- J1930 term: MFI
- Corvette term: Port Fuel Injection (PFI)
- Others: Electronic Fuel Injection (EFI)

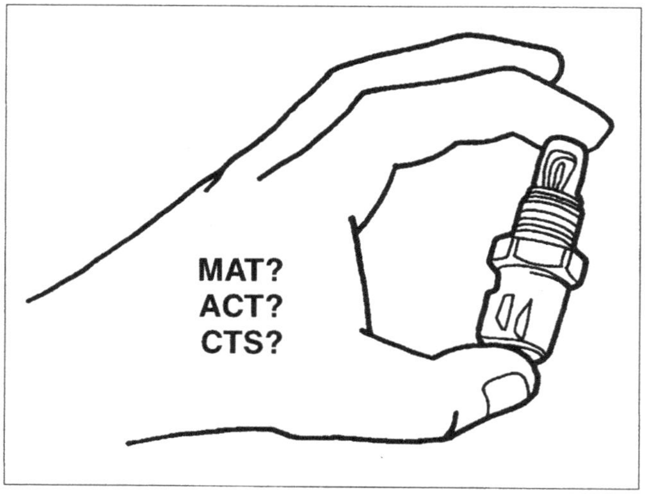

Fig. 1-9. What to call parts? Different corporations often use different names for similar parts. GM used the term Manifold Air Temperature (MAT) for it's temperature sensor. Ford used Air Charge Temperature (ACT). And Chrysler called it a Charge Temperature Sensor (CTS). The term *Engine Coolant Temperature* (ECT) is what is used in SAE J1930, the final word on terminology in automotive circles.

Sequential Fuel Injection

- J1930 term: SFI
- Corvette term: Sequential Electronic Fuel Injection (SEFI)

To reduce confusion, I will use the SAE terms.

- MFI, for PFI and EFI
- SFI, for SEFI
- Port Injection, to include both MFI and SFI
- TBI, for Cross-Fire Injection

In 1986, before SAE and the manufacturers began to bring these terms together, I discussed the nomenclature problem with the head of Service Training at GM. He explained that parts get named in the Engineering Department at the time the part is originated. By the time parts get to Training, perhaps two years later, he said, it's a done deal—a real mess to change them. Now, GM's Service Technology Group (STG) is in at the beginning, working toward common nomenclature for the same part in all manufacturers' cars. SAE J1930 "Electrical/Electronic systems Diagnostic Terms, Definitions, Abbreviations and Acronyms" spells it out, and believe me, it was *not easy.* I have worked with the SAE J1930 Committee, which has accepted several of my proposals for terms and abbreviations.

1.4 Metrics

Did you know that virtually all modern cars are manufactured with the metric system? In the current global economy, car components flow around the world. So even US cars are metric, since many of the parts are manufactured elsewhere. But for communicating with the service industry, in the U.S. only, metric dimensions are still often converted to English units.

The dashboard instruments of fuel-injected Corvettes offer you a choice between metric and English units. Most of the time I'll list both English and metric units. But I advise you, in the car business, the sooner you start thinking metric, the easier it will be for all of us. Here are a few simple helps to get comfortable with metrics.

- Hold a paper clip between your fingers—that's about 1 millimeter thick.
- Hold an aspirin tablet—that's about 1 gram.
- At sea level, barometric pressure is about a nice, even 100 kiloPascals (kPa).

Any technician knows the difference between metric hand tools and English-unit tools. When a 10-millimeter wrench doesn't fit, it's easy to reach for an 11-millimeter; but if a 5/8-inch doesn't fit, you have to think to reach for an 11/16-inch wrench. Most gauges read both metric and English units, so try metric measurements; you'll find they're really easier.

1

I produced several technical films on the Lunar Rovers—the "jeep" for the Apollo Lunar program. I even got to drive one! We had to be careful about referring to metric measurements. English units were used in the contracts, so some of our units in the films had to be English to relate to the contract. But the program was controlled by NASA, in Huntsville, and the director was a Dr. Werner Von Braun, the World War II director of the German ballistic missile (V-2) program. When the films were shown, he or his German space scientists kept asking, "How much is that in kilograms?" "How many meters is that?"

2. Basics of Fuel Injection

Before we talk about specific Corvette systems, you'll need to understand some basics about fuel injection.

2.1 Air/Fuel Mixtures

In gasoline-fueled engines, the air/fuel mixture is of critical importance, to the quality of combustion. The right air/fuel ratio is necessary to maximize engine power output and acceleration. Since the amount of air required by the engine varies with rpm and load, the required amount of fuel varies too.

Fig. 1-10. Carbureted engines need many more vacuum valves and hoses than engines with electronic fuel injection. The '80 LG4 Corvette is shown here.

The throttle of an engine regulates the airflow into the engine. The proportions of air and fuel are critical, but the throttle cannot meter the correct amount of fuel into the moving air. So the fuel metering system responds to throttle changes and adjusts to continuously supply the engine with a combustible mixture of air and fuel. Fuel injection is an accurate and sophisticated type of fuel metering system.

Gasoline engines use one of two basic types of fuel metering systems: carburetors or fuel injection. These systems both mix fuel and air, but in very different ways. To help you understand the differences in fuel injection, it might help to review how carburetors work.

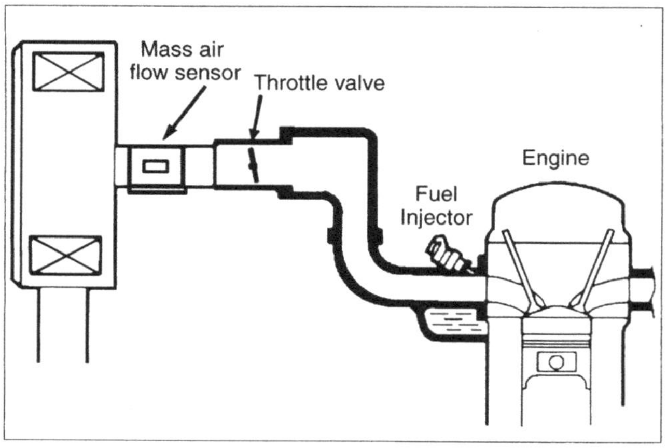

Fig. 1-11. The amount of air entering the engine is controlled by the throttle, and is measured by the airflow sensor. Measurement of air intake is sometimes done indirectly, by a Manifold Absolute Pressure (MAP) sensor. The throttle cannot meter the correct amount of fuel into the moving air.

2.2 Contrast with Carburetors

Carburetors take advantage of the *venturi* principle; the Italian physicist G. B. Venturi discovered that as more air flows through an opening, pressure decreases. (I know, it would seem like *more* pressure, but in a venturi, or throat, it's *less*.) So increased airflow through the carburetor throat, as the throttle opens, decreases pressure at the venturi. This decreased pressure pulls fuel from the carburetor float bowl into the intake airstream, where it's vaporized to form a combustible air/fuel mixture. Fuel is "metered" more or less in proportion to airflow.

Carburetor, strange word. Where did it come from? If you look it up in the dictionary you get : a device for carburetting—for mixing something (air) with hydrocarbons.

For many decades, a carburetor was a relatively simple way for metering fuel into the airstream entering an engine. Beginning in the 1980s, drivers demanded more performance and more driveability; they wouldn't accept the carburetor "stumbles" of the 1970s. At about the same time, the government mandated tighter control of emissions, and better fuel economy. The demands of drivers and the government necessitated more precise control of fuel metering, which carburetors could not provide, especially under extreme operating conditions, even with their complex set of fuel circuits, jets, air bleeds, chokes, and valves.

Until the 1980s, Europe's looser emission limits resulted in many European cars being built with carburetors, for delivery in Europe, but with fuel injection for the U.S. In 1981, in Paris, I spoke with the chief engineer of one of the largest European manufacturers. He was reluctant to change his engines to the more expensive fuel injection, for export to the U.S. He suggested that we in the U.S. were "a little paranoid about clean air."

Now Europeans also recognize the importance of clean air. Since the 1990s, European manufacturers have rushed to replace carburetors with computer-controlled fuel-injection systems, for cars sold in Europe.

2.3 Fuel Injection

Fuel injection systems deliver fuel by forcing it under pressure into the incoming airstream. These systems are more efficient than carburetors, for three reasons. First, directly or indirectly, they measure the amount of incoming air and use this measurement as a basis to meter fuel in precise amounts. Second, fuel delivery to the airstream under pressure permits more accurate control of its quantity. Third, unlike a carburetor bowl, which has a tendency to slosh fuel, fuel injection delivers the fuel more consistently, which provides more power in cornering and racing. Fuel metering can be precisely managed to meet the demands of different operating conditions.

2.4 Background

Fuel injection is not new, particularly to racing. Many engines built for Indy racing used a simple Hilborn mechanical system, injecting fuel continuously into each intake port. This was the basis for the Rochester mechanical system, adapted for some '57–65 Corvettes. On the street, it required work to be totally successful.

At Indianapolis, fuel injection was the winner. By 1970, the Bendix RS-11 system had taken over at Indy, and a fuel-injected car won first place. It used a mass airflow sensor—a way of measuring the mass, or weight, of air flowing into the engine. Based on airflow, the system mechanically controlled fuel delivery to the intake ports. While well suited to the near-constant high-speed operation of racing, it could not satisfy the changing needs of a street engine, which must start and run well from cold.

Electrojector—The Original EFI

Flashback to the mid-1950s, when a complete electronic fuel-injection system was being developed in another Bendix division, in upstate New York. The Electrojector system, as it was called, turned out to be ahead of its time—the first units used vacuum tubes which took 30 seconds to warm up before the engine could be started! As transistors became available, the electronic warmup problem disappeared, but the system cost twenty times as much as a carburetor and was less reliable. Manufacturers learned that using several carburetors was cheaper than one fuel-injection system, and Bendix shelved the electronic system in 1961. In the U.S., it was to stay on the shelf for twelve years.

Bosch Influence

Beginning in Germany in 1937, Robert Bosch GmbH adapted mechanical diesel-injection pumps to gasoline injection in aircraft, racing, and the famous 1955 Mercedes 300SL. But if the Bendix electronic system was expensive, the Bosch mechanical system was even more so, and lacked in metering accuracy and driveability.

By the mid-1960s, emission control added a new worry for the engineers. Emission-control concerns drove Volkswagen to ask Bosch for an electronic system for the VW Type 3 air-cooled engine. In 1966, a cross-licensing agreement between Bosch and Bendix enabled Bosch to quickly adapt the Bendix system. In 1967, VW brought out the world's first production electronic fuel-injection system; it came to be known as Bosch D-Jetronic.

Bosch/Bendix/General Motors

In a strange case of hands across the sea, some '76-80 Cadillacs were built with an electronic fuel-injection system sharing features from the U.S. Bendix Electrojector and the German Bosch D-Jetronic. It was called EFI. As in the Bosch D-Jetronic, the analog computer controlled only fuel injection— not ignition, and not idle rpm. One of the Cadillac engineers told me they had no technician training. Systems were tested with the infamous "Blue Box." When the box failed to indicate the problem, they sometimes replaced the whole fuel-injection system.

In 1980, Corvette was still using carburetors. But GM had a tested fuel-injection system in production: the Cadillac *Digital* Fuel Injection. (That's to say, "We're no longer analog.") Dave McLellan, former Chief Engineer of Corvette, told me that Cadillac DFI became the basis for the first Corvette electronic fuel injection, called Cross-Fire Injection. This Throttle Body Injection in '82-84 models provided Corvette with a rapid transition to electronic fuel injection, while the more sophisticated port injection was being developed.

3. Corvette Fuel Injection Systems

1

Corvette electronic fuel injection systems have changed over the years. All of the systems meter fuel in short pulses, rather than continuously, as in '57-65 Corvettes or in Bosch K-Jetronic systems. The principal differences include:

- whether fuel is delivered centrally or at each intake port
- whether port fuel injection is "group-timed" (delivering fuel to an entire bank of cylinders at once) or individually timed (delivering fuel to each cylinder sequentially in firing order)
- whether the load is measured (by mass airflow), or calculated (by speed density)

3.1 Throttle Body Injection (TBI)

Throttle Body Injection injects fuel above the throttle plate, about where carburetor fuel would be drawn into the air-stream. As in carbureted engines, the intake manifolds carry an air/fuel mixture, and are troubled with "wall wetting." The **L83** TBI is called "Cross-Fire" because each injector fires into a throttle body on one side of the engine with manifold runners crossing over to fire into the other cylinder bank.

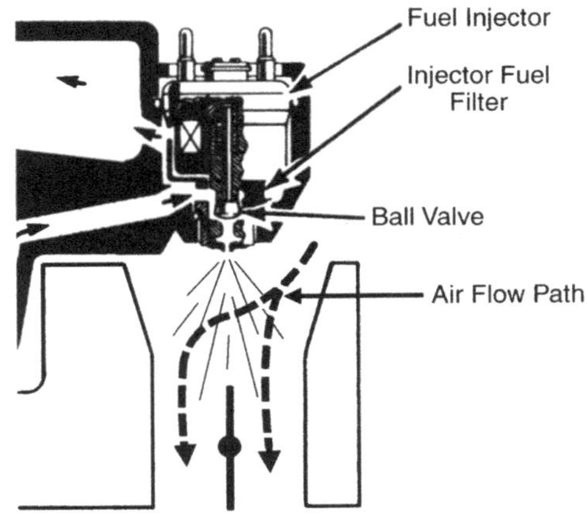

Fig. 1-12. L83: With TBI, injection pressure is low, near barometric, and is unaffected by manifold pressure.

GM calls the Throttle Body Injection system "Cross-Fire Injection," or CFI. Consider the usage of the term "CFI":

- **1980:** Central Fuel Injection (Ford's name for Throttle Body Injection)
- **1982:** Cross-Fire Injection (GM's name for Corvette Throttle Body Injection)
- **1993 and later:** Continuous Fuel Injection (J1930 term), formerly Bosch K-Jetronic (a type of Multiport Fuel Injection)

I hope I have de-confused you!

3.2 Tuned Port Injection (TPI)

As I've mentioned, **L98** Corvette port injection is known as Tuned Port Injection (TPI). In this system, the intake runners cross over from the plenum to the opposite cylinder bank. (See Fig. 1-13). They're "tuned", in the sense of being longer, to increase *mid-rpm* power by ramming more intake air into each cylinder. You'll find two types of "port injection": Multiport Fuel Injection and Sequential Fuel Injection.

Multiport Fuel Injection (MFI), used in Corvettes from **'85–93**, delivers fuel to multiple injectors at the intake ports, ganged as left bank/right bank.

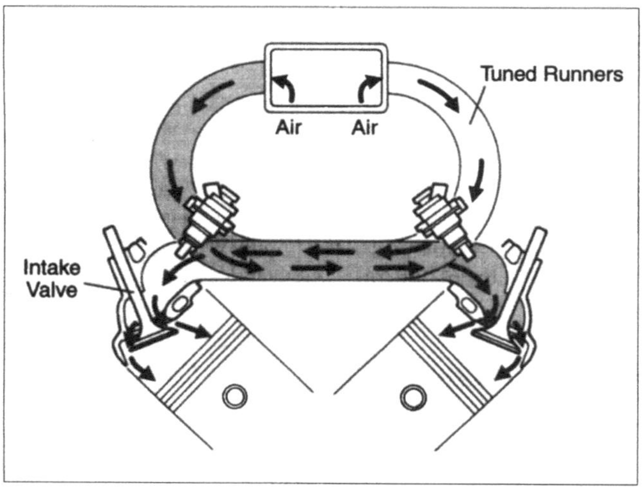

Fig. 1-13. Each bank of the Tuned Port Injection intake runners crosses over from the plenum to the opposite cylinder bank. Long runners increase mid-rpm power by ramming more intake air into each cylinder, but long runners tend to limit the maximum rpm; **L98** horsepower peaks at only 4,000 rpm.

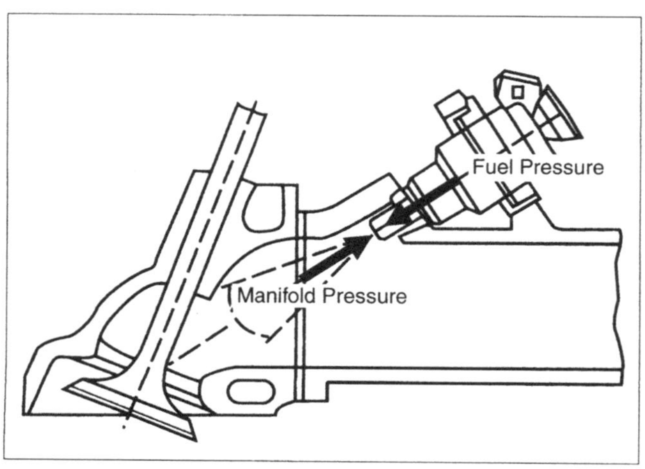

Fig. 1-14. L98, LT1, LT4, ZR1: These systems regulate injection absolute fuel pressure according to the manifold pressure.

Sequential Fuel Injection (SFI), used in Corvettes from 1994 on and in all **ZR-1** engines. Visually, it appears no different from the ganged MFI. The difference is inside the computer. Sequential systems deliver fuel to each individual cylinder in firing-order sequence. This provides less variation, between cylinders, than with ganged systems, and offers a smoother idle and acceleration, reduced emissions, and rpm limitation (because SFI can cut off one cylinder at a time).

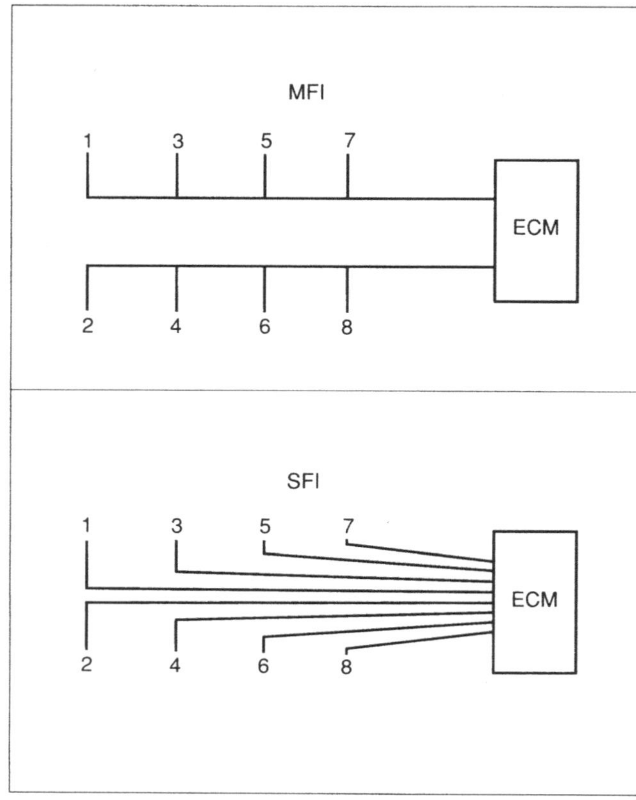

Fig. 1-15. The Corvette Multiport Fuel Injection systems deliver fuel to each individual cylinder. From **'85–93**, Corvettes pulse the eight injectors in two ganged banks of four. Sequential Fuel Injection pulses each cylinder separately.

3.3 Comparison—TBI Port Injection

Multiport systems deliver fuel at each intake port, opposite the intake valve. Multiport systems have several advantages over Throttle Body Injection systems:

- greater power, by operating with cooler, denser intake air
- improved driveability, by reducing the throttle-change lag that occurs while fuel travels from the throttle body to the intake ports
- increased fuel economy, by avoiding condensation of fuel on the interior walls of the intake manifold (called manifold wetting)

- simplified supercharger/turbocharger applications, because the compressor handles only air, instead of an air/fuel mixture

Throttle Body Injection does have a few advantages, but they're insignificant compared to the advantages of port injection:

- The cost for two injectors is less than eight injectors.
- Control can be simpler.
- The Throttle Body Injectors operate at lower fuel pressure than Multiport injectors, so the low-pressure fuel pump is less expensive, and the low-pressure lines are both less expensive and easier to seal.

4. Total Engine Control

Most people think of fuel injection as a replacement for a carburetor—a computer system controlling only the fuel delivered. Indeed, controlling only fuel was how it started in the early, mechanically operated Corvette Ram Jet systems (1957–65). Later, GM determined that total electronic engine control by computer had many advantages. Beginning with carburetors, GM called it Computer Command Control Ignition (C^3 I) to include the job of controlling ignition.

Using many of the same signals needed for computer control of fuel delivery, Corvette computers also control:

- ignition, or spark timing and dwell, eliminating most vacuum hoses and their servicing headaches
- idle rpm
- emission systems, improving driveability and again reducing the number of vacuum hoses

> Emissions are controlled by considering the entire powerplant and coordinating several elements:
> - the design of combustion chambers, valves, and manifolds
> - electronic control of fuel and spark timing
> - exhaust gas recirculation, air injection, and evaporative emissions
>
> This should change your thinking if you still regard emission control as separate, "hung-on" systems of EGR and air injection.

4.1 Ignition

Spark timing has always been important for power and economy. Recall the early automatic control, which was provided by flyweights, in the distributor, which mechanically advance timing for increased rpm and/or by vacuum diaphragms which retard timing for increased engine load. Correct spark plug timing depends on many of the same variables that control fuel metering, including engine speed, engine temperature, altitude, and, in most cases, whether the engine is knocking. Electronic ignition control uses these variables to compute timing (when the spark begins) and dwell (how long the coil is charged).

4.2 Idle-Air Control—Closed Throttle

Control of idle air, to regulate idle speed, contributes to fuel economy and reduced emissions. Using many of the same sensor signals already input, the control module adjusts idle rpm by varying the amount of air that enters the engine when your foot is off the accelerator.

The **L83**, **L98**, **LT1**, **LT4**, and **ZR-1** manage idle rpm by Idle Air Bypass Control. The **LS1** Throttle Actuator Control (TAC) manages idle rpm by motor control of the main throttle plate—there is no idle-air bypass. Electronic control reliability is insured by triple sensors on the accelerator, and double sensors on the throttle position.

4.3 Emission Control

Emission-control functions are usually included in the engine-control unit. For example, I'll describe those Corvette electronic systems that control EGR (Exhaust Gas Recirculation), AIR (air injected into the exhaust stream), and EVAP (fuel vapor control).

5. Applications

Following is a brief summary of **'82-01** Corvette changes. (Also refer to the Applications table inside the front cover.)

'82 L83: engine "Cross-Fire Injection"; (Double TBI) adapted from the Computer Command Control of carburetors, and from Cadillac fuel injection; a new 4-speed automatic transmission, but no manual; the last of the C3 (3rd generation) Corvette series; 200 horsepower @ 4000 rpm

'84 L83: all new body/chassis, designated C4 (4th generation); delivered 1.0 g cornering, with a stiff ride; minor changes to engine and (Cross-Fire) fuel injection/electronic management; new manual transmission, a 4-speed plus overdrive on the top 3 speeds is a no-cost option

'89 L98: a six-speed replaces the 4+3 manual transmission, still a no-cost option; Computer Aided Gear Shift (CAGS), otherwise known as "skip shift"

'90 L98: a change from Bosch MAF to AC Speed Density; analog tachometer and other gauges; digital speedometer; self-adjusting Throttle Position Sensor; first yaw control; 245 hp @ 4000 rpm

'90 ZR-1, LT5 engine: Dual OverHead Cams; 32 valves; SFI; Electronic Ignition (distributorless); 375 hp @ 5700 rpm

Fig. 1-16. '85 L98: You can recognize a Tuned Port Injection by the distinctive long intake runners.

'92 LT1: MFI Speed-density; Optispark ignition; electric AIR; engine computer is in the engine compartment; Mobil One synthetic oil eliminates need for oil cooler; Anti-Slip Regulation (ASR) traction control; 300 hp at 5000 rpm; red line at 5700 rpm

'93 ZR-1: 405 hp @ 5800 rpm—result of porting/polishing; linear EGR; Mobil One synthetic oil

'94 LT1: a change to SFI/Mass AirFlow; the first OBD-II, partial; the Powertrain Control Module (PCM) provides electronic control of the A/T; white instrument faces.

'95: the last **ZR-1**

'96: all Corvettes fully OBD-II compliant; the **LT1** has a new throttle body, and new fuel rail assembly; EVAP sensor; all LT1 models are A/T only

Fig. 1-17. The **'97 LS1** is more enclosed than previous engines. Notice the dual fuel lines—return and supply, The FPR is on the left fuel rail, with a crossover to the right fuel rail. The MAF is ahead of the radiator.

LT1, **LT4**, **LS1**, and **'93-01 ZR-1** Corvettes are factory-filled with Mobil One synthetic oil, which flows well cold and can operate at higher temperatures than mineral oils. The benefit to the Corvette bottom line: the powertrain can be built without an oil cooler. That's no problem, until you start adding power-increasing mods. (See Chapter 9.) I've been driving with Mobil One for the last twenty years, because an SAE paper convinced me that it flowed better than mineral oil when cold, and withstood high engine temperatures better. An extra benefit: a clean engine on the inside.

'96 LT4: a higher compression ratio and other tweaks deliver 330 hp @ 5800 rpm; engine control similar to **LT1**; no EGR; all **LT4** models are M/T only

'97 LS1: a totally new engine, still nominally 5.7L; increased power; 345 hp @ 5600 rpm; the throttle Actuator Control System (TAC) manages idle rpm by motor control of main throttle plate, with no idle air bypass; electronic control reliability is insured by multiple sensors for accelerator position and throttle position; the Fuel Pressure Regulator (FPR) is referenced to BARO; a new body, and chassis, designated C5 (5th generation); the transaxles are in the rear; A/T standard, M/T is an $815 option

'98 LS1: the first C5 convertible debuted; the EVAP canister tripled in size, to meet vapor storage requirements.

'99 LS1: a Semi-Returnless (Demand) system—the Fuel-Pressure Regulator is combined with the fuel filter in the rear near the left fuel tank, smaller injectors mean longer injection times to allow more precise control of injected fuel; a few engine changes, but it complies with the USEPA Low Emission Vehicle (LEV) standards, reducing HC 70%, NO_X 50%; still 345 hp @ 5600 rpm; the Onboard Refueling Vapor Recovery reduces escape of refueling vapors

'01 LS6: a high-performance engine with a Z06 option; Hardtop; M/T; 385 hp @ 6000; 385 lb-ft torque @ 4800; redline @ 6500; the compression ratio is 10:5 (**LS1** 10:1); many engine tweaks, including a 10% increased flow to the fuel injectors; a new intake manifold; plenum and runners increase the airflow by 10% in both the **LS1** and **LS6**; the **LS1** is still rated at 345 hp. "Yeah sure."

The **LT4**, a $1450 option, was offered for one year only. Why? It had been scheduled for an earlier model year but was delayed. It was finally ready for **'96**, but Corvette engineers realized that future emissions regulations would require an entirely new engine, the **LS1**.

Fig. 1-18. '99 LS1: A single fuel-supply line indicates a Semi-Returnless (Demand) system. The Fuel Pressure Regulator is combined with the fuel filter, in the rear near left fuel tank. Smaller injectors mean longer injection times, to allow more precise control of injected fuel.

In the Next Chapter:

Now you have a little background and a general idea about how the chapters cover electronic engine control. In Chapter 2, I'll tell you what Corvette electronic engine control must handle: the fundamentals.

My thanks to Karl Ludvigsen for details on the development of Rochester fuel injection in the early "fuelies", '57 to '65, from his book, *Corvette, America's Star Spangled Sport Car, the Complete History*.

Chapter 2

Engine Control Fundamentals

2

Contents

IN THIS CHAPTER, LOOK FOR:

- the basic factors of air/fuel ratio, spark timing, and idle-air control
- control of air intake as it relates to engine control
- the importance of different air pressures and fuel pressures to engine control
- several Closed Loop systems in engine control
- concepts of feedback and feedforward
- why various engine operating modes demand different air/fuel mixtures, spark timing, and idle-air control

1. INTRODUCTION

If you want to understand how your car's engine control works, you'll need to know the engine's requirements under different conditions. This will help you see why it needs the *sensors*, which send signals to the control module, and the *actuators*, which respond to the control-module output signals. I'll explain it this way for three reasons:

- Despite the different systems for different model years, the engine's systems are more alike than different.
- When you know what's behind these systems, you'll be much better equipped to go inside them for service or high-performance modifications.
- Electronic engine management operates by overlapping systems, simultaneously controlling parts with signals from many sensors. For example, you'll see how idle rpm is controlled by spark timing and Idle Air Control and how Exhaust Gas Recirculation changes injector pulse time and spark timing.

Fig. 2-1. Let's say the air in this car trunk weighs about 14 ounces; that is its mass. Under normal conditions, it takes this 14-ounce mass of air to burn the mass of fuel in this glass, about 1 ounce. We say that's an air/fuel ratio of 14:1. Mass, not volume, determines the air/fuel ratio.

Throughout this book, I'll follow generally accepted practice and discuss air/fuel ratios primarily in terms of mass. A Mass Air Flow sensor is an input sensor you'll hear about in **'85–88** and **'94-01** Corvettes. Some scan tools allow you to read intake air in terms of mass (grams/second). Or you may see carburetor flow rates expressed as volume, such as xx cubic feet per minute. In Earth's gravity, mass is effectively equal to weight, so for practical purposes we can consider the two to be essentially the same. Thinking of mass, or weight, is the best way to understand the basic factors that govern fuel delivery and combustion.

2. BASIC FACTORS

2.1 Air/Fuel Ratio

The main job of any fuel-metering system is to mix fuel with the incoming air in the proper ratio. Small variations in the air/fuel ratio can have major effects on power output, fuel consumption, and exhaust emissions.

The Basic Combustible Mixture

In round numbers, it takes about 14 parts of air to support complete combustion of one part of fuel—in other words, an air/fuel ratio of about 14:1. Gasoline air/fuel ratios higher or lower than this may still burn, but such combustion usually raises emissions and causes other side effects.

Fig. 2-2. Once upon a time, an air/fuel ratio somewhere close to 14:1 was good enough. But with today's finer tuned technology, we can get the more effective ratio, of 14.7:1.

As you read this chapter, you'll see why precise control of the air/fuel ratio has become so important. The relationship between air/fuel mixture and exhaust gas oxygen affects the electronic control of power, economy, and emissions. I'll also explore the new fuels and their effect on power and emissions.

Rich and Lean Mixtures

A rich mixture is one with a lower ratio of air to fuel. It has insufficient air (oxygen) to support complete combustion of the fuel. If you see black smoke pouring out the exhaust pipe, that's unburned fuel—I mean, lots of unburned fuel. Rich mixtures, such as 11:1 or 12:1, can cause:

- hard starting (flooding the engine)
- increased fuel consumption during cruise
- increased emissions of hydrocarbons (HC) from unburned fuel, and carbon monoxide (CO), the product of incompletely burned gasoline

Super-rich mixtures, say 10:1, can cause:

- reduced power
- increased carbon deposits
- fouled spark plugs
- diluted engine oil

Fig. 2-3. When all the air (oxygen) of a rich mixture is consumed in the cylinders, fuel is left over, so exhaust gas includes unburned fuel.

> Wait a minute! you say. When I pump in extra fuel, I may get less power? Holy cow! You'll find out more in Chapter

A lean mixture is one with a higher ratio of air to fuel. After complete combustion of the fuel, the exhaust contains leftover air. The fuel will burn completely, but more slowly and at a higher combustion temperature. Lean mixtures reduce power and run hotter, leading to detonation and knocking. That increases emissions of oxides of nitrogen (NO_x), the product of nitrogen in the air combining, or "burning," with oxygen. Lean mixtures lead to hard starting and lower performance. If the air/fuel mixture gets leaner than about 17:1, misfiring in the cylinder can lead to loss of power and backfiring, increasing hydrocarbons (unburned fuel) and carbon monoxide (partially burned fuel). In extreme cases, the high temperatures resulting from lean combustion will cause pre-ignition—violent combustion of the mixture that can cause serious engine damage.

The way the air/fuel mixture affects the exhaust gas is fundamental to engine control. Remember:

- Rich mixtures produce exhaust gases with little or no oxygen.
- Lean mixtures produce exhaust gases with extra oxygen.

Fig. 2-4. When all the fuel in a lean mixture is consumed in the cylinders, some air (oxygen) is left over, so exhaust gas also includes oxygen.

Stoichiometric (Ideal) Ratio

"Stoichiometric" means, roughly, "measuring the elements." I couldn't say it right until a friend pronounced it for me: "*stoy*-key-o-metric." A stoichiometric ratio is neither too rich nor too lean. It contains just enough fuel to burn all the oxygen in the air. This is the air/fuel ratio that provides the proper exhaust gas to operate the catalytic converter. For a warm engine running at part throttle, stoichiometric is a so-called "ideal" ratio, the one that offers the best compromise between power and economy: 14.7 units of air to 1 unit of fuel. (See Fig. 2-5.)

Beginning in the 1990s in many parts of the country, we began burning oxygenated fuels—fuels with additives containing oxygen. It takes less air to burn oxygenated fuels completely, because some of the oxygen comes from the fuel. Therefore, we need a fresh approach to our ideas concerning the traditional stoichiometric ratio of 14.7:1.

For a warm engine running at part throttle, stoichiometric may be an "ideal" ratio—the one that offers the best compromise between power and economy. But conditions can change, both in a fraction of a second and over a long period of time. To handle these changes, electronic engine control must change air/fuel ratios and spark timing with great speed and accuracy.

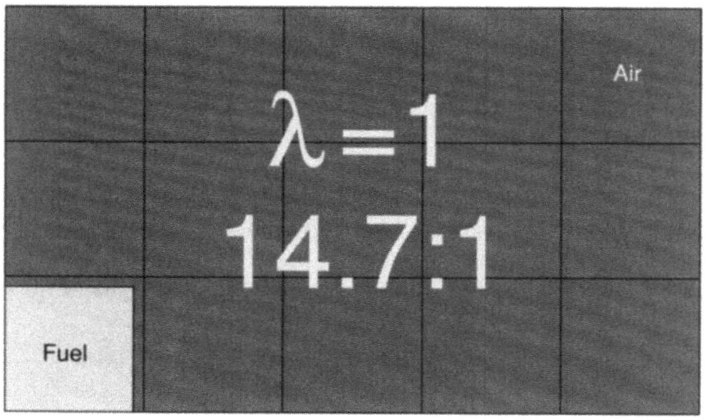

Fig. 2-5. A warm engine runs at part throttle most of the time. For this, an air/fuel ratio of 14.7:1 is nominal, or "stoichiometric." This is sometimes described in terms of λ ("lambda"), the Greek letter "L," for *Luft* ("air" in German). Bosch calls this the "excess air factor." When λ = 1, all the oxygen in the air combines with the fuel—that is, "no excess air."

Oxygenated fuels need less air than the traditional 14.7:1 because there is extra oxygen built right into the fuel.

Under the Oxygen Sensor's control, oxygenated fuel is metered to deliver an air/fuel ratio of about 14.2:1 at the injectors. The resulting exhaust is about the same as the exhaust from burning gasoline formulated without oxygen, the traditional 14.7:1.

When "gas was cheap and the air was dirty," carburetors usually delivered mixtures richer than 14:1, perhaps as low as 12:1. As the car aged and a little excess air leaked in around the gaskets of the carburetor or intake manifold, the engine still got a good combustible mixture. Also, because the carburetor was farther from the end cylinders than from the middle ones, a richer mixture was some compensation for unequal fuel distribution.

Fig. 2-6 shows another reason for running Throttle Body Injection and carbureted engines a little rich. Where the air/fuel mixture is less precise, variations in a rich mixture have only a small effect on power, while variations in a lean mixture affect power dramatically.

For the first emission controls, in 1968, lean mixtures reduced hydrocarbons (HC) and carbon monoxide (CO).

In today's fuel-injected engines, an air/fuel mixture richer than 7:1 will not burn, nor will a mixture as lean as 19:1. And we now know that fuel-injected engines generally cruise best at the stoichiometric ratio, usually 14.2 for oxygenated fuels, 14.7:1 for other gasoline. When other ratios are needed, Corvette electronic fuel injection changes the ratio for best performance.

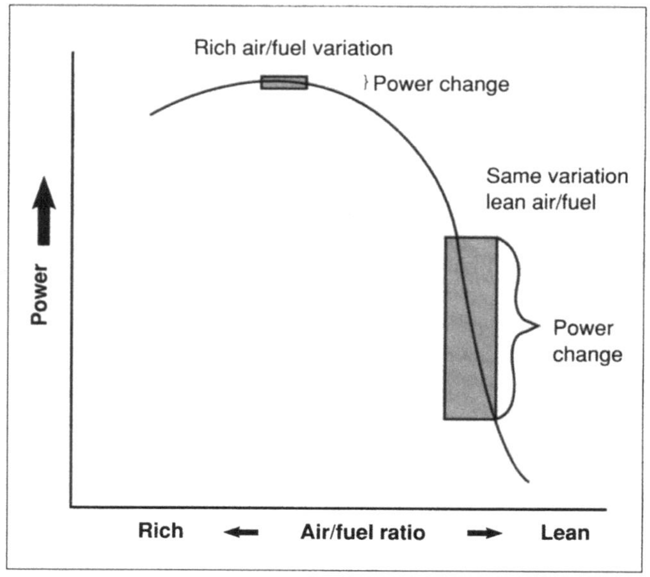

Fig. 2-6. Variation in a rich air/fuel ratio makes much less difference in power than the same variation in a lean ratio.

2.2 Spark Timing—Ignition Timing

Precise control of spark timing can improve performance. It is also important to engine control, for emissions, economy, and driveability. As the spark jumps across the plug gap, it takes, on average, about two milliseconds (thousandths of a second) to ignite the air/fuel mixture. The spark must fire at the proper millisecond so combustion pressure peaks just after the piston reaches Top Dead Center (TDC). (See Fig. 2-8.) After all, you want the combustion pressures to push down on the piston, and that's after TDC.

The three most important factors that determine elementary spark timing are engine speed, load, and temperature:

- During maximum-rpm operation, the piston may travel ten times as fast as during its slowest rpm. But the spark doesn't jump faster or ignite the mixture faster. To build combustion pressure after Top Dead Center, spark timing must be advanced more as engine rpm increases.
- During lighter-load operation, the air/fuel mixture is throttled, so less mixture is present in the combustion chamber, and it burns more slowly. Therefore, under part load, spark timing must be advanced.
- During lower-temperature operation, the air/fuel mixture is less likely to detonate. Therefore, spark timing is advanced.

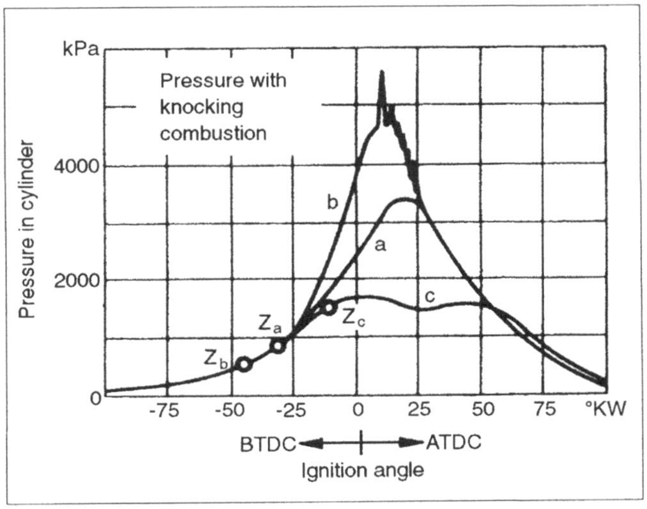

Fig. 2-7. Pressure in the combustion chamber varies according to spark timing advance.
- Line *b* shows that the spark timing (Z_b) is too advanced (early)—about 45° Before Top Dead Center (BTDC). If the pressure builds with the piston traveling upward, knocking in the combustion chamber may result. This costs power and is destructive to the engine.
- Line *c* shows that the spark timing (Z_c) is too retarded (late)—about 10° BTDC. The piston is already traveling downward; again, this costs power.
- Line *a* shows the proper timing (Z_a)—about 30° BTDC, producing maximum pressure shortly after the piston reaches TDC.

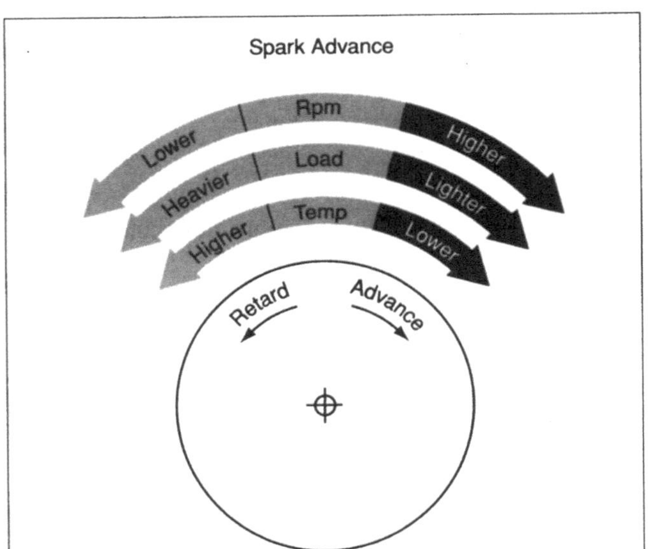

Fig. 2-8. Elementary spark timing is based on three engine factors:
- RPM—advance for higher rpm
- load—advance for lighter load
- temperature—advance when lower

Think of this as Distributor Reference—Corvette speak—even if there is no distributor. (See Chap. 4)

Engine control must also consider the effect of spark timing on emissions. Naturally, emissions are related to the air/fuel ratio and how it burns at different spark timings.

Variations in Spark Timing

All the above is based on that old qualifier, "All other things being equal." But variations exist cylinder-to-cylinder because of differences in manifolding and injectors, even in engines in good condition. Those variations increase with engine wear, carbon buildup in the cylinders, valve seating clearances and stem clearances, injector clogging, and deposits on intake valves. Compared to carburetors or throttle body injection, however, in which the air/fuel mixture (rather than just air) travels through intake manifold runners, port fuel injection tends to deliver a more consistent air/fuel mixture to the cylinders, and sequential injection is more consistent than multiport injection.

Some changes in operating conditions occur over long periods of time, years even. Others occur over shorter times, such as warmup. And some occur almost instantaneously, as in acceleration, deceleration, and starting.

Electronic control can combine all these factors, as well as idle smoothness and driveability, and deliver far more precise spark timing than simple vacuum/flyweight spark control.

Dwell Control

Dwell control refers to the instant the primary circuit is *closed*, beginning the saturation of the primary coil windings. Spark timing refers to the instant the primary circuit is *opened*, causing the coil to develop enough voltage to fire the spark plug.

Corvette high-compression engines require high-energy ignition systems. Yet in the interests of coil life and reliability, the dwell period must begin at the right millisecond. The coil must be charged just enough to satisfy the plug firing requirement, and no more. That means the engine-control system must control dwell.

2.3 Airflow, Fuel Metering, and Engine Load

Load measurements signal how hard the engine is working—how hard you're pressing the accelerator. Load increases with speed, hills, and headwinds. Load measurement is important for determining the amount of fuel to be injected, spark timing, emission control, and idle-air control. (See Fig. 2-10.)

When the control module needs load measurements, it's really asking, "How much air are we burning?" Or, more specifically, "What is the weight (mass) of the air we are burning?" Based on the weight of the air, the control module calculates the weight of fuel to be injected so that the air/fuel ratio is correct for the operating conditions. It must also consider that the density of air changes under two influences: temperature and pressure.

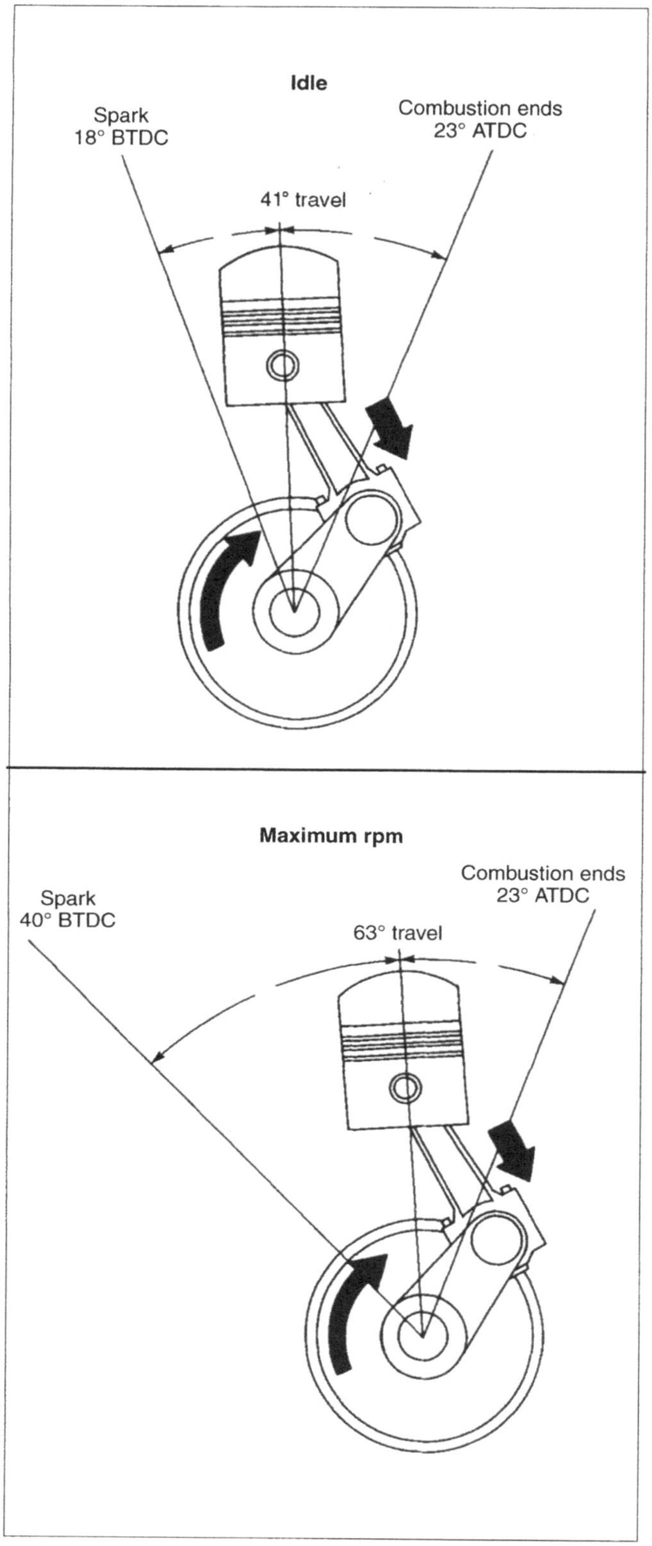

Fig. 2-9. To build combustion pressure after Top Dead Center, spark timing must be advanced as rpm increases.

Fig. 2-10. Load changes as a result of hills, headwinds, open windows, underinflated tires, dragging brakes,and top down. Anything which changes how much effort the engine is exerting. You get the idea.

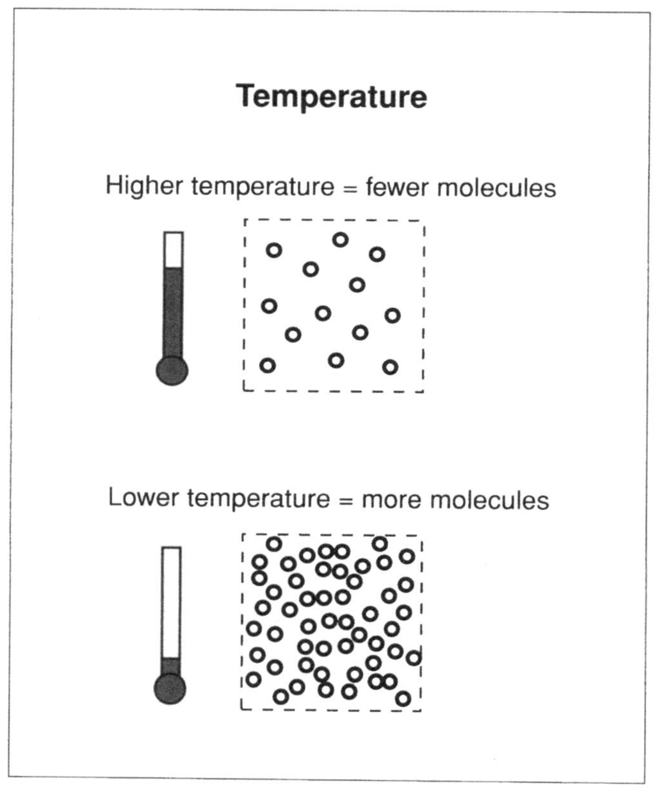

Fig. 2-11. The same *volume* of warm air weighs less than cool air, since warm air spreads out more, and less of it can fit into a given space. This means that less warm air can fit into your engine's cylinders. So you need load measurements to ensure that less fuel is injected.

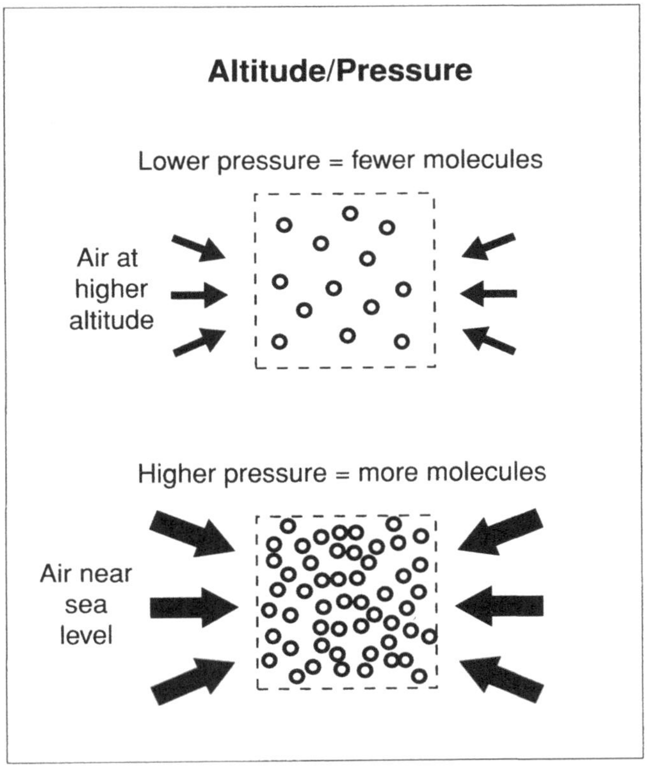

Fig. 2-12. At higher altitudes, the same volume of air weighs less than at sea level, because the lower pressure squeezes the air less, than under a higher pressure. Therefore, air pressure also has a direct effect on air/fuel ratios, and why load measurements are important.

3. Controlling Air Intake

You know that air is drawn into the engine with each intake stroke of each piston. The piston moving down on its intake stroke increases cylinder volume and lowers the pressure in the cylinder.

3.1 Throttle

The throttle restricts intake airflow. Opening the throttle increases manifold pressure. Driving on a level road, you can cruise along comfortably and maintain a desired speed with a relatively small throttle opening. When you come to a hill, you press down on the accelerator to maintain the same speed. The hill has demanded more work from the engine, created a higher load, and the engine needs more air and fuel to match that load. Load, and the resulting throttle openings, directly affects manifold pressure. Manifold pressure in turn affects airflow and, thus, fuel requirements.

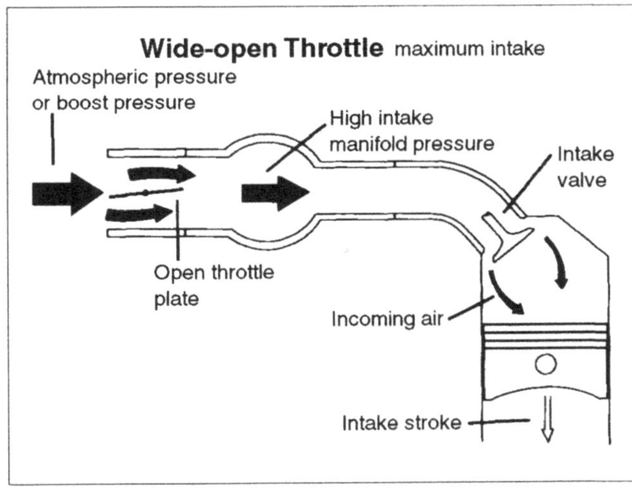

Fig. 2-13. When the throttle is open, the manifold pressure is greater than the intake pressure in the cylinders, so more of the air/fuel mixture moves into the cylinder as the intake valve opens.

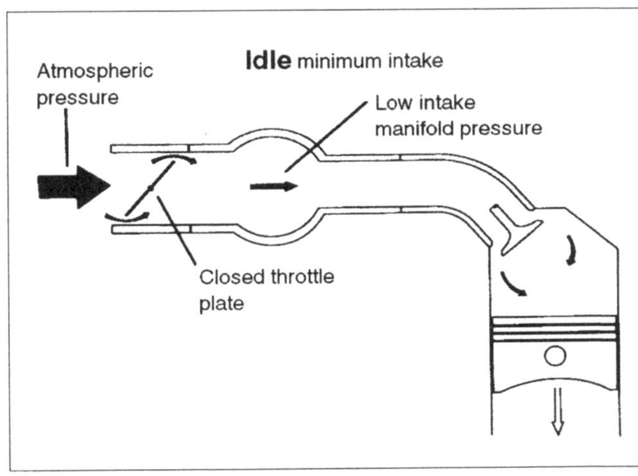

Fig. 2-14. When the throttle is closed, as in idling or coasting, atmospheric pressure has little effect on the manifold pressure. A low pressure-differential between the manifold and the cylinders results in low airflow.

Check your dictionary. You'll find one definition of "throttle" is "restricting flow". In the movies, when the bad guy squeezes the throat of the good guy to cut off his air supply, that's "throttling" him.

3.2 Ramming Intake Air

Getting more air/fuel mixture into the cylinders per stroke is the same as having a larger engine (more power). "Ram air intake" refers to the technique of ramming more air into the cylinders. As each intake valve opens and closes, each column of air resonates, or "bounces," back and forth in its runner. If the air column is bouncing toward the intake valve just as it opens, the increased pressure "rams" in extra air through each valve opening. The pressure also resists backflow caused by the rising piston on the first part of the compression stroke (while the intake valve is still open). The natural resonance depends on the *length* of the intake runner. In longer runners the air takes longer to bounce back and forth.

- Long runners increase torque at lower rpm.
- Short runners increase torque at higher rpm.

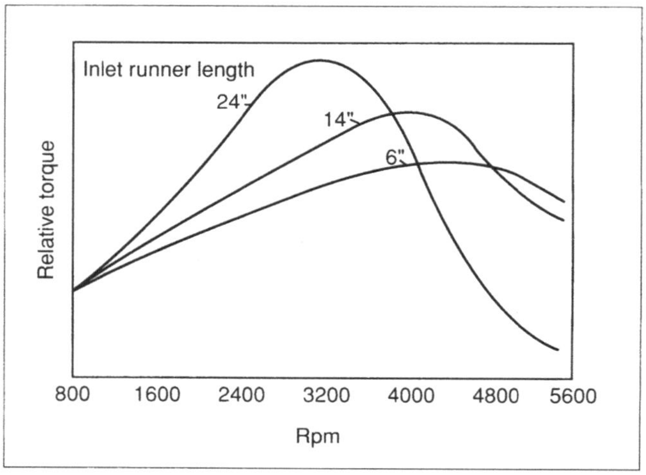

Fig. 2-16. Here are some examples of torque curves, for different inlet runner lengths. As you can see, the long runners increase torque at lower rpm, while the short runners have a higher torque with a higher rpm.

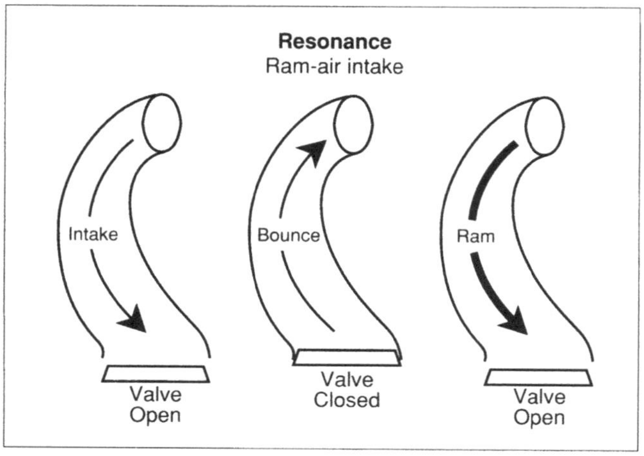

Fig. 2-15. As each intake valve closes, each column of air resonates or "bounces" back in its runner. If the air column is bouncing toward the intake valve just as it opens, the increased pressure "rams" in extra air through the valve opening.

Tuned Intake Runners

It doesn't take a rocket scientist to figure the effect of a single set of fixed-length runners. The natural resonance, or inertia effect, has a fixed rate, while the rate of intake valve opening changes from idle to maximum rpm, possibly by a factor of ten. In other words, the length is tuned for maximum resonance during a relatively small range of rpm, according to the probable use of the car. For the track, short runners would increase torque at higher rpm, but you're talking a major change in these manifolds or you're talking dual sets of runners, one set designed to add intake air at higher rpm, as in the **ZR-1**. Compare Corvette's intake runner length on different engines and you have clues to which engines rev faster.

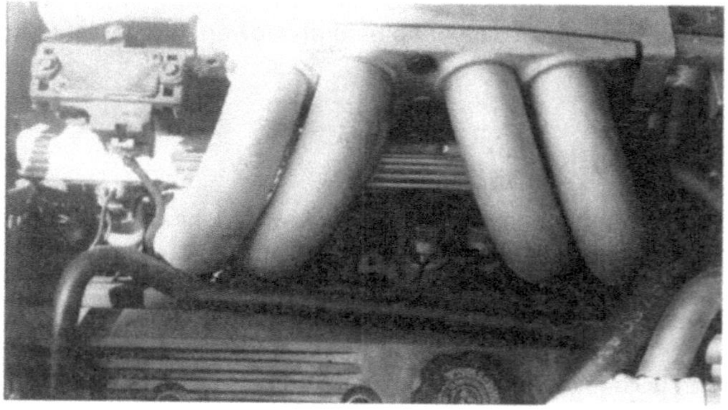

Fig. 2-17. L98: When you first see these runners, you think they're curving down into the same bank of cylinders below, but each runner actually feeds a cylinder on the opposite side of the engine. (See Fig. 2-18.)

With port fuel injection (**'85-01** Corvettes), the intake runners carry only air, instead of an air/fuel mixture. The intake system can therefore ram the intake air to improve engine efficiency without regard to fuel puddling or wall wetting in the runners. That's the idea behind Corvette's Tuned Port Injection and just about every port-injected engine.

Corvette engine control uses several methods to increase intake airflow, depending on the engine. These methods include the use of:

- fixed-length tuned intake runners
- dual tuned intake runners, long and short
- supercharging or turbocharging (aftermarket only)

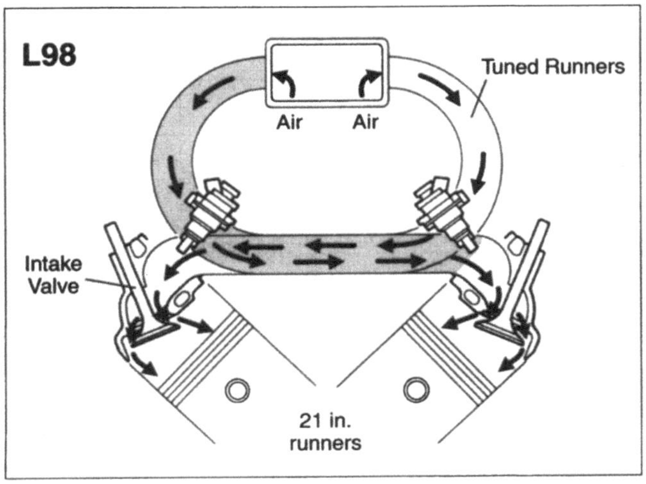

Fig. 2-18. L98: The right runners deliver intake air to the left cylinders and vice versa. This provides longer intake runners to ram in more air at mid-rpm. Each runner is about 21 inches (530 mm) from the plenum to the cylinder head.

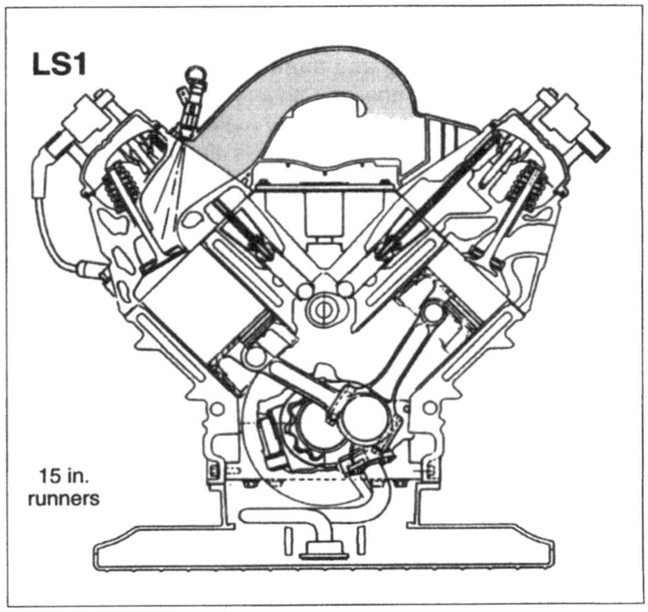

Fig. 2-20. LS1: The intake runners also cross over to the opposite bank. They are shorter than in the **L98** and longer than the **LT1**. Their length is right in between the others, at about 15 inches (380 mm).

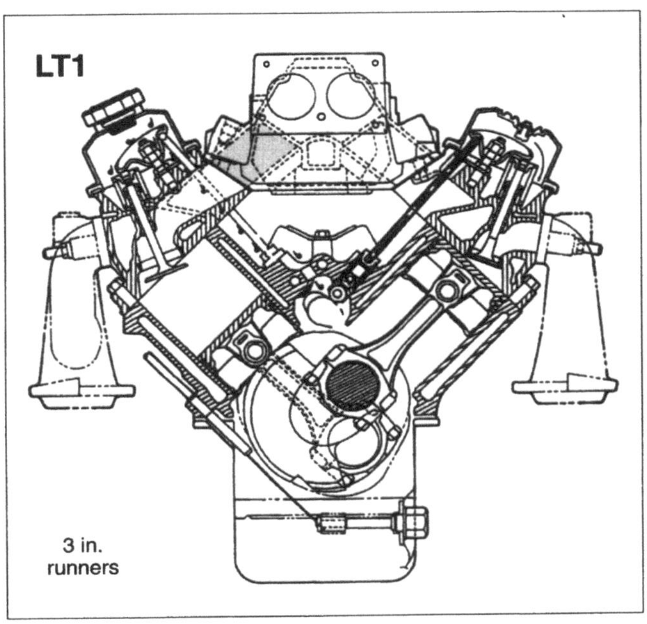

Fig. 2-19. LT1: The intake manifold passages are direct, delivering to the same side. The objective: torque bandwidth (torque at least 90% of maximum torque) peak horsepower of 300. The **LT1**'s runner length is only about 3 inches (75 mm).

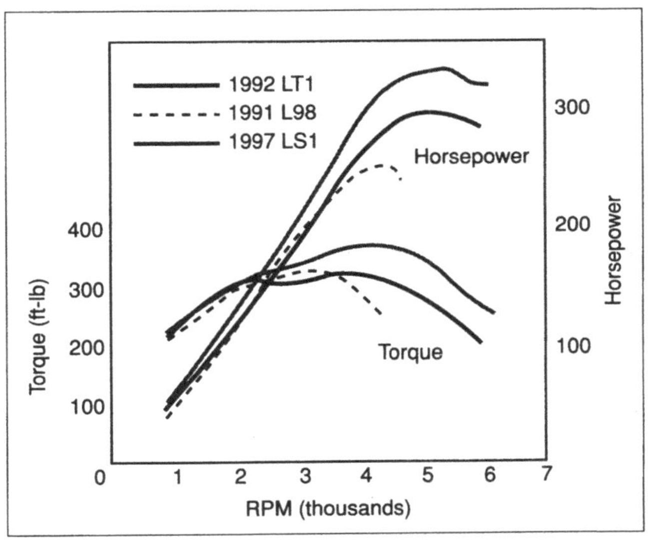

Fig. 2-21. LT1: Compared to the **L98** engine, with its long runners, the **LT1** gives up a bit of torque at about 3,000 rpm to get a much flatter torque curve, with more torque below 2000 rpm and above 3,600. That torque increase is part of the horsepower increase above 4,000 rpm.

LSI: With medium-length runners, power curves show increases over **LT1**. Similar torque curves show gains, mostly at higher rpms.

LS1, LS6: Increased intake air is made possible by low-restriction runners made of nylon. They have smoother interior surfaces to reduce drag and thermal insulation for a cooler (and therefore denser) charge. Other features include a single throttle plate, a 74 mm mass airflow meter, and no EGR, due to more efficient design. Exhaust flow is also increased—back pressure is less than 5 psi (34 kPa) at peak power.

For earlier engines, metal intake runners were used, and some aftermarket tuners polish them, to help reduce drag, as in the nylon versions.

On port fuel injection engines, look for a surge tank, or chamber, just downstream of the throttles. (See Fig. 2-22.) This separates the air column for each intake valve in its own intake runner, to dampen the effect of each cylinder on its neighbors. The **LS1** operates with a surge tank of about 11 quarts (10 liters), to insure good filling of each intake runner. The surge tank is low-restriction, starting with the peak-power of atmospheric pressure and dropping down to less than about 0.7 psi (5 kPa), at the rear of the intake manifold.

Compare the effect of tuned intake runners:

Table 2-1. Effect of tuned intake runners

Engine	Runner length (in.)	Peak power (hp @ rpm)	Peak torque (ft-lb @ rpm)	Fuel cutoff (rpm)
L98	21	230/250 @ 4300	330/340 @ 3200	5850
LT1	3	300 @ 5000	330 @ 4000	6200
LS1	10	345 @ 5600	350 @ 4400	6000

Fig. 2-22. ZR-1: Two sets of intake runners for each cylinder improve intake airflow.

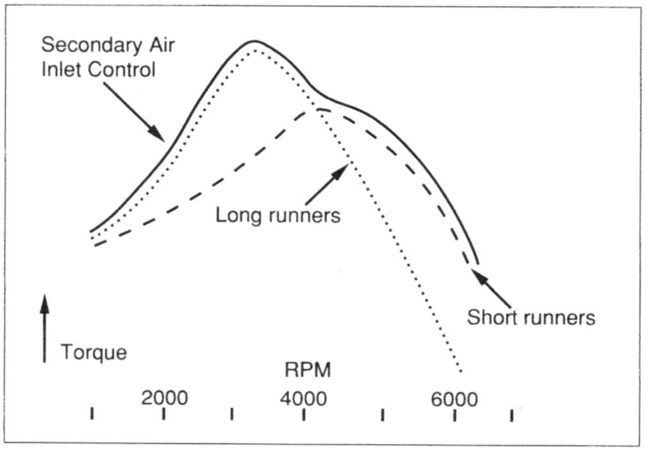

Fig. 2-23. ZR-1: Long runners increase airflow at lower rpm. With both runners open at higher rpm, short runners increase airflow to help provide a broader torque curve.

Dual Intake Runners (ZR-1)

Long runners are better at lower rpm, and short runners help at higher rpm; so why not use both? The **ZR-1** does just that, with entry controlled by the engine computer through throttle-plate-type valves.

At lower rpms, using only the longer intake runners, each air column takes longer to bounce, or resonate, between the surge tank and the intake valve. The slower ram effect is timed to the slower rate of intake-valve opening.

As engine rpm increases, the computer may add the shorter manifold runners, so both sets of runners are open. In Power Mode, the eight Secondary Port Throttles can open a separate set of shorter runners, which change the tuning to match higher rpm, above 4,000.

> You may hear the term Secondary Air *Inlet* Control used for the **ZR-1**'s Dual Intake Runners. But don't get that confused with Secondary Air *Injection* (AIR), which controls the air pumped into the exhaust for emission control.

Supercharging/Turbocharging

The third method of increasing intake air flow is actively ramming air by supercharging/turbocharging, as in aftermarket installations.

None of these Corvettes has a factory-installed supercharger/turbocharger. (For aftermarket info, see Chap. 9.)

Volumetric Efficiency

Engineers use the term *volumetric efficiency* to describe the efficiency with which the engine takes air into the cylinders. Taking 5.7 liters of air into a Corvette 5.7-liter engine is 100% volumetric efficiency. Most engines run only 70–80% volumetric efficiency, even at Wide Open Throttle (WOT), due to friction and flow restrictions. With a turbocharger or supercharger, compressing the intake air can raise volumetric efficiency to over 100%.

Working with a new scan tool, I was surprised to find a readout for volumetric efficiency. Of course, that's no problem for the engine-control computer: it knows the cylinder capacity, say 5.7 liters. Reading rpm, it counts two crankshaft revolutions (intake filling of all cylinders). Reading the Mass Air Flow sensor, it knows the air-mass intake, and knowing the intake air temperature, it calculates the volume. If the intake volume calculates to 4.0 liters every two revolutions, volumetric efficiency is 70%:

$$\frac{4.0 \text{ liters}}{5.7 \text{ liters}} = 0.70 = 70\%$$

Later scan tools read intake air mass in grams per second. "Why would technicians want to read volumetric efficiency in grams/second?" I asked myself. Hey, that could be a troubleshooting clue to a clogged air filter, deposits in the intake passages, even a clogged exhaust. It's also a guide to the effectiveness of performance modifications—larger passages, porting and polishing, increased valve lift, camshaft changes—really neat.

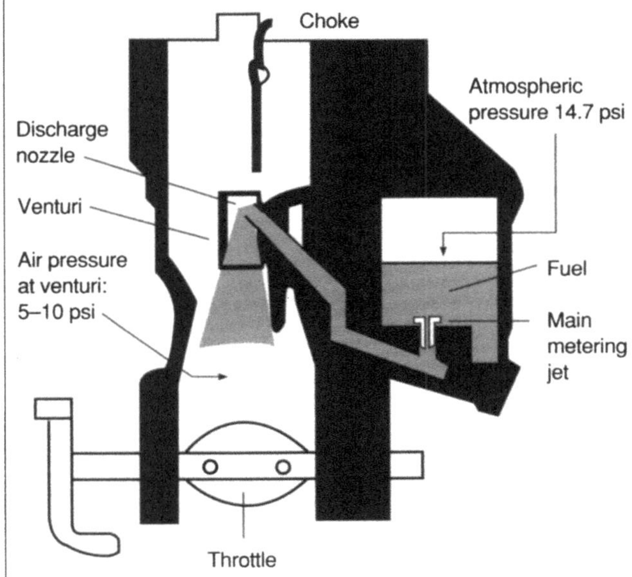

Fig. 2-24. The fuel-metering part of a carburetor operates with small differences in pressure. The atmospheric pressure on fuel in the bowl pushes the fuel into the reduced pressure area, which is called a "vacuum," at the discharge nozzle in the venturi.

4. Pressure

Fuel injection means fuel under pressure. When pressurized fuel is injected into the airstream, it vaporizes and mixes with the air into a good burnable mixture. Picture how a liquid is vaporized when you spray it under pressure from an aerosol can. In most Corvette fuel-injected engines, injectors operate at a *relative* fuel pressure of 47 psi (325 kPa). Relative pressure means pressure above manifold absolute pressure. (See injector pressures in Chap. 7.)

The fuel in a carburetor is normally at atmospheric pressure. A vacuum at the tip pulls gasoline from the fuel jets. When I use the term "vacuum," I'm talking about the difference between air pressure on the fuel in the carburetor bowl and air pressure in the venturi chamber. That difference is usually small. Fuel moves into the airstream because the moving air in the venturi of the carburetor is at reduced pressure, aided by the piston pumping actions. For a cold start, the carburetor choke restricts air entering the venturi, further reducing the air pressure, to pull in more fuel. But with injection, there is no choke, and no vacuum effect.

4.1 Pressure Measurement

As I get into the details of fuel injection systems, you'll see that many functions and relationships are defined in terms of pressure. They may be fuel pressure values in the fuel system, manifold pressure in the air intake system, or atmospheric pressure.

Barometric (Atmospheric) Pressure

If you watch the local TV weatherperson, you'll hear talk of "highs" and "lows." This refers to barometric readings which describe atmospheric pressure changes. Changing atmospheric pressure changes the density of the air. Denser intake air may slightly alter the air/fuel ratio and may affect how your engine operates.

We live in this "ocean of air," with the atmosphere pressing on all sides of us. It's *inside* too, so it balances out, and we don't get squished. We seldom feel it, except when it pops our ears in an skyscraper elevator or a rapid descent in an aircraft. Our engines do feel the pressure, though. When the engine pulls in air on the intake stroke, it reduces the absolute pressure below the atmospheric pressure and creates a vacuum. The engine control system must "feel" the barometric and manifold absolute pressures, and adjust accordingly.

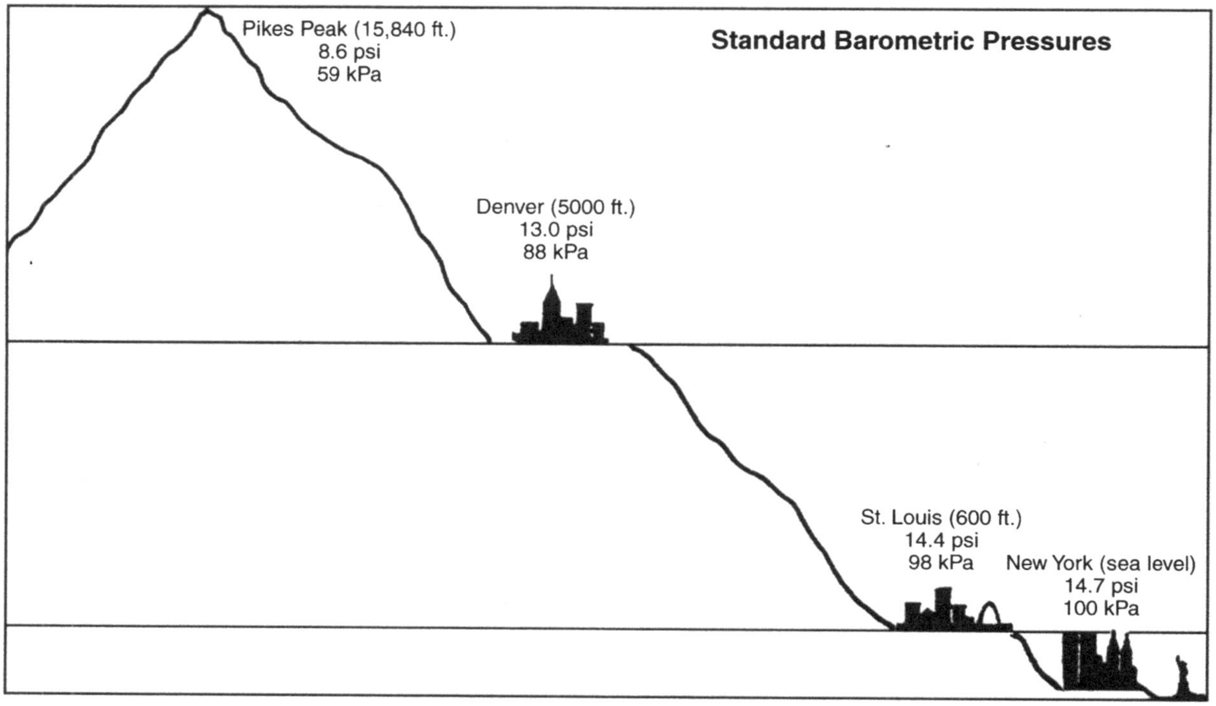

Fig. 2-25. Standard barometric pressure at or near sea level, such as in New York City, is 14.7 psi. Many drive at slight elevations, as in St. Louis (14.4 psi) or at higher altitudes, as in Denver (13 psi). The Corvette barometric sensor (BARO) measures the pressure, and the computer compensates.

Pressure is expressed in the following units. With the appropriate conversion factors, all are interchangeable:

- **kiloPascal (kPa).** SAE and most automakers use the metric unit for pressure. Many pressure gauges now read both psi and kPa. Atmospheric pressure at sea level is about 100 kPa—easy to remember.
- **pounds per square inch (psi).** Widely used in the U.S., this unit of pressure is defined as force in pounds divided by area in square inches. Atmospheric pressure at sea level is approximately 14.7 psi.
- **inches of mercury (in. Hg).** This originally referred to measurement of pressure with a mercury manometer. (*Hg* is the chemical symbol for mercury.) It's used to specify manifold vacuum. In these units, 29.92 in. Hg is the difference between standard atmospheric pressure at sea level and absolute vacuum. Indy racers refer to boost in terms of inches of mercury absolute, on a scale of 0–60.

Wait a minute, 14.7:1 is the ideal air/fuel ratio, and 14.7 psi is barometric pressure at sea level! How's that again? Pure coincidence, but that makes it easier to remember those two important numbers in engine control in U.S. measuring units.

Gauge Pressure

For many years, people traditionally thought about engine airflow and load in terms of vacuum—the vacuum created in the intake manifold by the intake strokes. Using atmospheric pressure as a baseline of zero, lower manifold pressures are expressed as a negative value, called a vacuum. That was convenient, because carburetors worked on the vacuum effect. (See Fig. 2-26.)

By the 1980s, the automotive industry was moving away from thinking in terms of vacuum. Widespread use of fuel injection under pressure called for positive thinking. Manifold Absolute Pressure (MAP) is measured from absolute zero and is always positive—whether it is below atmospheric pressure , as in idle or part throttle, or above, as in boost. (See Fig. 2-26.) In the world of fuel injection, vacuum measurements can be confusing.

Consider the piston on its intake stroke. Using absolute pressure as the reference point, the downward-moving piston is creating a low pressure in the cylinder. The absolute pressure in the intake manifold (MAP) is higher than in the cylinder. So intake air is pushed by the higher pressure toward the lower pressure. Your throttle controls its influence on the manifold pressure. As one factor in calculating load, manifold-pressure measurements are the key to "speed density" engine management systems.

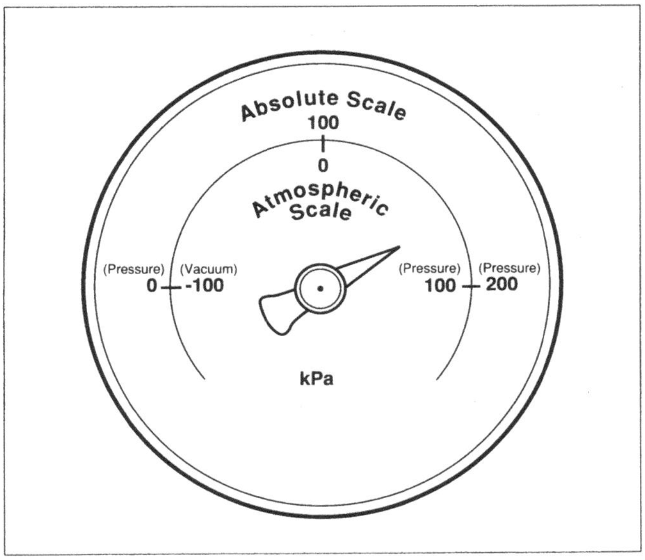

Fig. 2-26. *Absolute* gauges (*outer scale*) show all readings as positive. Zero to +100 kPa is a positive indicator of what is referred to as "vacuum."

Atmospheric gauges (*inner scale*) show pressure below zero as vacuum: minus 100 kPa to 0.

5. Control Systems

By now you are aware that computer control plays an overwhelmingly important part in maintaining the acceptable balance of power, fuel economy, emissions, and driveability. Corvette electronic engine control responds to measured inputs, precisely metering the appropriate amount of fuel, spark timing, and emissions for the conditions. A control system may operate one-way ("Open Loop") or two-way, receiving feedback that modifies its output ("Closed Loop").

5.1 Open Loop

Open Loop control takes information about operating conditions from various sensors and calculates signals to the actuators, including injector pulse time, spark advance, and idle-air control. In Open Loop, the computer does the best it can, based on inputs from sensors and information in its memories about what the engine should be doing. Accuracy of the Open Loop air/fuel ratio, spark advance, and idle rpm depends mainly on how well the system can predict the engine needs, based on its "knowledge" of operating conditions. But Open Loop mode has no feedback—it doesn't know the answer to the question, "How am I really doing? (See Fig. 2-28.)

Open Loop operation affects the Air/Fuel mixture during starting, warmup, W.O.T. acceleration, and deceleration.

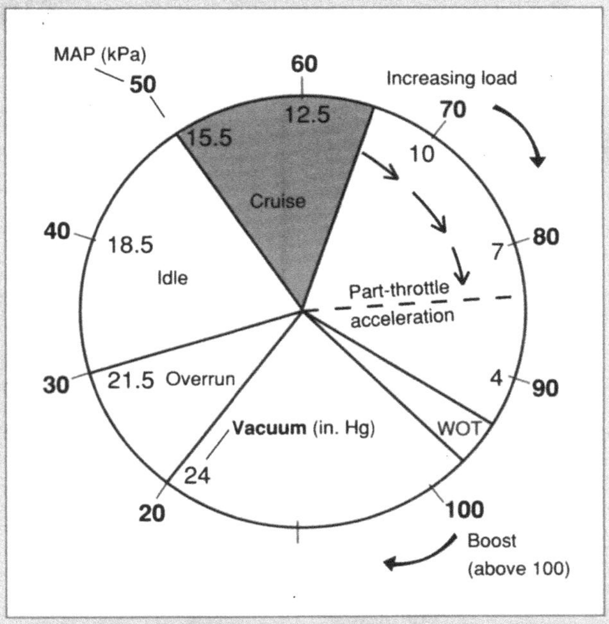

Fig. 2-27. For any midrange rpm, horsepower is roughly proportional to manifold pressure. Corvette MAP readings differ, depending on many factors, but the values shown are typical at or near sea level. Above approximately 1000 feet, expect lower numbers. The sectors indicate modern, tight engine MAP values.

Manifold Absolute Pressure (MAP) *increases* with load while the vacuum *decreases*. Actual Corvette values may vary, especially with engine modifications. Above about 1000 feet altitude, expect lower manifold pressures. The different values generally indicate the various engine states:

Overrun—during deceleration, MAP is at it's lowest, typically at 20-30 kPa

- **Idle**: 30-40 kPa (perhaps 50 with a wild cam), this indicates closed throttle loads, for example, when braking at a stop while in Drive (automatic transmission), or with the Air Conditioner compressor on
- **Cruise**: 50-65 kPa, light acceleration, and increased fuel economy
- **Part-throttle acceleration**: 65-90 (or more) kPa; Closed Loop operation runs, depending on conditions, and then usually Open Loop begins, if Throttle Position Sensor signals above 80% open.
- **Wide Open Throttle**: (W.O.T.) around 90-98 kPa; Open Loop runs; the higher this MAP, the higher the intake airflow, and the lower the resistance of the intake air ducting and air cleaner.

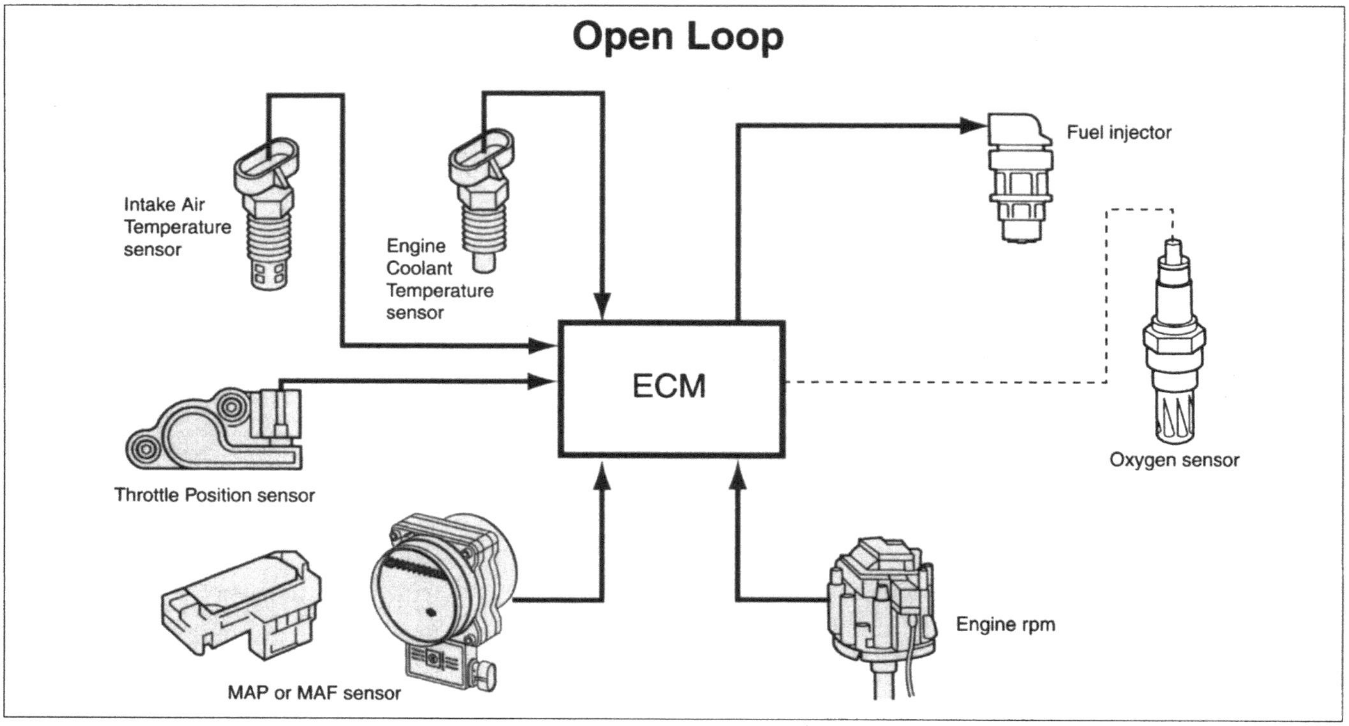

Fig. 2-28. During Open Loop, the computer determines the injector pulse times for what the engine *should* be doing, based on many sensors and memories. Oxygen sensor signals are part of the switch into Closed Loop.

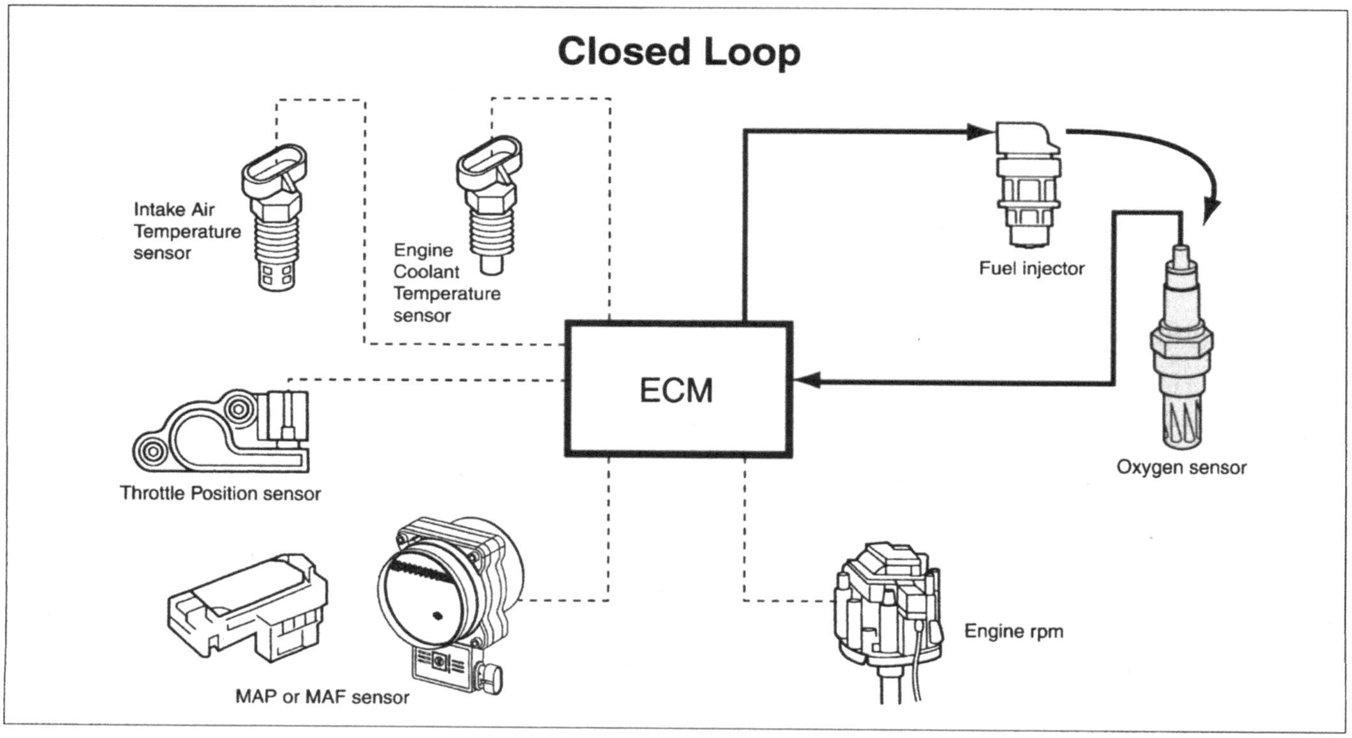

Fig. 2-29. During Closed Loop, the computer determines the injector pulse times based on oxygen sensor signals that sense what the engine *is* doing.

5.2 Closed Loop

Closed Loop control is normal for most operation. In a Closed Loop control system, information about whatever is being controlled is continuously fed back to the system as an input. Based on that feedback, the computer controls the actuators, to achieve the desired results.

When the air/fuel ratio control operates in Closed Loop (See Fig. 2-29), the system compares the Oxygen Sensor signal to the desired air/fuel ratio. Measurement is indirect, based on the Oxygen Sensor signal to the control module. The control module varies the injector pulse time to maintain the target air/fuel ratio.

Ask a hundred car guys about Closed Loop, and ninety-nine will respond "*Fuel* control," reflecting the usage in most manuals. That's unfortunate, because Closed Loop is more than fuel control. Several Corvette engine control systems, other than the Air/Fuel control, operate with Closed Loop.

Fig. 2-30. A home thermostat is a Closed Loop system, using feedback from the thermometer (sensor) to control the furnace or air conditioner (actuator).

On a thermostat, you set a target temperature—what you want the heating system to deliver. As the room temperature falls below the target temperature, the thermostat senses a lower temperature and automatically signals the furnace. When the temperature rises to the target setting, the thermostat senses the results and signals the furnace to cut back the heat—a Closed Loop system.

Knock control also operates in Closed Loop mode. (See Fig. 2-33.) Signals from the Knock Sensor cause the computer to retard spark timing (injectors may be cut off). When the knock signals stop, the actuators gradually return to normal. To prevent engine damage, the Knock Sensors are never operated Open Loop.

Idle-rpm control operates Closed Loop too, comparing idle rpm to target rpm. (See Fig. 2-34.) When the rpm is too low, idle-air is increased to prevent stalling. When the idle rpm is too high, idle-air is decreased.

Cruise Control uses Closed Loop for operation of the throttle. (See Fig. 2-35.) The computer compares the actual speed to the target speed and opens or closes the throttle accordingly. When the Cruise Control is disengaged, the throttle is operated in Open Loop, by the driver.

I like to describe Closed Loop as the equivalent of having a skilled technician riding under the hood, continually tuning the mixture and adjusting spark timing and other engine functions for best operation.

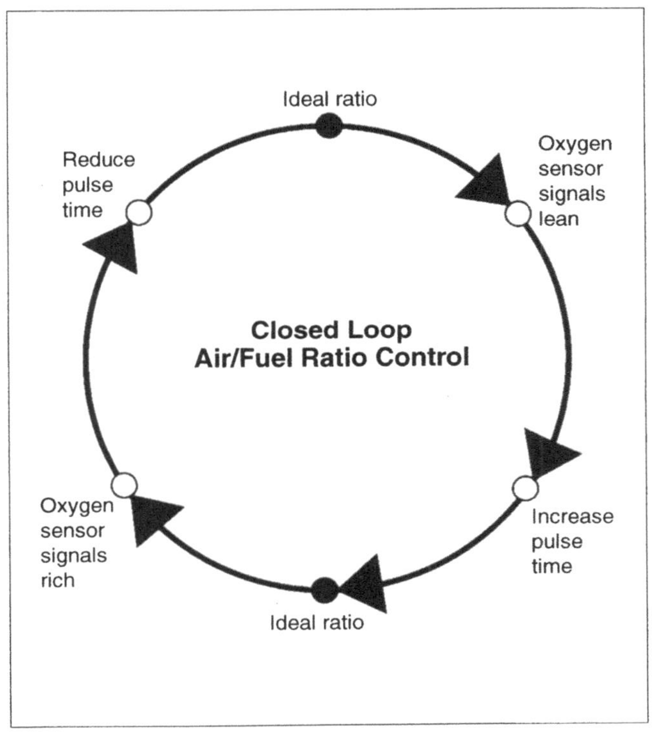

Fig. 2-31. During warm cruise, the computer controls the air/fuel ratio with Closed Loop, by sensing oxygen in the exhaust (an indirect measurement of the air/fuel ratio) and changing injector pulse time.

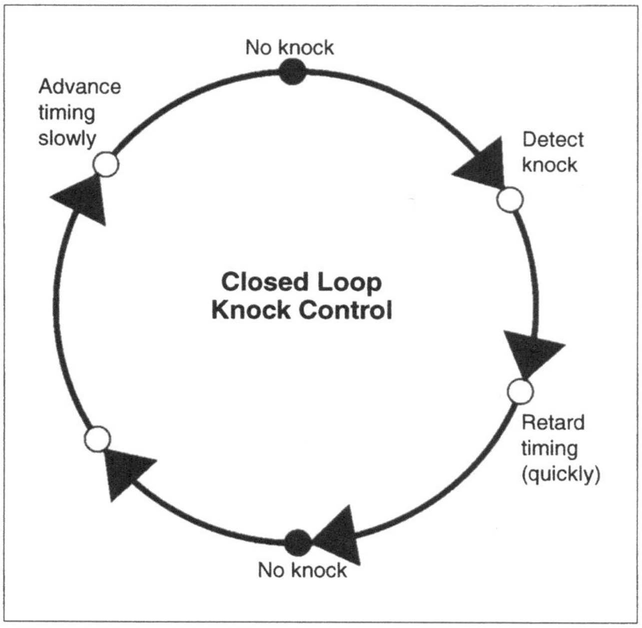

Fig. 2-32. Knock control operates in Closed Loop, to retard spark timing and cut off injectors.

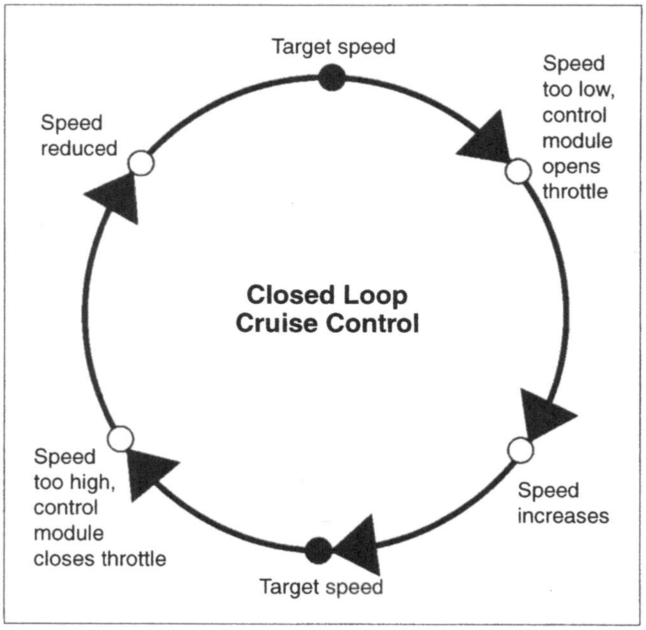

Fig. 2-34. Cruise Control operates in Closed Loop, comparing the actual speed (from the sensor) with target speed (from memory).

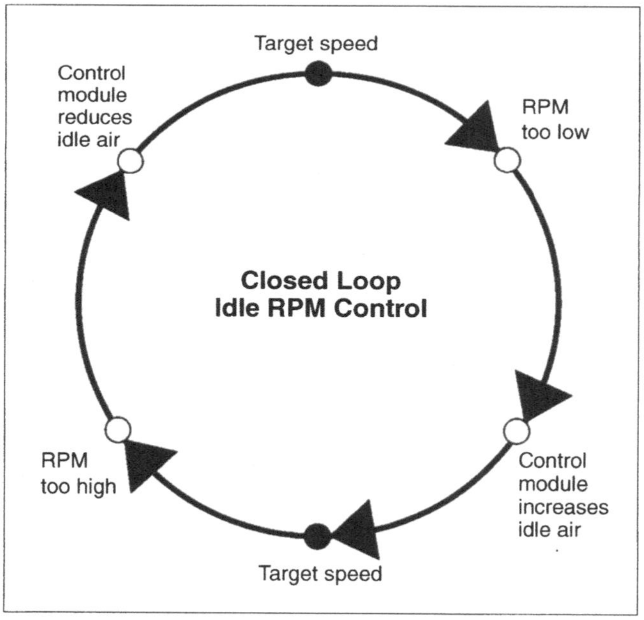

Fig. 2-33. Idle-rpm control operates in Closed Loop, comparing actual idle rpm to target rpm.

5.3 Feedback/Feedforward

Feedback is what the computer receives from the sensors. It looks back to monitor the results of its outputs—a sort of "How'm I doing?"

Feed*forward* is a relatively new term, the counterpart of feedback. Feedforward is when signals are sent to the computer which anticipate changes *about to happen*—a sort of "What should I be doing before X occurs?" Generally, feedforward signals relate to engine idle rpm, and therefore to idle-air control. Feedforward signals from engine accessories (air conditioning, power steering) allow the computer to anticipate a rpm drop caused by load, and prevent it instead of reacting to it. The result allows a lower idle engine rpm without stalling, saving fuel and reducing emissions.

6. Operating Modes—Strategies

If I were talking about the requirements of a stationary industrial engine, I'd expect it to operate under fixed conditions: constant rpm, constant load, nearly constant temperature, limited stops and starts, no acceleration, and no heavy-footed driver. Such an engine would operate quite nicely at a fixed air/fuel ratio and fixed spark timing. It could easily be tuned to optimize fuel economy and emissions and would require only the simplest of engine control systems.

Automobile engines are a different story. We expect them to perform under a wide variety of operating conditions. We also expect them to meet these demands effortlessly, at any time, and at the turn of a key. We expect "driveability", and a smooth, trouble-free performance.

Corvette engine management may control many systems within a millisecond.

The computer's primary management includes:

- the air/fuel ratio
- spark timing
- idle rpm
- emission control (Exhaust Gas Recirculation (EGR), air injection to exhaust (AIR), and fuel vapors (EVAP))

While you're at it, the computer manages:

- the automatic-transmission's shift speeds and timing *or* the manual-transmission's 1-4 upshift (**'90-01)**
- the engine cooling fan(s) and engine temperature
- the air conditioner
- engine overspeed
- vehicle speed (including tire speed limitation)
- engine diagnostics (remember any fault codes)
- **'90-92:** EBCM Electronic Braking Control Module
- **'93-01**: traction control
- **'94-01**: **LT1**, **LT4**, **LS1**: OBD-II data and performs an active diagnostic while running

And then it sends information to:

- the nonvolatile RAM (NVRAM)
- the malfunction Indicator Light (MIL/"CHECK ENGINE")
- the trip computer
- the digital instrument panel

Fig. 2-35. The Corvette engine computer must manage many systems at a time.

And finally it:

- keeps track of battery voltage
- remembers the short-term corrections and adapts for long-term ones

2

Corvette uses the term "mode of operation" to describe the different conditions affecting engine control. In this overview, to set the stage for the following chapters, I'll briefly describe the requirements of engine operating strategies—conditions as they affect fuel, spark, idle rpm, and emission control. (Look for more details of strategies in Chap. 8.)

You're already familiar with Closed and Open Loop operation, but there are others. Corvette lists the following strategies, called modes, which the engine control employs: Run (Closed Loop or Open Loop), Starting, Clear Flood, Acceleration Enrichment, and Fuel Cut-Off. Battery Voltage Correction is also considered a mode (though it's not a separate one, as it applies in all strategies). Later models include Fuel Backup, Deceleration Enrichment, Power Limitation, Catalytic Converter Overheat, and Overspeed Limitation. The following is a breakdown of some of the more prevalent modes which are used in your Corvette.

6.1 Closed Loop

When you're tooling along, the engine is operating under relatively constant conditions, but it can handle variations. The fuel quality may vary from one tank fill to the next. Weather and outside temperature may change. It may be dry or it may be pouring rain. Injector pulse time and spark timing may be slightly different for each of these conditions.

The computer can adjust to these changing conditions with little difficulty. Closed Loop can maintain an air/fuel delivery near the ideal ratio, and advance spark timing close to knock. It can satisfy the most important considerations of low exhaust emissions and fuel economy.

Closed Loop requires that the oxygen sensor be operating properly and be hot enough to keep switching the voltage between rich and lean. In addition, the coolant temperature must be above 105°F (40°C), and the engine must have operated for a specified time from ignition (generally 10-90 seconds), varying according to the engine control system. (See Fig. 2-30.)

6.2 Open Loop

Open Loop, visually, looks the same as Closed Loop, but operates quite differently. The most obvious use of Open Loop is at engine start. It also generally operates whenever Closed Loop just won't do the job. Unlike Closed Loop, Open Loop does not use Oxygen Sensor input. Rather, the computer calculates the fuel injector pulse width from the inputs of Intake Air Temperature (IAT), Engine Coolant Temperature (ECT), Manifold Absolute Pressure (MAP) or Mass Air Flow (MAF),

Throttle Position sensor (TP), and engine rpm. The following is a list of situations when Open Loop runs:

- start and warmup of a cold engine
- Clear Flood
- W.O.T. Acceleration
- Deceleration/Fuel Cut-Off
- Converter Overheat
- Overspeed
- Fuel Backup
- Traction Control

6.3 Start and Warmup of a Cold Engine

You expect the engine to start instantly when you turn the key, unaffected by temperature or time. Even if the car is at subzero temperatures, or is parked in the desert, you want it to run. And when it's been sitting for five weeks in the garage, you expect it to start right up. Open Loop helps this happen.

Cold crank means that the engine is cold. The engine temperature might be:

- –40° F (–40° C), sometimes called cold-cold
- 115° F (45° C), sometimes called warm-cold

The latter may seem hot to you, as a 98.6° (37° C) human, but compared to its normal operating temperature of about 205° F (95° C) your Corvette's engine still thinks it's cold.

In starting, the computer calculates the pulse width and timing for the fuel injectors, without input from the oxygen sensor. (See Fig. 2-28.) Also, of course, it calculates spark timing and dwell.

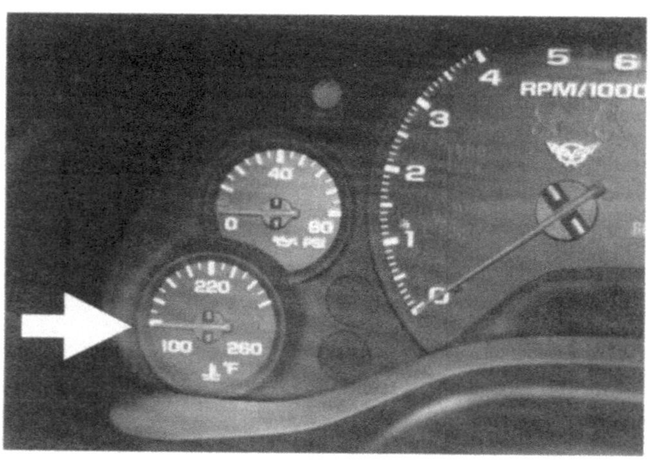

Fig. 2-36. 100° F means a *cold* engine. Savvy Corvette drivers know that if the engine has been off for an hour or more, fuel pressure may have fallen, increasing cranking time.

6.4 Clear Flood

Cranking enrichment quickly becomes a problem if the engine doesn't start right away. Prolonged cranking may produce flooding, an over-rich mixture at the intake manifolds and in the cylinders. Gasoline mixtures don't burn well if they're richer than about 10:1—or, for that matter, leaner than about 16:1. Spark plugs don't fire when they become fuel-fouled.

Foot-full-down cranking will usually clear a flooded engine by "sweeping it out" with air. When the engine catches, release the accelerator. If the rpm gets above 300-400 while you're cranking, the computer sees that as an engine start and switches to a run mode.

6.5 Wide Open Throttle Acceleration

The computer can provide full power without regard to economy or emission control. In Wide Open Throttle (W.O.T.) operation, Closed Loop switches to Open Loop. Fuel injection is enriched, and spark timing is advanced to increase power. With the rich mixture, extra advance can be programmed with less risk of detonation. Emission control stops—both EGR and air injection. The air conditioning compressor and cooling fan(s) are also turned off for 10 seconds.

Generally, depressing the accelerator up to 80%, to hold speed on an upslope or speed up some, continues Closed Loop operation, with the Oxygen Sensor functioning as the principal sensor. But pressing that last 20% says to the computer, "Don't bother with fuel economy. Forget emissions. Give me Power!"

Fig. 2-37. Wide Open Throttle acceleration is another Open Loop strategy. The computer provides full power without regard to economy or emission control.

6.6 Fuel Cut-Off

In Fuel Cut-Off mode, the computer may cut off one or more of the injectors. In later models, it may also retard spark timing or cut off spark to some of the plugs. This is, of course, Open Loop. The cutoff is a soft one, since a sharp one could throw you forward in your belt, like sudden braking does. Fuel Cut-Off occurs under the following conditions:

- engine overspeed
- car overspeed, often for tire ratings
- a coolant temperature over 255° F (125° C)
- rapid or long deceleration, such as coasting down a long hill, which prevents raw fuel from entering the catalytic converter(s)
- traction control, when the rear wheels slip under power
- no reference pulses sent to the computer
- key off, which cuts fuel to prevent after-run, sometimes called dieseling

6.7 Fuel Backup/"Limp Home"

When the computer, itself, is not operating normally, engine management changes to Fuel Backup, or "limp home" mode. This could be under any of the following circumstances:

- a computer voltage less than 9 volts
- a cranking voltage less than 9 volts
- missing or nonfunctioning Programmable Read-Only Memory (PROM)
- if the computer fails to issue Computer Operation Pulses (COP)

When one of these situations occurs, the computer runs on something called the CAL-PAK chip, operating the engine only on inputs from a few sensors: Manifold Absolute Pressure, Throttle Position, Engine Coolant Temperature, and rpm.

2

6.8 Catalytic Converter Protection

Catalytic Converter Protection, used in the **LT1**, **LT4**, and **LS1**, detects overheating, as computed from the Oxygen Sensor signals. The computer switches back to Open Loop, to enrich the mixture. With less oxygen to burn, because of unburned fuel (HC) and partially-burned fuel (CO) in the exhaust, the converters operate at a lower temperature.

6.9 Battery Voltage Correction

The engine computer is sensitive to low voltage, and the Driver Information Center in late-model Corvettes will warn of 10 volts or less, reading "Low Voltage." At 9 volts or less, the instrument panel and other systems may shut down. Continued operation under 9 volts may destroy the computer.

If the battery voltage falls too low, slower fuel-pump speeds may reduce the fuel delivery to the injectors. The computer increases the injector pulse time to compensate, and can raise the idle rpm to increase generator output. Electronic ignition in the **LS1** and **ZR-1** can increase dwell time, to improve

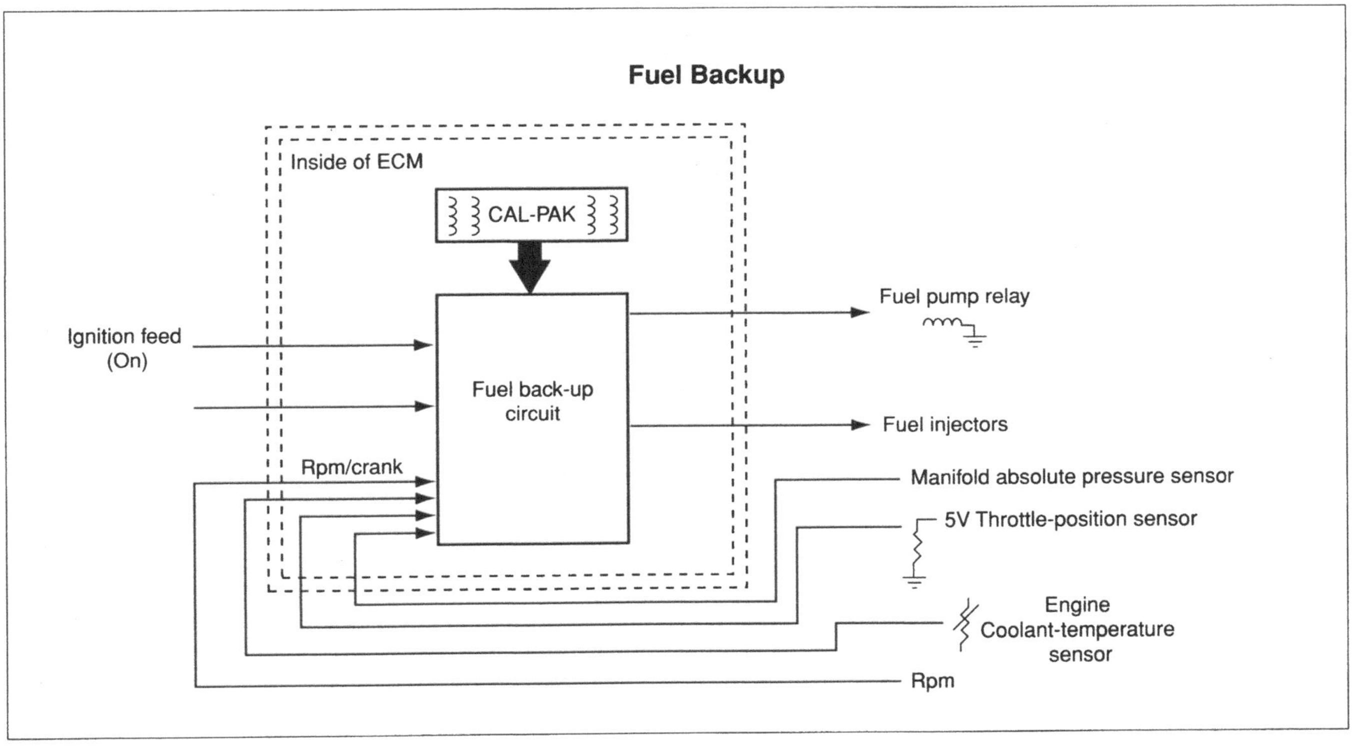

Fig. 2-38. When the computer does not operate normally, engine management changes to Fuel Backup, or "limp home."

coil saturation. Battery Voltage Correction is sometimes listed as a "mode," but it actually works in all modes.

All Corvettes since 1956 operate on 12 volts—that means all Corvettes covered in this book. Expect 42-volt systems in the future, to handle all the increased electronics.

6.10 Traction Control

Traction Control, used in the **LT1**, **LT4**, and **LS1**, is a strange combination of Wide Open Throttle and Fuel Cut-Off that reduces the tendency of slippage in the rear wheels, during W.O.T. acceleration. It retards the spark and/or closes the throttle, to decrease power to an amount the tires can handle. It may operate at Wide Open Throttle on dry road surfaces or even under medium acceleration on dirt, gravel, ice, snow, or wet roads.

7. Idle-Speed Stabilization

Usually, idle should be the lowest speed at which the engine runs smoothly. Decreasing idle speed reduces noise and fuel consumption. But the biggest obstacle to low idle speeds is the variation in load on the engine at idle. At idle, small changes have big consequences:

- Gross engine output at idle is about 5 horsepower.
- The power required to turn the engine, at idle in Park or Neutral, is about 5 hp.
- So, net output at idle ends up at about zero.
- But, the air conditioning compressor uses about 2 hp, that's 40% of the idle engine's output.

The result? If the air conditioning switches on, the engine will stall.

The load from the power steering pump can also affect idle rpm, especially during parking. With an automatic transmission, shifting into Drive or Reverse at idle increases load. The power required to operate the charging system varies with electrical load (headlights and rear-window defrost, for example). And friction loads change with temperature. But the computer monitors and constantly corrects idle speed, so it's not necessary to maintain a high idle, just to handle the variations in engine performance.

L83, L98: Even when the engine is warm, Closed Loop operation may switch to Open Loop, if the engine idles for too long. If you are working in Closed Loop for some service checking, and it goes into Open Loop, it won't send a signal, and you may not even realize that the modes have switched. **ZR-1**, **'94 -01 LT1** , **LT4**, **LS1** engines have *heated* Oxygen Sensors and they tend to stay in Closed Loop, even for long idles.

Table 2-2. The effect of added load on idle rpm

Condition	Automatic in Neutral/Park; Manual in Neutral or clutch out	Automatic in Drive or Reverse
Normal, AC off, power steering normal, straight or moderate turning	Lowest rpm for engine smoothness	Lowest rpm for minimum creep
Normal, AC ON, power steering normal, straight or moderate turning	Increased AC compressor rpm for required cooling	Increased rpm for AC cooling, but minimum creep
Only during A/C compresor switch-on, or power steering pressure increase, hard over	Increased rpm during AC compressor switch-on or power steering to prevent engine stall	Delayed idle speed increased to avoid unexpected surge
"Hot" ECT 110°C (225°F)	Increases idle rpm to cool engine	Delayed idle speed increased to avoid unexpected surge

8. Changes in Engine Conditions

The computer continuously adjusts the amount of fuel injected, to correspond to the amount of oxygen in the engine exhaust. The computer fine tunes the air/fuel ratio, as dictated by the MAP or MAF sensor. As the engine's mechanical condition changes with time, the feedback system can adapt, changing the air/fuel ratio. Perhaps a leaky valve is causing a change in combustion, so a bit more fuel is required in that cylinder. The closed loop fuel system will adapt to this change. It will also adapt to certain changes, in fuel quality or oxygen content, that may affect combustion.

8.1 Adaptive Control

The computers in **'85-01** Corvettes (beginning with the **L98**, with full implementation in the **'87** models) provide Adaptive Control for the air/fuel ratio. It's a Closed Loop function and operates two ways:

Short-Term Fuel Trim (STFT) (also known as "Block Learn") must operate quickly and continuously to maintain the air/fuel ratio close to the stoichiometric ideal. You might think of it as dealing with the momentary changes in engine conditions during driving, such as changing oxygen levels in the exhaust. But if the Short-Term Fuel Trim continues for more than ten seconds at more than ±10%, the computer will decide that the condition is not momentary, and initiates a long-term correction.

Long-Term Fuel Trim (LTFT) (also known as the "Integrator") compensates for changes in the engine over time. This includes STFT, at more than ±10% for more than ten seconds. If a valve is leaking slightly or an intake air leak is present, an Oxygen Sensor notices the change in combustion, and brings the system back within its limits. However, changes beyond the system's range can lead to driveability problems.

Disconnecting the battery usually means losing the memory that stores the adaptive learning. After disconnection, the learning process must begin all over, and your car's performance may change. To help the computer relearn, after the disconnection and reconnection, drive around under different conditions with the engine warm: at part throttle, in "crowd" (a little push above steady cruise), and at idle. How long does it take to relearn? Usually 5–10 minutes. If the computer were programmed to take longer, operation could be unsatisfactory. If it were programmed to relearn in 1 or 2 minutes, the computer could be adjusting itself for short-term changes that don't count. **'94-01** Corvettes, with the second generation On Board Diagnostic (OBD-II) and complex drive cycles, usually take longer.

9. Diagnosis

Trouble codes are stored in one of the computer memories, known as Random Access Memory (RAM). RAM is a volatile memory, it's not permanent and can be erased after serving its purpose.

If an engine-control problem arises, the computer is programmed to make do until the problem can be repaired. The memory in **'82–93** Corvettes includes a range of acceptable values from each sensor. An input signal outside that range may cause a Check Engine or Service Engine Soon signal from the computer, which will store the proper trouble code. The engine operates on a fixed "limp home" temperature until the sensor can be serviced. This means that a failure in the coolant temperature sensor may not even be noticed. The engine will operate quite well, so long as conditions do not change. You'll have a hard start if it cools off, and with fuel injection, as you know, no amount of accelerator pumping will change that. The stored trouble code will indicate proper replacement of the sensor or its wiring harness or connector.

Here's a warning, though: if you try to play games with the sensor inputs, you'd better know what you're doing, because the computer may be programmed to detect improper sensor input signals.

9.1 On Board Diagnostic (OBD)

The diagnostics in **'82–93** Corvettes came to be known as OBD-I, or On Board Diagnostic-*first* generation. A hard fault in an emission related system sets a code and turns on the SES/MIL light. A "hard" fault means it's not intermittent.

'94-01 Corvettes are a whole new ball game in diagnosis. OBD-II (On Board Diagnostic, *second* generation) looks at various inputs and outputs, so that emissions could be increased to 1.5 times the standard. In OBD-II, "system performance" is the key idea. This makes a huge difference, as you'll find in Chapter 10, Diagnosis.

In this chapter you have seen:

- the basic factors of air/fuel ratio, spark timing, and idle air control
- air intake control, as it relates to engine control
- the importance of different air pressures and fuel pressures to engine control
- several Closed Loop systems in engine control
- concepts of feedback and feedforward
- why various engine operating modes demand different air/fuel mixtures, spark timing,and idle-air control

In the Next Chapter:

Now you've got the fundamentals. In Chapter 3, I'll show you how Corvette engine control is affected by important considerations of emissions and fuels.

Chapter 3

Emissions and Fuels

3

Contents

In this chapter, look for:

- how engine emissions, exhaust, and fuel vapor create smog
- how engine controls reduce emissions while meeting performance expectations
- how your Corvette will be tested for emissions, particularly with new tests in the geographical areas that do not attain Clean Air standards
- how your Corvette will respond to changes in fuels related to starting, fuel performance, economy, and driveability

1. Introduction

In Chapter 2, you saw that the control of emissions is an important part of engine-control strategies, now we're going to talk about why. No matter how much performance we want out of a Corvette engine, if it's going to be street legal, it's got to be clean. All the electronic fuel-injected cars (that's **'82-01**, so far) were emission-tested at manufacture. With tighter limits and improved testing, emission control is becoming a key factor in engine design and maintenance.

Green Standards

"Green" is the worldwide term for an increasing concern for clean air and clean environment. Green concerns affect the Corvette you buy or modify, and the fuels you burn in it. Running a clean car is so important that I will emphasize emissions and economy in this book, even for performance modifications in Chapter 9.

Emission control can get pretty complicated, and you may be asking, "Why be Green?" Health impacts, for one thing.

- **Ozone**, an element of smog, can damage your lungs, making it harder to breathe.
- **Carbon monoxide** can reduce the oxygen flow to your brainand the rest of your body, which is obviously a bad thing.

Acid pollution is another concern. Exhaust gases contain nitric acid and sulfuric acid, leading to acid rain. In Los Angeles, the smog can be as acidic as lemon juice—imagine that in your lungs! Trees and other plants die, and lakes and streams become too acidic to support fish. Buildings are eaten away, especially marble.

2. Emission Control

With changing legislation and tougher regulatory standards, engine control systems have undergone significant changes. Emission standards have been tightened, but modern electronic engine control still provides the power and driveability demanded by Corvette owners.

The air we breathe, and the air the engine breathes, is about one-fifth oxygen (O_2). Oxygen, combined with the HC from the fuel, is what burns in the cylinders, which is how your engine is powered. But one-fifth isn't all that much. Actually, most of the air going into the cylinders, about four-fifths, is nitrogen (N_2).

Nitrogen is an *inert* gas—that is, it tends not to combine with other gases. Ideally, it would pass through the engine without change, just as it does through our lungs. But as you'll see, when it gets hot enough in the cylinders, nitrogen can burn with the oxygen, and produce one of the more troublesome pollutants in the exhaust—oxides of nitrogen. Think about it, when you heat bread just right, you get tasty toast. When you overheat bread, you get black stuff to scrape off. "Yecch!"

2.1 Combustion Byproducts

Combustion of the air/fuel mixture in the engine creates the gaseous byproducts that make up the exhaust. Some are relatively harmless, and some are definitely harmful. Traditionally, three exhaust gases have been controlled as the most harmful ones: Hydrocarbon (HC), carbon monoxide (CO), and oxides of nitrogen (NO_x). (See Fig. 3-2.)

Emission of these three gases is regulated by the Federal Clean Air Act of 1970. As revised in 1990, the regulations required reduced emissions beginning in 1994 and gradually tightened with new model years. Normal warm-engine operation has gotten so clean that one cold-engine start can cause

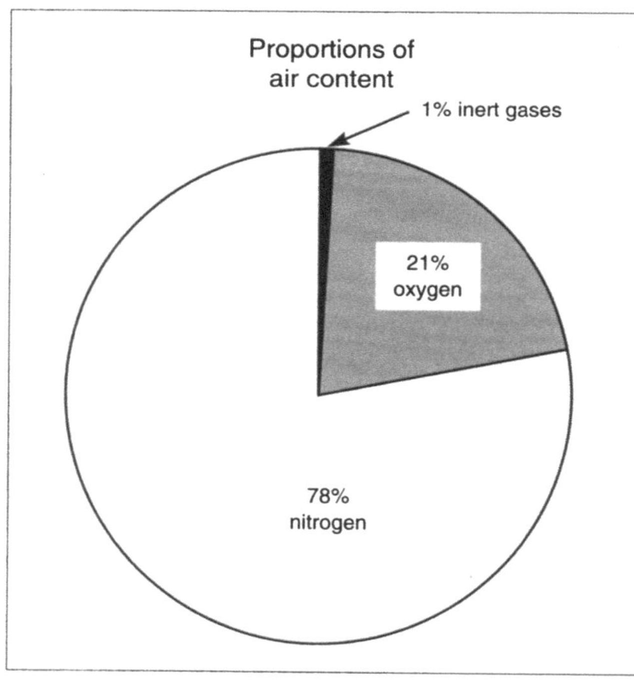

Fig. 3-1. Air intake includes about 21% oxygen, necessary for burning the fuel, and 78% nitrogen.

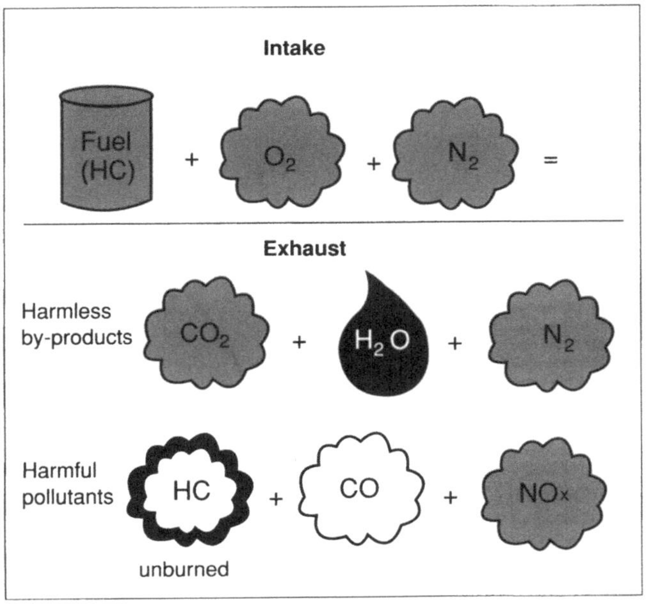

Fig. 3-2. (*Top*) The traditional emission equation: the engine takes in fuel (HC), oxygen (O_2), and nitrogen (N_2). (*Middle*) Combustion produces the relatively harmless byproducts: carbon dioxide (CO_2), water (H_2O), and nitrogen (N_2).
(*Bottom*) Some of the harmful pollutants produced are unburned hydrocarbon (HC), carbon monoxide (CO) , and oxides of nitrogen (NO_X).

more pollution than a lot of driving, so cold-start testing and emission control have become important. Also keep in mind that emissions depend on the motor vehicle *system*—both the powertrain/controls and the fuel being burned.

Exhaust gases are normally invisible. I know, you've probably seen "smoking" tailpipes, and mumbled something about all that pollution. However, what you see is probably not dangerous at all. It's usually just harmless steam, condensing from water vapor in the exhaust. (See Fig. 3-3.) I want to set you straight on pollution and tailpipes:

- A clear tailpipe is not a sure sign of a clean-burning engine. An estimated 10% of the cars are pumping out excessive *invisible* pollutants.
- A smoking tailpipe is not a sure sign of pollution. But if you see blue smoke, you're seeing an oil-burner, which is a polluter, but you won't fix it with the emission controls. The cure is engine overhaul.
- Significant amounts of HC can come from the fuel-vapor control (EVAP) system, even if the engine/catalytic converter(s) are running clean. Some HC can also come from refueling.

According to a recent SAE report I reviewed, the Environmental Protection Agency studied fifty high-mileage gross polluters, all running Closed Loop engine control. All failed the Inspection & Maintenance (I-M) test, some with ten times the allowable emissions. But none of these gross-polluting vehicles showed blue smoke at the tailpipe.

Fig. 3-3. Tailpipe steam is visible when water vapor condenses in cold air. Pollution is normally not visible, and what's visible is not necessarily pollution.

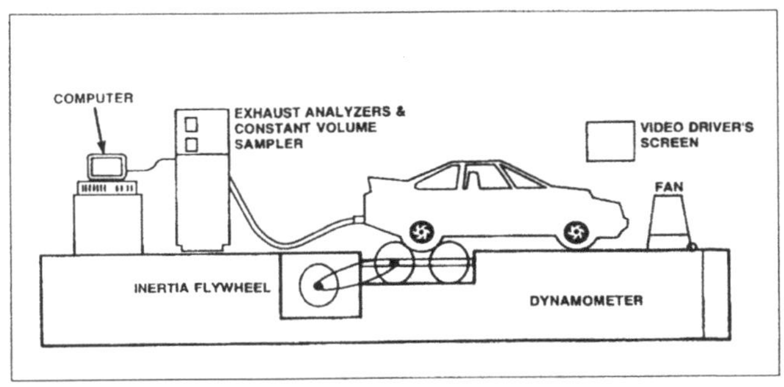

Fig. 3-4. For certification, emissions are measured under controlled operating conditions as the engine drives a dynamometer, a set of large rollers. Loading the engine is necessary to measure oxides of nitrogen (NO_x). An engine not under load produces no NO_x because the combustion is not hot enough.

Cars and light trucks sold in the U.S. must qualify under the Federal Test Procedure (FTP). Emissions are measured under controlled operating conditions on a dynamometer, or "dyno"—a set of large rollers, usually installed in the floor. (See Fig. 3-4.) Loading the engine by running the car on the dynamometer is necessary to measure NO_x. An engine operating without load produces no NO_x, because the combustion doesn't get hot enough. Such tests cannot be performed on the road, because variables in weather and driving can affect the results. (See Fig. 3-5.)

In the beginning, "cold" was defined as *engine*-cold, which is around room temperature, about 70–90° F (20–30° C). Beginning in 1996, the cold start is real winter cold. The car and engine are tested after being refrigerated down to about

15° F (–7° C). Other recent changes made, to reflect real conditions, include aggressive driving, air conditioner on, and "microtransient" throttle changes.

Here's the engine emissions story, one gas at a time (See Fig. 3-6.):

Hydrocarbon (HC): Unburned gasoline—that's HC. Gasoline is a mixture of compounds of hydrogen and carbon. Unburned gasoline—usually from misfire—comes out of the engine as HC. As the engine accumulates miles, other causes of HC include poor compression, vacuum leaks, incorrect timing or dwell, or a rich or lean mixture. As hydrocarbon limits tighten, HC out the tailpipe can also result from faulty engine design—combustion-chamber pockets, like the upper compression ring, where gasoline can "hide," preventing it from being burned.

HC emissions can result from a mixture that's either too rich or too lean. How's that?

- Rich mixtures cause HC, of course, from unburned fuel after the oxygen is used.
- Lean mixtures may cause misfires, which causes a similar situation to a rich mixture.

If the HC burns completely in the cylinders, it combines with oxygen to form water (H_2O) and carbon dioxide (CO_2). But it seldom burns completely. Catalytic converters can greatly reduce HC emissions.

Some concentrations are expressed as percentages and others as parts per million (ppm). What's the difference?

- Percentage is parts per hundred, and is used for CO, CO_2, and O_2.
- Parts per million provides a much finer readout, and is used for HC and NO_X.

Concentrations of these gases are so small that we have to look at a million parts to find them.

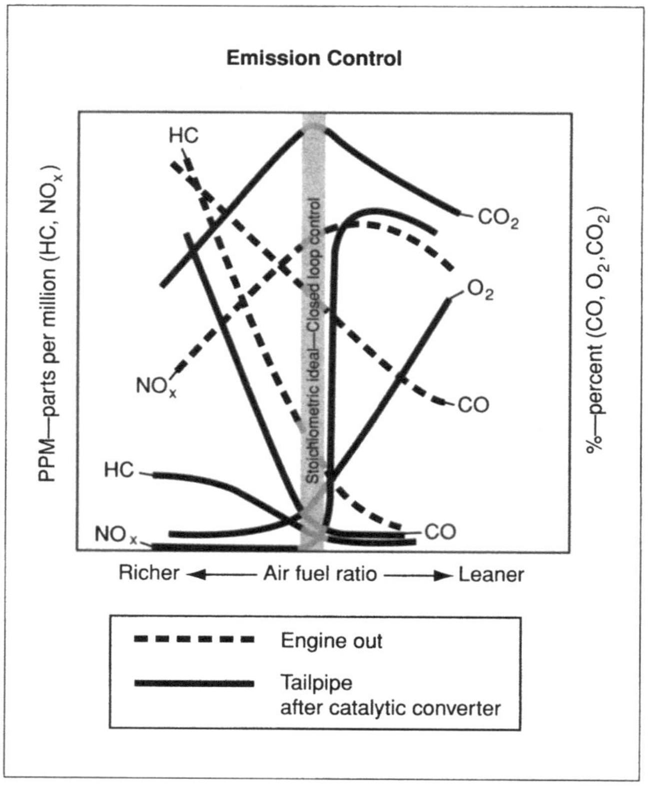

Fig. 3-6. The five gases show engine out, and tailpipe exhaust after the catalytic converter, (during Closed Loop, light load cruise, and with no EGR). If the engine runs rich, the exhaust shows the worst combination of HC and CO. Notice that high Carbon Dioxide (CO2) is a good indicator of stoichiometric combustion, combining C and O_2. Note: these readings are representative, not from any actual Corvette.

It pays to look back at this diagram during this section, as you consider each of the five gases.

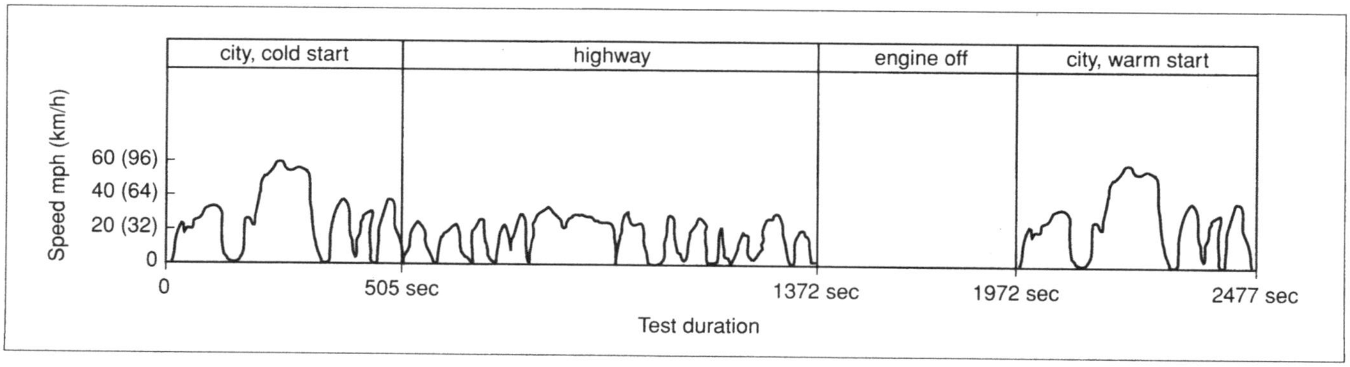

Fig. 3-5. The FTP test pattern was devised to duplicate a typical Los Angeles commute, including city and highway driving. The pattern remained unchanged from 1975–95, and obviously did not relate to cold weather.

City, cold start represents a typical drive (in LA), including stops and idling, as in traffic.
Highway is a longer run, still with stops, and more traffic.
Engine off kills the engine for 600 seconds.
City, warm start gets averaged with cold start.

Carbon Monoxide (CO): Partially burned gasoline produces CO, a poisonous gas resulting from incomplete combustion, generally from a rich mixture. CO is highest when the oxygen in the cylinders is low, because gasoline is only partially burned.

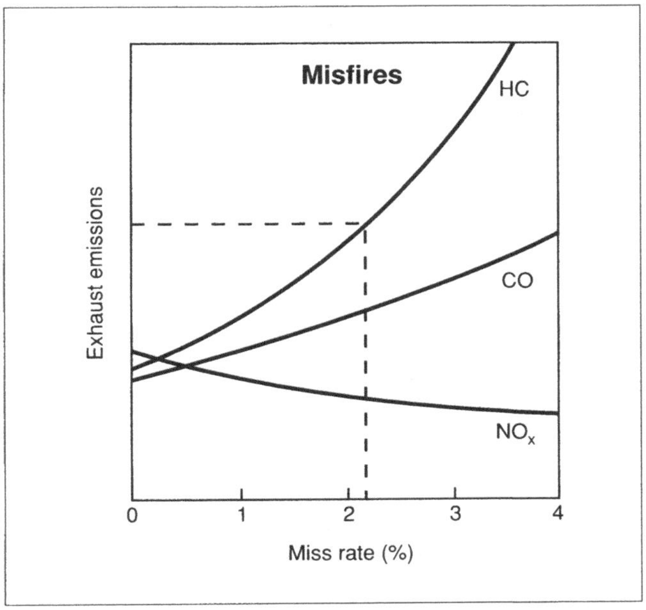

Fig. 3-7. In misfires, HC (unburned gasoline) and CO (partially burned gasoline) are high, even though NO_X (from high-temperature combustion) is down. That's why misfire detection is an important part of Corvette computer diagnostics.

Oxides of Nitrogen (NO_x): Nitrogen burns in the cylinders when the combustion chamber temperature exceeds about 2500° F (1400° C). (See Fig. 3-8.) What's measured is a mixture of oxides of nitrogen that we call "", it is a key element of smog. NO_x is worse with a slightly lean mixture, which burns hot. But, if the temperature in the cylinders stays low enough, nitrogen remains inert, passing through the engine and coming out the exhaust pipe as nitrogen.

> **NOTE —**
> "Idle" refers to *no-load* not rpm. Idle operation means low temperatures, so NO_X won't be apparent without load. Idle testing is restricted to HC and CO.

A five-gas analyzer sees two gases other than HC, CO, and NO_x. This data, read at the tailpipe, helps give readings on:

Oxygen: O_2 is low during the times that O_2 and CO combine to make carbon dioxide. Look for *low oxygen* in a healthy engine.

Carbon Dioxide (CO_2): CO_2 peaks when the air/fuel ratio is just right—not too rich, not too lean. Look for *high carbon dioxide* in a healthy engine.

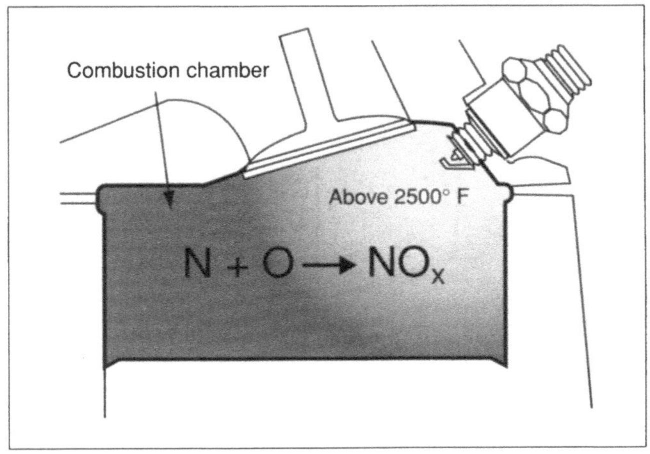

Fig. 3-8. When the temperature in the combustion chamber gets too high (above about 2500° F), the nitrogen also burns, combining with oxygen to form oxides of nitrogen.

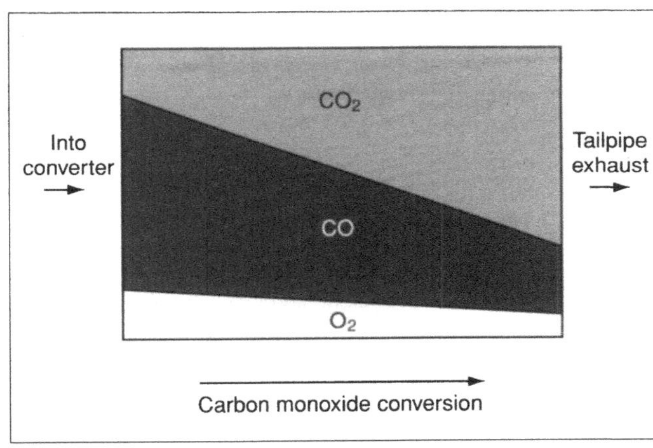

Fig. 3-9. Why do we measure CO_2, a so-called "harmless emission"? Because it's a good indicator of the conversion of CO and O_2.

Measuring Emissions

Before I go further, you should know that your car may be tested regularly with one of two different methods of measuring emissions: *concentration* , or *mass* related to distance. The method used can make a real difference in what you spend to keep licensing your car for street driving.

Limits have traditionally been in terms of concentration: What proportion of the exhaust gas is HC? What proportion is CO? But this doesn't tell us, in absolute terms, how much of each gas is present. Concentration tests favor large engines, since they put out lots of exhaust gases, of which only a small part is pollutants, it's more likely to test "clean" on a concentration basis.

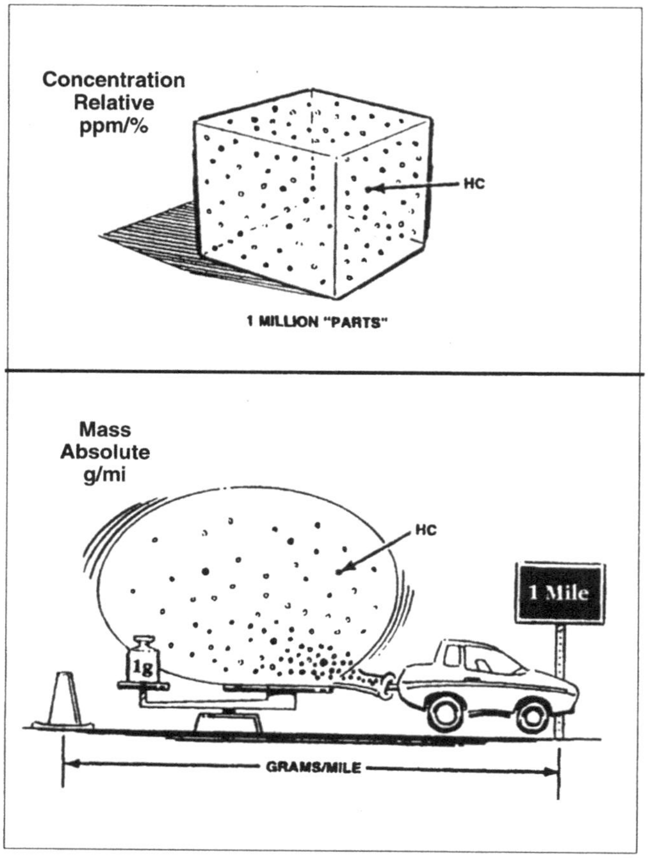

Fig. 3-10. Relative tests measure the concentration of the different exhaust gases, and are expressed as parts per million, or percent. (Top) The results will tell you what proportion is HC and what proportion is CO, for example.

Absolute tests measure the mass of the exhaust gas related to the distance traveled, and is expressed as grams per mile. (Bottom) The results will tell you exactly how much gas is emitted for each mile driven.

Some limits are defined as mass related to distance traveled: What is the actual mass, or weight, of HC? Of CO and CO_2? Of NO_X? Mass tests favor small engines, since they put-out less exhaust gas, but the concentration of pollutants can be larger and still be considered as "clean".

I think you can see the difference. Concentration is *relative*. We don't know how much the engine is producing—only what portion of the exhaust is HC and CO. We only know that the exhaust of car A is "dirtier" than that of car B, or that car C's is cleaner now than before its tuneup. Concentration measurements, the traditional tailpipe test, are chosen because we can make them with a simple exhaust gas analyzer. Concentration is usually measured with the engine *idling* (no load). So we can't measure the NO_X output of an idling engine, because none is being formed.

Mass is absolute and *distance* is absolute. We measure the mass, or weight, of each gas collected from the tailpipe. The test is done on a dynamometer, which measures how far the car would travel during the test, so the emissions can be measured as grams per mile. Each new car model is tested on a dynamometer for certification in g/mi (grams per mile), permitting comparisons in absolute terms. Beginning in the nineties, where the air is dirtiest—so-called "non-attainment" urban areas—cars and light trucks are also emission-tested this way.

Dynamometer testing is much more expensive and time-consuming than idle testing. Why is it necessary? The answer is the title of this book, "Fuel Injection—Electronic Engine Management."

- **'82-01** engines, with fuel Injection/electronic engine management, emit *significantly different emissions when under load.* With such modern cars, measurements at no load have little relationship to actual emissions under road-load.
- **Pre-'75** engines, with traditional carburetors and ignition, generally put out comparable HC and CO emissions at no-load (parked for emissions testing) as under road-load in actual driving. So a no-load test can be representative. But such emissions are much higher than allowed in later testing.
- **'75-81** engines with High Energy Ignition (HEI-1975) and **'80-81** engines, with electronic control of the carburetors, fall somewhere in between.

In other words, the two-speed idle testing, designed for older cars, is relatively meaningless in modern cars.

Loaded testing may include:

1. *Transients* of acceleration and deceleration are measured in absolute measurements. Dynamometer testing provides emission information and distance information (g/mi). Each test bay, known as an IM240, may cost $150,000-200,000, and is not likely to be found in your corner gas station.
2. *Steady-state* is measured on a concentration (percent or ppm) basis. Acceleration is simulated by steady-state speed-load combinations. The Acceleration Simulation Mode (ASM) is a lower-cost ($50,000-60,000 per test bay) alternative to transient testing. Steady-state tests include:
   - *25/25* tests are done at 25 mph, at 25% load, and are best for monitoring HC and CO.
   - *15/15* tests are done at 15 mph, at 50% load, and are best for monitoring NO_X.

   Imagine if you were held at a fixed speed of 15 mph, while applying 50% throttle; the engine control/emissions would simulate *acceleration*. 25/25 tests are similar but are more like a speed test.

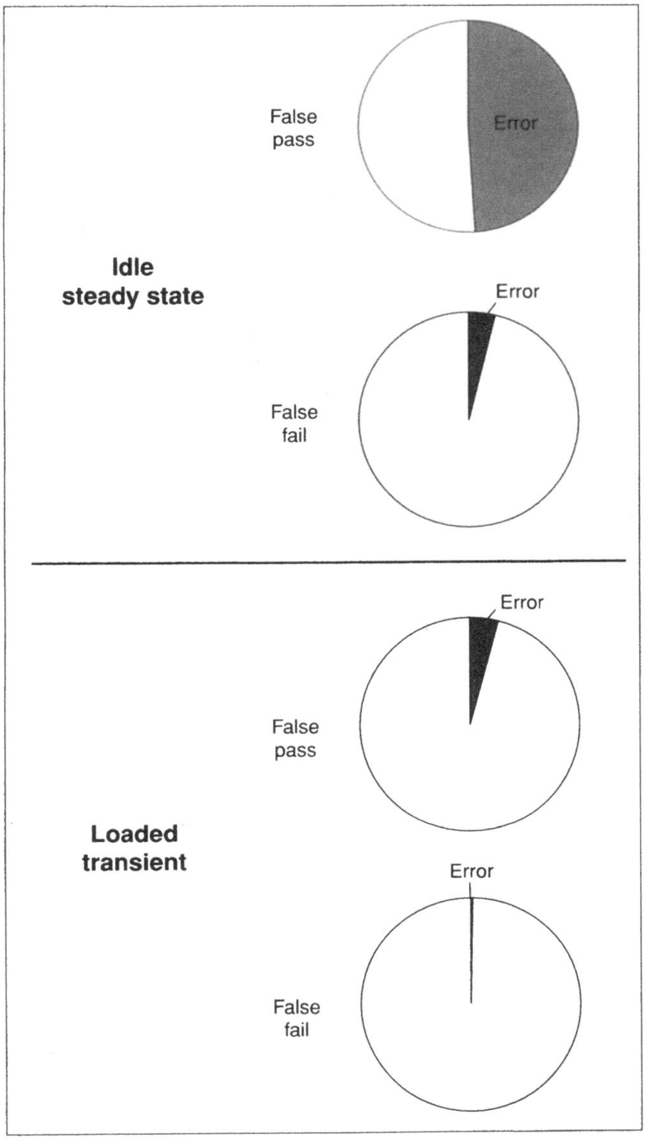

Fig. 3-11. Idle and steady state tests are likely to false-*pass* (certify a "dirty" car) about half the time, and false-*fail* a clean car about 4-5% of the time. Loaded, transient tests, such as the USEPA Federal qualification tests, show more accurate results. They false-*pass* about 4%, and false-*fail* almost never.

Harmful Emissions—Controlled

HC, CO, and NO_X are the classic emissions defined in the early 1970s. Recently, the discovery of how other gases contribute to pollution has produced a more precisely defined set of emissions. I'll cover those later in this chapter.

Recent measurements show that HC comes not only from the tailpipe but also from the fuel system. *As much as 40% of a car's HC emissions may come from the evaporative control system (EVAP)*, fuel tank, lines, and especially the filler pipe and cap. Obviously, we never see these emissions in *tailpipe* measurements.

Relatively Harmless Emissions—Not Controlled

CO_2 is important to emissions diagnosis because it is an indicator of complete combustion. When CO_2 is at its highest, you know that the air/fuel ratio is close to ideal (stoichiometric) and that the converter(s) are working.

Until recently, CO_2 was considered a harmless emission. But recent studies show it's accumulating in the upper atmosphere, trapping global heat, much as glass traps heat in a greenhouse. "Greenhouse gases" and carbon dioxide are now recognized by most international environmental scientists as real contributors to global warming.

Fig. 3-12. The underhood label, such as on this **'95 ZR-1** Corvette, identifies emission controls, and guides tune-up procedures. This is a 50-state OBD-I car. Other cars may be identified as California only. 49-state cars may be registered in California, but only under certain conditions.

The amount of CO_2 coming from the exhaust pipes is astonishing, even if invisible. Burning 1 gal (6 lbs) of fuel produces about 18 lbs of CO_2, or 750 cubic feet (at standard temperature and pressure)—that's about three times the volume of a Corvette! (See Fig. 3-13.)

The burning of fossil fuels—oil, gasoline, or coal—produces CO_2, and lots of it, as do wood stoves and forest fires. Unlike other pollutants, CO_2 cannot be treated to eliminate its polluting effects. Most experts consider that global warming of only a few degrees would have disastrous results, including rises in global temperature, successive heat waves, and iceberg melting, which could raise ocean levels and flood seaside properties.

What can we do in driving to reduce CO_2? Avoid unnecessary idling, for one thing. Turning off your engine when parked is a simple thing you can do. And of course, when you drive, better fuel economy means less CO_2.

Emission Limits

Why do government agencies keep tightening limits on cars and trucks? Haven't we done our share? Compared to pre-control, our modern cars emit 86% less HC, 92–96% less CO, and 76% less NO_X. But in that same time, Vehicle Miles Trav-

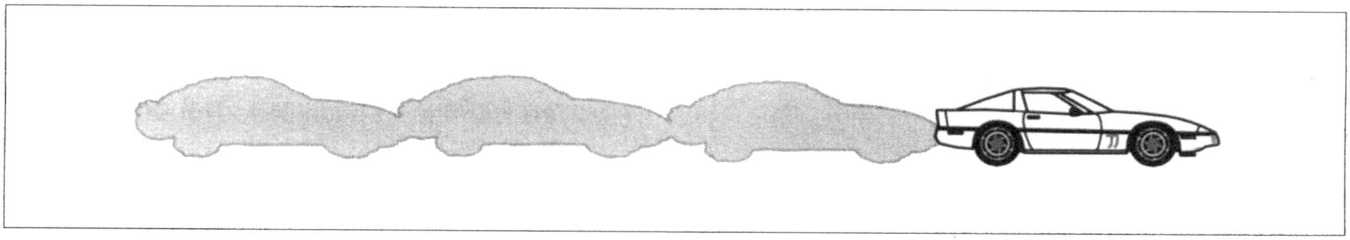

Fig. 3-13. With each gallon of gas you burn, you leave behind 18 pounds of CO_2, invisible clouds equal to about three times the volume of your car.

eled have more than doubled! This is for two reasons: there are more vehicles on the road, and we're driving them more often.

Tightening limits on these pollutants may be better for us than restrictions on driving, as in Mexico City, where each car must stay off the road one day a week. And although we don't hear much about it, controls are imposed on other sources of pollution: barbecue fire-starters, refineries, power plants, dry cleaners, bakeries (honest—gases from the yeast rising in the dough!). Basically, we've got to control these bad elements, one way or another.

2.2 Smog Formation

Smog refers to the "soup" of many gases, principally ground-level ozone, but including dust and particulate matter. Smog is cooked in the sunlight, with gases emitted from gasoline-powered cars. Smog is a fact of life that has been affecting our cars and trucks since the late 1960s. Increasingly, it will affect our driving and, indeed, how we can use our cars. Control has helped make the air cleaner than it was, but many metropolitan areas are still not safe for many days in the year—a problem we face at a time of increasing vehicle usage. The real costs of smog—damage to health, vegetation, property—are recognized as greater than the costs of reducing air pollution. You may not like smog control, but in most cases, the alternative is worse. Those of us who love cars and love to drive are coming to recognize what we can do to be responsible motorists.

Fig. 3-14. Smog began burning eyes in the 1960s (Los Angeles is shown here). Smog has become a threat in more major metropolitan areas, as Vehicle Miles Traveled increases. The real costs of smog—poor health, and damage to vegetation and property—are recognized as greater than the costs of reducing air pollution.

Ozone, O_3, has an extra atom of oxygen attached to an ordinary oxygen molecule, O_2. (See Fig. 3-15). Ozone's third atom gets stripped away from the O_2 in your lungs, and produces that burning sensation you get when you breathe smog. The longer hours of daylight, in the summer, cook the soup more, resulting in more ozone-heavy smog. In high concentra-

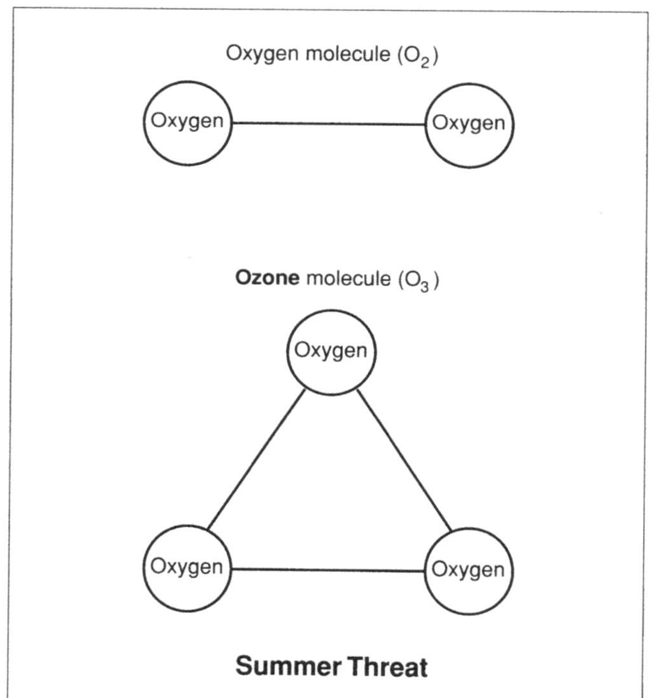

Fig. 3-15. In summer, ozone is the primary threat from smog. An oxygen molecule, O_2, contains two atoms of oxygen. O_2 is good stuff to breathe. Ground-level ozone, O_3, contains three atoms of oxygen. O_3 is *bad* stuff to breathe.

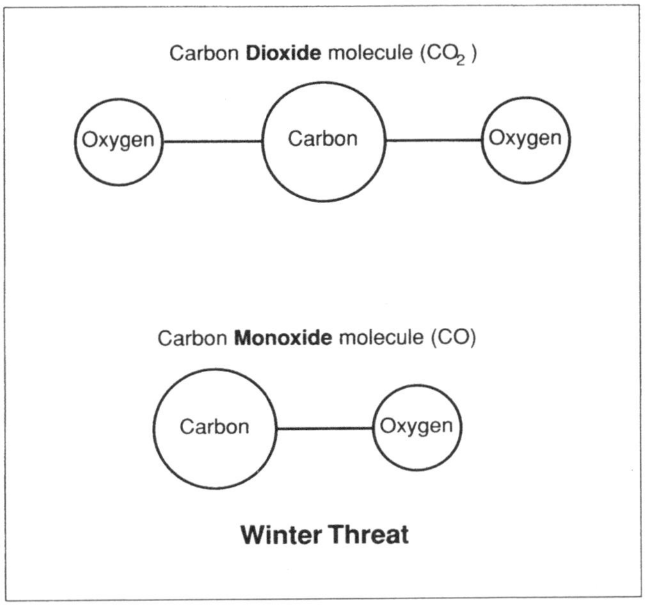

Fig. 3-16. In the winter, carbon monoxide is the primary pollution threat: more cold starts and longer warmups result in more partially burned fuel. Also, temperature inversions concentrate pollution in the lower atmosphere.

tions, ground-level ozone is hazardous to people and other growing things, such as trees and plants and other animals. As it happens, we need trees and plants because they use up some of the carbon dioxide in the atmosphere, making oxygen. "Woodman, spare that tree!" And we need all those other living things, too, for various reasons.

Carbon monoxide combines *one* atom of carbon with one atom of oxygen. (See Fig. 3-16.) CO is poisonous because it's an incomplete molecule, and seeks another atom of oxygen, to be complete carbon dioxide. Where do you suppose carbon monoxide gets oxygen? When you inhale it, it takes oxygen from your lungs. CO can kill you. Bummer!

Other Gases Emitted

Recent research shows that, in addition to the three major exhaust-gas pollutants, other emitted gases may increase smog formation at ground level. Some gases in the stratosphere also affect global warming, adding the following to our smog vocabulary:

Formaldehyde (HCHO): A compound of hydrogen, carbon, and oxygen, produced in trace quantities from the combustion of alcohol-based fuels, particularly with ethanol additives. It irritates the eyes, nose, and throat—for some people, even at minimal levels—and may increase cancer risk.

Methane (CH_4): A particular form of hydrocarbon found in natural gas. It is greenhouse gas, which contributes to global warming, along with CO_2.

When we redefine HC (because methane does not contribute to pollution in the same way as hydrocarbons) we get:

Non-Methane Hydrocarbon (NMHC): A federal (EPA) standard, beginning in 1994, measures non-methane hydrocarbon instead of HC.

Non-Methane Organic Gases (NMOG): These include both oxygenated and non-oxygenated HC. Beginning in 1994, California measures both NMHC and NMOG.

Volatile Organic Compounds (VOC): A broad category of gases, including NMHC, which vaporize at ambient temperatures (volatile) and are carbon based (organic).

> On a visit with CARB (CA Air Resources Board), I asked how they measure NMHC (Non-Methane HydroCarbons). "Easy," the engineer told me. "Measure the methane, measure the HydroCarbons. Subtract methane to get NMHC." In other words, Total HydroCarbons (THC) minus Methane = NMHC. Why didn't I think of that?

Added to our vocabulary are *oxygenates*—fuel additives containing oxygen. A gallon of oxygenated fuel (part oxygen) has less energy than a gallon of non-oxygenated gasoline, so it emits fewer HC and carbon monoxide than non-oxygenated gasoline. In engines with electronic engine control (all **'82-01** Corvettes), oxygenates affect emissions only in Open Loop, principally in the first four minutes of warmup, and during Wide Open Throttle. Oxygenates make more of a difference in pre-'80 Corvettes, without electronic engine control, because carburetors are more likely to run rich than fuel injection is. Oxygenates help correct this tendency. Beginning in '80 Corvettes, with Computer Command Control and Ignition (C_3I), and in **'82** models, with electronic fuel injection, Closed Loop operations result in stoichiometric operation, with or without oxygenates.

2.3 Effects of Air/Fuel Ratio on Pollutants

Just as variations in the air/fuel ratio change power output and fuel consumption, they also change exhaust emissions. As I described, the air/fuel ratio is a key to the complete combustion of fuel. It also affects the combustion temperature (and therefore the formation of NO_X) and exhaust-gas oxygen levels (and therefore the operation of the catalytic converter).

> During warm-engine cruise, air/fuel ratios must be controlled to a narrow range. How narrow? 14.7 ± 0.5%, or 14.63 to 14.77. That's narrow! (See Fig. 3-17.)

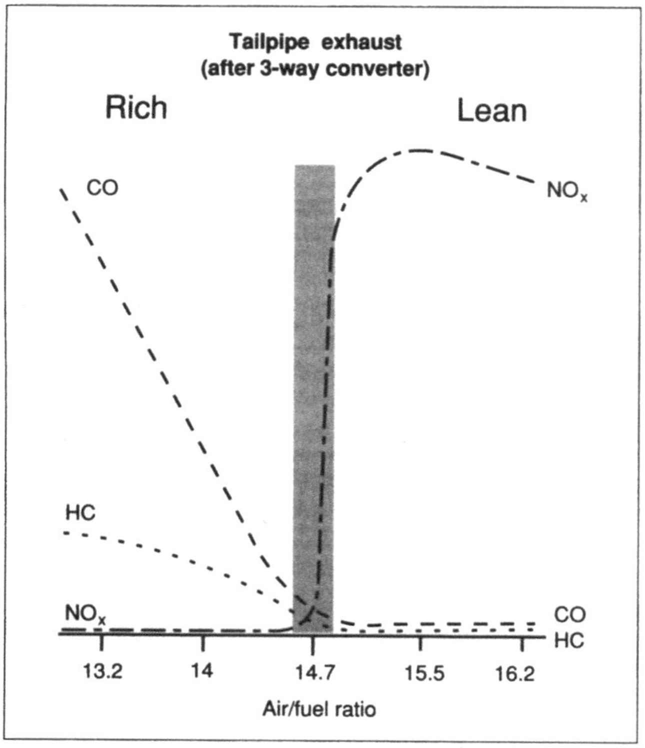

Fig. 3-17. Tailpipe exhaust, after the converter, shows the lowest emission of controlled gases when the air/fuel ratio remains in a narrow, 1% range around 14.7:1 at the converter. Richer mixtures result in more HC and carbon monoxide, while leaner mixtures result in more nitrogen oxides.

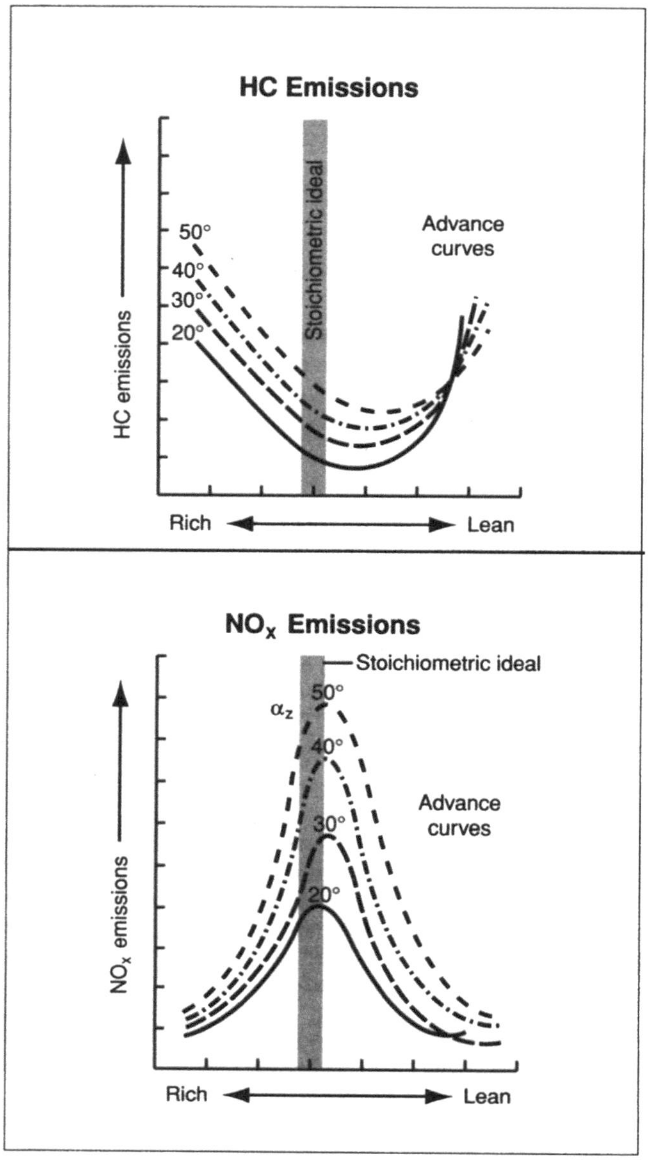

Fig. 3-18. The lowest HC emissions result from spark timing near Top Dead Center. As timing is advanced, say to 50°, HC emissions more than double. Emission of NO_X increases with spark advance above about 20° BTDC, because moderate spark advance helps lower combustion temperature.

2.4 Effects of Spark Timing on Pollutants

HC and NO_X emissions change according to the combined factors of the air/fuel ratio and spark timing. (See Fig. 3-18.)

2.5 Exhaust Gas Treatment

So far, I have been talking about the effects of the air/fuel ratio and spark timing, on engine-out exhaust. Most Corvette engines use Exhaust Gas Recirculation (EGR) to reduce the formation of NO_X during burning in the cylinders. All Corvette engines use Air Injection (AIR) to burn HC and CO in the exhaust manifolds, during early warmup. The **'82–95** engines also inject air into the catalytic converter, during later warmup.

Notice that EGR and AIR are *two different systems*; they act:

- on different pollutants
- at different places in the engine-exhaust system
- at different times in engine operation

Exhaust Gas Recirculation (EGR)

Exhaust Gas Recirculation reduces formation of NO_X. The EGR valve routes a small amount of exhaust gas (5–15% of the intake air) into the intake manifold and back into the combustion chambers. This dilution of the intake air lowers the combustion temperature, decreasing formation of NO_X.

Fig. 3-19. The EGR valve routes a small amount of the exhaust gas (5-15% of the intake air) into the intake manifold and back into the combustion chambers, in order to reduce the formation of nitrogen oxides. The design of the valve varies with some model years.

'90–93 ZR-1, LT4, LS1: These engines have no EGR system. Their internal design, valve overlap, and fuel metering minimize lean mixtures, which tend to produce NO_X, allowing them to pass certification tests without an EGR system. You may want to think about that before you lay out dollars for a new camshaft that may never pass a smog test.

Air Injection (AIR)

Adding air to the exhaust manifolds during warmup reduces emissions, by burning the HC and CO that result from the rich starting mixtures, as they exhaust from the cylinders. This also increases underhood heat. Exhaust-manifold burning has the additional advantage of heating the Exhaust Gas Oxygen sensor and catalytic converter, both of which must be hot to operate. With electronic engine control, air can be added to the oxidation converter part of the catalytic converter in the **L83** and **L98** Corvettes, as described below.

Most Corvettes use engine-driven pumps with control valves. These belt-driven AIR pumps waste engine power, because they are driven whenever the engine runs. On **LT1**, **LT4**, and **LS1** engines, an electrically driven AIR pump delivers air into the exhaust manifold for a maximum of four minutes after start. But these electric AIR pumps do not operate under *any* of the following conditions:

- the computer recognizes a problem and sets a diagnostic trouble code
- the computer switches to Closed Loop
- engine speed is greater than 2825 rpm
- Manifold Absolute Pressure (MAP) is less than 20 kPa (coasting)
- Manifold Absolute Pressure (MAP) is greater than 90 kPa with a vehicle speed over 60 mph (high-speed acceleration is detected by increased injection time)
- the Engine Coolant Temperature (ECT) is less than 14° F (–10° C).
- the converter temperature is excessively high, based on readings from the Oxygen Sensor(s).
- Short Term Fuel Trim (STFT) or Long Term Fuel Trim (LTFT) is outside normal range

3

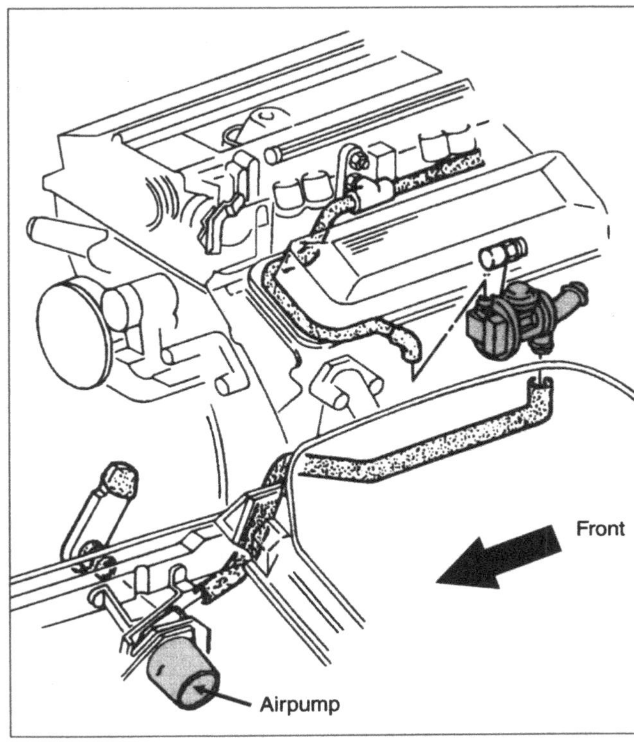

Fig. 3-20. LT1, LT4, LS1: An electrically driven AIR pump delivers air into the exhaust manifold for a maximum of four minutes, after start. The **ZR-1** electric AIR pump looks different, but operates the same way.

Fig. 3-21. L83, L98: The AIR pump is belt-driven, from the crankshaft, with computer control of the AIR switching and divert valves.

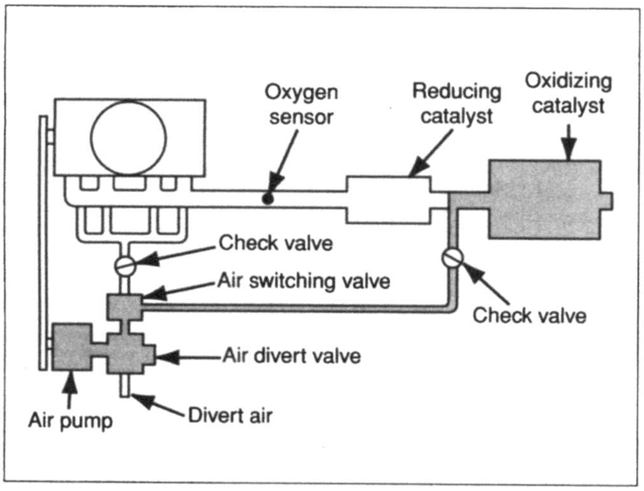

Fig. 3-22. L83, L98: AIR can be switched to the oxidation converter section of the catalytic converter, during warmup.

Catalytic Converters

A catalyst is a material that promotes a chemical reaction, without being affected itself. Catalysts in the converters include small amounts of rare and expensive metals like platinum, palladium, rhodium, and, lately, cerium. They convert the pollutants in engine emissions into less harmful substances, before they leave the car altogether.

Three-way catalysts (TWC), decrease NO_X as well as oxidizing CO and HC. The three-way converter adds oxygen, to oxidize the HC and CO into H_2O and CO_2, and it takes away oxygen, to reduce NO_X, separating it into N_2 and O_2.

Late model three-way converters utilize "oxygen storage" and don't depend on AIR injected into the converter. The base metal, cerium, acts as an oxygen sponge, soaking up oxygen and releasing it. Therefore, a good catalyst does both jobs:

- With less oxygen in the exhaust stream, as with a rich mixture, the converter releases oxygen, adding it to the HC and CO.
- With extra oxygen in the exhaust stream, as with a lean mixture, the converter soaks up the oxygen, removing it from the NO. For reduction (taking oxygen from NO_X) to match oxidation (adding oxygen to CO and HC), the proportion of gases in the engine exhaust must be closely controlled. That means the intake air/fuel ratio must always be near stoichiometric.

Converters must be hot to do any converting, so the faster they warm up, the sooner they become active. Therefore, they're located as close to the engine as possible. Some are right under the exhaust manifold; others are further down, under the car. The conversion process also raises the temperature of the converter, increasing its efficiency. The design and location of Corvette's converter changes with some model years, depending on engine design, emission limits, and/or the chemistry of the catalysts, as well as the amount of space under the car and the space and temperature underhood.

Even a few misfires can send raw fuel into a converter. When it burns there, it can cause permanent damage, so don't drive with a misfire, as from a bad plug or plug wire.

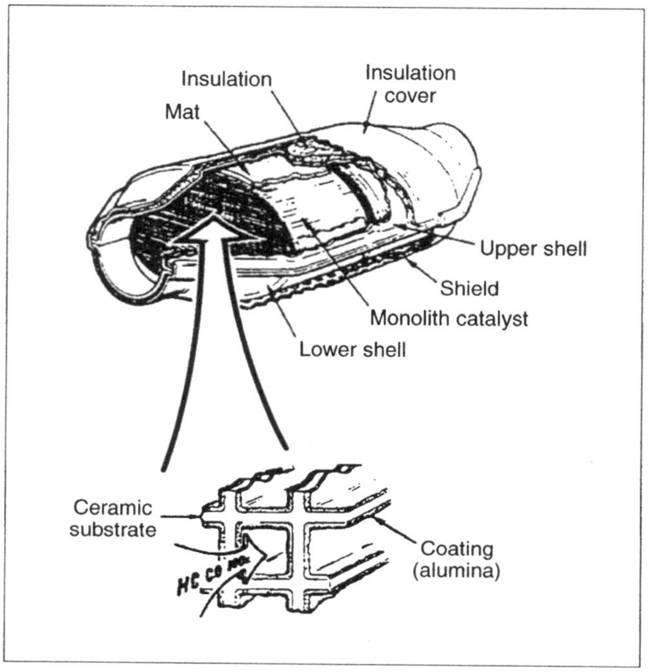

Fig. 3-23. The catalytic converter is a complex structure, designed to expose exhaust gases to catalytic materials.

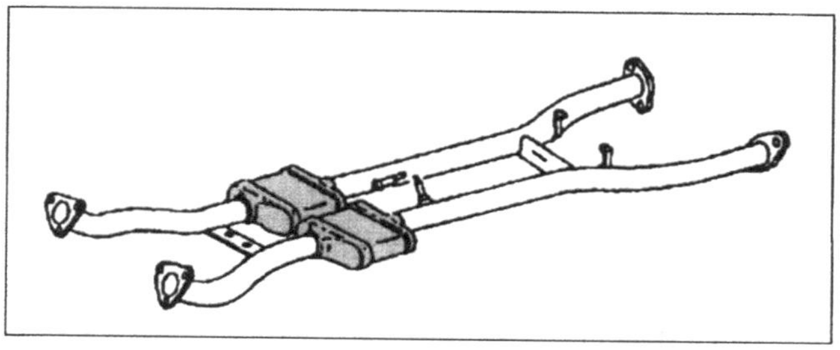

Fig. 3-24. LS1: The twin three-way converters are located under the car.

Underhood converters warm up perhaps a minute sooner than undercar converters. Don't *underhood* converters raise underhood temperatures? Yes, but remember, they operate only during the first four minutes or less, when the engine is cold, and so are underhood temperatures. After this, the air/fuel ratio is controlled by the oxygen sensor.

Table 3-1. Corvette catalytic converters

Engine	No. of converters	Location
L83	1	Undercar
L98	3	2 underhood 1 undercar
ZR-1, LT1, LT4	2	Underhood
LS1	2	Undercar

2.6 Canister Purge—Evaporative Fuel Vapor

Unburned HC can pollute the atmosphere, if it evaporates from the fuel system or escapes during refueling. A canister draws in the fuel vapor that forms over the liquid fuel in the tanks. The canister is purged when the vapor is released into

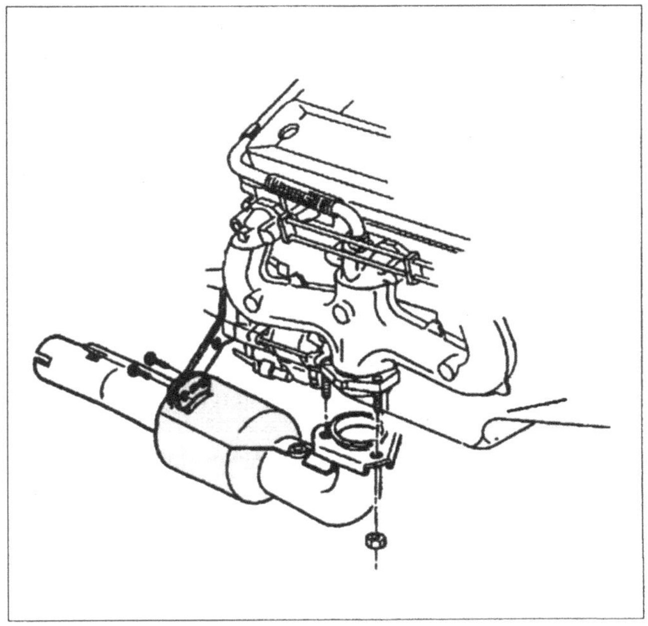

Fig. 3-25. ZR-1, LT1, LT4: Many late-model converters are located close to the exhaust manifold. It's a simple design, which requires no AIR pump delivery to converter.

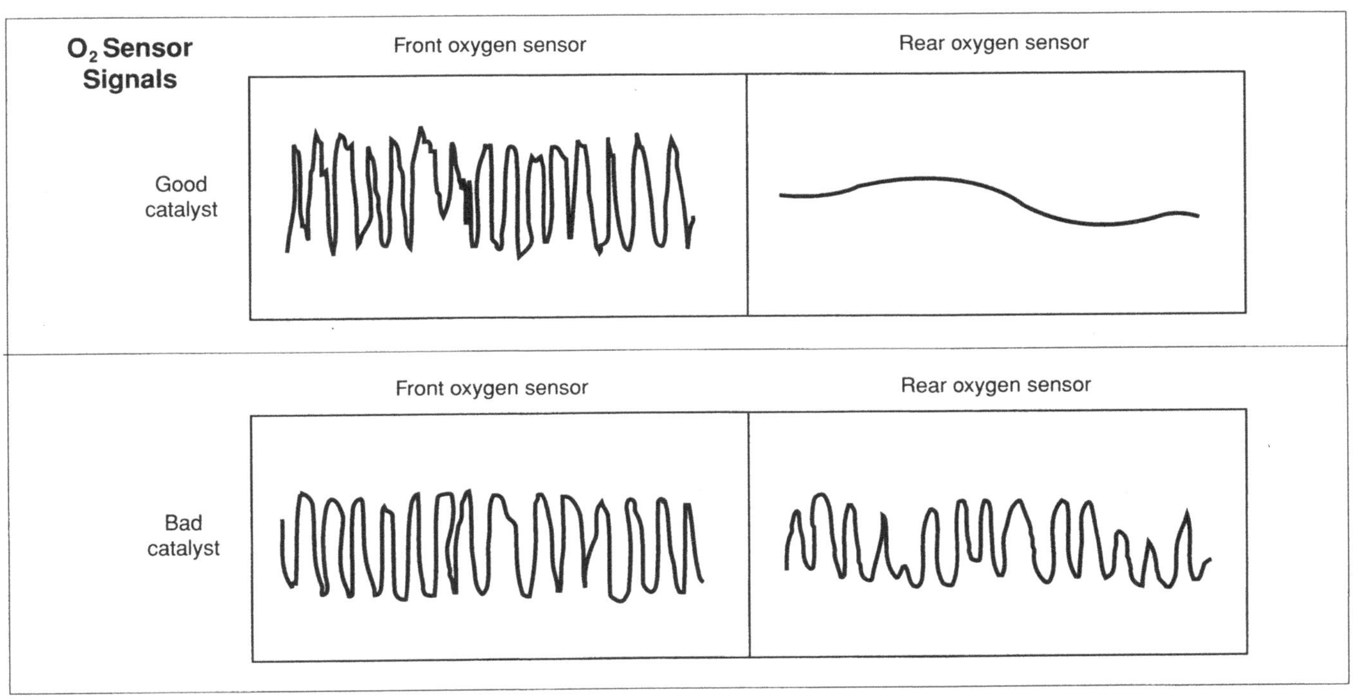

Fig. 3-26. '94-01 LT1, LT4, LSI: For the "Good catalyst," compare the voltage signals from the front oxygen sensor (ahead of converter) to the signals from rear sensor (after converter). The oxygen content of the exhaust is consistently low, unlike with the failing catalyst (*bottom*), which has diminished oxygen-storage capacity.

the engine. Computer control also cuts back on the injection pulse time, to maintain proper air /fuel ratios.

The more you know about emissions, and the limits placed on them by legislation, the more you'll understand how fuel-injection and engine-management systems work, and therefore what performance modifications are worth investing in.

2.7 Conflicting Demands on Engine Control

Corvettes are often bought for their power, but modern engine control systems face additional demands:

- fuel economy—increasing concern over cost and availability of gasoline
- exhaust emissions—environmental concerns and resulting legislation
- driveability—drivers' demands for quick starting and smooth, trouble-free performance under any operating conditions

Each of these factors places different demands on the engine control system. The design engineer must consider tradeoffs. Adjusting the system for maximum power also means increasing fuel consumption, and may compromise driveability. Minimizing fuel consumption means sacrificing power and, potentially, driveability. Choosing either maximum power or minimum fuel consumption may mean increased emissions. The modern fuel-delivery system must be able to maintain strict control of the air/fuel ratio, to meet these conflicting demands in the most acceptable way.

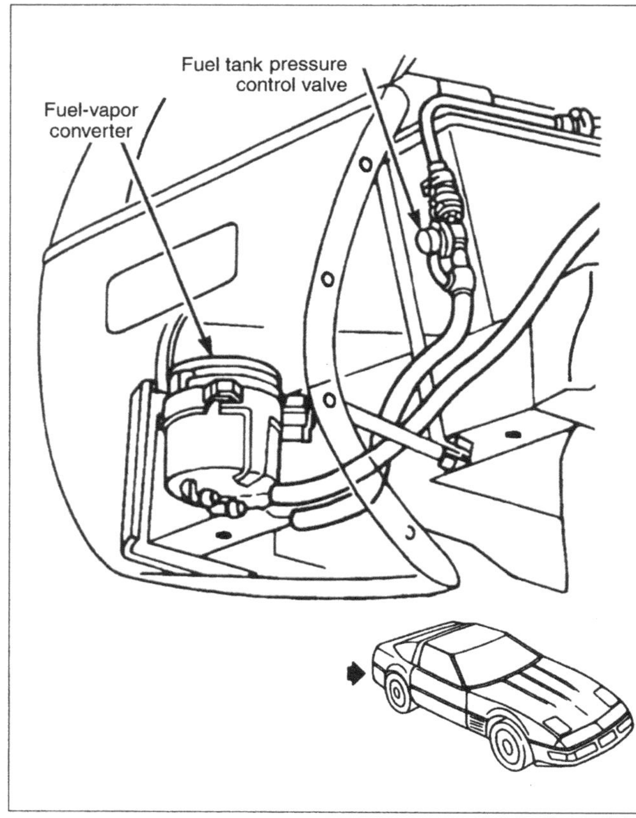

Fig. 3-27. ZR-1, LT1, LT4, LS1: The fuel-vapor canister is mounted behind the right rear wheel, near the fuel tank(s). In the **L83** and **L98** the canister is mounted underhood.

Table 3-2. Typical Corvette USEPA test data

	LT1[a] **4-speed A/T**	**LT1[a]** **4-speed A/T**	**LT1[b]** **6-speed M/T**	**LT1[b]** **6-speed M/T**	**ZR-1[c]** **6-speed M/T**	**ZR-1[c]** **6-speed M/T**
Vehicle ID	CP2Y 011-FEC	CP11Y 003	C11Y 004	C11Y 004	C19Y 118 ➞	C10Y 003A
Displacement (cu. in.)	350	350	350	350	349	349
Horsepower	300	300	300	300	375	375
Test weight (lb)[d]	3625 ➞	3750	3625 ➞	3750	3750	3750
Axle ratio	2.59* ➞	3.07*	3.45	3.45	3.45	3.45
City mpg	18.7	16.4	18.9	18.9	17.1	17.8
Highway mpg	31.6	27.7	32.3	32.3	30.6	32.5
Adjusted ratings[e]	17/25	17/25	16/25	16/25	17/25	17/25

➞ indicates compared tests
[a] Two different **LT1** vehicles, different test weights and axle ratios
[b] Same **LT1**, different test weights
[c] Two different **ZR-1** vehicles, same test weight and axle ratios, showing variability of tests, similar cars
[d] Test weight = curb weight + 300 lb
[e] Test mileage is adjusted to represent the actual driving variances from the published ratings:
city mpg = 90% of test data; highway mpg = 78% of test data

3. Fuel Economy—CAFE Fines and Gas-Guzzler Taxes

From the emission tests, fuel usage can be calculated, leading to the widely-publicized EPA fuel economy ratings.

- The **City** rating is the fuel used during both the first 505 seconds, in the cold-start/warm city cycle, averaged together with the last 505 second warm city-cycle.
- The **Highway** rating is based on the 867 seconds following the first city pattern.
- The **Composite** rating is an average, based on 55% city and 45% highway.

Fuel economy ratings *do* affect your Corvette.

CAFE (pronounced "cah-fay") stands for Corporate Average Fuel Economy. *Corporations* are charged a fine for any model year in which the average rated miles per gallon doesn't meet federally mandated fuel economy ratings. These are averaged separately for GM domestic cars (including Corvettes), import cars, and light trucks.

> It is GM policy to maintain its CAFE ratings high enough so the corporation avoids paying any CAFE fines. Corvette fuel economy is part of GM/CAFE.

A **Gas-Guzzler tax** is levied on each *purchaser* of a car which is rated at fewer miles per gallon than specified by law, typically about 22.8 mpg. This tax is paid at the time of purchase, and ranges from about $1,000–7,000.

> It is also GM policy to maintain its fuel economy ratings high enough so that no Corvette buyer—indeed, no buyer of any GM car—has to pay a Gas-Guzzler tax.

CAFE

The CAFE fine is $5 for each 0.1 mpg for which each car sold fails to meet the CAFE standard, for the total of all domestic cars sold that year. Typically, GM sells about 5 million cars a year. For purposes of the law, the CAFE figure is the Federal Test Procedure's composite test for miles per gallon. If GM misses its CAFE, the fine will be $5 times 5 million cars, or $25 million, for each 0.1 mpg for that model year! Luckily, for GM, credits are allowed to be carried forward or backward a few years. (For more information, see the sidebar on Page 54.)

Remember, these are *ratings*. When you hear statements like, "The mileage you'll get is (quote from these ratings)," you can say, "No, that's not what drivers get, that's the *rated* mileage." The mileage you actually get depends on many factors: weather, temperature, altitude, road condition, tire inflation, engine condition, and, most important, how you drive your Corvette. Perhaps you understand the EPA qualifier, "*Your mileage may vary*." Unless you make your fuel-economy-rating measurements with the same precision, your mileage really will vary.

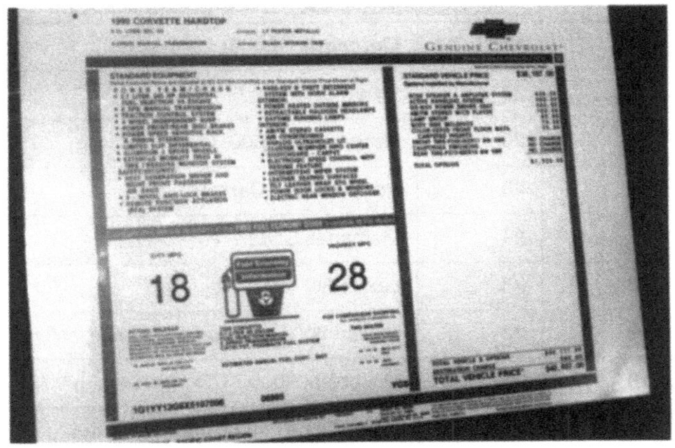

Fig. 3-29. This manual-transmission Corvette is rated at 18 mpg city and 28 mpg highway. These mpg ratings are derived from the test data, but have been adjusted by USEPA by certain factors to make them closer to actual usage.

GM must consider both customer demand for Corvette performance, and CAFE standards. Corvette's highway mileage ratings are much better than some competitors. "Blame" it on Corvette's Computer Aided Gear Shift (CAGS), or 1–4 upshift, and the Gas-Guzzler tax. Let's face it, CAGS may be a necessary nuisance but the **LS1** is an improvement over previous manual transmission cars. Of course, any performance driving bypasses the 1–4 upshift control and permits all manual-transmission gears.

> The official economy/emission runs are made only after a break-in of at least 4,000 miles. One EPA engineer told me, "We don't test any brand-spanking-new cars. *They vary too much just off the production line*." He went on, "April and October are the right months to check your mileage on the road—no winter-cold operations and no air conditioning." So to test your own Corvette mileage, break it in if it's new, and measure in the spring and fall.

The **'90-01** Corvette manual transmission 1–4 upshift, also known as the Computer Aided Gear Shift (CAGS), is matched to the mild accelerations of the federal emissions/economy test. This improves the Corvette rated composite miles per gallon (and emissions). (See Chap. 6 and 9.)

Table 3-3. Typical published economy ratings (not test data)

City mpg	Highway mpg	Model
18	28	Corvette*
For comparison:		
9	**14**	**Lamborghini Diablo**
13	**17**	**Mercedes SL600**
14	**22**	**Viper**
16	**20**	**Mercedes SL500**
16	**23**	**Porsche 911 Carrera**
17	24	Mercedes SL320
18	23	Lexus SC 400
18	23	Nissan 300ZX (automatic transmission)
18	24	Mitsubishi 3000GT/Dodge Stealth

*Manual and automatic transmissions usually rate about the same.
BOLD models: Based on test data, car/engine combinations subject to Gas Guzzler Tax.
Mercedes-Benz pays the U.S. Treasury the largest CAFE fines: $18 million (1992), $13 million (1993), $11 million (1994). BMW: $7 million (1993).

4. New Fuels

Gasoline, refined from petroleum, is one of the most concentrated forms of liquid energy. Yet its worldwide use helps foul the air and increases U.S. import of what could become a scarce resource. It's important to consider the engine/fuel combination.

Beginning in 1995–96, in all states, fuels have been changing, to help reduce automobile pollution. In general, the objectives, for providing cleaner gasoline, are to reduce smog-forming emissions (by reducing aromatic HC), reduce cancer-causing benzene, and maintain catalytic converter efficiency (by reducing sulfur).

> **Why New Fuels?**
>
> Changing the engine *hardware* only affects new cars. And it can take up to twenty years to turn over 95% of the fleet—even longer with Corvettes, because owners care for them for decades. Estimates are that 90% of the Corvettes built are still running, even if some are driven only to show. On the other hand, changing *fuel* affects almost all vehicles on the road immediately.

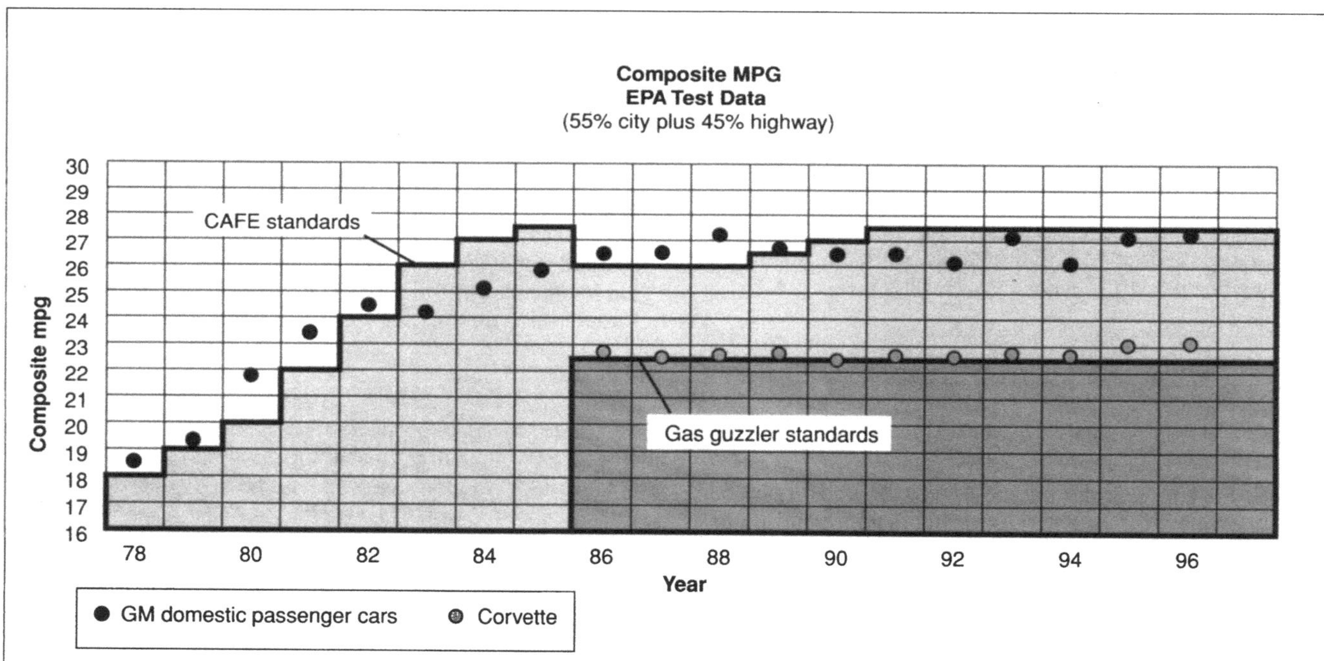

Fig. 3-28. Each manufacturer must consider two related, but different, fuel-economy standards, in order to avoid fines. Many of the cars GM manufactures don't actually meet the standards, but, fortunately for GM, the sales of their more efficient models are higher, which counterbalances the sales of their less efficient ones. It all averages out OK in the end, so they don't get fined.

If you're considering adding *lead* to your fuel, be aware that since 1996, *it's illegal to burn leaded fuel in any vehicle on U.S. public roads.* Of course, lead will quickly poison your catalytic converter(s) and Oxygen Sensor(s)—and it'll cost a bunch to get street-legal again. Manganese additives, like MMT, have been illegal in California since 1977. *MMT is not GM-recommended in any state.*

You may find Manganese additives in states which are not as concerned about low emissions. Ask your service station operator. If you burn fuel with MMT, your Corvette may experience:

- reduced spark-plug life
- reduced emission-control system performance
- lighted MIL (Malfunction Indicator Light)

The composition of the fuel you buy depends on the smog problems of the part of the country in which you're driving, as measured daily by nonattainment of either *ozone* standards (usually *summer*) or *carbon monoxide* standards (usually *winter*), or both. The solutions to ozone are different from solutions to carbon monoxide, and if you have both, that's different again.

Measurement of automotive pollutants is affected by

- local weather, including temperature inversions (warm air above cool air)
- prevailing winds
- terrain
- concentration of vehicles, and traffic congestion

Before I look at what you'll be burning in the foreseeable future, let's define our terms. (See Table 3-4.)

Octane

Octane is a measure of a fuel's resistance to knock. The higher the octane, the less the knock. All Original Equipment Manufacturer (OEM) engines must be able to operate on regular 87 octane fuel, although the **LS1** owner's manual recommends premium. Quoting the **LS1** manual: "Use premium unleaded gasoline of 91 or higher octane rating for best performance. You may use middle grade or regular unleaded gasolines, but your vehicle may not accelerate as well."

How is that? The answer is, when the Knock Sensor detects ping, the system retards the timing. Most Corvettes show improved performance with 89 octane, and even more with 92–93 octane. And high-mileage engines tend to build up carbon, raising the effective compression ratio and the tendency to knock with regular fuel. Corvette engines for **'82-01** are equipped with Knock Sensor(s), to satisfy the requirement that they operate on fuels rated as low as 87 octane, generally with computer-retarded spark timing.

3

Table 3-5. Factors affecting octane requirement

Design/operation	In use
Compression ratio	Barometric pressure/altitude
Ignition timing	Temperature
Air/fuel ratio	Humidity
Combustion temperature: Inlet air temperature Coolant temperature	Combustion chamber deposits
EGR rate	
Combustion chamber design	

Volatility

Volatility is a measure of how easily gasoline vaporizes, usually on a scale of 8–15. The higher the number, the greater the volatility. Seasonal blends usually need higher vapor pressure in winter than in summer. Fuel injection aids vaporization, even at a lower vapor pressure, because it delivers fuel under pressure (compared to carburetors, which depend on vacuum to aid vaporization). (See Fig. 3-31.)

Table 3-4. Fuel terms defined

Term	Definition	Variable	Effects
Octane	Anti-knock index	87 regular 89 mid-grade 92–93 premium	Reduces tendency to knock, up to certain limits
Volatility	Tendency to vaporize	Reid Vapor Pressure	*Winter*: improved cold start and warmup *Summer*: worse VOC emissions (even with reduced vapor pressures)
Oxygenate	Oxygen added to fuel	ethanol, MTBE*	Reduces open-loop CO emissions

* MBTE is being phased out—there is increasing evidence that MTBE is leaking from underground fuel storage tanks.

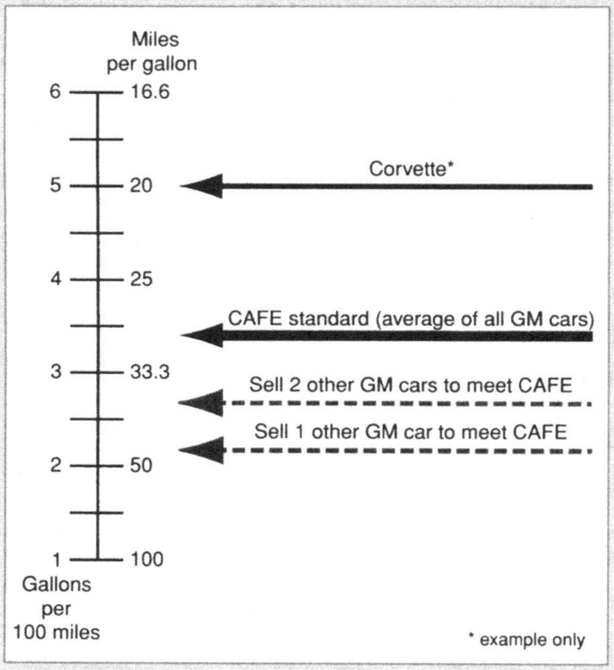

Fig. 3-30. To meet CAFE standards GM needs to balance the effects of their less efficient models by selling more of the cars that exceed the standards.

Maybe you don't care about Corvette rated fuel economy, but Corvette engineers do. Their calculations, like the CAFE standard, are based not on miles per gallon but on how many gallons are burned while traveling a specific distance—say, 100 miles. What counts is the amount of gas used, not the distance traveled.

What's the result? For a hypothetical example, let's say that Corvette's weighted average, of city and highway ratings, is 20 mpg. Let's also say that the current legal CAFE requirement is 28 mpg. But CAFE isn't measured in mpg, it's measured in gallons per 100 miles. So we have to convert the mpg numbers.

A CAFE standard of 28 mpg works out to be 3.6 gallons per 100 miles (100/28). A Corvette rated at 20 mpg would use 5 gallons per 100 miles (100/20). 1.4 gallons more than the CAFE rating.

To balance the Corvette's effect, for each Corvette sold, GM must sell another domestic car rated at 1.4 gallons less than the CAFE rating, which would be a rating of 2.2 gallons per 100 miles. A car that burns only 2.2 gallons per 100 miles would be rated at 45 miles per gallon (100/2.2).

In other words, to balance a 20 mpg Corvette, at 8 mpg below a CAFE standard of 28, GM must sell *one 45 mpg car* (17 mpg above CAFE), or *two 36 mpg cars* (8 mpg above CAFE). And since it's easier to make a car rated at 36 mpg, GM usually chooses to go the latter route. Additionally, they usually underprice their smaller cars, so that they sell better. This way they avoid the expensive fines.

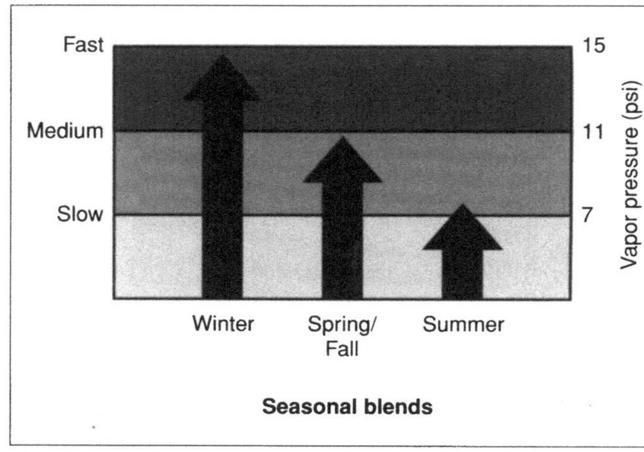

Effects of gasoline volatility on vehicle performance

Volatility too low	Volatility too high
Poor cold start	High evaporative emissions, from canister overload and purge
Poor warmup performance	
Poor cold-weather driveability	Fuel economy may deteriorate
Increased deposits: Crankcase Combustion chamber Spark plugs	
Unequal fuel distribution, in Throttle Body Injection vehicles	

Fig. 3-31. Gasolines must strike a seasonal balance in their volatility.

Oxygenates

Oxygenates are fuels or additives that contain Oxygen (O_2), added by the refinery, to assist in complete combustion during Open Loop operation. By oxidizing more of the HC in the fuel, they decrease HC and CO, especially after cold starts. Oxygenates have little effect on combustion during Closed Loop operation, though.

Oxygenates reduce the amount of energy available in fuel. MTBE has 19% less energy than gasoline, so the total effect in a 15% MTBE blend, is 2.9% less energy (.19 x .15 = .029). Ethanol has 34% less energy than gasoline, so the total effect, in a 10% ethanol blend, would be 3.4% less energy.

The principal advantage of oxygenates, in fuel-injected cars, is during cold start and warmup, in Open Loop. With heated Oxygen Sensors (**'94-01**), cold-start, Open Loop operation may last less than one minute. They're also useful for Wide Open Throttle operation (another Open Loop). In general, oxygenates are more important for carbureted cars, particularly those without computer control (before 1980 in California, and 1981 elsewhere).

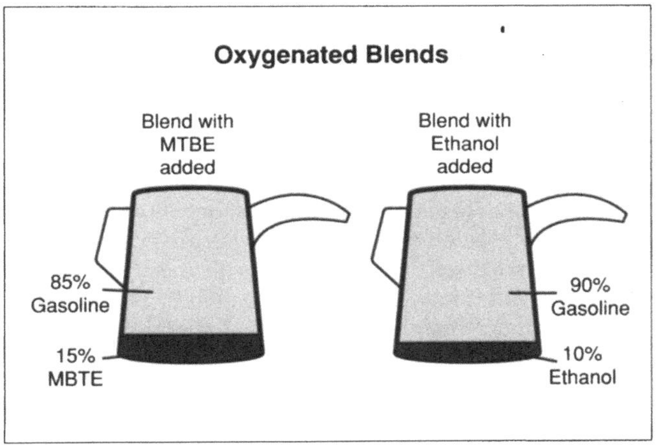

Fig. 3-32. MBTE blend, 15%, means that about one gallon of MTBE is mixed with every six gallons of gasoline. Ethanol blend, 10%, means that one gallon of Ethanol is mixed with every nine gallons of gasoline.

4.1 What Fuel Do You Burn in Your Corvette?

The fuels you'll burn will depend on where you're driving in the U.S. What this means for Corvette owners is that fuels can help clean up earlier Corvettes, without the need for added smog equipment.

Conventional Gasoline

Conventional gasoline has a low vapor pressure (volatility), to reduce HC emissions, and has additives (detergents), to reduce engine deposits. It is sold in places where air quality is satisfactory, which includes most of the nation's "open-space" area. However, most cars are concentrated in metropolitan areas, where conventional gas is beginning to be eliminated, in favor of some of the more environmentally friendly kinds of fuel.

Winter Oxygenated Gasoline

Winter oxygenated gasoline is sold in many states (but not California), generally October through May. It's required where air-quality measurements show significant CO non-attainment days, in other words, days when Carbon Monoxide measurements are excessively dangerous.

In the 21st century, if you drive coast-to-coast or even cross-country, you might be burning several kinds of gasoline. All are different from the fuels of the early 1990s, including what was then called "reformulated gas". The fuels you burned in 1997 probably differed from those in 1998, and 2000, and so on. In most parts of the country, summer fuels differ from winter fuels.

Winter gas has oxygenates. It also has increased vapor pressure, to assist Open Loop operation, as in cold starts, warmup, and Wide Open Throttle acceleration. Expect fuel economy losses of 2–3% compared to conventional gas, plus the usual winter losses of 5–15%.

Reformulated Gasoline, Federal (RFG), Phase II, 2000

Phase II reformulated gasoline is sold in many states (but not California). It is required where air-quality measurements show significant ozone non-attainment days (often in the summer, and in the city). The differences from earlier gasolines include:

- reduced toxic chemicals, like benzene (a carcinogenic aromatic), other aromatics, and olefins
- greatly reduced sulfur, a substance that can mix with moisture to create sulfuric acid, which damages lung tissue among other things
- reduced summer vapor pressure, to decrease Volatile Organic Compounds (VOC) in refueling, tailpipe emissions, and fuel-vapor system losses, (compared to the previous level of 9.0 psi, the Phase II VOC level for southern states is 7.1 psi, and for northern states it's 8.0 psi)
- increased deposit-control additives, to reduce deposits on the injectors and valves
- added oxygenates, year round (oxygenates have limited effectiveness in reducing Closed Loop tailpipe VOC, or toxic emissions, particularly in fuel-injected engines, but they're still required by federal law)

Expect fuel economy losses of 2–3%, compared to conventional gasoline, plus the usual winter losses of 5–15%. This gasoline will reduce VOC by about 28%, and reduce NO_X about 7%. Expect increased prices at the pump, based on the increased refinery costs of 2–5 cents per gallon.

California Reformulated Gasoline (RFG), Phase II, 2000

California reformulated gasoline is sold only in California and is required in the entire state. The differences from other gasolines (limits by volume) are:

- reduced toxic chemicals (some carcinogenic):
  - benzene
  - formaldehyde
  - 1,3-butadiene
  - polycyclic organics
  - other aromatics
  - olefins, to reduce the reactivity of fuel-vapor losses and reduce NO_X
- reduced sulfur
- reduced summer vapor pressure (limit 7.0 psi), to reduce VOC in refueling, tailpipe emissions, and fuel-vapor system losses, compared to the previous California limit of 7.8 psi

- increased deposit-control additives, to reduce deposits on injectors and valves
- oxygenates, all year (oxygenates have limited effectiveness in reducing Closed Loop tailpipe emissions, particularly in fuel-injected engines, but they're still required by federal law)

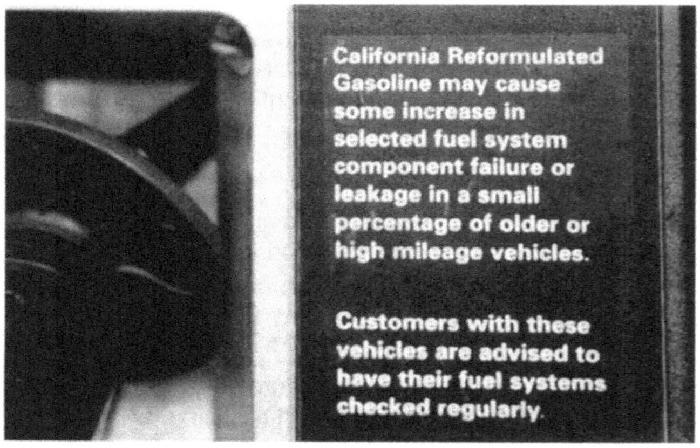

Fig. 3-33. The California reformulated gasoline label on a gas-station pump warns of the possibility of fuel-system problems, generally for cars that are ten or more years old, or that have been driven for 100,000 miles or more.

Expect fuel economy losses of 1%, compared to California's previous winter gas, plus the usual winter losses of 5–15%. This gasoline will reduce CO by about 11%, VOC (NMHC), tailpipe, and fuel-vapor emissions by about 17%, NO_X by about 11%, and sulfur dioxide by about 80%. Expect increased prices at the pump, based on the increased refinery costs of 2–5 cents per gallon.

Consider the following, additional advice from the '97 Corvette owner's manual on California RFG fuel. Paraphrasing from p. 6-4, if your underhood emission-control label indicates that your Corvette is a California car, then it has been designed to operate on California RFG fuel. If California RFG is not available, your engine can be operated on non-California fuel. Even though the engine will perform adequately on non-California fuel, operation of emission control systems may be adversely affected. You may fail a smog test, and the Malfunction Indicator Light may be lit on the instrument panel. Repair of any damage caused by non-specified fuel may not be covered by the warranty.

4.2 Bottom Line

Many of these fuels may not have existed when your Corvette was manufactured. And they may have different effects on your system, than the fuel that it was designed for. But, the bottom line is that automakers participated in the testing of the new fuels, and they approve of their use. In general, they don't have any serious negative effects on Corvette engines. Fuels do vary, though, from one refinery to the next. If you have trouble, try changing brands. As the folks at GM say: Usually, all a Corvette needs is a good-quality gasoline.

Some cautionary words on *supplemental* fuel additives:

From GM's Research and Development: All major (fuel) brands contain detergent additives to keep injectors clean, and other additives to minimize intake deposits. Some 92–93 octane fuels have more additives than 87 octane. GM does not recommend continued use of supplemental fuel additives. Octane boosters, like toluene, can improve performance only up to engine capability. Overuse—adding more than 20% of the fuel—may cause adverse effects or damage original fuel-system components.

With **'85–88** Bosch injectors, do not use cleaning additives in the gas tank. Fuel-system cleaners are not recommended, by GM, because of the kinds of materials used in the fuel tank bladder, fuel pump O-rings, and fuel line hoses. If cleaning is required, use the GM specified procedure. On the other hand, Multec injectors, in the **'89-01** models, do not need cleaning, and doing so could actually damage them.

Also, Gordon Killebrew, of *Late-Model Corvette Solutions,* and former Corvette Action Center expert, told me that in **'89–91** Corvettes, with Multec injectors, using any kind of supplemental fuel additives can cause a breakdown in the O-ring that separates the coil from the housing. This allows fuel to enter the injector coil, causing driveability problems, such as engine misses, and roughness. The **'92-01** injectors resist deterioration from these types of fuel additives, though.

5. Summary

In this chapter you have seen:

- how engine emissions, exhaust and fuel vapor form smog.
- how engine controls operate to reduce emissions while meeting performance expectations
- how your Corvette will be tested for emissions, particularly with new tests in the areas that do not attain Clean Air standards
- how your Corvette will respond to changes in fuels—starting, fuel performance, economy, and driveability

IN THE NEXT CHAPTER:

You have the basics under control. In Chapter 4, I begin with the "nerves" of the system—the sensor signals which tell the control module what your engine is actually doing, as well as what you would like it to do.

Chapter 4

Sensors—the Nerves

Contents

In this chapter, look for:

- what each sensor looks like and how it operates
- the different kinds of input signals which sensors send to the computer
- the importance of each sensor, under different operating conditions

1. Introduction

1.1 Terminology

Beginning in 1993, many of the names for sensors, control modules, actuators, and other parts were changed, as mentioned in Chapter 1. SAE standard J1930 provides common terms for the same general part, throughout the automotive industry. This chapter uses the terminology applicable for the years 1993 to present.

1.2 Overview

The sensors are the "nerves" of the system. They send input signals, generally about "how much" (how hot is the coolant?), or about "when" (when did a crankshaft or camshaft reach a Top Dead Center reference point?). Though, some sensors simply answer questions with a "yes" or "no", by switching a signal on or off.

Many of the sensors send input signals that are important to control outputs, or actuators, for several subsystems. For example, the Engine Coolant Temperature (ECT) sensor is important to the control of fuel, spark timing, idle rpm, and emission control. It is also important during several operating conditions, including cranking, warmup, cruise, idle, and acceleration. At the end of this chapter you'll find detailed schematics, with locations of the various sensors and actuators.

1.3 Types of Sensors

You'll find six different types of sensors, classified according to their input signals:

Thermistors are temperature-sensitive variable resistors that operate on a 5 v. reference voltage (VREF). Resistance changes with temperature, changing the signal voltage going to the control module.

Potentiometers are mechanically variable resistors that also operate on VREF. Signal voltage changes with position or rotation.

Signal generators create a changing signal directly, without VREF. Some send voltage signals; others send frequency signals.

Hall effect devices create digital pulses, signals which change as the shaft rotates.

Magnetic pickups create an alternating current (AC) signal that changes as a tooth, or notch, passes a stationary magnet.

Hot-wire sensors create a voltage or frequency signal that changes according to the mass of air passing them.

Sensor signals operate on low power, so sensor currents are measured in *milliamps* (one mA = 1/1000 ampere). That means that input signals are sensitive to the effects of corrosion, particularly at the connectors. In contrast, actuator signals may be relatively high, since they actually power things, rather than just sending information. Currents in actuators can get up to two amperes.

Most Corvette sensors operate on a 5 volt *reference voltage* (VREF), which is stepped down from B+ (battery voltage). A few sensors (**LS1**'s Crankshaft position and CaMshaft position sensors) operate on a 12 volt reference voltage.

1.4 Types of Switches

Some switches just say "yes" or "no" to the control module, about conditions that could affect the engine. These are referred to as *discrete* sensors. In GM-speak, the signals they send are either Logic high (yes) or Logic low (no).

Fixed current-limiting resistors reduce the current flow to the control module. A typical 2,000-ohm resistor reduces the current flow to a tiny 6 mA. (From Ohm's law, 12 volts ÷ 2,000 ohms = .006 A, or 6 mA.) For example: an air conditioner request from the ignition switch.

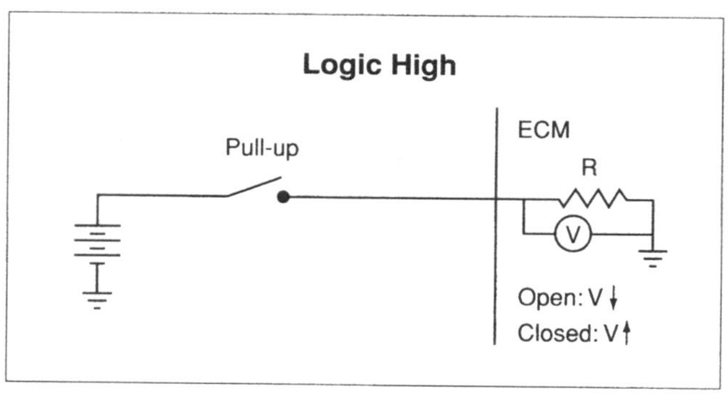

Fig. 4-1. In Logic high, a pull-up operates on a power source *outside the computer.* Notice the battery symbol outside on the left. A closed switch allows the external-source voltage to flow, "pulling up" the signal to the computer. An open switch would send a low-reference signal.

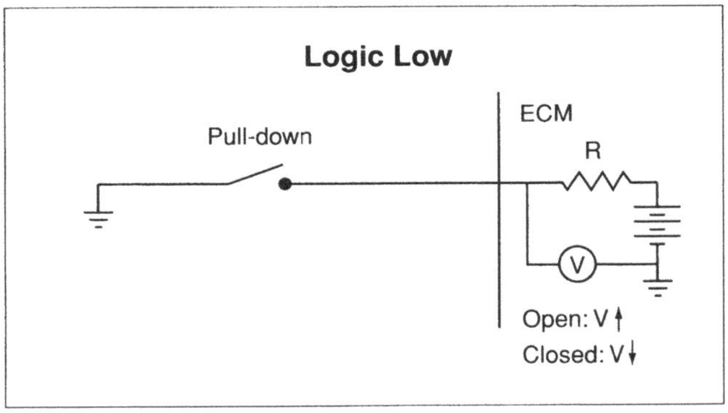

Fig. 4-2. In Logic low, pull-down operates on a power source *inside the computer*, usually 5 volts. A closed switch grounds the source voltage, pulling down the low reference signal to the computer. An open switch would send a high reference signal. For example: the Park-Neutral switch.

When you see the term "EI"—Electronic Ignition—look out!

- In the seventies, "Electronic Ignition" *was* the term applied to the system that replaced points and condensers.
- Now, "EI" applies to electronic systems that control distribution through the computer, eliminating the distributor.

I argued with the SAE J1930 committee against this new meaning ascribed to "EI," proposing instead "Electronically Distributed Ignition (EDI)."

Bottom line: I lost. **LS1** and **ZR-1** engines use Electronic Ignition (EI), formerly referred to as a Direct-fire Ignition System or Distributorless Ignition System (DIS).

At least we got rid of "Distributorless." Nobody calls fuel injection "*carburetorless*", anymore.

4

2. Distributor Reference

2.1 Engine rpm, Crankshaft Position, and Cylinder Identification

You'll find several different "Distributor Reference" (Corvette-speak) systems, which provide input signals for rpm, crankshaft position, firing order information, and cylinder identification. These systems include:

L83, L98: a pickup coil and core in the distributor, and High Energy Ignition (HEI)

LT1, LT4: a CaMshaft Position (CMP) sensor, Opti-Spark (also known as Distributor Ignition, DI), and the CranKshaft Position (CKP) sensor (added in **'96** models, for OBD-II)

LS1: a CranKshaft Position (CKP) sensor for rpm and crankshaft position, a CaMshaft Position (CMP) sensor for cylinder-stroke identification, and Electronic Ignition (EI)

ZR-1: a CranKshaft Position (CKP) sensor, CaMshaft Position Sensor (CMP), and Electronic Ignition (EI)

Distributor reference signals determine the speed of the crankshaft's rotation (rpm). These signals also indicate the position of the crankshaft (expressed as an angle from 0–360°). Crankshaft position and cylinder identification signals are important to fuel injection, spark timing, idle rpm, and emission control. The signal for engine rpm is probably the most important. Without it, the engine will not run. And with it, you can usually limp home, even if signals from most other sensors are lost.

2.2 L83, L98

Crankshaft rpm and position signals, which regulate spark timing and fuel-injector timing, are based on the High Energy Ignition (HEI) system, located in the distributor.

The magnetic field becomes stronger as the teeth on the timer core move toward the pole-piece teeth. As the air gap gets smaller, the pickup coil signals the ignition module. This turns the primary current on. (See Fig. 4-3.)

As the timer-core teeth move away from the pole-piece teeth, the magnetic field becomes weaker. And as the air gap

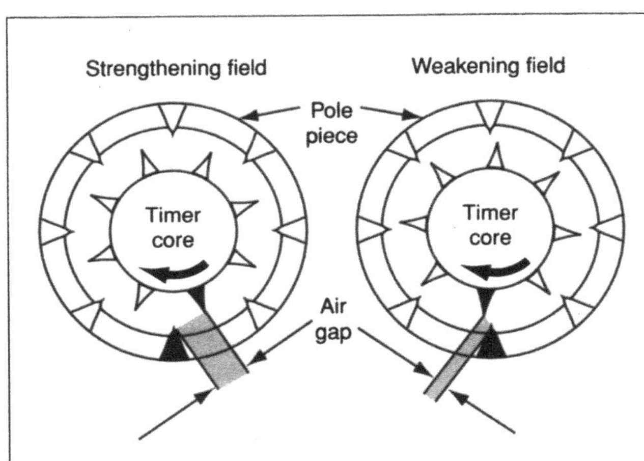

Fig. 4-3. **L83**, **L98**: Electronic Spark Timing (EST) begins with a rotating *timer core* inside a *pole piece* in the distributor. There are eight points, one for each of the eight cylinders. The relation between the timer-core teeth and the pole-piece teeth determines the firing of the spark plugs.

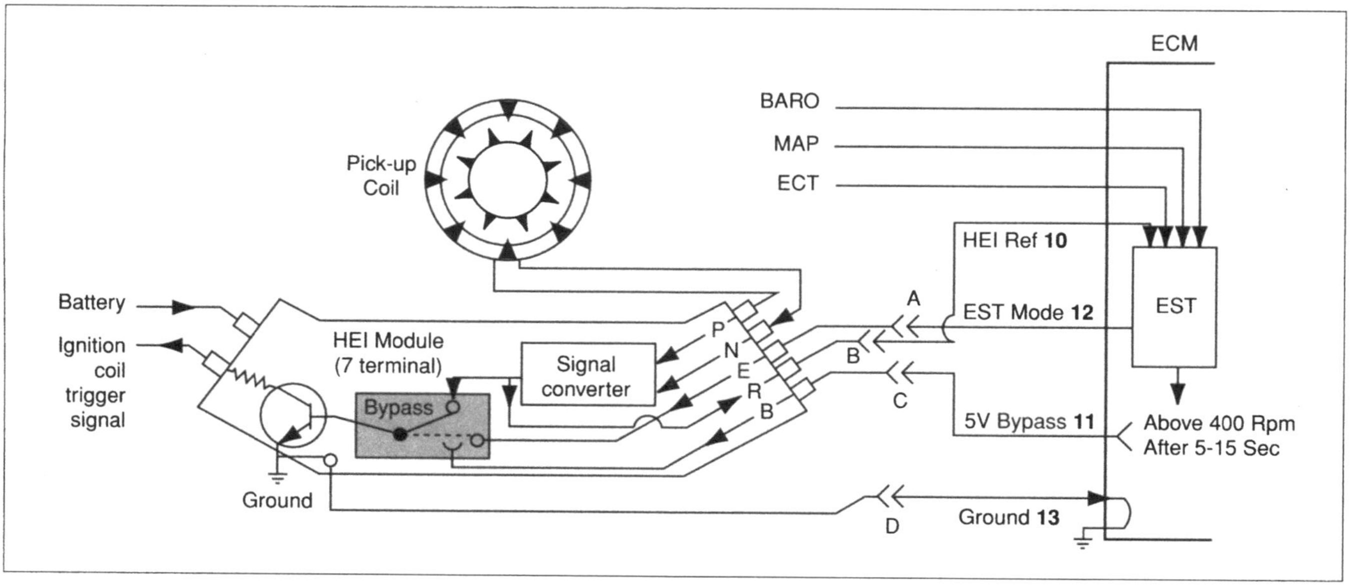

Fig. 4-4. L83, L98: The pickup coil and HEI module in the distributor send HEI reference signals (#10) to the engine computer, which responds with EST (#12) and bypass (#11) signals.

gets larger, the pickup coils signal the ignition module to turn-the primary current off. This induces a high voltage current in the secondary coil, and the distributor fires one spark plug.

During engine cranking, *bypass* timing is fixed at, or close to, base timing. As the name suggests, this timing bypasses any modifications from the computer. Above 400 rpm, or after 5–15 seconds, the 5-volt bypass circuit is opened. This switches spark timing from the HEI module to the engine computer, so that timing is variable, instead of being fixed at base. (See Fig. 4-4.)

The terms Electronic Spark Timing (EST) and Electronic Spark Control (ESC) can be confusing. I'll explain ESC later on in this chapter.

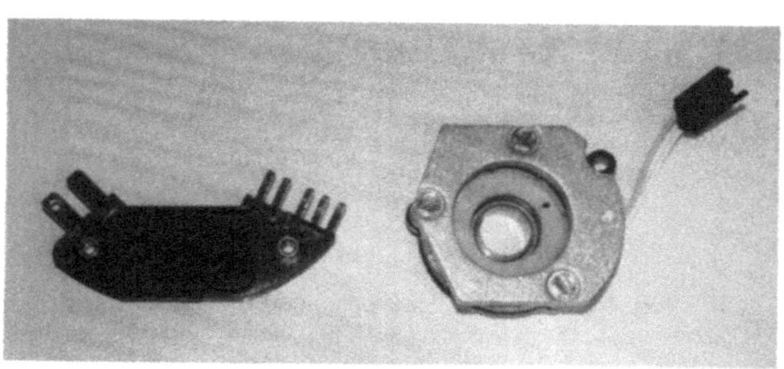

Fig. 4-5. L83; L98: The HEI module and pole piece ihave been removed from the distributor.

2.3 LT1, LT4

CaMshaft Position (CMP) Sensor

In the **LT1** and **LT4** engines, the CaMshaft Position (CMP) sensor indicates crankshaft rpm and position (for cylinder identification). That's in contrast to most engines, where crankshaft signals are sent by a CranKshaft Position (CKP) sensor.

The CaMshaft Position (CMP) sensor is in the distributor, mounted directly on the end of the camshaft. This system is known as:

- Distributor Ignition (DI), in SAE J1930-speak
- Opti-Spark, in GM-speak

It operates optically by a pair of Light-Emitting Diodes (LEDs). These LEDs send light through two sets of slots, in a rotating sensor disc, to the Optical Position Sensor, a pair of photo-electric diodes.

When a slot allows light from the LED to fall on a diode, a signal is sent to the computer and Ignition Control Module. And when the slot stops allowing the light through, another signal is sent. The inner ring of eight slots is the low-resolution track. The outer ring of 360 slots is the high-resolution track.

On the inner ring, the signal when the LED light is blocked indicates the Top Dead Center for each cylinder, and provides crankshaft rpm and position, as in a CranKshaft Position (CKP) sensor.

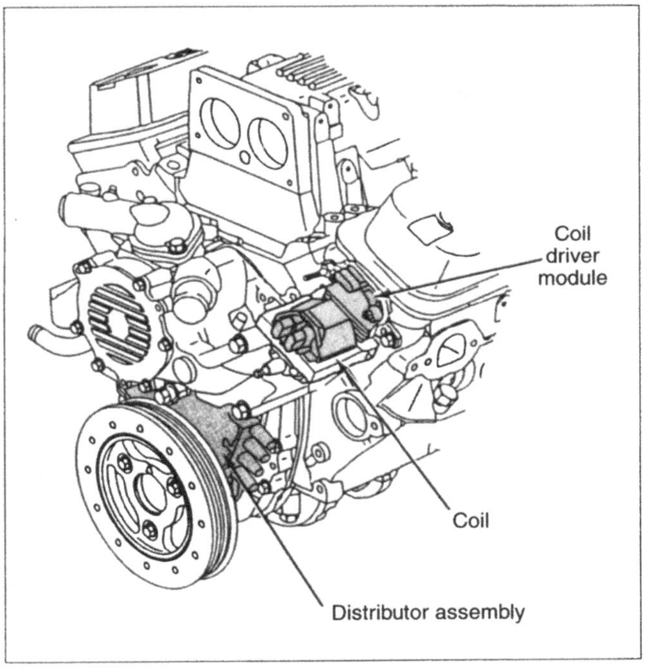

Fig. 4-6. LT1, LT4: Crankshaft rpm and position are signaled by a distributor assembly, mounted directly on the end of the camshaft. This improves timing accuracy, compared to HEI, which is a gear drive from the camshaft to the distributor.

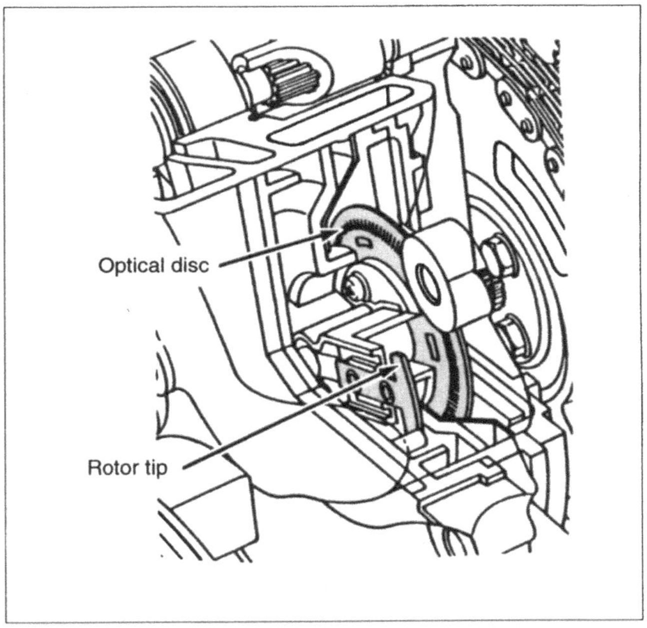

Fig. 4-8. LT1, LT4: Inside the distributor are the optical disc and rotor.

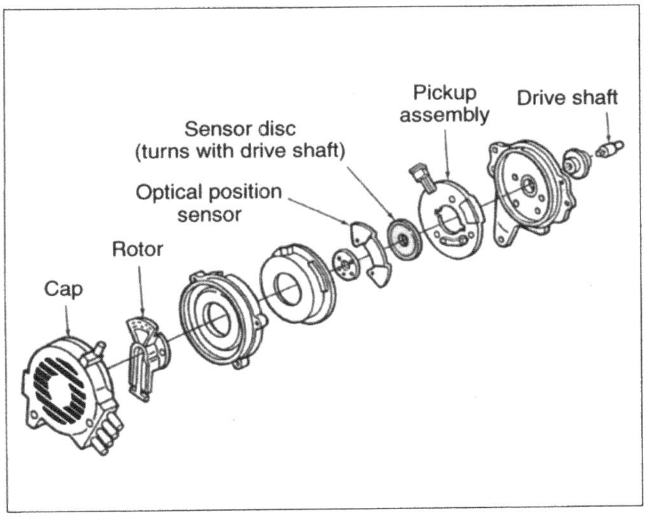

Fig. 4-7. LT1, LT4: Inside the distributor are a stationary pickup assembly and a sensor disc, that rotates with the camshaft. Each revolution of the disc fires all of the eight cylinders, in two crankshaft rotations.

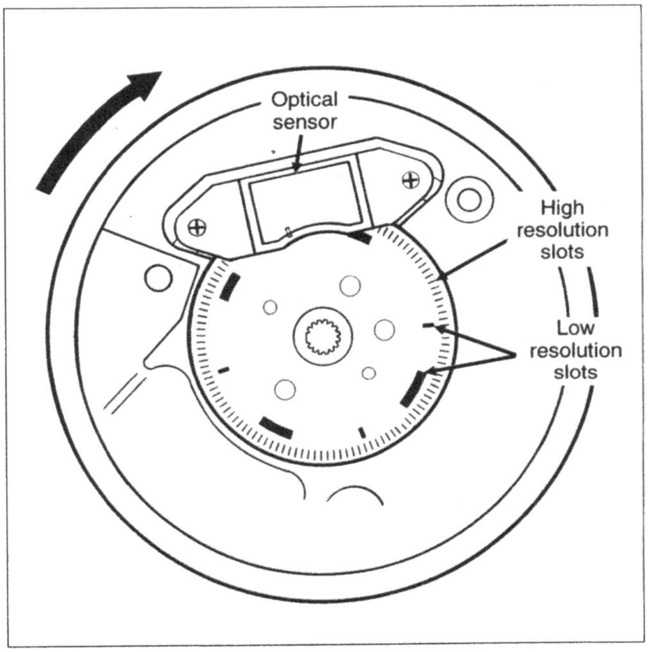

Fig. 4-10. LT1, LT4: In the outer ring of the sensor disc, look for 360 high-resolution slots, one for each degree of camshaft rotation. Look for an inner set of eight low-resolution slots, one for each cylinder. Viewed from the front of the engine, the disc rotates clockwise.

4

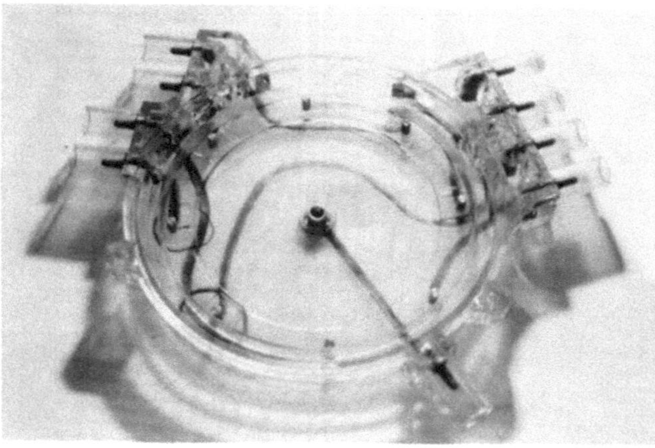

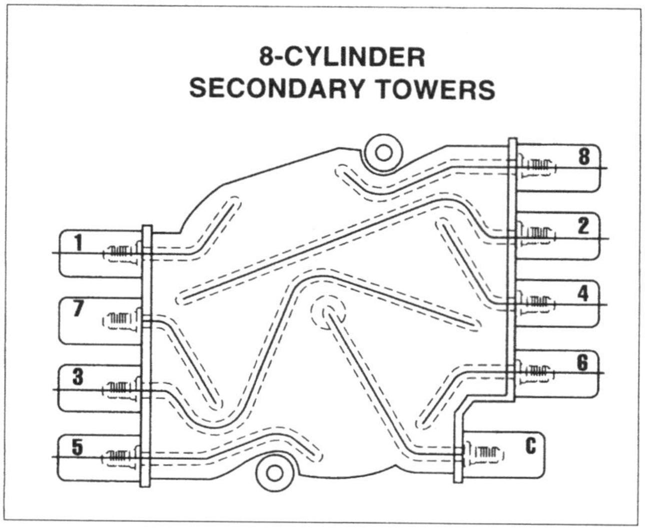

Fig. 4-9. LT1, LT4: The distributor cap shows the secondary leads for the coil and plugs. The leads cross over to the left for 1,3,5,7, and to the right for 2,4,6,8, so that the firing order is correct.

Fig. 4-11. LT1, LT4: You can clearly see the slots, and their different sizes, in this photo of a sensor disc.

The inner ring also sends out cylinder identification signals. The slots for cylinders 8, 3, 5, and 2 are not the same size—they increase in width—so the size of the slots identifies the individual cylinders:

Cylinder 8: 7°
Cylinder 3: 12°
Cylinder 5: 17°
Cylinder 2: 22°

The sizes are shown as angles, since that's how a technician measures them, but the computer actually times the light coming through the slots. So the signal when the light is allowed through indicates cylinder identification. (See Fig. 4-11.)

Wait a minute, you might be saying. That all works fine for identifying cylinders 8,3,5, and 2, but how does the computer find the others? Easy— cylinder #1 immediately follows #2 (the 22° signal), and then #4 follows #8, and so on, in spark firing order. This system can identify the cylinders in about a quarter of a crankshaft revolution, reducing start-up pollution. Some earlier engines need one or even two full crankshaft revolutions. Earlier starts reduce start-up pollution, and please Corvette drivers, who want their start *now*.

In total, the sensor disc's lo-res ring sends out sixteen separate pulses (eight on, and eight off), per camshaft revolution. These pulses measure rpm, provide injector pulses, calculate spark timing, and energize the fuel pump. This makes the lo-res signal a must. If its pulses are missing, the engine will crank but won't start.

OK, so what does the outer-ring signal do? Well, its hi-res track consists of 360 equally spaced slots, each switching the signal on and then off. So the photo-electric diode sends a total of 720 pulses to the control module, for each rotation of the disc on the camshaft. These pulses signal spark timing.

Of course, one rotation of the camshaft equals *two* rotations of the crankshaft. And with the 720 pulses from the outer ring that gives one pulse for each 1° rotation of the crankshaft. We're talking about 1° here. Precision spark timing is increasingly important for emission reduction and fuel economy, and the high-res signal helps provide that fine tuning.

Improved Spark Control—Starting and Acceleration

Opti-Spark operates at all times, under computer control, with sensor inputs. During starting, timing can be modified for several factors, such as Intake Air Temperature (IAT). This is in contrast to the fixed starting timing of bypass ignition systems, like HEI on **L83** and **L98** engines.

Spark Scatter: Opti-Spark, with one signal for each degree, reduces spark scatter from the ordinary time-based spark delivery to angle-based delivery:

- Time-based means that spark timing is the same, for both low-speed operation as for maximum rpm.

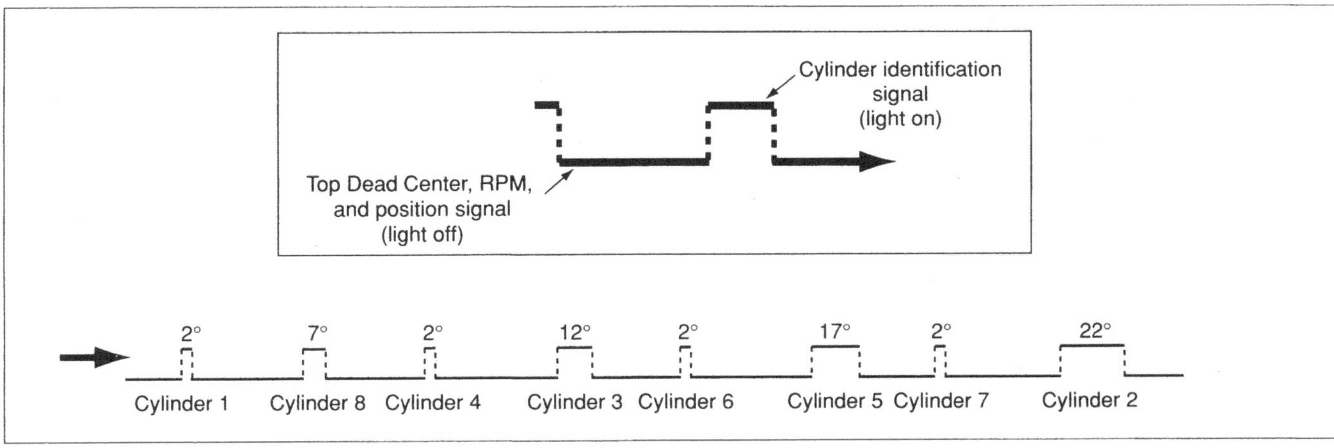

Fig. 4-12. LT1, LT4: The computer receives the signal, from the sensor disc's inner ring, as a straight line, with the open slots allowing the light through to the diode. When the diode does not see the light, the crankshaft rpm and position signals are indicated. When the light is allowed through the slot it hits the diode, which sends the cylinder identification signal. The length of the alternating signals is what is important.

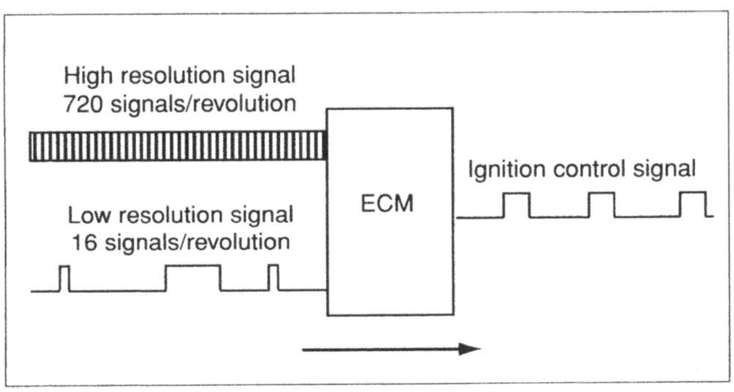

Fig. 4-13. LT1, LT4: Compare the 720 hi-res signals for fine tuning with the 16 lo-res signals. The computer continuously compares the hi-res signals with the lo-res.

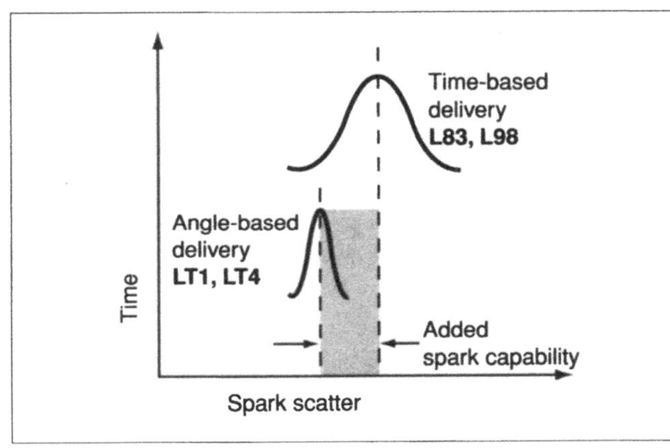

Fig. 4-14. LT1, LT4: Opti-Spark, with one signal for each degree of crankshaft rotation, reduces spark scatter, compared to the ordinary time-based spark delivery.

- Angle-based means that the rotation of the distributor disc indicates spark timing. At higher rpm, the faster disc rotation shortens the time, reducing the scatter. This term gets its name from the angle of the rotating disc.(See Fig. 4-14.)

This allows the computer to add more spark advance before engine knock occurs. The results are better control, during acceleration, transmission shifts, and coasting, as well as added power, and improved fuel economy. And you don't even need to adjust timing, with Opti-Spark. **LT1** and **LT4** distributors are called "net-build", and are so accurate, no messing around is necessary.

Distributor Vent

The distributor vent system in the **'95–96 LT1** and **LT4** prevents buildup of ozone, caused by sparks jumping in the distributor. The ozone and air are pulled from the distributor cap through a check valve, to be burned in the combustion. Fresh air is picked up from ahead of the throttle body and delivered to the base of the distributor. (See Fig. 4-15.)

Misfire Detection '96 LT1/LT4, LS1

Detection of misfires has become increasingly important for emissions. As a result of the *counterforce* pulses of compression and power, the crankshaft decelerates and accelerates slightly in each revolution. (See Fig. 4-16.) This effect is dampened by the flywheel. The computer times the regularity of these minor speed changes and looks for the momentary drop in crankshaft speed, as each cylinder fires.

In a misfire, one or more of the power impulses is missing, changing the regularity of the crankshaft's twisting and untwisting. Measuring such an irregularity detects a misfire—

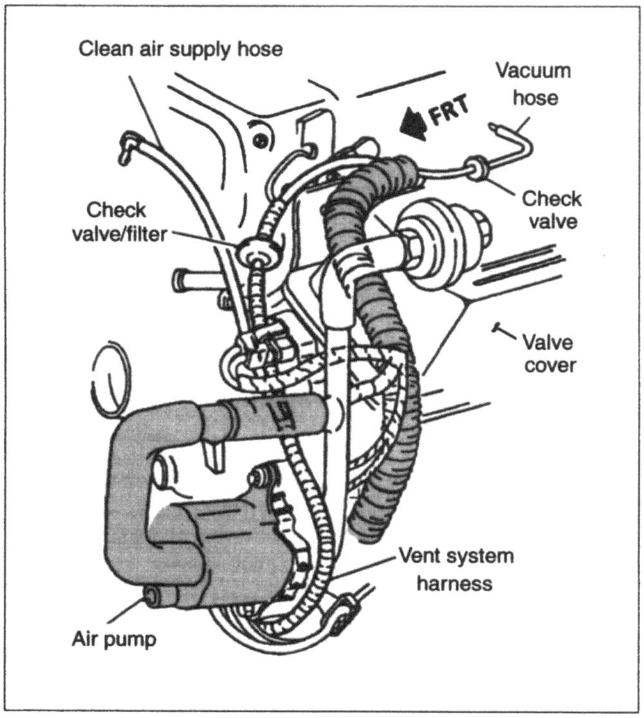

Fig. 4-15. '95–96 LT1, LT4: Fresh air is picked up from ahead of the throttle body and delivered to the base of the distributor. The distributor vent system prevents buildup of ozone caused by sparks jumping in the distributor.

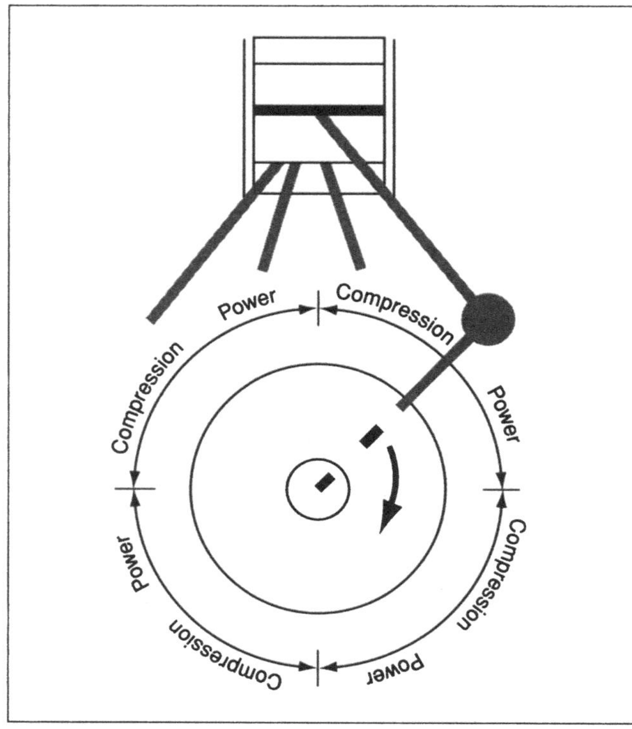

Fig. 4-16. For each crankshaft revolution, the four cylinders produce four compression strokes, which twist the crankshaft counterclockwise. And four power strokes twist the crankshaft clockwise. Under these counterforces, the crankshaft decelerates and accelerates in each revolution, dampened by the flywheel.

yes, even one cylinder! These variations cannot be sensed by a sensor driven by the camshaft. Why? Because the camshaft generates its own impulses, caused by the chain drive from the crankshaft.

In engines with On Board Diagnostic, second generation (OBD-II), misfires are sensed, and stored, in the computer. Just a few misfires can result in raw fuel getting into the catalytic converter(s), causing overheating of the converter core, raw fuel coming out of the tailpipe, and hydrocarbon pollution.

A rough road can also cause irregularity of crankshaft twisting, induced by the movement of the propellor shaft. In the engine computer, special CranKshaft Position (CKP) software is able to distinguish the pattern of most rough roads. As a backup, the AntiLock Braking System sensors can detect rough-road variations in rear-wheel speed.

- With manual transmission, rough-road signals require careful analysis to distinguish them from misfire signals, because the crankshaft is usually solidly connected to the drivetrain.
- With automatic transmission, the torque converter clutch is disconnected when a misfire is suspected. This lets the computer concentrate on crankshaft irregularities, unaffected by the rough road.

Boy, what these engineers do, so that you won't worry!

So in the **LT1** and **LT4**, Opti-Spark provides the distributor reference signals for timing of fuel injection and ignition. And in the **'96 LT1** and **LT4**, the CKP provides misfire detection.

2.4 LS1, LS6

CranKshaft Position Sensor (CKP)

The **LS1**'s CranKshaft Position Sensor provides accurate information on rotation, including engine rpm, crankshaft position, and cylinder identification. The CKP can sense misfires from the momentary variations in crankshaft rotation. This does not depend on rotation speed, which means that the CKP is capable of sensing at nearly zero rpm. Sensing crankshaft location improves timing accuracy, as compared to a CaMshaft Position (CMP) sensor, because slack in the camshaft chain drive has no effect. Locating the sensor near the flywheel also minimizes any possible effect of crankshaft twisting, as the different cylinders fire.

The engine will start, and run, with only the CranKshaft Position (CKP) sensor inputs, but starts even quicker with the additional input of the CaMshaft Position (CMP) sensor.

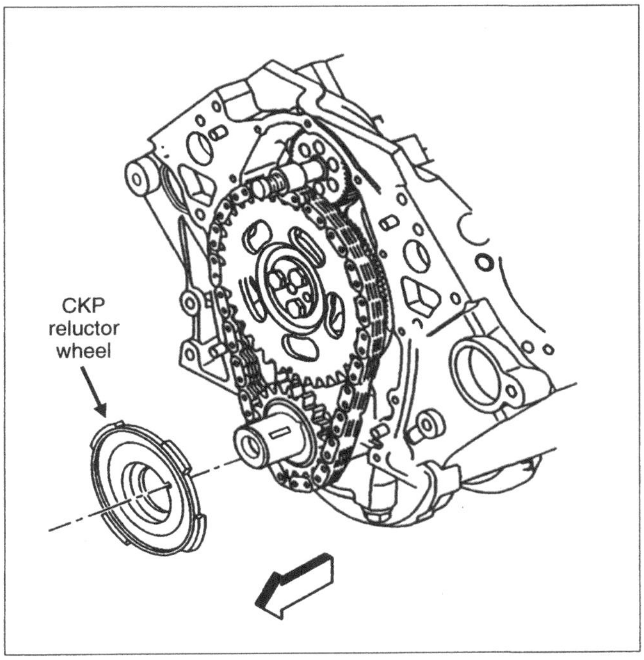

Fig. 4-17. '96 LT1, LT4: This CranKshaft Position (CKP) sensor has the sole job of misfire detection. The four teeth of the reluctor wheel correspond to the four cylinders that fire on each crankshaft revolution. The computer compares the CKP sensor signals, to the crankshaft position signals, from the distributor, to identify each of the eight cylinders.

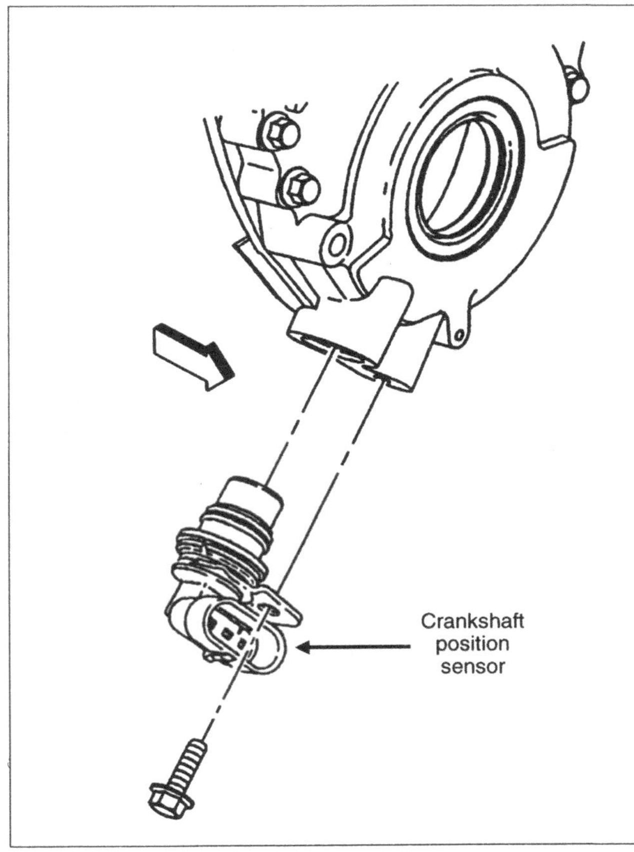

Fig. 4-18. '96 LT1, LT4: The CranKshaft Position (CKP) sensor detects misfires for OBD-II, by sensing minor variations in crankshaft rotation.

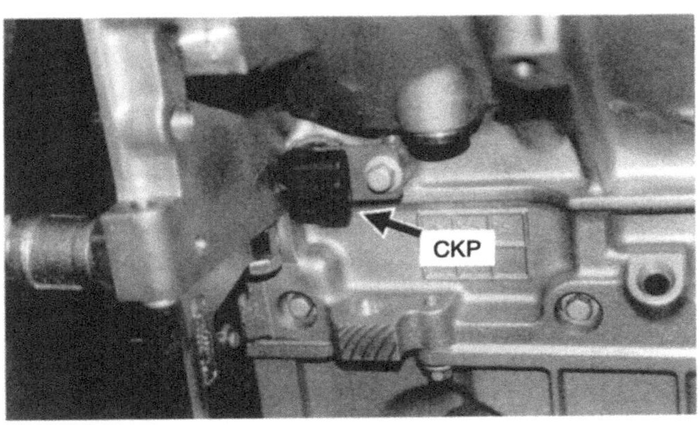

Fig. 4-19. LS1: Look for the CranKshaft Position (CKP) sensor in the right rear of the block, below the starter.

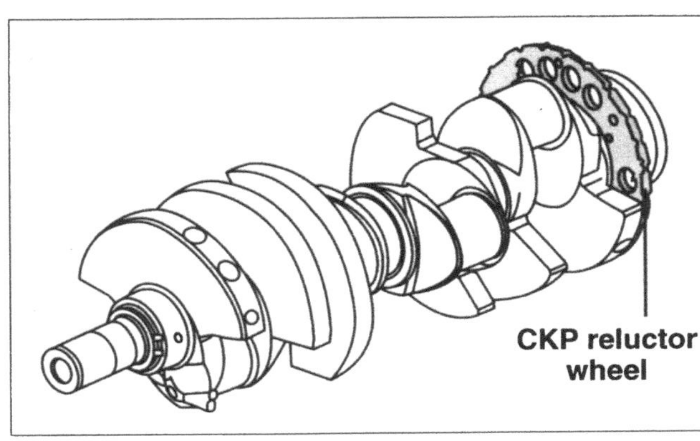

Fig. 4-20. LS1: A dual reluctor wheel is mounted on the rear of the crankshaft, in what is known as the "quietest deflection point." The sensor is mounted close to the reluctor wheel, with the minimum of variation in the air gap.

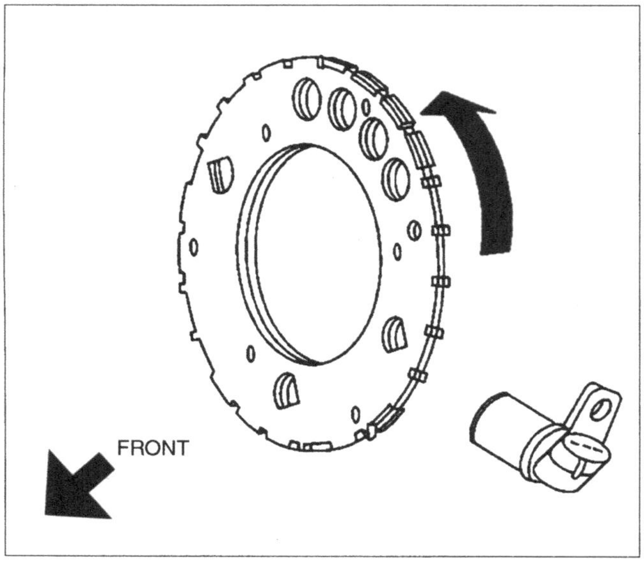

Fig. 4-21. LS1: Notches on the dual reluctor wheel permit quick identification of the individual cylinders, for precision in injection timing and spark deliver. "Quick Start" reduces starting emissions.

Fig. 4-22. LS1: The CKP reluctor wheel is on the rear of the crankshaft.

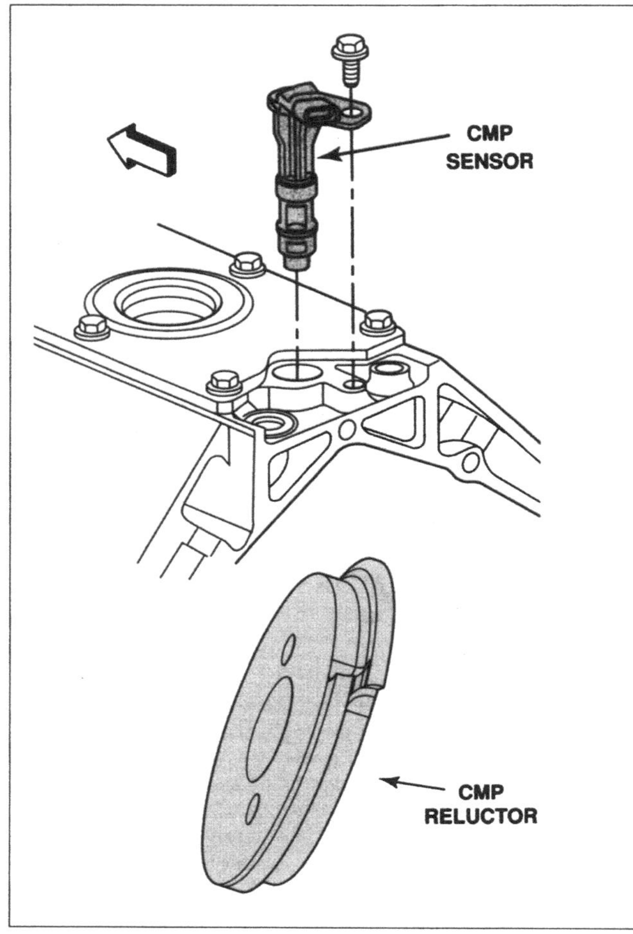

Fig. 4-23. LS1: Look for the CaMshaft Position (CMP) sensor on top of the block, at the rear of the valley cover. The dual-track reluctor wheel is directly below, on the *cam*shaft.

Reluctor wheel, a new term. Reluctance is the resistance, in a circuit, that is caused by a magnetic field. Notches in the rim of the reluctor wheel cause variations in the magnetic field. The pickup senses these changes, altering its voltage output to the computer.

CaMshaft Position (CMP) Sensor

The CaMshaft Position (CMP) sensor is mounted on top of the engine block, at the rear of the valley cover. It reads signals from a *reluctor wheel*, with one notch (1X) mounted on the rear of the camshaft. The computer combines the CaMshaft Position and CranKshaft Position signals, to identify whether a cylinder is on its firing or exhaust stroke.

If the camshaft signal is missing, the computer attempts a start based on the firing stroke. If the Mass Air Flow (MAF) signal increases, the computer has made a good guess—the cylinder was on its firing stroke. If airflow doesn't increase, and the computer has made a bad guess, it resynchronizes, according to the intake stroke, the opposite cam position. This results in slightly longer cranking. So the CaMshaft Position (CMP) sensor allows quicker starts but is not necessary for starting or running the engine.

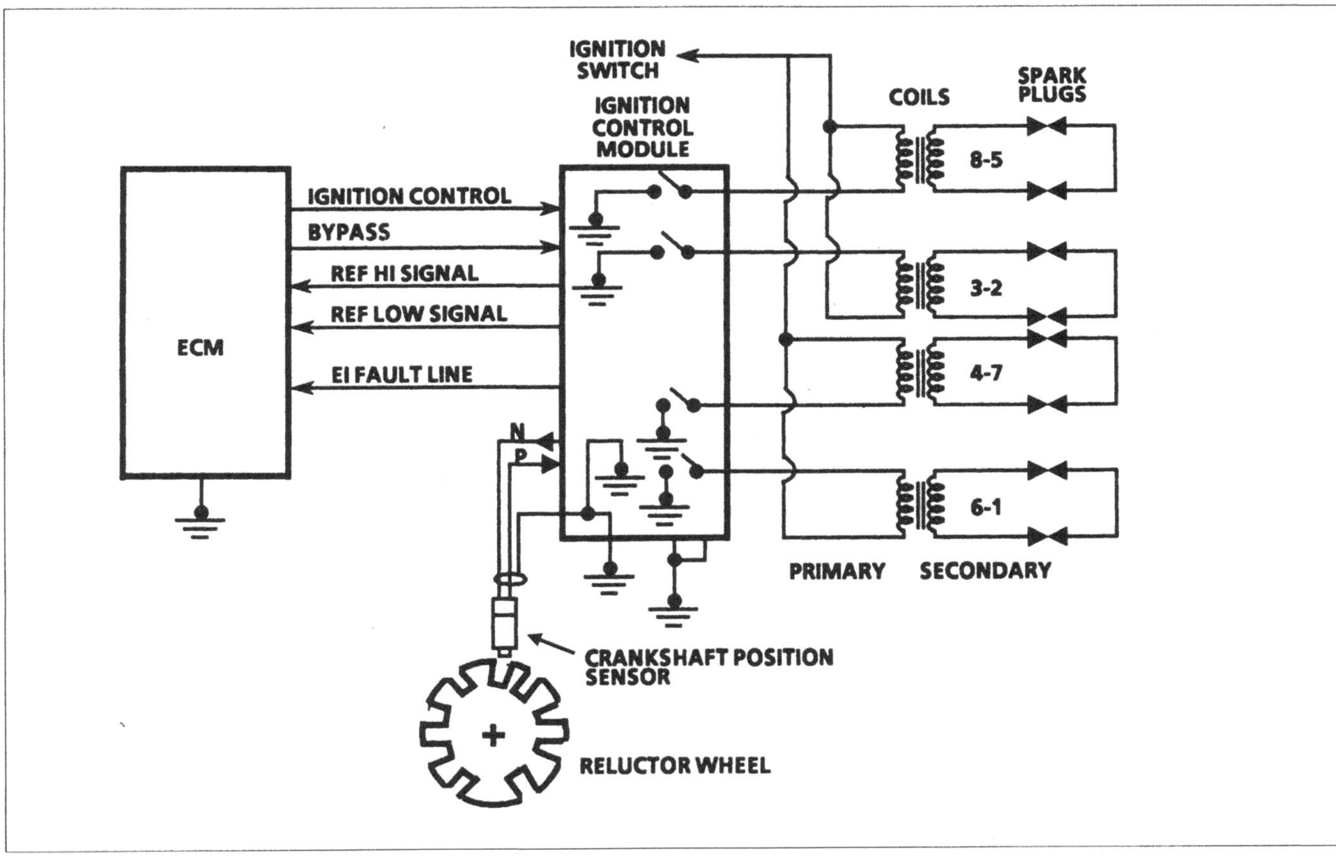

Fig. 4-24. **ZR-1:** The CranKshaft Position (CKP) sensor signals crankshaft rpm and position. Notice that the eight cylinders are paired, to operate from four coils. The Ignition Module is separate from the Engine Control Module.

2.5 ZR-1

CranKshaft Position (CKP) Sensor

The crankshaft-based sensor in the **ZR-1** can send more accurate signals, than those resulting from a camshaft-distributor-gear system, as in the **L83** and **L98**, or a camshaft-chain drive, as in the **LT1** and **LT4**. The CKP signal provides timing through the Ignition Module and the computer, for the ignition system. With the **ZR-1** winding to speeds as high as 7,000 rpm, accuracy of these timing signals is critical.

Aside from eliminating chain/gear inaccuracies, the engine takes its timing from the rear of the crankshaft, to reduce the effect of crankshaft twisting during power strokes. That's close to the flywheel, and the "quietest deflection point." The increased timing accuracy, for injection and spark timing, means more power and acceleration, aids the built-in engine diagnostics, and reduces emissions.

In the **ZR-1** engine, a notched or slotted wheel is machined from the solid part of the crankshaft. The eight slots on the wheel produce changes, in the pickup's magnetic field, which send pulses to the computer, one for each pair of cylinders. Based on these pulses, the computer can determine both engine rpm and crankshaft position.

RPM: The computer counts the slots to determine rpm, as well as determining top dead center (TDC) for each pair of cylinders. In your mind, picture the wheel as it rotates past the CKP. Numbered slots (at 90°) pull the reference voltage high. Unnumbered slots (at 45°) pull the reference voltage low.

Crankshaft position: The ninth slot sends an odd pulse, known as the synchronization (sync) pulse, for the crankshaft position, so the computer can identify each individual cylinder of each pair.

Referring to Fig 4-25 and 4-26, follow the sequence, rotating the CKP wheel in your mind counterclockwise past the CKP sensor. Remember that the CKP wheel and crankshaft rotate twice. During crank, the 4/7 coil is the first to be charged, after the sync pulse:

- slot 4/7 pulls VREF to high

- 45° slot (upper left) pulls VREF to low, firing the 4/7 coil
- slot 3/2 pulls VREF to high
- 45° slot (upper right) pulls VREF to low, firing the 3/2 coil
- slot 6/1 pulls VREF to high
- 45° slot (lower right) pulls VREF to low, firing the 6/1 coil
- slot 8/5 pulls VREF to high
- 45° slot (lower left) pulls VREF to low, firing the 8/5 coil

Repeat for the second revolution.

The AC voltage from the sensor can change, from as low as 1/3 volt at cranking speeds, to 180 volts (ouch!) at high rpm. With the sensor fixed in the crankcase, and with the reluctor wheel a solid part of the crankshaft, there is no timing adjustment. This EI is called "net-build" timing.

You can see that the CranKshaft Position (CKP) sensor signals crankshaft rpm and position for *ignition* firing. But turning twice for eight cylinders, it can't signal which is the next cylinder on any rotation for injection. That is the job of the CaMshaft Position (CMP) sensor.

CaMshaft Position (CMP) Sensor

CaMshaft Position (CMP) sensor signals are sent by a semiconductor known as a "Hall effect" switch. This switching transistor has no moving parts, which means no wear or adjustments. It consists of a solid-state switching element and a permanent magnet, separated by an air gap.

The Hall effect switch is triggered by a reluctor wheel. In the **ZR-1**, the wheel is machined with one slot, not the usual Hall effect blade or vane. As long as the reluctor wheel cuts off the magnetic field from the Hall element, reference voltage is pulled high—12 volts. When the reluctor wheel slot allows magnetic flux to flow through the Hall element, reference voltage is pulled low. (See Fig. 4-30.)

When does the system fire the injector?

- Each numbered slot fires the injector for one cylinder. Remember, the **ZR-1** can fire both of its *two injectors per cylinder*, under certain high-power conditions.
- Each unnumbered slot fires *ignition* (VREF to low).

The signals are precise, and are essentially digital.

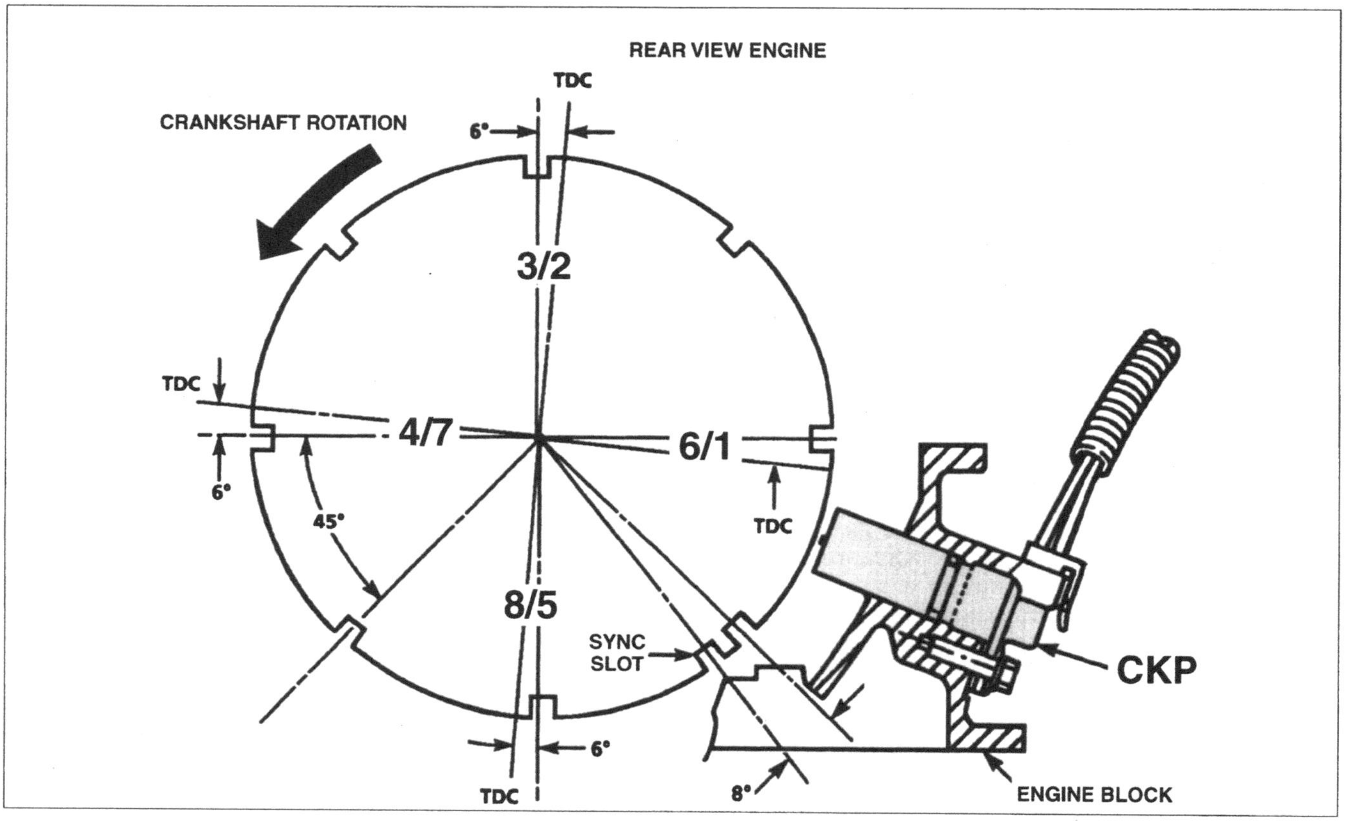

Fig. 4-25. ZR-1: On the crankshaft, an eight-slotted wheel sends pulses to the computer. A ninth slot sends an odd pulse, known as the synchronization (sync) pulse. Remember, you're looking at the wheel from the rear of the crankshaft, so its rotation arrow is counterclockwise.

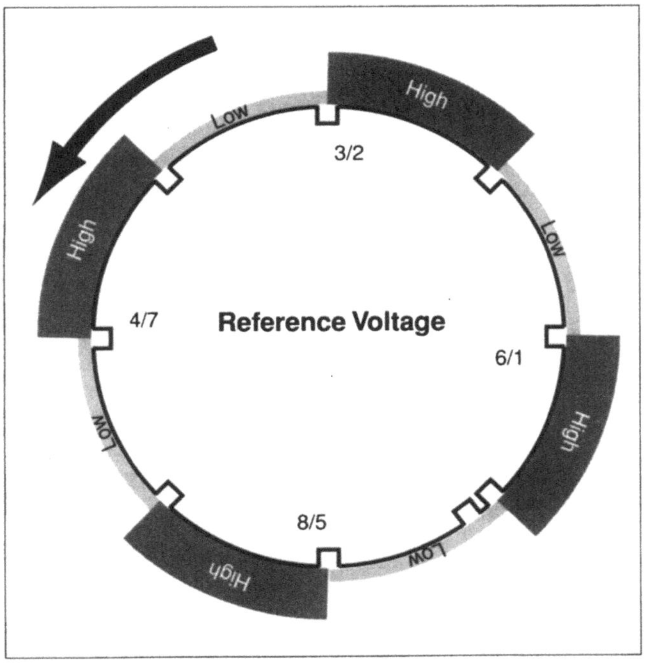

Fig. 4-26. ZR-1: As the wheel rotates, each slot passing the sensor causes a voltage-change signal. Numbered slots signal the computer to pull reference voltage high. Unnumbered slots signal the computer to pull reference voltage low.

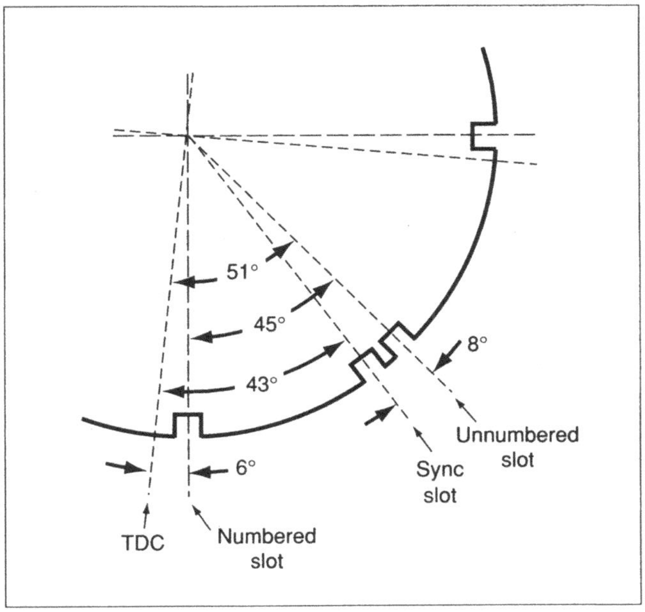

Fig. 4-28. ZR-1:
Each of the regular slots is 45° from the next. TDC is 6° after the numbered slots. So each TDC is 51° after an unnumbered slot.
The *sync* slot is 8° after one of the 45° slots. So the next TDC is 43° (51 – 8 = 43) after the odd slot
The odd slot provides base timing 43° before the next TDC.

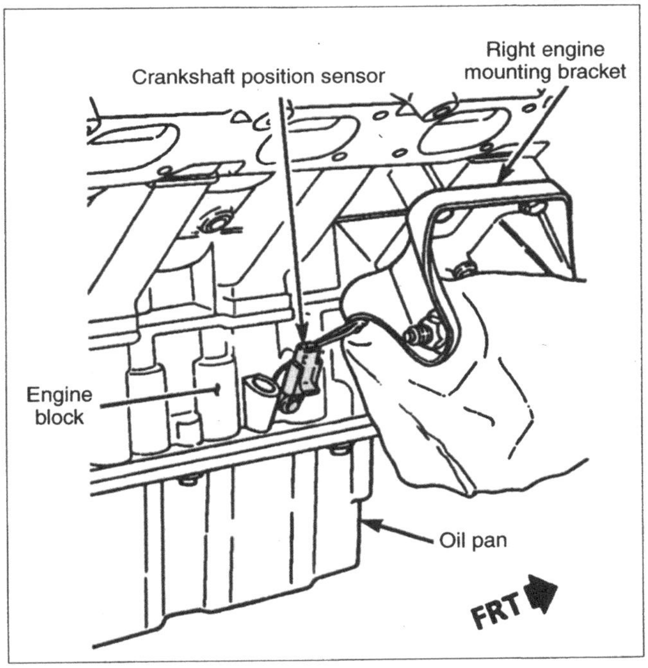

Fig. 4-27. ZR-1: As long as the reluctor wheel cuts off the magnetic field from the Hall element, reference voltage is pulled high. When the reluctor wheel slot allows magnetic flux to flow through the Hall element, reference voltage is pulled low.

Don't think of the Hall sensor as a "hallway" but as a device named for a Dr. Hall. He discovered a way to accurately switch a small current flow with no contacts and no wear. In the **ZR-1**, the openings in the reluctor wheel allow the magnetic flield through, but in Ford's Hall device, a rotating blade allows the magnetic current to flow, instead of cutting it off.

In **ZR-1** engines, the CMP sensor and CKP sensors work together, to provide the Distributor Reference signals and the injector-timing signals. The Ignition Control Module (ICM) determines its own spark timing, during starting (CKP below 400 rpm), bypassing the spark advance signals from the computer.

2.6 Summary RPM, Crankshaft Position, Cylinder Identification

In summary, you'll find different ways to measure crankshaft rpm and position, to identify the cylinders, and in the **'96 LT1/LT4** and **LS1**, to detect misfires for OBD-II. Electronic-fuel-injection Corvettes may operate with one or two different Crankshaft Position and Cylinder Identification sensors, depending on the model year.

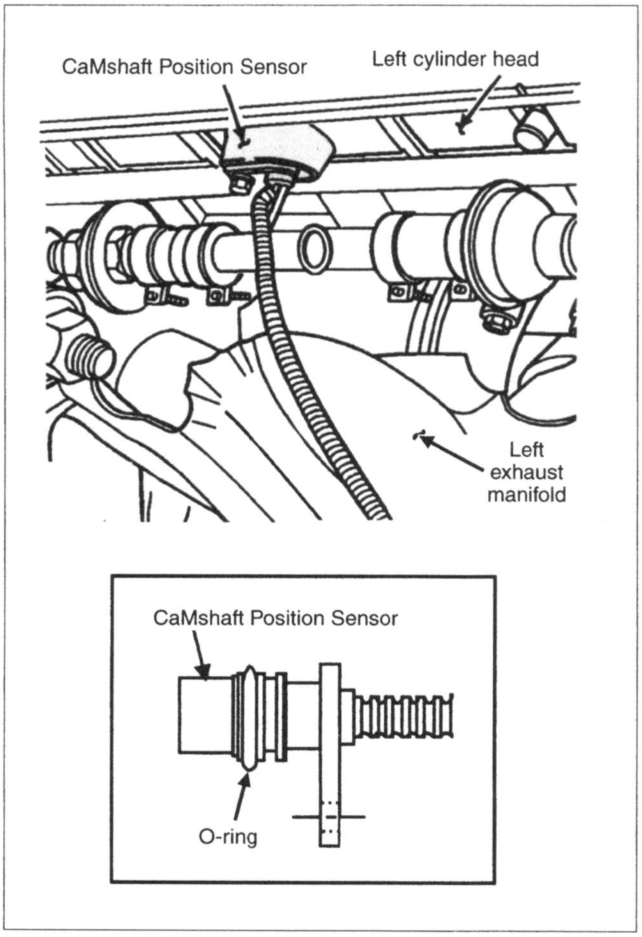

Fig. 4-29. ZR-1: Look for the CaMshaft Position Sensor in the left camshaft cover. Think of the CMP as signalling which cylinder gets injection.

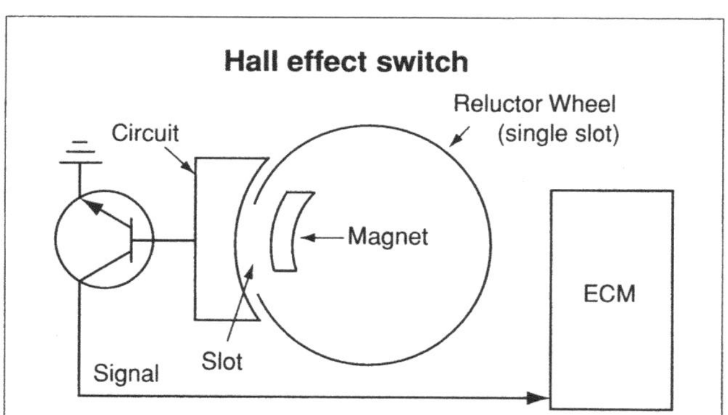

Fig. 4-30. ZR-1: As long as the reluctor wheel cuts off the magnetic field from the Hall element, reference voltage is pulled high. When the slot in the wheel allows magnetic flux to flow through the Hall element, reference voltage is pulled low.

3. Engine Load

Engine load (along with rpm) determines the amount of air/fuel mixture that must be taken in and burned. Depending on the model year, Corvette uses different sensors—speed density or Mass Air Flow—to measure how hard the engine is working. Sometimes you'll find one or the other—and, recently, both. Compare speed density and MAF.

Speed Density

Speed density systems calculate engine load using:

- engine speed (Distributor Reference)
- air density, by sensing the Manifold Absolute Pressure (MAP) along with other factors, to calculate a mass airflow rate. Measuring load by MAP is indirect, requiring several computer calculations.

Mass Air Flow

Mass Air Flow directly measures the mass (or weight) of air entering the engine. Measuring load by MAF is direct. In model years **'85-89** and **'94-01**, Corvette engines measure load with MAF. (See Fig. 4-32.) (Note that the **ZR-1** did not change from MAF to MAP in **'94** because the model's engines had already been built.)

MAF does a better job of maintaining accuracy during dynamics, including acceleration and deceleration. From mass airflow and rpm, the computer determines the weight of the fuel, to provide the proper air/fuel ratio. Remember, the engine burns the weight (mass) of fuel, not the volume.

MAF sensor signals in hertz (Hz) can be read directly, using a frequency meter. They can also be read with a scan tool through the computer, which uses the frequency outputs to calculate Mass Air Flow in grams per second. Many scan tools show the weight, or mass, of airflow in grams per second. Typical flow, depending on altitude is 5–9 grams per second, at idle, and 20–26 grams per second at 2500 rpm, with no load (in neutral or park).

Don't be confused by the similarity of the terms "MAP" and "MAF." They sound similar and are spelled almost the same, but the systems both look and operate differently. You may also get confused because some Corvette engine control systems use MAP, some MAF, and some both, depending on the model year. (See Table 4-1.) In fact, **'94-01** engines measure load, for fuel control, with MAF, but also use MAP as backup. In addition, MAP is measured, and compared with throttle angle and rpm. If MAF is out of range for the expected pressure, the computer sets a Diagnostic Trouble Code (DTC), and switches to MAP.

In addition, MAP is the primary input for spark timing and for altitude corrections.

Table 4-1. Corvette load sensors

Model Year	Normal engine models	High-output engine models
'82–84	**L83** MAP (pressure)	None
'85–89	**L98** MAF (flow)—Bosch	None
'90–91	**L98** MAP (pressure)	**ZR-1** MAP (pressure)
'92–93	**LT1** MAP (pressure)	**ZR-1** MAP (pressure)
'94–95	**LT1** MAF (flow) MAP backup	**ZR-1** MAP (pressure)
'96	**LT1** MAF (flow) MAP backup	**LT4** MAF (flow) MAP backup
'97–01	**LS1** MAF (flow) MAP backup	**LS6** MAF (flow) MAP backup

In Corvette engines, the RPO, or "L" number, won't tell you which load sensor is used as the primary and which is the backup, so refer to Table 4-1 to find your application.

When you consider aftermarket modifications, remember that the MAF sensor compensates for intake or exhaust modifications that increase air throughput. But the MAP sensor doesn't.

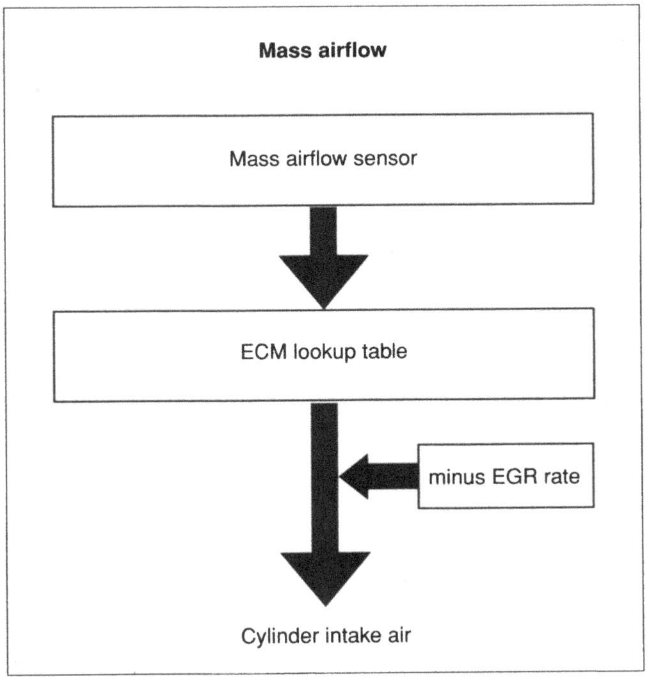

Fig. 4-32. Measuring load by Mass Air Flow is direct. The computer handles any compensations, based on sensor input. The **'85-89 L98** and **'94-95 LT1** computers also compensate for the Exhaust Gas Recirculation rate. In the **LT4** and **LS1**, there is no EGR.

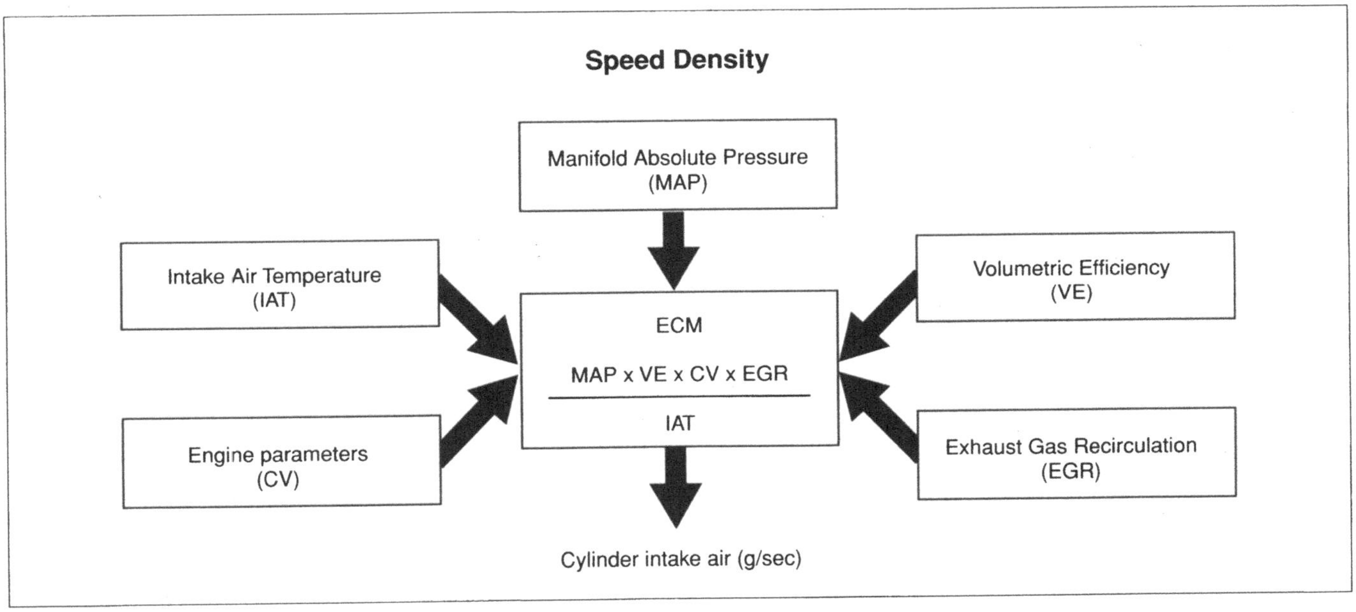

Fig. 4-31. With speed density, the computer calculates the engine load by sensing Manifold Absolute Pressure. It combines the MAP with the Intake Air Temperature and other engine parameters, like volumetric efficiency and EGR rate, to calculate the mass of the intake air.

4

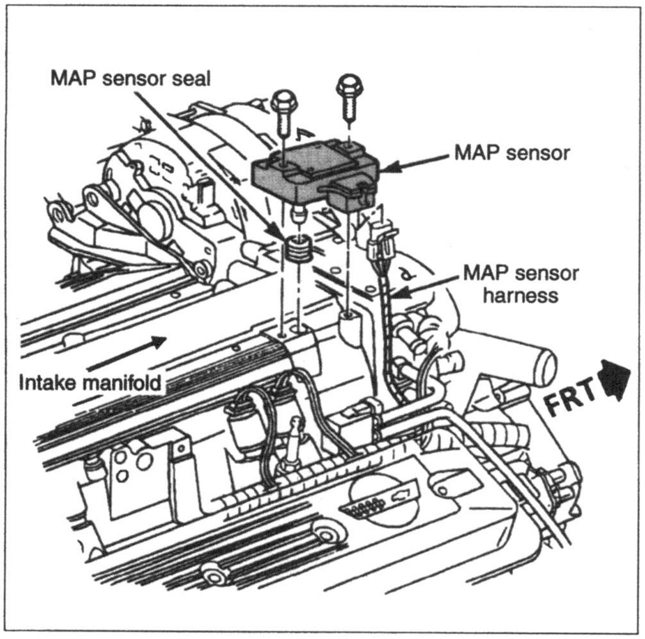

Fig. 4-33. L98, LT1, LT4, LS1: The MAP sensor is ruggedized to operate directly on the intake plenum. The **'96 LT1** is shown here. In the **L83** the MAP sensor is mounted on the firewall.

3.1 Manifold Absolute Pressure (MAP) Sensor

The Manifold Absolute Pressure (MAP) sensor measures increases in manifold pressure, above absolute zero.

Absolute sensors measure the Manifold Absolute Pressure or barometric pressure (BARO). Some MAP sensors can be switched, from measuring MAP, to momentarily check on BARO. In other engines, separate sensors continuously measure absolute pressures, BARO, and MAP.

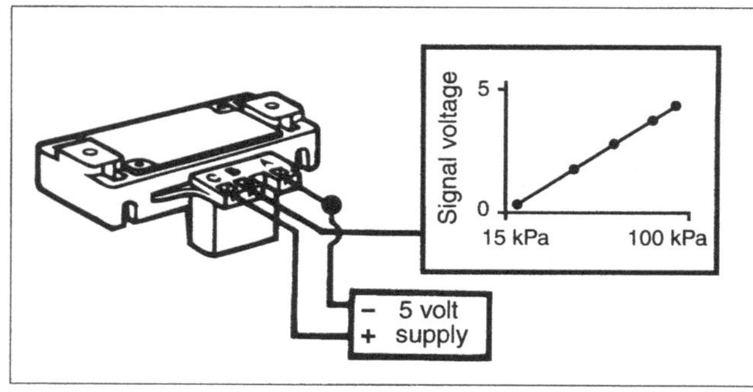

Fig. 4-35. The Manifold Absolute Pressure sensor operates on a VREF of 5 volts. Low MAP (idle, 15–30 kPa), produces low voltage, about one volt. High MAP (Wide Open Throttle, about 95 kPa) produces high voltage, about 4–5 v.

The **LT1**, **LT4**, and **LS1** use BARO (absolute) sensor signals for spark timing, as well as for MAP speed density backup to MAF load signals.

Design and Operation

Inside the MAP sensor (and in the BARO sensor, too) there is a diaphragm, a small silicon wafer, approximately 3 millimeters square, which acts as a strain-gauge. As a result of pressure changes, it flexes and produces a change in resistance which affects the signal voltage to the computer. (See Fig. 4-34.)

Barometric Pressure (BARO) Sensor

In some Corvettes, the control module uses barometric pressure to help control fuel injection, spark timing, and emissions. Barometric pressure is particularly important when the vehicle is driven at altitudes significantly above sea level. Most engines sense barometric pressure using the MAP sensor, when the engine is off, or during Wide Open Throttle. Though some engines use a separate barometric pressure sensor.

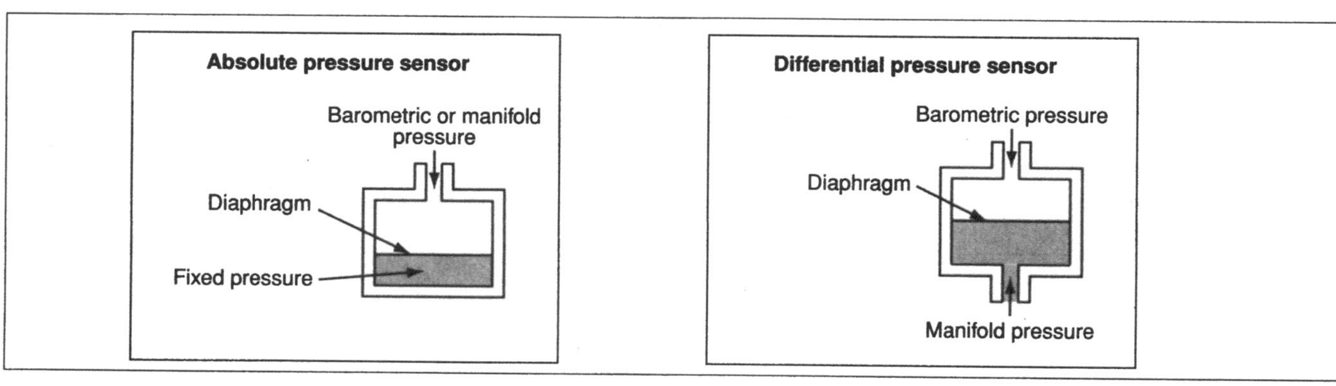

Fig. 4-34. An absolute sensor (*left*) sends signals comparing barometric, or Manifold Absolute Pressure, to a fixed reference pressure sealed inside—a very low pressure. In contrast, a differential sensor (*right*) sends signals comparing manifold pressure to barometric pressure.

3.2 Mass Air Flow (MAF) Sensor

The Mass Air Flow sensor provides a direct measurement of the mass, or weight, of intake air. Particularly during transients, mass airflow has a more direct correlation to actual engine load than Manifold Absolute Pressure does. In addition, the MAF sensor:

- measures air mass, so it requires no correction for density changes, related to higher altitudes
- has no moving parts, which means mechanical simplicity and quick response
- signals adapt to changes in air mass in one to three milliseconds

You'll notice Corvette's continuing adoption of MAF, since the **'94 LT1**, **LT4**, and **LS1**. MAF isn't used in the **'90-93** models, where the load is sensed by MAP.

Fig. 4-36. L98: The Bosch MAF sensor is located midway between the air cleaner and the intake plenum. This helps insure good laminar flow through the sensor.

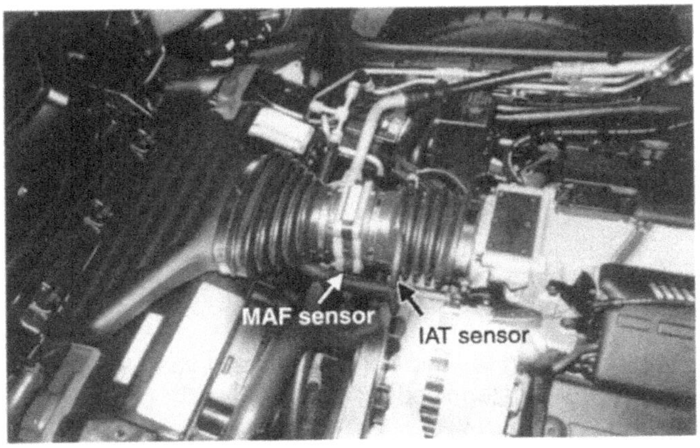

Fig. 4-37. '94-96 LT1: The AC MAF sensor is clearly visible between the throttle body and the intake air duct. The IAT sensor is just behind the MAF sensor.

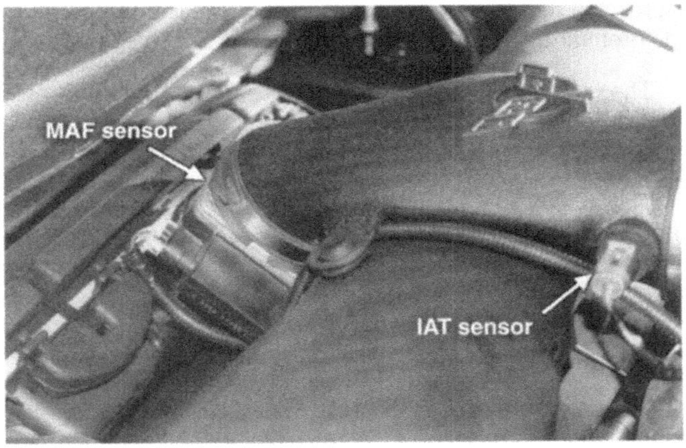

Fig. 4-38. LS1: The MAF sensor is almost invisible, ahead of the radiator. The IAT sensor is at the right.

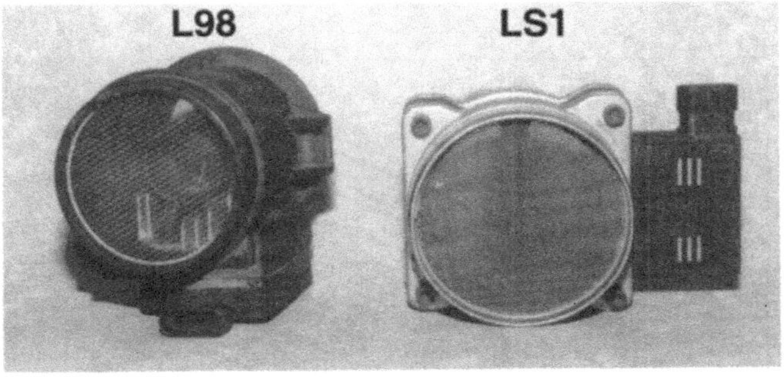

Fig. 4-39. Compare the **L98** Bosch MAF *(left)* with the **LS1** AC/Delphi MAF *(right)*. The AC/Delphi sensor has a 74 mm opening (almost 3 in.)—a "nominal throat internal diameter," in Corvette-speak. Does that baby flow!

Design and Operation

Corvettes use one of two different MAF sensors, either Bosch or AC/Delphi, depending on the model year. Both depend on the measurement of current, flowing through heated wires, to measure airflow. Picture a fan blowing across an electric heater. With a small mass of air moving past the heated wires, the cooling effect is small. The more air that flows past the heated wires, the greater the cooling effect. This is the principle of MAF sensors.

A MAF sensor has a very fine heated wire, and another, cooler, wire that provides a reference temperature. As the air flows past the heated wire, it gets cooled. And when this happens, internal controls add heat, to maintain a specific temperature differential between it and the cooler wire. Measuring the heating current used, to maintain that differential, provides a MAF output signal.

The Bosch hot wire is heated by the MAF control circuits, to 100° C (212° F) above the ambient air. The cold wire measures the temperature of the incoming air. The temperature differential is generally 100° C (180° F), in Bosch sensors.

The hot wire on AC/Delphi MAF sensors operates at a higher temperature differential than on Bosch units, at 200°C above the ambient air, but it works on the same principle.

The **'85–89 L98** Bosch MAF sensor outputs an analog signal, of varying voltage, typically .04–5 v. Accurate measurements depend on the mesh entry screen, used to improve laminar flow. ("Laminar" means "smooth layers.") If you remove the screen, you might improve airflow by a smidgen, but it may disturb your fuel injection, particularly during acceleration.

On the other hand, the **LT1**, **LT4**, and **LS1** AC/Delphi MAF sensor outputs a digital signal, also of varying frequency, typically 32–150 mHz. This signal is converted from the voltage variation in the hot-wire circuit.

The greater the airflow cooling the wire, the greater the signal—voltage or frequency.

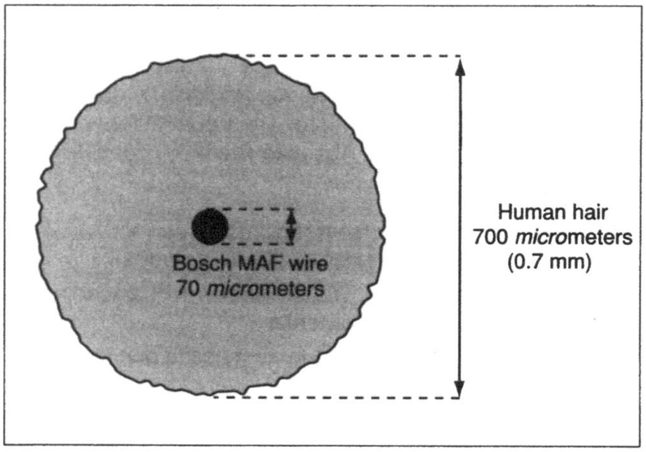

Fig. 4-41. The Bosch wires are fine, 70 *micro*meters in diameter; a human hair is about ten times that size. You can see the wires only when the light is just right, and against a good background.

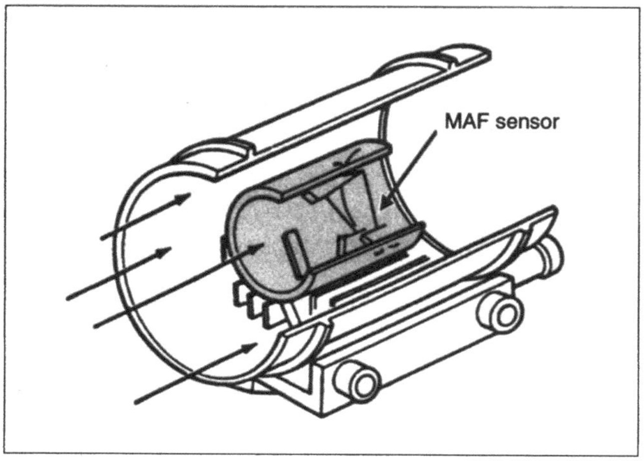

Fig. 4-40. '85–89 L98: This Bosch MAF sensor measures airflow through the smaller tube, sampling the flow. Accurate measurement depends on laminar flow, which is established by a mesh entry screen.

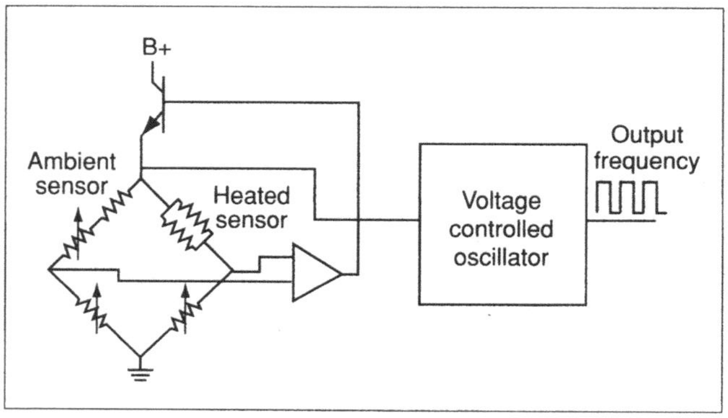

Fig. 4-42. '94-01 LT1, LT4, LS1: the MAF sensor's circuits compare temperature, and the voltage drop across the ambient sensor (the cold wire), to the voltage drop across the heated sensor (the hot wire). The **'85–89** Bosch MAF sensor's output is an analog signal, while the **94'-01** AC/Delphi MAF output is a digital frequency.

Burn-Off (Bosch Only)

L98 '85-89: Shortly after you shut off the engine, the control system heats the Bosch MAF sensor wires up to about 1000° C (1800° F), for about four seconds, long enough to burn off any dirt that may have accumulated.

The AC/Delphi MAF sensor, on the other hand, has no burn-off system, so it's sensitive to dirt. Change your air filter! A contaminated MAF can cause a lean mixture, resulting in stumbles, and possibly a lit Malfunction Indicator Light (MIL).

In normal operation, the AC/Delphi sensor's hot wires do not glow. Actually, they don't look hot at all, even though they normally operate at around double the temperature of the Bosch version.

So, according to your Corvette's model year, the engine-load sensor is determined by:

'82–84: MAP for speed density, an indirect measurement

'85–89: Bosch MAF, a direct measurement

'90–93: MAP for speed density, an indirect measurement

'94-01: AC/Delphi MAF, a direct measurement; and MAP as a backup, for speed density, which is indirect

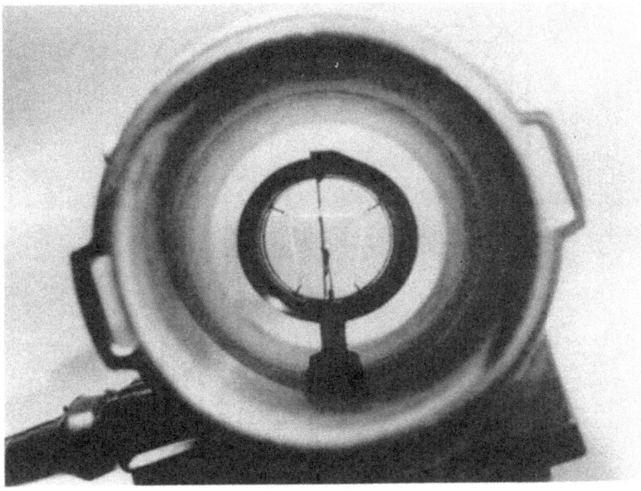

Fig. 4-43. '85–89 L98: Shortly after you shut off the engine, the control system heats the Bosch MAF wires way up, to remove any dirt that may have snuck in. You can see the glowing wires in this photo, in the center of the sensor, in an upsidedown "U" shape.

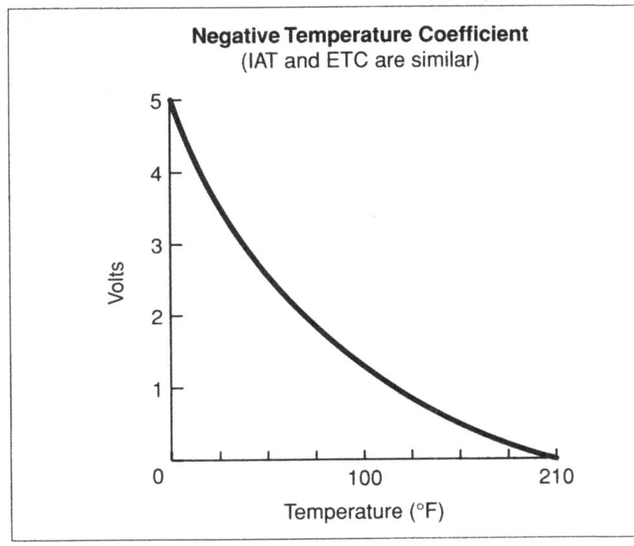

Fig. 4-44. The solid-state IAT and ECT sensors have less resistance when warm, than when cold—indicating a negative temperature coefficient. This is the opposite of most resistors, which increase resistance when warmer. The IAT sensor operates on VREF, 5 volts.

4. Intake Air Temperature (IAT) Sensor

As I discussed in Chapter 2, air density affects the air/fuel ratio. Temperature is an important factor in air density. For speed density operation, the control module calculates air mass, using pressure and temperature signals. When the Intake Air Temperature (IAT) is too cold, less than 5° C (40° F), the it signals the computer to restrict EGR flow. (See Fig. 4-44.)

The Intake Air Temperature sensor has a thermistor (a thermal transistor). Through the computer, the IAT sensor causes modification of spark advance and enrichment, during acceleration. (See Figs. 4-31 and 4-32.)

5. Engine Coolant Temperature (ECT) Sensor

The Engine Coolant Temperature (ECT) sensor is also known as the Coolant Temperature Sensor (CTS). (See Fig. 4-45.)

Engine temperature, sensed as coolant temperature, is one of the most important factors in the computer's calculations. The principal job of the Engine Coolant Temperature sensor is to help the computer modify the air/fuel ratio—making it richer when the coolant is colder. But the ECT affects a lot more: spark timing, idle rpm, knock sensing, the cooling fan, emission control (including EGR and EVAP), and Torque Converter Clutch (TCC) lockup.

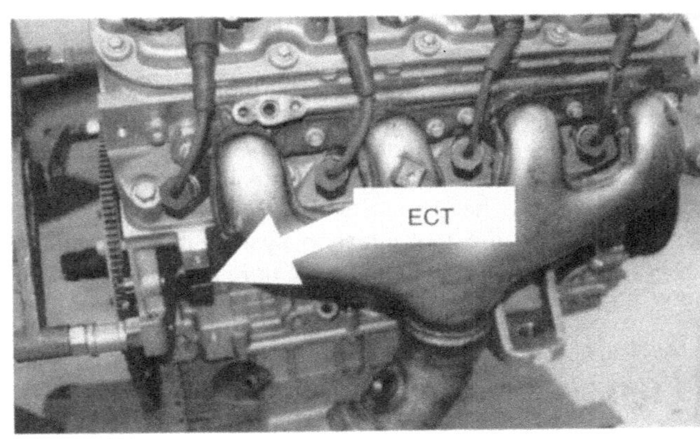

Fig. 4-45. LS1: The Engine Coolant Temperature (ECT) sensor is in the right cylinder head. In other Corvette engines, it's usually in the heater outlet fitting. in the engine. The ECT sensor is similar to the Intake Air Temperature (IAT) sensor but is designed to operate in the coolant instead of the air.

ECT voltage signals vary according to the temperature. The unique two-step signal shunts the circuit at 50° C (120° F), increasing sensitivity, by providing a greater range for the signal. (See Fig. 4-46.)

If your engine has not run for several hours, ECT, IAT, and Engine Oil Temperature (EOT) should read similar.

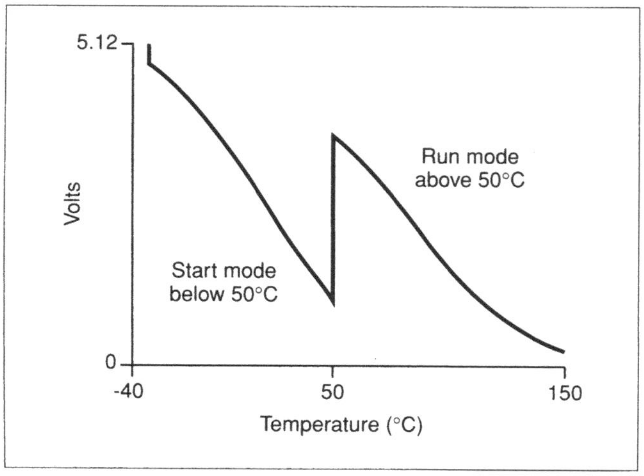

Fig. 4-46. Late-model ECT sensor voltage signals vary according to temperature. As you can see by the sharp spike, the two-step signal shunts the circuit at 50° C (120° F). This provides a wider signal range. In effect, the left curve is engine start, and the right curve is engine run.

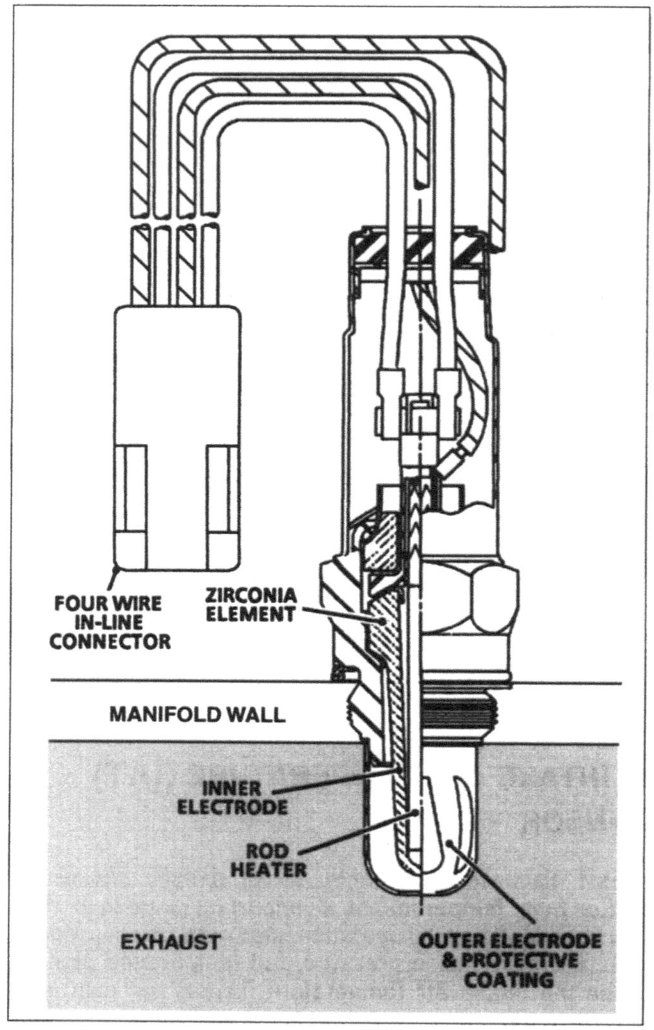

Fig. 4-47. The O_2 sensor compares the oxygen content of the air inside the tube, to the oxygen outside the tube. Late models (with heated sensors) use a four-wire connector.

6. Oxygen Sensor (O_2S) and Heated Oxygen Sensor (HO_2S)

Engine-management systems normally maintain the air/fuel ratio, for best emission control, by sensing exhaust-gas oxygen. The control module monitors Oxygen Sensor signals, so it can adjust the injector pulse times, to maintain the ideal air/fuel ratio. The Oxygen Sensors provide the key feedback signals, so the fuel injection system can operate in Closed Loop. Oxygen Sensors affect air/fuel ratios and emissions, but have no effect on spark timing or idle-rpm control.

6.1 Oxygen Sensor Design

An Oxygen Sensor (O_2S) is something like a small chemical generator. It generates a voltage signal, based on the chemical difference between the oxygen content of exhaust gas and the oxygen content of ambient air. But its signals are good only when it's hot, and I mean hot! That's why the **ZR-1**, **'94-'96 LT1**, **LT4**, and **LS1** Oxygen Sensors are electrically heated.

> This sensor has many names:
> **Oxygen Sensor (O_2S):** Corvette- and SAE-speak. ("O_2" is the chemical symbol for oxygen.)
> **Exhaust Gas Oxygen (EGO) sensor:** Ford-speak
> **Lambda sensor:** Bosch-speak. ("Lambda" is the Greek letter "L"; the German word "*Luft*" refers to "air" in the exhaust.)
> Manufacturers are agreeing to use SAE-speak.

The O_2 sensor generates voltage when the oxygen content of ambient air, inside the tube, is greater than the oxygen content in the exhaust, outside the tube. The outside of the sensor operates in the manifold, in the exhaust stream. There is a hollow zirconia-element, so that ambient air reaches the inside of the tip. Both sides of the tip are covered with metal conductive layers, one a positive electrode, the other negative.

The O_2 sensor only generates voltage when the zirconia element is hotter than about 315° C (600° F), and when the exhaust has little or no oxygen, the result of a rich mixture.

6.2 Oxygen Sensor Operation

Ordinary Oxygen Sensors are warmed up by heat from the exhaust. Heated Oxygen Sensors (HO_2S) are heated electrically to warm up sooner, and to stay warm during extended idle.

In an Oxygen Sensor, cold zirconia is solid. And, as you can imagine, it is difficult for electrons to move between the positive and negative electrodes, to generate electricity (voltage signal). Warm zirconia is less solid, making it easier to generate voltage signals.

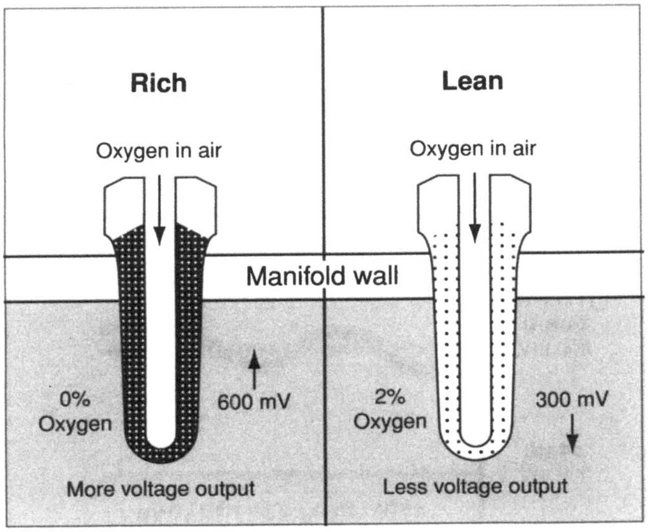

Fig. 4-48. (*left*) In burning fuel, a rich mixture uses up all the oxygen, so none remains in the exhaust stream, outside the sensor. But with a lean mixture, there is leftover oxygen in the exhaust, and the sensor generates less voltage. (*right*)

The air/fuel ratio changes the Oxygen Sensor voltage, which is related to exhaust oxygen. A rich mixture means less exhaust oxygen, and a higher voltage output. A lean mixture means more exhaust oxygen, and a lower voltage output. An O_2 sensor also tends to generate less voltage as it ages.

Corvette's engine computer applies a reference voltage (bias voltage), usually of 450 mV (0.45 V). This reference signal takes the place of the Oxygen Sensor output, during Open Loop operation. In Closed Loop, the computer compares the reference voltage to the actual voltage, generated by the Oxygen Sensor. Voltage below 450 mV is considered lean; above 450 mV is considered rich.

Here are two tips:

- The newer the O_2 sensor, the more the voltage changes, swinging from as low as 10 mV, to as much as 1000 mV. As a sensor ages, voltage changes get smaller and slower—"lazy sensor" voltage changes may lag behind changes in the exhaust gas oxygen.
- The higher the rpm, the faster the voltage changes—more exhaust gas is flowing across the sensors.

'82-93 Oxygen Sensors can be good indicators of catalytic converter efficiency, by signaling how quickly the sensor voltage changes from low to high. This is called "cross count," (crossing 450 mV) from rich to lean. Some scan tools can measure cross count in a period of 1.25 seconds. (You're seeing the measure of the activity of the converter.) Cross counts are an important diagnostic indication.

Beginning in **'94** models, the rear Oxygen Sensor, which is downstream of the converter, is a better indicator of converter efficiency than sensor activity.

4

Oxygen Sensor Heating

Oxygen sensors in the **L83**, **L98**, and **'92–93 LT1** are warmed by exhaust heat.

In the **ZR-1**, **'94–96 LT1**, **LT4**, and **LS1**, Oxygen Sensors are heated electrically, and by exhaust. This warms them up sooner for Closed Loop operation, to deliver fewer emissions and to keep them hot during extended idle. Closed Loop operation is important during extended idle, particularly during diagnosis and servicing. For **'97-01** models, the heater's power is increased to 18 W, to shorten the time to Closed Loop.

During warm up, the sensor is still inside a ±150 mV threshold, and is not yet ready to control. Open Loop continues as long as voltage swings within the threshold. During warm-up from exhaust gases (and the **LT1 '94-96**, **LT4**, **LS1** O_2S heater), the Oxygen Sensors begin changing the voltage, as the internal resistance decreases.

Within a few minutes, or around 30-90 sec. with heated O_2S, the sensor voltage swings exceed ±150 mV (from the 450 mV bias voltage). The fuel control can go Closed Loop, provided that the computer's clock-time exceeds 30 sec. (70 sec. with unheated O_2S), and the ECT rises above the programmed temperature. At about the same time, the catalytic converter gets hot enough to "light off". Now do you see the importance of heating the Oxygen Sensors, when it comes to emission control?

As emission limits tighten, heated O_2 sensors and more sensors are needed, particularly in '94 and later, for OBD-II.

6.3 Oxygen Sensor Locations

Oxygen sensor locations"$1HO_2S2$" and "$2HO_2S1$" may seem confusing (see Table 4-2.), but just remember:

- The first number defines the engine bank. As you sit in the driver's seat, 1 = left, 2 = right.
- The last number defines the position of the sensor. 1 = before the converter, 2 = after the converter.

These are sometimes written with both numbers following: HO_2S11, 21, 12, or 22.

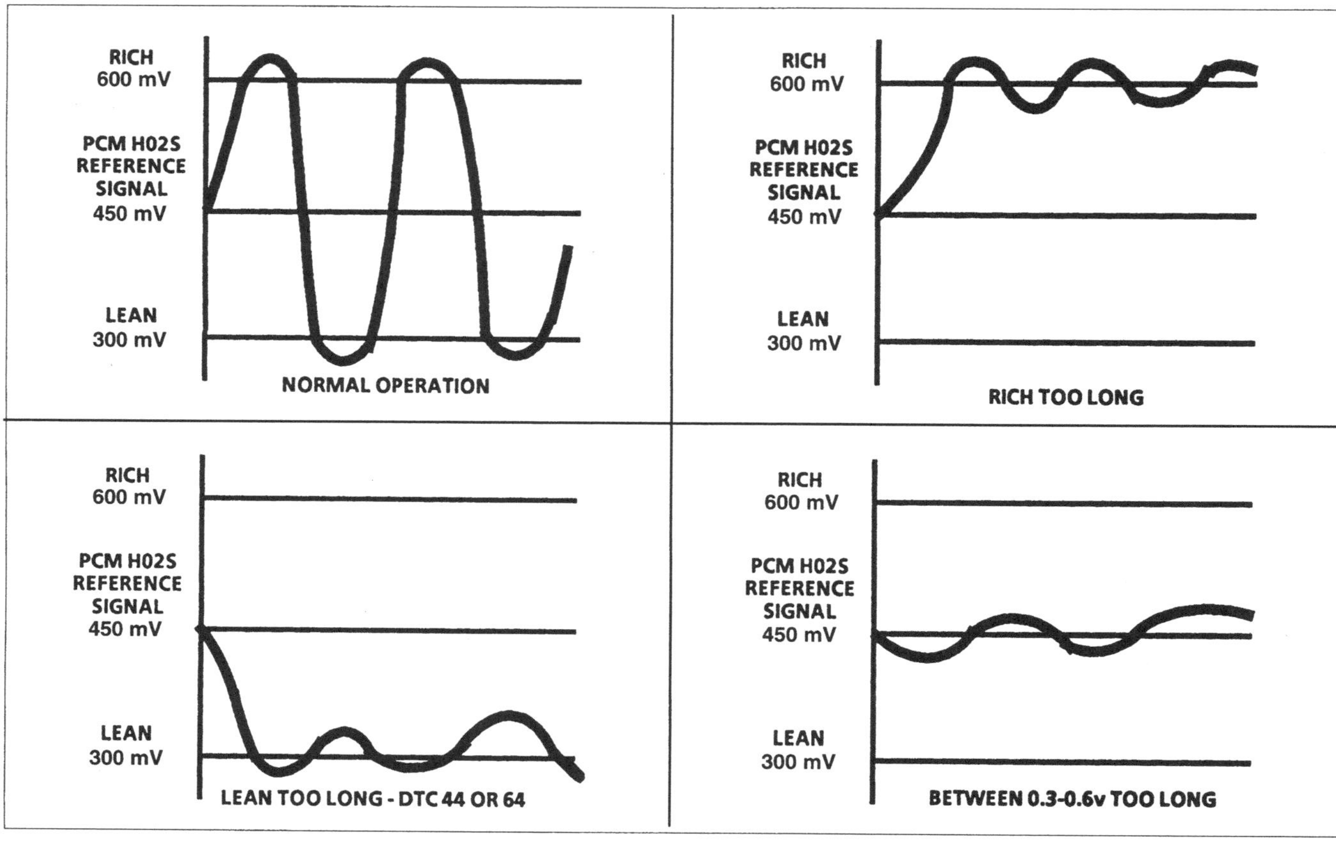

Fig. 4-49. Compare normal voltage trends (*upper left*) with:
- (*lower left*) a lean mixture (extra O_2) causes low voltage, near 300 millivolts
- (*upper right*) a rich mixture (little O_2) causes high voltage, near 600 millivolts
- (*lower right*) a contaminated sensor varies little, from the 450-millivolt reference signal (this is known as a "lazy sensor")

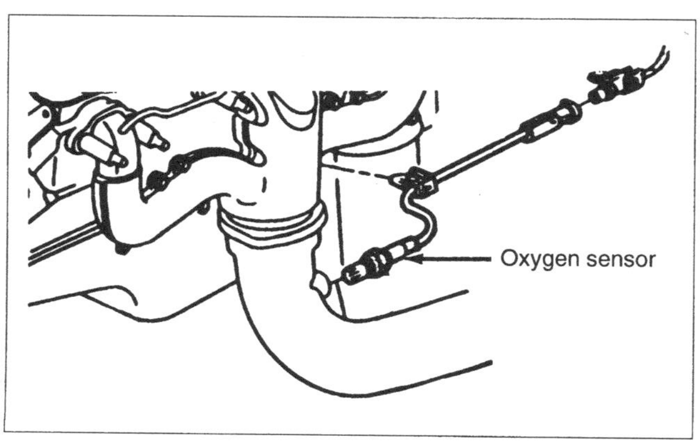

Fig. 4-50. '85–90 L98: Look for the Oxygen Sensor in the left downpipe, under the exhaust manifold.

Table 4-2. Corvette Oxygen Sensors

Model/s	Left exhaust (up-stream)	Right exhaust (up-stream)	Left converter (down-stream)	Right converter (down-stream)
L83 **'85–90 L98**	O_2S	No	No	No
'91 L98 **'92–93 LT1**	O_2S	O_2S	No	No
ZR-1	HO_2S	HO_2S	No	No
'94–95 LT1	$1HO_2S1$	$2HO_2S1$	No	$2HO_2S2$
'96 LT1; LT4; LS1	$1HO_2S1$	$2HO_2S1$	$1HO_2S2$	$2HO_2S2$

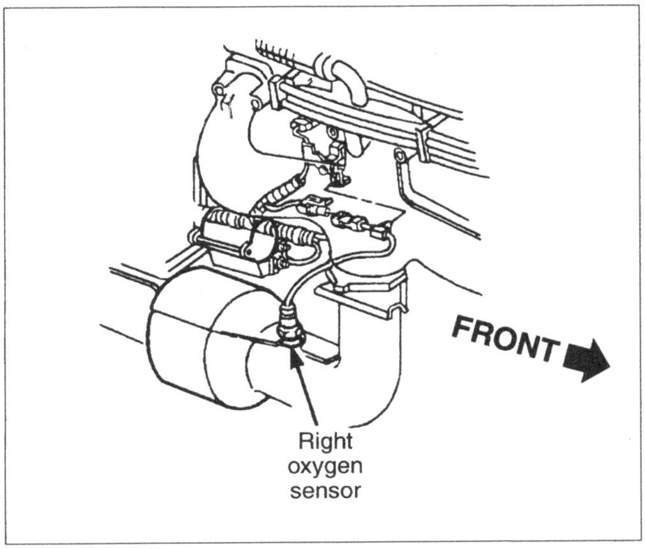

Fig. 4-51. '91 L98, '92–93 LT1: Two underhood O_2S measure hot gases directly from the exhaust manifold. So there is no need for an after-converter measurement. The right one is shown, the left is similar.

The upstream sensors for all model years send signals of the *engine* exhaust as primary feedback to trim the air/fuel ratio. The downstream sensors for '94-and-later send signals of the *converter* exhaust for OBD-II catalyst diagnostics.

Closed Loop Control—Fuel

In Chapter 2, I discussed Open Loop/Closed Loop fuel injector systems. Closed Loop is the normal operation of Corvette air/fuel ratios, with Oxygen Sensors signaling the control module about oxygen content in the exhaust. And with this information, the computer continually adjusts the air/fuel mixture, by changing fuel-injector pulse time.

If the air/fuel mixture is lean, during normal warm operation, an Oxygen Sensor generates a lower voltage. This signals the control module to increase injector pulse time, to enrich the mixture. And when the mixture is rich, the Oxygen Sensor voltage increases, and the control module reduces pulse time, to lean the mixture. This happens continuously, back and forth, since the mixture is essentially never absolutely perfect.

This oscillation is small and rapid, with only about a 0.5% change in the air/fuel ratio, sometimes occurring several times per second.

- For gasoline without oxygenates, which have a stoichiometric ratio of 14.72:1, the oscillation range is about 14.67–14.77 air to one part fuel.
- With oxygenates, the air/fuel ratio is lower, like 14.2:1, so the range might be 14.15–14.25 air to one part fuel.

Such small changes are not noticeable in overall engine performance.

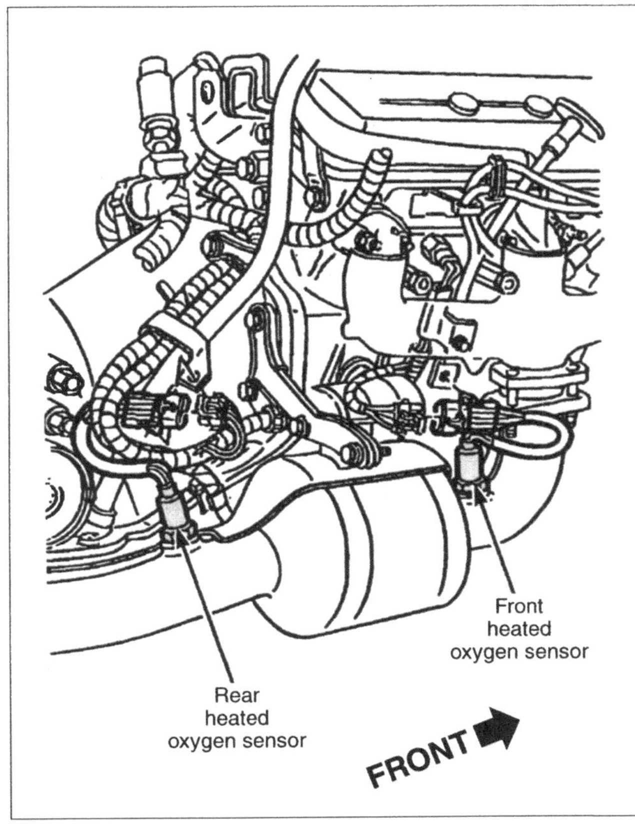

Fig. 4-52. '96 LT1, LT4: Four underhood O_2S measure hot gases directly from the exhaust manifold, upstream and downstream of the converter, for OBD-II. The right two are shown, the left are similar.

Short Term Fuel Trim (STFT), or Closed Loop air/fuel ratio control, must operate quickly and continuously, to maintain the air/fuel ratio close to the stoichiometric ideal. While Long Term Fuel Trim (LTFT) can compensate for changes in the engine over time. For example, if a valve is leaking slightly or if there is an intake air leak, an Oxygen Sensor senses the change in combustion, and brings the system back within its design limits. Changes beyond the system's range, though, can still lead to driveability problems.

Referring to Table 4-3, fuel trim values help identify rich and lean problems resulting from the fuel system or the Oxygen Sensor(s) being tricked by other, possibly mechanical, problems.

Failure of the Oxygen Sensor(s) or circuit(s) is one of the leading causes of failure to pass emission tests.

7. Throttle Position (TP) Sensor

L83, L98, LT1, LT4, ZR-1 The Throttle Position (TP) sensor signals the computer just what you would expect—throttle position, a typical feedback signal. It also signals what the driver expects the engine to do—either change speeds, or stay the same.

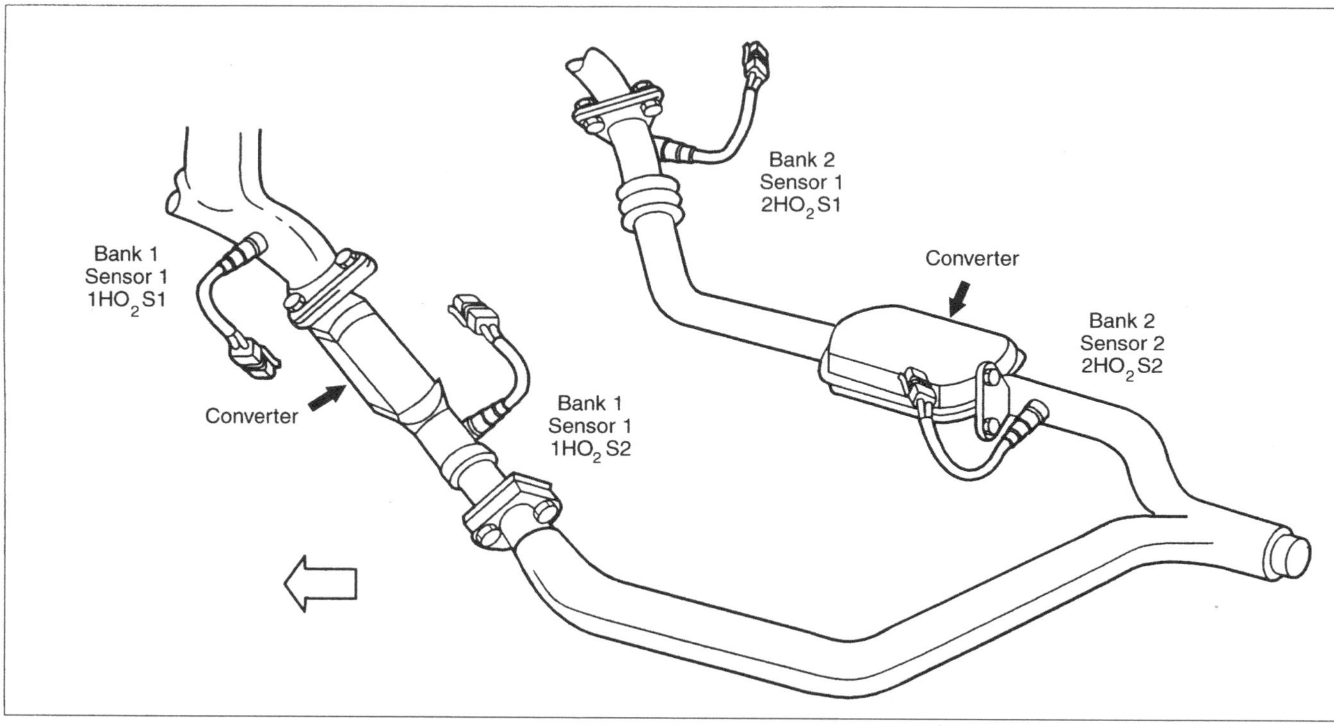

Fig. 4-53. LS1: There are separate O_2 sensors and converters for each bank. On bank one, the HO_2S1 is farther downstream, because there is limited underhood space. Each HO_2S2 is behind the converter.

Table 4-3. Lean/rich diagnosis

	Oxygen Sensor reads Lean (low O_2 volts) **Short or Long Term Fuel Trim is high (0–100% or *above* 128).**	**Oxygen Sensor reads Rich (high O_2 volts)** **Short or Long Term Fuel Trim is low (minus 100–0% or *below* 128).**
Vehicle operation indicates a lean mixture: **high NO_x** **stumbles, stalls** **surges** **poor performance**	Oxygen Sensor is correctly verifying lean operation Cause: fuel system malfunctioning Check for low fuel pressure vacuum leak, intake manifold leak lean injectors MAF/MAP	Oxygen Sensor being "tricked" rich Cause: not enough air in the exhaust Check for too much EGR Oxygen Sensor ground loose or dirty no star washer Oxygen Sensor poisoned
Vehicle operation indicates a rich mixture: **high HC, CO** **black smoke** **catalytic converter odor** **fouled plugs**	Oxygen Sensor being "tricked" lean Cause: too much air in the exhaust Check for air to the heads all the time cracked or leaking exhaust manifold cylinders not firing, sending unburned air/fuel mixture into the exhaust	Oxygen Sensor is correctly verifying rich operation Cause: fuel system malfunctioning Check for high fuel pressure restricted fuel return line leaking injectors canister purge continuously purging low coolant or IAT values restricted air cleaner Oxygen Sensor contaminated fuel-contaminated oil MAF/MAP

7. THROTTLE POSITION (TP) SENSOR

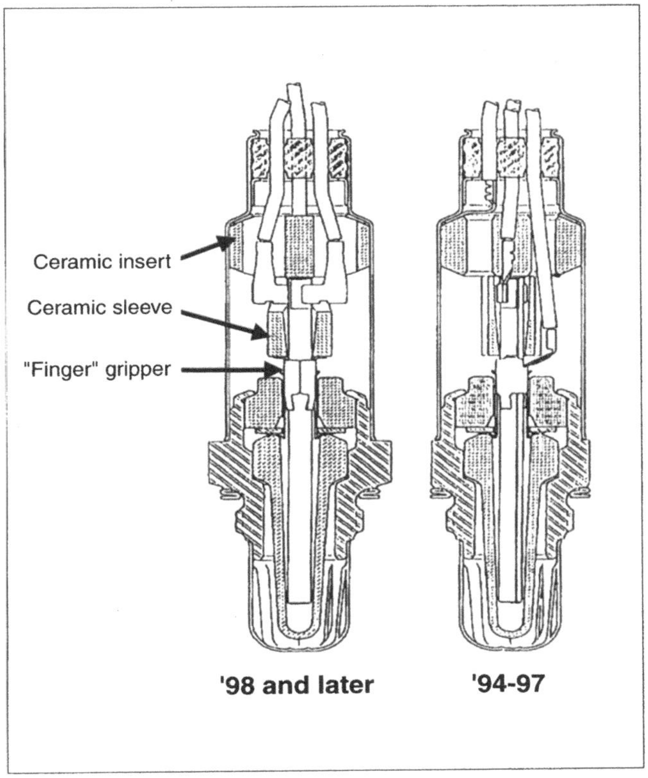

Fig. 4-54. The more robust second-generation (**'98-01**) heated Oxygen Sensor (*left*) has "strain relief", which supports internal components using two ceramic parts and an improved gripper. The appearance and part numbers are unchanged from previous sensors (*right*).

Fig. 4-55. L83, L98, ZR-1, LT1, LT4: The Throttle Position sensor is important, because it signals the computer about driver input, through the accelerator pedal. **L98** shown. The **L83**, **LT1**, **LT4**, and **ZR-1** are similar.

When you step on the accelerator, the Throttle Position signal increases before the Manifold Air Pressure (MAP) or Mass Air Flow (MAF) signal is increased. So the TP signal can cause an enriched mixture, as the throttle is being opened, something like the accelerator pump of a carburetor. In a way, it contributes to calculations about load, as signaled by the MAF or MAP sensors:

- Feedforward signals indicate that the airflow will be increasing soon: the TP sensor senses this.
- Feedback signals indicate that the airflow has already increased, or is X% open: the MAF or MAP sensor senses this.

The control module interprets signals from the TP sensor in four ways:

4

Amount of throttle opening—how much is the accelerator being depressed? This is important to cruise strategy, and transmission torque-converter clutch lockup.

Rate of throttle opening—how fast is the accelerator being depressed? This is important to the Acceleration Strategy.

Closed throttle position—is the engine either idling or decelerating?

Wide Open Throttle position—if the throttle is more than 80% open, the sensor signals the computer for acceleration enrichment, spark advance, EGR and purge shutoff, air conditioning compressor cutoff, and to de-choke for flooding, during crank.

In Open Loop operation, Throttle Position signals combine with signals from other sensors, like the Manifold Absolute Pressure (MAP) or Mass Air Flow (MAF), and the Engine Coolant Temperature (ECT) sensors, to provide information for injector pulse time calculations, for the best air/fuel ratios.

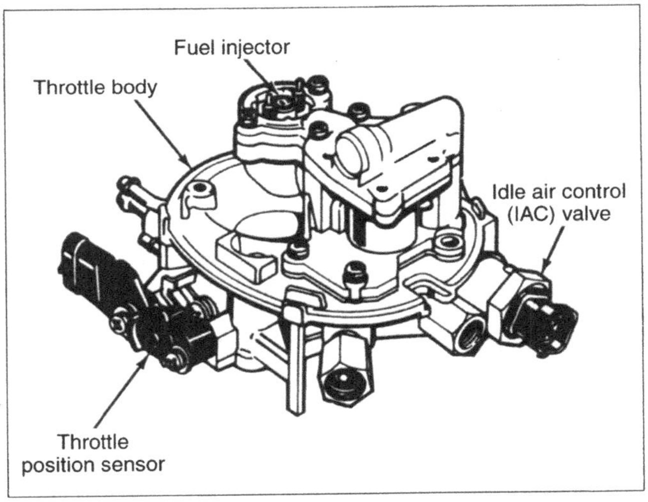

Fig. 4-56. L83: Throttle Body Injection operates with the TP sensor only, which is located on the rear of the throttle body.

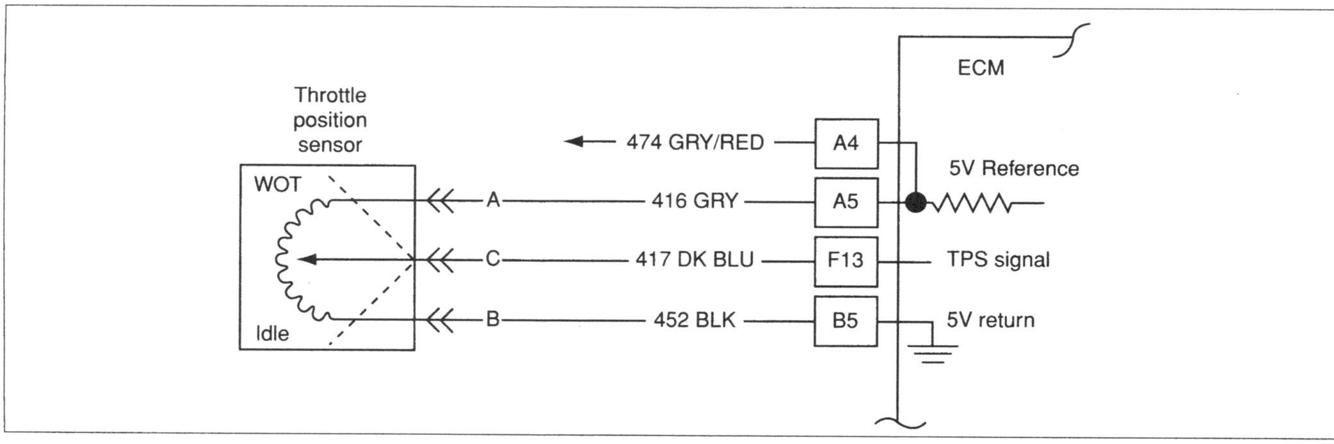

Fig. 4-57. L83, L98, LT1, LT4, ZR-1: The computer supplies VREF to the grounded resistor. Signals range from about .5 V at idle, to 4.5–5 V at W.O.T.

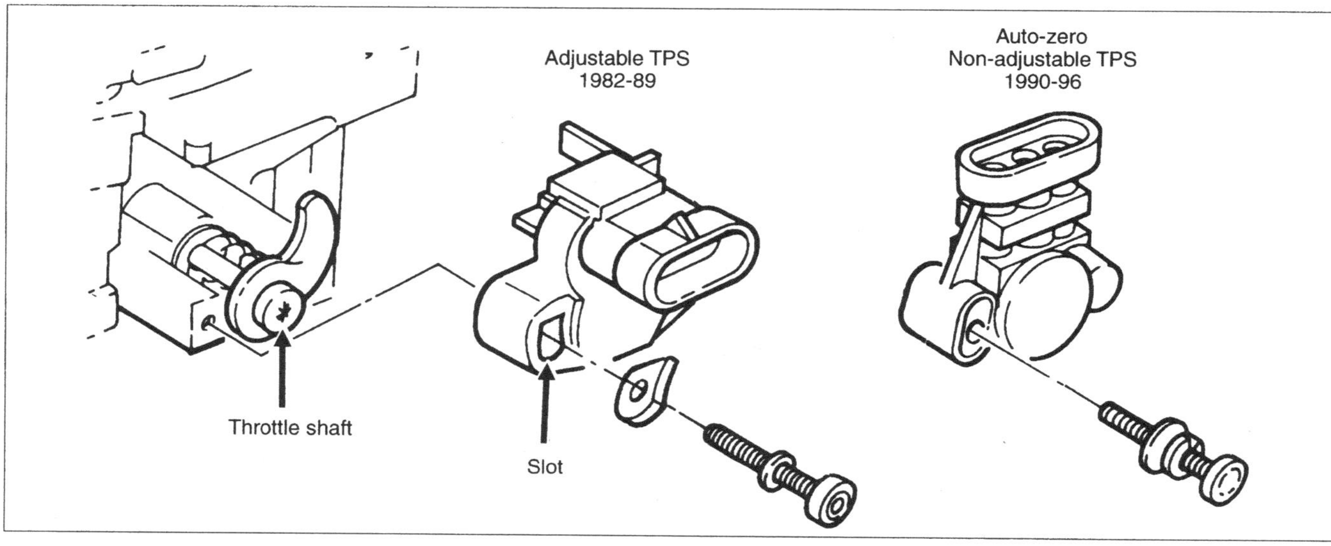

Fig. 4-58. '82-89: Adjustable TP sensors require a zero signal from the closed throttle, an important adjustment.
'90-96: Non-adjustable TP sensors are auto-zero'd in the engine computer.

The Throttle Position potentiometer operates from a 5 V VREF. The TP signals affect injected fuel, spark timing, idle rpm, evaporative emissions, the torque-converter clutch, and air conditioning. So the TP sensor is an important one.

Adjustable and Nonadjustable

The **'82–89** Throttle Position sensors require a zero signal from the closed throttle, important for idle. The sensor is mechanically adjusted to set the closed-throttle position. For **'90–96** models, the sensor is electronically calibrated in the computer (called "auto-zeroing"), and does not need adjustment. **'97-01** Corvettes use Throttle Actuator Control (TAC), described in the next section.

8. LS1: Throttle Actuator Control (TAC)

Throttle Actuator Control (TAC) is also known as Electronic Throttle Control (ETC). But I prefer TAC, since ETC can be easliy confused with ECT (Engine Coolant Temperature sensor).

Note that the previous section, on adjustable Throttle Position sensors, applies to throttles in Corvette models before **'97**, which are operated by a mechanical cable, from the accelerator pedal, or by cables from the Traction Control or Cruise Control. But beginning in **'97**, with the **LS1** models, Throttle Actuator Control (TAC) became of great interest to Corvette owners, who have always had a mechanical cable between the accelerator pedal and the throttle(s) in the intake.

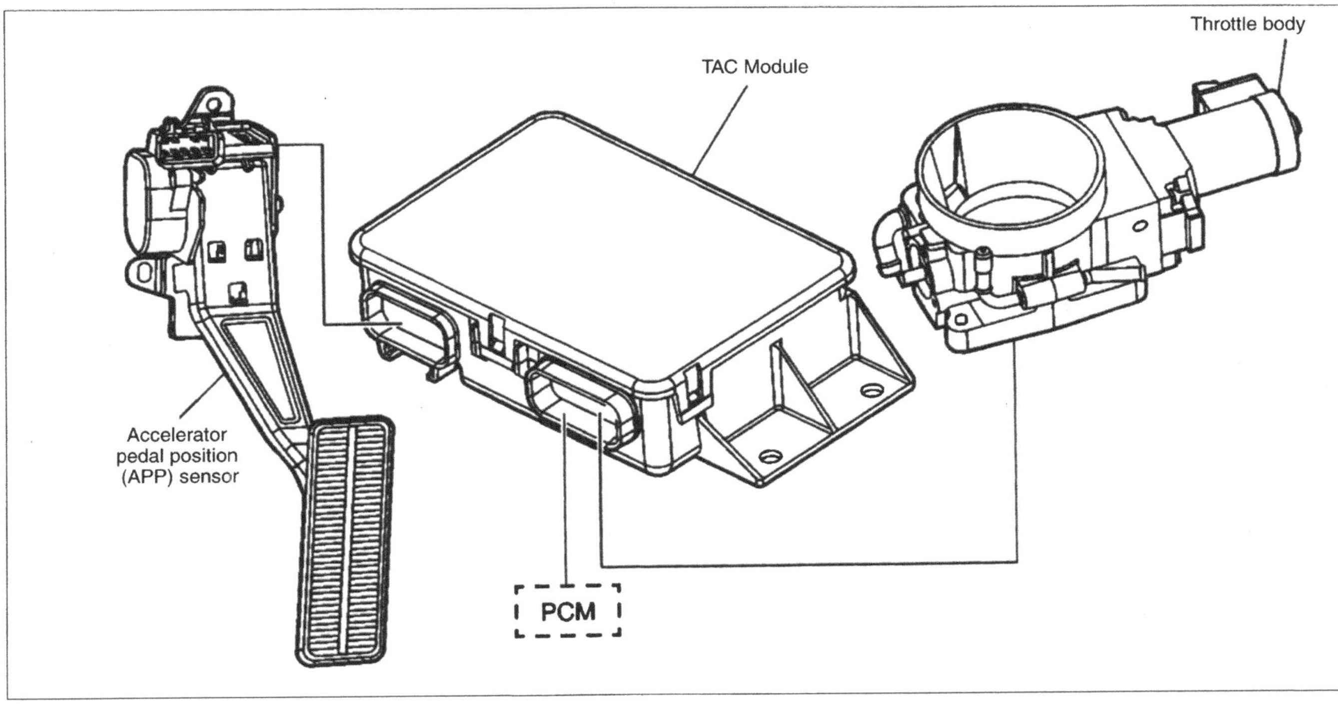

Fig. 4-59. LS1: The Accelerator Pedal Position (APP) sensor (*left*) signals the TAC module (*center*) which sends signals the the Throttle Body (*right*) and the PCM (Powertrain (engine) Control Module) .

The TAC system replaces all those earlier mechanical cables, because it operates on signals from the Accelerator Pedal Position (APP) sensor, through the Powertrain Control Module (PCM). On the throttle body, an electric motor drives the throttle to the desired positions, with feedback from the Throttle Position (TP) sensors. Some other advantages of TAC, include:

- It provides a better launch feel, and it can be calibrated to respond to changes in engine power.
- It is lighter, simpler, and more user-friendly. It controls the throttle with more than the driver's toe. Through computer inputs, TAC regulates the throttle for the Electronic Brake and Traction Control (EBTC) and Cruise Control. It also eliminates the traditional Idle Speed Control (ISC).
- It uses multiple redundancy, for safety. Several sensors and actuators cross-check each other, something like a Boeing 747 airliner with its three autopilots.

8.1 Accelerator Pedal Position (APP) Sensor

In determining the throttle opening, based on how you press the accelerator pedal, the computer considers:

- Engine Coolant Temperature (ECT)—Is the engine too hot, or too cold?
- Electronic Brake and Traction Control (EBTC)—is the engine delivering more torque than is really necessary, resulting in excessive slippage in the drive wheels?
- Knock—how badly is the engine knocking, and for how long?

Three APP sensors signal the computer as you move the accelerator pedal from 0 to 100%. Sensor #1 sends an increasing voltage signal, while #2 and #3 decrease, but at different rates.

Accelerator pedal position and movement sends three different APP signals to the computer. Don't the three different voltage changes drive the TAC computer crazy? No. The TAC is programmed to sort out and translate those signals into similar terms, so they can be compared:

- Each APP signal is compared to range specifications.
- The computed angles of all three are compared to one another.

Each of the three APP sensors operates differently ("has unique functionality"—Corvette-speak). It's designed that way so that an error is less likely to affect all three sensors.

When all three APP sensors agree, within certain limits, full power is available. When the computer runs diagnostics on the three sensors and one fails, the computer stores a Diagnostic Trouble Code, but the system continues to operate normally. If two of the sensors fail, the TAC will operate the engine at reduced power.

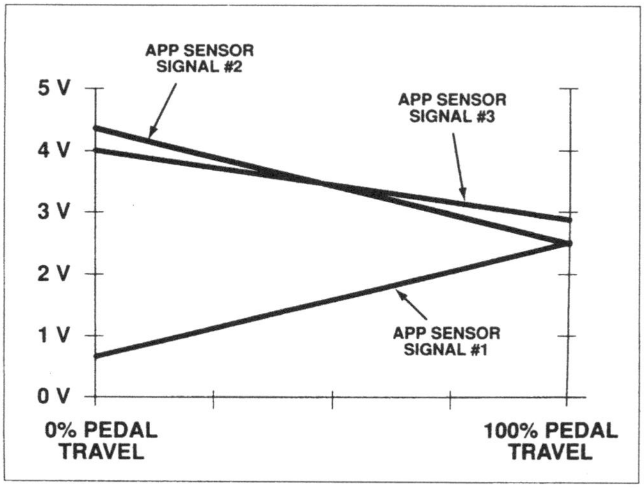

Fig. 4-60. LS1: Three APP sensors signal the computer, as you step on the accelerator pedal from 0-100%. Sensor #1 sends an increasing voltage signal, while #2 and #3 decrease, but at different rates.

Nobody wants a mistake in any of the electronics that replace the throttle cable, so the signal is triple-redundant. So you can mash the pedal as fast as you want. Unlike previous Corvette accelerator pedals, it does not push back when Traction Control reduces power to maintain traction. Compression springs and friction discs provide a good feel, relating displacement to your foot force. With the **LS1**, we don't say, "Step on the throttle." We say, "Step on the accelerator", since the computer will 'step on' the throttle, probably better than we can.

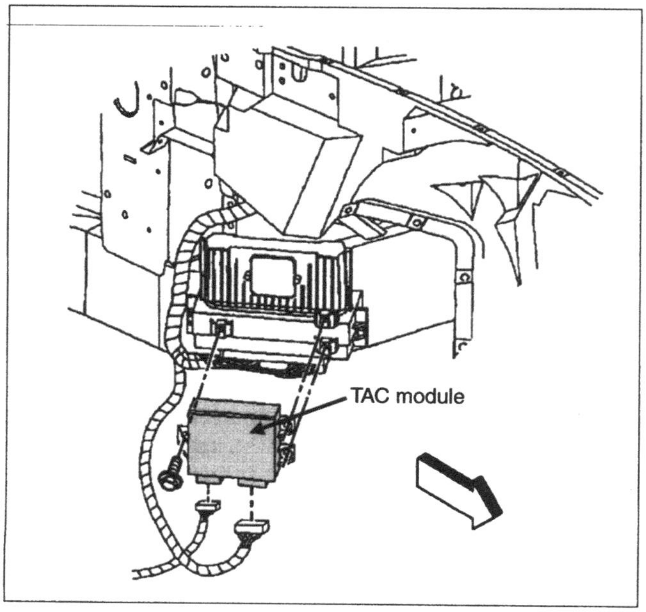

Fig. 4-61. LS1: Look for the TAC module beneath the engine computer, under the hood.

8.2 TAC Module

The TAC module and the computer communicate through a dedicated, redundant data circuit.

- Dedicated: this is its sole purpose; no other signals move through that data circuit.
- Redundant: it has two paths, either of which does the job.

Reliability is a prime consideration.

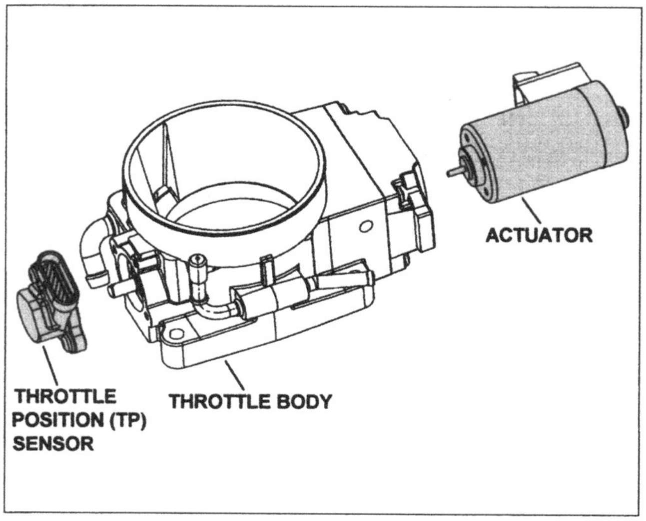

Fig. 4-62. LS1: The actuator motor, on the throttle body assembly, operates the throttle. A reduction gear provides precise, sensitive action. The dual Throttle Position sensor provides feedback, signaling the TAC Module about the results of its actuating signals to the motor.

8.3 Throttle Position (TP) Sensors

On a scan tool, the **LS1** computer displays:

- TP desired angle—what is it supposed to be?
- TP indicated angle—what is it, currently?
- TP sensor #1 angle—does it agree with TP sensor #2?
- TP sensor #1 voltage—it should increase with an increase in the throttle opening.
- TP sensor #2 voltage—it should decrease with a decrease in the throttle opening.

Throttle Actuator Control redundancy lets you keep driving with a TAC systems code, while warning you with the dash light. If several TAC codes are set at the same time, the PCM will default to a reduced power mode, and switch to a more tightly restricted control, for limited safe operation. In the unlikely event of several failures in the TAC system, it may command the engine to shut down, entirely.

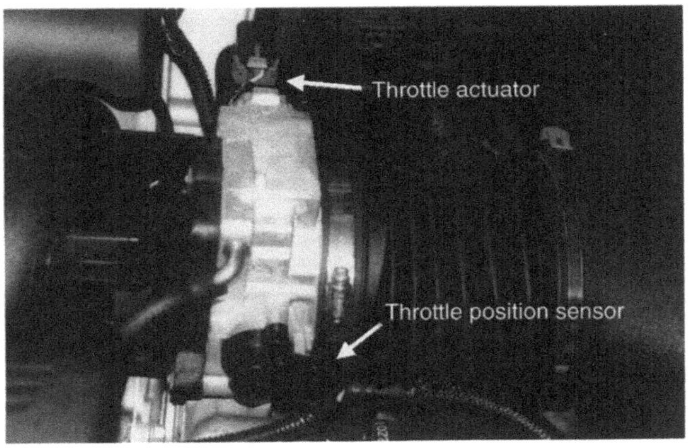

Fig. 4-63. LS1: The Throttle Position sensor is actually two compact sensors, mounted directly on the throttle shaft.

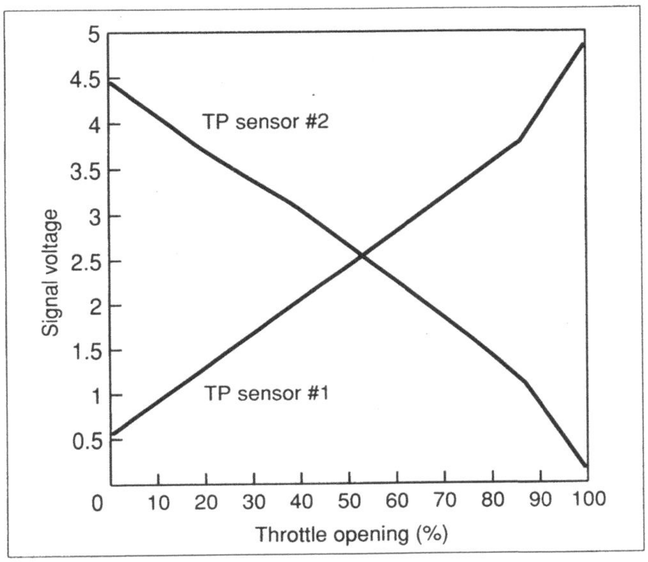

Fig. 4-64. LS1: Two TP sensors operate with opposite voltage changes, for the same open/close action. As the motor opens the throttle, Sensor #1's voltage signal increases, pulled up to 5 volts, while #2's voltage decreases, to ground.

Surprisingly, when you consider eliminating many of the previous systems of throttle control, including cruise control and Electronic Brake and Traction Control, the **LS1** Throttle Actuator Control is actually much simpler than the previous systems.

9. Knock Sensor (KS)

For years, knowledgeable drivers have known that higher-octane fuel generally did not add to engine power, unless spark timing was advanced at the distributor. The **'82-01** Corvettes, all of which have electronic fuel injection, operate with Electronic Spark Control (ESC).

Knocking limits depend on many factors: intake air temperature, engine temperature, engine deposits, combustion-chamber shape, mixture composition (air/fuel ratio and stratification), fuel quality, air density, altitude, and even the weather.

9.1 KS/ECS and Fuel

Spark advance, and the resulting power output, is usually related to the octane rating (anti-knock index) of the fuel being burned. The higher the octane rating, the greater the resistance to knock. When knocking occurs with the lower-octane fuels, the retarded timing causes a power loss. Indeed, Corvette engineers reported a horsepower loss of 10% with regular gasoline, on the **L98**.

The Knock Sensor is a Closed Loop feedback signal used to control spark timing—by as much as 20° retard. You might think of it as an insurance policy, allowing spark timing to be more advanced, with less concern that harmful knocking will destroy the engine.

The Knock Sensor's design and location are carefully matched to the engine characteristics:

- What are the frequencies associated with destructive knocking?
- What location is the most useful, for detecting these frequencies?

The answers do not come easily, considering all the other vibrations from all those moving parts in the engine, such as valves opening and closing. Ensuring that the KS is tightened to the block is important, for optimal sensitivity to the knocking frequencies.

The Knock Sensor operates in Closed Loop. Typically, some cylinders begin knocking before others. In the **ZR-1**, **LT1**, **LT4**, and **LS1**, the control module can identify and retard timing for just the cylinders that are knocking.

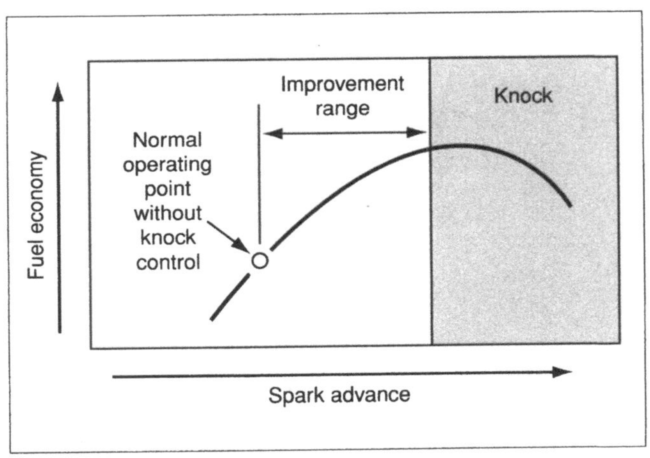

Fig. 4-65. Power and fuel economy increase with spark advance. With knock control, spark advance can be increased with more accuracy, based on the ESC protection from serious detonation.

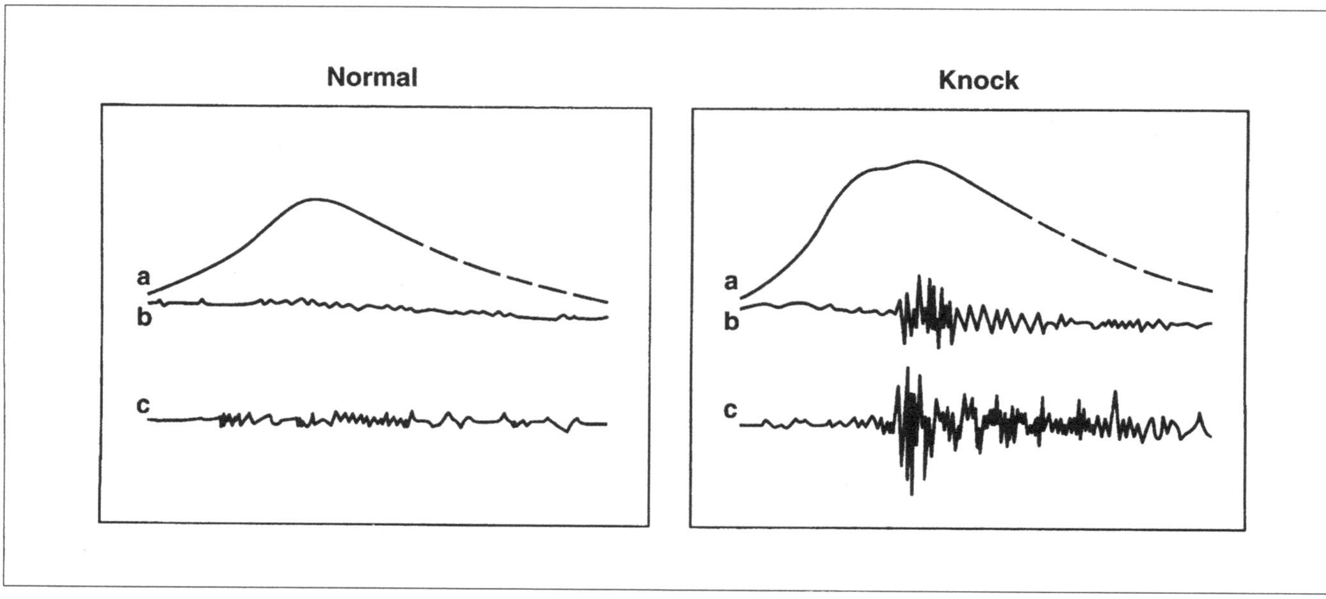

Fig. 4-66. At normal spark firing (*left*), combustion pressure rises (*a*), the Knock Sensor senses normal minor vibrations (*b*), and sends signals (*c*) that the Knock Sensor module interprets as "No problem." When self-ignition produces knock (*right*), the higher cylinder pressures (*a*) produce large, irregular variations in the pressure signal (*b*); the Knock Sensor signals the computer (*c*), "Help. Retard timing now!"

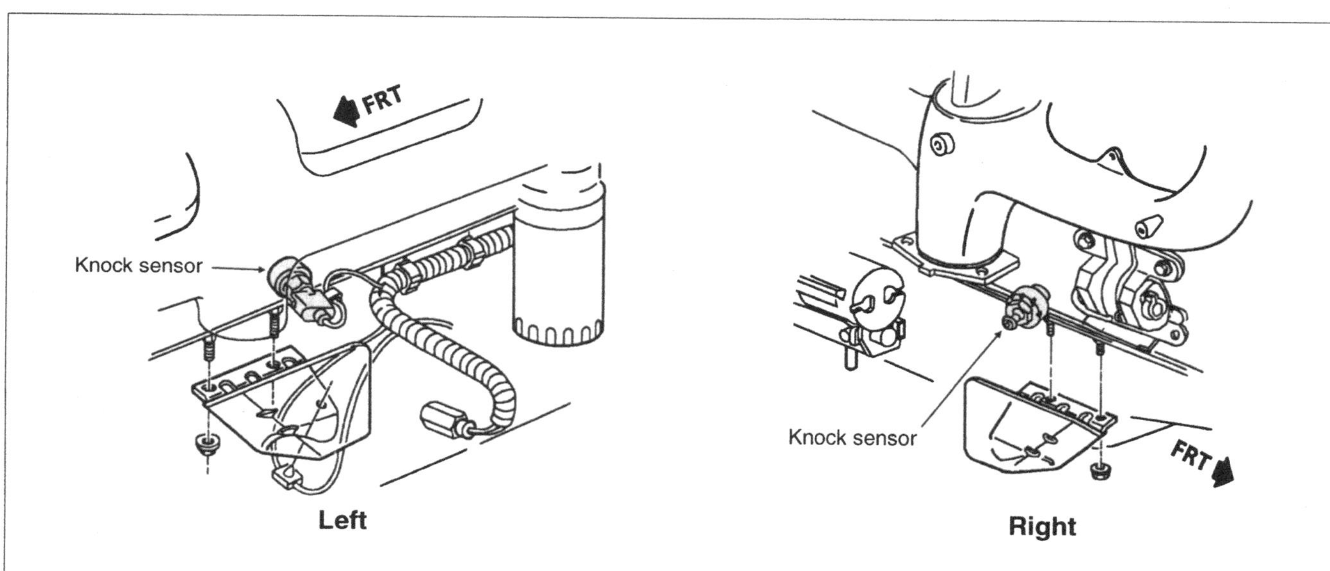

Fig. 4-67. LT1, LT4: (*shown here*) These engines have two Knock Sensors on the lower part of the block, one on each side.
L83, L98: (*not shown*) These earlier models have a single Knock Sensor, near the starter. The **ZR-1** is similar.

You may hear advice like "If it doesn't knock, burn the cheapest fuel." That was yesterday, before knock sensors were designed to prevent knock on any fuel, from regular to premium. When the modern ESC senses knock, the electronic engine management retards timing. Of course, that reduces power. So with premium fuel, you'll usually get more power.

9.2 Design and Operation

A Knock Sensor is a tuned vibration sensor, mounted on the engine block. It's similar to a tuning fork, in that it vibrates most at a certain narrow band of frequencies. But when a Knock Sensor vibrates, its crystal generates a small voltage (about one volt) which increases with the severity of the engine vibrations associated with knocking.

For example, the **'87–88** Knock Sensor (with a blue cover) has a design frequency of 5200 hertz, with a range of 4800–5600 hertz. Based on engine dynamometer tests, those specific vibrations are chosen because they're known to indicate engine knocking, for that engine type. So the Knock Sensor module is programmed to ignore any other vibrations. Retard occurs 4° at a time, up to 20°, but no lower than basic timing. If, after a few seconds, retard (or relaxed accelerator pressure) reduces or eliminates knock, then the control module again advances timing, in smaller 2° steps. (See Fig. 4-68.)

L83, L98: The Knock Sensor control is cleverly fail-safe—a lack of knock, plus normal operation, advances spark timing to normal. With no detonation, the ESC module sends a voltage signal to the computer, for normal advance. When the Knock Sensor detects detonation (or if the sensor fails), the ESC module sends a signal to the engine computer, retarding spark timing.

4

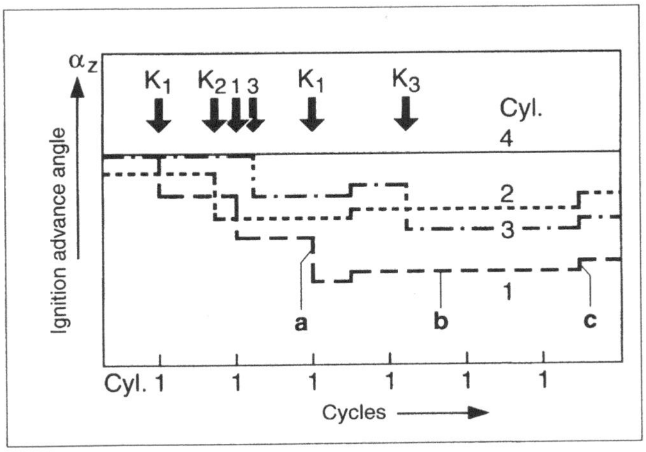

Fig. 4-68. LT1, LT4, LS1: This is a diagram representing Closed Loop knock control. Adaptive Knock Sensor processing is cylinder-selective, based on the tendency of one or more cylinders to knock before the others. One four-cylinder bank is shown, with knock in cylinders 1–3 (no knock in cylinder 4). (*a*) ignition retard; (*b*) step width for advance; (*c*) advance.
Knock at cylinder 1 (K1) causes cylinder 1 retard. Knock at cylinder 2 (K2) retards cylinder 2, independently of cylinder 1. On the next cycle, K1, still knocking, further retards cylinder 1. Knock at cylinder 3. (K3) retards cylinder 3 timing, independently of 1 and 2. K1, still knocking, retards cylinder 1 further. Reduced knock at cylinders 1, 2, and 3 causes timing to advance on each of the cylinders, and so on.

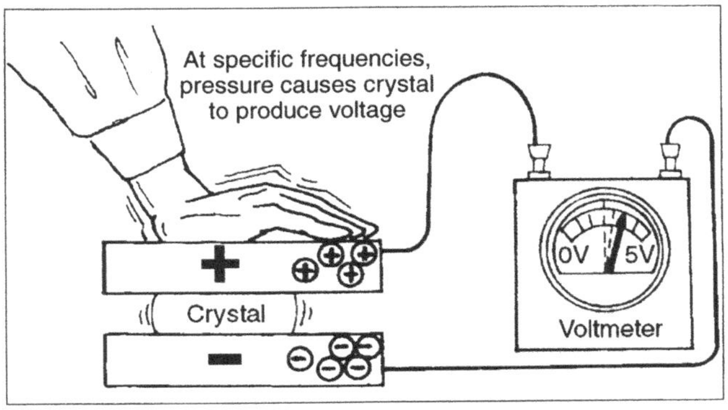

Fig. 4-69. Engine vibrations cause the Knock Sensor's crystal to vibrate, generating voltage signals.

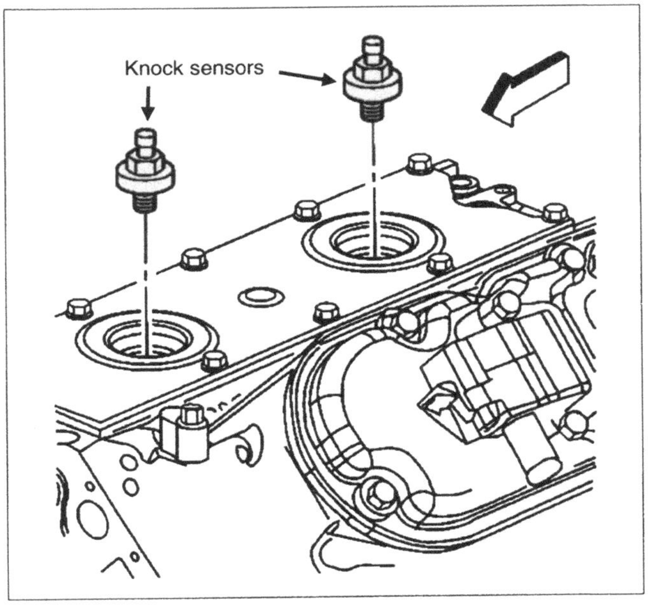

Fig. 4-70. LS1: The combination of front and rear Knock Sensor signals, in the center valley, allows the computer to distinguish individual cylinder knocking.

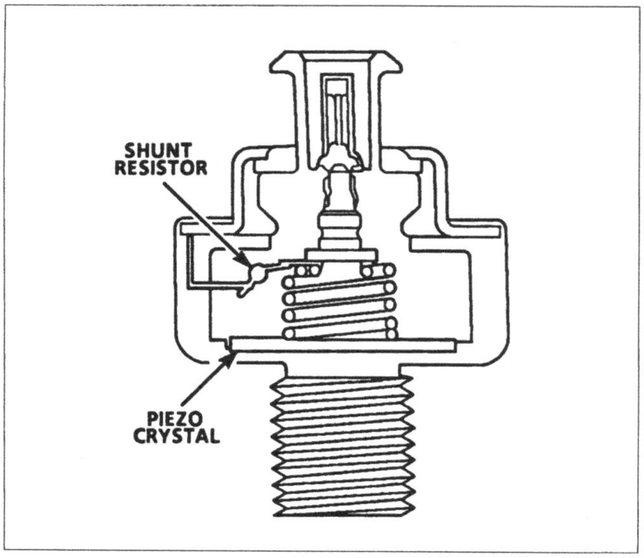

Fig. 4-71. The KS is designed to generate a voltage at only a specific frequency, That voltage signal, usually at about 1 V, increases with the severity of engine vibrations.

LS1, LT1, LT4: The adaptive Knock Sensor processing is cylinder-selective, based on the tendency of one or more cylinders to knock before the others.

LT1, LT4, ZR-1, LS1: With no detonation, and the system otherwise functioning normally, the module pulls the 5 V, VREF, down to about 1.5 V.

- If the Knock Sensor picks up detonation, the knock signal voltage rises to VREF, retarding the spark timing up to 20°.
- If the Knock Sensor fails, voltage rises to VREF, retarding the spark timing.

Late-model Corvettes use two methods for calibrating the Knock Sensor(s):

- When knocking is absent, the computer looks at the Knock Sensor's vibration signals. These vibrations are deemed "OK". The computer then assumes that any other kinds of vibration signals are knocking.
- At Wide Open Throttle, with the engine warm, the computer momentarily advances spark timing to actually cause knocking. When it detects the KS signals, it stores the resulting frequencies, as being associated with knocking. During this test, if the system signals knock, all is well. If none is detected, the computer sets a trouble code.

Reverse-flow cooling in the **LT1** and **LT4** reduces the tendency to knock, compared to the **L98**. It provides cooler cylinder heads, valve seats, and spark plug bosses, and warmer cylinder bores for reduced lower-ring friction. The tendency to knock in these engines is also reduced by the greater timing precision, from the Opti-Spark ignition and reduced spark scatter.

10. Other Sensors

Several other sensors provide inputs that influence engine management.

Vehicle Speed Sensor (VSS)

The Vehicle Speed Sensor (VSS) is essentially a magnetic pickup, mounted on the transmission to signal the control module. It operates with two inputs, one low-resolution and the other high.

VSS is important during Cruise Control and Torque Converter Clutch lockup, and for the engine cooling fan(s), speedometer, and odometer. It also controls the speed-chime signal and radio volume (which increases with vehicle speed).

Wheel Speed Sensors

The wheel speed sensors, in the Antilock Braking System (ABS), are used for Anti-Slip Regulation (ASR)/Torque Control (TC). There is one sensor for each rear wheel, since each may operate at a different speed. If there is any slippage, the wheel speed sensors signal to the Electronic Brake and Traction Control (EBTC) module, causing the computer to apply the brakes selectively, retard timing, and reduce fuel injection. Don't confuse the wheel speed sensors with the Vehicle Speed Sensor (VSS), which measures the speed of the entire car.

Engine Oil Switches

Engine Oil Temperature (EOT) is measured differently, depending on engine model:

- **'92-95 LT1**: The EOT switch signals when the engine is too hot. The computer fuel-pump control is related to the operation of the oil cooling system.
- **'93–95 ZR-1:** The EOT signals when the oil temperature is "hot enough" for high-power operation, activating of the secondary injectors and intake air.
- **'96 and later**, There is no EOT switch or gauge.

The **LS1** uses an Engine Oil Pressure (EOP) switch, instead. The EOP switch sends signals to the computer, and sets low- and high-pressure codes, for OBD-II analysis. (See Fig. 4-74.)

Feedforward Switches

Generally, feedforward switches are important to the control of idle rpm. As I mentioned in Chapter 2, idling at the lowest

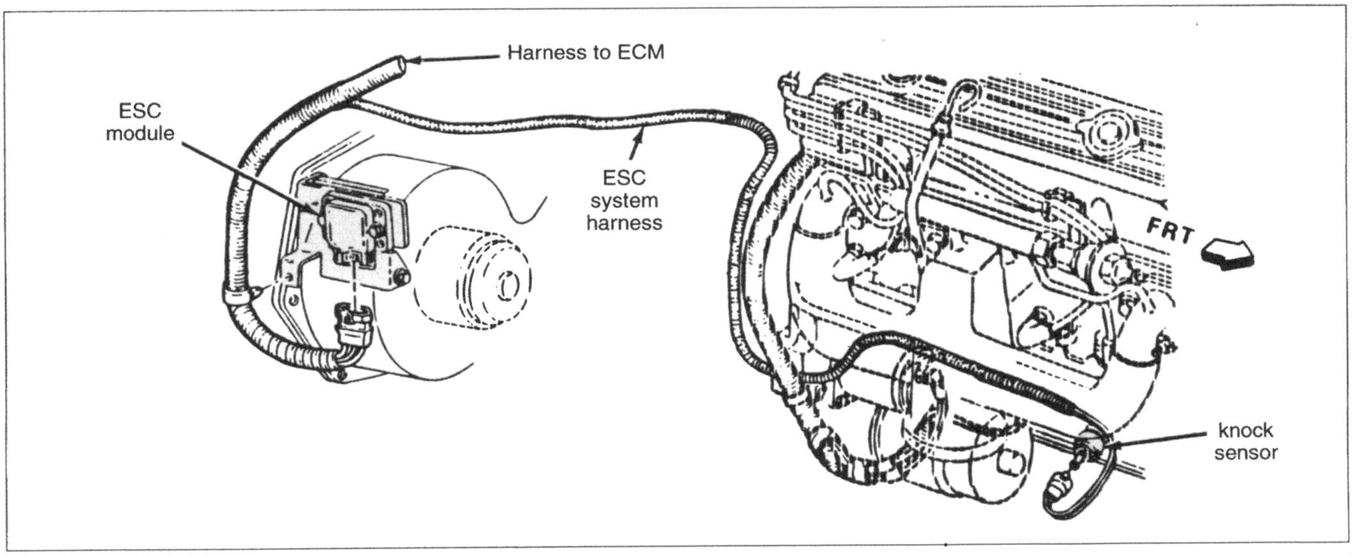

Fig. 4-72. L83, L98: The Knock Sensor, on the engine block, signals a separate ESC module. Through the engine computer and HEI module, spark timing is retarded when reference signal is grounded, or "pulled low."

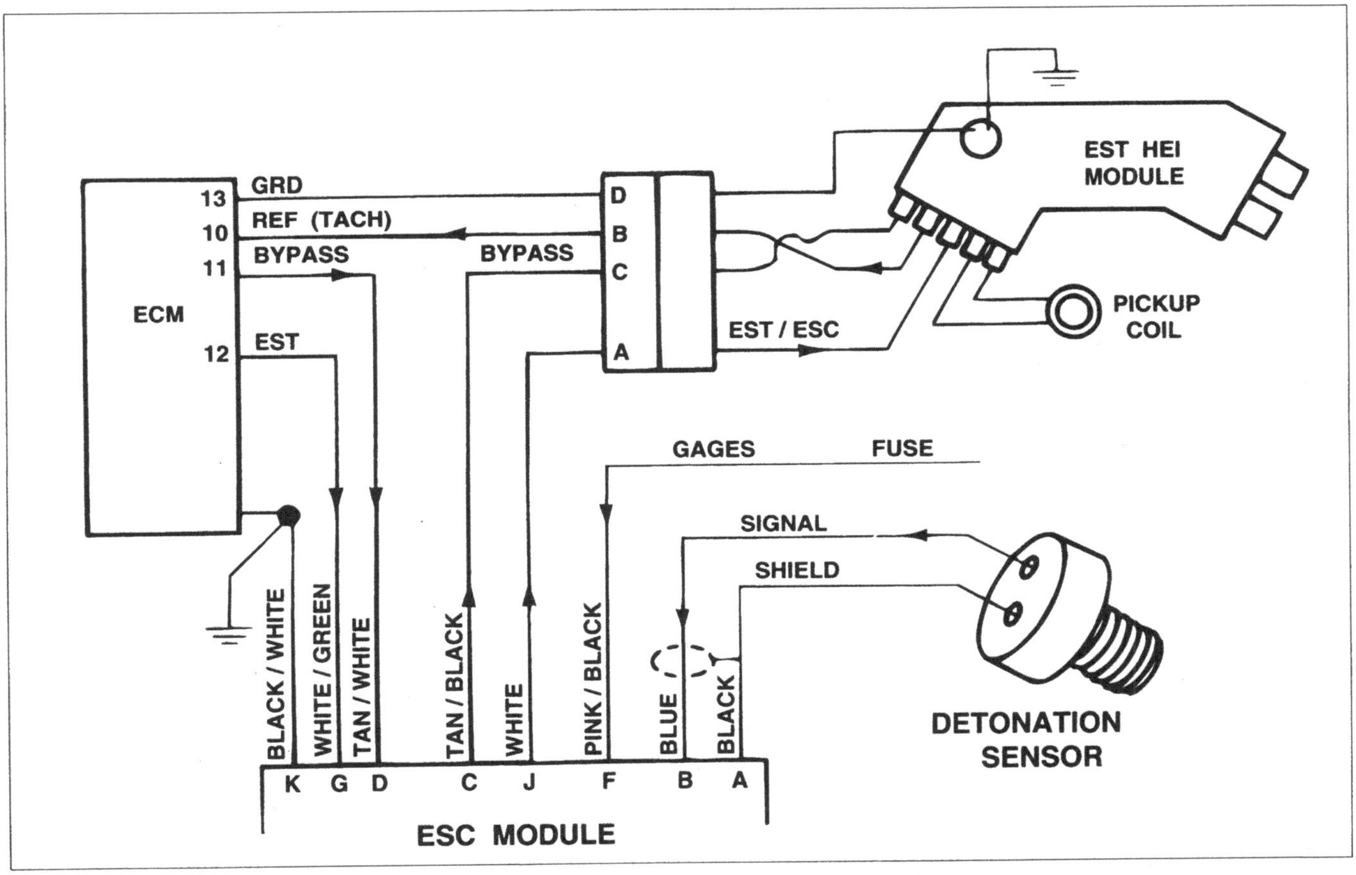

Fig. 4-73. L83, L98: A separate ESC module processes signals from the Knock Sensor.

Fig. 4-74. LT4, LS1: Corvette's factory-installed Mobil One can run much hotter than mineral-based oil, eliminating the need for an oil cooler. So, these models have an Engine Oil Pressure switch, instead of an Engine Oil Temperature switch.

rpm is important for fuel economy and emissions, so feedforward switches prevent stalling when load is added.

The following are examples of feedforward switches:

- the **air conditioner request signal** (which advises the control module if the air conditioner clutch is about to be engaged)
- the **air conditioner refrigerant pressure switch**
- the **air conditioner pressure cycling switch**
- the **brake switch** (which cancels Cruise Control and Torque Converter Clutch lockup)
- the **Power Steering Pressure Switch (PSPS)** (which signals the control module when power steering is operating at high pressure, as from a sharp turn in parking, usually with the engine idling; when the PSPS closes, it anticipates the power steering load and calls for an increase in idle-rpm)

10.1 Transmission Specific Sensors

Manual Transmission (M/T)

In the **L83** and **L98**, the overdrive is driver-switchable, depending on the:

- Engine Coolant Temperature (ECT)
- rpm
- Vehicle Speed Sensor (VSS)
- Throttle Position (TP)
- Mass Air Flow (MAF) or Manifold Absolute Pressure (MAP)

In the **ZR-1**, **LT1**, and **LS1**, the 1–4 upshift depends on the:

- Engine Coolant Temperature (ECT)—above 50° C
- Vehicle Speed Sensor (VSS)—12–19 mph
- Throttle Position (TP)—less than 35%

> The "hated" 1–4 upshift (skip shift) ensures that Corvette drivers avoid the federal Gas-Guzzler tax. In the EPA tests, the original drive cycle is so mild that second and third gears would be utilized, decreasing fuel economy. But the gear blockout can be avoided by shifting out of first gear above 19 mph, or by double-clutching with 35% throttle opening.

Automatic Transmission (A/T)

Automatic transmission is controlled by the:

- Torque Converter Clutch (TCC)—the solenoid valve initiates the TCC apply and release
- Pressure Control Solenoid (PCS)—under control of the engine computer, the PCS controls the line pressure
- Vehicle Speed Sensor (VSS)—which signals the engine computer
- 1–2 and 2–3 shift solenoids—which control upshifts and downshifts
- Fluid Pressure switch—which signals the manual valve position and the fluid temperature, to the engine computer
- 3–2 Shift Solenoid—which ensures positive 3–2 downshifts
- TCC Pulse-Width Modulator (PWM)—which controls the rate of apply and release

Torque Converter Clutch lockup depends on the:

- Engine Coolant Temperature (ECT)—above 65˚C
- Vehicle Speed Sensor (VSS)—above various values, according to gear
- Fourth, third, and second gear switches—which vary lockup according to speed, and retard spark timing in higher gears (**L83** and **L98** in fourth gear only)
- Throttle Position (TP) sensor—which unlocks the clutch during braking, and some acceleration and coasting
- **LT1**: Transmission Oil Temperature (TOT) switch—which applies the Torque Converter Clutch (in fourth gear only) to reduce transmission oil temperature, when it is above 155˚C

Automatic transmission signals are sent to and from the engine computer from the:

- Vehicle Speed Sensor (VSS)
- transmission switches, 4/3, 3/2

The following engine signals are important to transmission control:

- Throttle Position (TP) sensor—for downshifting when the driver wants more power
- Engine Coolant Temperature (ECT) sensor—for restricting torque converter lockup until the engine is warm

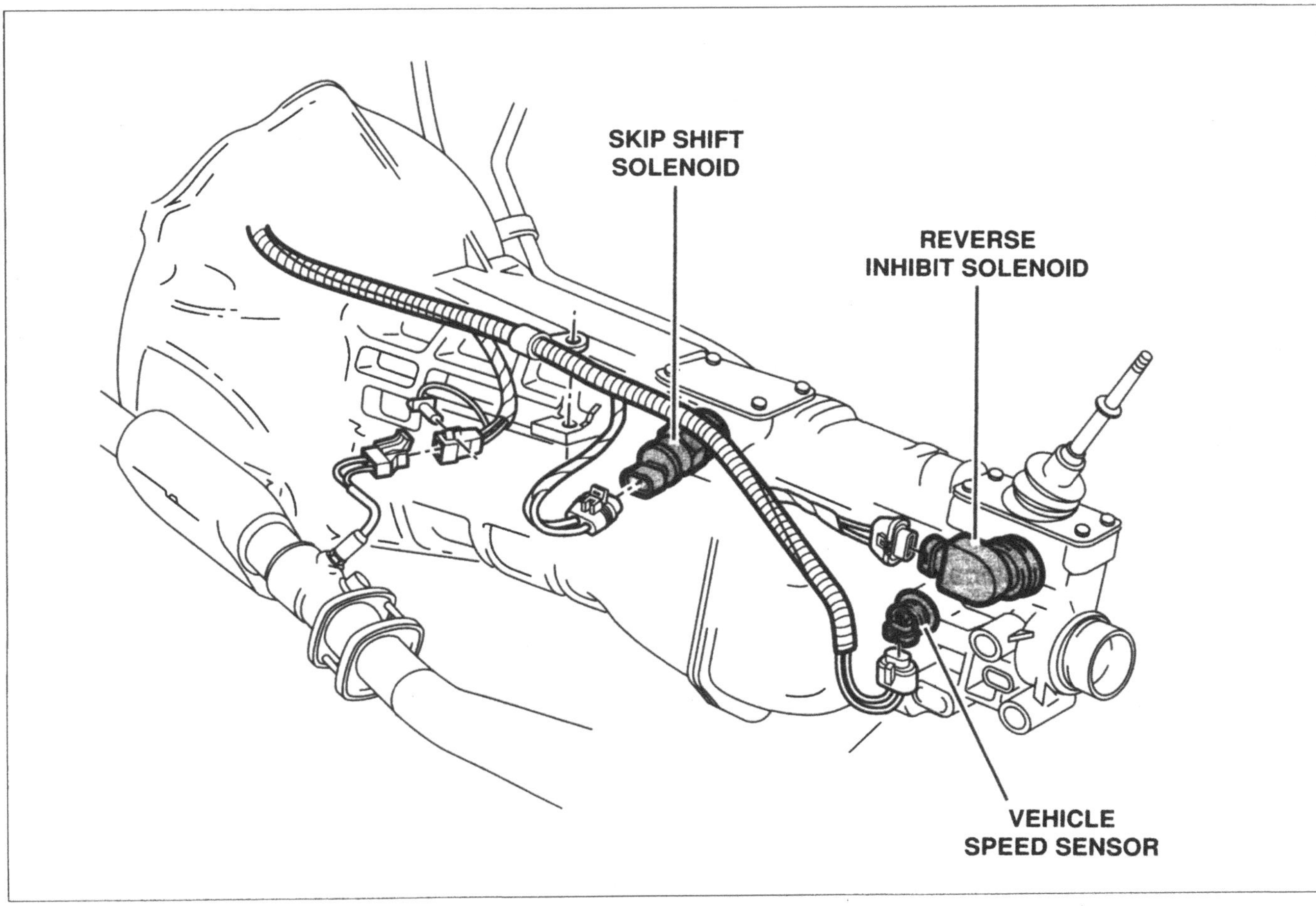

Fig. 4-75. LT1, LT4: (manual transmission) The electric PCM controls include:
- a skip-shift solenoid
- a reverse-inhibit solenoid—which blocks reverse shift above 5 mph
- a Vehicle Speed Sensor

11. Conclusion

You've seen six types of sensors, which monitor conditions and send input signals to the control module. Many of these are returned as signals referenced to VREF. You've seen typical switches, many of them feedforward, which signal to the control module to anticipate loads that might cause an idling engine to stall.

When you look at late-model electrical schematics, or at the large number of pins in the connector to the control module, you'll begin to appreciate the large number of sensor signals going to the computer.

Sensors provide inputs to the control module. Compare the diagrams on the following pages, to see how inputs differ in the various Corvette engines from different model years.

12. Sensor and Actuator Locations

On the following pages are some diagrams which indicate the locations of the various sensors and actuators in various Corvette engines.

In the Next Chapter:

Using the sensor inputs, the engine's computer checks its memory and does computations, to decide what needs to be done. In Chapter 5, I show you just what goes on in the computer—also known as the Powertrain Control Module.

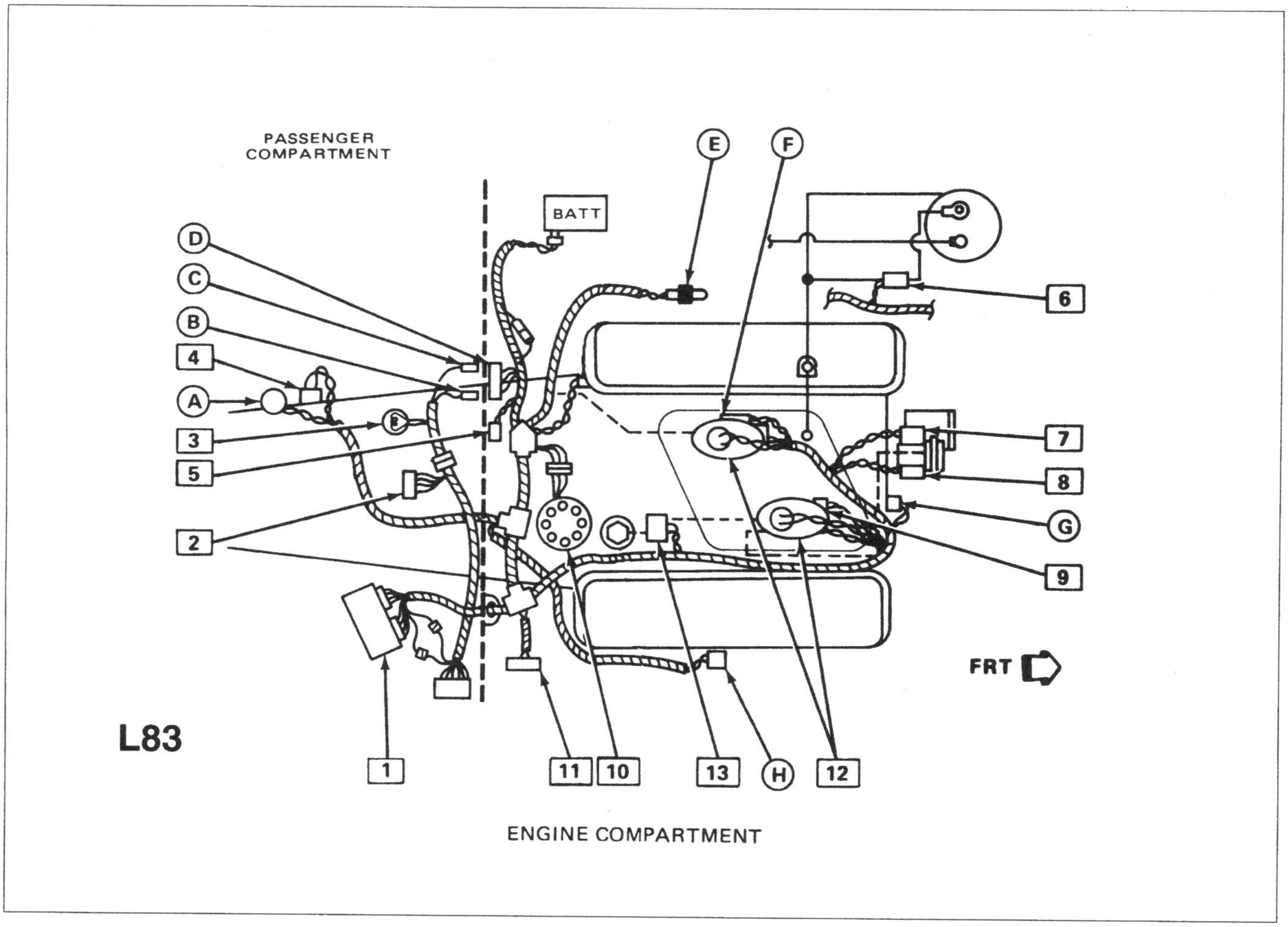

Fig. 4-76. L83: Throttle Body Injection with MAP

Sensors (circles)
A Vehicle Speed Sensor (VSS)
B Brake switch
C Park/Neutral Switch (A/T)
D Manifold Absolute Pressure (MAP) sensor
E Oxygen Sensor (O_2S)
F Throttle Position (TP) sensor
G Engine Coolant Temperature (ECT) sensor
H Knock Sensor (KS)

1 ECM
2 ALCL connector (DTC)

Actuators (squares)
3 Check Engine light
4 Transmission Converter Clutch (A/T)
5 Fuel Pump relay
6 EVAP canister purge solenoid valve
7 AIR control solenoid valve
8 AIR switching solenoid valve
9 Idle Air Control (IAC) valve
10 Electronic Spark Timing (EST)
11 Electronic Spark Control (ESC)
12 Throttle Body injector
13 EGR solenoid valve

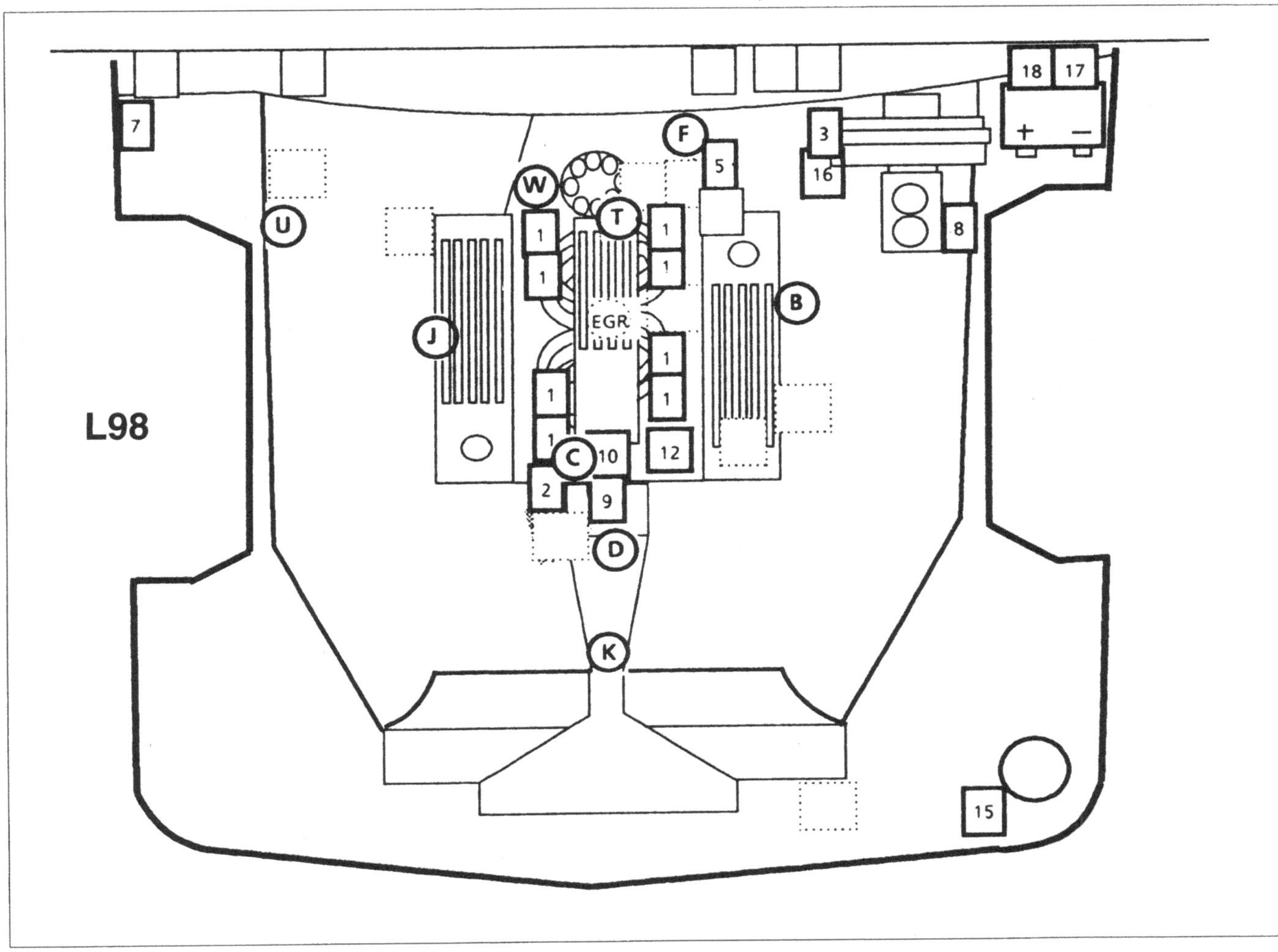

Fig. 4-77. '85–89 L98: Multiport Fuel Injection; the Bosch MAF requires burnoff
'89–91 L98: Multiport Fuel Injection with MAP

Sensors (circles)
B Oxygen Sensor (O_2S)
C Throttle Position (TP) sensor
D Engine Coolant Temperature (ECT) sensor
F Vehicle Speed Sensor (VSS)
J Knock Sensor (KS)
K Mass Air Flow (MAF) sensor ('85-89)
T Intake Air Temperature (IAT) sensor
U A/C pressure switch
W EGR diagnostic temperature switch

Actuators (squares)
1 Fuel injector
2 Idle Air Control (IAC) motor
3 Fuel Pump relay
5 Transmission Converter Clutch (TCC) connector
7 Throttle actuator
8 Engine fan relay
9 AIR divert valve
10 AIR switching valve
12 EGR vacuum solenoid
15 EVAP canister purge solenoid valve
16 Transmission overdrive relay M/T
17 MAF sensor power relay ('85-89)
18 MAF sensor burnoff relay ('85-89)

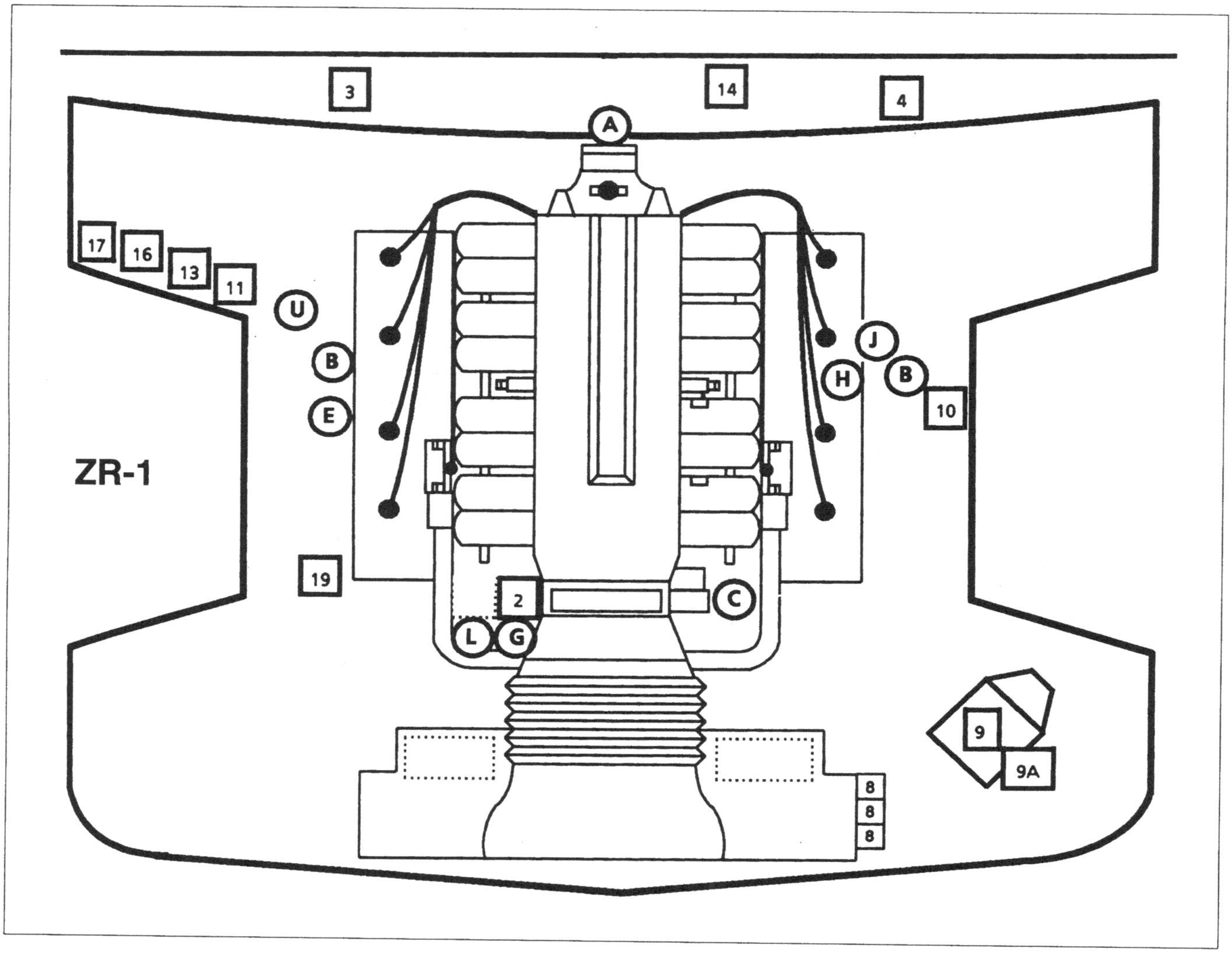

Fig. 4-78. ZR-1: Sequential Fuel Injection. The **'95** model is shown, others are similar. Note the extra sensors for DOHC, Secondary Air Inlet Control, and secondary injectors.

Sensors (circles)
A Manifold Absolute Pressure sensor
B Heated Oxygen Sensors (HO_2S)
C Throttle Position (TP) sensor
E CranKshaft Position (CKP) sensor
F Vehicle Speed Sensor (VSS) (not shown)
G Intake Air Temperature (IAT) sensor
H CaMshaft Position (CMP) sensor
J Knock Sensor (KS)
L Engine Oil Temperature (EOT) sensor
U A/C cooling fan switch

Actuators (squares)
2 Idle Air Control (IAC)
3 Fuel pump relay (primary)
4 Fuel pump relay (secondary)
8 Cooling fan relay
9 AIR pump
9A AIR pump relay
10 Air bypass valve
11 2-3 blockout relay
13 A/C clutch relay
14 2-3 blockout solenoid
16 Secondary SFI control module #1
17 Secondary SFI control module #2
19 Linear EGR valve

4

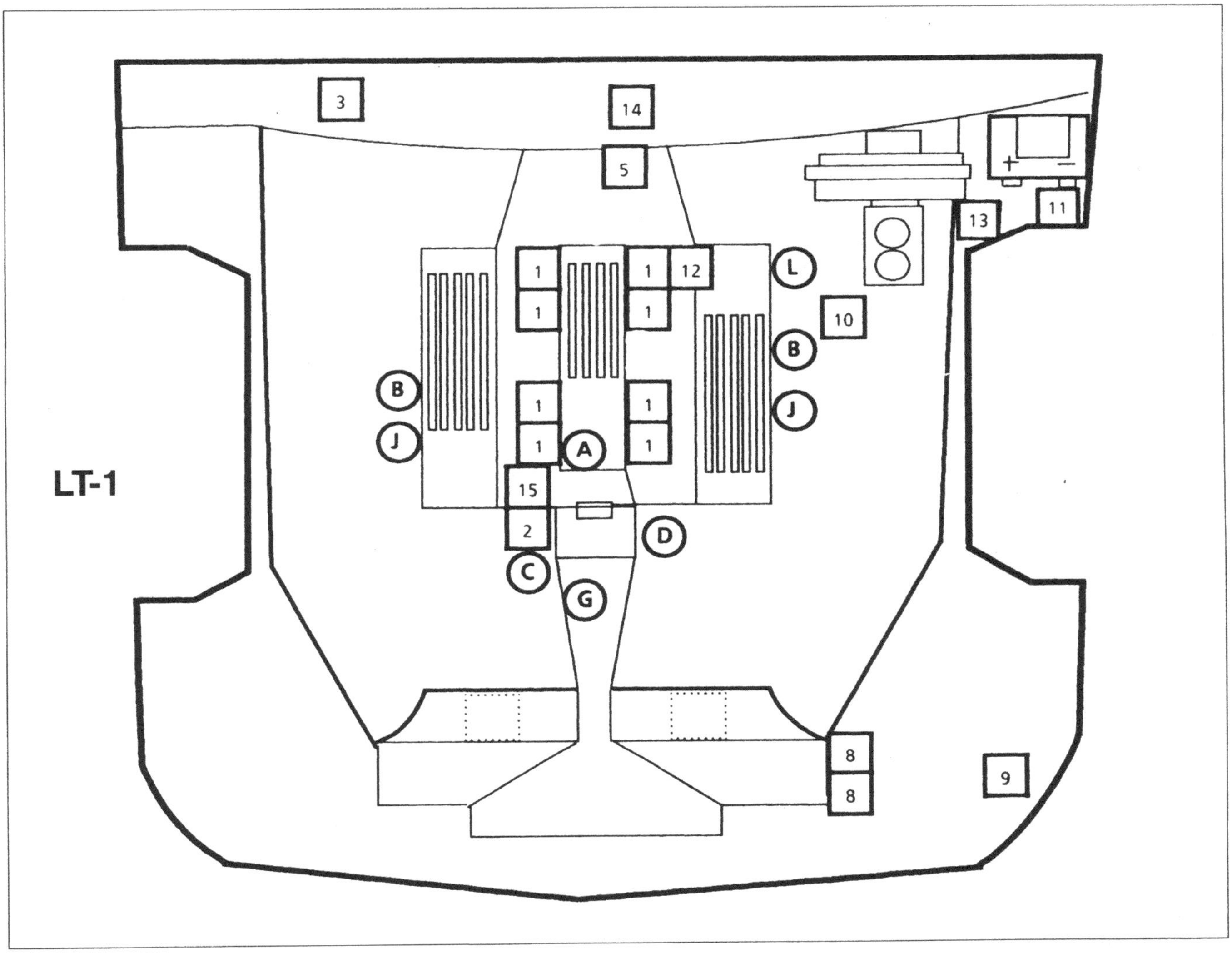

Fig. 4-79. '92–93 LT1: Multiport Fuel Injection with MAP

Sensors (circles)
A Manifold Absolute Pressure sensor
B Oxygen Sensor (O_2S) (two)
C Throttle Position (TP) sensor
D Engine Coolant Temperature (ECT) sensor
G Intake Air Temperature (IAT) sensor
J Knock Sensors (KS)
L Engine Oil Temperature (EOT) sensor

Actuators (squares)
1 Fuel injector
2 Idle Air Control (IAC) valve
3 Fuel Pump relay
5 Torque Converter Clutch (TCC) connector
8 Cooling fan relay
9 AIR pump
10 AIR bypass valve
11 2-3 blockout relay M/T
12 EGR solenoid valve
13 A/C clutch relay
14 2-3 blockout solenoid M/T
15 EVAP canister purge solenoid valve

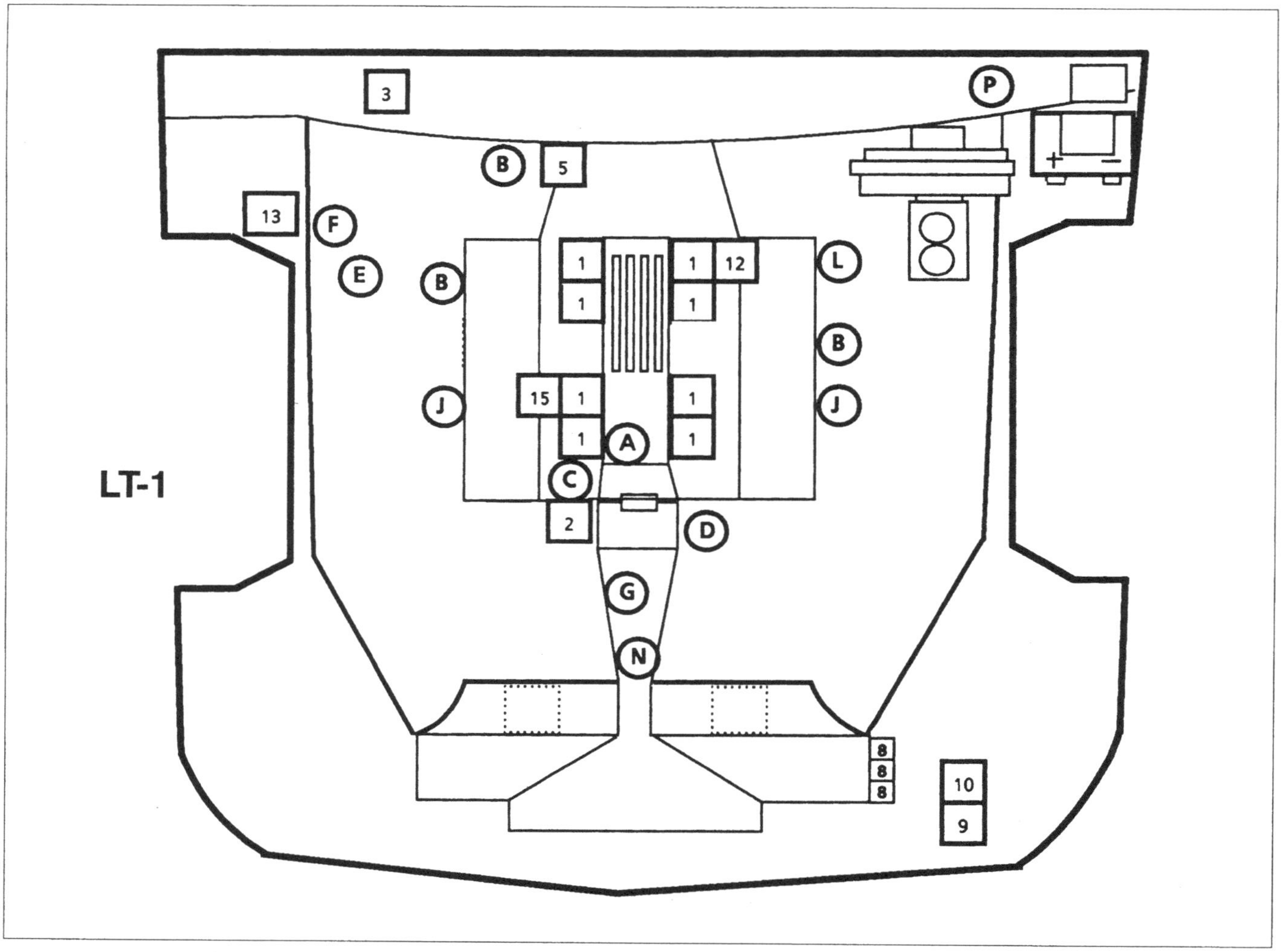

Fig. 4-80. '94–95 LT1: Sequential Fuel Injection with AC/Delphi MAF

Sensors (circles)
A Manifold Absolute Pressure sensor
B Heated Oxygen Sensor (HO_2S)
C Throttle Position (TP) sensor
D Engine Coolant Temperature (ECT) sensor
E A/C refrigerant pressure sensor
F A/C pressure cycling switch
G Intake Air Temperature (IAT) sensor
J Knock Sensor (KS)
L Engine Oil Temperature (EOT) sensor
N Mass Air Flow (MAF) sensor
P Brake switch

Actuators (squares)
1 Fuel injector
2 Idle Air Control (IAC) valve
3 Fuel Pump relay
5 Transmission electrical connector
8 Cooling fan relay
9 AIR pump
10 AIR pump relay
12 EGR vacuum control signal solenoid valve
13 A/C clutch relay
15 EVAP canister purge solenoid valve

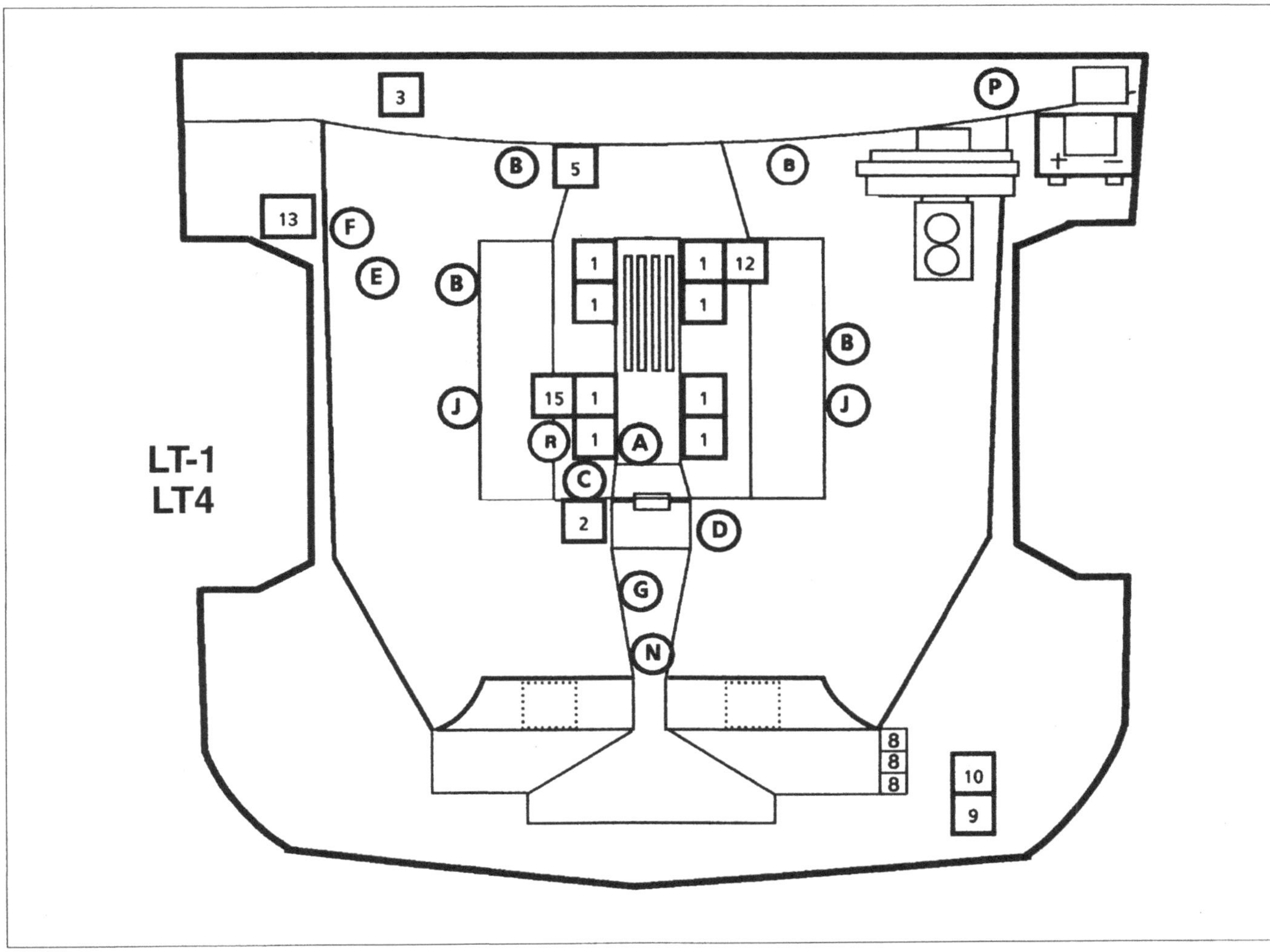

Fig. 4-81. '96 LT1; LT4: Sequential Fuel Injection with Delco MAF

Sensors (circles)
A Manifold Absolute Pressure (MAP) sensor
B Heated Oxygen Sensor (HO_2S)
C Throttle Position (TP) sensor
D Engine Coolant Temperature (ECT) sensor
E A/C refrigerant pressure sensor
F A/C pressure cycling switch
G Intake Air Temperature (IAT) sensor
J Knock Sensor (KS)
N Mass Air Flow (MAF) sensor
P Brake switch
R EVAP vacuum switch

Actuators (squares)
1 Fuel injector
2 Idle Air Control (IAC) valve
3 Fuel Pump relay
5 Transmission electrical connector
8 Cooling fan relay
9 AIR pump
10 AIR pump relay
12 EGR vacuum control signal solenoid valve
13 A/C clutch relay
15 EVAP canister purge solenoid valve

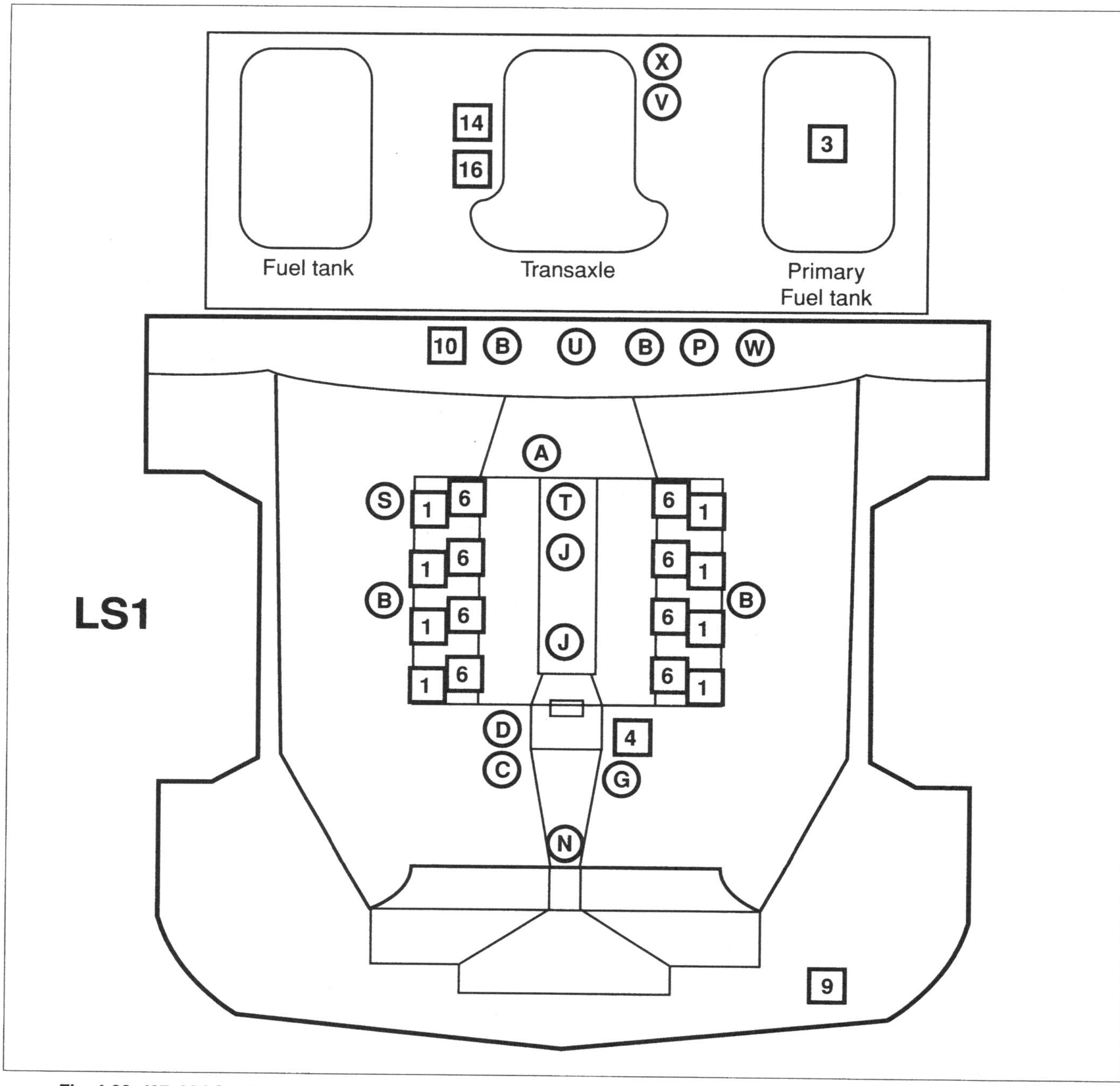

Fig. 4-82. '97–02 LS1: Sequential Fuel Injection with MAF

Sensors (circles)
A Manifold Absolute Pressure sensor
B Heated Oxygen Sensors (HO_2S)
C Throttle Position (TP) sensor
D Engine Coolant Temperature (ECT) sensor
G Intake Air Temperature (IAT) sensor
J Knock Sensors (KS)
N Mass Air Flow (MAF) sensor
S CranKshaft Position (CKP) sensor
T CaMshaft Position (CMP) sensor
U Accelerator Pedal Position (APP) sensor
V Park/Neutral Switch (PRND P) A/T
W Clutch switch M/T
X Vehicle Speed Sensor (VSS)

Actuators (squares)
1 Fuel injectors
3 Fuel Pump relay
4 Throttle actuator
6 Ignition coils
9 Air pump
10 Air pump relay
14 2-3 blockout (skip shift) solenoid M/T
16 Reverse-inhibit solenoid M/T

Chapter 5

Control Module—The Brain

Contents

In this chapter, look for:

- the purposes of the engine's computer—called the Engine Control Module (ECM), or the Powertrain Control Module (PCM) if it also controls the transmission.
- input conditioning—how the computer handles the different kinds of input signals, analog and digital
- calculation—how the Central Processing Unit (CPU) uses input signals, and its several memories, to calculate output signals
- output drivers—how the computer signals actuators in different ways

Fig. 5-1. The control module is about the size of a hardcover book, but it can cost as much as twenty books. It should—it contains more information than twenty books, plus the processing power of a good desktop computer!

1. Introduction

In Chapter 4, you saw the sensors, the "nerves" of the system. The sensors provide the input of information, monitoring conditions in the engine and powertrain. Now I'll talk about the brain of the engine, the different engine control modules that:

- process the information
- calculate the necessary engine operations
- signal the actuators to operate the engine
- perform diagnostic checks of the system
- store trouble codes and alert the driver

When you finish this chapter, you'll be able to explain in simple terms what happens in the computers, so they're no longer the mysterious "black boxes." And you'll understand the changes in computers that are required for improved control of power and emissions.

2. Control Modules

I'll use the term "control module," "engine computer," or just "computer" to cover the different types of control that you'll find in the different model years. Beginning with electronic fuel injection in '**82**, you may find the SAE terms:

- **'82–93:** Engine Control Module (ECM)—which handles just the engine, and has a separate Transmission Control Module (TCM)
- **'94-01**: The Powertrain Control Module (PCM)— which handles both the engine and the transmission, (including shift schedules and engagement for automatic, or the 1-4 upshift for manual transmission)

The computer is based on a digital processor, a cousin of the desktop personal computer. And, like a home computer, the Central Processing Unit (CPU) operates with chips, microprocessors which do the calculating and the memorizing. But instead of getting input from a keyboard, it receives the input from the sensors. And instead of delivering output to a screen or a printer, it delivers the output to the engine's actuators. Besides sending output to the fuel injectors, spark timing, and idle-air control, it diagnoses the system. The computer can:

- recognize problems
- alert the driver, with a malfunction light (Malfunction Indicator Light, Check Engine light, or Service Engine Soon light)
- store codes, to assist in troubleshooting (for example, code 55 for an engine computer failure, and code 51 for a PROM failure)

See, the computer can even check itself!

2.1 Types and Locations

As in most new technologies, Corvette's engine computers are growing in capabilities. Nearly every year, since the first Corvette electronic units came out in 1982, they do a little more, and they do it a little faster. Because of this the units tend to have different locations. Early computers, **'82-91**, are sensitive to electrical noise, as well as to heat and vibration.

L83, L98: Look for the computer in the passenger compartment. The first, in the **'82**, is located behind the driver's seat. (See Fig. 5-2.)

LT1, LT4, ZR-1, LS1: The computers are hardened enough to be located inside the engine compartment. The advantages are shorter cables to the engine, and fewer connectors through the firewall. The disadvantage is that backprobing is not allowed!

PROM

L83: The first electronically fuel-injected Corvette operates with the simplest computer, called the Engine Control Module (ECM), usable in many GM cars. The removable Programma-

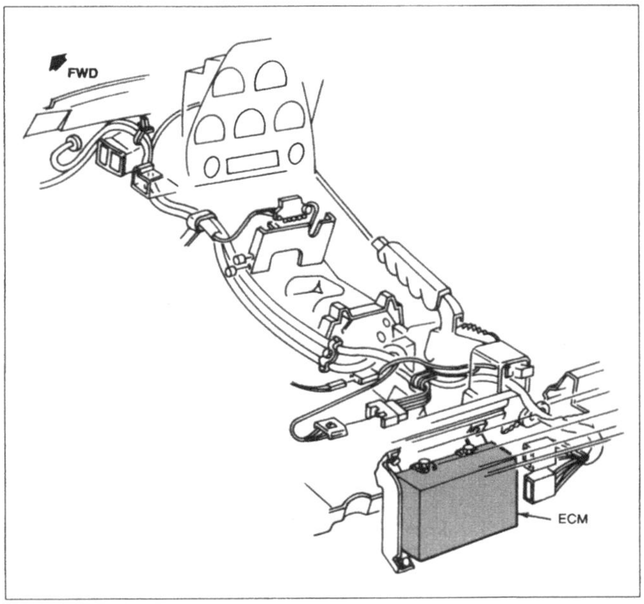

Fig. 5-2. L83 '82: Look for the computer behind the driver's seat, inside the battery compartment. The longer cables were subject to interference.

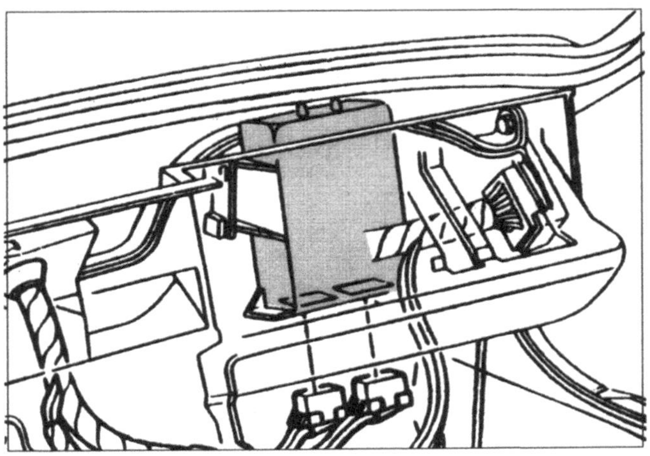

Fig. 5-3. L83 '84: Look for the computer under the right side of the dash, behind the glove compartment.

ble Read-Only Memory (PROM) calibrates the control module according to the car's engine, transmission, weight, axle ratio, and several other factors. Manufacturers designed the control module and PROMs separately, to allow wide interchangeability of control modules throughout manufacturing and servicing. Some think that changing the PROM is the first step to increasing performance, but it's often not really the best way. (See Chapter 9.)

The Memory and Calibration (MEM-CAL) assembly includes the PROM and Calibration Pack (CAL-PAC). And the CAL-PAC includes the backup fuel control, to help you "limp home" if the engine computer goes out.

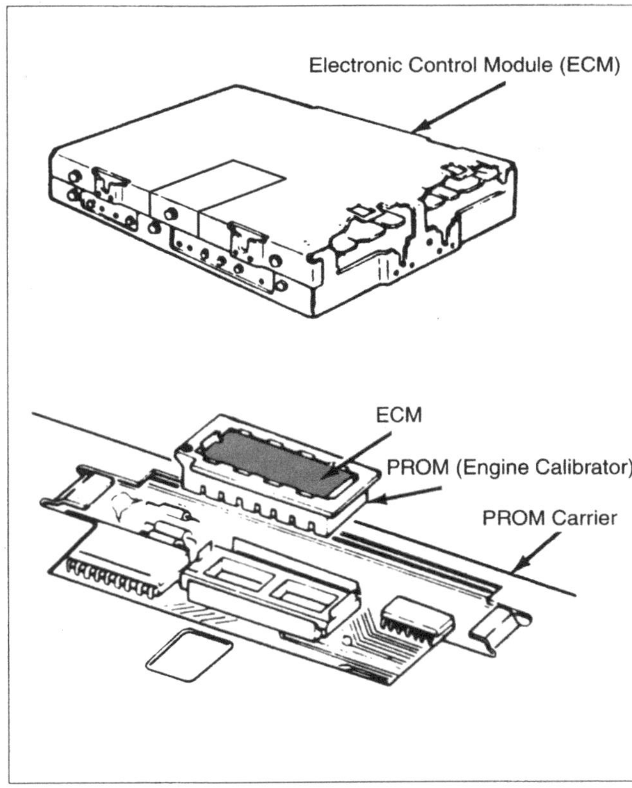

Fig. 5-4. L83 '84: The PROM carrier allows you to easily transfer the PROM from one ECM to another. Changing PROM is a popular method of changing performance, but it's not always the best method.

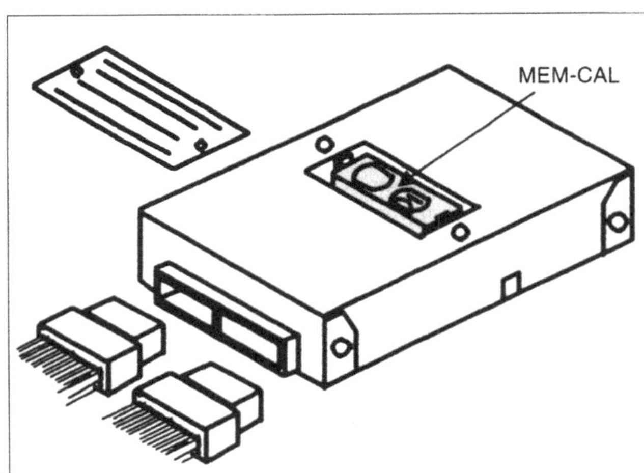

Fig. 5-5. L98 '86–89: The engine computer includes the control module and (under the access cover) a single Memory and Calibration (MEM-CAL) unit.

5

Warning:

If you are considering replacing a PROM, make sure that it is the proper one for your engine. Never install a new PROM just because the box fits or the connectors seem to match. The cases, and even the connectors, may look similar on the outside, but variations in the engine models, and years, require very specific programming to function properly.

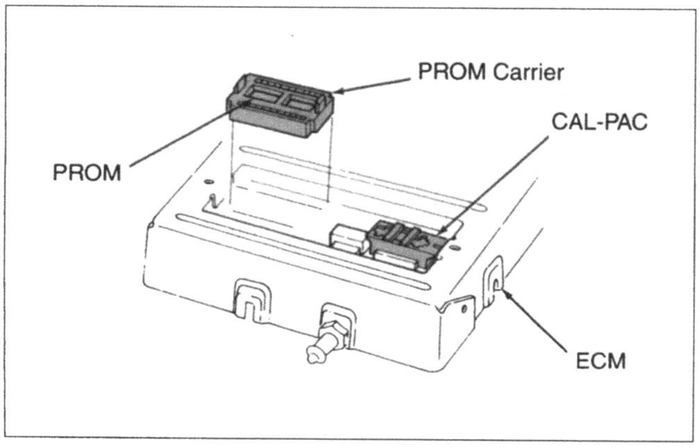

Fig. 5-6. '87 L98: PROM is separate from CAL-PAC. Header connectors (not shown) replace edgeboard connectors.

L98: The GMP4 control module needs a fast processing speed, for the port injection and for better engine control.

'94-96 LT1, LT4, LS1: The Electrically Erasable PROM (EEPROM) is a big difference from PROM. EEPROM is permanent memory, it's not removable and it can only be changed electronically. Reprogramming calibrations can be input by GM CD-ROM dealers, using a scan tool plugged into the Data Link Connector (DLC).

In addition to controlling the engine, the Powertrain Control Module in **'94-01** Corvettes controls the manual transmission 1–4 upshift, or the separate hydraulic control for automatic-transmission shifting.

Among the features in the **LS1**'s new 32-bit processor and Modula (GM speak), are improved spark timing, and improved filtering of knock signals, resulting in fewer "false retards". The new software includes 375 kilobytes of code, 75 kilobytes of ROM, for calibration purposes, and 13.5 kilobytes of RAM. Among its capabilities are predicting airflow and "wall wetting," for transient fuel requirements—useful when you step on it hard.

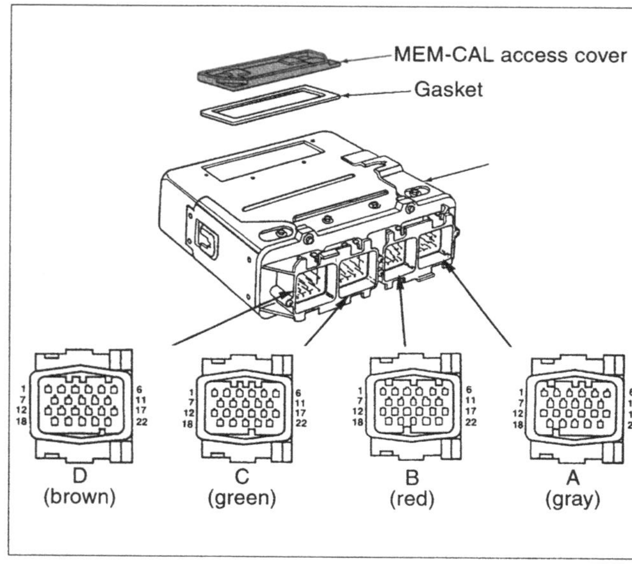

Fig. 5-7. LT5, '92–93 LT1: The PROM, part of the MEM-CAL assembly, stays with the car, even if the computer is changed. Connectors on the **'90–91 L98** look similar to one on the **'92–93**, but, of course, they have different pins and circuits.

Fig. 5-8. LT1, LT4: New heat-resistant chips and components, along with cooling fins, permit computer operation even in underhood heat. This computer is located behind the left front tire. The computer in the **ZR-1** is similarly located but uses different components.

Knock Sensor (KS)

Fuel-injected Corvettes include knock sensing as part of Electronic Spark Control (ESC). In the **L83** and **L98**, a lack of signal, or "pull down," indicates normal operation, and normal spark timing. This "fail safe" system tends to protect the engine from destructive detonation. (For details on the Knock Sensor, see Chapter 4.)

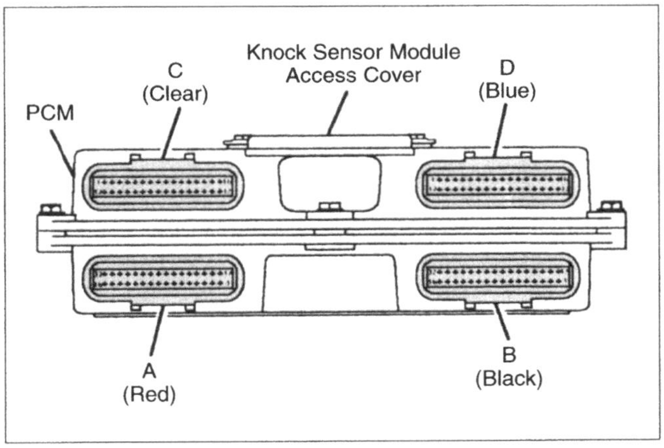

Fig. 5-9. '94–96 LT1, LT4: These connectors handle the increased number of leads to the computer, as required by OBD-II. Notice the color-coding of the four connectors.

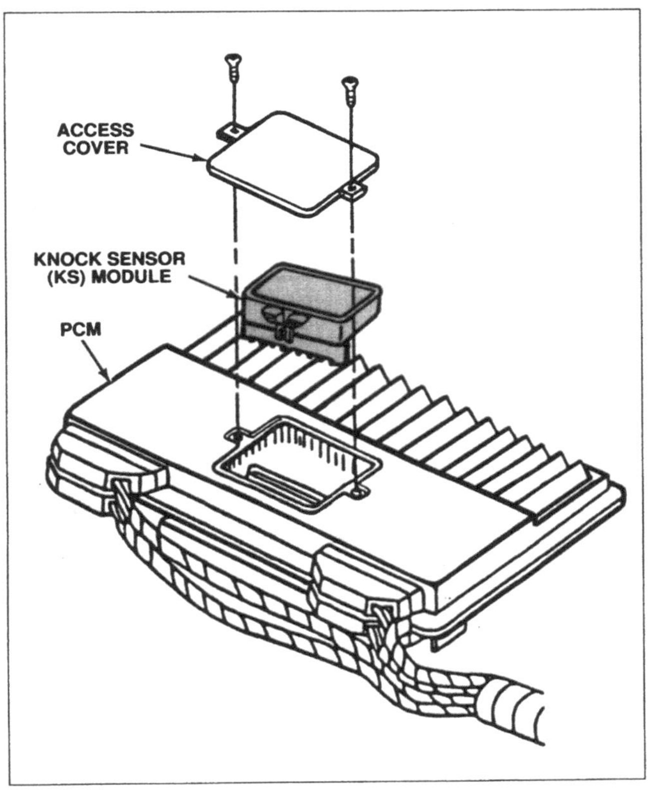

Fig. 5-10. '94-01 LT1, '96 LT4: The computer operates with a removable Knock Sensor module, under an access cover.

CAL-PAC

The computer goes into backup ("limp-home") mode if the system isn't functioning properly. (See Chapter 2.) This backup mode operates with the CAL-PAC, using inputs from only four sensors: MAP or MAF, Throttle Position, Engine Coolant Temperature, and rpm. You'll get home or to the shop, but without full engine control.

2.2 Transmission Control

Automatic Transmission

In **'94-01** engines, the PCM controls both engine and transmission. In the **'82-93** the automatic transmission (A/T) is controlled by the Transmission Control Module (TCM). In this book I'm concentrating on just engine controls, but you should recognize the close link between engine control and transmission control. For example:

5

- Signals needed for automatic transmission control come from many of the sensors for engine control, including throttle position and barometric pressure.
- Fuel cutoff, during upshift, reduces torque for smoother shifting and reduced load on the clutches.
- Ignition retard, during downshift, again, provides for smoother shifts, and reduces the load on the clutches. (In manual transmissions, the PCM controls the 1–4 upshift, or skip shift.)

Manual Transmission

In the **'91-01** manual transmission engines, there is no TCM. The engine computer controls the 1-4 Upshift Control (Skip Shift) See Chap. 4: Chap. 9, CAGS (Computer Aided Gear Shift).

2.3 What a Computer Does

The jobs which are handled by the seven different parts of the control module are:

- monitoring how many instructions are processed each millisecond (even the "slow" **'85** engine computer processes 300 each millisecond)—done by the internal clock
- receiving the signals from the sensors—done in the input interface
- processing the information—done in the CPU
- outputting signals to the actuators—done in the output interface
- storing the programed memory—done in the ROM
- storing temporary values—done in the RAM
- storing calibration—done in the PROM or EEPROM

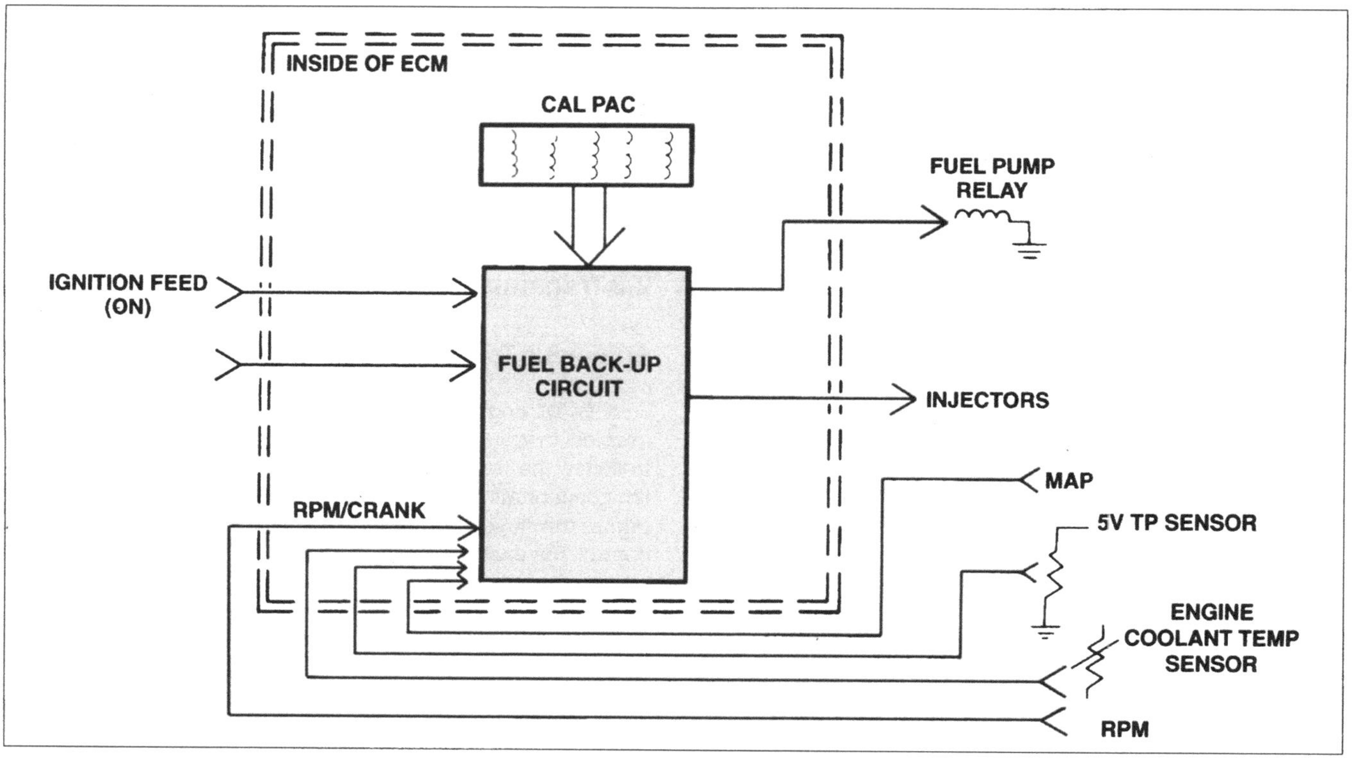

Fig. 5-11. The CAL-PAC works as the computer's safety net, if there is a system failure.

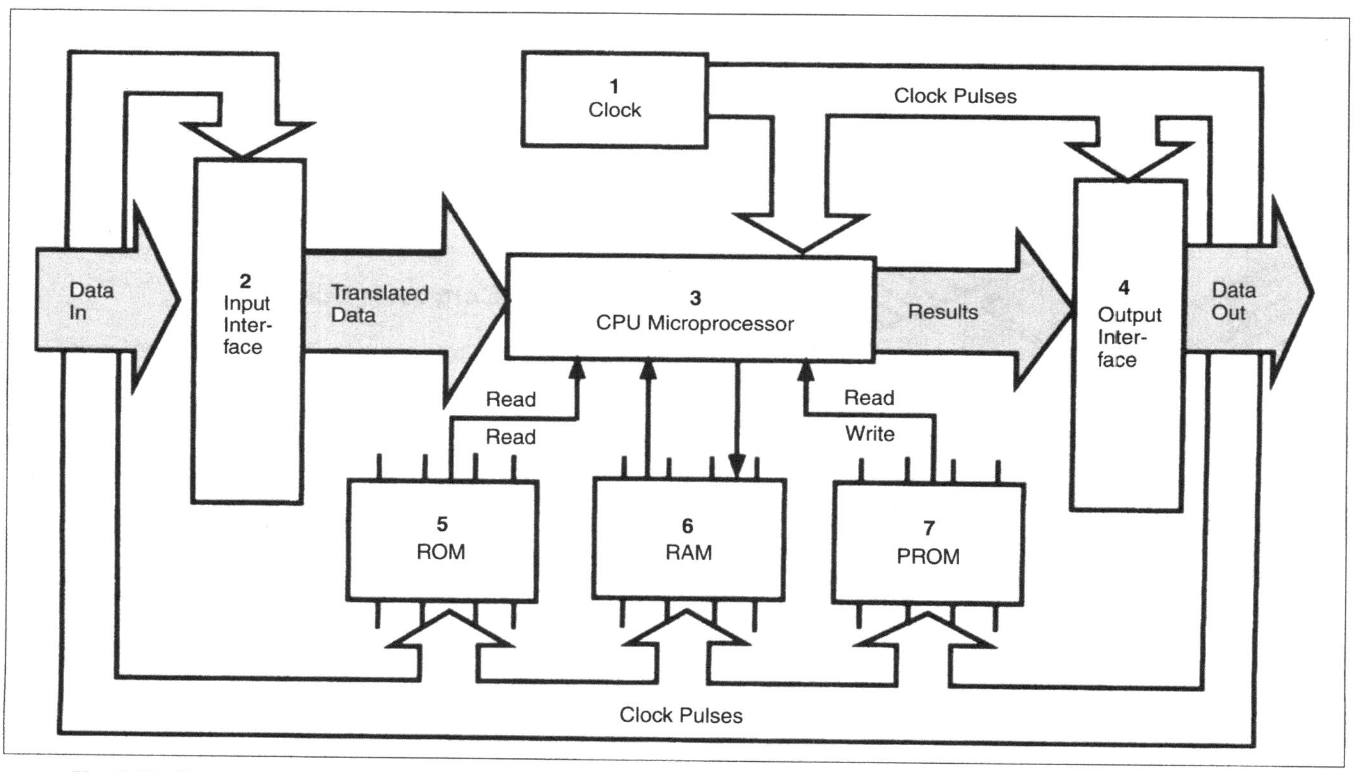

Fig. 5-12. The computer's jobs are: "data in" (from the sensors); "processing" in the Central Processing Unit; and "data out" (to the actuators). The CPU is also called the Microprocessor Unit (MPU).

3. Input Conditioning

Some of the sensors send signals that are analog, while the engine computer uses signals which are digital. From most of these signals, the analog data must be translated by the input interface, before the microprocessor can use it. Some people refer to this process as language translation. But don't get confused—we're not talking about computer languages. We're talking about translating (conditioning) signals into the right basic form, for processing or actuating. It's really more equivalent to changing the actual alphabet of a language.

> In 1963, I studied "Computers" at a Communication Cybernetics seminar given by the Air Force Office of Scientific Research. Back then, computers were room-sized mysteries, operated in an air conditioned, super-clean environment, by computer specialists (in white coats, of course). When we needed computer output, we brought our input data, knocked on the door, handed it over to the specialists, and waited a day for the output!

3.1 Analog vs. Digital

Analog? Digital? What's the difference? A digital voltmeter displays numbers, like a calculator, while an older analog voltmeter has a swinging needle. "Numbers" and "needles"—is that the difference between digital and analog? Don't you believe it. It's important to know the real difference: level steps versus continuous change. Digital measures in steps, jumping from one step to the next, providing no information in between the steps. Analog measures continuously, providing all the information available.

For example, the steps in most digital clocks are one minute in length. They can read no more accurately than one minute. Whereas most analog clocks show the hands moving continuously around the dial, allowing you to see where the hand is when it is between minutes. An analog measurement can be more accurate than a digital measurement that uses large steps. But by increasing the number of steps—measuring in seconds, or even hundredths of a second—digital accuracy can be just as useful as analog. (See Fig. 5-13.)

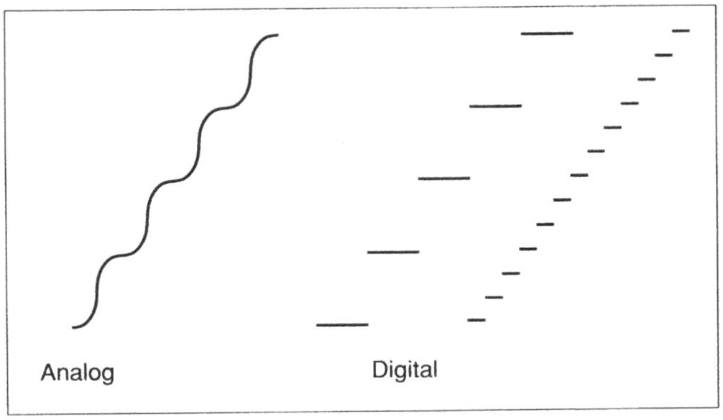

Fig. 5-13. Analog signals flow continuously, while digital signals jump from one level to the next. Digital signals can get more accurate with more (and smaller) steps.

In a computer, like the one in your Corvette's engine, the steps and the empty spaces between them, are expressed as ones and zeros (indicated by either "on" or "off" signals). So, digital data is especially useful to the engine computer, because with only two possible kinds of signals, it is less likely to be corrupted by small changes in the current, as the computer's wires run around all the harnesses and connectors.

3.2 Amplification and Conversion

5

The following signals come into the computer as analog:

- the Engine Coolant Temperature (ECT) and Intake Air Temperature (IAT) sensors (variable resistors)
- the Throttle Position (TP) sensor (a potentiometer)
- the Oxygen Sensor (O_2S) (a signal generator)

Most input signals have a very small voltage. This means that they must be amplified. These are then converted, or "conditioned", into digital signals and sent off to the CPU. Some sensors, like those from signal generators or switches, send signals already in the digital format, and they need no amplification or conversion. In future engine-control systems, you'll see more sensors which condition their own signals; these kinds of sensors are called "smart sensors".

4. Central Processing Unit (CPU)

The microprocessor, or the chips in the Central Processing Unit, are the heart of the computer. Each chip is a Large-Scale Integrated (LSI) circuit, with hundreds or thousands of transistors. With different Corvette engines and transmissions, axle ratios, and wheel sizes, each PROM chip is specific to each application.

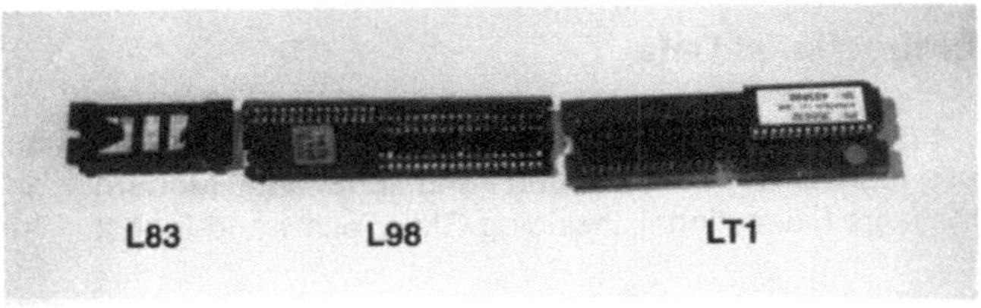

Fig. 5-14. The chips in the Central Processing Unit, are really the heart of the computer. These chips are two to three inches long.

4.1 Processing Speed

'94-01 LT1, LT4, LS1, ZR-1: For an eight-cylinder SFI engine, running at 6,000 rpm, the microprocessor must sense, process, and output the individual cylinder injection times and spark timing four hundred times every second—that's once every two and a half milliseconds! Remember, the crankshaft revolves twice for every eight cylinder cycle.

In this example, the calculations work out like this:

$$6000\text{rev/min} \div 60\text{sec/min} = 100\text{rev/sec}$$

So one revolution fires four of the cylinders in 0.01 sec, or ten milliseconds. Divide that by the four cylinders, and you get an injection and spark in each cylinder every 2.5 milliseconds.

Fortunately, the engine runs under transient conditions only some of the time. At other times, the chip has a few milliseconds to tend to other matters, like idle rpm, emission control, diagnostics, relation to the transmission, vehicle speed, and trip computer. That's a lot of processing! And it must deliver it for the 3,000 to 5,000 hours that the average engine runs.

Serial Data

Serial data is another term for "data stream". Several of the engine control system's parts communicate with each other using digital voltage signals, pulsing between high and low, or on and off. Each pulse is known as a bit. A series of eight bits equals one byte, or one "word". (Here's where the idea of a computer language really comes in.) These "words" help the computer carry information that is more complicated than just "yes" or "no". Serial data just indicates that the computer is using this, more complicated, kind of signal.

Baud Rate

Baud (named after French inventor Jean Baudot) is the term applied to the speed of data transmission, in bits per second. Compare these baud rates (GM Class 2 data stream):

- **'82–85:** 60 bits per second
- **'86–95:** 8,192 bits per second
- **'96-01:** 10,400 bits, or 10K (bytes), per second

Bidirectional Data

Late-model computers are known as bidirectional—you can use them to both receive and send serial data. Modern scan tools are bidirectional, including GM's Tech 1 and Tech 2.

Interrupts

The CPU chip is able to interrupt its computation for some of the less time-critical events, in order to tend to fuel injection and spark timing. When you stomp on the accelerator, or release it suddenly, the chip quickly responds. When it does this, it stores, in memory, the program which it was working on, which it will return to when matters are less pressing.

5. Memories

The control module contains three kinds of memories. I'll introduce them and then look at each in detail.

Read Only Memory (ROM): This is a permanent body of data about how the engine control operates. It can be read but not changed. It doesn't need applied power, since it is essentially hard-wired, so it retains data when the key is off.

Random Access Memory (RAM): This is short term memory, which changes, as in ECT, MAF, MAP, and VSS. RAM is like a scratch pad, where you write down a telephone number, use it, and then throw it away. Some RAM lasts until you turn off the key, after which it is erased. **Nonvolatile RAM (NVRAM)** is mid-term memory—it stores Diagnostic Trouble Codes (DTCs) even when the key is off, but loses the data when the car's battery is disconnected.

Programmable Read Only Memory (PROM): This is long-term data about a specific application. It is used in Corvettes from **'82–93**. Data may be specific to the axle ratios, wheels and tires, and other engine/vehicle variables. PROM is usually removable: you exchange the PROM to change the data. It also doesn't need applied power, so it retains data when the key is off. **'94-01** Corvettes have **Electrically Erasable PROM (EEPROM)**. It's permanently soldered to the circuit board, so you can't change the chip—but you can change the data. Data can be erased and reinstalled, using the Data Link Connector.

Among the many jobs of ROM is storing lookup tables. Among the many jobs of RAM is keeping track of fuel trim data. Some people (myself included) have confused these two because they're illustrated with similar tables. But I want to caution you to keep them separate even though they may look similar on paper.

> PROM is sometimes considered the universal "key to performance," but see Chapter 9 for reasons why it may not always be.

5.1 Read Only Memory (ROM)

Lookup Tables

Lookup data tables are actually located in the cells of the computer's memory chips. To determine precise data for the lookup tables, engineers test each variation of engine. They find the best air/fuel ratio and spark timing, for each condition of speed, load, and other variables: in heat and cold, on the dynamometer, and in the mountains. They look for the best power and economy, all the while meeting emission limits. If you are considering a replacement chip, check the validity of the development work. Chip replacement seems so promising and so simple. But they're not, and the parts dealer is likely to be the one to do the most "promising".

Blocks

Many of the values are stored in blocks (that's data blocks, not engine blocks). This is where the old GM term "Block Learn" comes from. The separate blocks contain similar data, for example, where the air/fuel ratio may be the same for many values of load and rpm. You can see that there can be literally thousands of combinations of load and rpm, far more than can be stored, and far more than necessary. So storage in blocks is a practical solution, something like the different folders in a filing cabinet. Bosch calls its lookup table a "map," (not to be confused with Manifold Absolute Pressure).

The computer can interpolate from the data block information: for example, if the ECT is at 100° F (halfway between 50° and 150°, which it has data for), the computer can calculate that the air/fuel ratio at high load should be 11:1 (halfway between the 9:1 and 13:1 ratios indicated by the lookup table). (See Fig. 5-15.)

In **'94-01** Corvettes, the EEPROM allows you to change the stored values, without changing the chip. Reasons for reprogramming in the EEPROM include:

- adapting the computer to new sensors, which may be introduced during production
- selecting features according to customer options, or for later changes (for example, tire or wheel size)
- making improvements based on field experience

Strategies—ROM

Strategy ROM holds the plans, created by engine designers, for the timing and control of the engine control systems. For example, the normal fuel-control strategy is to maintain a stoichiometric air/fuel ratio, while the cold-cranking fuel-control strategy is to enrich the mixture.

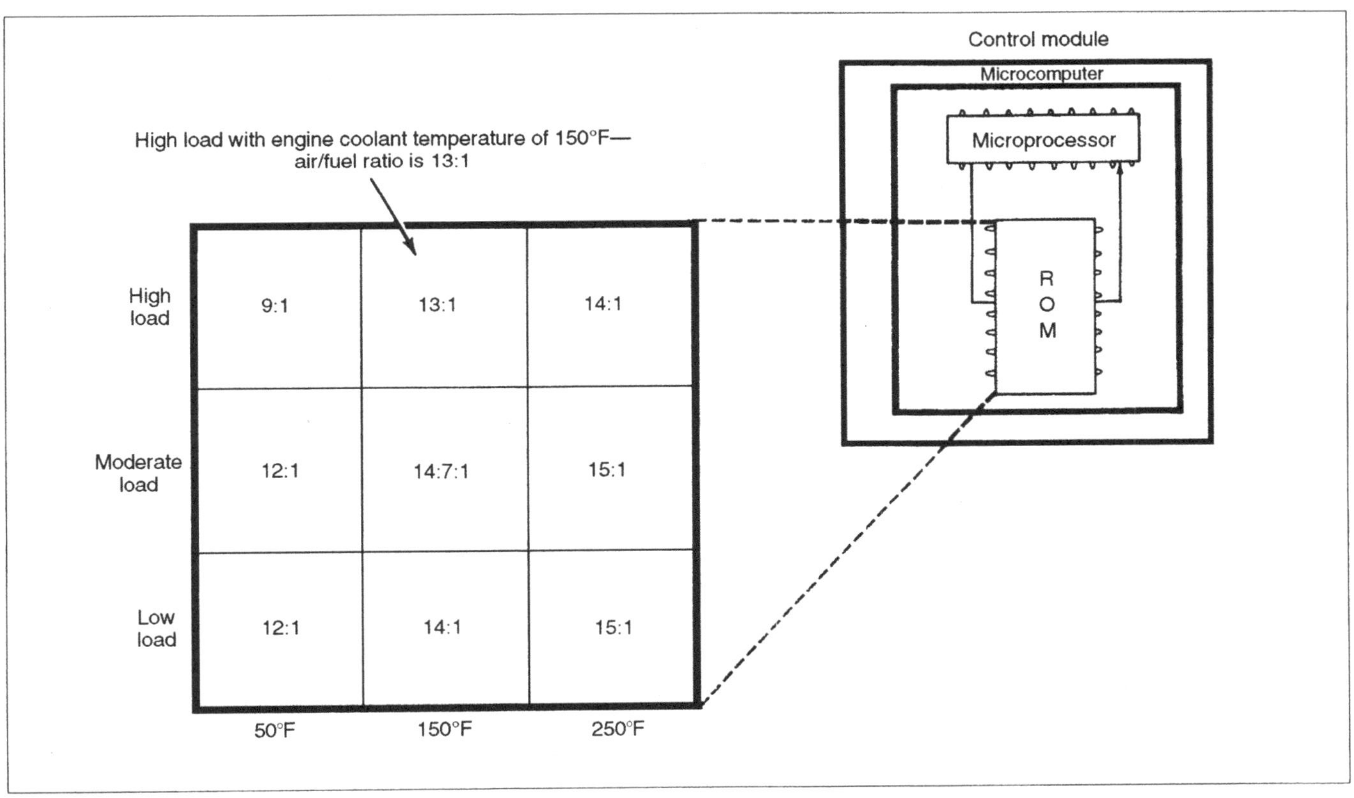

Fig. 5-15. Here is a ROM lookup table, with the air/fuel ratios for different conditions of load (MAF or MAP) and for Engine Coolant Temperature: richer for high loads, at a cool ECT; leaner for low loads, at a hot ECT. What does this table say about guys who want to sell you a low-temperature thermostat to boost power? See Chapter 9. (These values are for example only, and are not actual.)

In contrast to the lookup tables and strategies stored in ROM and also in PROM, fuel-trim data is stored in RAM, changing as you drive.

> Corvette, the government, and most of the aftermarket dealers and customers, are interested in EEPROMs with proper safeguards built in. The first two want to be sure the authorized service facility can reprogram (within proper engine control and emission limits) but performance guys *cannot*. But many Corvette owners depend on aftermarket products, so aftermarket servicing is even more important for them. You might say that the EEPROM in **'94-01** Corvettes has put the slip-in-a-new-chip guys out of business, while creating a new business, for those who can change the data on the chip.

5.2 Random Access Memory (RAM)

Random Access Memory stores short term information, as needed, which may be for only a few milliseconds or for several hours, until the key is turned off. For example, on earlier-model Corvettes (**L83** and **L98**), when you turn on the ignition, the control module takes a few milliseconds to collect the BARO pressure signal and store it in the RAM. It may update that data occasionally, collecting and storing new BARO data. But, when you turn off the key, RAM says, "Forget it. Don't store that old data, because barometric pressure changes with the weather. I'll store new data when the engine is restarted." On later models, BARO is measured continuously, and not stored in RAM.

You can think of Nonvolatile RAM (NVRAM) as a special kind of RAM, powered directly by the battery, instead of from the ignition key. NVRAM's main purpose is to store Diagnostic Trouble Codes (DTC), also known as Service Codes. NVRAM also stores adaptive corrections.

> The earlier term for Nonvolatile RAM was "Keep Alive Memory," or KAM. They both mean that RAM will continue to store the data even after the key is turned off. NVRAM is data to be remembered for a while, then forgotten, and erased. For example, Diagnostic Trouble Codes remain in memory even with the key off. In the **'94-96 LT1**, **LT4**, and **LS1**, disconnecting the battery for 30 seconds erases NVRAM, except for the OBD-II codes.

Adaptive Strategy

Adaptive strategy regularly shifts the base calibration, to compensate for changes in Barometric Pressure, Intake Air Temperature, fuel composition, and small drifts in sensor or actuator signals. "Wait a minute," you say, "the control module is already measuring those." Yes, but the strategies stored in the ROM are based on how the original test engine responded to sensor inputs. Adaptive strategy looks at how your specific engine is responding—and further, how your engine is responding to how you are driving it, under the current conditions of temperature, humidity, altitude, and fuel (such as the amount of oxygenate it contains). The adaptive system "learns" those modifications, and then teaches them to the computer control. I'm talking about an intelligent car, which adjusts itself to its own needs as well as the the driver's. Some **'85 L98** Corvette control modules are adaptive; more in **'86**. From **'87-01**, all are adaptive.

Adaptive strategy is long-term correction, based on short-term corrections which have been repeated. Short-term feedback applies corrections in milliseconds, as with a Knock Sensor and idle rpm, or in a second, as with an Oxygen Sensor. Long-term adaptation applies corrections over several minutes.

For example, basic (short-term) feedback from the Oxygen Sensor is relatively slow—perhaps 100–1000 milliseconds (and must be, for stability). Basic feedback corrects for steady-state errors, caused by aging and failures. But suppose the oxygen sensor keeps sending rich-mixture (go lean) signals as short-term correction, under certain rpm/load conditions. The computer notes these repeated short-term corrections and shifts the base calibration for that rpm/load combination toward lean. The computer has "learned" that this engine needs less fuel than the test engine, under the same conditions. Adaptive strategy is limited in its correction, to prevent shifting calibrations to unsafe, or improper, fuel injection or spark timing.

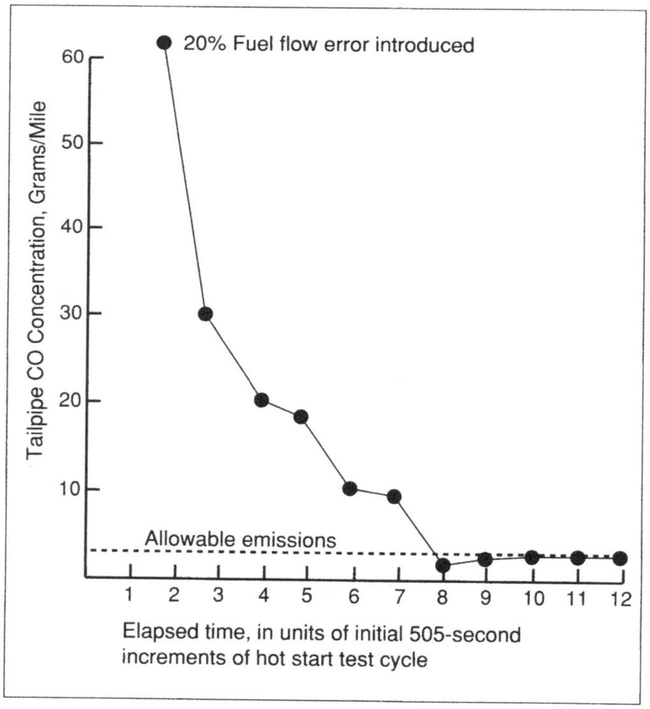

Fig. 5-16. In one test (not Corvette), the computer's adaptive strategy corrected an engine which had purposely been set 20% rich. Carbon monoxide fell from over 60 grams per mile to nearly zero.

In one legitimate engine-control development test, a system was calibrated properly for carbon monoxide emission of 3.4 grams per mile, then set 20% rich from Open Loop fuel injection. At 20% rich, the carbon monoxide emission was over 60 grams per mile, seventeen times the limit, at the time. Within one hour's driving (long-term) with many accelerations and decelerations, as during normal urban driving, the computer had adapted. It had "learned" well enough to reduce carbon monoxide even below the 3.4 limit, to almost zero.

> I heard one story, going around Detroit, that adaptive strategy could recognize the conditions when a car was on the dynamometer doing an EPA emissions test. When test values matched the EPA standard test, the computer could "adapt", shifting its calibrations to insure emissions/economy test success, even if those calibrations might be poor for performance or cold-cold starting. Once the test car was off the EPA dyno, adaptive control could shift back to "the good stuff". The story goes on that EPA heard of this and threatened to program EPA computers to recognize when the test-car engine computer had adapted!
>
> When I queried an EPA friend, the engineer grinned and said, "Sure, the engine computer could recognize the fixed initial engine-starting temperature and certain fixed acceleration rates of the Federal Test Procedure." But, he went on, "That's a good story, but there's no truth to it."
>
> But a manufacturer's powertrain engineer (not Corvette) smiled at the story and told me, "It was easier than that. The hood switch [the underhood light-switch connection to the computer] didn't last long. When else would the car be driving with the hood up, except during an EPA test on the dynamometer?"

NVRAM Storage

NVRAM stores adaptive strategies. That means the computer adapts to your engine and your driving—but only as long as no one disconnects the battery. If the car is serviced or loses battery power, for example during installation of a theft alarm or cellular car phone, look out. You can expect the system to lose the adaptive RAM, along with settings for the electronic radio and clock. To prevent this, some technicians plug in an auxiliary power supply. A simple 9-volt battery clipped to the battery connectors will do the trick. Or, after disconnecting a battery, good technicians often drive the car to restore the adaptive values in the NVRAM. Some say ten minutes is enough, but OBD-II cars may take up to an hour.

If the car drives strangely after being serviced electrically, the engine may have lost its adaptive memory in NVRAM. Adaptive strategy needs time to work, after replacement of any part of the engine control system, and when the car is new.

> Drivers who occasionally drive their adaptive-system cars on the drag strip are often puzzled about their high elapsed times. The system adapts to street driving. When they run the strip a few times, the computer readapts to the strip, and their ETs improve.

Fuel Trim

Fuel trim is a typical adaptive strategy. It's a number indicating a correction to the injector pulse width, set up in two parts: a volatile short-term RAM memory registers, and a set of NVRAM cells are arranged by load and rpm. The control module is constantly updating short-term fuel trim, responding to changes in operating conditions, like load, rpm, and oxygen sensor voltage.

The Short Term Fuel Trim (STFT), previously known as the "Integrator", varies the base pulse width, a Closed Loop factor. STFT is a single number, quickly updated, perhaps many times each second, as conditions change. You'll recall that the Oxygen Sensor voltage changes more or less continuously, as the exhaust crosses over from rich to lean to rich to lean.

The STFT numbering system was changed for the **'94-01** engines, but the principle is the same for all GM cars with adaptive control.

'86–93: The Short Term Fuel Trim is known as the "Integrator". Zero correction is 128, on a scale of 0–255. 0 is the richest/go-lean signal; 255 is the leanest/go-rich signal. The normal operating range for Integrator/Short Term Fuel Trim is 115–141 (±10%).

'94-01: Zero correction is 0 on a scale of –100% to +100%. –100% is the richest/go-lean signal; +100% is the leanest/go-rich signal. The normal range for Short Term Fuel Trim is ±10%.

Thinking about fuel-trim corrections as "plus", "zero", and "minus", makes good sense and is consistent with OBD-II.

> Let me warn you of confusion potential:
>
> - The "Go lean" command is the result of being rich.
> - The "Go rich" command is the result of being lean.
>
> We argued this repeatedly in developing fuel-injection training: "Did you say 'Rich'? Or 'Go rich', meaning "lean condition"? A simple, less confusing, answer is to say "Add fuel" instead of "Go rich"; "Subtract fuel" instead of "Go lean."

5

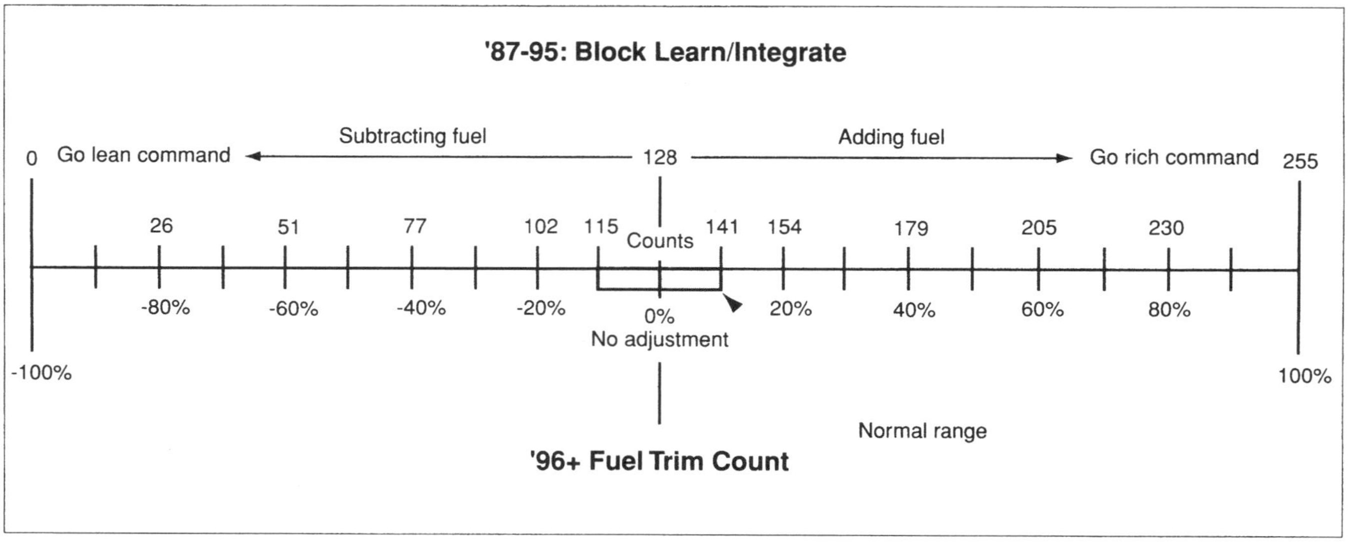

Fig. 5-17. Fuel trim is a number which indicates a correction to the injector pulse width. Block Learn/Integrate (**'87-95**) is based on counts, with zero correction as 128. The **'96-01** STFT is based on a straight percentage, with zero as no adjustment.

Long Term Fuel Trim (LTFT) was previously known as "Block Learn", because it referred to the relearning process of the blocks. LTFT is a series of cells which provide a long-term fuel correction factor, for each combination of load and rpm. The computer monitors the STFT, and if it is beyond the normal variation of ±10%, the data in the LTFT is switched from one block to another. Continued monitoring of the STFT, indicated by Oxygen Sensor signals, indicates whether the STFT is still out the of normal range. The computer continues to change the LTFT data, until the STFT is accaptable.

For LTFT, the maximum fuel correction is ±100%, added to the ±STFT maximum of ±100%, in the same direction. If an Oxygen Sensor indicates that the air/fuel ratio is still wrong, the computer sets a Diagnostic Trouble Code. During power enrichment, as in Wide Open Throttle, the computer freezes the STFT, and the LTFT is prohibited from correcting for enrichment.

Tables 5-1, 5-2, and 5-3 are simplified examples of STFT memory cells, with **'96-01** LTFT numbers first, in bold, followed by the **'86–95** Block Learn numbers.

In Table 5-1, for any load/rpm combination, the STFT is 0 (Integrator is 128), so no LTFT/Block Learn correction is required.

Table 5-2, indicates minor corrections, based on Oxygen Sensor signals. In a fraction of a second, increasing Oxygen-Sensor voltages can reverse fuel correction numbers to less than 0%/128. As long as the STFT numbers are within normal, within ±10%, no LTFT/Block Learn corrections are required.

Table 5-1. No correction required

0%/128	**0%**/128	**0%**/128	**0%**/128
0%/128	**0%**/128	**0%**/128	**0%**/128
0%/128	**0%**/128	**0%**/128	**0%**/128
0%/128	**0%**/128	**0%**/128	**0%**/128

Load →
Rpm →

Table 5-2. Lean exhaust with STFT less than 10% requires short-term enrichment, but no long-term fuel correction

0%/128	**+2%**/130	**0%**/128	**+1%**/129
+1%/129	**+2%**/130	**+2%**/130	**+3%**/131
+5%/133	**+6%**/136	**+1%**/129	**0%**/126
+5%/135	**+4%**/132	**+1%**/129	**+2%**/130

Load →
Rpm →

Table 5-3. Lean exhaust with STFT more than 10% requires long-term fuel correction

Load ↑				
	+20%/154	+15%/148	+15%/148	0%/128
	+15%/149	+12%/144	+14%/146	0%/128
	+13%/145	+13%/145	+13%/145	0%/128
	+11%/142	+13%/145	+11%/142	0%/128

Rpm →

For some load/rpm combinations, as in Table 5-3, STFT varies more than ±10%, so LTFT corrections are required—in this case, going rich. A lean exhaust calls for variations when the STFT numbers are *a*bove +10%/141, requiring added fuel. In contrast, a rich exhaust will call for blocks below −10%/115, subtracting fuel.

During step-on-it acceleration, the computer strategy commands power enrichment, but it also freezes the STFT at 0%/128. Why? Because during acceleration, you don't want to compensate for a too-rich mixture. You really do want the power enrichment. And then, after acceleration, the computer will dutifully return you to where you were, before you stepped on it.

The computer adapts every ten minutes or so. Why ten minutes? And why is this important to you? Because, if it adapted in less than ten minutes, it would be "chasing its own tail"—sometimes trying to adapt to changes which have already been adapted for. And if it adapted in more than ten minutes, it might altogether miss adapting to important changes. So, It seems like ten minutes is about right. Remember: your engine performance will be affected if you change parts, like the Throttle Position Sensor or air filter, or even if you disconnect the battery.

> When I first learned of adaptive strategies, I talked with one of GM's trainers. I asked why I could find nothing about "learning adaptation" in GM training. "Not yet," he said, "we're worried it'll blow their minds." Later, when I attended the **LT1** course, we learned about adaptive Short- and Long-Term Fuel Trim (Integrator/Block Learn). And, while it might be hard to keep straight, it didn't blow any minds.

6. Power

As you would imagine, the engine computer requires power: significant power through the ignition switch, when on, and small power from the battery, for the NVRAM (usually about 2 milliamps). In about fifty hours, the computer drains around one amp-hour from the battery.

Voltage Control

The computer supplies a controlled reference voltage,either 5 or 12 V, for the sensors and actuators. This provides for some stability, even as the battery voltage changes, due to load or temperature. Typical voltages are over 14 V during charging, and under 9 V during cold cranking.

Reference voltage is especially important for three-wire sensors, which depend on the resistance division of a potentiometer. These include the Throttle Position sensor, as well as sensors that translate changes into voltage input signals.

7. Output Control

Output control is necessary, since the relatively large-current circuits are controlled by very small-current computer signals. The computer's inputs deal in milliamps, but it takes full-size amperes to drive some of the solenoids and injectors. It will help you in your diagnosis if you keep this difference in mind:

- small currents use milliamps for input
- large currents use amperes for output

Output Drivers

The computer provides a ground circuit for the actuators. Small voltage signals from the computer cause the transistor output drivers to complete the ground circuit of the actuator. You can think of the driver as a solenoid.

L98: The Quad Driver II (QDR II) is protected by current limiting. Quad Driver Modules (QDMs) protect against circuit faults.

'92–93 LT1: The QDMs protect against circuit faults, and a fault line provides a feedback signal to the computer. Trouble codes may be set and/or readouts sent.

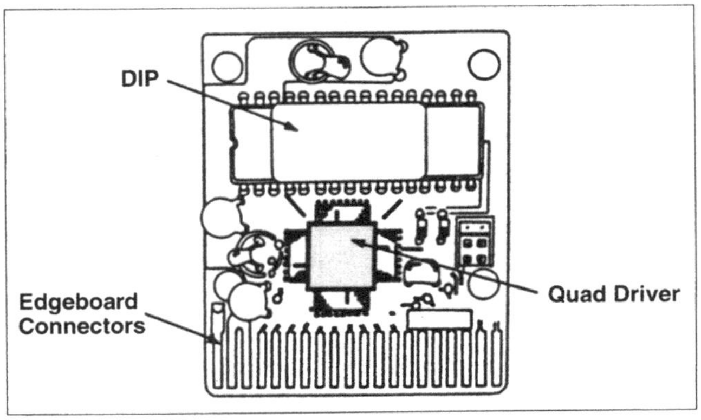

Fig. 5-18. L98: Quad drivers (QDRs) deliver output signals that control a group of four ("quad") actuators. Most actuators are powered, by the battery, whenever the key is turned to On or Crank.

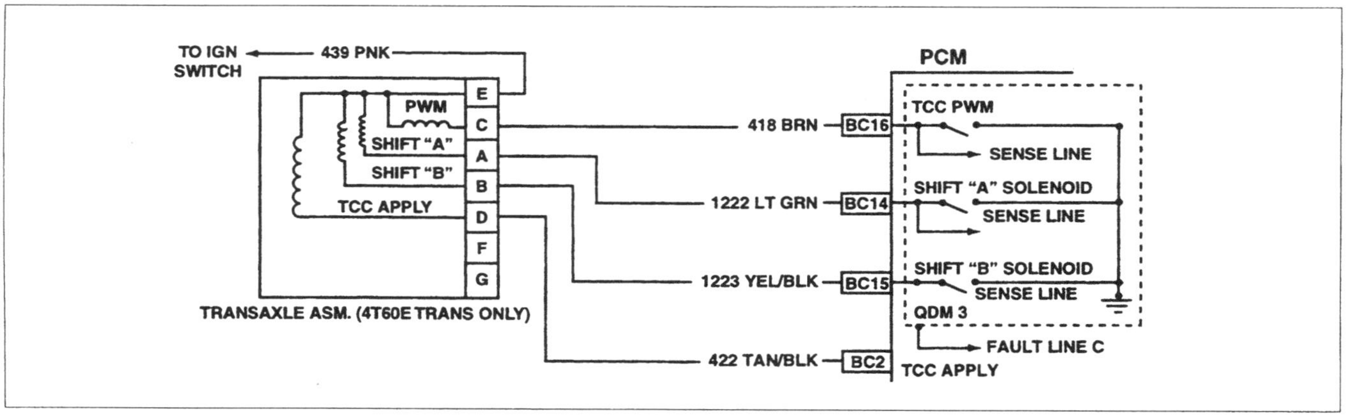

Fig. 5-19. LS1: In the Powertrain Control Module, the Torque Converter Clutch-Pulse Width Modulation (TCC PWM) controls the application of the TCC. Solenoids "A" and "B", in the PCM control, shift engagements of the transaxle.

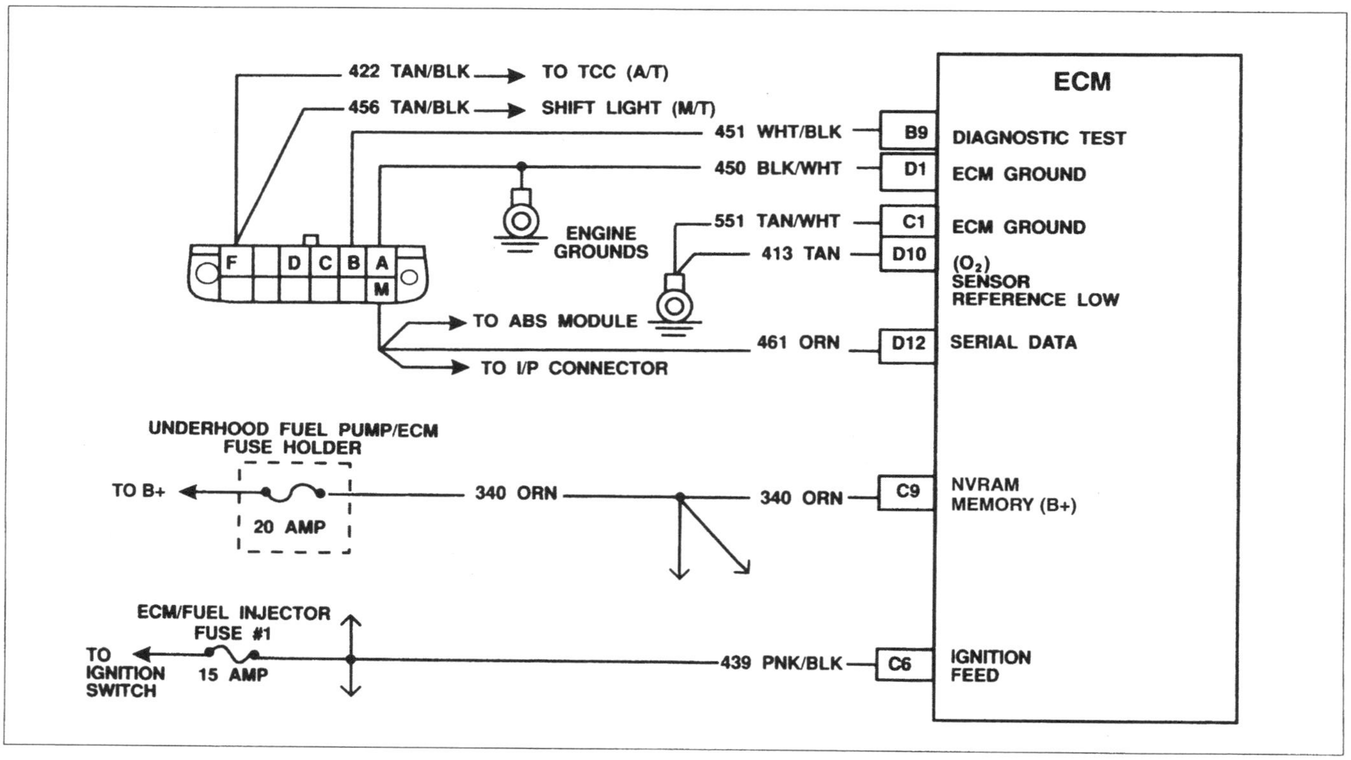

Fig. 5-20. The Engine Control Module requires power, to run all of its various chips. Some of the power is switched on via the ignition key, while other power is routed straight from the car's battery.

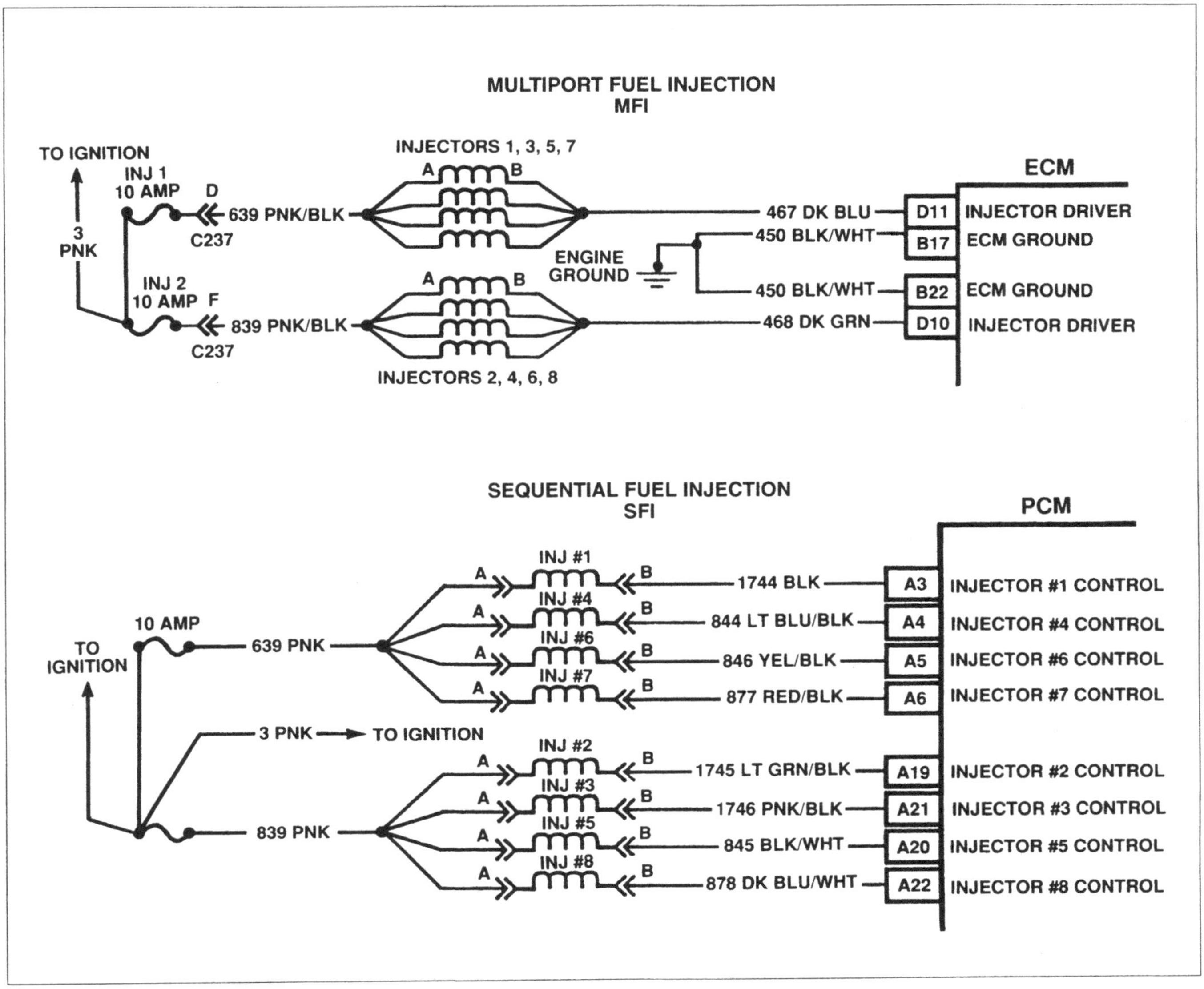

Fig. 5-21. This is injector output control. The injectors are battery powered when the ignition is on—actuated by output drivers grounding each circuit in the computer.
L98, '92–94 LT1: MFI—four of the injectors are grounded at the same time, with fuel being injected into either the left bank or right bank.
LT5, '94-01 LT1, LT4, LS1: SFI—each injector has its own output control, with fuel being injected into one cylinder at a time.

5

LT5, '94-01 LT1, LT4, LS1: The Output Driver Modules (ODMs) can control up to seven circuits, and are current limited. They have fault detection, signaling the computer which output is at fault.

Pulse-Width Modulation (PWM)

Computer outputs are usually digital. And, because most circuits are powered, they're completed by grounding in the computer, or in computer-controlled relays:

- yes/on = grounding for current flow
- no/off = no current

These output signals can operate a solenoid, relay, or solid-state readout. To accommodate the different input-voltage requirements of these devices, Pulse-Width Modulation (PWM) provides a signal that is modulated, or adjusted, to the correct voltage, somewhere between the system voltage and zero.

Think of Pulse-Width Modulation as duty cycle. It acts on a series of digital pulses from the output drivers. The computer varies the on-time/off-time (duty cycle) of the pulses to control an actuator (for example, the idle-air bypass motor), by turning the circuit on and off rapidly. A 50% PWM, of a 12-volt signal, produces a 6-volt signal.

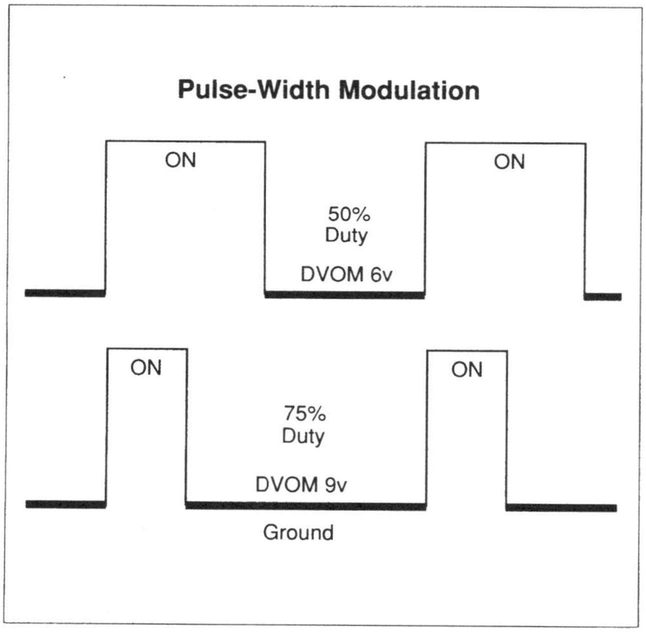

Fig. 5-22. Pulse-Width Modulation acts on digital pulses from the output drivers. The computer varies the on-time/off-time (duty cycle) of the pulses to control an actuator. The current flows only when the circuit is grounded.

When you measure the on/off ratios, you're measuring the percentage of time that the current is flowing, through a grounded circuit—remember that most output drivers ground the circuit to control output.

- When the pulses are off (grounded) for 50% of the time, the circuit passes 50% of the current it would with a closed (completed) circuit.
- When the pulses are off (grounded) for 75% of the time, the circuit passes 75% of the current it would with a closed (completed) circuit.

With a battery current of 12 volts:

- 50% duty cycle averages out on a digital voltmeter to read 6 volts
- 75% duty cycle averages out to 9 volts

People have asked me, "How can a digital voltmeter read duty cycle?" And the answer is, it does just what the circuit does, it averages the voltage-on, voltage-off times.

8. Trouble Codes (Service Codes)

(For specific codes, see Chapter 10.)

8.1 System Self-Test, Diagnostic Self-Test

One of the programs, stored in the ROM, tests the operation of the computer. Every 100 seconds, this program samples the various input and output signals, and compares them to normal ranges. When it senses an improper signal, it stores a Diagnostic Trouble Code (DTC), also called a Service Code. If the indicated trouble is not serious, a "soft" code is stored, for later readout, but does not signal the driver. With more serious trouble, a "hard" code is stored, and the Malfunction Indicator Light (MIL), on the instrument panel, is lit. OBD-II systems indicate emission trouble codes, in particular.

If a fault is only indicated during part of a pulse, the fault won't be stored. This means that intermittent symptoms aren't often stored as trouble codes. The self-test program also won't store a service code unless the same fault has happened several consecutive times. This helps prevent false trouble codes, but, again, it means that intermittent driveability problems won't be indicated by the DTC's.

8.2 Limits of Sensor Inputs

When a Check Engine light appears, you begin diagnosis by reading the service code. You can do this with a scan tool, an analog VOM (**'82–93**), or even by observing the flashing pattern of the Check Engine light (**'82–93**). Corvette models from **'94-01** require an OBD-II scan tool. Once you've determined the offending circuit, you can probe it with your VOM tool. You can then determine what caused the trouble—a sensor, a connector, an open or short in the circuit, or some other problem. (More about that in Chapter 8.)

When trouble codes first appeared, scoffers said, "Whoa, how stupid can they be, asking a computer to check itself? What happens if the computer itself is in trouble?" Well, it turns out that the computer:

- primarily checks the sensors, which provide external signals
- checks the PROM separately, reporting any such problems as a code 51
- doesn't fail as often as people had expected, while the external sensors, connectors, and harnesses fail more frequently

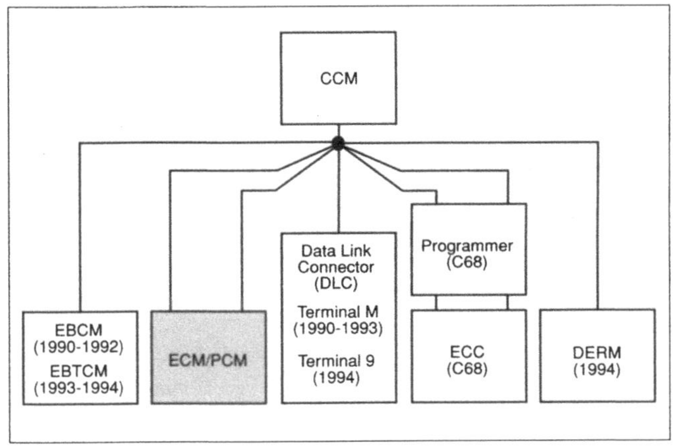

Fig. 5-23. The ECM or PCM is part of a larger control module system, the Central Control Module (CCM).

8.3 On Board Diagnostic (OBD)

Beginning in the **'91** California models, and then spreading to the rest of the country, the engine computer monitors more components than it had previously. In addition, engine operating information is stored in a "freeze-frame" for several seconds, before and after a Check Engine light signals a malfunction. The diagnostics stored in the NVRAM are becoming more valuable for troubleshooting and servicing Corvette engines. Beginning in the **'94** models, the second generation of the On Board Diagnostic (OBD-II), stores even more information, particularly when it comes to intermittents.

5

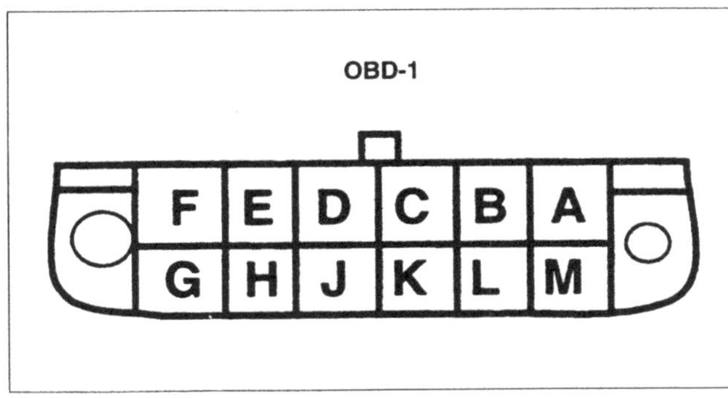

Fig. 5-24. The OBD I, 12-pin Data Link Connector

Pin	Identification (typical)
A	Data Link Connector ground
B	Diagnostic request (ECM)
C	AIR solenoid
D	Malfunction Indicator Lamp
E	Serial data link (160 baud)
F	Torque Converter Clutch solenoid
G	Fuel pump
H	Antilock Braking System
J	Entertainment and comfort
K	SIR diagnostic request
L	Serial data loop (8192 baud)
M	Serial data link (8192 baud)

8.4 Data Link Connector (DLC)

It started with the idea that the car's electronics could be checked on the assembly line, before the car was shipped. So, near the end of the line, a factory inspector could plug a tool into the connector to check everything out. Then the technicians realized that the serial data, accessible through the connector, could help diagnosis, too. In the early eighties, the connector was known as the Assembly Line Communications Link (ALCL). Later, it was known as the Assembly Line Data Link (ALDL). And, by the late eighties, through SAE, the industry agreed on the term Data Link. To see this new piece of technology for yourself, look under the dashboard, in between the steering column and the center of the dash.

Compare the 12-pin OBD-I Data Link Connector (see Fig. 5-24) to the 16-pin OBD-II connector (see Fig. 5-25), used on all cars from **'96-01**. Pin #2 is a Class 2 data line; pins #4 and #5 are ground circuits; pin #16 is battery power. Some of the pins are reserved for use by individual manufacturers.

9. Other Control Modules

The ECM or PCM is part of a larger control module system within the Central Control Module (CCM). The CCM's main jobs include the PASS-key (theft protection), HVAC, liquid

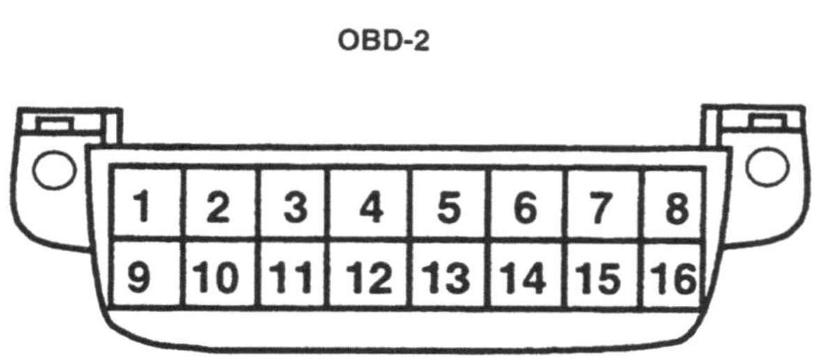

Fig. 5-25. The OBD II, 16-pin Data Link Connector

Pin	Identification (typical)
1	Secondary UART 8192-baud serial data (ckt 800), secondary Class B (ckt 710) or 160-baud serial data (1995 only) (ckt 461)
2	J1850 bus and L line on 2-wire systems, or single-wire (Class 2) (ckt 1807)
3	Ride control diagnostic enable (ckt 1826)
4	Chassis/body ground pin (ckt 150)
5	Signal ground pin (ckt 451)
6	PCM/VCM diagnostic enable (ckt 448)
7	K line for International Standards Organization (ISO) application
8	Keyless entry enable (ckt 1455) or MRD theft diagnostic enable (ckt 477)
9	Primary UART (ckt 1061)
10	J1850 bus line for J1850 2-wire applications
11	Electronic Variable Orifice (EVO) steering (ckt 1294) or Magnetic Steering Variable Assist (MSVA)
12	ABS diagnostic (ckt 799) or CCM diagnostic enable (ckt 555)
13	SIR diagnostic enable (ckt 326)
14	Entertainment and comfort bus (ckt 835)
15	L line for International Standards Organization (ISO) application
16	Battery power from vehicle unswitched (4 amps maximum)

crystal displays (LCDs: the clock), and interior lighting. Because I've chosen to concentrate on the ECM/PCM, other system modules are not discussed in this book. But I'll list some of them here, for your reference:

- **'90–92:** The Electronic Braking Control Module (EBCM) relates the engine computer to the Antilock Braking System (ABS)
- **'93-01:** The Electronic Braking and Traction Control Module (EBTCM) relates the powertrain (PCM) to ABS/TC (Traction Control).
- the programmer
- The Diagnostic Energy Reserve Module (DERM) provides for air-bag deployment, even if battery power is lost in a vehicle crash.

In the Next Chapter:

- Now you know what happens to the input signals. In Chapter 6, you'll see how engine operation is controlled by signals output to the injectors, and other actuators.

Chapter 6

Actuators—the Muscles

Contents

IN THIS CHAPTER, LOOK FOR:

- what happens to each actuator as it receives signals from the control module for the air/fuel ratio, idle-speed control, spark timing, and emission control
- how the control module manages relays and controls, and provides information signals

1. INTRODUCTION

Actuators are what it's all about. The reason that we have sensors and computers is to control the actuators. The actuators are the parts which "make things happen". I like to think of the actuators as the muscles of the system.

1.1 Terminology

Beginning in 1993, many of the names for the sensors, control modules, and actuators were changed. SAE standard J1930 provides common terms for the same general part, throughout the automotive industry. This chapter uses the terminology applicable for the years 1993 and later (**LT1**,**LT4**, **LS1**).

2. AIR/FUEL(A/F) RATIO

The air/fuel ratio is a key factor in power and emissions. This ratio is regulated in the injectors, according to the sensor inputs for various engine operating conditions.

'82, '84: Central injectors (Throttle Body)

- **L83:** the A/F ratio is constant at the injectors, except for minor variations in barometric pressure.

'85-01: Port injectors

- **L98, LT1, LT4, ZR-1:** A/F computations are based on a constant, relative fuel pressure.
- **LS1:** A/F computations are based on a constant, absolute fuel pressure. As the MAP changes, the computer electronically compensates, changing the A/F ratio by changes in pulse time.

2.1 Injectors

Injector solenoid valves are electrically "hot" (using B+ power) whenever the ignition is on. For fuel delivery, the control module grounds the injectors. Thus, in case of a short in the injector lead, it will shut off instead of flooding.

Operation

Injectors operate on similar principles, even though the shapes may vary. They are precision actuators, metering fuel with accuracy—sometimes less than one drop per actuation!

Throttle Body Injectors—Crossfire

L83: Each of the two Throttle Body Injector (TBI) units holds a separate injector. The A/F mixture is delivered, through long runners, to the opposite bank of cylinders, hence the name Crossfire Injection. To achieve the same amount of air intake, the two TBI units must deliver the same amount of fuel as eight port injectors. The Throttle Body Injectors operate at the relatively low constant fuel pressure of 9-13 psi (60-90 kPa).

Port Injectors

Beginning in the **'85** models, and continuing up to the current engines, you'll find port injection systems in all Corvettes. Compared to the TBI systems, Port injection provides superior fuel-delivery control, for improved power and smoothness, as well as reduced emissions.

Fuel pressure is supplied from the rail. When an electrical pulse is delivered to the injector coil, the armature lifts the valve at the tip, allowing fuel to flow through. This pulse is very short, and is measured in milliseconds.

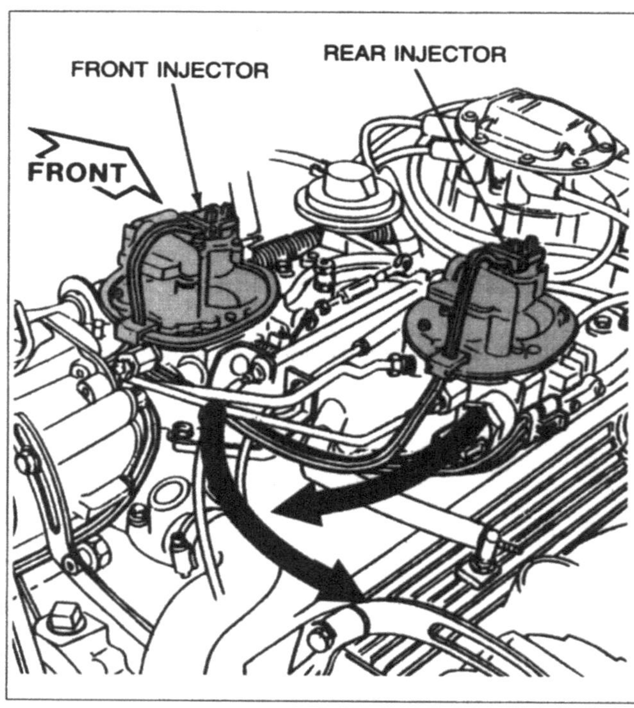

Fig. 6-1. L83: the front injector delivers fuel into the manifold for the right bank of cylinders; the rear injector goes to left bank.

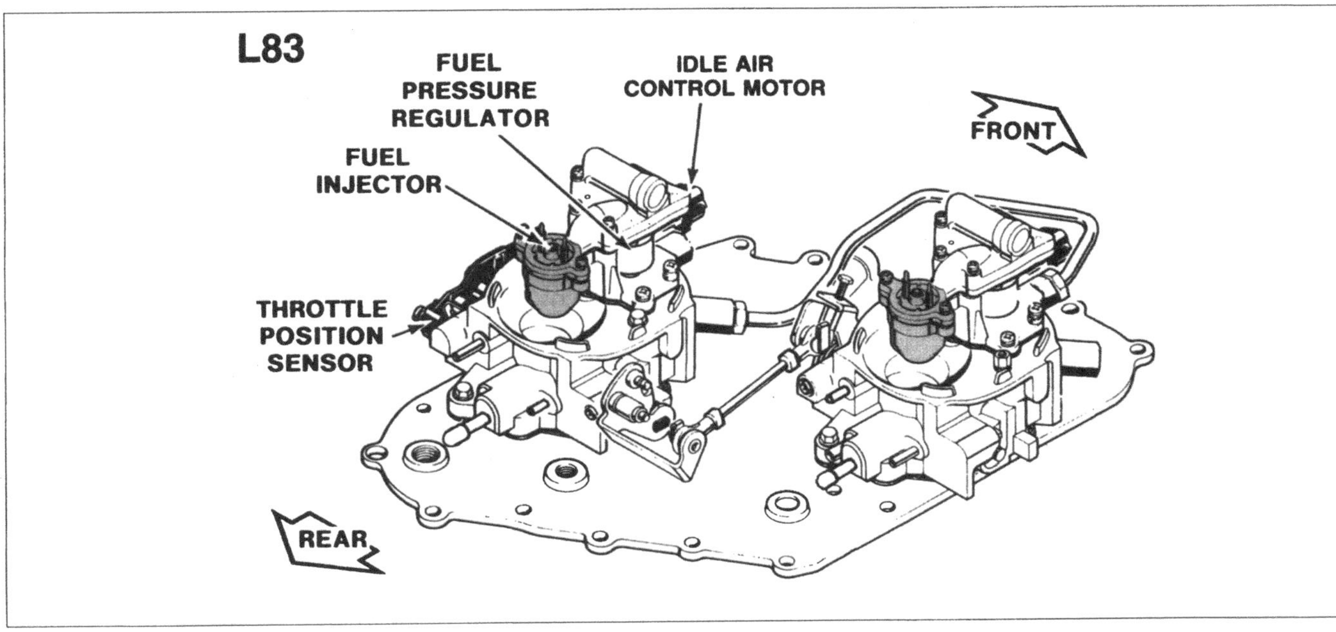

Fig. 6-2. L83: The injector and filter are mounted in the fuel-metering body. The Throttle Body Injectors are larger than the port injectors, and deliver four times the fuel, compared to the eight port injectors (four per bank).

6

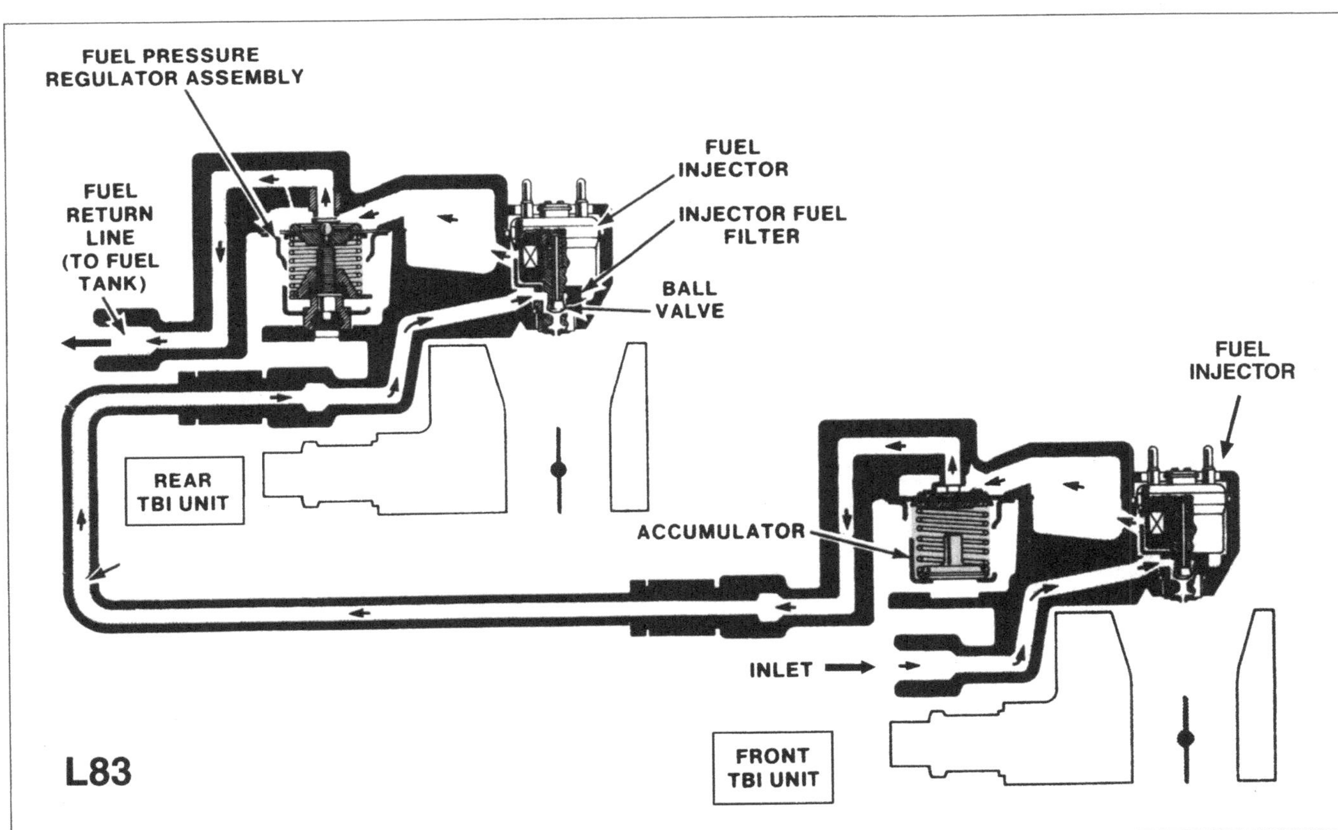

Fig. 6-3. L83: The left injector unit's Fuel-Pressure Regulator manages both injectors. The right injector accumulator stabilizes the pressure as each injector opens and closes. Fuel enters the right unit first, then goes to the left, and finally returns to the tank.

Injector types are not necessarily interchangeable. Check the part numbers. Some model year Corvettes used different injector part numbers on different cars. *Do not intermix part numbers*. You can use either part number, but make sure to use the same number for all of the injectors on your engine.

Know that improving performance is not a simple matter of replacing the OE injectors with larger flow-rate injectors.

Controlling the Amount of Fuel Injected

For any given amount of intake air, two factors usually govern amount of fuel injected:

- pulse-time—the amount of time open, usually just milliseconds (see Figs. 6-10 and 6-12)
- relative fuel pressure—the difference in the pressure at the spray tip, and the opposing Manifold Absolute Pressure (MAP) (see Fig. 6-13)

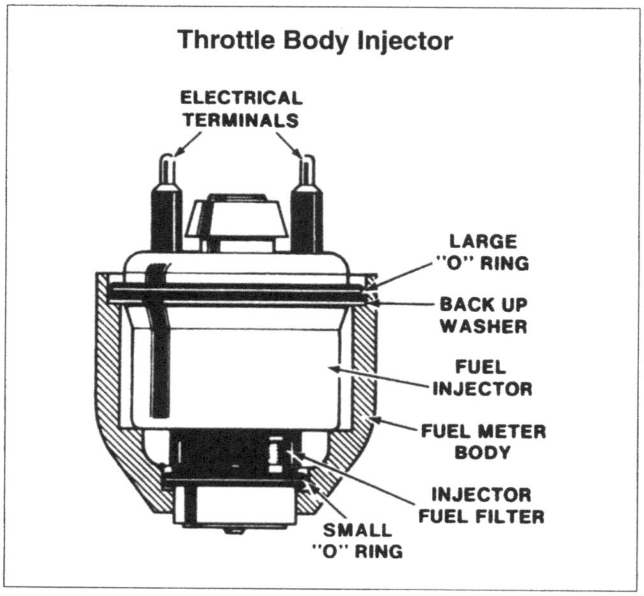

Fig. 6-4. L83: Each top-feed injector sits in its body with two O-rings.

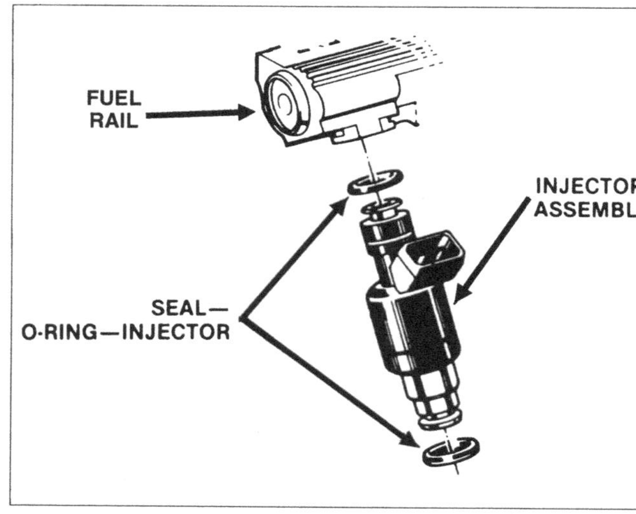

Fig. 6-6. The port injectors are held by two O-rings. The upper ring seals the injectors to the fuel rail. The lower ring seals the injector to the manifold, and partially insulates the injector from the engine's heat and vibration.

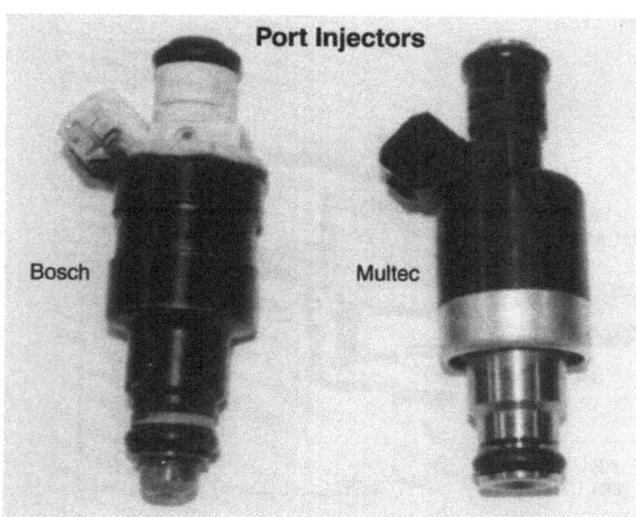

Fig. 6-5. The Bosch injector (**'85-89**) is on the left, and the Multec injector (**'90-96**) is on the right.

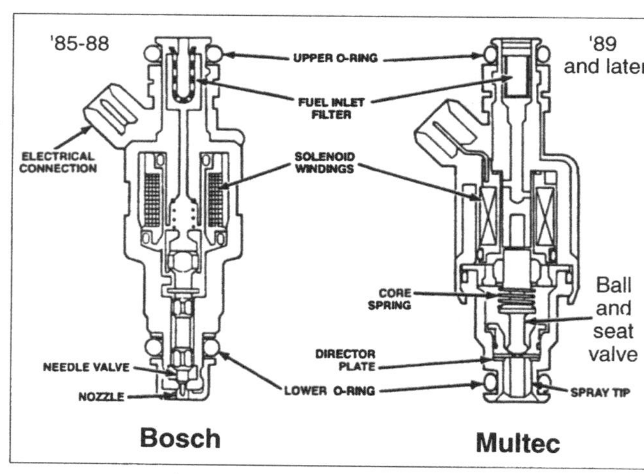

Fig. 6-7. '85-88 L98: The Bosch injector has a needle valve, also known as a pintle.
L98, '89-91 LT1, LT4, ZR-1: The Multec injector uses a ball and seat valve, which is less sensitive to dirt.

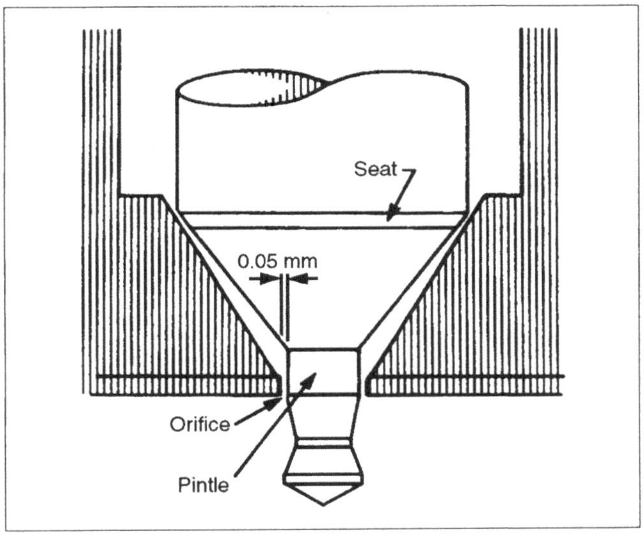

Fig. 6-8. '85-89: Bosch injector tolerances are very fine, and are subject to fuel-delivery variations from dirt in the fuel.

Variations in those two factors, pressure and time, are important for your understanding of electronic fuel injection, as applied to different model years, and to increased performance.

Pulse Time

Pulse Time, sometimes called pulse-width, is measured in milliseconds (ms). It is the time from the instant that the open signal is received to the moment that the close signal is received.

> Millisecond, that's one-thousandth of a second. Some wit has defined a millisecond as: "The time between when the stoplight changes to green, and the guy behind honks!" We'll say a millisecond is a very short piece of time.

6

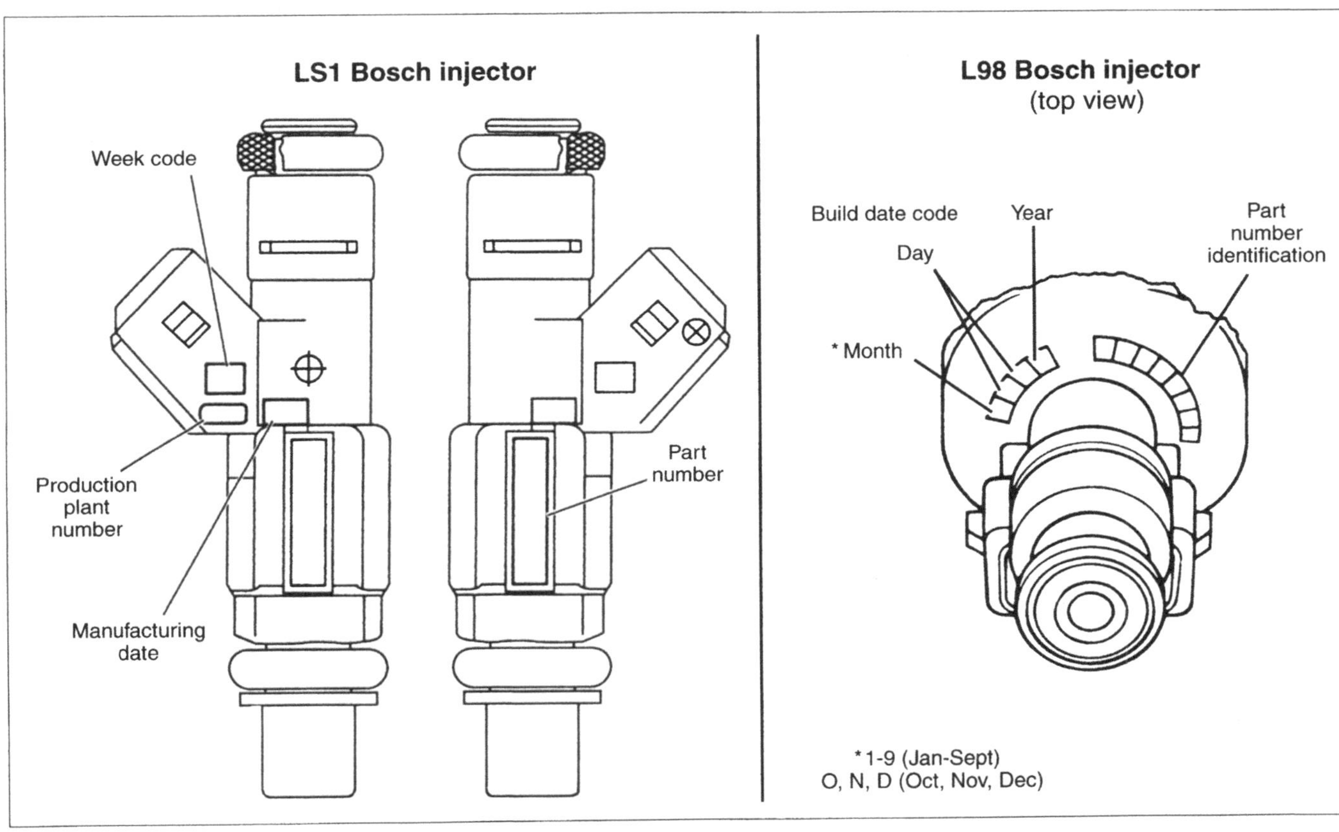

Fig. 6-9. Injectors include a part number, a product plant number, and a manufacture date.

Fuel Pressures—Port Injection

Port injectors operate downstream of the throttle, so changes in manifold pressure usually require varying the operating injector pressure. The manifold pressure changes frequently, with a corresponding "back-pressure" effect on the injected quantity. But not all engines are affected by the manifold pressure:

- The **L83** has a cross-fire system, and there is no effect on the TBI injectors.
- **LS1:** Injector pressures operate only at absolute pressure, and do not change with the manifold pressure. (In the **'97-98 LS1,** fuel pressure does change slightly, as the barometric pressure changes.)

L98, LT1, LT4, ZR-1: Computations are based on a constant, relative fuel pressure. The fuel pressure regulator changes the injector's absolute fuel pressure, according to the MAP, to maintain that constant relative fuel pressure.

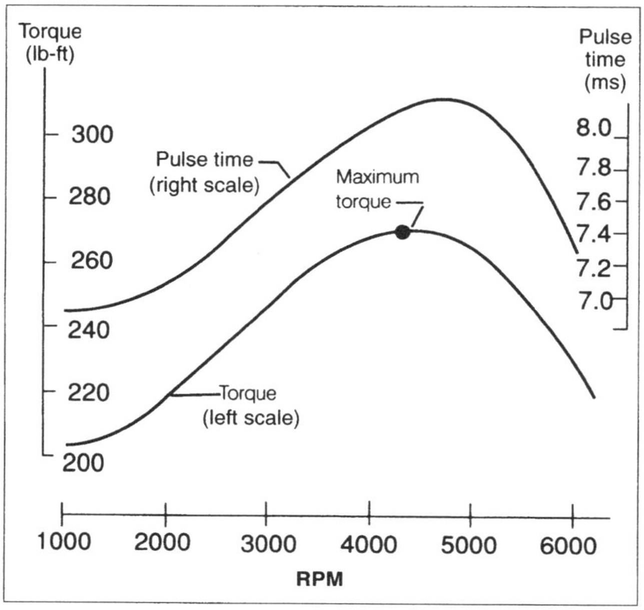

Fig. 6-10. This graph shows a close relationship between the W.O.T. pulse-time and torque. The W.O.T. injector pulse-times shown are representative, rather than actual, and vary with the engine and calibration.

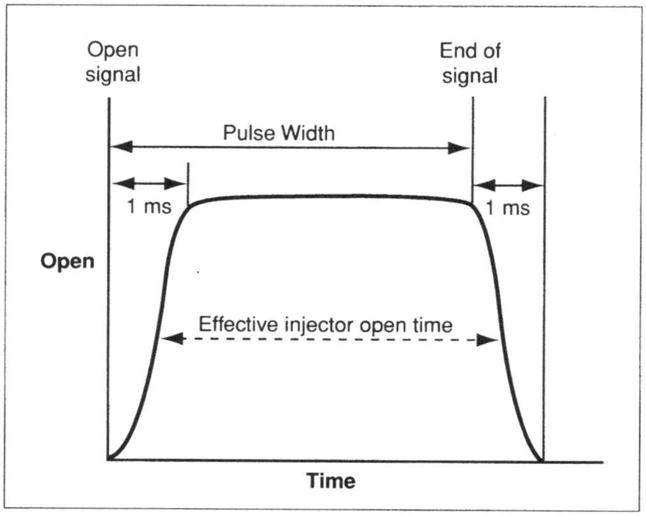

Fig. 6-12. The injector valve takes about one millisecond to open, gradually increasing delivery, if you can call one millisecond gradually. Opening time is counted in the pulse width, while closing time is not. But because the two are virtually the same, the pulse width averages out to equal the effective injector-open time.

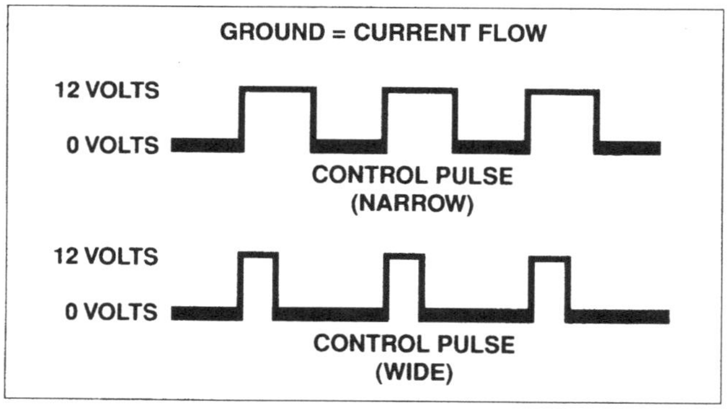

Fig. 6-11. Fuel is delivered only when an injector is grounded. (These kinds of diagrams can be misleading. It may appear that lines which go up are pulses. But this isn't the case. Only lines heading down (towards zero volts) are grounded, and are indicative of pulses.)

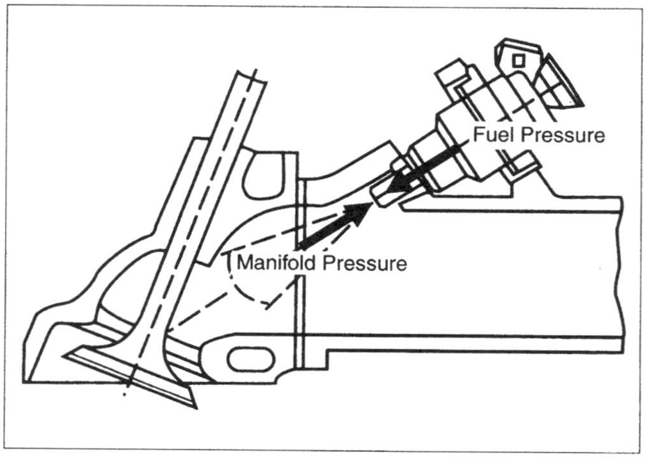

Fig. 6-13. L98, LT1, LT4, ZR-1: As the MAP changes, the port injector's fuel pressure changes correspondingly.

Metering Fuel Injection

In electronic fuel injection, the needle lifts a tiny amount, for a short time, and delivers a small amount of fuel. What is the tiny amount of lift? In the **L98**, for example, the Bosch port-injector lift is about 0.15 mm (0.006 in.), about the thickness of two pieces of paper. This lift is fixed, the same every time.

Suppose you want to add more fuel than the original calibration specifies. Perhaps you've cleaned up the intakes to admit more air. For "street" cars with Mass Air Flow (MAF) sensors, the control module will usually increase the fuel injected, with no modification on your part. Remember, with MAF, the engine is measuring the air flow and injecting fuel to match. In contrast, Manifold Absolute Pressure (MAP) systems affect the air intake, so with most modifications, you will need a new chip.

Perhaps you want to run richer for the track. Then you would increase the fuel pressure, or increase the injector-open times, or maybe both. You must consider the injection time—how long the injectors are open under different conditions, and the pulse period—how much time is available in the engine cycle, for injector delivery.

Time Factors

The pulse signal in the control module grounds the injectors for shorter or longer times, in order to vary the amount of fuel injected. Injection time may be as short as one millisecond.

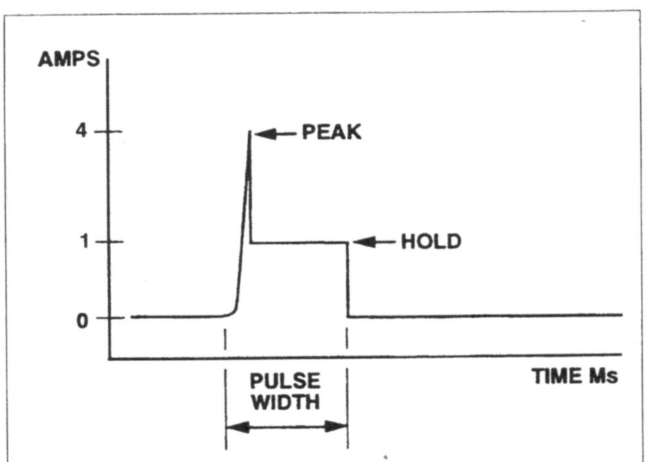

Fig. 6-14. This is an injector current from relatively low resistance injectors (1-2 ohms). The current flow peaks, to open the injector quickly, for precision metering. The injector drive circuit includes a current-limiting device, to hold the injector open without overheating the solenoid.

The typical Corvette injector is driven open, but as the current reaches four amps, resistance in the current-limiting circuit pulls current down to one amp, this is known as "hold." With this kind of driver, the injector opens and closes more quickly. It gives more precise control of fuel metering, over a broad range of rpms and fuel delivery, while minimizing injector heating.

Port Injectors—MFI, SFI

Corvette port injectors are top fed. Fuel, under pressure, enters through the injector filter, and exits in a conical spray pattern. The injectors are held by O-rings, aided in some engines by clips. If the O-rings crack, ambient air may enter, and the mixture may go lean, increasing idle rpm.

A port injector operates perhaps 500 million times in the life of the car—with a precision measured in *micro*seconds (millionths of a second), this is equivalent to:

- as often as 100 times per second, or every 10 *milli*seconds (thousandths of a second)
- as often as once for every foot a Corvette races (at 150 mph).

6

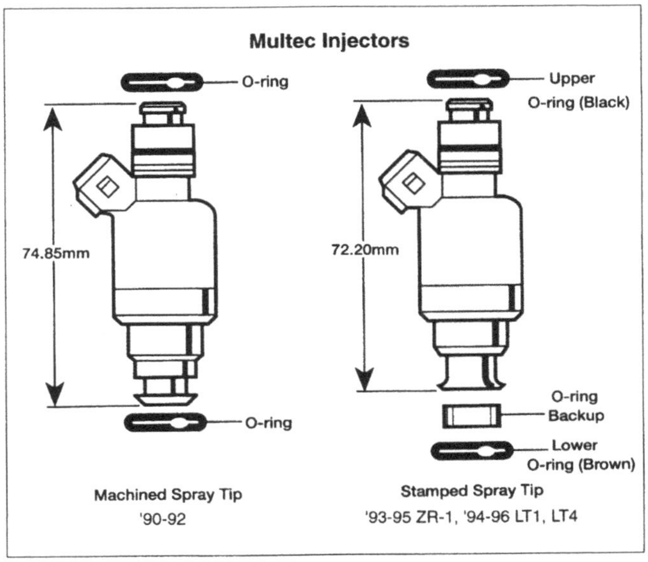

Fig. 6-15. '89-92: The injectors (*on left*) are longer, and are mounted with two black O-rings.
'93-95 ZR-1, **'94-96 LT1**, **LT4**, **LS1:** The injectors (*on right*) are shorter, requiring a lower O-ring Backup, and a different lower O-ring, colored brown.

With a larger bore, a stamped spray tip delivers a better spray pattern, for improved air/fuel mixing. Look for the stamped spray tip Multec in the **'93 LT5** and in the machined tip in the **'94-01** Corvette engines. Obviously, the two kinds of spray tips are not interchangeable.

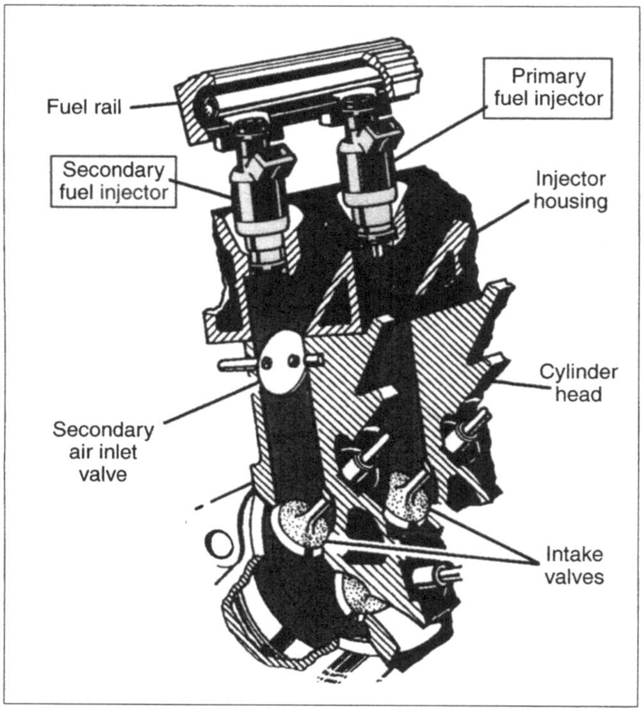

Fig. 6-16. The **ZR-1** operates with two injectors for each cylinder. The primary injector delivers fuel during all operations. While the secondary injector adds fuel only during high power and rpm, (near W.O.T.) and only when the secondary air inlet valve is open.

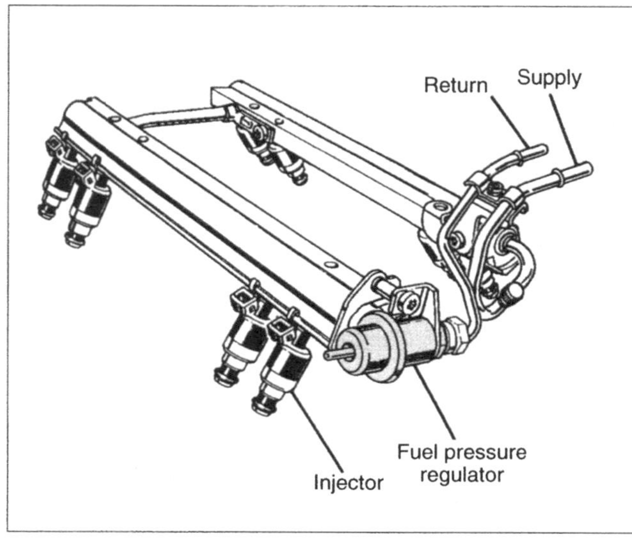

Fig. 6-17. The Fuel Pressure Regulator (FPR) maintains approximately constant relative fuel pressure (constant pressure differential) as the manifold pressure changes.(For more, see Chap. 7.) The **LT1** is shown, others are similar.

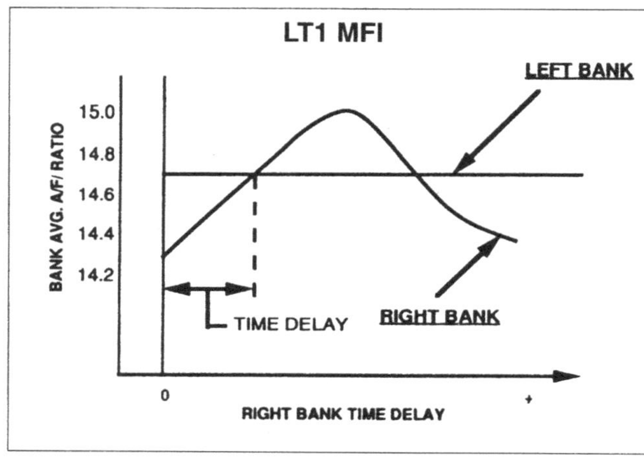

Fig. 6-18. '91-92 LT1: MFI group fires by bank. The right bank delay compensates for the closer location of the Fuel Pressure Regulator. This reduces bank-to-bank variation.

Operation

You might think that an injector is pulsed open just as the intake valve opens. And with SFI, it is. But with MFI it's not that simple. (See Figs. 6-17, 18, and 19.)

Cold-Start Valve (CSV), also known as Cold Start Injector (CSI)

'85-88 L98: The MFI includes a Cold-Start Valve, or injector, mounted on the engine. When you turn the ignition on, the CSV injects vaporized fuel into the manifold, in addition to the fuel injected by the port injectors. The CSV operates with a temperature/time switch, independent of the computer control. If you are familiar with Bosch systems, that's what you're seeing on **'85-88** Corvettes. (See Fig. 6-20.)

The Cold-Start Valve's bimetal switch opens when heated, whether by the coolant or by the heating element. The switch grounds at or below 35° C (95° F). The heating element then opens the switch, to stop CSI delivery. This delivery is:

- up to eight seconds, at –5° C (–20° F)
- one second, at 35° C (95° F)

'89-01 Corvettes eliminate the CSV. The port injectors operate at increased pulse times, which provides enough cold-start enrichment.

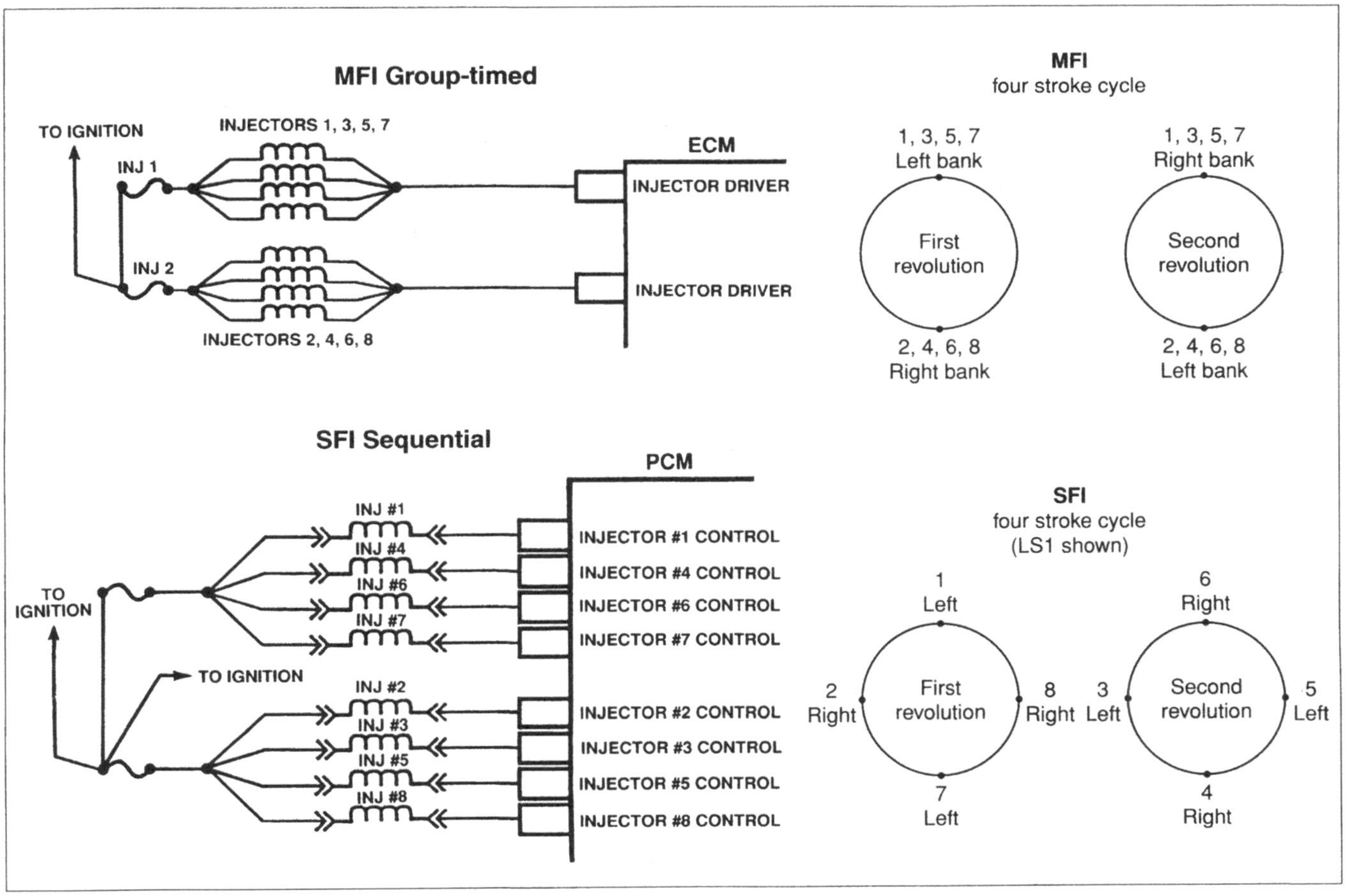

Fig. 6-19. '85-91 L98, '92-93 LT1: MFI is group timed. Each bank operates with a single injector driver, capable of handling four cylinders at a time. The injector fires once for each crankshaft revolution. This is also sometimes referred to as "bank firing."

'94-96 LT1, LT4, ZR-1, LS1: SFI is sequential. Each SFI injector is pulsed open separately. This indicates the firing order, once every other crankshaft revolution.

3. Idle RPM

The computer control maintains the lowest practical idle rpm. At idle rpm, even 25-50 rpm can make a big difference in smoothness, as well as in emissions. Idle rpm can be affected by various auxiliary loads, including power steering, air conditioning, and generator loads. The IAC depends on sensor inputs: the rpm (modified by engine load), MAF or MAP, TPS, ECT, battery voltage, and the AC clutch signal.

The computer compensates for any air added to the intake manifold, such as the Positive Crankcase Ventilation flow, and the vacuum losses through the accessories. Any vacuum accessory which leaks into the intake manifold will add air, increasing idle speed. EGR is not a factor because there is no exhaust gas admitted during idle.

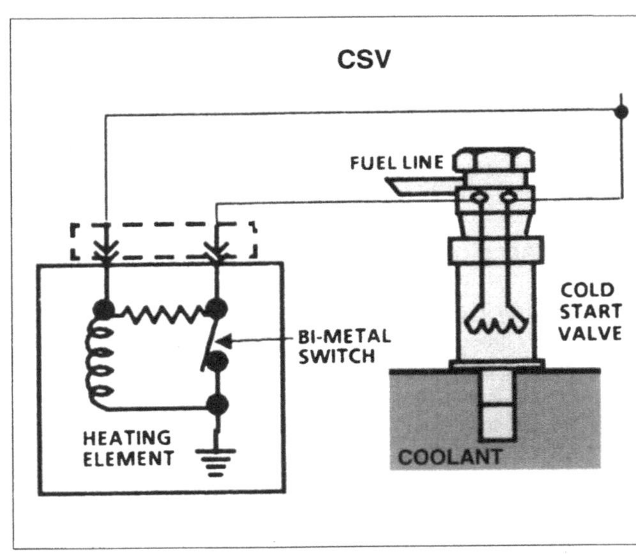

Fig. 6-20. '85-88 L98: The CSV is powered by the starter solenoid, and is grounded through the Cold-Start Valve switch, which is a coolant thermistor with a heating element. The CSV has no connection to the Engine Control Module.

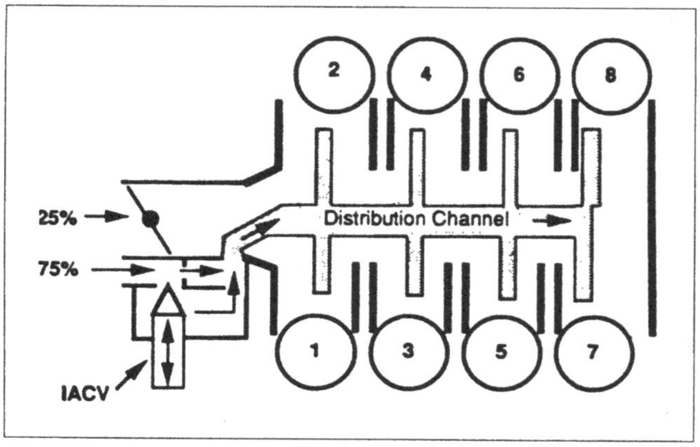

Fig. 6-21. **L98, LT1, LT4:** To control idle rpm, the IAC bypass channel can provide 75% of the air intake at idle. The other 25% enters around the closed throttle plates.

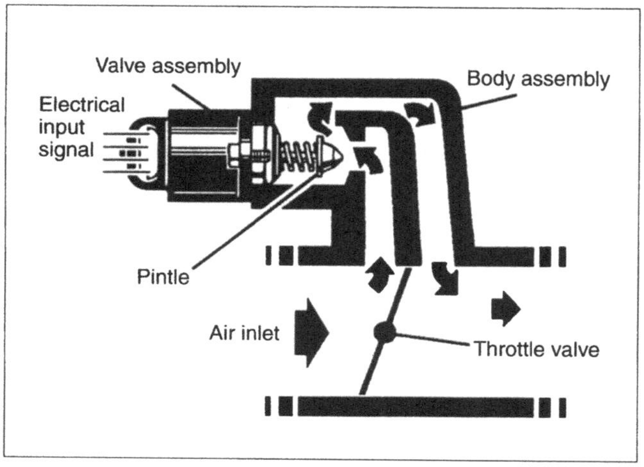

Fig. 6-22. **L98, LT1, LT4:** Air flows around the closed throttle valve, through the IAC, and into the manifold. The amount of air is determined by the position of the IAC pintle valve, as it opens and closes to increase or decrease rpm.

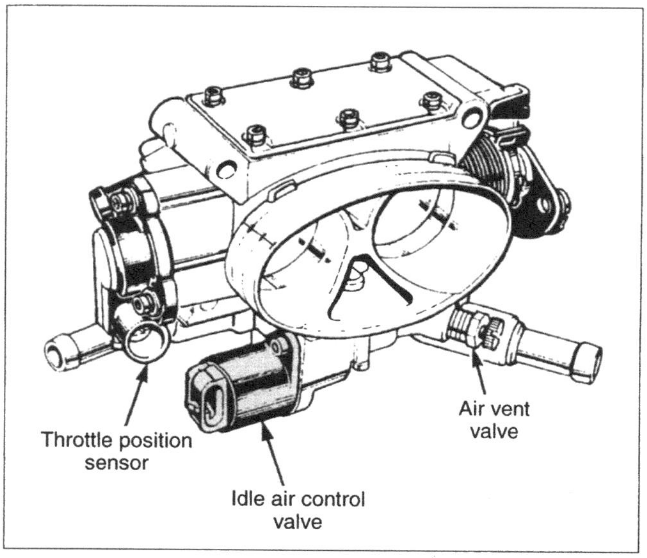

Fig. 6-23. The **LT1**, **L98,** and **LT4** are all similar. Look for the Idle Air Control under the throttles, on the same side as the Throttle Position Sensor.

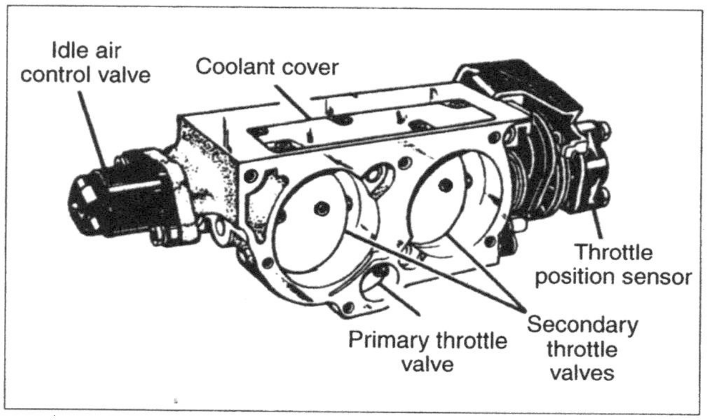

Fig. 6-24. **ZR-1:** The Idle Air Control operates only on the primary throttle, because the secondary throttles do not operate at idle.

3.1 Idle Air Control

The Idle Air Control (IAC), in the throttle-body assembly, bypasses a closed throttle, adding or subtracting bypass air to control engine idle speed. (See Fig. 6-21.)

For the IAC to regulate bypass air, it must start from somewhere near the middle.

- One set of coils drive the pintle closed, to reduce rpm.
- The other coils drive the pintle open, to increase rpm.

During idle, the computer calculates the proper position of the IAC valve (partly open), for the desired idle rpm. As the other sensor inputs vary (TPS and ECT), the IAC moves to change the rpm. The engine cannot start with a closed IAC valve, so, at engine shutdown, the IAC resets the pintle to the proper position for the next start.

The computer moves the pintle valve in small steps, which can be measured on a scan tool, plugged into the Data Link Connector (DLC)—formerly ALDL.

ZR-1: The primary throttle, with a 22 mm (0.9 in) diameter, handles idle, and most operation up to about 70 mph. The dual secondary throttles, with 59 mm (2.25 in.) diameters, handle maximum power, above about 80% primary throttle opening.

LS1: There is no IAC, and idle rpm is controlled with the Throttle Actuator Control (TAC). (Described in Chap. 4.)

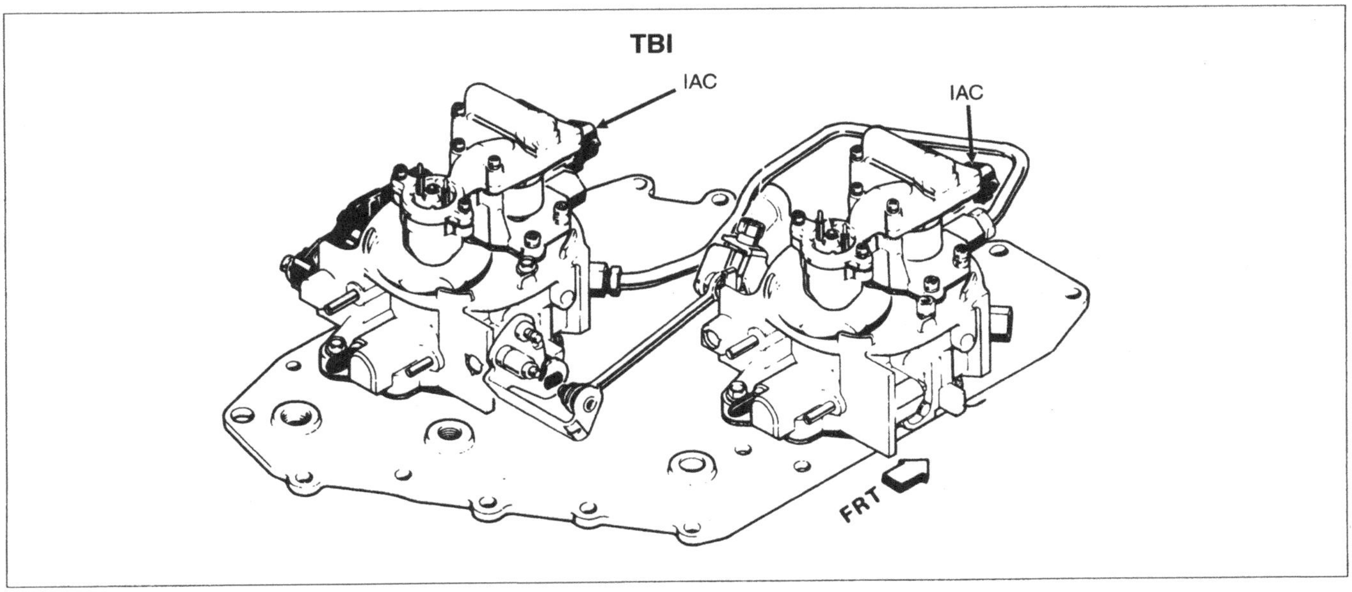

Fig. 6-25. L83: Look for two IACs, one on each throttle body.

6

Pintle shape

The pintle shape is related to the movements programmed in the computer. *If you replace Idle Air Control, be sure to use the correct part numbers.* They differ by pintle shape, and also by spring tension.

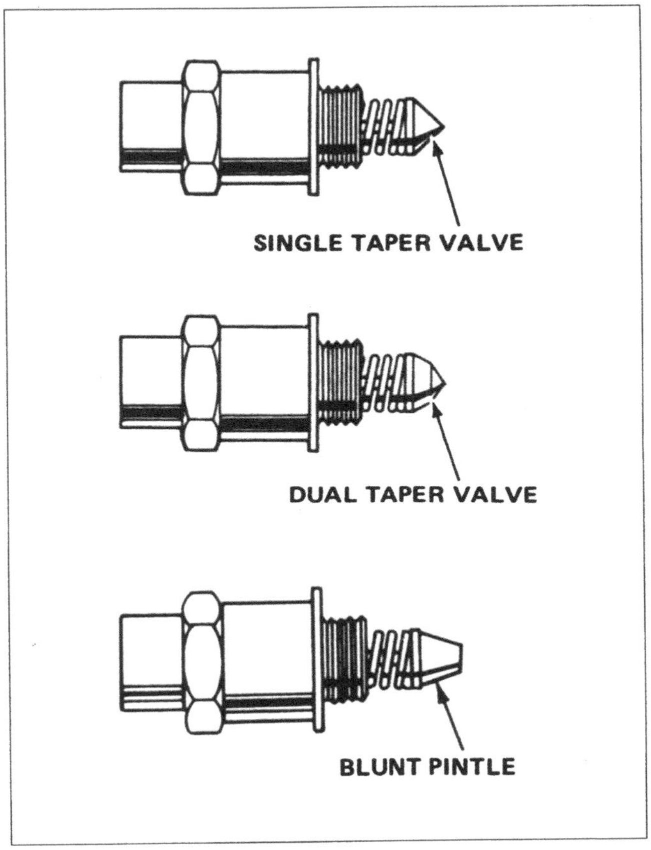

Fig. 6-26. As the pintle moves in and out, the valve affects the bypass air. Different pintle shapes have different effects. **LT1**, **ZR-1**, **LT4** and late **L98** engines use the dual taper valve, which provides the best control of idle rpm.

Keep straight the following differences:

- **idle air:**
  Idle Air Control (IAC) meters idle air—sometimes known as Idle Speed Control (ISC) (**'82-96**).
- **air pumped to the exhaust:**
  Air Injection Reaction (AIR), a belt-driven pump routes air to the exhaust ports to complete the combustion of unburned fuel, during warm-up (**'82-91**). During cruise, air is sent to the catalytic converter, to assist in conversion.
  Secondary Air Injection (also AIR). An electric air pump is used instead of a belt-driven pump. The pump injects air into the exhaust, but does not affect idle rpm (**LT1**, **LT4**, **LS1**, **ZR-1**).
- **extra air intake:**
  Secondary Air Inlet Control (SAIC)—The eight port throttles admit additional air into to the secondary injectors at full power. There is also no effect on idle rpm (**ZR-1**).

Anti-Slip Regulation/Throttle Actuator Control

Using the ABS Wheel Speed sensors, the Traction Control System senses any undesirable acceleration in the drive wheels. If either of the rear wheels is rotating faster than the front wheels, the Anti-Slip Regulation acts to restore traction and regulate slip. It does this through:

- spark retard
- disabling cruise control
- unlocking Torque Converter Clutch (in A/T)
- individual rear brake application, through ABS, at speeds below 50 mph

ASR was replaced, in the **'97 LS1**, by the Throttle Actuator Control (TAC), which handles idle rpm and slip regulation. There is no separate system for IAC or ASR. (I describe this fully in Chap. 4, "Sensors".)

Idle Speed Control by Spark Timing

Idle speed control is assisted by the Electronic Spark Timing. The EST increases or decreases the spark-angle retard, to reduce any idle-rpm variations. Being an electronic change, the EST can respond in milliseconds, faster than Idle Air Control, which is a mechanical change.

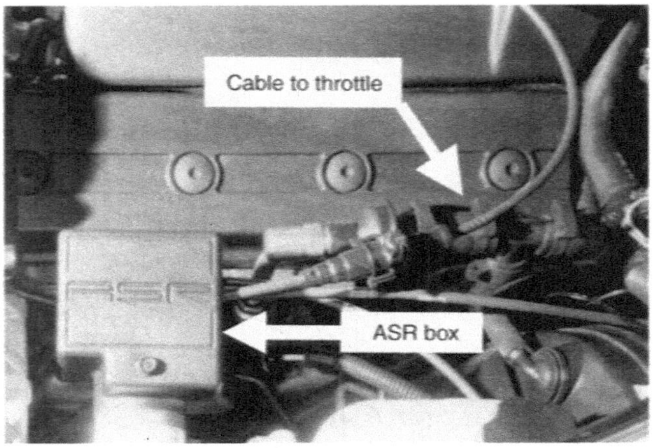

Fig. 6-28. LT1, LT4: The Anti-Slip Regulation (ASR) Adjuster assembly includes inputs from the four ABS wheel-speed sensors, and can suspend the cruise control, and relax the throttle cable. Control is through the engine computer, and either the Electronic Brake Control Module (**'92-93**), or the Electronic Brake/Traction Control Module (**'94-96**).

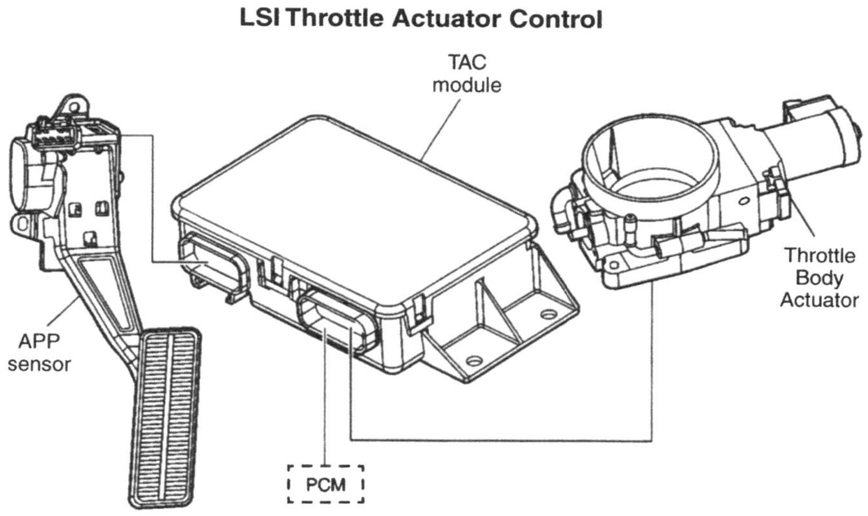

Fig. 6-27. LS1: Beginning in **'97**models, the Accelerator Pedal Position (APP) Sensor sends signals to the Throttle Actuator Control Module. The TAC Module sends signals, through a dedicated serial link, to the Throttle Body.

4. Secondary Air Inlet Control (SAIC) ZR-1

The **ZR-1**'s Secondary Air Inlet Control (SAIC) manages air delivered to the secondary intake valves. For normal operation, the SAIC closes off the shorter set of tuned intake manifold runners. So, even though all sixteen intake valves open, the engine operates through the primary intake runners, with only a single intake valve per cylinder. With no air and no fuel delivered to the secondary intake valves, economy and emissions are improved during normal speed, and normal load.

With intake air delivered to both of the intake valves, for each cylinder, more fuel, and more power, is available on demand.

Why an electric pump? (See Fig. 6-31.) Well, under full-throttle conditions, how much intake manifold-vacuum is available to operate these SAIC throttle valves? Mighty little. So, if the pressure in the reservoir drops below 12 in.Hg (40 kPa), the vacuum sensor turns on the electric vacuum pump.

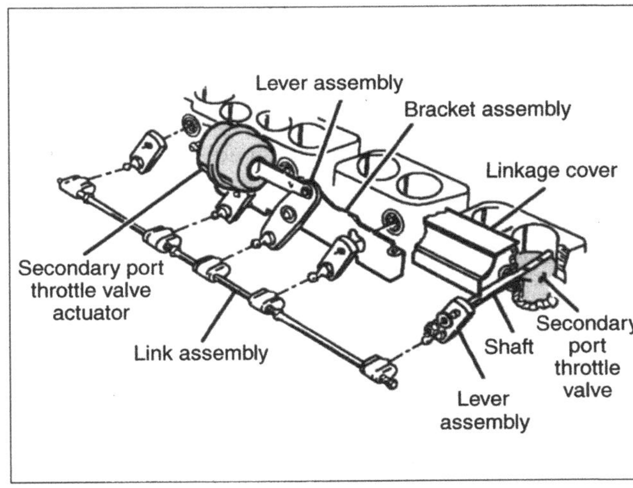

Fig. 6-30. ZR-1: When secondary air (and injectors) is called for by the computer, a vacuum is supplied to the port throttle valve actuator, for each cylinder bank. The link assembly operates the four secondary port throttle valves in that bank.

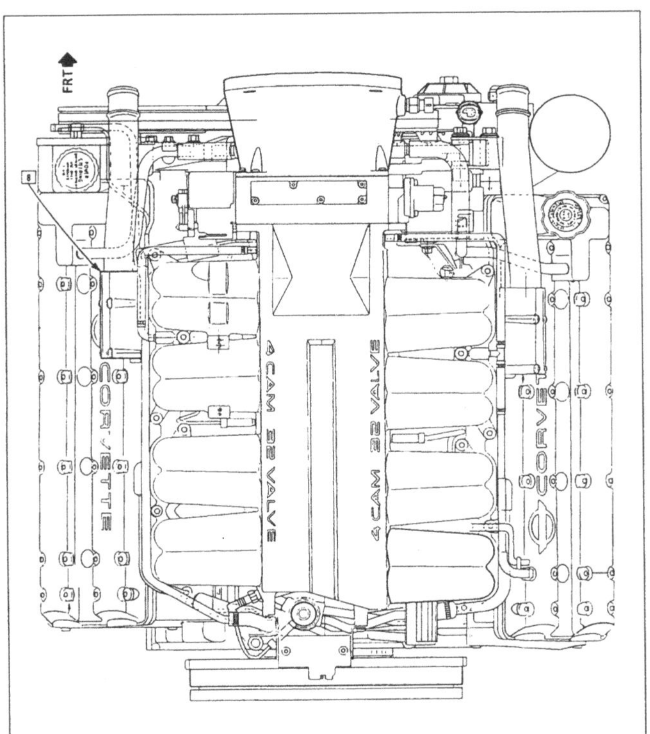

Fig. 6-29. ZR-1: For most operation, intake air is provided only through primary runners. For high-power output, the secondary runners provide additional intake air, through the secondary intake valves.

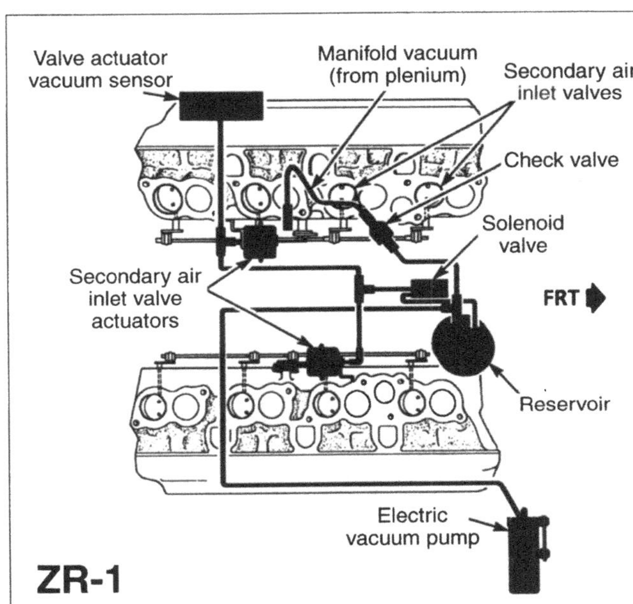

Fig. 6-31. ZR-1: For high power, the solenoid valve applies a vacuum to the two SAI Valve Actuators, one for each cylinder bank. And the eight secondary air-inlet valves open. The manifold vacuum is provided by the intake manifold and the electric vacuum pump.

Operating Conditions for *full* power—with Secondary Air Inlet Control, injectors and fuel pump:

- power switch to FULL
  Note: this is switchable up to 4,000 rpm
- Engine Oil Temp: 20-150° C (68-300° F)
- Engine Coolant Temp: 20-150° C (68-300° F)
- B+ above 10v.
- no current trouble codes

The Power switch is sometimes called the Valet Switch—just like the person who parks your car lends you an extra hand, the power switch lends your car an extra hand, so to speak.

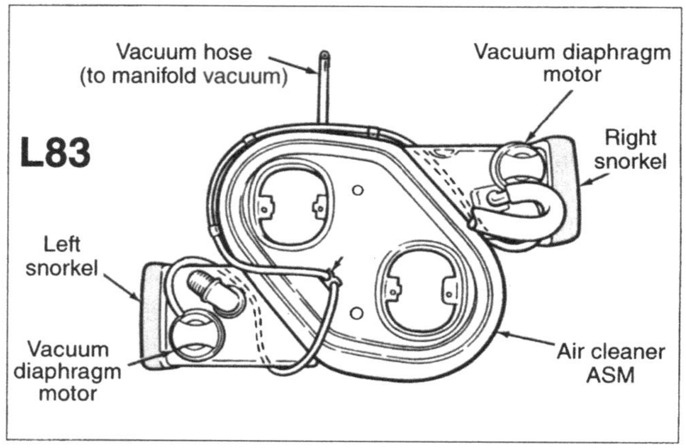

Fig. 6-33. L83: Snorkels on the front and rear units take in air from around the exhaust manifold, as needed. The sensor controls the amount of heat. Notice that the thermac is strictly mechanical—there is no connection to the computer.

5. Thermac, '82, '84 L83 TBI

L83: The Thermostatic Air Cleaner (thermac) heats the TBI intake air, as necessary, to insure that the vaporized fuel, from above the throttles, reaches the intake ports. Obviously a thermac is not necessary for port fuel injection, which passes air, without fuel, through the intake manifolds.

When you floor it, for maximum acceleration, the manifold pressure falls to near zero, so the spring overcomes the vacuum motor. (See Fig. 6-32.) The damper door then closes, allowing full outside air to fill the intake manifolds. Outside air is colder, and heavier, which gives you more power, and the rich acceleration mixture does not need any more heat. This describes operation of the left (rear) unit. The right (front) unit normally draws in hot air only, though at full throttle, it also switches over to full outside air.

Here we go again, with some potential confusion: GM's thermac heats the intake air, while Ford's "thermactor" is responsible for sending air into the exhaust manifolds and catalytic converter.

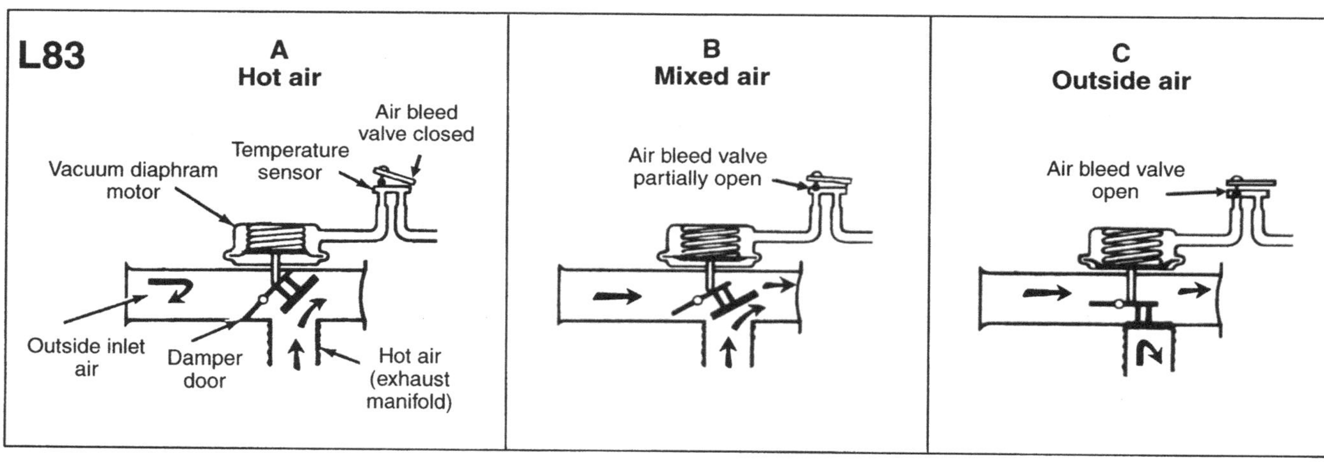

Fig. 6-32. L83:
(A) When the temperature sensor measures cold air, below 85° F (30° C), the vacuum motor closes the damper door, to shut off the outside air and allow hot air in.
(B) As the temperature increases, the vacuum motor opens the door, to allow some cooler air in.
(C) When the temperature rises above 130° F (50° C), the motor closes the damper door, so only outside air enters the air cleaner.

6. Spark Control

L83, L98, LT1, LT4 Corvettes use one of two distributor-ignition (DI) systems:

- **L83, L98:** High Energy Ignition (HEI) is a bypass system—the Ignition Module processes crank reference signals in a base mode, and sends rpm reference signals to the computer, for ignition-control timing.
- **LT1, LT4:** The Optispark system controls all ignition-coil timing with the computer, through the module. The sole function of the module is to turn the coil on and off.

ZR-1, LS1 Corvettes use Electronic Ignition (EI), previously called "Distributorless Ignition", i.e., complete computer control.

All Corvettes use Knock Sensor(s) (KS) with Electronic Spark Control (ESC) to reduce or prevent spark knock, known as detonation.

Two kinds of ignition controls have quite similar names but do different kinds of jobs:

- Electronic Spark Timing (EST), used in HEI systems, provides spark timing, without the traditional flyweights or vacuum retard/advance. If you see the big distributor, you're talking HEI with EST.
- Electronic Spark Control (ESC) is Corvette-speak for the Knock Sensor system, which retards timing, to reduce spark knock (detonation). ESC is used in both HEI and Optispark systems.

(And then there is "ESC"—Electronic Suspension Control. Where, oh where is SAE J1930 when we need it?)

6.1 DISTRIBUTOR IGNITION (DI)

High Energy Ignition (HEI) L83, L98

High Energy Ignition (HEI) systems provide the Electronic Spark Timing (EST).

Fig. 6-34. L98: Remove the cover of the HEI distributor to access the coil on top. The crankshaft position and rpm signals are sent to the computer, to control spark timing. Other sensors provide load, engine temperature, and barometric pressure.

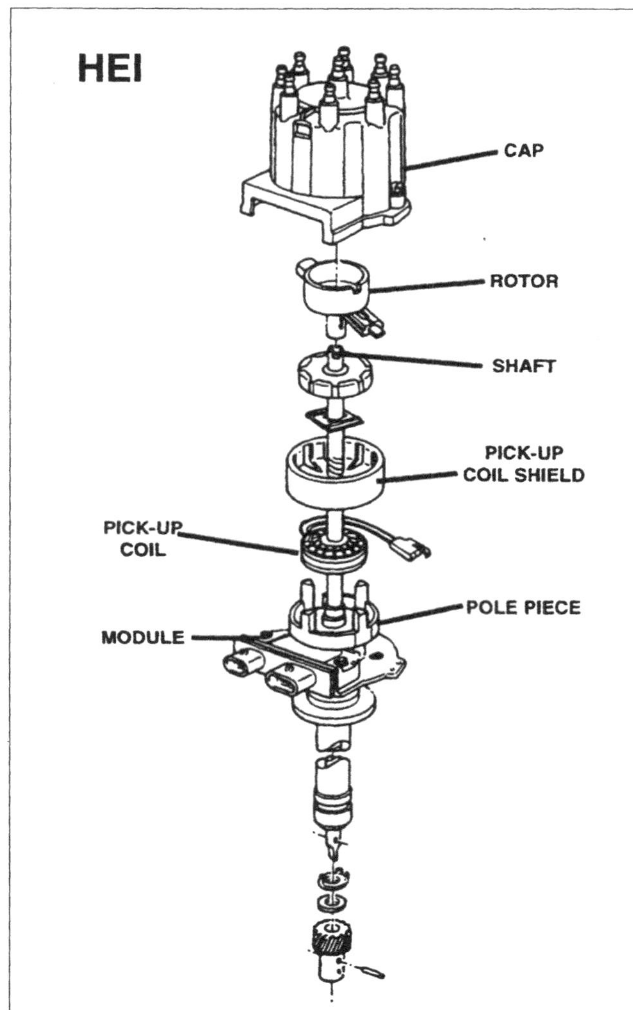

Fig. 6-35. The distributor assembly includes a rotor, pick-up coil, pole piece, and module.

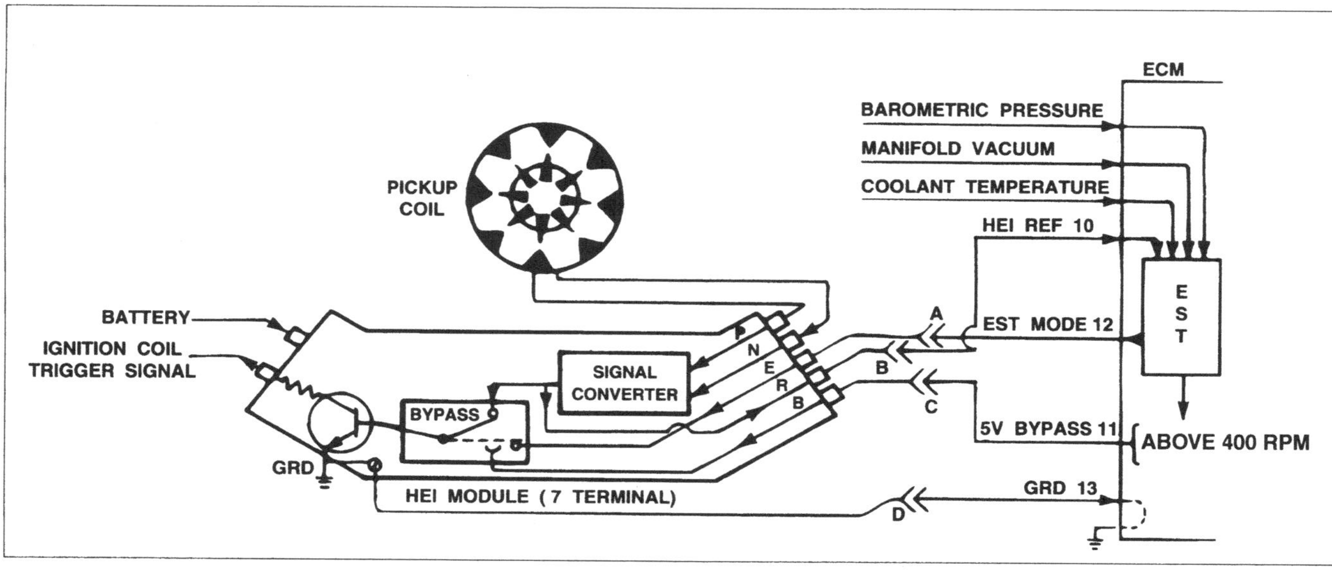

Fig. 6-36. **L98, L83:** The distributor module uses the computer's output to advance or retard spark timing. The inputs which influence spark timing include BARO, MAP, and ECT.

Refering to Fig. 6-36, the four connectors are responsible for the following signals:

- B) The distributor reference sends rpm and crankshaft position signals, and the base signal for EST and injector triggering.
- D) The reference ground ensures that the ground circuit has no voltage drop, so that the computer and distributor have the same ground plane, providing the same voltage levels to both.
- C) At about 400 rpm, The bypass terminal switches from by-pass timing in the module, to run timing in the computer.
- A) The EST terminal triggers the distributor module, to advance or retard timing from the reference signal.

The distributor reference sends signals to the computer, to operate the fuel pump and the injectors. The EST monitors engine load, MAF or MAP (including the AC compressor and Power Steering pump), TPS, rpm, ECT, TCC, and KS. The EST calculates a spark-timing signal for each crankshaft revolution.

The **LT1**, **LT4**, **LS1** have no Bypass circuit. The computer controls the timing in all modes.

Dwell Control

Closed-Loop dwell control reduces the current, and wattage consumed. This reduces heating in the coil. (See Fig. 6-39.)

Current limiting occurs just at the point of discharge. The result: reduced heating of the coil.

Optispark system

LT1, **LT4** Opti-spark is so named because of the way in which the distributor uses light beams to sense the engine rpm, crankshaft position, and cylinder identification. (See Chap. 4) The compact distributor, on the front of the camshaft, reduces the engine bulk, allowing the lower hood line of the typical Corvette bodystyle.

The distributor's direct-drive, from the camshaft, reduces distributor-gear backlash, typically found in most distributor right-angle drives. The Ignition Control Module, an actuator, operates similarly to Electronic Spark Timing (**'82-91**). The Ignition Control Module operates through the computer, to control spark timing and dwell, during all operating conditions.

Because of the high resolution, spark timing in the **LT1** and **LT4**, more closely follows the needs of the engine during transients. Advancing the spark, without causing knock, increases power and fuel economy. (See Fig. 6-42.)

Ignition Control Module (ICM)

The Optispark Ignition Control Module (sometimes called the Ignition Coil Module) operates the same way during crank as during run. There is no bypass circuit, as used in HEI.

We all know that the distributor rotor does not fire all four cylinders in order, in one bank, and then all four in the other bank. Leads go to 4, 6, 8, 2 from one side of the distributor to the right bank, and 5, 3, 7, 1 come from the other side and go to the left bank. (See Fig. 6-44.) How *do* they do that? Inside the cap, conductive ink tracings carry a secondary current, from

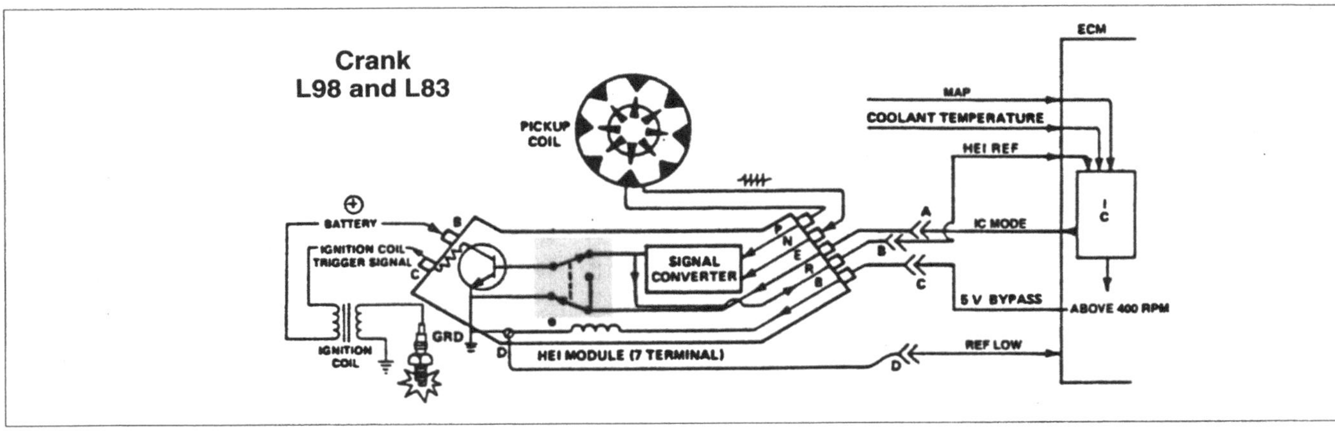

Fig. 6-37. L98, L83: Crank (By-pass) mode—think of base timing as being controlled by the ignition module. The base timing is not necessarily fixed, but may be advanced by the ignition module circuits.

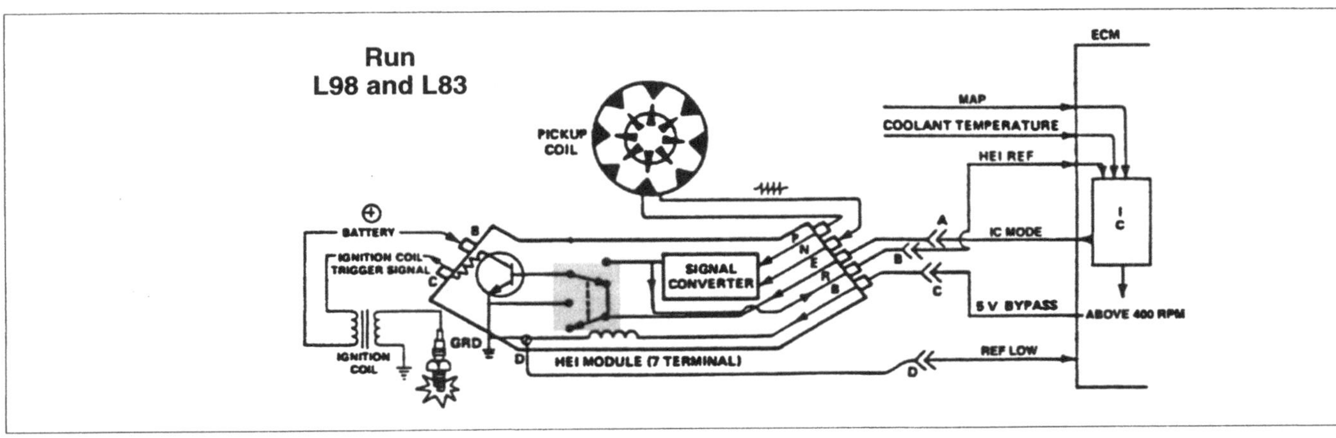

Fig. 6-38. L98, L83: Run mode—EST spark timing is controlled by the computer, based on sensor inputs.

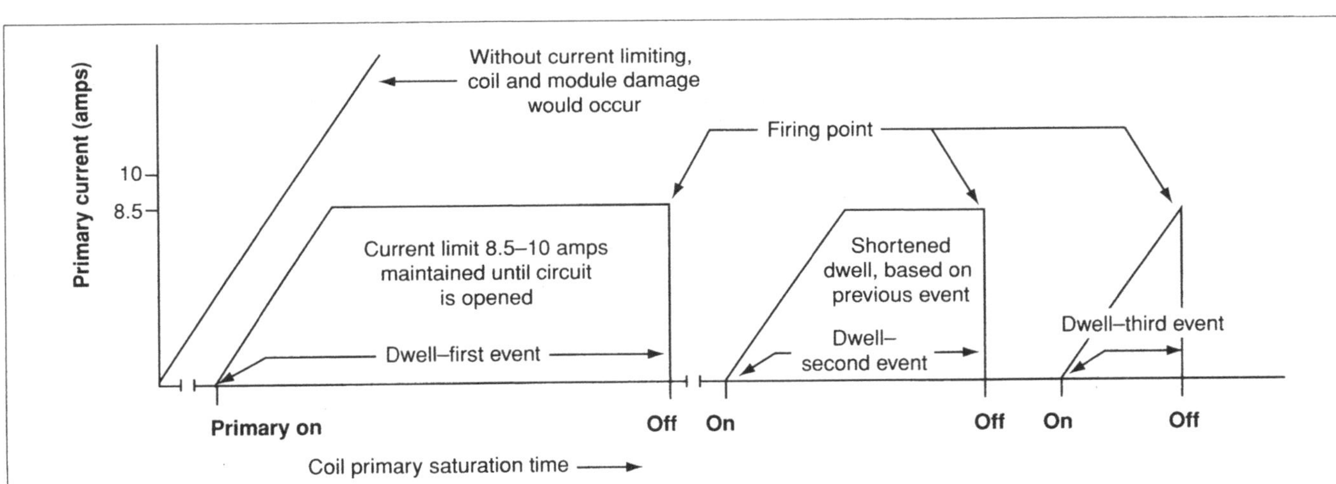

Fig. 6-39. L83, L98: The Ignition Control Module monitors each time the coil builds up, measuring the maximum current obtained. Dwell time is lengthened or shortened, as necessary, until reaching the maximum current (8.5-10 amps), at the instant of firing.

the rotor to the proper plug wire. These tracings are protected by a thick layer of moisture-resistant material. But, as it turns out, they contribute to an ozone problem.

Ozone Venting in the Distributor

Ozone and other harmful contaminants are caused by a secondary ignition-voltage arcing between the rotor and cap terminals. These contaminants are removed by the distributor vent system.

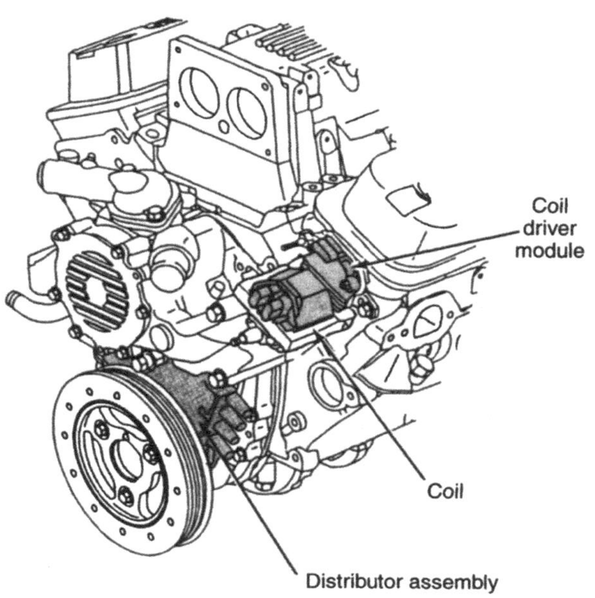

Fig. 6-40. LT1, LT4: The distributor is tucked away. You don't need to adjust th e timing or replace the usual rotor, points, or condenser. In fact, you can't mess with it at all.

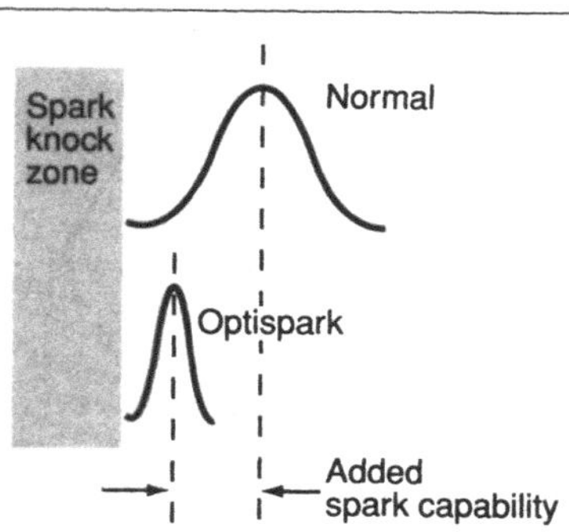

Fig. 6-42. LT1, LT4: Spark scatter is defined as the difference between the spark angle, as calculated by the computer, and the actual spark delivered to the cylinder (shown as "Normal"). Optispark signals are sent with less scatter. The added spark capability from the improved (360X) resolution provides greater precision during acceleration.

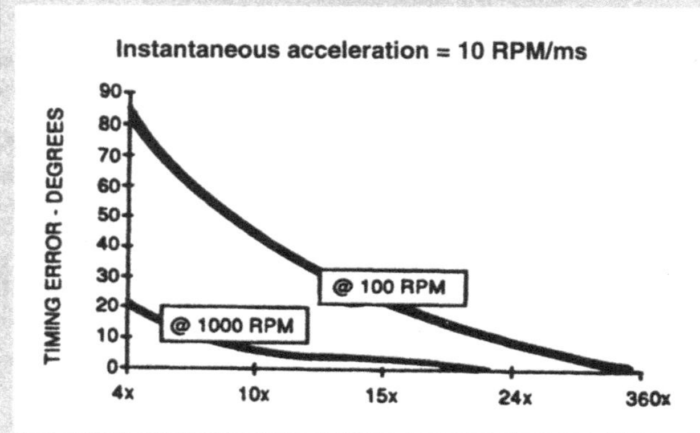

Fig. 6-41. Compare the ignition resolution at 1000 rpm and 100 rpm.

LT1, LS4: These curves show the Ignition Resolution (accuracy). Check the 360X hi-res signal (one degree of rotation) on the right of this curve. 360X provides almost continuous rotation data to the computer, even at 100 rpm—slow cranking. This accuracy reduces errors in spark timing. Why is the circular motion of the crankshaft not steady? For two reasons: 1) transients, from Corvette acceleration, for example, and from very slow engine speeds, such as cold weather cranking; and 2) the changing opposing forces from the compression and power strokes of each of the eight cylinders, even at a steady rpm. These plots are based on transient accelerations/decelerations of 10 rpm/ms, that is, in one millisecond, the crankshaft changes by ten rpm.

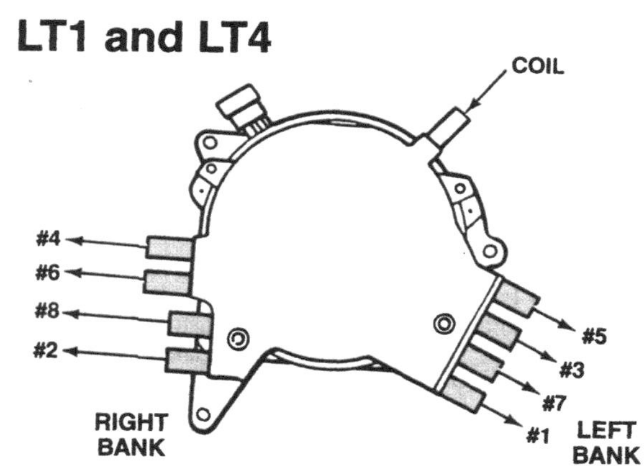

Fig. 6-44. LT1, LT4: You may wonder how the Optispark distributor's rotor sends ignition sparks to the proper cylinder, when the timing alternates banks, 1L-8R-4L-3R-6L-5R-7L-2R. Good wondering! Conductive tracks lead the high-voltage ignition power to the proper side, 1,3,5,7 on the left side and 2,4,6,8 on the right side.

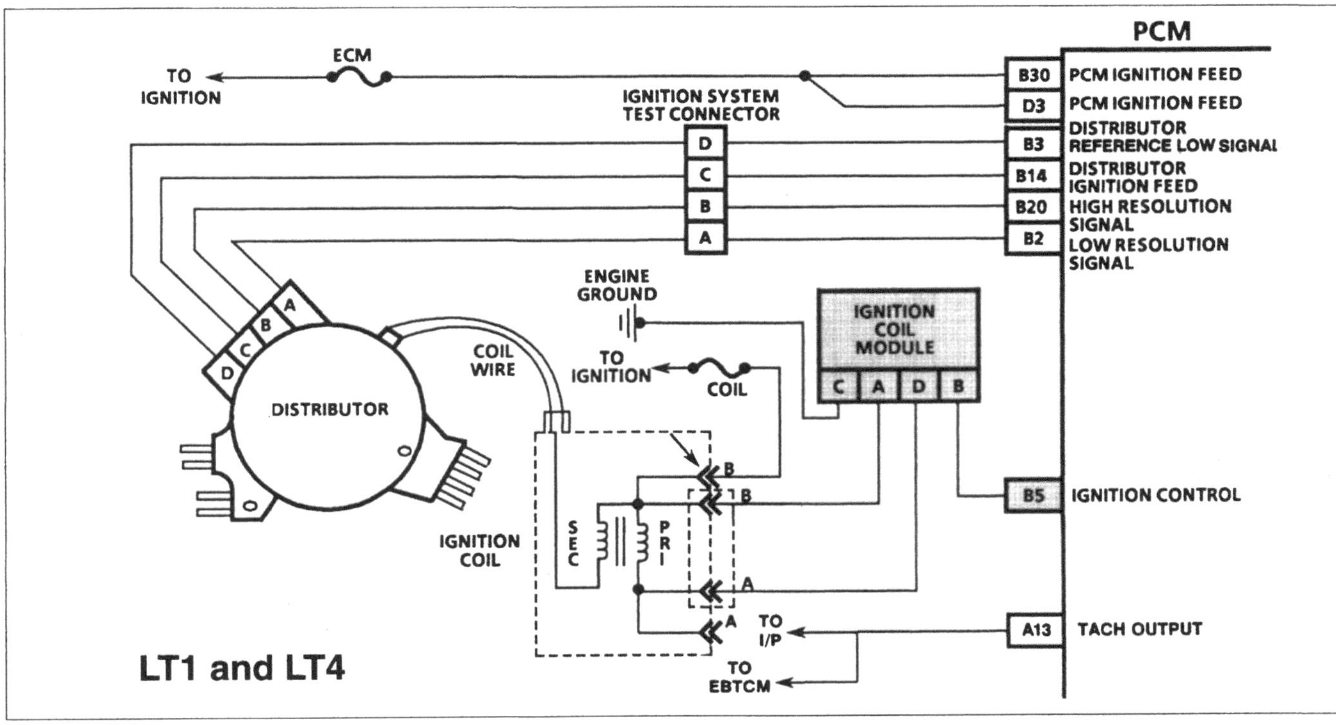

Fig. 6-43. LT1, LT4: The engine computer's Ignition Control (connector B5) drives the Ignition Coil Module (connector B). Connectors A and D control the current flow to the primary circuit, and activate the secondary circuit, which fires the proper plug. These are the distributor connections to the computer: B3 is the distributor reference low signal; B14 is the distributor ignition feed; B20 is the high-res signal; and B2 is the low-res signal.

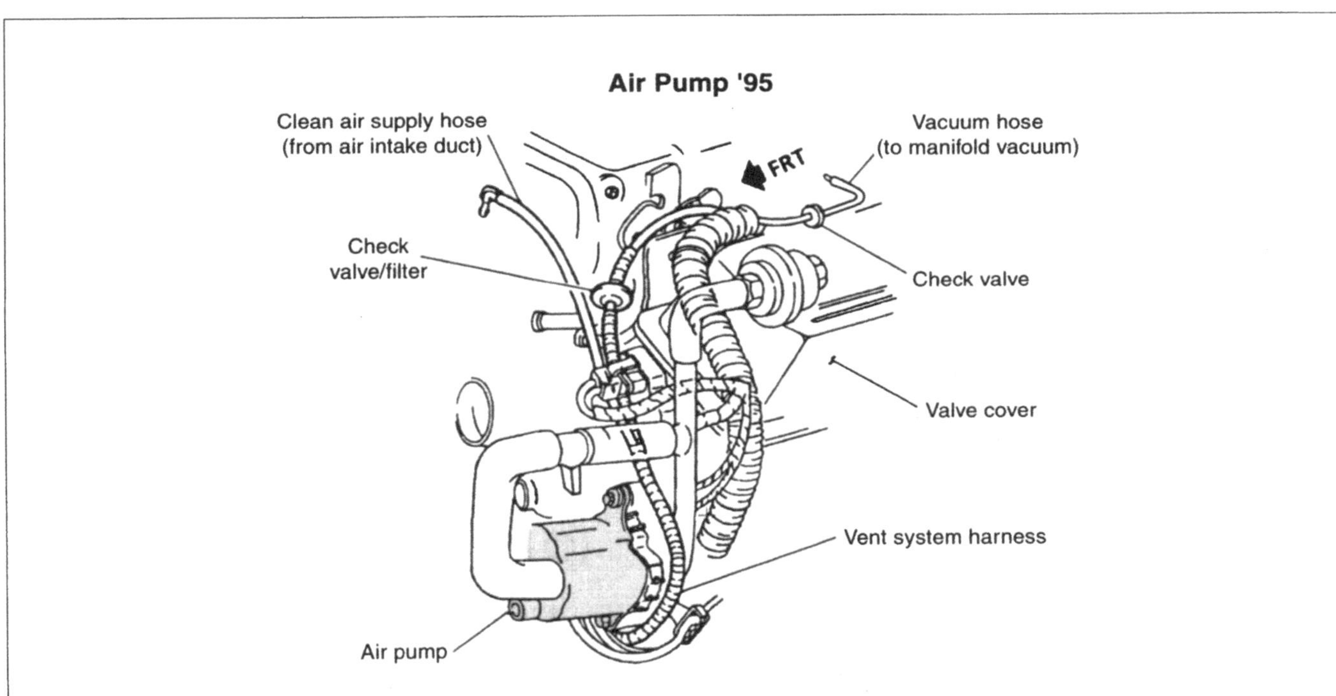

Fig. 6-45. LT1 '95: The air pump pulls clean air from ahead of the throttle body, through the distributor. Air and ozone are pumped through the check valves and hose, to the intake manifold.

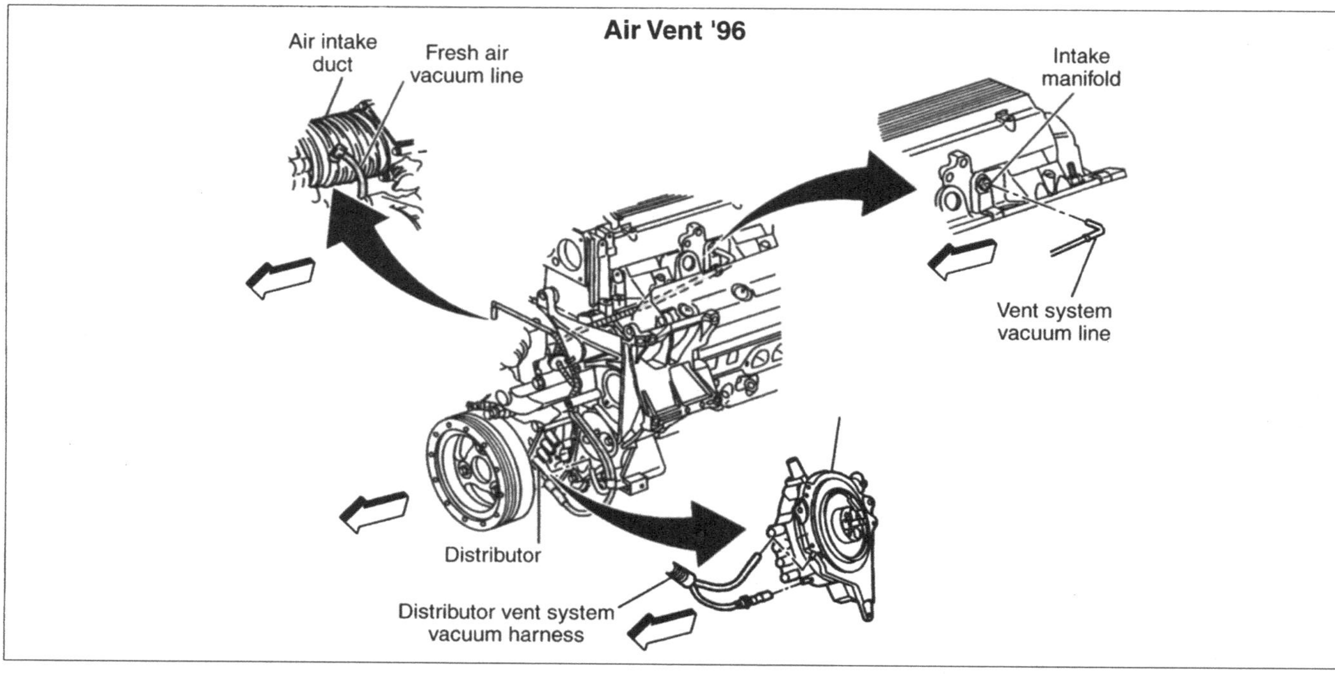

Fig. 6-46. LT1 '96, LT4: The air venting here is similar to the **'95**, except that it operates on a manifold vacuum without an air pump.

6.2 ELECTRONIC IGNITION (EI)

Corvette first used Electronic Ignition (EI) in the **ZR-1**, and continued with a different EI in the **LS1**.

- **ZR-1:** fires plugs in pairs—there are four coils
- **LS1:** fires each plug with its own coil—there are eight coils

ZR-1 Electronic Ignition (EI)

In the "waste-spark" system each coil fires, and:

- The system delivers most of the energy where it's needed most, into the cylinder on compression.
- The system also delivers a minimal amount of energy into the other cylinder.
- Every other firing, the high-energy delivery changes from one of the pair's plug to the other.

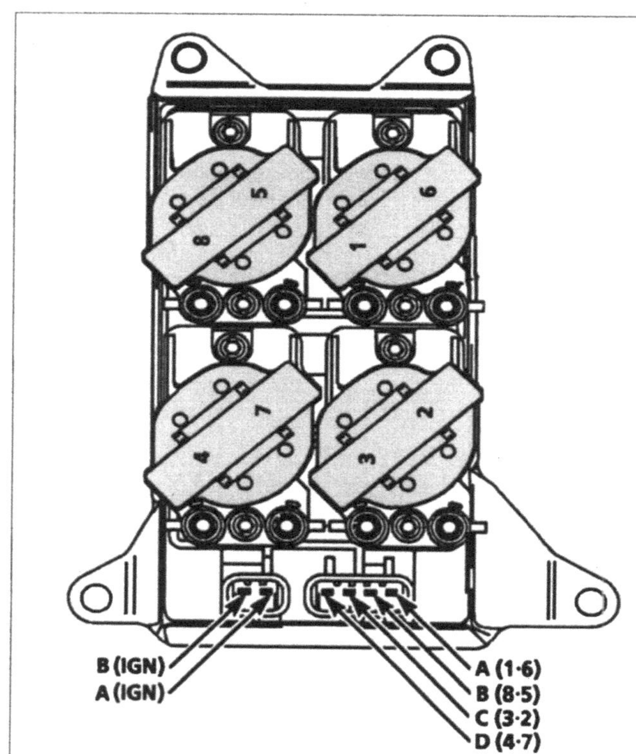

Fig. 6-47. ZR-1: Four coil packs are controlled by the computer. Each coil pack fires a pair of plugs at the same time, one in each bank.

How do the plugs know which cylinder gets the high energy spark? It may seem like magic, but here's the secret: Resistance in the plugs. The amount of current flow remains the same throughout the series circuit, but the plugs use different amounts of energy when they fire. While one plug is on its compression stroke, the other plug is on its exhaust stroke. On the compression stroke, cylinder pressure is high, which requires a higher voltage to fire the spark, against the resistance of the compressed mixture. On the exhaust stroke, cylinder pressure is low, so the waste spark only requires a low voltage to fire through the low resistance.The power spark

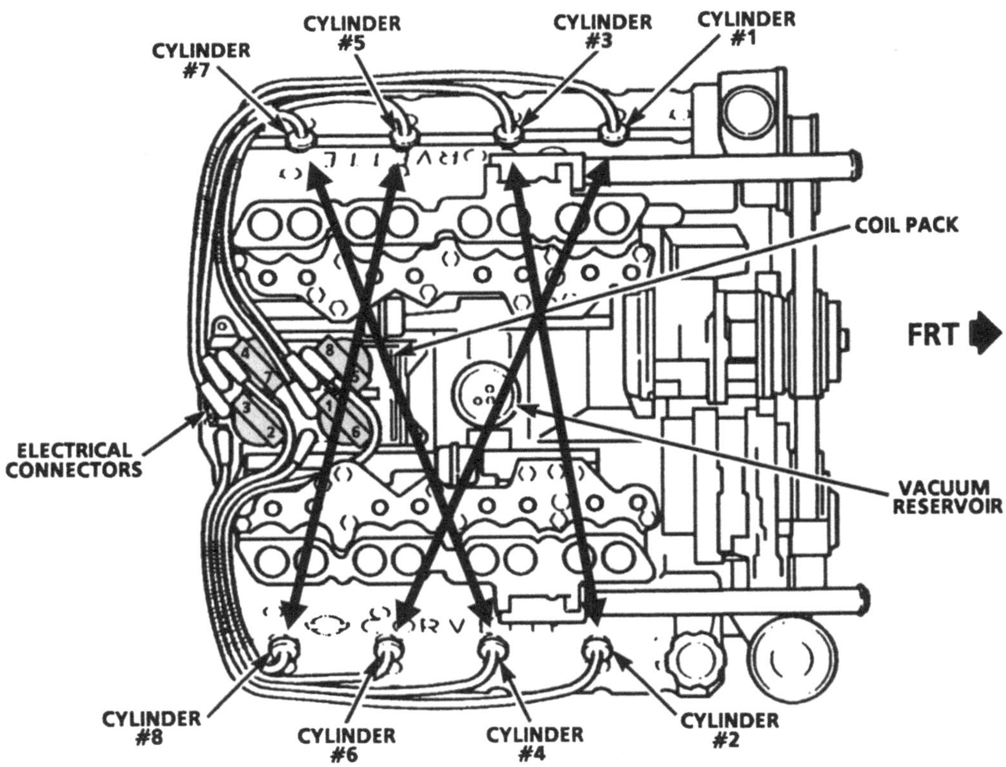

Fig. 6-48. ZR-1: The EI system is based on the fact that each pair of V-8 cylinders comes to TDC at the same time, 7-4, 5-8, 3-2, 1-6 (indicated by the arrows across the banks).

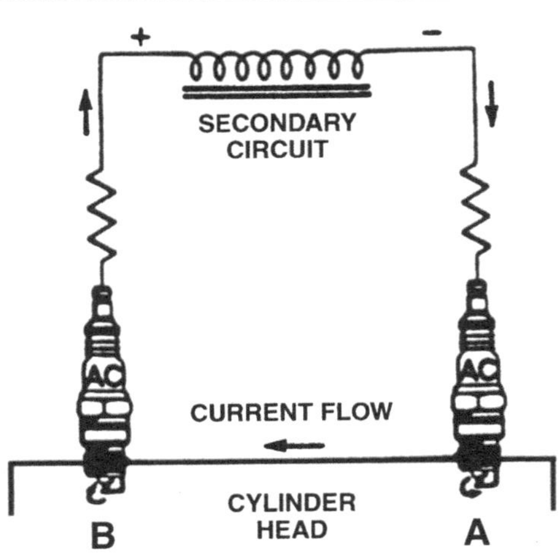

Fig. 6-49. ZR-1: From the secondary coil , current flows to plug A and jumps from the center electrode to the side electrode. Crossing through the head to plug B, the current jumps from the side electrode to the center one, completing the series circuit through the plug cable to the coil.

goes to the proper cylinder (under compression) and the waste spark goes to the other cylinder (during exhaust), based on cylinder pressure and the associated resistance. How about that! It is not magic!

Ignition Control Module (ICM)

ZR-1: Each Ignition Control Module (ICM) sends a reference-high signal, a 5 V square wave. Look at Fig. 6-50, the sensor wheel. As each of the "numbered" slots crosses the sensor, the reference voltage switches from low to high. When the next slot (at 45°) crosses, the reference voltage toggles low. The ICM sends signals to the Engine Control Module about rpm.

OF FURTHER INTEREST —
ZR-1: "reference high" is different from the "high resolution" of the **LT1** and **LS4**.

The engine computer also needs to know the crankshaft position, so the sensor signal has an extra pulse, caused by the extra slot, which follows the TDC 6/1 slot. This is known as the sync pulse, and it enables the ICM to identify the cylinder pairs. Actually, after the sync pulse, the first coil to be charged is not 5/8, but 4/7.

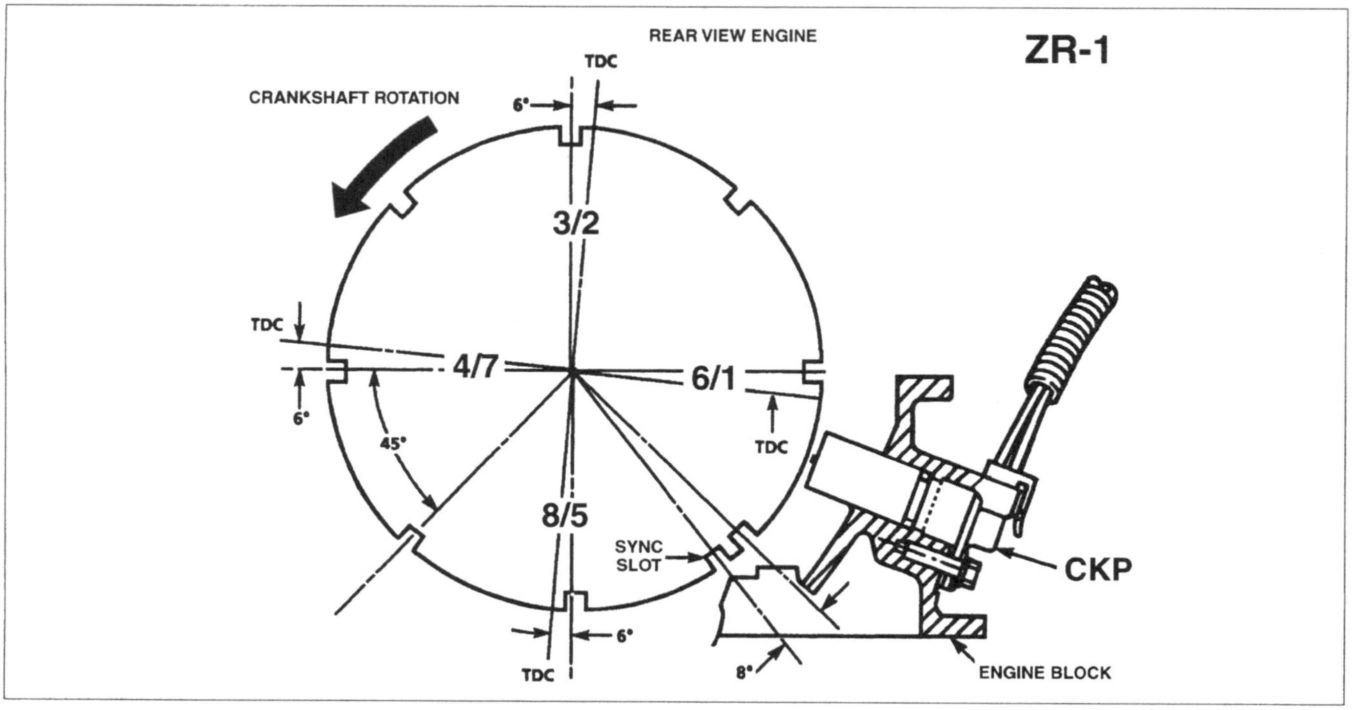

Fig. 6-50. ZR-1: Notice the four slots labeled as cylinder pairs: 4/7, 3/2, 6/1, 5/8. Recall the cylinder pairs identified with each 90° slot, and the alternate slots at 45°. Also notice the odd slot, which is the sync slot.

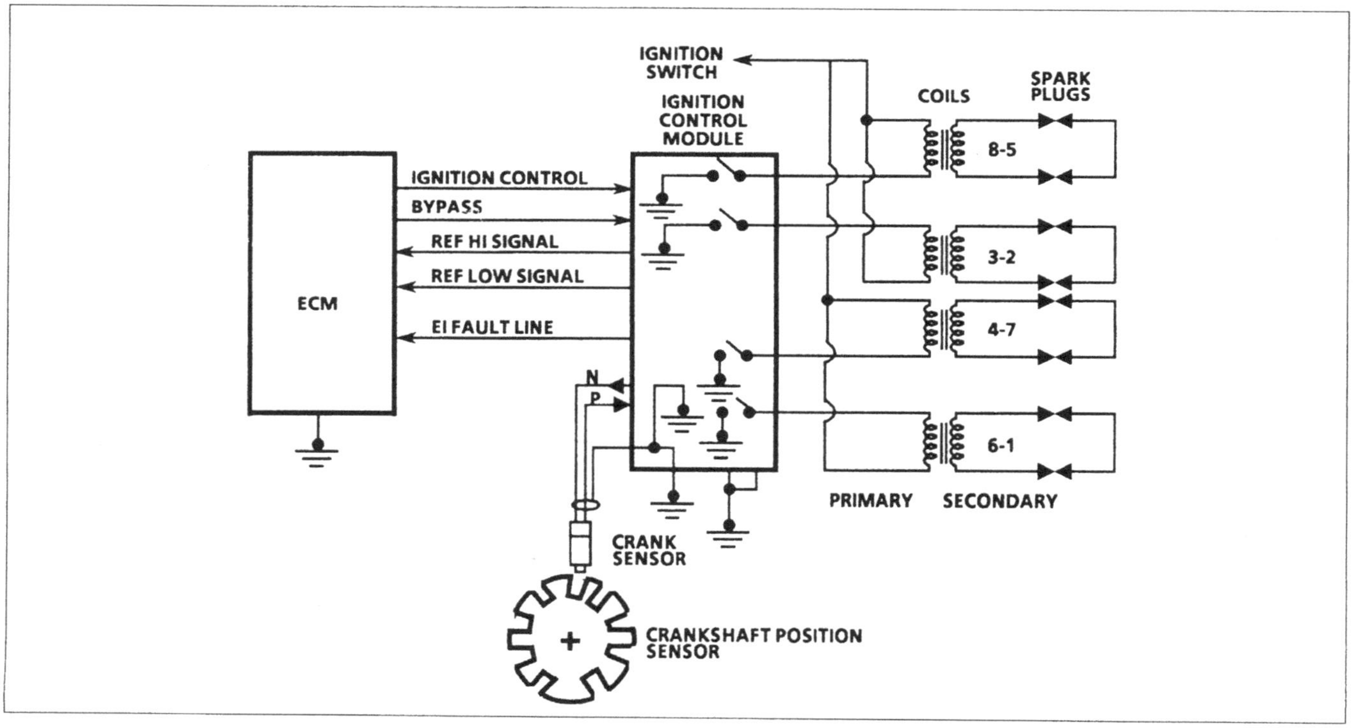

Fig. 6-51. ZR-1: The Ignition Control Module uses the CranKshaft Position signal, modified by the Electronic Spark Timing, in the computer, to turn on and turn off each coil pair.

Still refering to Fig. 6-50, note that reference switching is caused by the eight slots at 45° angles, pulling the signal high at the 6/1 slot, then falling to low at the next 45° slot, and then back high at the 5/8 slot, and so on.

Bypass

ZR-1: The Electronic Ignition operates in two modes, using EST during run, and bypass during crank. During crank, the CKP sensor sends pulses, indicating less than 400 rpm, to the Ignition Control Module (ICM), which signals the computer. This causes the computer to send a low voltage bypass signal back to the ICM. The result:

- The ICM determines the spark timing from the engine computer, bypassing the Electronic Spark Timing.
- The ICM sends a current to the proper coil, for a fixed timing of 6° BTDC, using the slot 51° BTDC. (What's 51°? Add the 45° from the right slot to the 6° to the sync slot.)

The **ZR-1** Ignition Control Module operates at over 100 watts of power. It requires a heat-sink with dielectric grease. 100 watts! We're talking 8 amps at 12.5 V. Remember, you don't hold your hand on a burning 100 W light bulb! The sensor voltage signal varies with crankshaft rpm, ranging from 1/3 V, at slow cranking to 180 V, at high rpm. That could spoil your DVOM!

LS1 Electronic Ignition

The eight individual coils with ignition modules provide:

- increased time for coil saturation, while the other seven coils fire
- increased cool-down time between firings

CAUTION —
LS1: *The secondary coil may operate at voltages above 40,000. So never pierce a plug wire or boot for testing, or you'll be guaranteed future problems.*

The computer controls all of the ignition elements (timing/dwell and sequence) based on inputs from the sensors which are related to fuel-injector sequence and pulse-time: the CMP sensor and 1X Camshaft Reluctor Wheel, CKP sensor and 24X Crankshaft Reluctor Wheel, MAF/MAP engine load, ECT, IAT, MAP for BARO, and KS. (See Chap. 4, Sensors, for more info.)

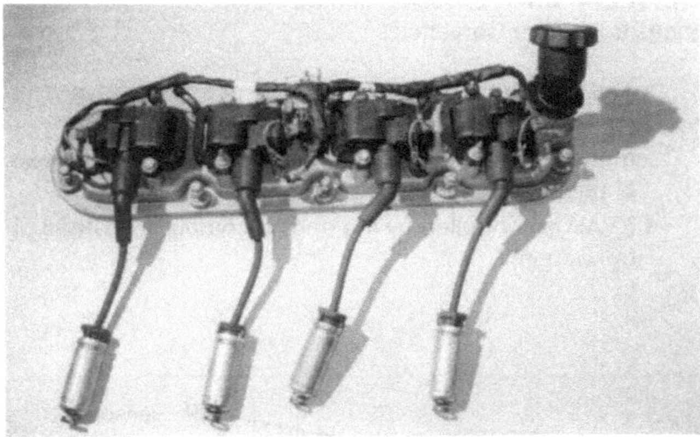

Fig. 6-52. LS1: Each plug has its own coil/ignition module, known as the "coil near plug". This provides a short lead, for minimum energy loss to the plug-wire's resistance. It also helps minimize the engine width, where a coil *on* the plug, wouldn't. The left bank coil/plug assembly is shown here.

6

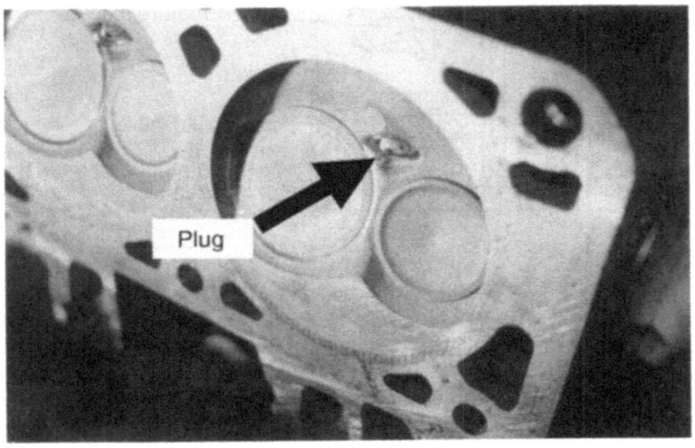

Fig. 6-53. LS1: A "piston-eye" view of the cylinder head. The plug tip is between the larger intake valve and the exhaust valve. A plug life of up to 100,000 miles reduces the need for easy access to change plugs.

7. Emission Control

Emission control is sometimes thought of as a "system". That was true in the 1970s, when most manufacturers hung EGR, AIR, and EVAP parts on the engine. With today's limits, though, emission control is the job of the entire powertrain design, including the internals of the engine, and even the fuels burned. But as a measure of what can be done today with engine design, consider that the **LS1**, **LT4** and **'90-92 ZR-1** certify to the tight limits of the 1990's with no separate EGR control.

LT1, LT4, ZR-1: Emission control varies, with major differences from other Corvettes:

- The AIR pump is electric, instead of engine-driven.
- Catalytic converters are simpler, and are close-mounted directly under the exhaust manifolds, instead of under the car.
- EVAP is controlled by the engine computer, instead of by vacuum.

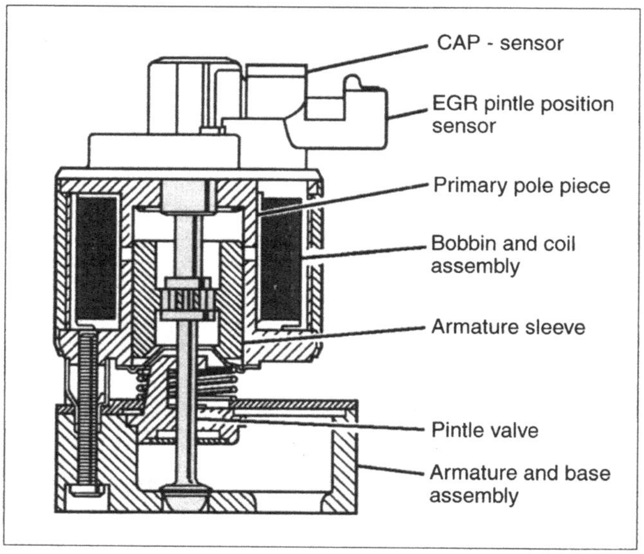

Fig. 6-54. '93-95 ZR-1: The linear EGR valve is computer-controlled, with no vacuum. Pulse-Width-Modulated signals vary the magnetic field of the coil. The Pintle-Position Sensor helps adjust the pintle position ten times faster than a vacuum-operated EGR valve.

7.1 Exhaust Gas Recirculation (EGR)

EGR lowers combustion chamber temperatures, to minimize the formation and emission of Nitrogen Oxides (NO_X). The computer changes the amount of fuel injected, and spark timing, considering EGR flow. That's because EGR causes some loss of power (exhaust gas reduces intake of air).

Negative Backpressure EGR

The control module operates the EGR solenoid valve only when it receives all of the following input signals (varies with model year), within the design limits: engine coolant temperature, intake air temperature, throttle position, engine load (MAP or MAF), Park/Neutral Switch, and vehicle speed.

Fig. 6-56. LT1, L98, L83: The negative backpressure solenoid valve controls the intake vacuum which operates the valve. The amount of EGR is controlled by rapidly opening and closing the vacuum valve, this is Pulse-Width Modulation.

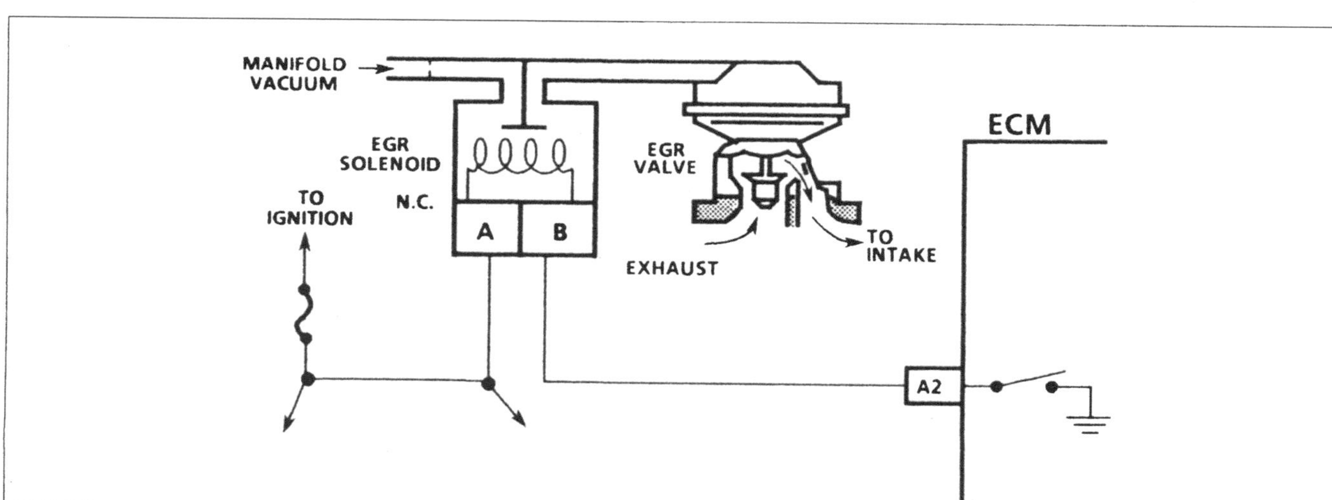

Fig. 6-55. '82-96: The EGR flow is controlled by a combination of the intake manifold vacuum, and negative-exhaust backpressure.

Fig. 6-57. L98: Look for the EGR control solenoid near the throttle body.

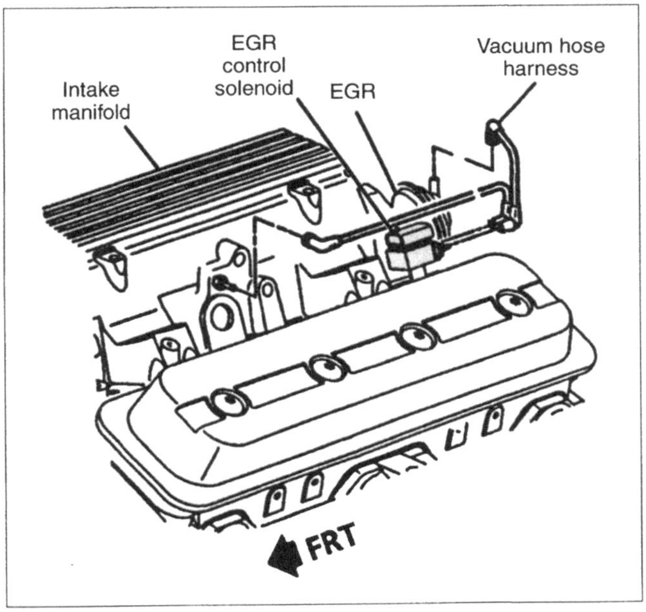

Fig. 6-58. LT1: Look for the EGR control solenoid at the rear center of the engine.

EGR is shut off at idle and during warm up, because:

- NO_X formation is minimal
- EGR could cause rough idle

EGR is shut off during Wide Open Throttle (W.O.T.) operation because it causes a loss of power (W.O.T. is usually only used for short periods).

7.2 Secondary Air Injection (AIR)—previously Air Injection Reaction

Secondary Air Injection (AIR) adds air, and oxygen, to the exhaust during warm-up, to promote "afterburning". This also heats the unheated Oxygen Sensors. In contrast to EGR and intake systems, AIR deals with exhaust gases after they have left the engine, and has nothing to do with the engine combustion.

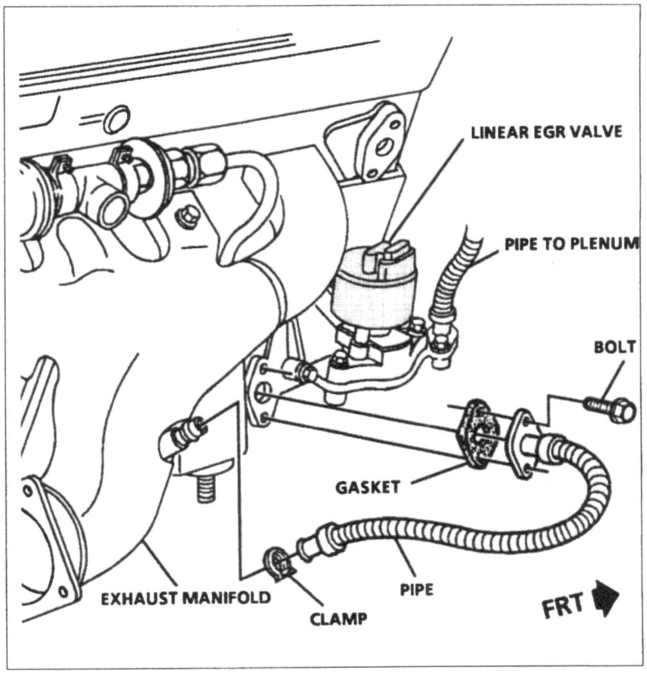

Fig. 6-59. ZR-1, '93-95: Look for the Linear EGR valve under the right bank. There is direct PCM control with no solenoid valve.

Why is Secondary Air Injection sometimes identified by the acronym AIR? Why not SAI? The SAE J1930 committee decided to keep the long-used and accepted term, which originally stood for "Air Injection Reaction".

L83/L98: An engine-driven pump delivers air to the exhaust manifold, or switches to an oxidizing catalytic converter.

LS1, LT1, LT4, ZR-1: An electric pump delivers air to the exhaust manifolds, but not to the catalytic converters.

L83, L98 AIR

The air switching valve sends air to the two-way conversion section, at the rear of the converter, during Closed Loop. The air diverter valve gets rid of the pumped air by dumping it back into the air cleaner, during rich operation, deceleration, or high-rpm operation, or whenever a trouble code is set by the computer.

LS1, LT1, LT4, ZR-1 AIR

The electric pump operates during cold-engine starts, but only if the ECT signals above 15° C (60° F). The computer stops the pump and closes the valve when the fuel system goes Closed Loop. The maximum run-time limit is 240 seconds.

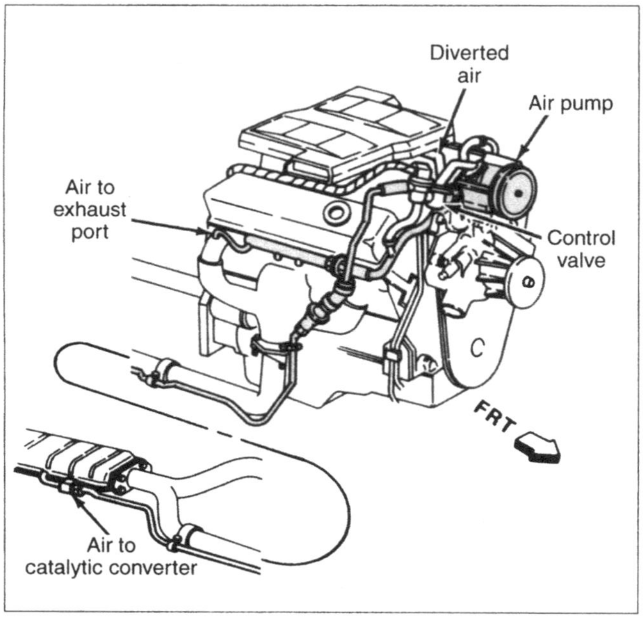

Fig. 6-60. L98,L83: The engine-driven pump runs whenever the engine runs. The control valve switches air to the exhaust manifolds during warm-up, or any time the engine is Open Loop.

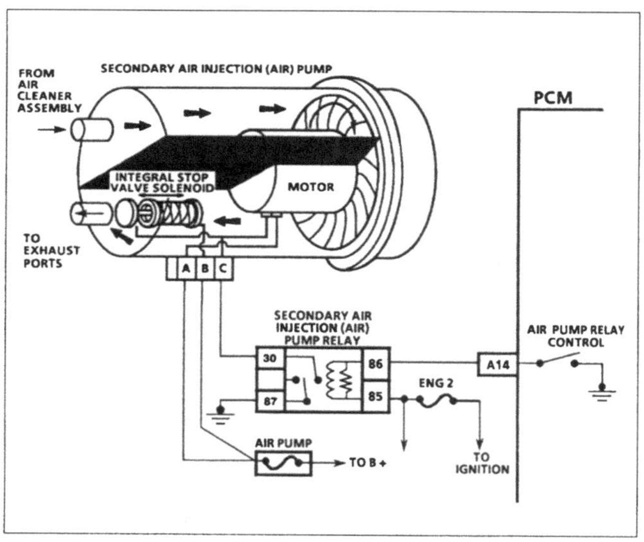

Fig. 6-63. LS1, LT1, LT4: The computer grounds the relay, to operate the pump and the integral stop-valve solenoid. The **ZR-1** is similar.

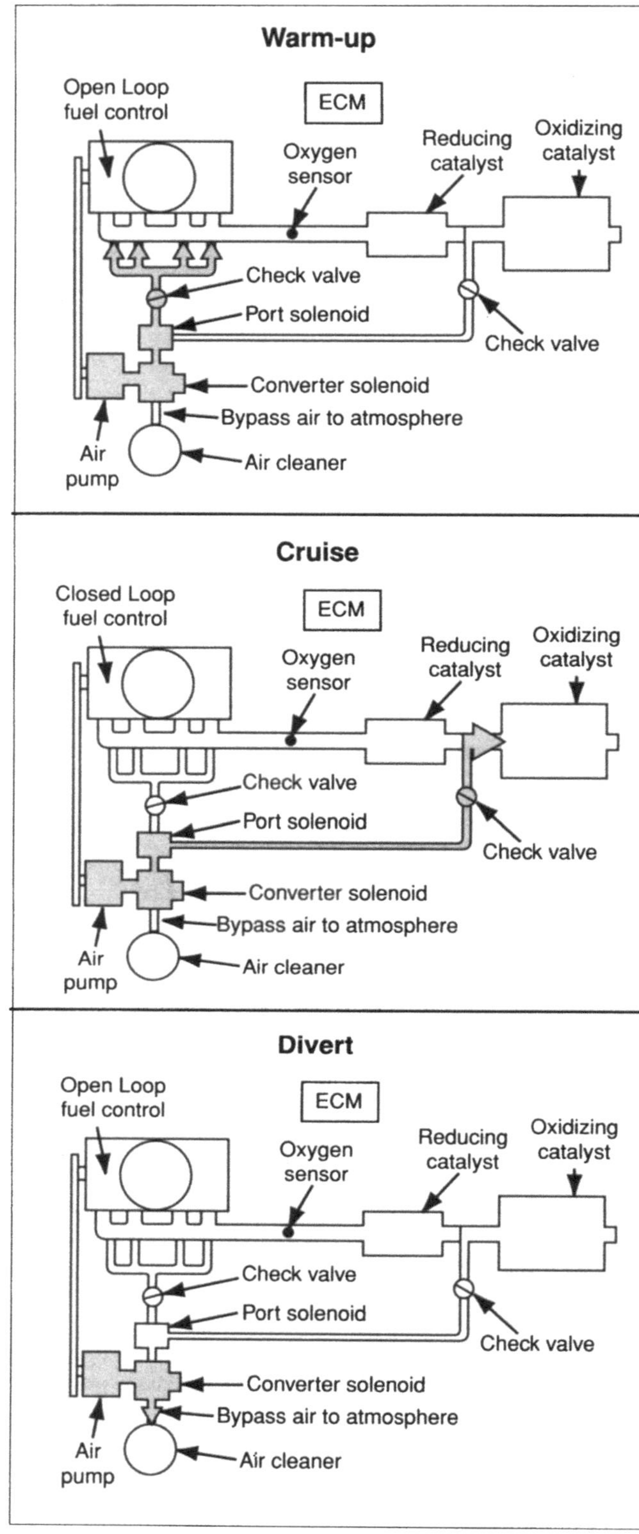

Fig. 6-61. Air flows during:
1. warm-up—air goes to the exhaust manifolds
2. cruise—air goes to the oxidizing catalyst
3. divert—air is "dumped" into the air cleaner

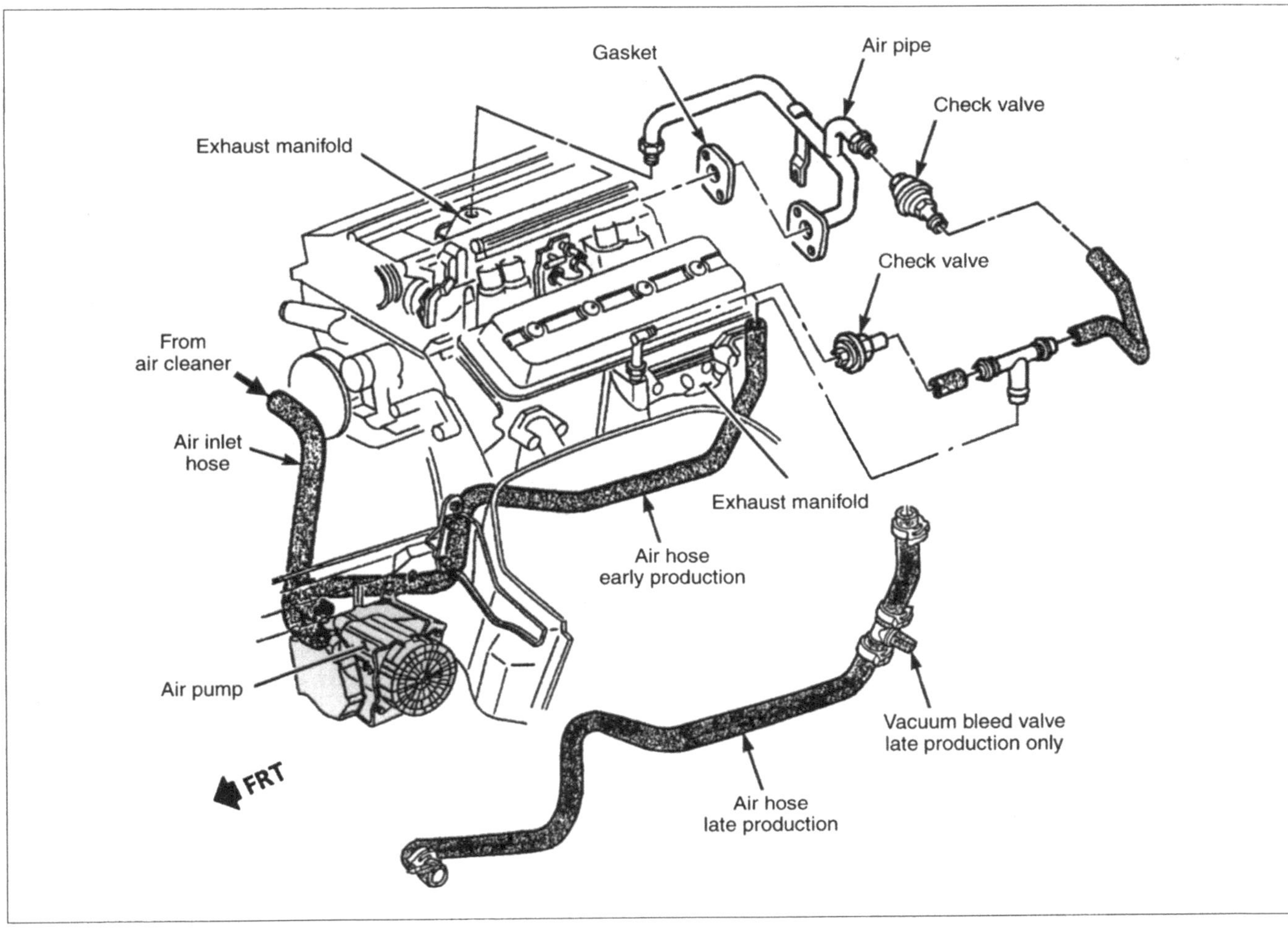

Fig. 6-62. LS1, LT1, LT4: The electric pump delivers air to the exhaust manifolds during warm-up. There is no need to switch the air to the converter, because the converters have no separate two-way oxidizing section. The **ZR-1** is similar.

A late-production change in the **'95** model provides a vacuum-bleed valve. This valve, plus running the electric pump during crank, helps prevent vacuum buildup.

(For the Evaporative Emission Control System (EVAP), see Chap. 3 "Emissions", and Chap. 7 "Fuel Systems".)

8. Information Controls

Electronic Engine Management systems actuate various signals for the driver.

Driver Information

You'll find several versions of the so-called "Check Engine" light. Consider how the indicator has changed meaning, from early **'82**. (See Fig. 6-64.)

Manual Transmission

ZR-1: You'll see the "SHIFT" light at:

- W.O.T., maximum acceleration
- maximum power for gears 1-5
- 5800 rpm

Why "5800", when specs say power peaks at 6000? Actually, the power curve is quite flat above about 5600. Under maximum acceleration, the engine rpm actually is 6000 when the tach reads 5800. The slight lag is built into the tach to help readability.

OF FURTHER INTEREST —
ZR-1: the shift light operates in contrast to the One To Four light, which is programmed to operate at low rpm/low power, indicating shift for economy.

Fig. 6-64. • **'82-85** has a "CHECK ENGINE" light. This wording caused over-concern of some drivers.

• **'86-92** has a "SERVICE ENGINE SOON" (SES) light, similar, but less threatening.

• **'93** has a "SERVICE (engine pix) SOON" light.

• **'94-00** Look for two lights:
The "SERVICE (engine pix) SOON" light is for all engine problems which are not related to emissions.
The *Malfunction Indicator Light* (MIL) is for emission-related problems only. SAE J1930 required MIL nomenclature only for malfunctions that could cause emissions to exceed 1.5x original limits of certification. A flashing light indicates a misfire condition, which may cause damage the the converter. A steady light indicates an Emission Control System malfunction, which requires service, checking the stored codes.

Fig. 6-65. LT1, LT4, ZR-1, LS1: In the manual transmission engines, the One to Four shift lamp will show when conditions require upshift from gears 1 to 4, or to put it the other way, to prohibit shifting into gears 2 or 3.

Power Switch & Indicator—ZR-1

Secondary injectors can be activated and secondary-air intake valves may open. Such *full* power operation requires satisfactory engine oil temperature (EOT), throttle position (TP), manifold absolute pressure (MAP) and engine rpm.

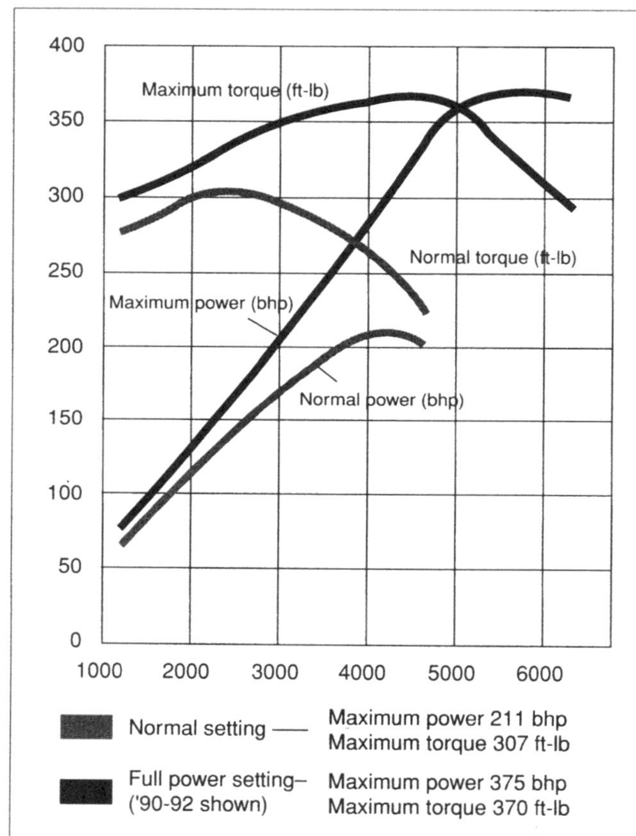

Fig. 6-66. ZR-1: When the power key is turned to the "FULL" setting, you can expect power increases, depending on model year:
• maximum HP, up 75%
• maximum torque, up 20%, and much flatter

9. Relays and Controls

For diagrams detailing specific engine model relays and controls refer to Chap. 12—Schematics.

9.1 Fuel-Pump Relay(s)

The fuel-pump relay closes when the ignition is turned on, and remains closed for two seconds, or until the engine is cranked for starting. If the computer does not receive any engine-turning signals (ignition reference), it opens the relays, cutting off the fuel pump(s). Look for the Fuel-Pump Relay(s) inside the passenger compartment under the dash. The **ZR-1** uses two relays to control the two fuel pumps.

9.2 Air-pump Relay

LS1, LT1, LT4, ZR-1: The air-pump relay closes, to operate the Electric AIR pump, when the computer commands the operation of AIR. It also energizes the Electric Switching Valve (ESV) to direct pump air to the exhaust ports. (For details, see Chap. 3.) Look for the relay on the electric pump, ahead of the left-front wheel, and the ESV on the engine itself.

9.3 Cooling Fan Relay(s)

'96 LT1, LT4, LS1 have three Cooling Fan Relays. Relay #1 turns both cooling fans in a series, so that both operate on low. Relay #1 is commanded by the computer when any of the following conditions exist at idle:

- certain engine computer trouble codes are set
- CTS is above 104° C (219° F)
- EOT is above 132° C (270° F)
- A/C head pressure is above 139 psi

When more cooling is needed, the computer commands both relays, #2 and #3, to turn both fans in parallel, for high speed operation.

'92-95 LT1, LT4, LS1 have two Cooling Fan Relays, one for each fan. The Primary Cooling Fan Relay turns on the primary cooling fan under conditions similar to those for the #1 relay on the later engines (as detailed above). The Secondary Cooling Fan Relay turns on the secondary fan with higher temperatures and/or AC head pressures.

9.4 Computer Aided Gear Shift (CAGS)—or "ONE TO FOUR" or "SKIP SHIFT"

'90-01: The Computer Aided Gear Shift (CAGS) is part of 6-speed manual transmissions. During mild acceleration, as defined by the engine computer as low throttle position, and low speeds, CAGS restricts use of the second and third gears, forcing the shift from first to fourth. Why?

- Because Corvette's great available power and broad torque curve provide satisfactory acceleration without requiring 1-2-3-4. The one-to-four shift is fine for mild acceleration. You've got a lot of "go" in fourth.
- Shifting 1-2-3-4 during mild acceleration wastes fuel. You wouldn't think that it would be much, but during the mild accelerations of EPA emission/fuel economy testing, such shifting 1-2-3-4 could waste enough fuel to shift Corvette into "Gas-Guzzler" territory.

In owner's-manual-speak: "It's possible, and in some cases preferable, to skip a gear or two."

Table 6-1.CAGS upshift

Model Year	CAGS speed range	CAGS maximum throttle
'90–96	between 12 and 19 mph	35%
'97-00	between 15 and 19 mph	21%

6

CAGS only controls your shifting when the engine is warmer than 120° F (49° C) and when you use light throttle. If you want to avoid the 1-4 shift, simply apply a tad more throttle, or shift into second gear before you reach the indicated speed range.

10. Summary

You've seen the actuators, or the "muscles", controlled by output signals from the computer, the "brain", based on inputs from the sensors, the "nerves". You've also seen the:

- A/F ratio, Idle-Air Control, and **ZR-1** Secondary Air Inlet Control
- Spark Control for Distributor Ignition, and for Electronic Ignition
- Emission Control for EGR and AIR
- Information Control
- relays and controls

IN THE NEXT CHAPTER:

Fuel injection is great, but only as long as the fuel injectors receive clean fuel at the right pressure. Find out how in Chapter 7, "Fuel Delivery—Circulatory System".

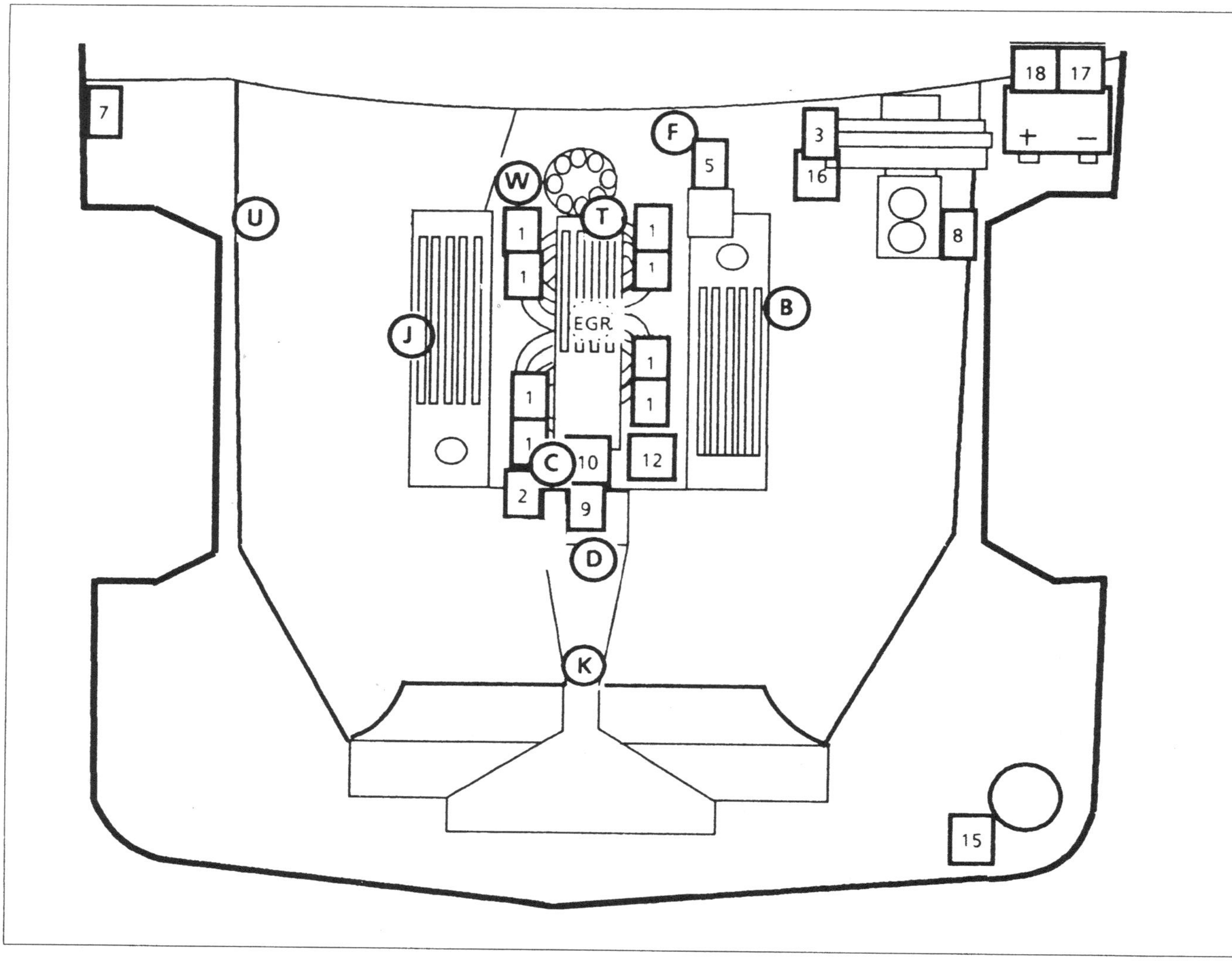

Fig. 6-67. '86 L98: The actuators and sensors (the **'85** is similar). Corvette refers to MFI and "Tuned Port Injection". Compare this simple system to the growing complexity of the following years.

Sensors (circles)
B Exhaust Oxygen sensor
C Throttle Position (TP) sensor
D Engine Coolant Temperature (ECT) sensor
F Vehicle Speed Sensor (VSS)
J Knock Sensor (KS)
K Mass Air Flow (MAF) sensor
T Manifold Air Temperature sensor
U A/C pressure switch
W EGR diagnostic temperature switch

Actuators (squares)
1 Fuel Injector
2 Idle Air Control (IAC) valve
3 Fuel Pump (FP) relay
5 Transmission Converter Clutch connector
7 Electronic Spark Control module
8 Cooling fan relay
9 AIR injection divert valve
10 AIR injection switching valve
12 Exhaust Gas Recirculation vacuum solenoid
15 Fuel vapor canister solenoid
16 Transmission overdrive relay (M/T)
17 MAF sensor power relay
18 MAF sensor burn-off relay

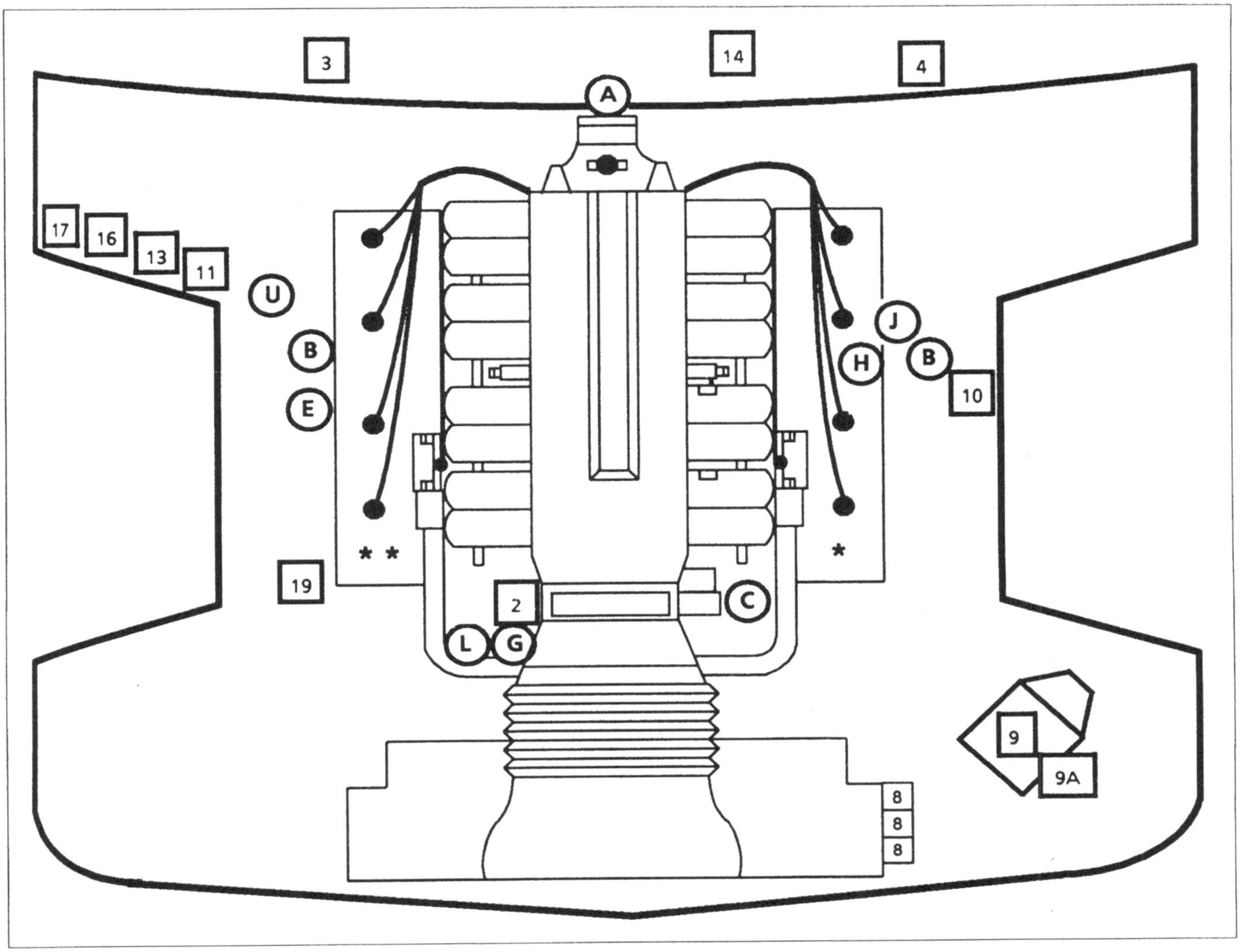

Fig. 6-68. ZR-1: There is added complexity, necessary to handle the Secondary SFI. The **'95 ZR-1** differs from the previous **ZR-1** by adding a third cooling fan relay. The primary and secondary fuel injectors are hidden by the plenum.

Sensors (circles)
A Manifold Absolute Pressure (MAP) sensor
B Heated Oxygen Sensor (HO_2S)
E Crankshaft Position sensor
F Vehicle Speed Sensor (VSS)
(not shown, on transmission)
G Intake Air Temperature (IAT) sensor
H Camshaft Position sensor
J Knock Sensor (KS)
L Engine Oil Temperature (EOT) sensor
U A/C cooling fan switch

Actuators (squares)
2 Idle Air Control (IAC) valve
3 Fuel Pump (FP) relay (primary)
4 Fuel Pump (FP) relay (secondary)
8 Cooling fan relay
9 Secondary Air Injection AIR pump
10 AIR bypass valve
11 2nd and 3rd gear blockout relay
13 A/C clutch relay
14 2nd and 3rd gear blockout solenoid
16 Secondary SFI control module #1
17 Secondary SFI control module #2
19 Linear EGR valve

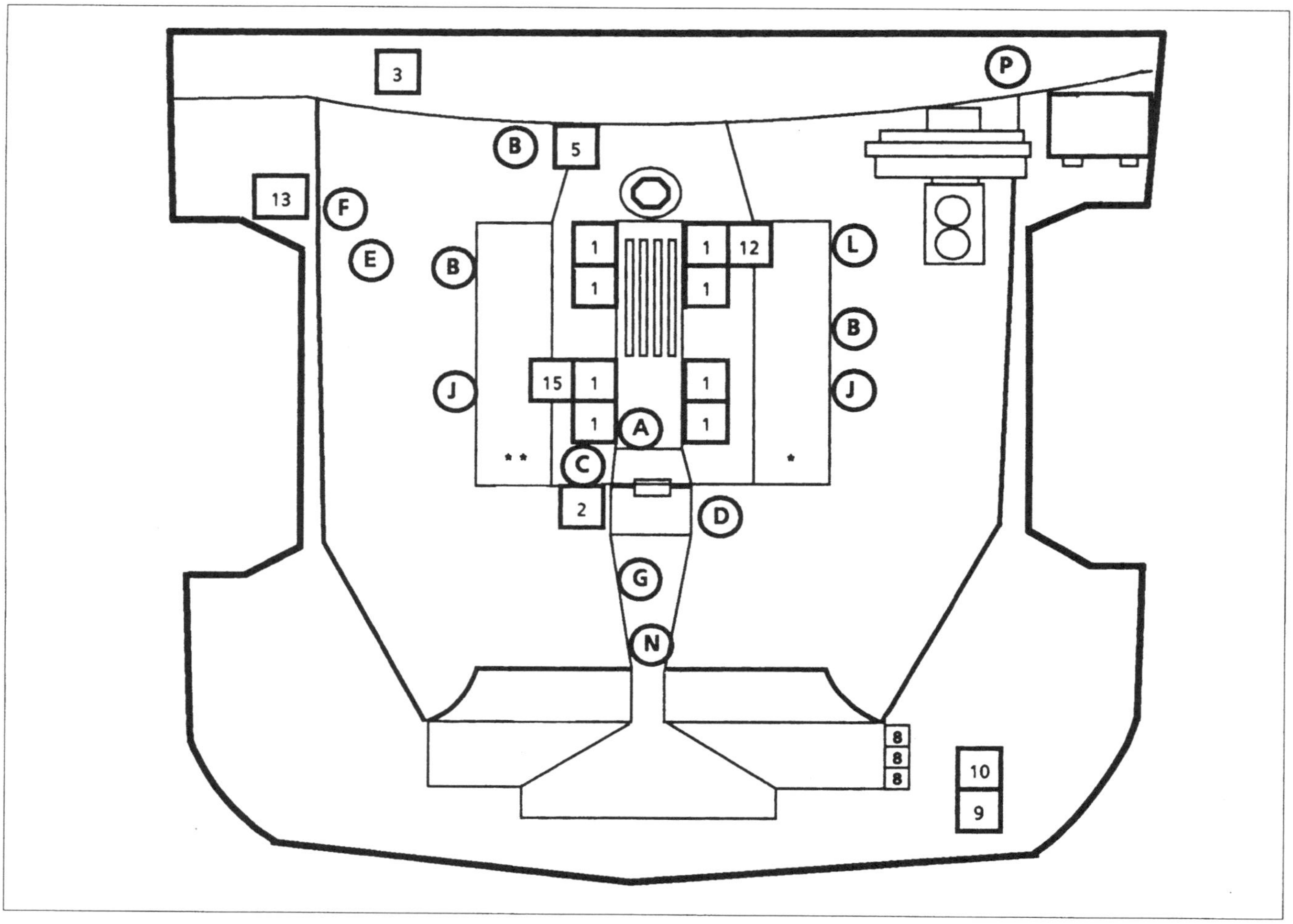

Fig. 6-69. '94-96 LT1, LT4: The actuators and sensors. In this system, the actuators are controlled by signals from the computer for the A/F ratio (eight individual injectors for SFI), Idle-Air Control, spark-timing, and emissions. This **'96 LT1** controls the cooling fans, electric air pump, and transmission controls. The **LT1** shows the EGR valve, while the **LT4** has no EGR.

Sensors (circles)
A Manifold Absolute Pressure (MAP) sensor
B Heated Oxygen Sensor (HO_2S)
C Throttle Position (TP) sensor
D Engine Coolant Temperature (ECT) sensor
E A/C refrigerant pressure sensor
F A/C pressure cycling switch
G Intake Air Temperature (IAT) sensor
J Knock Sensor (KS)
L Engine Oil Temperature (EOT) sensor
M Vehicle Speed Sensor (VSS)
(not shown, on transmission)
N Mass Air Flow (MAF) sensor
P Brake switch

Actuators (squares)
1 Fuel Injector
2 Idle Air Control (IAC) valve
3 Fuel Pump (FP) relay
5 Transmission electrical connector
8 Cooling fan relay
9 Secondary Air Injection (AIR) pump
10 AIR pump relay
12 Exhaust Gas Recirculation vacuum solenoid
13 A/C clutch relay
15 EVAP canister purge solenoid valve

Chapter 7

Fuel Delivery—the Circulatory System

Contents

IN THIS CHAPTER, LOOK FOR:

The four jobs of the Corvette fuel system

- deliver clean fuel under pressure to the injectors
- compensate injector fuel-delivery for varying MAP
- control fuel-pump relays and safety circuits
- control fuel vapors with EVAP system

1. INTRODUCTION

You've seen how sensors deliver input signals to the control module, which delivers output signals to the actuators. Now I'll look at the fuel systems, delivering fuel to the fuel rails at the proper pressure and in sufficient quantity. The fuel system is something like the *circulatory system* in our bodies, with a pump delivering fluid under controlled pressures. Owners of fuel-injected Corvettes want nothing to do with fuel starvation, no matter the cornering, no matter the acceleration.

LT1, LT4, ZR-1, L98: Excess fuel in the rails, not delivered through the injectors, flows back to the tank through the regulator and the return line.

LS1 '97-98: Excess fuel in the rails, not delivered through the injectors, flows back to the tank through the regulator and the return line. Fuel that is heated by the engine heat but not burned is recirculated back to mix with cooler fuel in the tank.

LS1 '99-01: There is a Fuel Pressure Regulator in the tank and a single supply line to the rails.

Fuel Tanks

Fuel tank(s) are *lightly pressurized*—about 1 psi (7kPa) to reduce the evaporation of HC and other pollutants into the atmosphere. What pressurizes the tanks? Fuel vapor-pressure, the natural build-up of fuel vapor in a tank. It's a closed system, not normally open to atmospheric pressure.

Improper fuel pressures can be caused by extreme performance driving that may starve the pump(s) in a cross-car tank ('82-96). This results in insufficient fuel pressure, even though the pump and the Fuel-Pressure Regulator may be operating properly. The key factor is a tank less than half full.

LS1 narrow, tanks are designed to feed under severe G loads. The design target is to maintain fuel pressure under one G force, accelerating, decelerating or cornering.

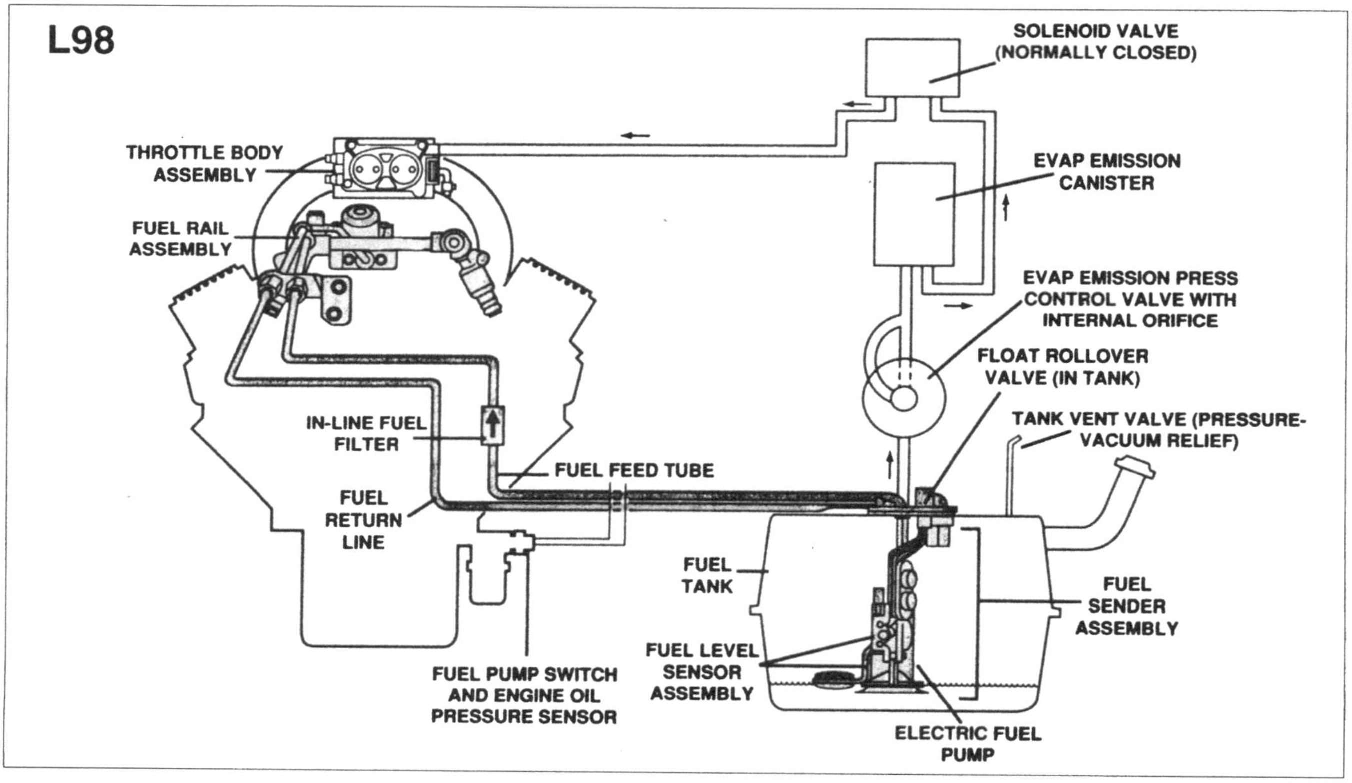

Fig. 7-1. L98: (LT1, LT4 are similar) the typical Corvette fuel system includes a fuel tank, fuel sender assembly (electric fuel pump with fuel-level sensor), fuel-feed tube with in-line fuel filter, fuel-rail assembly with fuel-pressure regulator, EVAP system with canister, and a tank vent valve.

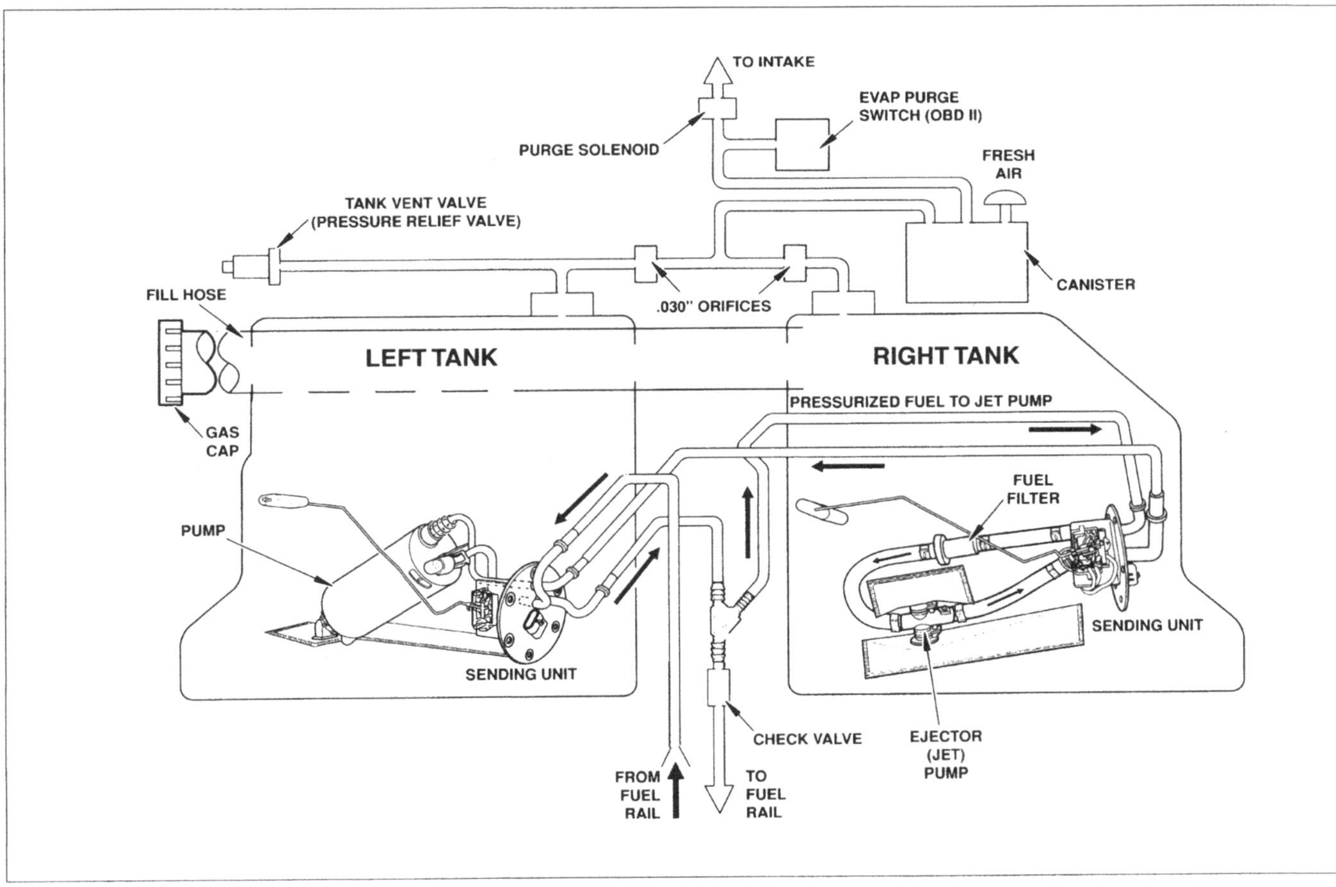

Fig. 7-2. '97-98 LS1: The fuel-delivery system returns excess fuel directly from the regulator so that all of the fuel entering the rails is burned. The Fuel Pressure Regulator is on the rail assembly.

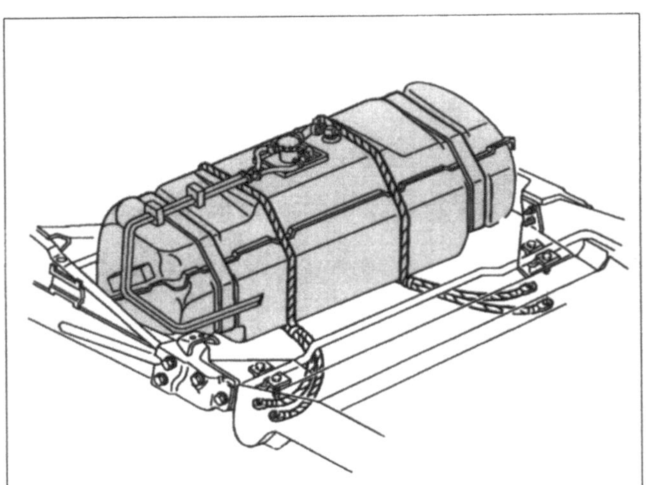

Fig. 7-3. '82-96: With traditional full-width tank shape, fuel can slosh to one side during hard cornering or acceleration. Pressure may drop off, causing lean mixtures, with the possibility of loss of power, surging or misfire.

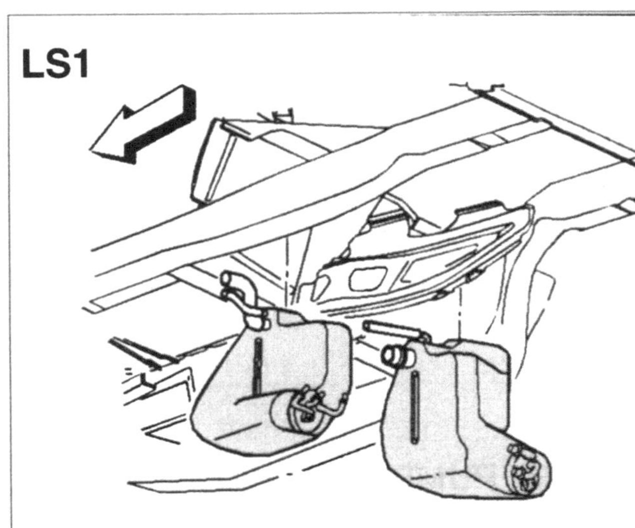

Fig. 7-4. LS1: A vertical tank shape tends to reduce sloshing under high G manuevers.

7

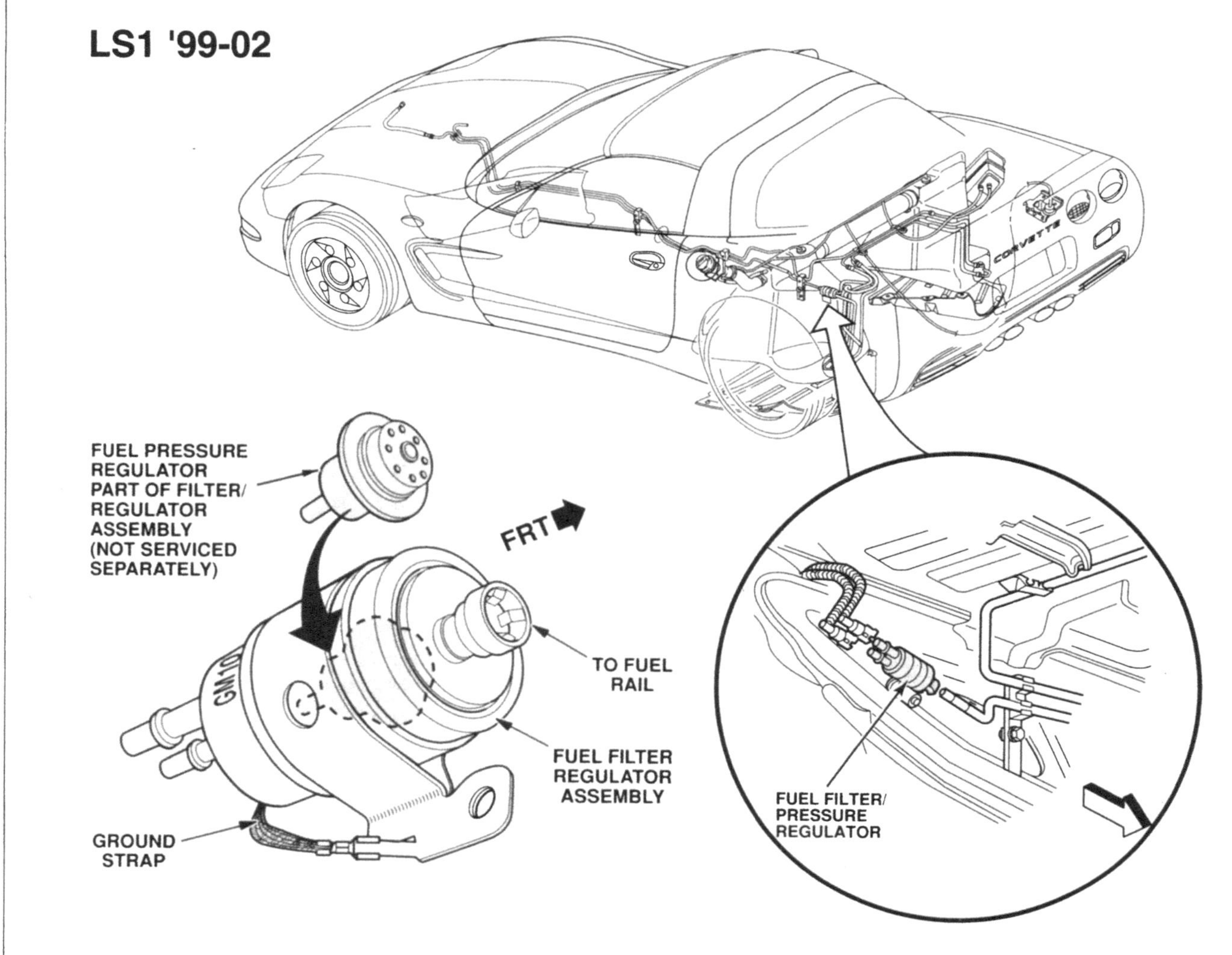

Fig. 7-5. '99-01 LS1: The Fuel Pressure Regulator/filter assembly is located near the left fuel tank. There is a short supply line and a return line between the left tank and the FPR/filter assembly (*top figures*). The fuel pressure *damper* is on the rail (*bottom figure*). The **LS1** has an unusual fuel system which includes interconnected *dual* 9-gallon tanks to accomodate the Corvette frame structure and to provide space for the rear-mounted transaxle. (See also Fig. 7-19.)

2. FUEL PUMP/ SENDER ASSEMBLIES, FILTERS

A high-pressure pump in the tank delivers fuel, reducing vapor lock. In addition to the in-tank strainer, look for an in-line filter. In-tank sensor assemblies include a strainer and a pump. The **ZR-1** uses two pumps.

LS1: You might think that 2 fuel tanks would need 2 electric pumps, but no. The left tank has the one electric pump, effectively delivering from both tanks. (See Fig. 7-2.)

The **LS1** left tank electric pump also delivers a small amount of pressurized fuel to the right tank *siphon-jet pump.* This pressurized fuel creates a venturi action in the siphon-jet pump that draws fuel from the right tank to the left tank. Remember, there is no electric pump in the right tank.

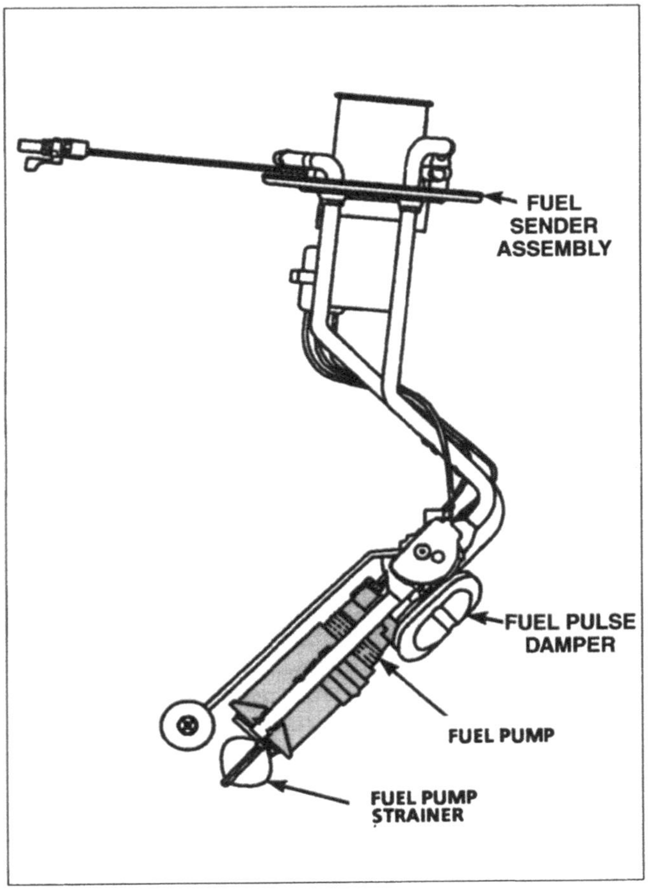

Fig. 7-6. LT1, LT4: (**L83**, **L98** are similar) Fuel Sender Assembly includes a fuel-pulse damper, fuel pump, and fuel-pump strainer. The pulse damper reduces noise generated by the fuel pump. The strainer is self-cleaning.

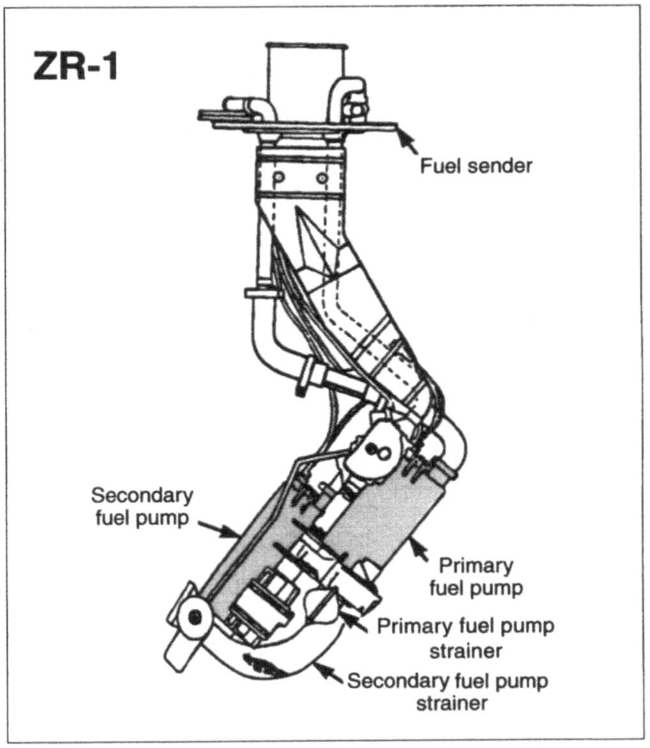

Fig. 7-7. ZR-1: The Fuel Sender Assembly includes two fuel pumps. The secondary fuel pump runs during *full* power operation as well as other conditions, (see page 155). **ZR-1** has no pulse damper.

7

Old-timers may remember when we siphoned gas from a fuel tank for another car that ran out of gas. We put in a hose and sucked to pull the fuel up to our mouth. When fuel filled the hose, if we were lucky, we squeezed our end of the hose and dropped it into the empty tank before we got a mouthful of gas. Ugh! Today, we know that gasoline is so toxic that hiking to the gas station might be better for us than such fuel siphoning. But it's OK for the car, itself, to siphon gas inside the tank.

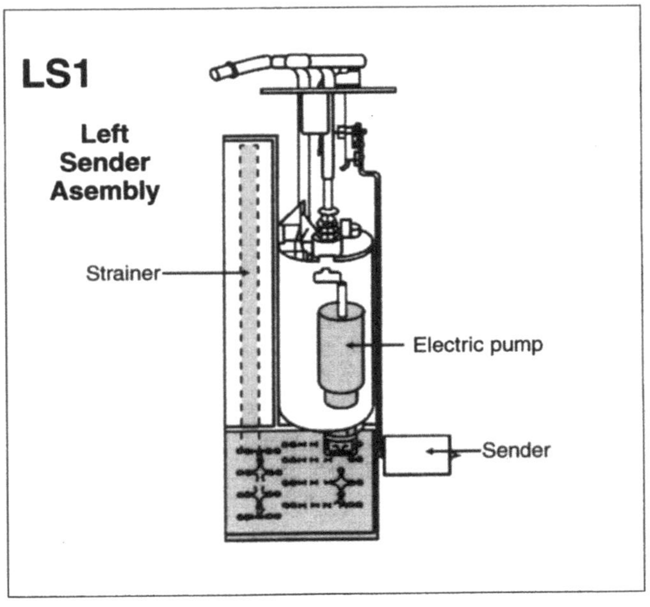

Fig. 7-8. LS1: The *Left* Fuel Sender Assembly's *electric* pump delivers fuel to the engine at a specified flow and pressure. Fuel delivery is at a constant rate even during aggressive maneuvers. The strainer is self-cleaning. The sender signals the amount of fuel in the *left* tank to the gauge.

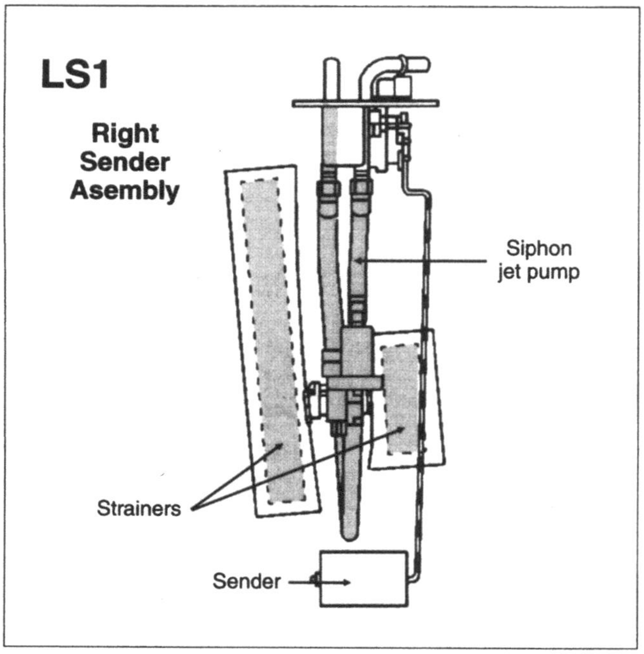

Fig. 7-9. LS1: The *Right* sender assembly's *siphon-jet* pump sends fuel to the left tank. Note the two strainers, which are also normally self-cleaning. The sender signals the amount of fuel to the gauge.

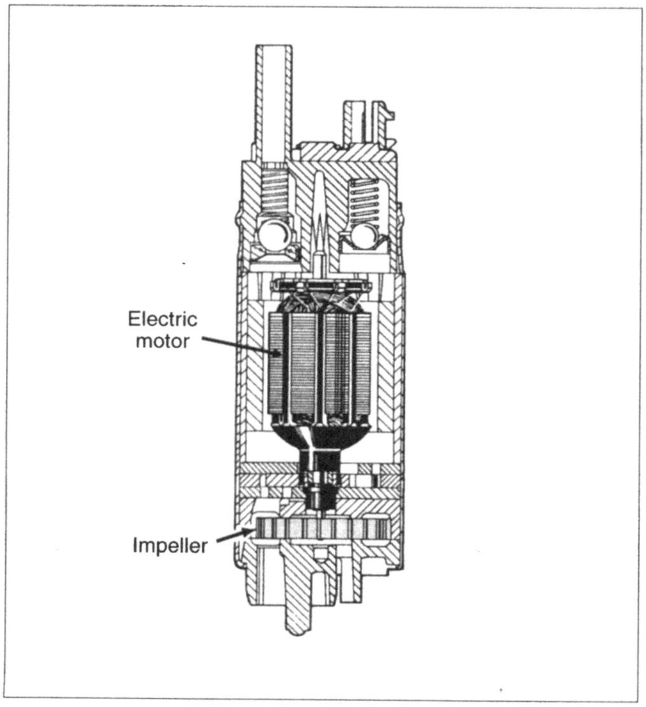

Fig. 7-10. Fuel pumps are known as roller-vane pumps. The motor operates in the gasoline in the pump housing. The **L83** is shown, others operate on a similar principle. The **LS1** uses an upgraded Gerotor pump.

2.1 In-line Fuel Filter

Although many cars specify periodic replacement of *in-line fuel filters*, Corvette has no specified service interval for replacement. If the fuel keeps flowing, then replacement is not necessary.

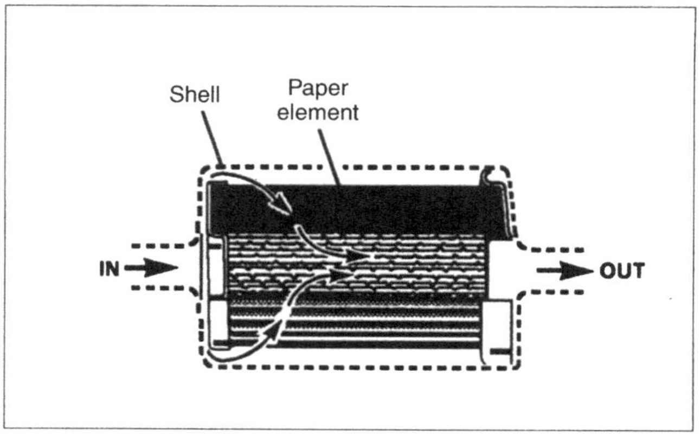

Fig. 7-11. The In-Line Fuel Filter is a simple steel shell with a paper element. The cleaner fuel your fuel is, the less often you need to replace the filter.

3. Fuel-pump Electrical Circuit

Fuel pump circuits are designed to keep the pump running. **L83**, **L98**, and **LT1** engines use an oil-pressure switch to back up the relay to limp home, if necessary, such as if the relay fails.

Note that the *computer supplies battery power* (B+) to the fuel pump with its engine ground. If this seems opposite to other computer circuits (those grounded in the computer), there's a safety reason. Consider a short to ground: the pump would operate even if the engine were not running. Whoops!. That kind of dumped fuel could spoil your whole day!

Engine Oil-pressure Switch

LT1, L98, L83: The oil-pressure switch can turn on the pump directly, bypassing a failed fuel-pump relay. But closing the oil-pressure switch may require long cranking times, especially with a cold engine.

3.1 Relays

LT1, LT4, L98, L83, LS1: Compare the relay shown in Fig. 7-12 with the two relays required by the two pumps in **ZR-1,** Fig. 7-14.

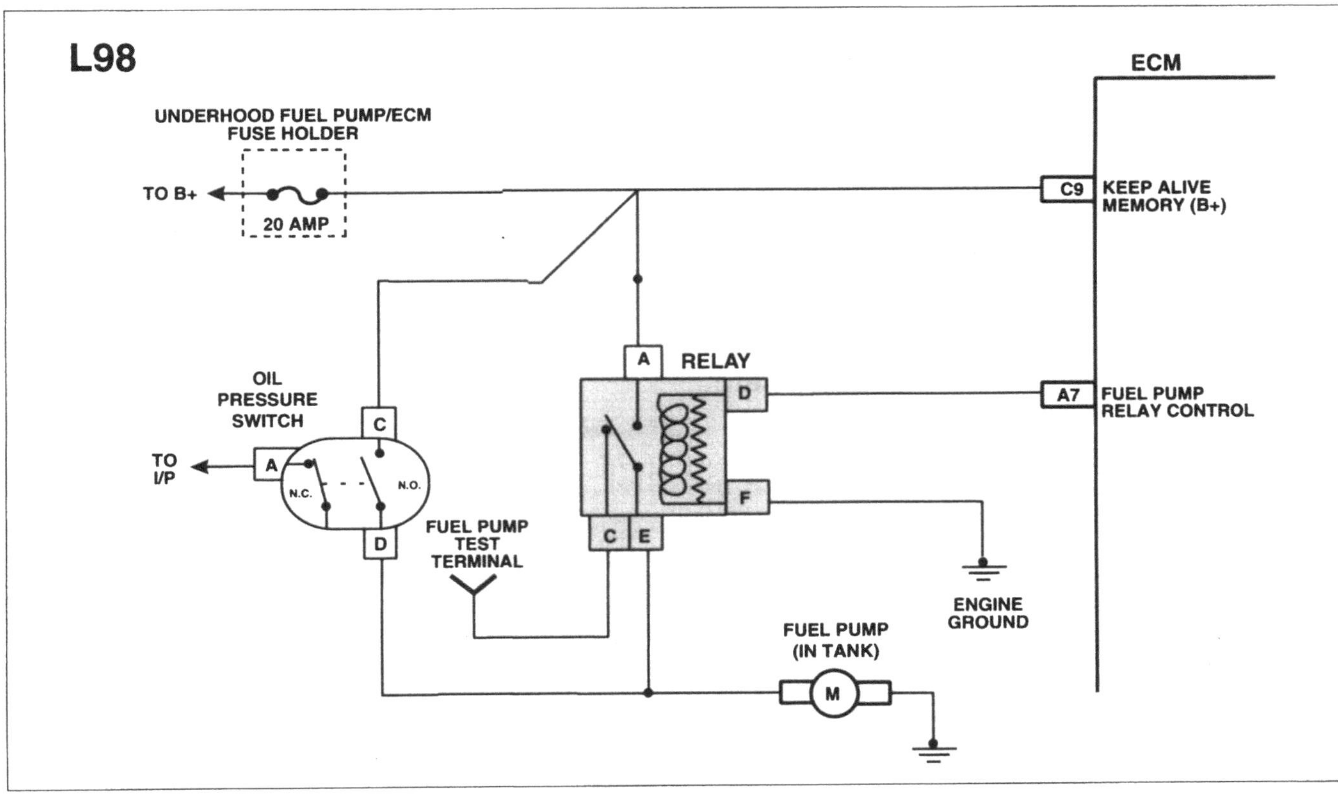

Fig. 7-12. L98 : The fuel pump relay is energized by the computer to power the fuel pump. The fuel pump is also energized by the oil-pressure switch, bypassing the relay. This keeps the fuel pump operating even if the relay or computer signal is lost. **LT1**, **L83** are similar.

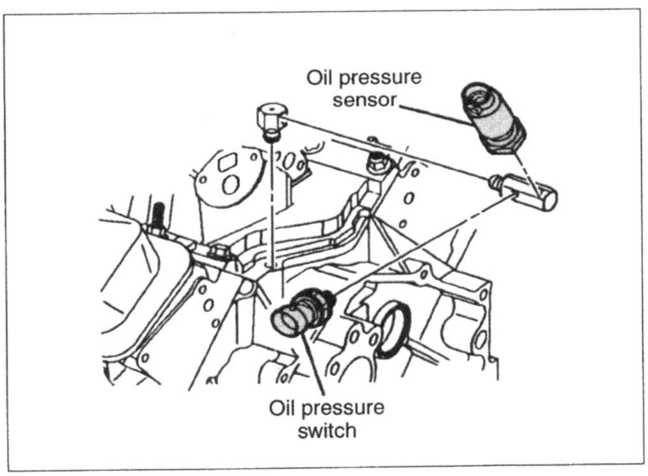

Fig. 7-13. LT1, L98, L83: The oil-pressure switch, in parallel circuit to the fuel-pump relay, closes after oil pressure builds to 28 kPa (4psi), indicating that the engine is turning over.

ZR-1

The *primary* pump relay operates similarly to **L83**, **L98**, **LT1**, **LT4**, and **LS1**. (See Fig. 7-14.) The pump is normally on, since the relay is open:

- any time reference pulses are being received
- during *crank*
- when the computer's voltage monitor indicates low voltage (less than 12 v. on circuit B-8 for 2 seconds after the ignition is turned on).

The *secondary* pump relay operates its pump only:

- during *crank* and *warm-up*—ECT and EOT are below 20° C (68° F)
- when the engine power is switched to *full*—TPS is above 80%, and ECT and EOT are below 150° C (300° F)
- when the rpm is below 4,000

The secondary fuel-pump relay shuts off its pump if:

- either HO_2S indicates lean (low voltage, less than 450 mv.) during *full* power, as at W.O.T.
- any DTC sets

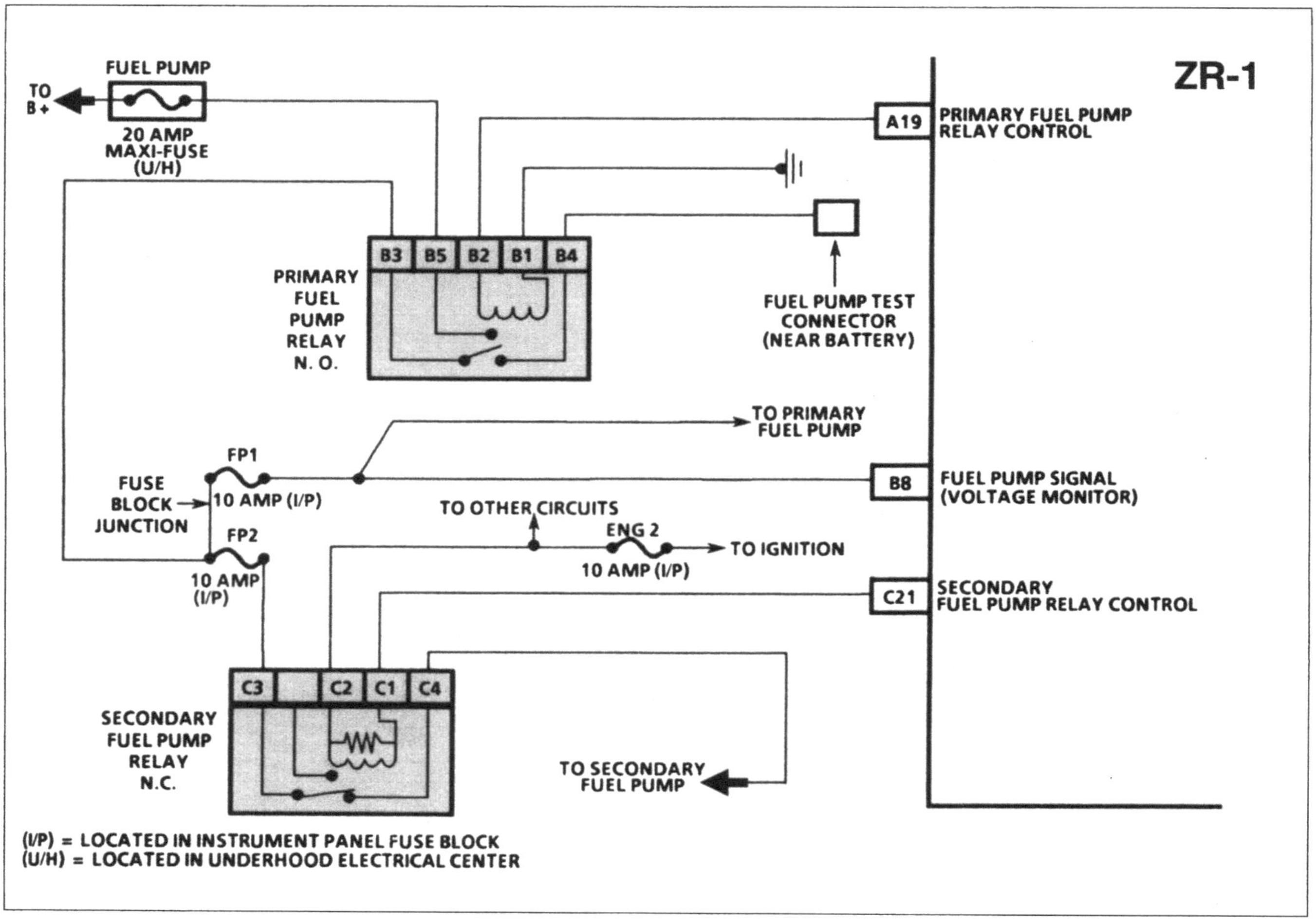

Fig. 7-14. ZR-1: The *primary* pump relay (normally *open*) compensates for low system voltage. The *secondary* fuel-pump relay (normally *closed*) insures pump operation for adequate fuel pressure during *full* power, and several other conditions.

4. Fuel Rail Assembly

Each fuel rail distributes fuel to the port injectors in that bank. Consider one important difference in fuel-rail assemblies:

L98, LT1, LT4, ZR-1, '97-98 LS1: Look for the Fuel Pressure Regulator (FPR) on the rail assembly. You'll find a fuel supply line and a *fuel return line.* (See Fig. 7-16)

LS1 '99-01: You'll find a fuel supply line but *no* fuel return line. The Fuel Pressure Regulator (FPR), combined with fuel filter, is near the left tank. (See Fig. 7-4)

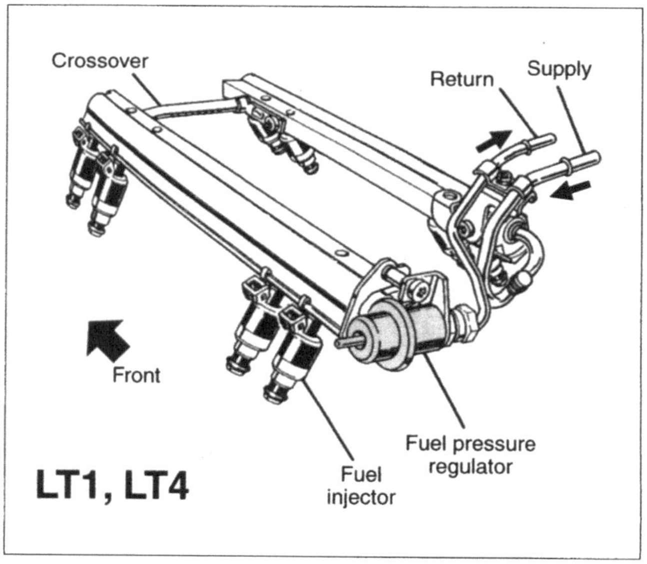

Fig. 7-15. LT1, LT4: The Fuel Rail Assembly includes a rail for each bank with a *crossover* tube, and the Fuel-Pressure Regulator has supply and return lines. Model-year details vary slightly. **L98** is similar.

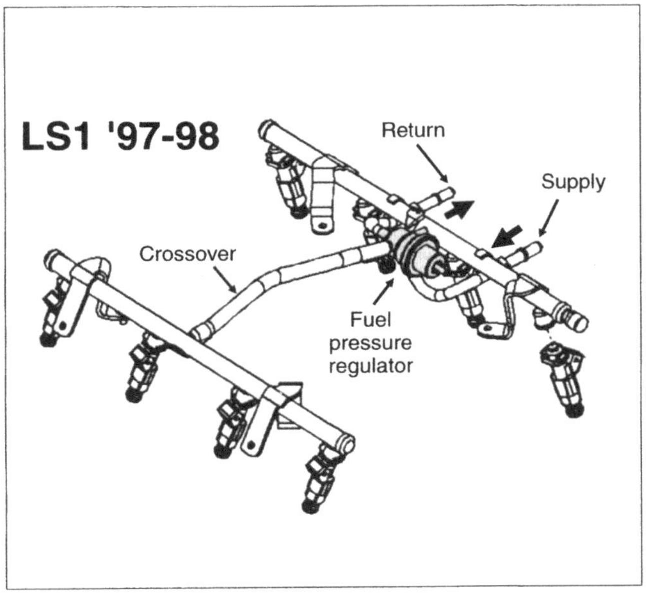

Fig. 7-16. '97-98 LS1: The Fuel Pressure Regulator has a central crossover, and supply and return lines.

LS1: The *symmetry* of the Fuel Rail Assembly layout tends to equalize the pressure at all of the injectors. The commanded injection is the same for all of the cylinders. This provides smoother combustion, reducing engine vibration.

'97-98 LS1: Fuel recirculates from the rail-mounted FPR rather than from the rail assembly as in previous systems. Fuel passing into the rails is completely consumed by the engine, which means that less fuel is heated in the engine. The reduced fuel temperatures in the lines and tank tend to reduce evaporative emissions.

'99-01 LS1: Fuel recirculation is eliminated from the rail assembly. Pressure is regulated near the tank so only one line is needed to feed the rails and injectors. All fuel delivered to the engine is consumed, there is no return. (See Fig. 7-21.)

4.1 Fuel-pressure Regulation

The Fuel Pressure Regulator is important for the precise metering of fuel from the injectors.

- If the fuel pressure is too low, the resulting lean mixture may cause power loss, surging, or misfire.
- If the fuel pressure is too high, the resulting rich mixture may cause hard starting, black exhaust smoke, and an overheated catalytic converter.

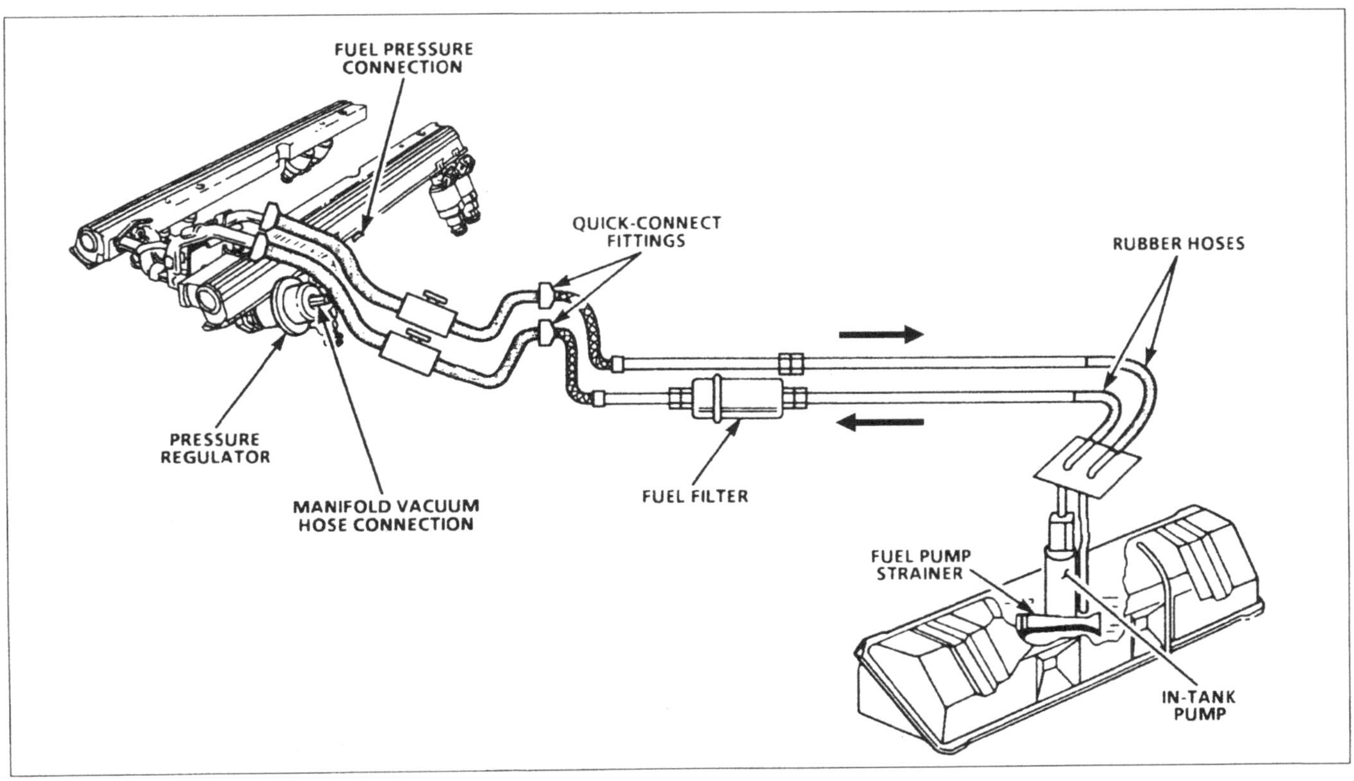

Fig. 7-17. LT1, LT4: The pump in the fuel tank supplies fuel to the fuel rails through the Fuel Pressure Regulator.

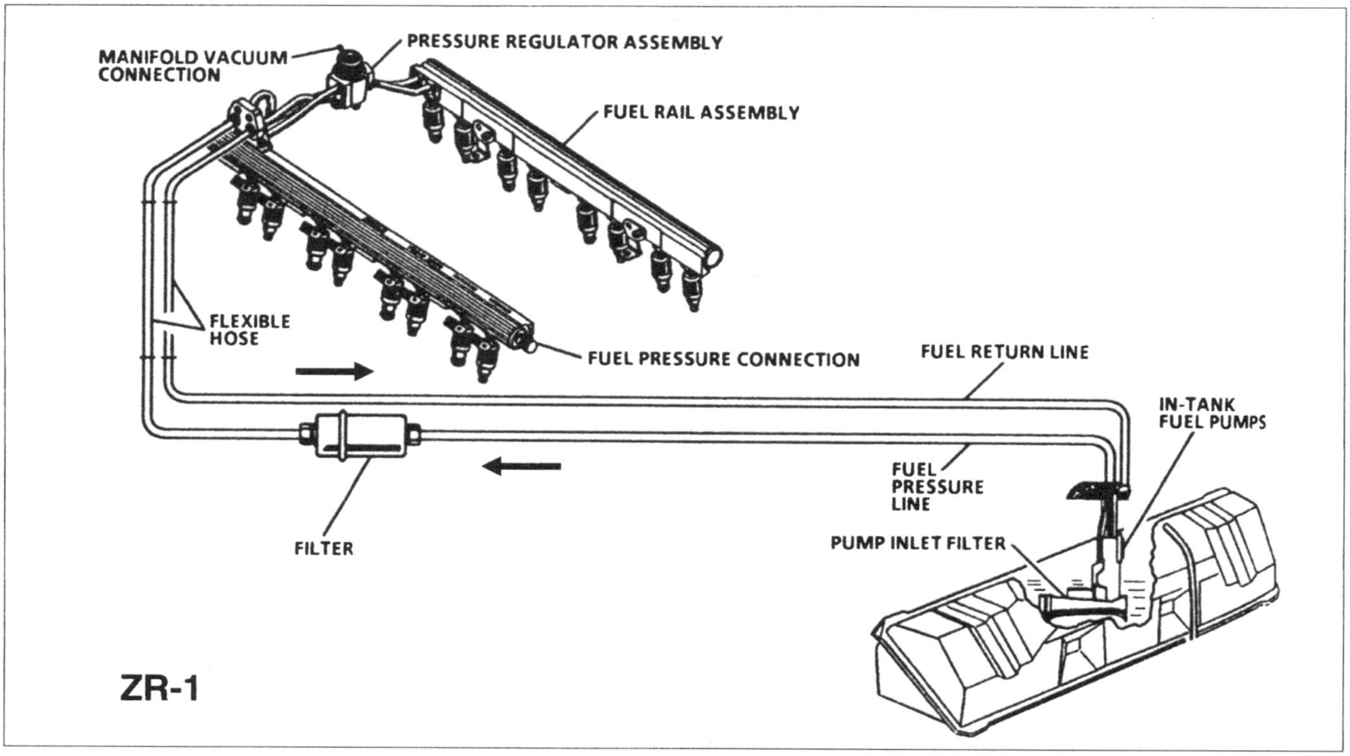

Fig. 7-18. ZR-1: Each fuel rail supplies four primary and four secondary injectors. The added second pump delivers fuel into both rails under certain conditions, such as *full* power. **LT1** and **LT4** are similar.

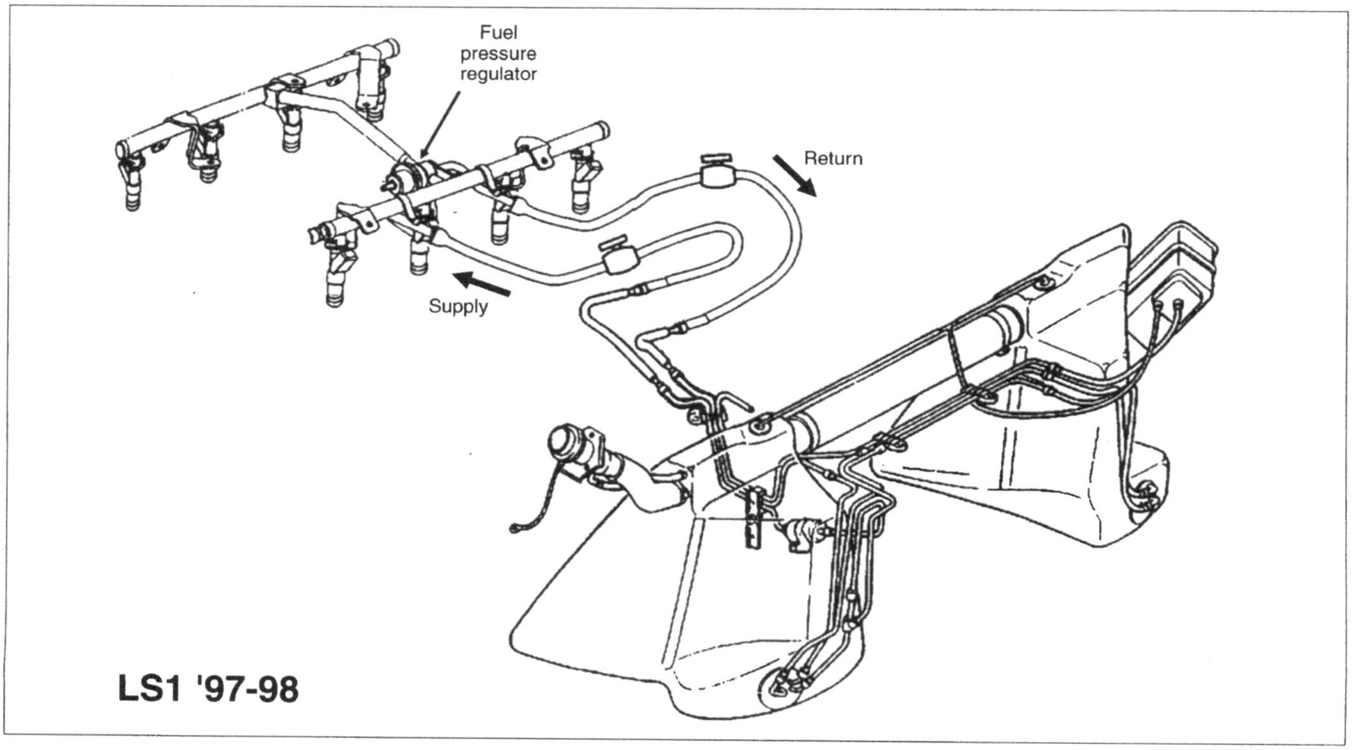

Fig. 7-19. '97-98 LS1: Fuel that passes into the two fuel rails is injected and burned. Recirculation is direct from regulator, not from fuel rails.

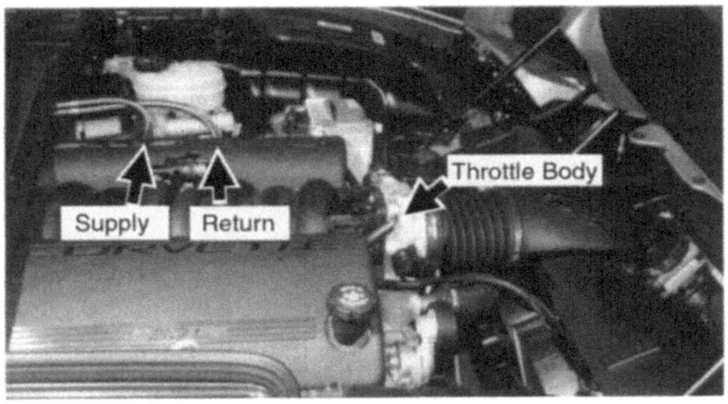

Fig. 7-20. '97-98 LS1: There are *two* fuel lines, one for supply and one for return. Compare to the **'99-02 LS1**.

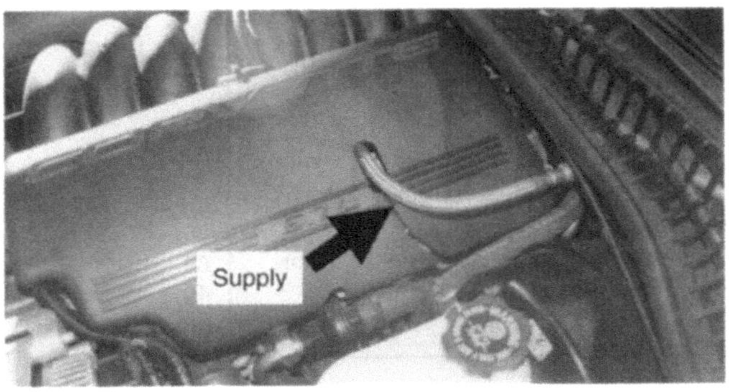

Fig. 7-21. '99-02 LS1: A *single* line delivers fuel to the rails, on the engine, from the Fuel Pressure Regulator/filter assembly, located near the left fuel tank.

Recall from Chap. 3, that the effect of improper mixtures on the functioning of the catalytic converter and on emissions: too lean—higher NO_X; too rich—higher HC and CO. In Chap. 9, "Tuning for Performance", you'll see that fuel-pressure regulation is a major consideration.

> It's an old tale that increased fuel pressure automatically means increased performance. *It may mean less.* Experienced tuners generally consider increasing fuel pressures only if other mods increase airflow.

Relative Fuel Pressure

Start with the idea that the system must increase the amount of fuel injected as you open the throttle, increasing the load and the manifold pressure. As manifold pressure

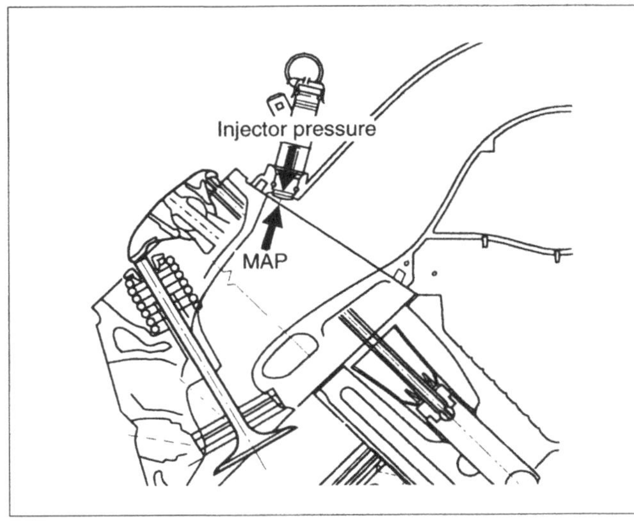

Fig. 7-22. *Relative* fuel pressure is the difference between the injector pressure pushing fuel *in*, and the manifold pressure pushing *out*, resisting delivery.

increases, injecting the right amount of fuel can be controlled in two ways:

7

- Method #1. Maintain constant *relative* fuel pressure by changing *absolute* fuel pressure. This is the method you're familiar with in **L98**, **LT1**, **LT4**, **ZR-1** and most other fuel injection systems. (See Fig. 7-23.)
- Method #2. Maintain constant *absolute* fuel pressure. Change the injector *pulse-time*, the computer increases the pulse time, relative to the throttle opening. Look for this in **LS1**. (See Fig. 7-24.)

L83 TBI injects above the throttles and is not affected by changing manifold pressures. So I'll concentrate on *port-injection* fuel pressures.

Comparison: Changing Absolute Fuel Pressure/Changing Pulse Time

For the following comparison, I'll use similar fuel pressures to illustrate the principles. Even though we both know that the two systems operate at different pressures, the principles are the same.

4.2 Fuel Pressure Regulator

The job of the Fuel Pressure Regulator (FPR) is to insure that fuel delivery per injection is not affected by changes in manifold pressure. When you understand this principle of *relative* fuel pressure (injection pressure versus manifold pressure) you will understand the checking of fuel pressures. You'll also understand how greater delivery for off-road performance operation may be increased by increasing fuel pressures, or it may be *decreased*.

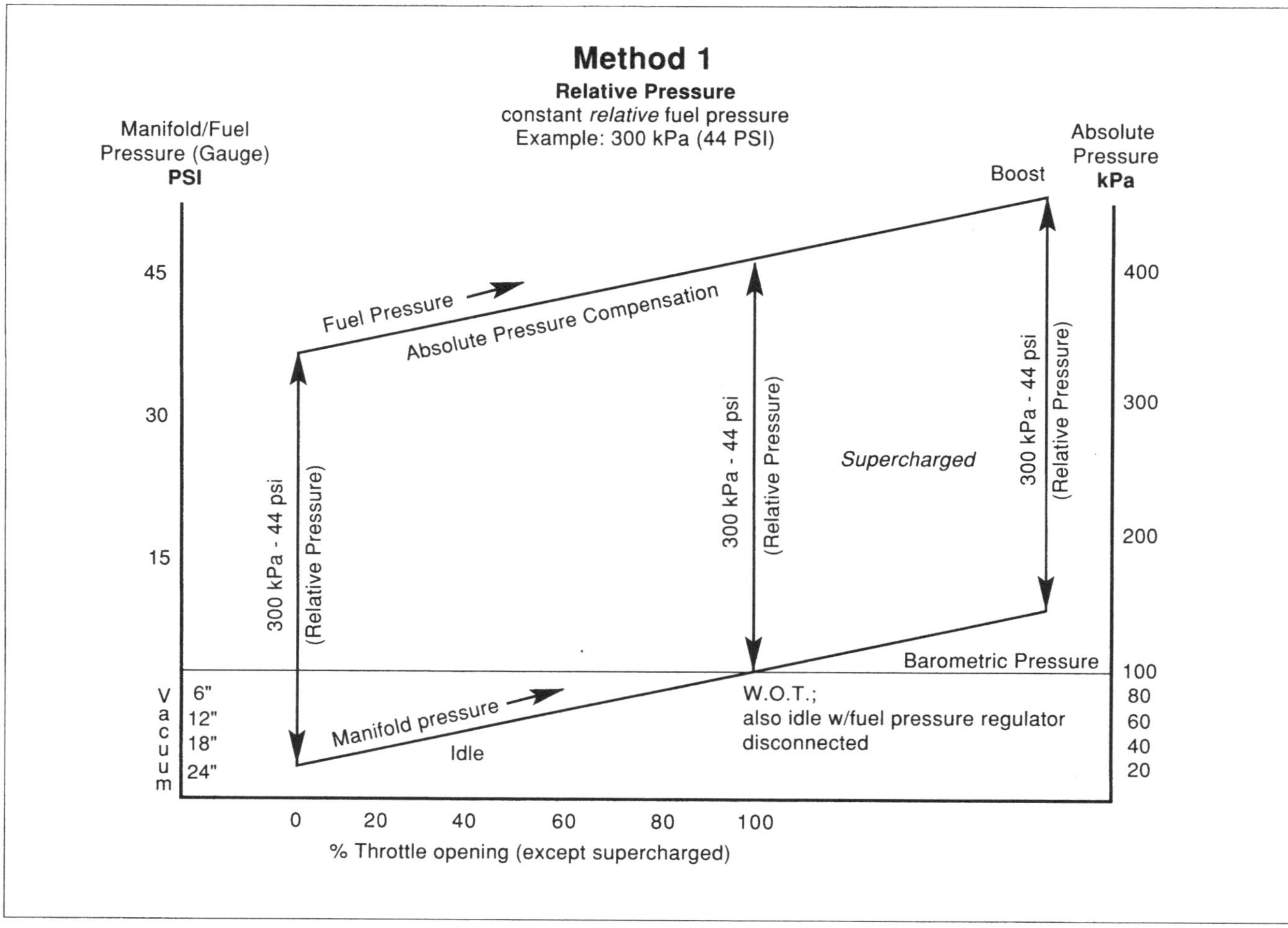

Fig. 7-23. L98, LT1, LT4, ZR-1: The Fuel Pressure Regulator (FPR) increases *absolute fuel pressure* as *manifold absolute pressure* increases with throttle opening. Notice that the *relative* pressure is constant, so the amount of delivery changes with the absolute fuel pressure.

L98, LT1, LT4, ZR-1: To keep relative pressure constant as the throttle is opened and closed, the FPR is connected to the intake manifold. Manifold pressure acts on the diaphragm to hold the relative pressure constant.

- When the relative fuel pressure is too high, fuel pressure resists the combination of spring pressure and Manifold Absolute Pressure. The fuel pressure raises the diaphragm, allowing fuel to flow through the return line to the tank.
- When the relative fuel pressure matches the combination of spring pressure and Manifold Absolute Pressure, the spring and relative fuel pressure press down on the diaphragm so that no fuel returns to the tank. (See Fig. 7-26.)

When the engine is off, spring pressure keeps the regulator valve closed, maintaining pressure in the system. **L98, LT1, LT4, ZR-1:** If the relative fuel pressure were constant for all manifold pressures, then at low engine loads, with the throttle partly closed, reduced manifold absolute pressure would increase fuel delivery. *Note*: This does not apply to **LS1**, where changing the MAP does not change rail pressures but causes computer-compensation of pulse times.

Think positive. Keep in mind the difference between barometric gauge pressure and absolute pressure, usually about 100 kPa at or near sea level. Your fuel pressure gauge reads *gauge* pressure—that is, fuel pressure *above barometric.*

LS1: Pulse-time changes compensate electronically much faster than *mechanical* FPR. Pulse-time changes are especially valuable during *transients*, as during acceleration, MT shifting, and deceleration to reduce emissions.

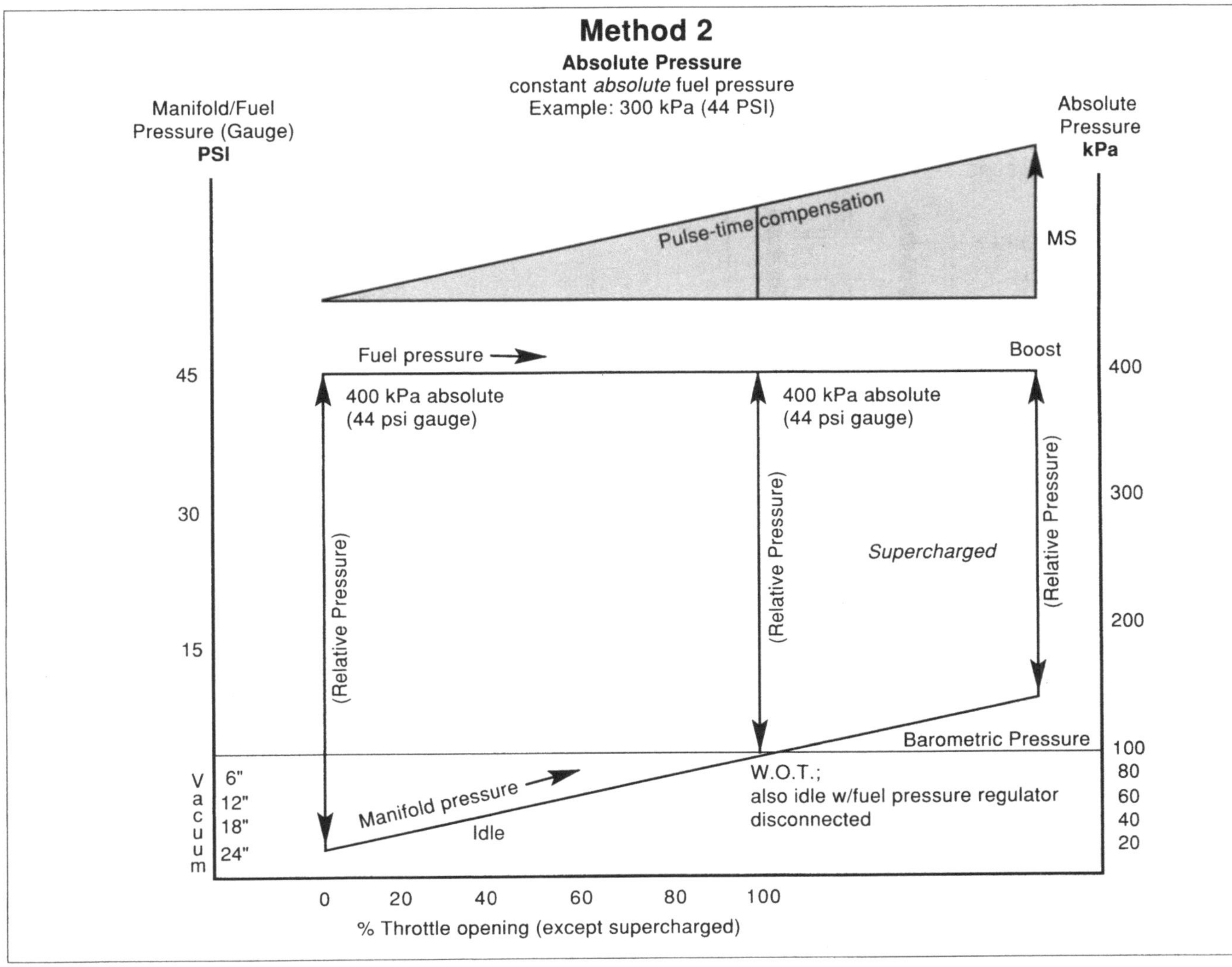

Fig. 7-24. LS1: The FPR keeps *absolute* fuel pressure constant. As throttle opening changes manifold pressure, the computer compensates by *adding pulse-time* (shaded area).

L98, LT1, LT4, ZR-1: The Fuel Pressure Regulator maintains constant *relative* fuel pressure, controlling injector pressure against manifold pressure.

- Insufficient fuel pressure that drops off during hard cornering or acceleration may cause a lean mixture. Results: loss of power, surging or misfire.
- Excess fuel pressure may cause rich mixtures that could cause hard starting, and show up as black exhaust smoke or exhaust odors.

As you have seen, for each millisecond of injector pulse time, the amount of fuel delivered through the injector tip depends on the relative pressure.

'99-01 LS1: The pulse damper tends to smooth out any variations in the rails as the injectors open and close. Fuel pressure variations are dampened by the spring.

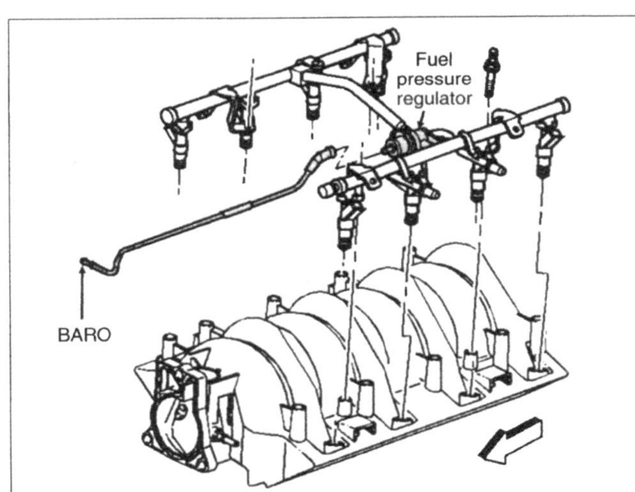

Fig. 7-25. '97-98 LS1: The pressure regulator line senses *barometric* pressure, not manifold pressure. *Note*: the BARO line senses barometric pressure ahead of the radiator rather than engine compartment pressure itself.

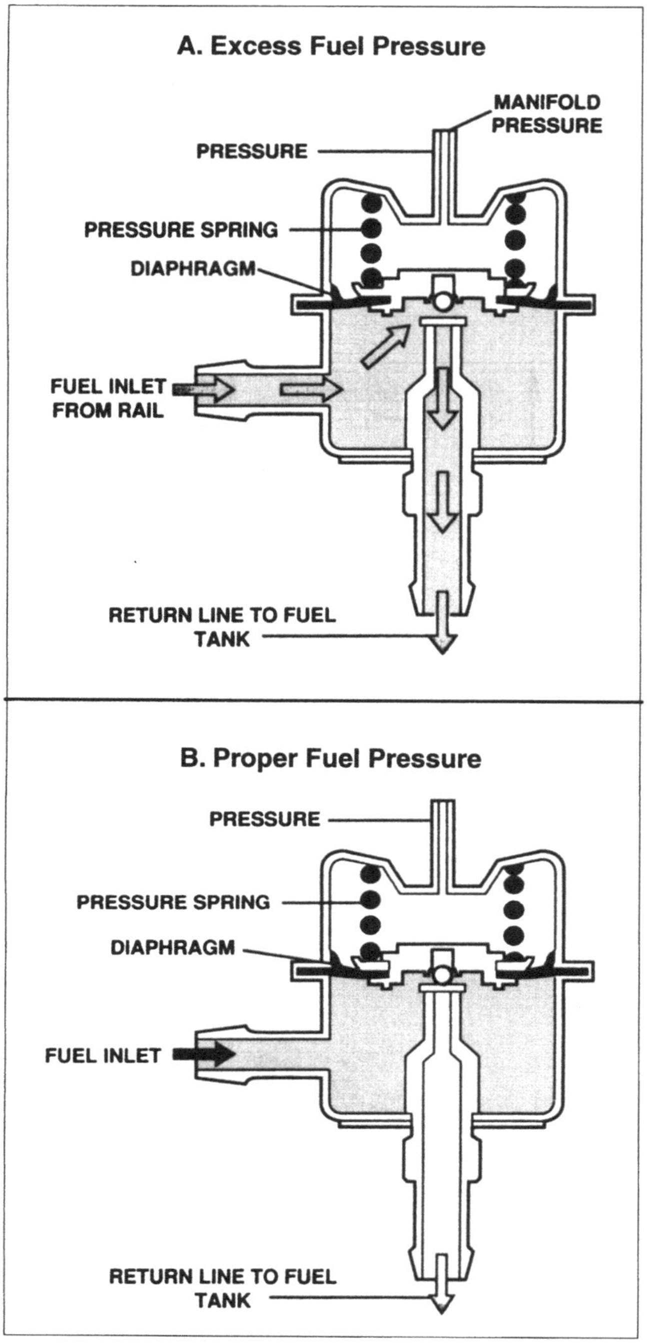

Fig. 7-26. L98, LT1, LT4, ZR-1:
A. When the relative fuel pressure is too high for the manifold pressure, the diaphragm rises to return fuel to tank.
B. When relative fuel pressure matches the manifold pressure, the diaphragm shuts off the fuel return.

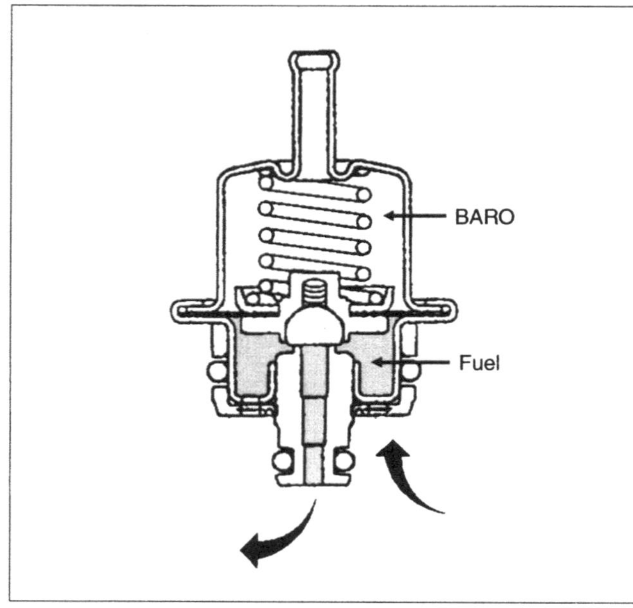

Fig. 7-27. LS1, '97-98: The Fuel Pressure Regulator (FPR) has an opening, above the spring, which is connected by a hose to the atmospheric pressure ahead of the radiator. This is equivalent to BARO. The fuel pressure does *not* change with the changing manifold pressure.

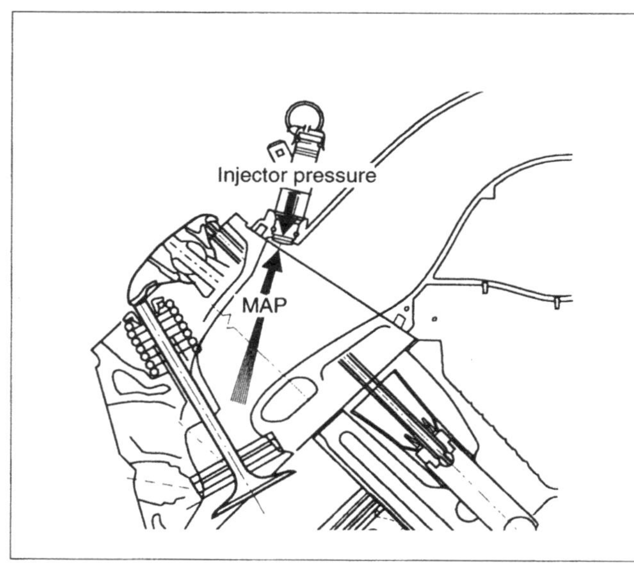

Fig. 7-28. LS1: The Manifold Absolute Pressure sensor signals the computer to change the *pulse-time* to handle MAP changes. The absolute pressure remains constant. (See Fig 7-24.)

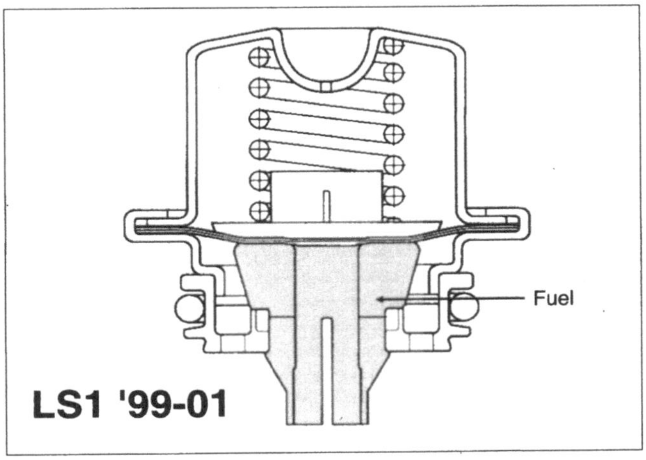

Fig. 7-29. '99-02 LS1: The *Pulse damper* is located on the rail. It looks like a Fuel Pressure Regulator but has a closed top. Fuel pressure variations are dampened by the spring.

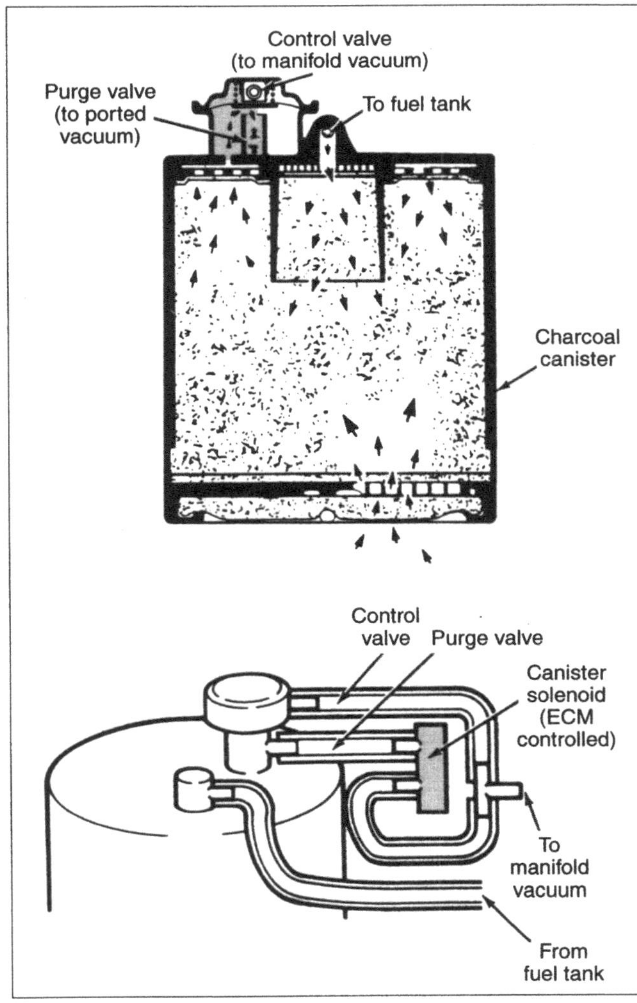

Fig. 7-30. L98: Atop the underhood canister, the computer-controlled solenoid control valve is normally OFF. Under specified conditions, it is open to admit vacuum to the purge valve. This purges the canister by delivering vapors to the intake manifold. Purge is prevented under conditions of low ported vacuum, such as deceleration.

5. EVAP SYSTEMS

Evaporative Fuel Vapor

Unburned hydrocarbons that evaporate from the fuel system can pollute the atmosphere unless they are contained. *Evaporated hydrocarbons may add up to 40% of the emitted polluting HC.* While the engine is not running, the EVAP canister contains fuel vapors that form over the liquid fuel in the tank. The tank must not be sealed, or the fuel pump could not draw fuel when the engine runs. Fuel vapor is purged from the canister under the following conditions (this is **L98** specific, others are similar):

- engine warm, ECT above 60° C (140° F)
- more than 1 minute after engine start
- above specified road speed, 2 mph
- above specified throttle opening, above 3%, off idle
- cruise, acceleration, but not deceleration or idle

The canister is filled with activated charcoal which absorbs and stores the vapors until they are drawn into the engine. The canister is open at the bottom. The fuel vapors, which are lighter than air, tend to rise to the top of the canister.

LT1, LT4, ZR-1: The Tank Pressure Control Valve closes off the fuel tank until fuel vapor pressure rises above 5 kPa (0.7 psi). Under this pressure, the valve opens, allowing vapor to flow to the canister.

LT1, LT4, ZR-1: Fuel vapors are drawn into the engine but only under certain engine conditions similar to **L98**: the engine is warm, after running a specified time and above a certain road speed and throttle opening—that is, when the engine can draw in and burn the vapors.

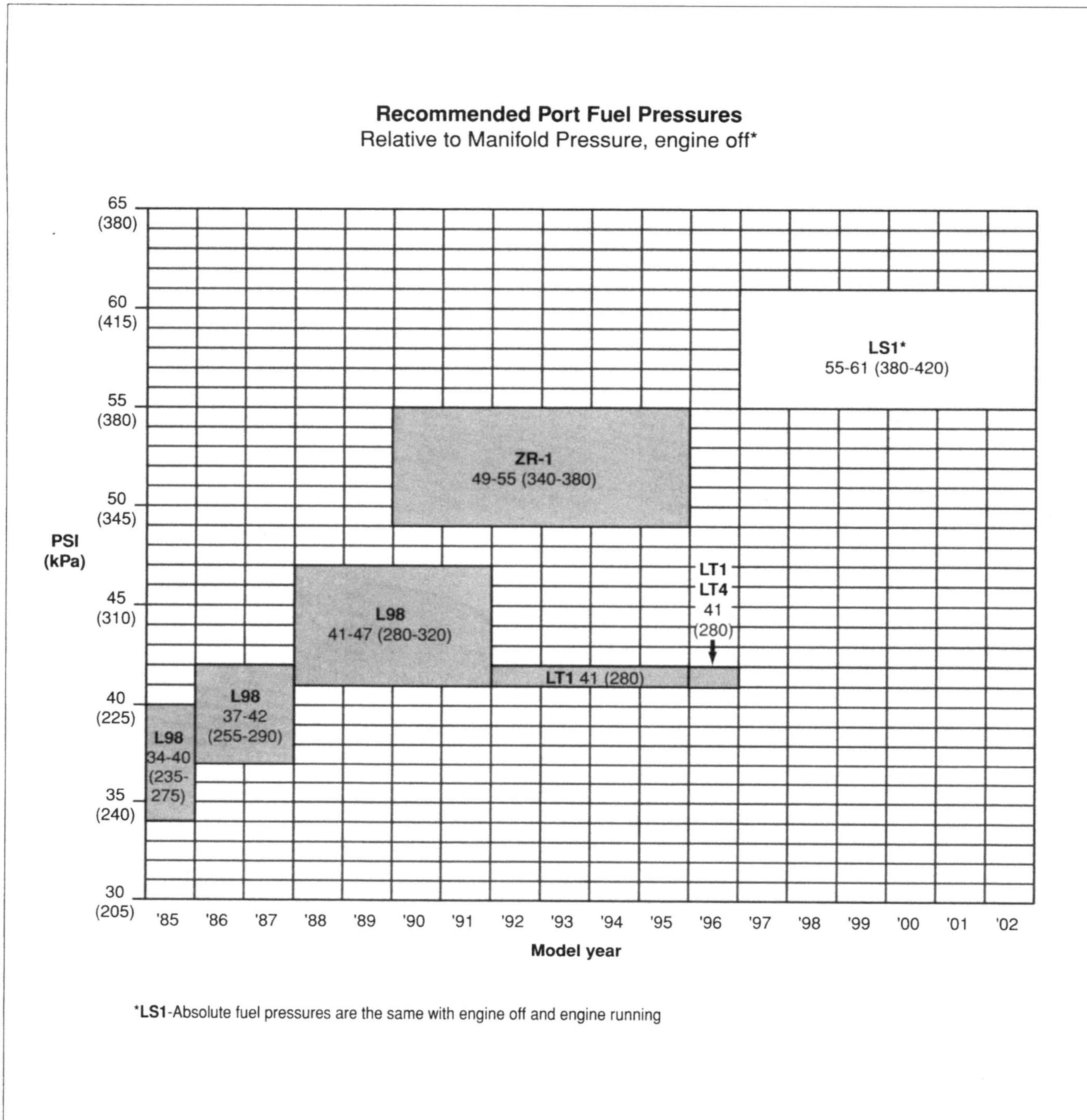

Fig. 7-31. Recommended port fuel pressure ranges tend to increase with later model years.
•**LT1, LT4** have a single value rather than a range.
•Most are relative; **LS1** is absolute.
Most fuel pressures are specified as a *range*–anywhere in here is ok. **LT1** and **LT4** specify a *single value* because these engines are sensitive to fuel pressure.

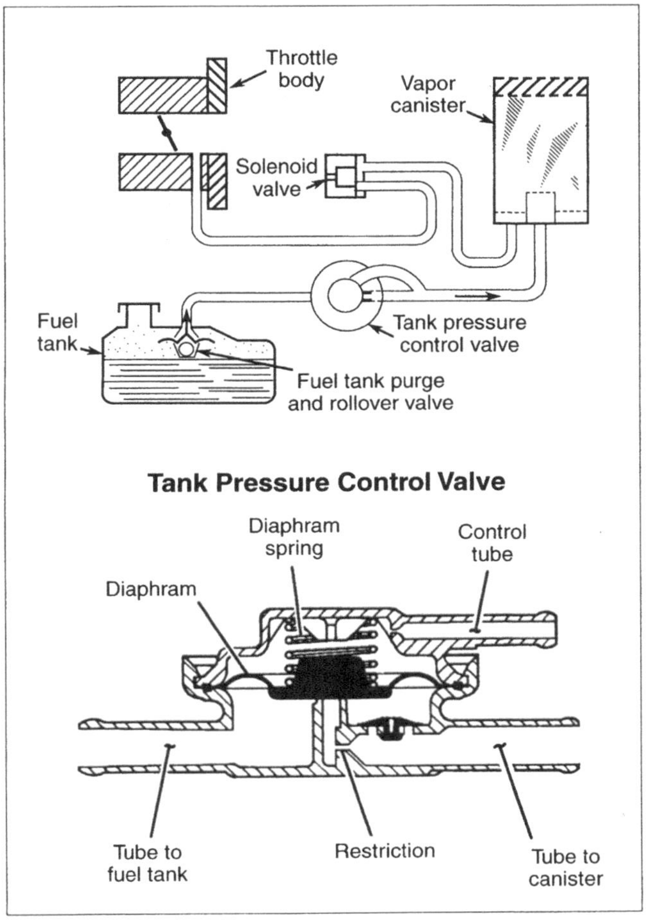

Fig. 7-32. LT1, LT4, ZR-1: When the tank pressure rises above 5 kPa (0.7psi) the Control Valve opens to allow vapors to be drawn into the canister. The internal restriction relieves pressure slowly, preventing a buzzing sound. The canister stores fuel vapors while the engine is not running.

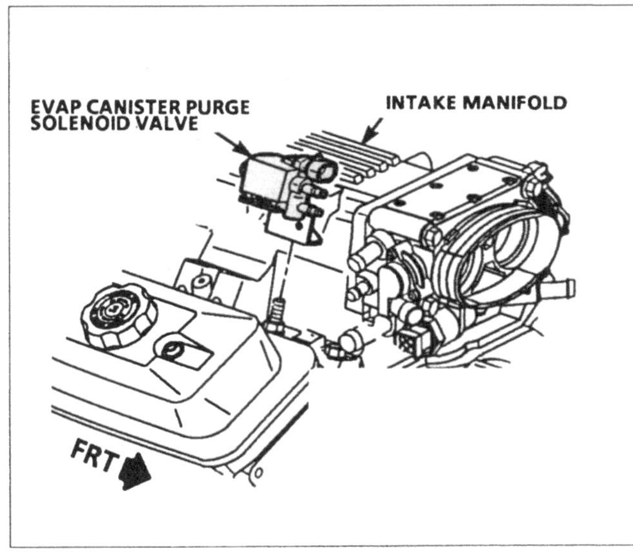

Fig. 7-33. LT1: The canister purge solenoid valve is located on the intake manifold. The valve feeds vapor from the rear-mounted canister to the manifold. **LT4** and **ZR-1** are similar.

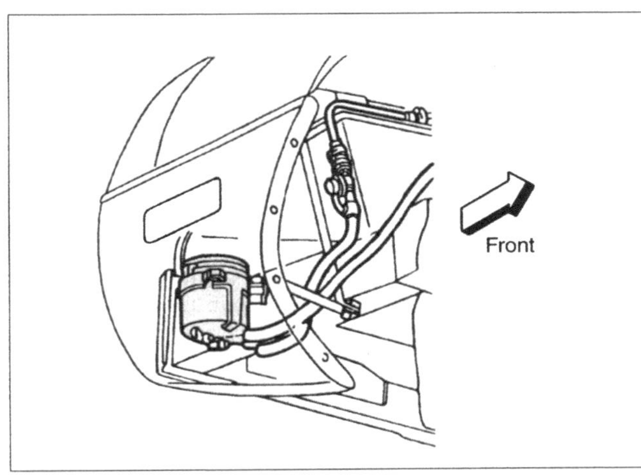

Fig. 7-34. LT1, LT4, ZR-1: Look for the cylindrical canister near the fuel tank, behind the right rear wheel. The canister is purged by the canister-purge valve in the engine compartment. The **LS1**'s canister is also in in rear, but it is a rectangular, instead.

7

Enhanced EVAP

EVAP systems, including fuel-filler caps, have come under close scrutiny in newer emission-control programs. Recent studies show that a damaged filler cap may release HC to the air, almost as much as HC as from the tailpipe of a well-regulated car. Leaks in EVAP canisters, hoses or connectors are also sources of HC pollution.

- **'99-01 LS1** use Enhanced EVAP to control EVAP system vapors under requirements of OBD-II.
- All other Corvettes use a simpler system, now known as Non-Enhanced EVAP.

LS1, '99-01: Enhanced EVAP operates a much tighter system necessary to meet *OBD-II's* tighter limits of fuel vapor for HC pollution. Computer diagnostic controls can momentarily seal the system and detect leakage, setting a code if excessive. Enhanced EVAP system controls vapors with computer management of a Pulse-Width Management (PWM) valve. The canister is rectangular and larger, 1850 cc (~2 quarts).

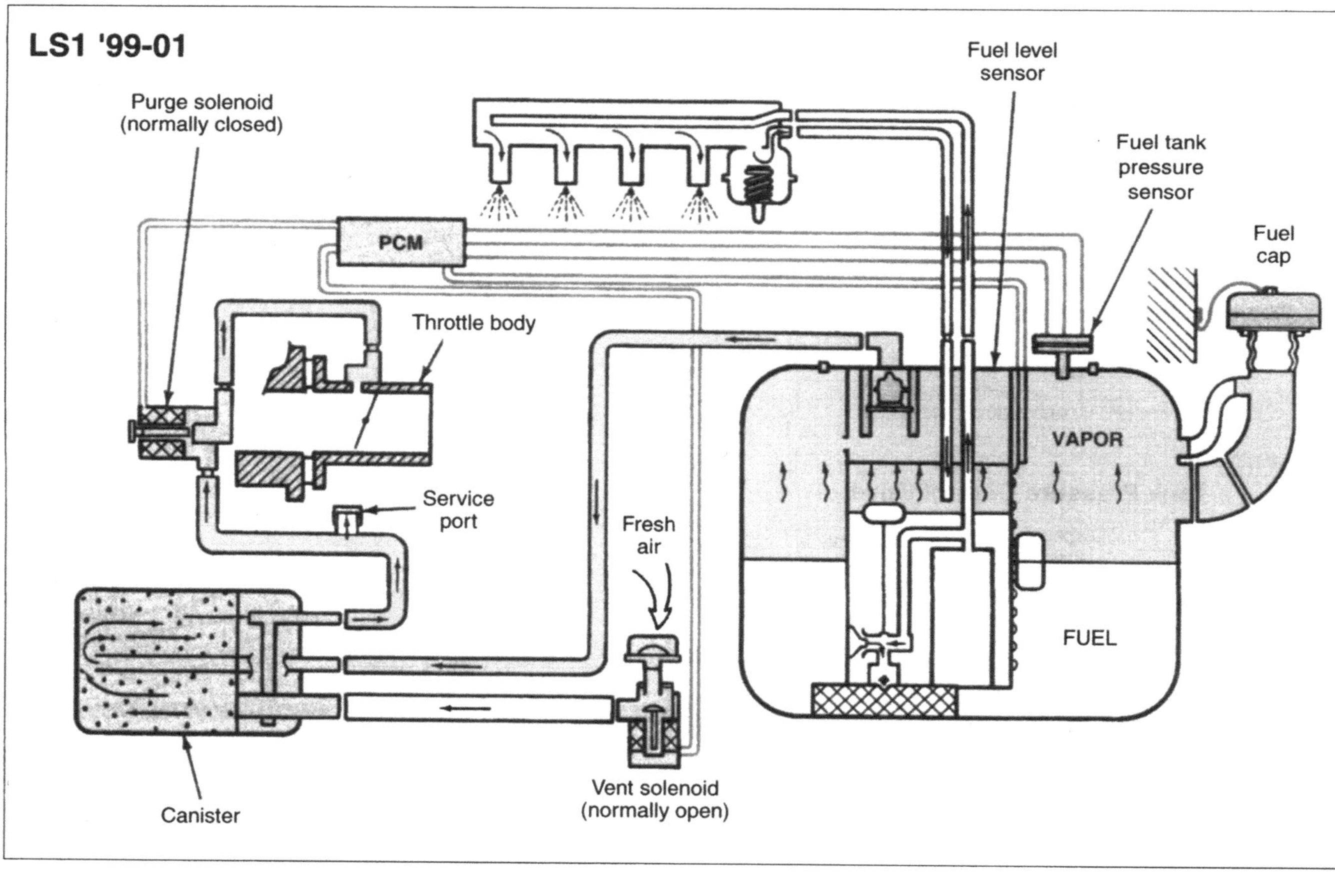

Fig. 7-35. LS1 '99-01: This *enhanced* EVAP system meets the requirements for OBD-II. The fuel-tank pressure-sensor, purge switch, purge solenoid, and vent solenoid are new.

The **LS1** controls the EVAP flow *to* the canister, and monitors the purging *from* the canister, as well as computer-controlled diagnostics during driving. (See Chap. 10.) When the purge valve is closed, the vacuum purge switch signals that purge has stopped.

The purge valve is powered whenever the ignition relay is energized, and is grounded in the computer to purge the canister. After 25% PWM of the purge valve, the EVAP Vacuum Purge Switch opens, signaling the computer that the purge system is operating.

In the **'99-01 LS1** the Fuel Tank Pressure Sensor (FTPS) momentarily checks the integrity of the fuel system. During this checking for leaks, the computer:

1. opens the purge valve
2. closes the vent valve, creating a vacuum in the tank as fuel is used
3. closes the purge valve to seal the tank

Then, as part of OBD-II self-diagnostics, the computer looks at *rate of change* of air pressure in the EVAP system. This

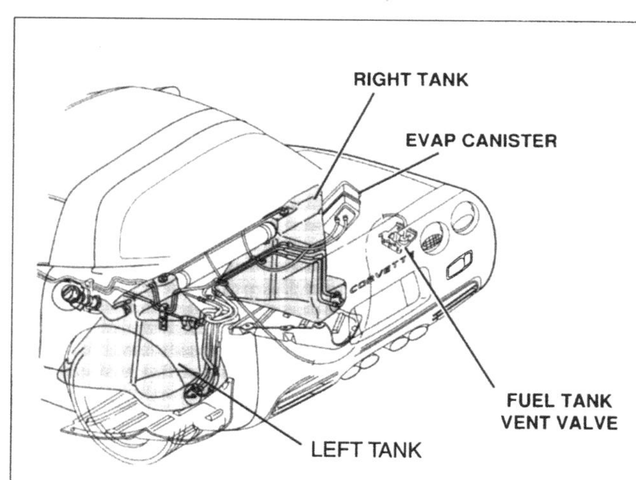

Fig. 7-36. The EVAP canister is near the right tank. The vent valve can be momentarily closed by the computer to seal the tank for diagnosis.

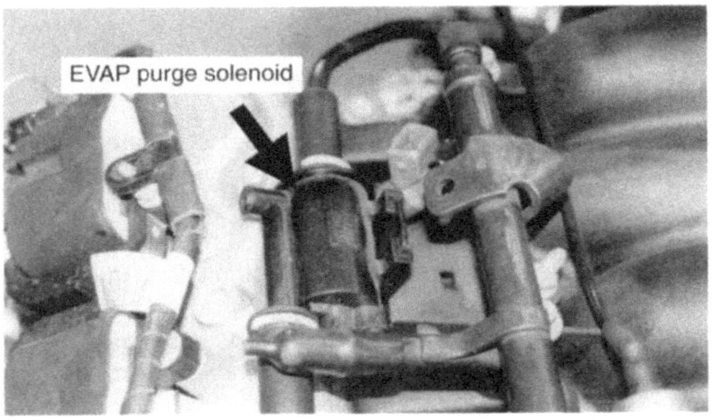

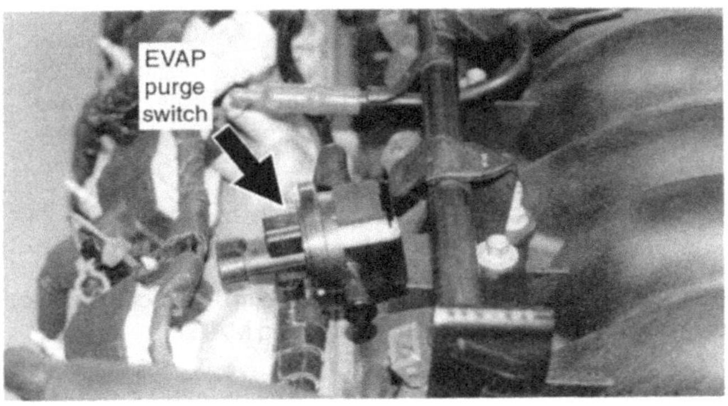

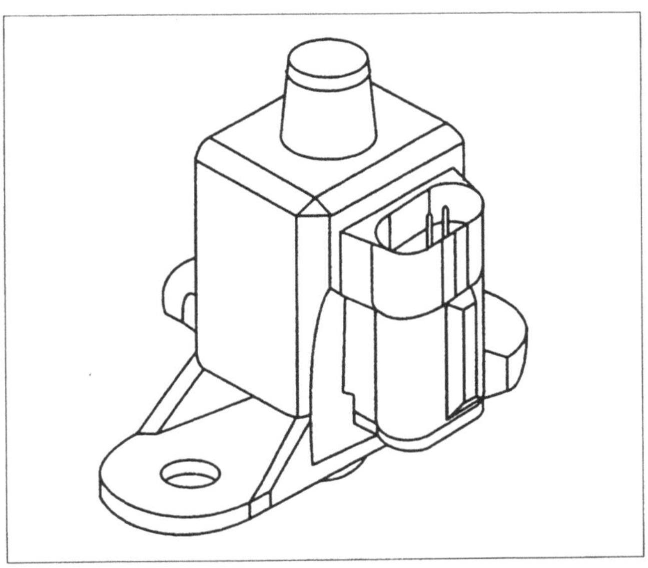

Fig. 7-37. '99-01 LS1: The purge solenoid (on the manifold) controls the purge valve (on the canister) allowing vapor to flow from the canister to the engine. Pulse-Width-Modulated (PWM) flow provides precise control. During OBD-II diagnostic tests, the computer opens the purge valve to create a vacuum in the tank, and then closes it to seal the system.

diagnostic will only run when the tank is between 15% and 85% full. Fuel level affects the volume of vapor and therefore this rate of change.

Normally closed, the purge valve is opened only under all of these conditions:

- Delayed after engine start, 150 seconds when cold—ECT below 50° C (120° F); 45 seconds when cool/warm—ECT above 50° C
- TPS greater than 3 degrees, off idle
- VSS greater than 10 mph (16 km/hr)

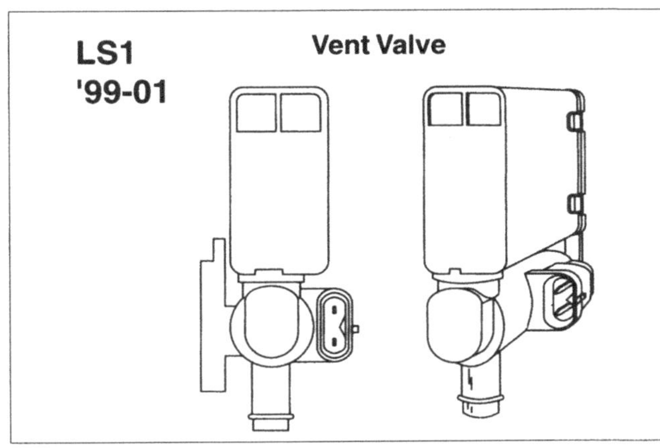

Fig. 7-38. '99-01 LS1: The vent valve near the canister allows fresh outside air into the canister during purge, similar to the fresh air vent of previous systems. During diagnostics, the computer closes the vent valve to briefly create a vacuum in the fuel tanks as fuel is pumped out.

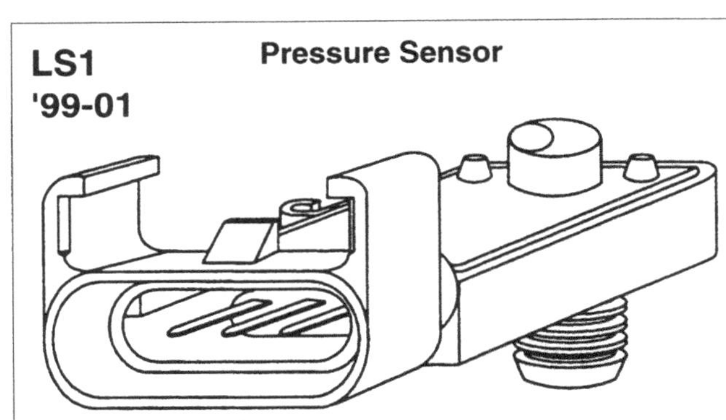

Fig. 7-39. '99-01 LS1: Look for the Fuel Tank Pressure Sensor (FTPS) on top of the left fuel-tank sender unit. The FTPS measures the difference between the pressure/vacuum *in*side the tank and the air pressure *out*side, using a three-wire strain gauge, much like a MAP sensor.

Onboard Refueling Vapor Recovery (ORVR)

'99-01 LS1: Onboard Refueling Vapor Recovery (ORVR) reduces escape of vapors during refueling. Of course, you have smelled those gasoline vapors while you stood over the refueling process. None of us wants to breathe those hydrocarbon (HC) vapors; none of us should want to add HC to the atmosphere to form more smog. With tighter limits, *refueling HC* must be controlled. Onboard RVR systems are better than the refueling vapor system built into the gas station filler nozzles, as in California and many other states, and much better than those states which permit gas stations to use filler nozzles with no vapor recovery.

OF FURTHER INTEREST—
Before the **LS1** the location of the fuel filler on top of the rear deck tends to carry the refueling vapors up and away into the atmosphere. In the **'97-98 LS1**, the filler is just behind the driver's door, you can get a good dose in your nose.

The fuel-filler hole is small, just big enough to take the smaller nozzles required since the end of leaded fuels. See #3 below.

1. Insert the nozzle.
2. The breather valve will open the EVAP canister refueling vent
3. Begin fueling. The venturi filler causes a gasoline seal around the nozzle, trapping vapors.
4. The tank fuel level rises, forcing vapors into the EVAP canister
5. As fuel level rises in filler breather bottle, the float valve closes the refueling vent valve to the canister
6. Pressure causes fuel to rise in the filler neck.
7. Rising fuel causes the gas station pump to shutoff
8. Remove the nozzle.
9. The breather valve closes the EVAP canister refueling vent
10. The breather valve opens a smaller EVAP vent line.
11. The remaining vapors flow to the canister.

They really don't want those refueling fuel vapors released into the atmosphere. You don't want them up your nose.

Umpteen years ago, it was determined that too much HydroCarbon was being released into the air during refueling. Remember the signs, "Don't smoke during refueling." That's because there's a lot of inflammable vapor in the air from refueling. How should the HC be controlled?

- The *gas-station* people said, "Control refueling HC vapors in the cars. It will cost too much to change all the pumps."
- The *automakers* said, "Control HC at the gas stations. We're talking millions of cars, far more than the number of gas pumps."

California decided, "Control vapors at the gas pumps." But the double hoses were bulky and hard for owners to handle (most stations were "self-service"). Few states followed. Actual smog measurements showed that we still have problems controlling HC. So, with Onboard Refueling Vapor Recovery, we're moving toward controlling refueling vapors at *both pump* and *car.*

6. SUMMARY

Corvette port systems operate at high pressures compared to carburetors.

L98, LT1, LT4, ZR-1: Engine-running *absolute* fuel pressures are mechanically-controlled by the Fuel Pressure Regulator. Fuel pressures are modified by FPR to maintain a constant *relative* fuel pressure at the injector tips to compensate for changing MAP.

LS1 maintains near-constant fuel pressure, changing injector pulse-times to compensate for changing MAP.

Most Corvettes use a single electric in-tank pump. Exceptions:

- **ZR-1:** Dual electric pumps in single tank
- **LS1:** Electric pump in left tank, siphon pump in right tank

Remember the port-injection fuel rails, with the regulator and fuel-pressure test point. Exception: **'99-01 LS1** fuel-pressure regulator is *near* the tank. There is no return from the engine.

In this chapter, you've seen four jobs of Corvette fuel systems:

- deliver clean fuel under pressure to the injectors
- compensate injector fuel-delivery for varying MAP
- control fuel-pump relays and safety circuits
- control fuel vapors with EVAP system

IN THE NEXT CHAPTER:

You've now seen all the Engine Control components in Chapters 4-7. In Chapter 8, I'll show you how Corvette Electronic Engine Management brings them together, using strategies to control engine operating conditions.

Chapter 8

Strategies

Contents

In this chapter, look for:

- the different engine-control strategies for different Corvette fuel-injection systems as they have changed over the years
- how the sensors, control module, and actuators operate during each of the strategies
- control of fuel, spark, idle air, and emission systems during each of the strategies

1. Introduction

I'll describe the operation of the electronic engine management/fuel injection systems during each of the strategies that I outlined in Chapter 2. I'll discuss the three basics of engine control, fuel, spark timing, and idle air.

Closed Loop is the way your Corvette manages fuel control while you are cruising. Emissions are evaluated by the oxygen sensor which has direct control of the fuel injector.

Open Loop uses a variety of sensors to regulate fuel control but does not provide any feedback for evaluating efficiency. This strategy operates during times of change, such as starting, Clear Flood, heavy acceleration, Wide Open Throttle, and fuel backup.

L83, L98, '92-93 LT1, ZR-1: Strategies are stored in the *MEM-CAL* (MEMory CALibration) assembly in the control module. To change strategies, replace the MEM-CAL.

'94-96 LT1, LT4, LS1: Strategies are stored in the *EEPROM* (Electronically Erasable PROM). To change strategies, electronically erase existing information and record new information.

I'll use the terms "computer" and "CM" to indicate the control module, whether I'm referring to ECM (*Engine* Control Module) for **'82-93**, or PCM (*Powertrain* Control Module) for **'94-01**.

2. Closed Loop

Closed Loop is the simplest of strategies because conditions are relatively stable, and because the engine operates without some of the special controls applied to a cold engine. Closed Loop strategy is designed for moderate power, maximum fuel economy, and minimum emissions. No matter how much you enjoy Corvette performance, your engine operates Closed Loop most of the time. You may be surprised to learn that level cruise generally draws 15-30 horsepower, a fraction of the max output of a Corvette engine.

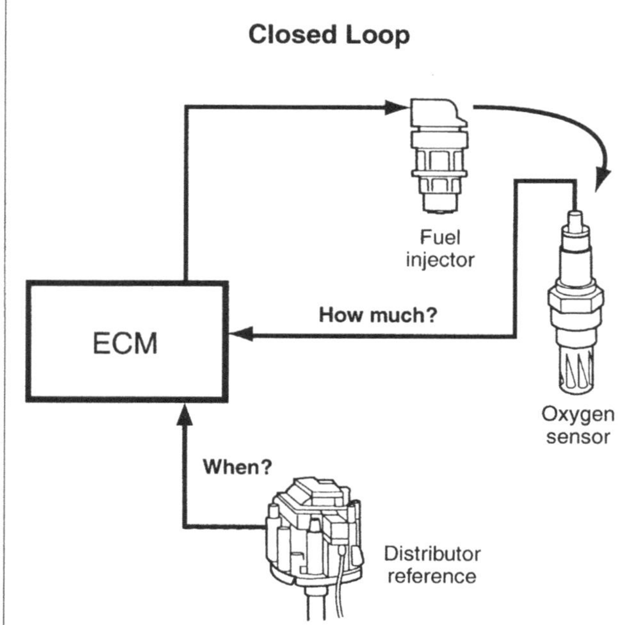

Fig. 8-1. Closed Loop is the baseline operating condition for light-moderate loads with a warm engine. The oxygen sensor is the primary input, determining "how much" fuel. A rotating sensor, such as the distributor or CranKshaft Position sensor, determines "when" fuel is injected.

Start with the two inputs that signal fuel injectors:

- The oxygen sensor *in*puts a signal to the CM which *out*puts to the fuel injectors. The signal tells "how much" fuel is needed to control the Air/Fuel ratio. Each fuel injector receives a pulse signal according to the continually changing oxygen sensor voltage signals.
- The distributor/CranKshaft Position sensor *in*puts allow the CM to signal the fuel injectors "when?"

Oxygen sensor emission control is most effective during cruise conditions. Most engine operating hours are at warm cruise so emission control has a great total effect on emitted gases. The computer manages emission systems such as Exhaust Gas Recirculation (EGR), Canister Vapor Purge, Air Injection (AIR) systems, and Three-Way Catalytic Converter (in the **L83** and **L98**). Computer control strategies must balance operations:

- When is emission control right for the *engine*? For example, in Cruise, the engine easily handles EGR up to 15%. The engine loses a little power because the exhaust gases don't mix with fuel and burn as air does.

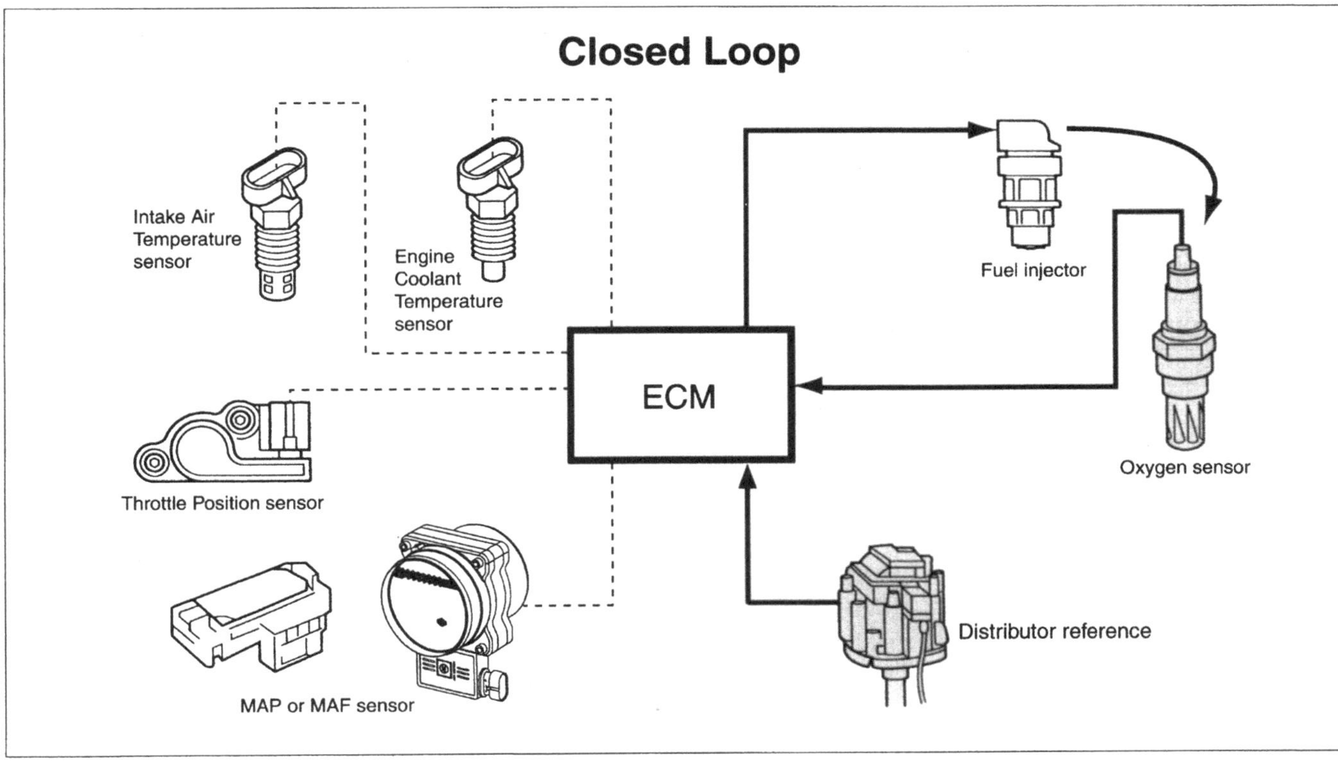

Fig. 8-2. Oxygen sensor and distributor reference are primary inputs. Other sensor inputs are part of computer determination that the engine is ready for Closed Loop, including IAT, ECT, MAF/MAP, TPS.

8

But cruise is part-power anyway, and you tend to take care of that with your accelerator foot, so it doesn't matter.

- When is it right for the *catalytic converter*? For example, during Cruise, the converter is hot enough to reduce emissions, but a misfire may introduce extra fuel that could burn in the converter, causing overheat that could destroy the converter.

2.1 Oxygen Sensors (O_2S) and A/F Ratio

The oxygen sensors and distributor references are *primary* inputs. Secondary inputs include sensors (IAT, ECT, TPS, MAF/MAP) that advise the computer to enter or remain in Closed Loop, or switch to Open Loop. (See Fig. 8-2.)

The amount of fuel injected is continuously fine-tuned by signals from oxygen sensors. O_2S signals advise the computer of the oxygen content of the exhaust. Remember:

- When a sensor *observes oxygen* (as little as 2%) in the exhaust, it generates *low* voltage, signaling a lean mixture. The computer adds fuel by slightly *in*creasing injection pulse time.
- When a sensor *observes little or no oxygen* in the exhaust, it generates *higher* voltage, signaling a rich mixture. The computer slightly *de*creases injection pulse time.

'91 L98, '92-93 LT1: MFI engines left and right oxygen sensors adjust A/F ratio by *bank*. Computer determines the necessary adjustment to each bank of injectors.

'94-96 LT1, ZR-1, LT4, LS1: Sensors of sequential fuel-injected (SFI) sense individual-cylinder oxygen-content-changes by timing in the exhaust stream. With the cylinder-identification sensor, the computer determines the necessary pulse-time adjustment to each individual injector.

I've discussed variations in oxygenated fuels in Chap. 3, "Emissions and Fuels", and the operation of the O_2S (HO_2S) in Chap. 4, "Sensors". Do the oxygen sensor signals cause control of the A/F ratio to be 14.7:1? Well, they do cause control of A/F ratios to stoichiometry for the fuel being burned. But *14.7:1 is a nominal value*, not stored in the CM memory.

Corvette **LT1** SFI engine, non-oxygenates gas and fuel, 3000 rpm at 60 mph, 20 mpg (round numbers):

Closed Loop	Per minute	Per second
Air burned	4.5 lb (2 kg)	0.075 lb (3.3 g)
Fuel burned (8 injectors)	0.3 lb (.13 kg)	0.005 lb (.22 g)
Pulses (8 injectors)	12,000	200
Pulses (each injector)	1,500	25
Fuel burned (each injection)	.0001 lb (.000045 kg, .045 g)!	

We're talking a fraction of a drop per SFI injection pulse!

OF FURTHER INTEREST—
Think of Closed Loop as Warm Cruise, but it may also include Warm Idle, particularly after cruise, or at a stop light.
•**L83, L98, '90-92 ZR-1, '92-93 LT1:** Unheated oxygen sensors may cool down enough to go Open Loop, such as when you're checking Closed Loop operation at idle.
•**'93-01 ZR-1, '94-01 LT1, LT4, LS1:** With heated oxygen sensors there is no such problem.

2.2 Engine Speed/Injector Timing

In all modes, the computer uses input signals signaling crankshaft RPM for injector timing, answering the question "When?" **'94-96 LT1**, **LT4** and **LS1** measure both crankshaft RPM and Position answering "Which cylinder?"; they also monitor both input and output signals for OBD-II.

CranKshaft RPM/CaMshaft Position and High Energy Ignition signal engine rpm/crankshaft position for injector timing.

Consider injections for 4-stroke cycles:

- **L83 TBI**: For every two revolutions of the crankshaft, each Cross-Fire injector opens/closes twice, alternating left bank, right bank. Call this "RPM dependent". The computer knows how fast the crankshaft is turning, but not where it is in relation to TDC of any specific cylinder.
- **L98, '91-93 LT1 MFI:** For every two revolutions of the crankshaft, each gang of 4 injectors opens and closes twice, alternating left bank (four cylinders), right bank (four cylinders). GM-speak: "Double Fire." Again, this is RPM dependent.
- **LT1 94-96, LT4, ZR-1, LS1 SFI:** For every two revolutions of the crankshaft, each sequential injector opens/closes once for its one cylinder. This is RPM and crankshaft position dependent; the computer knows when cylinder #1 is at TDC, and therefore knows the *crankshaft position* for each of the other cylinders.

2.3 Spark Timing

Closed Loop spark timing is advanced to about 30° BTDC, close to knock. Excess knock is signaled by the KS in Closed Loop: (See Chap. 4, Sensors)

- **L83, '85-90 L98:** A single Knock Sensor adjusts timing for *all cylinders.*
- **'91 L98, LT1, LT4, ZR-1:** Two Knock Sensors adjust timing by a *separate bank* of cylinders.
- **LS1:** Two Knock Sensors adjust timing by *individual cylinders.*

If all those input signals match the values stored in the memories for Closed Loop Strategy then the computer switches to a Closed Loop mode.

2.4 Emission Control

AIR

L98, L83: During Closed Loop cruise, the AIR system delivers secondary air to the catalytic converter to assist in oxidizing the HC and CO.

LS1, LT1, LT4, ZR-1: AIR normally operates during warm-up only, up to 240 seconds maximum. AIR is unnecessary during other modes.

Fig. 8-3. L83, L98: The AIR pump is crankshaft driven, turning whenever the engine runs.

EGR

In EGR, exhaust gases replace some of the intake air (5-15%). As a result, the cylinders need less fuel to burn. The computer calculates EGR intake and reduces the injector pulse time to account for the reduced intake air that is dis-

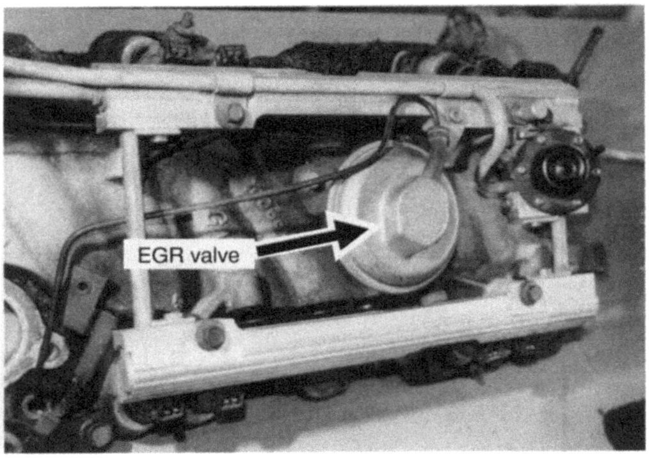

Fig. 8-4. EGR minimizes NOx by reducing in-cylinder temperatures. Engine operation must be in the proper range before the computer turns on EGR:
1. *ECT* is neither too cool nor too hot
2. *TPS* is part throttle, not idle, not W.O.T. L98 shown.

placed by EGR flow. The flow of exhaust gas is controlled by the computer: either the EGR Vacuum Regulator (EVR) solenoid or the **'93-01 ZR-1** Electronic Linear EGR Valve control.

Effect of Exhaust Gas Recirculation

EGR affects Closed Loop fuel-injection. The EGR valve is opened by Pulse-Width Modulation (PWM). This provides the proper opening to recirculate the best amount of exhaust gas into the intake manifold, minimizing the output of NO_X.

> **LS1**, **LT4**, and **'90-92 ZR-1** have no EGR because the engine was designed to reduce NO_X emissions through valve overlap and cooling, and through calibrations of fuel injection and spark timing.

EGR flow during Closed Loop affects the spark timing because the EGR slows the burning of the air-fuel mixture. The computer changes its signal to advance the spark timing by a few degrees, in order to ignite the fuel earlier, allowing for the effect of the EGR.

EVAP

When a canister purge sends in extra fuel vapors to be burned, the computer reduces the injector pulse time. This maintains the oxygen sensor signals at around 450 mv. bias, even with the added fuel from the EVAP system. (See Chap. 4 "Sensors", Chap. 5, "Control Module".)

3. Open Loop

Several strategies are similar in that O_2S does not control A/F ratios. In Open Loop, computer monitors oxygen content to switch over to Closed Loop.

Open Loop applies to several operating conditions:

- start and warmup of cold engine
- Clear Flood
- heavy acceleration
- fuel cut off
- fuel backup

And, on late models, it also operates during converter overheat and power and speed limitation.

3.1 Start/Warmup of Cold Engine

Signals from cold oxygen sensors are not suitable for controlling the injector pulse time and are not used in Open Loop. MAP/MAF signals are low and irregular during cranking and do not provide good load signals. Rpm signals are slow and irregular but are used to determine injection timing.

- **'94-96 LT1, LT4, LS1:** IAT is sensed.
- **'95-96 LT1, LT4, LS1:** MAP checks air density for two seconds at ignition, operating as the BARO for altitude corrections.

Starting Outputs

8

The fuel pump(s) is on for two seconds, continuing with the crank signal.

- The computer determines the amount and timing for the injection output. Normal injection is one pulse for each rpm reference pulse, this is known as *synchronous* pulses.

- At an ECT below –40°, long injector pulses may produce mixtures at the intake valves of A/F ratio of 1.5:1

> **L98, LT1, LT4, LS1:** During crank, don't step on the accelerator even a little. Above a 3% TPS signal the computer would increase the pulse time and perhaps flood the engine. In the **L83** the crank mode operates at up to 50% TPS signal. Of course, pumping is a no-no—if you pump the accelerator, those TPS signals really upset the computer, it thinks "What does the driver want?" If you have flooded the engine, let the starter cool for about fifteen seconds, and then crank with the accelerator full down to clear the flood. In the **ZR-1** if there is no start after three seconds, try pushing the accelerator pedal about a third of the way down. If it still doesn't start after 15 seconds of cranking, then use the full throttle method.

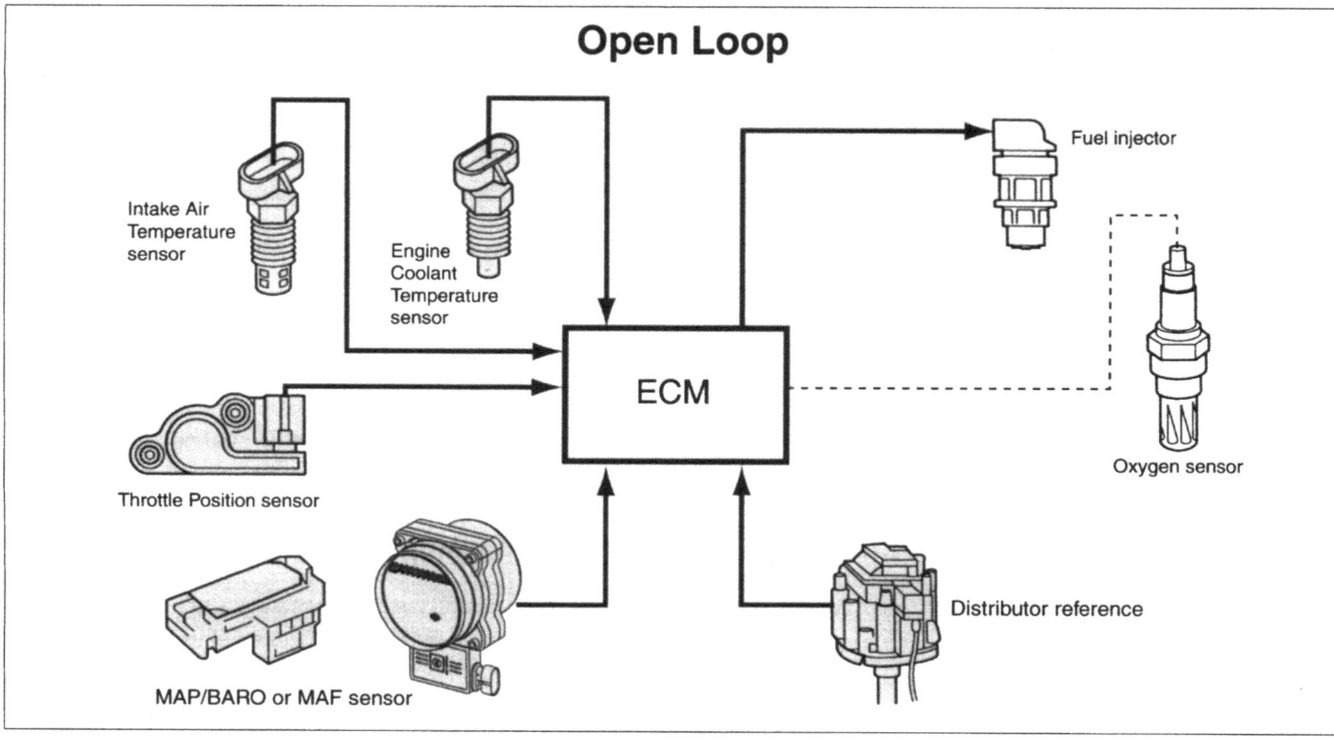

Fig. 8-5. Open Loop strategy is based on inputs from the IAT, ECT, MAP/MAF, TP, and distributor reference. The oxygen sensor is not controlling, but it, along with the timer, are still monitored.

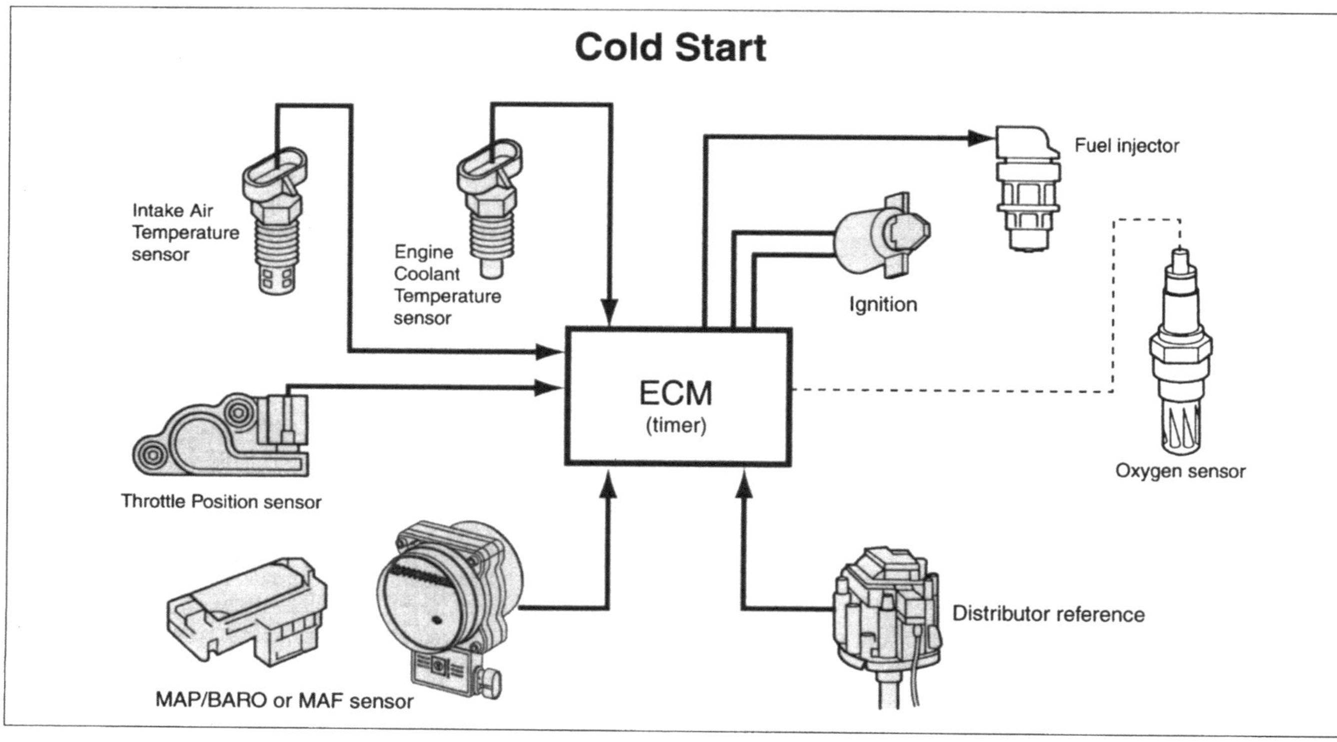

Fig. 8-6. Cold Start is still Open Loop. The inputs are the same, but also include the ignition and the timer in the computer, as well, The ECT is perhaps the most important here. This applies only to **'94-96 LT1, LT4, LS1**.

- As temperatures rise from –40° to 94° C (200° F), injector pulses shorten accordingly in order to produce decreasingly rich A/F ratios.
- With a warm engine at an ECT of 94° C (200° F) the computer commands synchronous short pulses for normal A/F. In the **'94-96 LT1, LT4, LS1** the computer also includes cylinder identification for SFI injection.
- The computer may add extra asynchronous fuel injection pulses based on time instead of on crankshaft revolution. They may be 80 pulses per second (12.5 ms) or 160 times per second (6.25 ms) depending on calibration.

With steady rpms indicating engine start, the CM changes the fuel controlled Look-Up tables from *Cold Start* to *Begin Warm Up*. This reduces additions to the basic injection pulse-time. Pulse time is gradually reduced as the engine warms up. The IAT input signals add fuel injection pulse time as needed. Engineers must devise a warm-up strategy for a balance in enrichment:

- A richer mixture means: better combustion in cold cylinders, improved cold idle, and a reduced chance of a cold stall. Furthermore, extra fuel delivered to the catalytic converters burns sooner and heats them up sooner.
- A leaner mixture means that the engine will have cleaner emissions.

Cold Start Injector and Cold Start Valve

'85-88 L98: Starting mixtures are enriched by the Cold Start Injector (CSI) from Bosch, also know as the Cold Start Valve, as well as by the computer controlled pulse times. The Cold Start Fuel Injection Switch, also called the Thermal Time Switch, controls the cranking enrichment based on time and temperature as follows:

- The CSI is activated for a maximum of eight seconds when the switch is at it's coldest. As the switch warms it gradually reduces the CSI time.
- The CSI is off when the switch is above 95° F (35° C).

Note: The CSI is independent of computer control, there is no Coolant Temperature signal involved in CSI.

L83, '89-91 L98, LT1, LT4, LS1, ZR-1 have no Cold Start Injector. Remember, the computer provides starting enrichment by lengthening the pulse time of the principle injectors (in the **L83** it's TBI; in the others it's the port injectors).

Engine Start

Rpm signals rising above about 400 (300 in some engine systems) indicate engine start to the computer.

- **LS1, LT1, LT4:** The computer controls the spark timing in all modes, there is no bypass timing,
- **ZR-1:** The computer controls the spark timing during crank. The first bypass mode is at 6° BTDC. Above 400 rpm, spark timing switches to another bypass mode, timed only by engine rpm. Above 600 rpm, steady running causes a switch over to full computer control, through the Ignition Control (IC).
- **L98, L83:** There is similar computer control during crank, but the bypass mode is fixed at 10° BTDC. Above 400 rpm, spark timing switches to the computer.

During cranking, the computer also starts a timer of its own. After 20 seconds, if the engine has not fired, the computer reduces injector pulse times to prevent flooding.

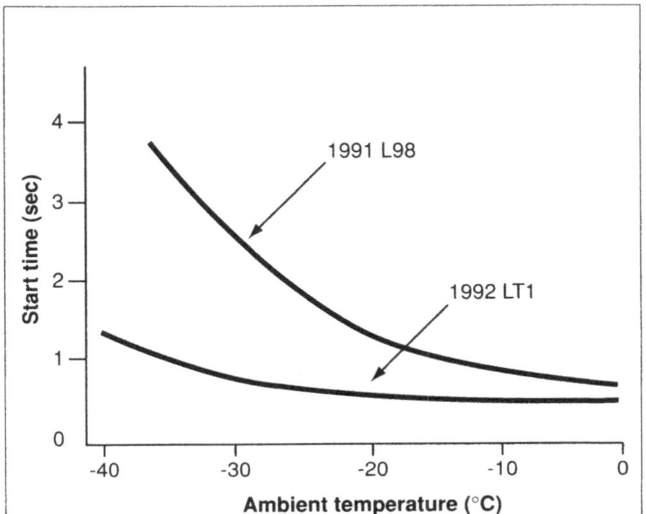

Fig. 8-7. LS1, LT1, LT4, LS1: Cold-start improves with precise spark timing. Individual cylinder timing is delivered within 1° of crankshaft rotation. Even a cold engine may start in less than one crankshaft revolution if there is still fuel pressure holding in the rails.

Idle Air Control

Idle rpm is also controlled in a Closed Loop mode. The computer compares the rpm signal to each different target idle rpm and modifies the output signals to the Idle Air Control (IAC) (or the throttle, in the **LS1**), to maintain the target idle rpm.

L83, L98, LT1, LT4, ZR-1: The IAC is partly open during starting, allowing intake air to bypass the closed throttles. Each time the engine is shut off, the computer signals the IAC to fully close the valve, then to reset to the start position.

LS1 has no IAC. The throttle plate is positioned for starting by the TAC motor, based on output signals from the computer. Starting air is determined by inputs from ECT and IAT.

8

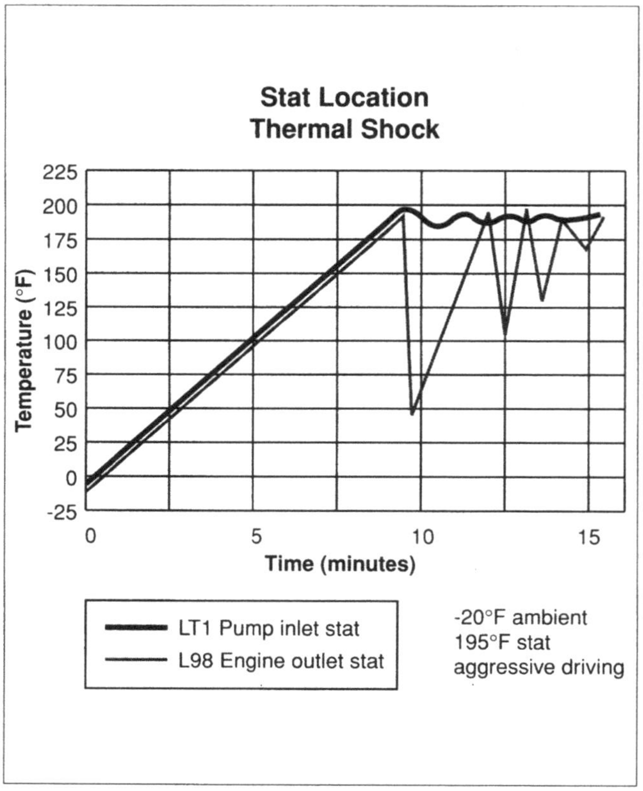

Fig. 8-8. LT1: The thermostat controls the coolant *entering* the engine .This produces less thermal shock than in the **L98**, where the thermostat controls the temperature *leaving* the engine. In both, a typical warmup with aggressive driving, takes:
- from 0° F, about 10 minutes
- from 100° F, about 5 minutes

Warm-up

As the engine warms, and the Engine Coolant Temperature (ECT) and Intake Air Temperature (IAT) increase, the idle rpm is gradually reduced toward a normal idle rpm:

- **LT1, LT4, ZR-1, L98, L83:** The IAC gradually closes.
- **LS1:** The electronic throttle gradually closes under the control of the TAC.

Warm-up Open Loop operates for a relatively short time compared to the rest of an emission control test, which runs Closed Loop. Many engines fail the emission qualification tests during the first minute, the start/warm-up. That is one reason for the **LS1** quick start, based on quick cylinder identification (90° crankshaft revolution).

> Electronic fuel injection eliminates the fast-idle cam of a carburetor, so you do not need to "kick it off the cam". Notice too, that warm-up cold-idle rpms are much lower with fuel injection than with carburetors.

Switch Open Loop to Closed Loop

Three conditions must be met before the Computer switches from Open Loop, as in Warm Up, to Closed Loop:

1. the O_2S must be switching back and forth around 450 mV, from above 600 mV, to below 300 mV
2. the ECT must be warm enough, typically above 70° C (158° F)
3. a certain amount time has elapsed on the computer's timer.

Spark Timing—Warm-up

Warm-up strategy changes the spark timing ROM tables. This advances the base spark timing according to the rpm and MAF/MAP inputs. Base timing may be further advanced depending on the ECT signals because a cold engine is less likely to knock. Spark timing advance increases combustion chamber temperatures. This warms the oxygen sensors and catalytic converters.

Emission Control—Warm-up

Emission control is especially important during Warm-up. Unheated oxygen sensors (**'82-93**) are not warm enough to send satisfactory signals, so the A/F ratio is decided by the computer, which processes several other emissions-related inputs.

Diverting AIR to the exhaust manifolds results in less pollution, as the addition of air (containing oxygen, i.e., O_2) into the manifold tends to:

- burn or oxidize HC into H_2O and CO_2
- burn CO into CO_2

Also, the burning A/F mixtures give off hot gases which help to heat unheated oxygen sensors (O_2S), and catalytic converter(s). HO_2S are warmed electrically.

The AIR pump is driven differently in different engines:

- **LS1, LT4, '95-96 LT1:** An *electric* AIR pump relay is energized during crank, to prevent a vacuum
- **'92-94 LT1, ZR-1:** An *electric* AIR pump relay is energized after the engine starts, provided ECT is above 60° F (15° C).
- **L98, L83:** The AIR pump is *engine-driven*. A computer

controlled diverter valve switches open until the engine starts.

EGR:

- **LT1, '93-95 ZR-1, L98, L83:** During starting, the computer shuts off EGR, which would only get in the way. The other engines have no EGR.

EVAP: During warm-up, the computer shuts off the canister purge to prevent it from interfering with the A/F ratio.

3.2 Clear Flood

Corvette Clear Flood mode either provides a lean A/F mixture of 20:1, or completely cuts off the fuel flow so that the overrich mixture is swept out. To use Clear Flood when engine does not start, turn the ignition while holding down the accelerator.

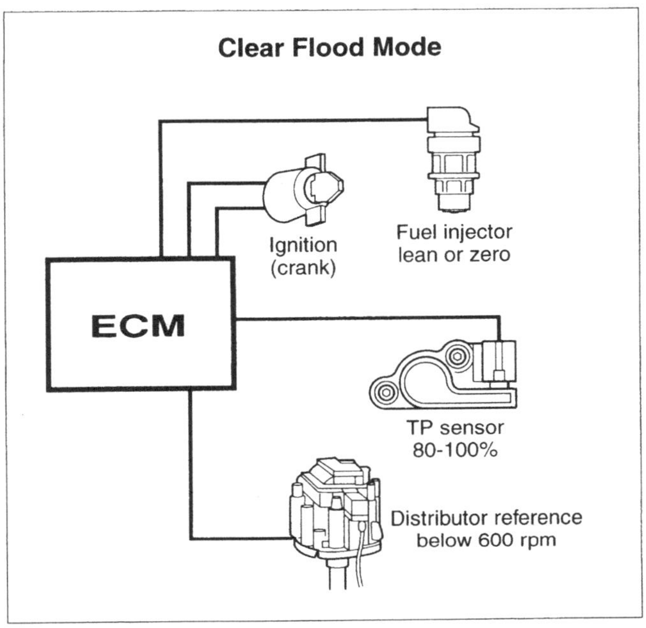

Fig. 8-9. Clear Flood is signaled by three elements:
- a crank signal from Ignition KEY
- TPS more than 80%
- RPM less than 600

Applying Clear Flood will usually sweep out a flooded engine with air. Of course, no other sensors affect the computer control of the injectors during this mode. When the engine catches, release the accelerator. When you ease off on the accelerator, the computer switches out of Clear Flood.

3.3 Acceleration Enrichment

When you floor the throttle, intake air and manifold pressure suddenly increase. This would normally tend to make the A/F ratio go lean, but the TPS and MAP/MAF signal the computer to increase the pulse times to prevent lean stumble.

Fuel injection is enriched by longer injector pulse times. **LT1, LT4, ZR-1, L98, L83** add asynchronous pulses timed between the regular synchronous pulses. During Acceleration Enrichment, the computer also makes the following adjustments:

- Short Term Fuel Trim drops to 0%. It is frozen so that there is no STFT effect on Long Term Fuel Trim(Block Learn).
- Spark timing is advanced to increase power. With the rich mixture, extra advance can be programmed with less risk of detonation.
- Emission control stops effecting the fuel(EGR, AIR).
- The air conditioning compressor and **LT1, LT4, ZR-1, LS1** (electric) engine cooling fans are turned off for 10 seconds.
- The computer sets pulse times to 160 times per second (6.5 ms)

Closed Loop Acceleration

Generally, you can ease the accelerator down to about 80% to speed up, or to hold speed on an up slope. The engine will continue in it's efficient Closed Loop mode, with the Oxygen Sensors as the principal sensors, though MAP/MAF and TPS signals may cause enrichment during the acceleration downpress, preventing lean-stumble. It's that last 20% press, however, that says to the computer, "*Go Open Loop. Don't bother with fuel economy or emissions. Just give me power!*" So, if you want fuel efficiency, keep it under 80%. If you need immediate power, then floor it.

8

3.4 Fuel Cut/Spark Control

The computer may order Fuel Cut/Spark Control during several modes:

- deceleration
- power/torque management
- overspeed
- converter overheat

In addition, fuel is cut off if computer does not see any reference pulses. Turning the ignition off also cuts off fuel to prevent after-run, sometimes called dieseling.

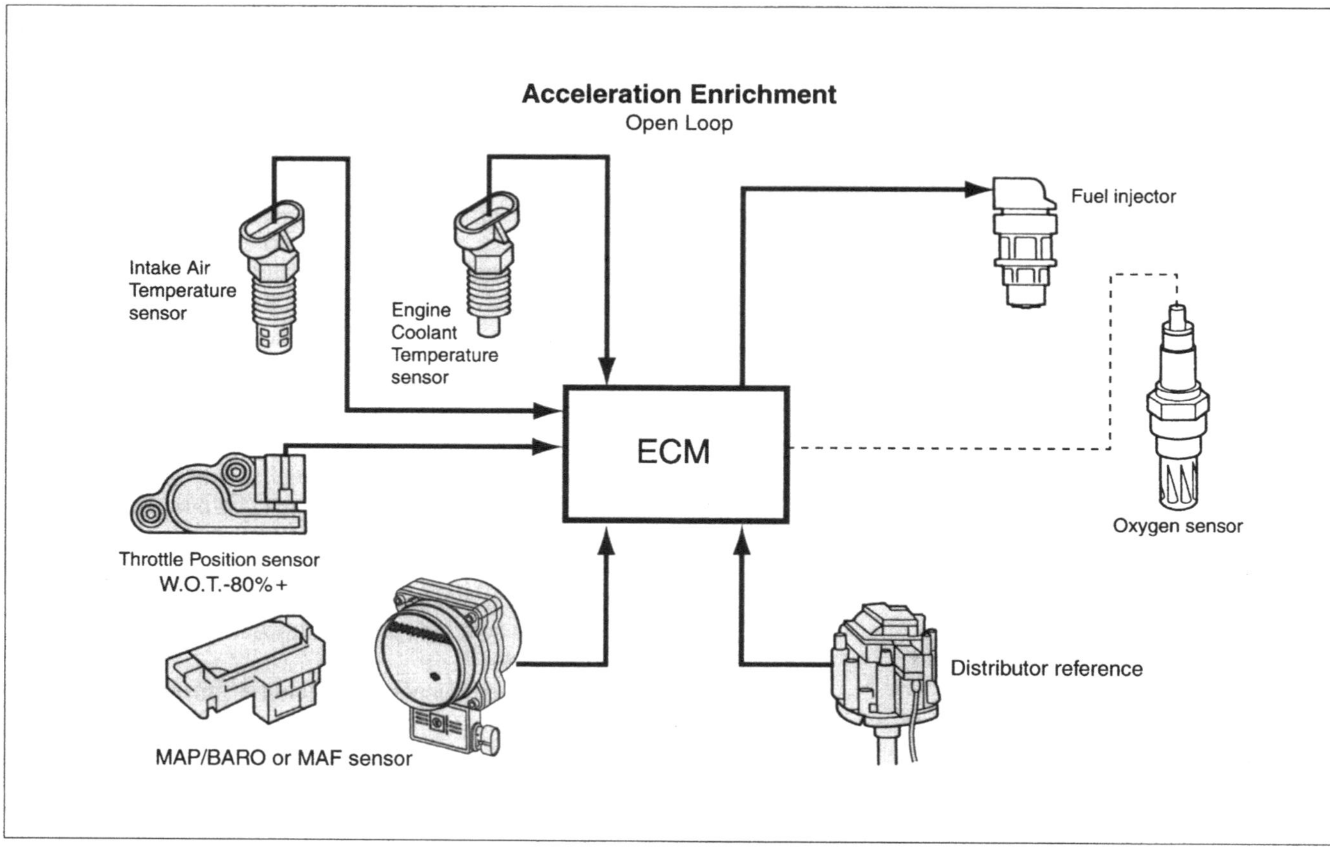

Fig. 8-10. In Wide Open Throttle acceleration, the TPS signals more than 80% in run mode. This strategy provides full power without regard to economy or emission control.

Deceleration

Consider two kinds of deceleration, signaled by the TPS and MAP:

1. Cut *Back* is a gradual ease off to slow down. It causes shorter injector pulse times which reduce power. It is Closed Loop operation. Light-throttle deceleration is an example of this.
2. Cut *Off* cuts off fuel entirely. Rapid or long deceleration or coasting down a long hill will cause this. It prevents raw fuel from entering catalytic converter. It is Open Loop, of course. As engine speed nears idle rpm, the computer restarts fuel injection to prevent stalling.

During deceleration, spark timing may be retarded, or cut off, from some plugs, depending on fuel cut-off; we wouldn't want raw fuel delivered to the converter(s).

L83, L98, LT1, LT4, ZR-1: During cruise, the IAC retracts, depending on vehicle speed. When you snap your foot off the gas, closing the throttles for deceleration, the IAC admits more air, making the mixture go lean. This reduces the HC and CO emissions.

Overspeed

Fuel Cut strategy produces a "soft cut" instead of a sharp one, which, when you're accelerating with W.O.T., might feel like sudden braking.

Car overspeed, a different term, comes from the Vehicle Speed Sensor, and is often used for tire ratings.

In overspeed, crankshaft rpm signals cut off fuel supply at:

- **LT1:** 5800 rpm
- **LS1:** 5800 rpm
- **LT4:** 6400 rpm
- **ZR-1:** 7100 rpm

LS1 '98-01: Torque Management retards spark and cuts back fuel injection during:

- heavy acceleration from a standing start
- brake application (more than 40%) with moderate to heavy throttle (with Traction Control on)
- abusive maneuvers, such as shifting into a gear at high throttle angles, or shifting from reverse to drive and back to reverse, to rock the car
- certain Throttle Actuator Control faults
- all A/T upshifts and downshifts

ZR-1: You'll feel engine stumble at 7100 rpm. Overspeed will cut fuel and power, reducing the rpm to 7000. If you do not ease off throttle, the engine will accelerate to 7100 and keep cycling against the rev limiter until you back off. This is referred to as "bumping 7100-7000."

Note: Maximum power at is 6000-6500.

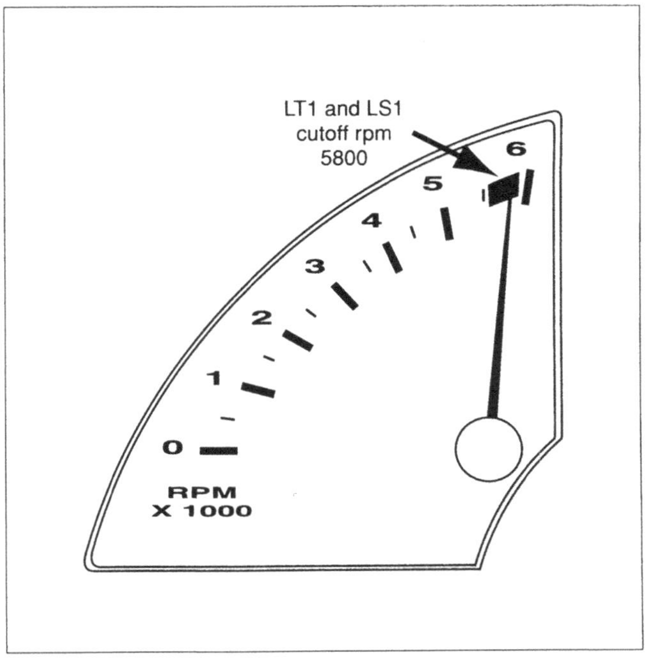

Catalytic Converter Protection

LS1, LT1, LT4: Catalytic Converter Overheat is computed from the oxygen sensor signals. If the converter indicates overheat, the computer switches to Open Loop and increases pulse width to enrich the mixture. With less oxygen, the converters will operate at lower temperatures. When temperatures return to normal, Closed Loop returns.

Torque Converter Clutch Lockup (A/T)

Most of the time, the computer's strategy calls for lockup of the automatic-transmission Torque Converter Clutch (TCC).

- ECT indicates that the engine is warm enough.
- VSS indicates car speed, usually above 30 mph.
- TP indicates a mid-range, not heavy, acceleration or deceleration.

Also, the computer can control A/T shift speeds and firmness according to engine operating conditions.

The TCC is unlocked under several special conditions:

- during brake pedal application
- when the Transmission Fluid Temperature (TFT) is above 263° F (128° C)
- during misfire in the **'94 LT1**and **LS1**, to free the crankshaft from a lockup to automatic transmission, which could dampen crankshaft speed fluctuations for OBD-II

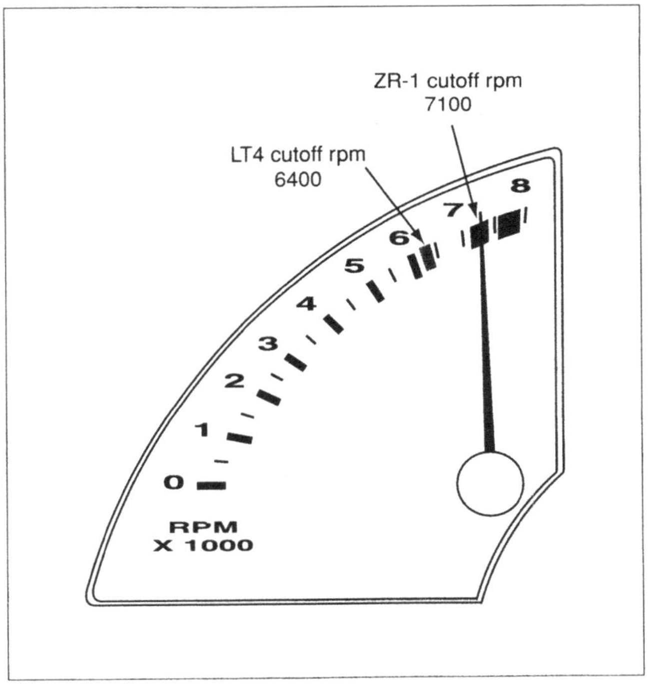

Fig. 8-11. Engine overspeed cuts off the fuel supply when the rpm gets too high. The **ZR-1** and **LT4** can handle higher rpms than the **LT1** and **LSI**.

8

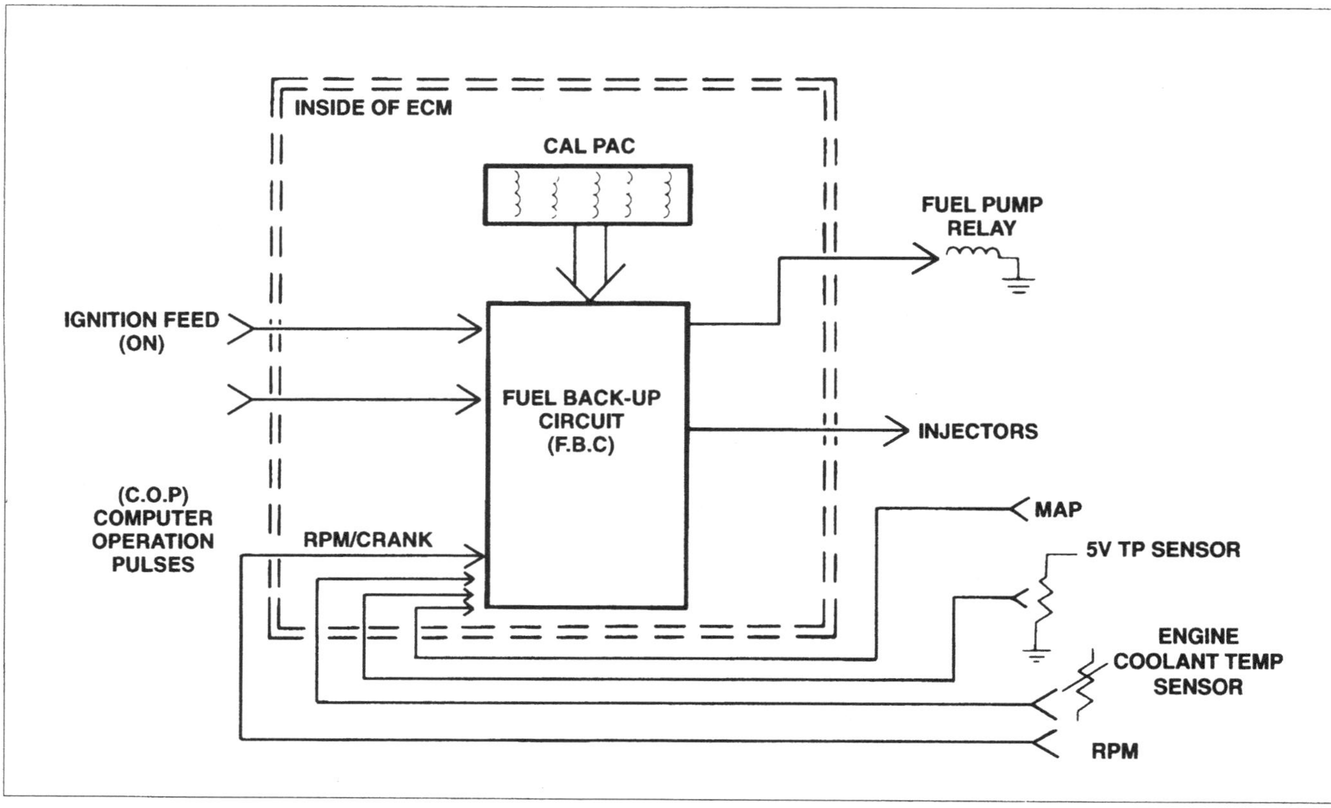

Fig. 8-12. In Backup or limp home mode, the computer runs on the CAL-PAK chip and on sensor inputs from the TP, MAF/MAP, ECT, and rpm speeds (distributor reference). This is also Open Loop mode.

3.5 Fuel Backup—"Limp Home"

Most Corvette engines seem to run well in backup mode, while some other car engines may operate quite erratically.

L83, L98, LT1, LT4: When the computer does not operate normally, engine management changes to Fuel Backup, or "limp home" mode. Backup mode could be needed under any of the following:

- computer voltage is less than 9v.
- cranking voltage is less than 9v.
- PROM is either missing or not functioning
- the computer fails to issue any Computer Operating Pulses

4. Battery Voltage Correction

In any mode, if the battery voltage falls too low, the computer compensates in several ways:

- the injector pulse time is increased, to deliver the necessary fuel
- idle rpm is increased, to raise generator output
- **LS1, ZR-1:** dwell time is increased by the the electronic ignition, to improve coil saturation

5. Continuous Test—Trouble Codes

Trouble codes may light the MIL/SES during operation and be stored. *Note*: **L83** does not provide continuous test.

- **L98, LT1, LT4, LS1:** During all normal operation, the computer continuously tests inputs. One at a time, each circuit is tested for shorts and opens.

- **LT1 '94-96, LT4, LS1:** The OBD-II control module notes any illogical readings, such as an Engine Coolant Temperature sensor changing from warm to cold to warm. The first time an error is observed, it is counted but not registered, because the system only cares about repeat offenders. After an error occurs several times, it is stored as a trouble code in the memory in the Non-Volatile RAM (NVRAM).

6. SUMMARY

In this chapter, you have seen how the ECM strategies manage fuel injection/engine control, and, in late-model Corvettes how the PCM manages both engine and transmission, A/T or M/T. You've also seen:

- different engine control strategies as they have advanced with model years
- how sensors, control modules, and actuators operate during different strategies
- how the computer controls fuel, spark, idle air, and emissions during each of the strategies

IN THE NEXT CHAPTER:

You now know how the production engines operate. But If there's anything a Corvette owner really wants, it's learning how to modify or tune for greater performance. I'll talk about that in Chap. 9.

Chapter 9

Tuning

Contents

In this chapter, look for:

- legal issues involving warranties, tampering, and legislation
- your decisions, street or track?
- high-performance basics
- bolt-ons and mods
- questionable tricks
- a list of suppliers

1. Introduction

I hope you've read the first eight chapters, because understanding the detailed workings of Corvette electronic fuel-injection/engine-control systems will help you to decide what you can do to make your Corvette faster. Now I'll take a look at the next step, for some enthusiast's, modifying engine-control systems for high performance. Whether you are a DIY proponent (Do It Yourself), or a DIFM (Do-It-For-Me) fan, you will want to know at least enough to ask questions, and to understand the trade-offs that usually accompany modifications. What your engine-control system needs depends on what you want to improve. You may be:

- track racing—maximum power and acceleration for extended time periods
- drag racing—maximum W.O.T. power for short periods (the shorter the better!)
- a driving enthusiast—performance is important, but so are emissions, cruising, long life, and cold starting

What can you get from the Chevrolet Performance Group, and from the aftermarket tuners and parts suppliers? Corvette has revolutionized stock-block racing, based on several hundred-thousand fuel-injected "small blocks". These engines have the potential for substantial power gains through modifications under the stock fuel-injection systems.

I'll also describe legal issues in this chapter. They could affect what you can do with your Corvette as far as emissions, and warranties are concerned.

> I'll concentrate on performance related to the subject of this book, electronic fuel injection/engine control for small-block (known as "**SBC**" for "Small Block Chevrolet.") I know you can do a lot with exhaust systems, headers, cams, and pistons, not to mention suspensions, brakes and the looking-good stuff, but that's a whole other book.

You'll find many roads to improving the performance of your car. Those of you familiar with Bosch fuel injection have a head start, particularly with the **'85-88 L98**, with it's Bosch-based Corvette fuel injection, and GM computers. You don't always need to start with fuel-injection modifications; the original systems have a lot of unrealized potential.

With the C5R, GM entered racing officially. The specially-built 5.98L **LS1** engines deliver 600 hp at 7200 rpm, and 495 lb-ft of torque at 5600 rpm. And they're designed to do that for 24 hours, at LeMans.

The Chevrolet Performance Group provides strong support to aftermarket performance shops and owners for track and for street. The small-block fuel-injected engine is famous for, and well suited to, hop-up for the strip. Its large production volumes attract many aftermarket suppliers of performance gear.

OF FURTHER INTEREST —
Horsepower ratings are measured in two ways:
Gross: is a bare engine on a dyno, with super-premium fuel (103 octane), pre-1973 ratings
Net: is the complete engine with water pump, generator, air filter

Fig. 9-1. For many, modifying your car for better performance is one of the great pleasures of owning a Corvette. Burnout (on top) and front-end lift (bottom) seem to spell "performance."

Fig. 9-2. The **LS1** engine is well-designed for performance, but the tuners seek and find ways to improve it.

2. Legal Issues

OK, legal issues are not the most fun, but they are realistic and must be faced. Before modifying an engine, consider the issues of warranty and tampering. More and more, you'll want to answer the question, *"Are these modifications certified?"* In this chapter, and in tuner company advertisements, you'll find reference to certification and Executive Orders (E.O.) by California Air Resources Board (CARB). California has: a) some of the worst pollution in the nation, b) more vehicles than any other state. Do California E.O.s affect you in the other 49 states as well? You bet, for two reasons:

- CARB engineers test modifications extensively and verify manufacturer testing. CARB certifies which modifications do not increase vehicle emissions. They maintain vehicles within the emission limits specified for the model year, as certified for OEM engines, transmissions, and axle ratios, and yes, even for wheel/tire diameters which can affect emissions. The other 49 states, and USEPA, usually accept CARB E.O.s, unless the Federal cars differ extensively from California cars
- Most aftermarket manufacturers are certified to obtain E.O.s from CARB in order to protect their street-legal market. In most states, you gotta be *street legal* to license your car. It's test clean or no plates.

Consider the performance products with Executive Order numbers:

- Some products are acceptable on the highway. On executive order lists, these are indicated by dash-numbers following the overall E.O.#.
- Other products are suitable only for off-road applications. Some manufacturers offer both kinds of products, indicated by their literature and advertising.

The E.O.# list changes as more manufacturers test more of their products for certification. When you see a manufacturer's number in this book, you know that manufacturer is working with CARB on certification. Check with the manufacturer, or CARB to verify the latest status of the particular mods you are interested in.

OBD-II and You

You should know that tuning of **'94-01** Corvettes is affected by On-Board Diagnostics, 2nd generation, known as "OBD-II." Does OBD-II affect performance mods? You bet! OBD-II computers keep a much closer watch over the engine management and can set well over 100 Diagnostic Trouble Codes (DTC). See Chap. 11 for DTC lists.

Each car manufacturer has set up a couple of "OBD-II leaders". GM chose the **'94** Corvette **LT1** as a phase-in, to test a small set of cars in the customers' hands. The **'94-95 LT1** has the OBD-II 16-pin connector and some OBD-II trouble codes, but are not full OBD-II. **ZR-1** doesn't even think about OBD-II. **ZR-1** stopped engine production for the **'94-95** after 1993. By **'96**, all Corvettes comply fully with OBD-II; that makes a big difference to the aftermarket.

Some tuners report that some performance mods (apparently triggering some problems in compatibility with OEM powertrain computer) can light the "Check-Engine" indicator, and cause surging, as if the computer were adding and subtracting fuel and/or ignition timing. Many tuners are taking longer to design, test and certificate their mods to avoid such problems. And if the tuner has done the proper job of testing for certification, the mods will probably cost more. So don't plan on just throwing parts at your **'94-01**.

9

"Street-legal" is a widely used term, generally applied to performance modifications that do not cause excess emissions. Corvette and most aftermarket performance modifiers are designing mods that run clean and certify with an E.O.#, often labeled "street legal". If you want to be sure, stick to certified parts.

Does the Gas-Guzzler Tax Affect You?

GM corporate policy amounts to: GM will not produce any car subject to Gas-Guzzler tax, not even the **ZR-1**! See Chap. 3.

Fig. 9-3. Serious tuners, such as Lingenfelter, check out their engine mods on the dyno and on the road. They test for power and torque, driveability. For street-legal engines, they test emissions.

Computer-Aided Gear-Shift (CAGS)

Also known as Skip Shift 1-4.

The Gas Guzzler Tax is the reason that **'90-01** M/T Corvettes have that "#@*$" 1-4 upshift. Officially, it's Computer-Aided Gear-Shift (CAGS). Under certain limited conditions, CAGS blocks any shift up to second or third gear. It fits the original EPA gentle drive schedule. You can easily drive to avoid the skip shift, but it can be a nuisance.

Fig. 9-4. ZR-1, LT1, LT4, LS1: The dash light (with M/T) tells you that, based on car speed, throttle position and engine temperature, your next shift will be "one to four."
'90-91 L98 models also include a six-speed M/T CAGS but there is no dash display.

You can, actually, disconnect CAGS, if you feel it necessary. You may want certain "short", or performance, axle ratios, for example. Corvette will not furnish these on your new car because it would fall short of Federal fuel-economy standards and require Corvette to add Gas Guzzler tax to the price on that car. But, as I understand, the Gas Guzzler tax applies:

- only when the Corvette is sold new by the dealer, not to non-dealer sales
- not to after-sale mods installed six months after new-car delivery

In other words, six months after a new car has been registered, performance mods do not involve the Gas Guzzler tax. However, the law may change and/or be stricter in your state. Check with a service station which conducts emissions tests in your area for local regulations.

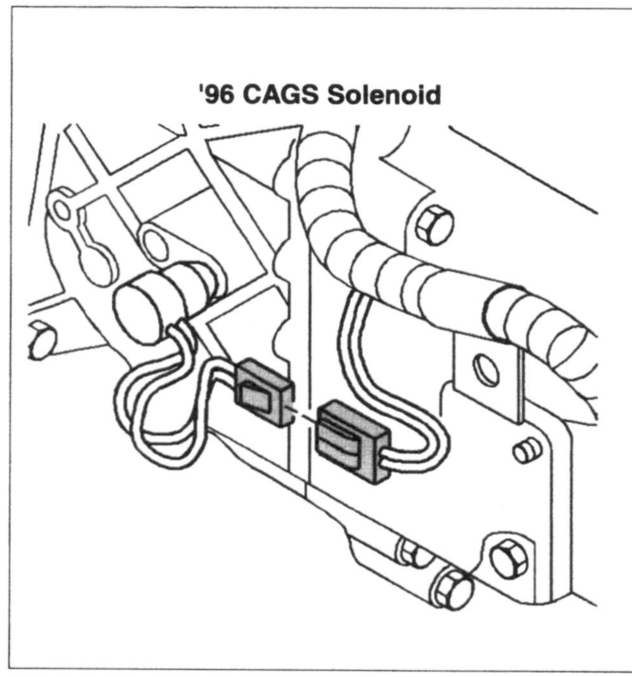

Fig. 9-5. '90-96: The Computer Aided Gear Selection (CAGS) connector looks like this, on a typical C4 manual transmission. CAGS forces the 1-4 shift only between 12-19 mph, and at light throttle openings.

To avoid CAGS, remove the electrical connector from the CAGS solenoid (after raising the vehicle properly). You can even avoid setting a trouble code by using a CAGS plug-seal kit, such as available from Lingenfelter Performance Engineering for less than $50. Two connectors prevent moisture from corroding the otherwise-open contacts. Jason Haines, project engineer from Lingenfelter, advised me that, if you decide to change back to reconnect CAGS, power up and exercise the solenoid; some have been known to stick in gear.

'97 CAGS Solenoid

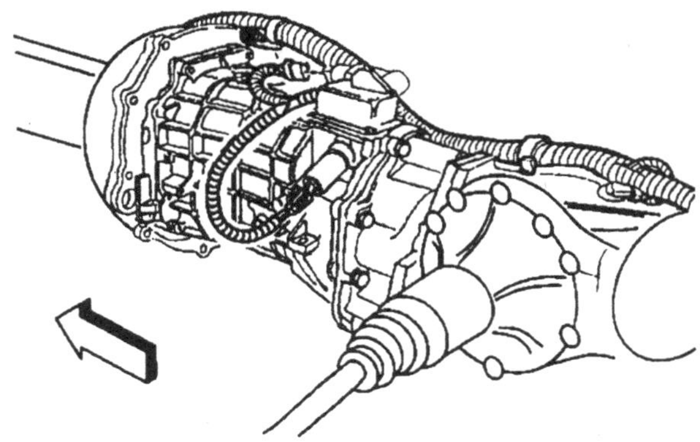

Fig. 9-6. '97-01: The Computer Aided Gear Selection (skip-shift) connector is on the manual transaxle, on a C5. CAGS is less noticeable on these M/Ts, because it operates only between 15-19 mph, and at a throttle opening less of than 21%.

Fig. 9-7. For a frequently-updated list of CARB Executive Orders from SEMA (Speciality Equipment Manufacturers Association) look at their website www.sema.org

3. Modifications and Future Legislation

Things are a-changing. In the near future:

1. Cars will be *one or the other*—really street-legal, or track, nothing in between. No more "street legal, wink, wink." The fine print will get larger and enforced, "for use only off-road" or "not legal for sale or use in *pollution-controlled vehicles*"—and that's all **'82-01** Corvettes, all are electronic fuel-injected.
2. There will be less exchanging of chips. If you open the computer, you forfeit the warranty. And, beginning with **'94** models, the chip is an EEPROM, it's permanent. You don't replace it, but you can reprogram it.
3. There will also be less changing of cams, rockers, and pistons, as Corvette and other manufacturers reach for better performance while remaining clean and licensable.

One tuner told me something like; the more the chips control performance, the more we find that reprogramming takes more engineering, but a lot less parts changing.

Frank Bohanon, Technical Director of *Speciality Equipment Manufacturers Association* (SEMA), maintains a list of Corvette and aftermarket products with CARB Executive Order Numbers (E.O.#s). SEMA provides it to smog-test stations and referee stations conducting visual inspections. The E.O.# list continues to grow, perhaps 5-10 a month. So always check with your aftermarket manufacturer regarding the status of his product. And if you flunk a visual inspection during an emissions test, get that list. (See also end of this chapter.)

Until recently, smog tests were run at "idle"— with no load. But where air quality is worst, in parts of 37 states, new tests measure emissions *under load* on a dynamometer (see Chap. 3), and that may affect which cars/engines/mods pass. Also, USEPA is now testing more for the "real world", colder "cold-starts", and at higher speeds and more realistic accelerations. Idle tests have little validity with the Corvette engines in this book, fuel injected/electronically-managed.

3.1 Tampering

Fuel-injection systems are considered to be part of emission-control equipment. Modifying the fuel-injection system is, therefore, modifying the emission-control system, and that raises some questions relating to federal laws as well as state laws.

> You may have heard that only the professional technician is subject to a fine if he alters emission-control equipment, but the owner Is exempt from such restrictions, right? Once upon a time, yes. Today, wrong-o!
>
> Now, under a 1990 amendment to the Clean Air Act, the $2500 fine extends to owners and individuals. Regardless of who does it, tampering is a Federal offense. But if your mods are on the E.O. list, that is *not* tampering.

Remember there are several forms of modification:

- "Strictly for the track" means that it is never to be driven on public roads.
- "Street legal" means that the government has certified it as not affecting emissions, and it is fine for any car.
- "Convertible" (that's what I call it) means that you can make changes after you arrive at the track, and then change back for the drive home. Some people may change the chip, or (this is really hard) change the

EEPROM calibration, and then change back. At the track, some change the stock MAF to a larger MAF, and then back to stock for the drive home.

Practically speaking, if the car is to be registered and driven on public roads, then the laws in your state and the legality of your modifications are more important than ultimate performance. You may think that you won't get caught, but you don't want to be identified as a "gross polluter" do you?

> From "CHEVROLET POWER", Official Factory Performance Guide:
>
> SPECIAL PARTS NOTICE
>
> This part has been specifically designed for Off-Highway application only. Since the installation of this part may either impair your vehicle's emission control performance, or be uncertified under current Motor Vehicle Safety standards, it should not be installed in a vehicle used on any street or highway. Additionally, any such application could adversely affect the warranty coverage of such an on-street or highway vehicle.

The above is a typical disclaimer for off-road products. It does not apply when the owner can show a CARB Executive Order number.

4. Street or Track?

4.1 When are Parts Legal?

Some Corvette performance cars/engines, and those from independents such as Doug Rippie Motorsports, TPIS (Tuned Port Induction Specialties), and Lingenfelter Performance Engineering are street-legal even though they may use parts listed as not street legal. That seemed strange to me until they explained the difference between certified-legal cars/engines, and certified-legal parts.

- Changing one part, such as a throttle body may be street legal or it may not, depending on many other factors, such as camshaft timing and transmission.
- Changing one part without considering the rest of the engine and its control system could ruin idle quality or damage the catalytic converter.
- You can buy a street-legal complete engine, or a complete car that the modifier specifies as certified, with an E.O. number. Under government rules (USEPA, CARB) the engine or car can be certified as a complete system. In such a case, the tuner/manufacturer has demonstrated that the *modifications work properly with each other.*

Even without an E.O.#, modification parts may be acceptable:

- in California, if it is considered a *replacement* part
- in 49 states, if the manufacturer represents that it meets the requirements of USEPA Memo 1A or is not otherwise in violation of anti-tampering provisions of the Clean Air Act.
- if it is certified as part of a complete package

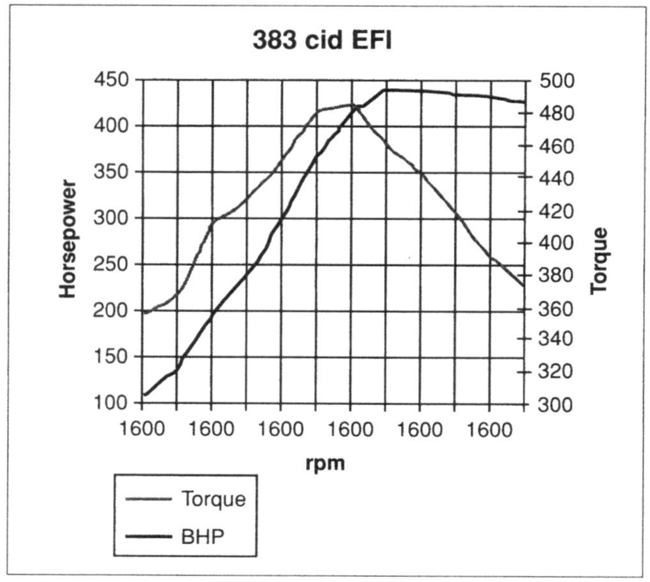

Fig. 9-8. You can buy a complete **LT1** engine, about $6600 and up, shown with an optional SuperRam intake system. Depending on the camshaft, it is certified 47/49-state street legal, E.O. #333. Individual parts may not be street-legal, but as a whole the engine performs within emission limits.

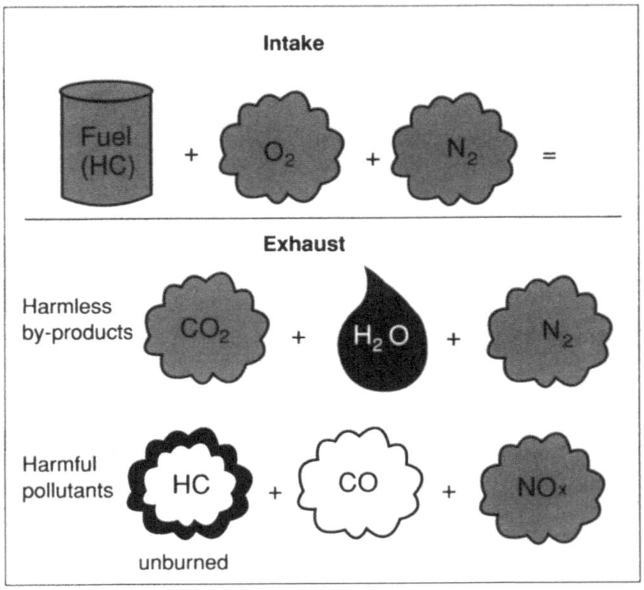

Fig. 9-9. Three exhaust pollutants are controlled by the Federal and California test procedures. But an *idle* test measures only two harmful pollutants. NO_X cannot be measured, because at idle there is no NO_X to measure.

Sometimes, the question of street-legal can get complicated. The **LT1** kit shown in Fig. 9-8 is a "crate engine", defined as follows:

- 50-state legal when installed by *Lingenfelter*
- 49-state legal (not CA) when installed by someone other than Lingenfelter in a **'92-92 LT1**, which is considered an OBD-*I*. CA law says something like, "No E.O. without a complete Lingenfelter installation manual."
- 47-state legal (not CA, NY, or MA) when installed by someone other than Lingenfelter in a **'96 LT1**, an OBD-*II*. Again, it says something like, "Not legal without a complete Lingenfelter installation manual."

4.2 Mods and Emissions

Your Corvette may require a smog test, usually every one or two years. Depending on the state, and the smog levels of the metropolitan area, your test may be one of these:

- a simple 2-speed idle test: HC and CO
- a dynamometer test under load: HC, CO, and NO_X
- a **'96-01** OBD-II check of the Diagnostic Trouble Codes

Emissions, and their effects on our health and environment, are becoming increasingly serious. Tighter regulations are increasingly being demanded, in response to continuing evidence of danger. Many more states now operate under the standards set by California. Such cars may be labelled: "CA/Northeast states", or "CA/NLEV," (California/National

Fig. 9-10. In the presence of the "tree," or a waving green flag, the whole scene changes. Some exciting examples of Corvette fuel injection mods can be found on tracks and strips.

Low Emission Vehicle). For 2001, the **LS1/LS6** are labeled "NLEV", National Low Emission Vehicle (all 50 states).

4.3 Planning for Performance

At first glance, it may seem that I'm out to discourage modifications to the fuel-injection system, but that is hardly the case. Time spent under the hood, investigating and experimenting with fuel-injection modifications is fascinating and educational. My intent is simply to help you:

- avoid wasting time and money
- avoid some of the more common and costly mistakes
- avoid getting your car labeled as a "gross polluter"
- recognize the OBD-II differences in calibrations and modifications

It seems that too many owners change a chip or throw parts at the engine, and then complain of driveability problems. You need basic knowledge to build the best running fuel-injected engine for your needs. For many applications, the stock fuel system is generally the best system, so my recommendation is to minimize your changes to your fuel-injection system unless you know they are really going to pay off.

If you haul your car to the track, legality is only an issue in terms of the applicable rule book. Emissions are not usually a factor at all in racing, although some tracks require catalytic converters for the backpressure. **ZR-1** is particularly sensitive to backpressure.

Consider this; before automotive fuel injection ever found its way into a passenger car, it was used on racing cars, by manufacturers seeking a more controllable and efficient alternative to carburetors. Later, the first passenger car applications of fuel injection began in the fifties on Corvettes and Mercedes. The point is this: Precise control of fuel delivery results in more complete combustion and a more efficient engine. Efficiency is the one of the basic building blocks for optimum horsepower, clean exhaust, and maximum fuel economy.

4.4 Driveability

Driveability is the delivery of smooth-running performance under a wide variety of operating conditions. The *cornering advantage* is evident in even the first mechanical fuelies of 1957-65. Fuel under pressure in the system goes through corners a lot better than fuel sloshing free in a carburetor bowl. A fuel-injection system measures many of the factors which affect performance, engine temperature for example, and compensates for different conditions to deliver the appropriate air-fuel mixture.

Any stock fuel-injection system is a compromise design that balances power output against concerns of driveability, fuel economy, and exhaust emissions. When you plan to modify fuel injection to deliver more performance, or to match engine modifications, you will be making tradeoffs. Many people overlook the tradeoffs. The important point is this; not all modifications will retain the driveability, safety or fuel economy of the original, unmodified system.

Fuel-injection driveability concerns are quite different from those with carburetors. For example, if the camming is too aggressive, fuel injection can be a total pain. Extreme fluctuations in manifold air flow and manifold pressure can send false signals to the computer. Mods that are suitable for **'85-93** engines may set codes in **'94-01** (OBD-II) Corvettes.

LS1, LT4, '90-92 ZR-1: Consider that these engines spare you EGR—I mean, that really is an achievement. Running to current emission standards is the payoff from the proper balance of valves, cam lifts,and profiles with combustion chamber shaping. Performance camshafts with large overlaps tend to introduce exhaust gases, tending to reduce NO_X as EGR does. Before you replace a camshaft, be satisfied that the car will pass an emission test if you want to use the car on the road.

The message of this section is this: For registered cars, driven on public roads, emission-control laws are a necessary consideration and the tests are getting tighter. You can stay ahead of the game by adding only certified modifications according to the laws of your state.

5. HIGH PERFORMANCE BASICS

For street use, I'll describe the general implications of feedback control found on all electronic fuel-injected Corvettes. What happens when you modify these systems? Remember, they all operate with oxygen sensors and catalytic converters to maintain stoichiometric A/F ratios for most driving—except warm-up, deceleration and W.O.T.

5.1 Principles of Air Flow and Volumetric Efficiency

Traveling downward on its intake stroke, each piston creates a pressure, lower than barometric, called a "vacuum". Barometric pressure pushes air into the cylinders mixed with fuel, where it burns to produce power. If you increase the air flow through the engine, MAF injection systems are programmed (within limits) to add more fuel and produce more power. For MAP systems, see below.

Volumetric Efficiency

If you know the volumetric efficiency, you can directly measure the pay-off of your performance mods which affect air intake. Some of the newer scan tools read volumetric efficiency, (see Chap. 2). In theory, the amount of air that is taken in by an engine is determined by displacement and rpm. In practice, other factors reduce the theoretical maximum:

- Valve timing, valve lift and manifolding limit the amount of air that can be taken in on the intake stroke; exhaust systems limit how much can be pumped out on the exhaust stroke. When we say, "volumetric efficiency", we use the term to describe how well the engine pumps air—the true value as compared to the theoretical 100%.
- In the real world, automotive engines are not very efficient air pumps. Volumetric efficiency is reduced on the intake side by the air filter, the air-flow sensor (very little), the throttle valve, the EGR spacer, and the intake manifolds and ports. They impede the free flow of air into the combustion chambers. Volumetric efficiency is further reduced by the restrictions of the exhaust system—exhaust manifolds, catalytic converters,

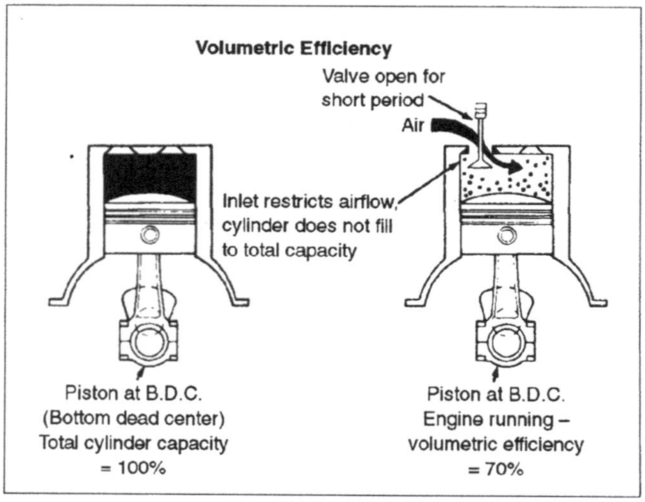

Fig. 9-11. Volumetric efficiency is measured as a percentage of total cylinder volume. If the cylinder takes in only 70% of its swept volume, the Volumetric Efficiency is 70%.

mufflers, and tailpipes. But believe me, Corvette engineers work hard on designs to move more air into and out of their engines—that's part of what they sell.

The **LS1**'s intake is designed for volumetric efficiency, and that means low restriction, less than a 5 kPa pressure drop from the BARO to the rear of the intake manifold. Glass-filled nylon intake runners are thermally insulated to maintain cooler intake air, and they are polished on the inside. All this takes in more air for more power.

Speed-density/MAP Systems are Affected by Mods

L83, '90-91 L98, '92-93 LT1 , ZR-1: When you increase the intake air flow, does the MAP system know enough about the increased flow to add fuel properly? The answer to that is: *seldom*. Believe me, the difference between *MAP* and *MAF* is far more than one letter. When it comes to mods, MAP and MAF measurements of engine load are *two different ball games*. Remember, the MAP is the primary sensor in **'90-91 L98**, **'91-93 LT1**, and all **ZR-1.** Beginning in **'94**, MAP is secondary to the primary MAF.

Speed-density systems (MAP) do not measure airflow directly. Instead, the computer *estimates* the airflow, based on stock cams, valves, and intake air manifolds. It starts with the MAP and Intake Air Temperature (IAT), modifies it for barometric pressure (BARO), and adds data from the computer memory. Many performance mods—not only ones for the engine, but also exhaust and axle ratio changes—compromise the accuracy of those calculations. They change the real air flow to be different from the estimated one. In contrast, the MAF measures the real air flow. For more, see Chap. 4, "Sensors".

Increasing Air Flow

Owners of MAP-equipped cars face a good deal of tuning, a lot more than required for MAF-equipped engines. What we're talking about here is changing calibrations. That's why some tuners will sell you a chip *designed for your car*, based on their own dyno testing of your car.

Nearly all the performance mods, from the widely-used and simple K&N air filter to the costly supercharger, have one common goal, to increase air flow through the engine. The gains may be tailored to the middle rpm range to improve torque, or to the high rpm range to maximize peak horsepower, but the idea is the same. Higher lift and longer duration cams, larger valves, ported cylinder heads, aftermarket intake manifolds, exhaust headers, and cat-backs all have the same job. They reduce air-flow restriction. The **LS1** intake system, with its large 74 mm MAF and intake manifolding, is well suited to increased outputs, up to 500 horsepower.

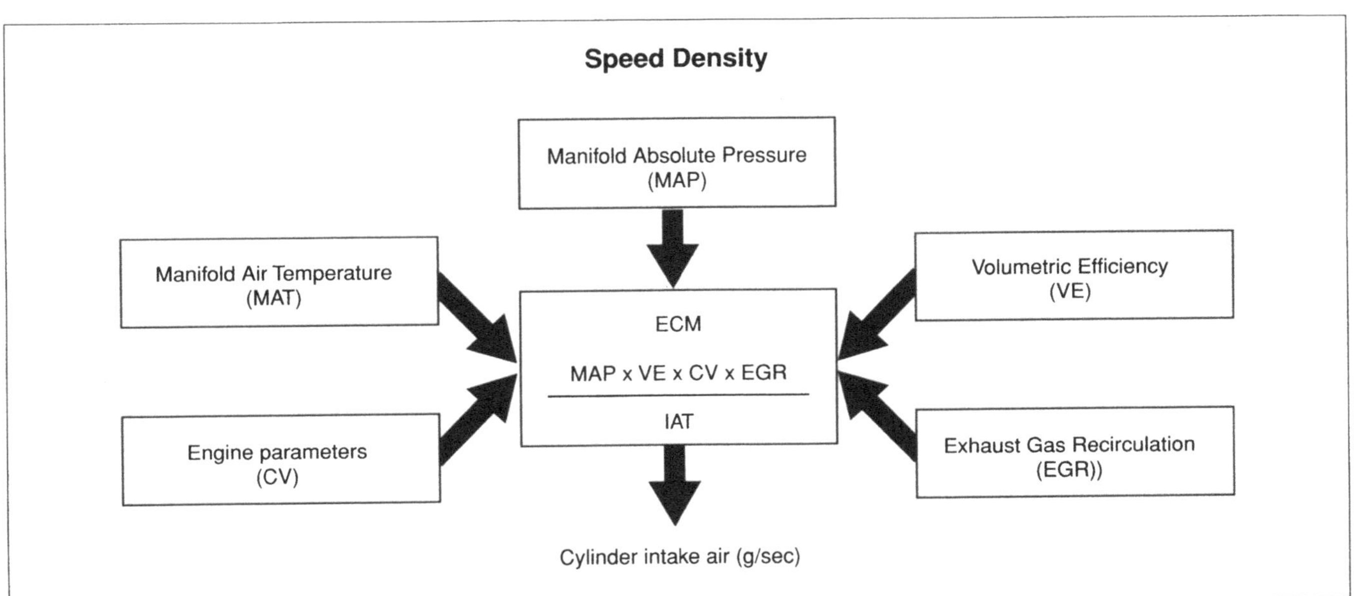

Fig. 9-12. Speed density systems combine Manifold Absolute Pressure signals with Manifold Air Temperature (Intake Air Temperature-IAT). The engine's computer considers engine parameters, like Volumetric Efficiency and EGR rate, to calculate the estimated mass air-flow rate.

5.2 Principles of Fuel Metering

Naturally, when engine improvements allow increased air flow through the engine, the fuel system must compensate. It must deliver a proportionally greater amount of fuel to maintain the proper air-fuel ratio, or the engine will run lean. In practice this may not be a problem, as stock Corvette systems are quite flexible and are able to compensate for some impressive air-intake increases. The air flow and fuel delivery capabilities of the stock fuel-injection system have been chosen to correspond with the performance demands of the engine—its volumetric efficiency. Beyond a certain point, modifying the engine changes the volumetric efficiency so much that you will need to modify the original fuel-injection system; for example, with higher-flow injectors.

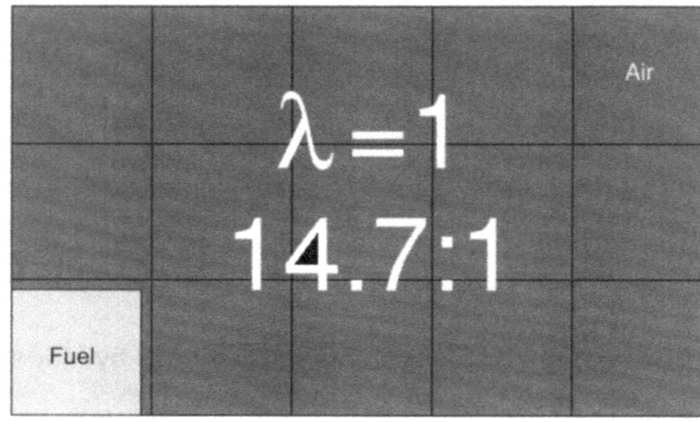

Fig. 9-13. Delivering exhaust gas in the A/F stoichiometric ratio (between 14.63 and 14.77:1) is required to operate the catalytic converter. Stoichiometric is also known as "lambda = 1", and is the correct amount of air needed to burn all of the fuel.

5.3 Principles of Air-Fuel Ratio and Performance

Under normal, part-throttle running conditions, operating in closed-loop, the systems precisely adjust the air-fuel mixture for minimum exhaust emissions. Corvette fuel-injection systems recognize full or near-full throttle acceleration as a special condition with special requirements.

The systems also provide a richer mixture to meet the brief demand for maximum power. At or near Wide-Open Throttle, during heavy acceleration, during rapid throttle opening:

- The computer increases the injector pulse-times, ignoring O_2S signals. The computer says to the O_2S: "*I know, I know, we're running rich, and that's the way I want it! Not stoichiometric now; that's for warm cruise*."
- The Fuel-Pressure Regulator increases the injector pressures to match the increased manifold pressures. Exception: the **'97-98 LS1** increases *pulse times* to match increased manifold pressures.

And that's before you add any performance mods. Emissions increase at W.O.T., but on the road the tradeoff is acceptable, because of the short periods of time spent at full throttle.

Power and specific fuel consumption both vary with the air-fuel ratio. The stoichiometric ratio, though, is not necessarily the optimum ratio for peak power or for minimum fuel consumption.

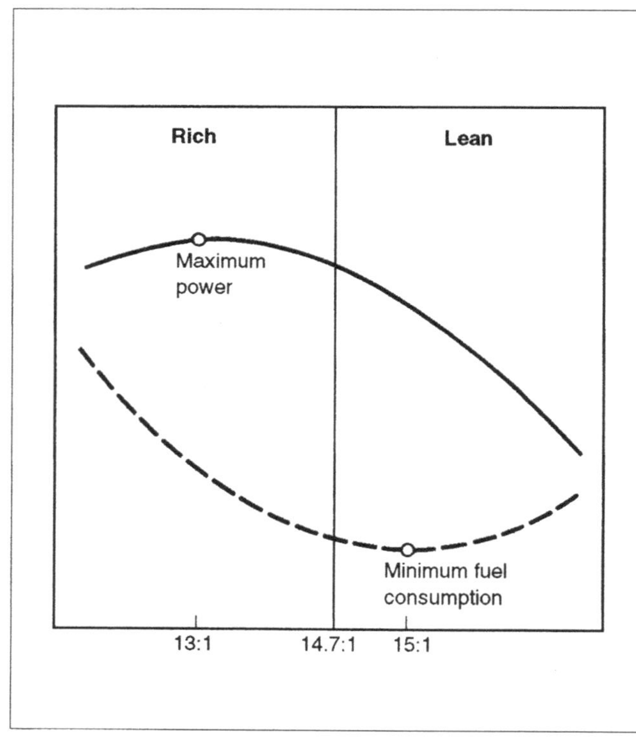

Fig. 9-14. Peak power is achieved with a slightly richer air-fuel mixture, approximately 13:1. Minimum fuel consumption is achieved with a slightly leaner mixture, about 15:1. Emissions are another matter entirely.

5.4 Principles of Adding More Fuel

First of all, except perhaps for track racing, you'll want to add more fuel only if it can combine with more oxygen, to provide more power. In theory, it is possible to fine-tune the air-fuel mixture, either to maximize power or to minimize fuel consumption. This is tempting, of course, but take a closer look.

- The computer already provides mixture enrichment at or near W.O.T. so some of what you might hope to gain by optimizing the mixture for maximum power is already there.

- Reconsider the curves in Fig. 9-14. In the areas of interest—near the maximum power point and the minimum fuel consumption point—those curves are relatively flat. Even if the system can be adjusted to deliver the perfect mixture (just at the point of maximum power), any gain promises to be pretty small. No huge amounts of horsepower to be unlocked here. And if you miss and go too rich, its easy to end up de-tuning instead of improving!
- On one hand, your engine might benefit from adding fuel at the maximum rpm, or on the other hand, at the maximum torque, probably at 3.000-4,000 rpm.
- Adding fuel at idle may trouble you with surge or a lumpy idle and poor tip-in.

You can add fuel at Wide Open Throttle, but weigh the pros and cons. An increase in the *injection pressure* results in:

- *Plus*: it is relatively simple and low cost
- *Minus*: it requires an adjustable Fuel Pressure Regulator, overfueling is cancelled by the O_2S in Closed Loop, higher pressures may lock up injectors, there is limited power for the injector solenoid coil, there is a possibility of erratic operation, and it may add fuel improperly at some rpm ranges

Installing *larger capacity injectors* (OEM 24 lbs/hr to 30 or 36 lbs/hr.) results in:

- *Plus*: it is more reliable
- *Minus*: in **L98, LT1 '92-93** there is little or no gain with a batch-fire injector system (it works better with SFI, **'94-01**), it requires recalibration of the PROM or PCM, too big a jump may cause problems at idle or at specific points on the curves, it costs more

Consider trying both, increase the pressure (but by less) and install larger injectors (but not by as big a jump from the OEM injectors).

Note that **'99-01** Corvettes have *smaller injectors* with longer injection pulse times, for better computer control.

Of all the methods that try to optimize the air-fuel mixture for peak power, the ones that provide more gain than pain are likely to have extensive dynamometer testing and road testing behind them.

5.5 Principles of Closed Loop Systems and Performance

Corvette fuel-injected cars are equipped with up to four oxygen sensors and one or two catalytic converters. The relatively clean engine exhaust, resulting from combustion at the stoichiometric ratio, is necessary for proper operation of a catalytic converter. Any significant deviations from stoichiometric A/F ratios will increase engine-exhaust emissions dramatically; if the catalytic converter cannot convert then the tailpipe emissions will go off the charts. As shown in Chap. 3, Fig. 3-17, as the mixture becomes rich, hydrocarbons (HC) and carbon monoxide (CO) go up. As the mixture becomes lean, oxides of nitrogen (NO_X) increase very rapidly.

What's more, closed-loop systems are affected by misfires, such as a bad plug or loose plug wire. **'94-01** OBD-II Corvette computers treat misfires very seriously, and record even one or two misfires. With or without OBD-II, I recommend that you, also, treat misfires seriously, for there is the danger of the other 7 cylinders running lean.

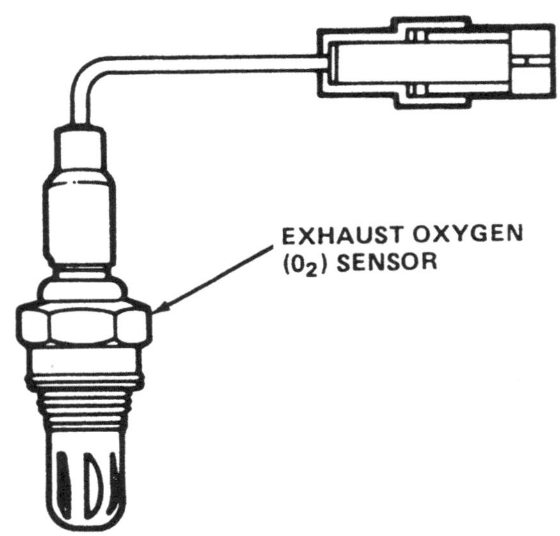

Fig. 9-15. The HO_2S (Heated Oxygen Sensor) may improve the ability to stay in Closed Loop, even if reduced backpressure, turbochargers, or other mods fail to maintain exhaust temperatures above 600°F (315°C).

'82-93: With reduced backpressure, unheated oxygen sensors (O_2S) may not get hot enough to get into—or stay in—Closed Loop. You may see a flashing SES light, or repeated switching between Open Loop and Closed Loop. If you do, you may want to change your unheated O_2S to *Heated* or *HO_2S*. Of course, that involves wiring for power to the O_2S heater.

LT1 '94+, LT4, LS1, ZR-1 engines generally remain in Closed Loop with HO_2S.

9

'94-01 LT1, LT4, LS1, LS6, ZR-1 can show Closed Loop in the Field Service Mode (with the diagnostic terminal grounded, engine running, and the parking brake on).

Table 9-1.Field Service Mode

	DVOM Swings	SES Flashes
Open Loop	Rapid about 2.5 per second	Rapid
Closed Loop	Slow about 1 per second	Slow
Lean Exhaust	Low Less than 0.45 v.	Off most of the time
Rich Exhaust	High Higher than 0.45 v.	On most of the time

On the street, you want the catalytic converter for reasons of emissions and inspection. In addition, these feedback systems provide driveability and fuel economy advantages. It's important that the system continuously fine-tune the air-fuel mixture to match different conditions, even different fuels. What's more, the computer maps for air-fuel ratios and timing are set for the exhaust-system back-pressure of the stock exhaust with its converters.

Remember, even if you modify the system to make it capable of providing more fuel, in most part-throttle (Closed Loop) operation, the oxygen-sensor system will still do just what it was designed to do. It will continuously adjust the air-fuel mixture to approach the stoichiometric ratio, the narrow range around 14.7:1 (with gasoline). In short, no gain, except maybe in open-loop operation at full throttle.

Problems may come when fuel system modifications force the system to the limits of its normal range of adjustment.**'87-91 L98, LT1, LT4, ZR-1, LS1:** With Adaptive Control, the system may constantly sense an over-rich mixture. When it tries to adjust to be more lean in Short Term Fuel Trim (STFT), formerly Block Learn, it reaches the limits of its adjustment range. In such a case, the modifications and the resulting rich mixture will override the oxygen sensor's ability to correct, and will hurt driveability and exhaust emissions. It may even hurt performance. Overrich mixtures can burn out catalytic converters, and may even cause fires.

Instead of working within a fixed range of feedback-control parameters, adaptive systems are able to accumulate data during operation. Long Term Fuel Trim (LTFT), formerly Integrate, *adjusts the center of the operating range* according to the needs of the individual car, to this individual driver, even to the fuel of the current fill-up.

5.5 Principles of Ignition Control

Ignition timing is another performance factor that goes hand-in-hand with fuel delivery. The manufacturer's conservative *fuel*-mixture control is done in the name of low exhaust emissions. This may also be accompanied by OEM's conservative *ignition* timing curves, primarily because of concerns over the variable quality of available gasoline. Many tuners of fuel-injected cars advocate revised ignition timing specifications to further unlock the performance potential of revised fuel control.

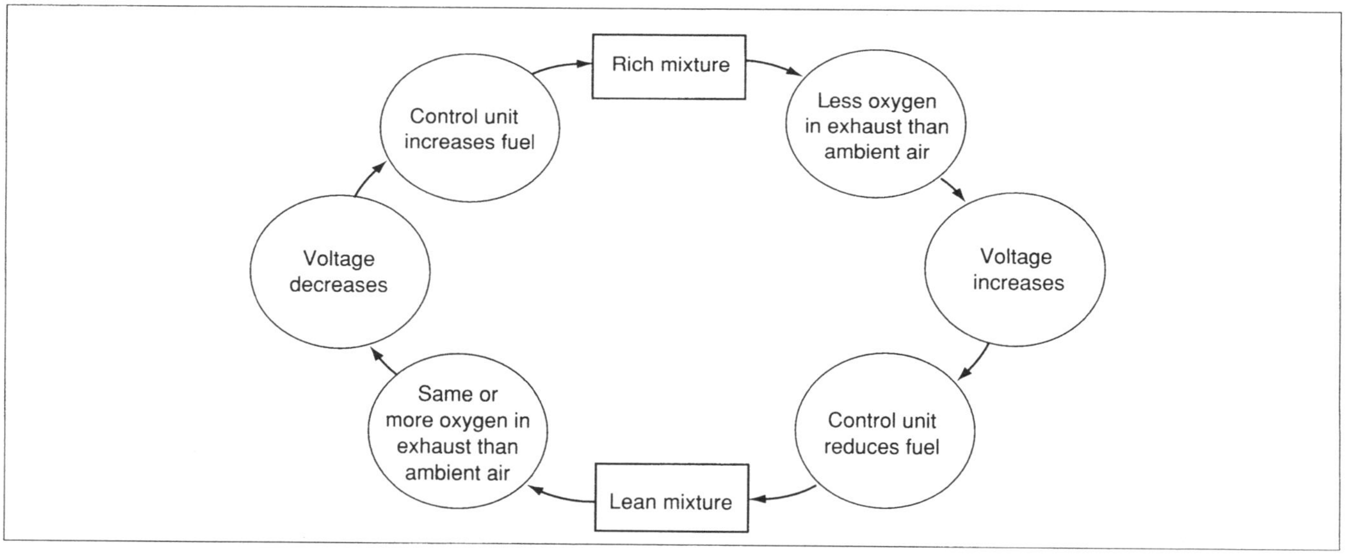

Fig. 9-16. In closed-loop operation, the oxygen-sensor system will continuously try to correct the air-fuel mixture to a stoichiometric value—no matter what modifications (up to a point) have been made.

By law, all *OEM engines must operate on regular (87 octane) fuel*, even when the Owner's Manual recommends premium. For Corvette and many engines, that means Knock Sensors, and it means conservative OEM spark timing. The 87 octane operation requirement *does not apply to aftermarket mods.*

To quote driver's Manuals:

LS1: "Use premium unleaded gasoline rated at 91 octane or higher for best performance. You may use middle grade [88 octane], or regular [87], but your vehicle may not accelerate as well."

ZR-1: "Mild knock or detonation can briefly occur on 87 octane fuel in severe conditions, such as hard acceleration uphill from low rpms. This is normal and will not harm the engine. Consistent detonation should not occur using 87-93 octane fuel."

For a number of good reasons, including that it is a Federal offense to burn leaded fuel on the highway, and that lead in your fuel will 'lunch' your oxygen sensor(s) and your converter(s) in a few tankfulls, do not use leaded fuel, "racing fuel", or fuel additives.

For street and most driving:

- Spark timing is better when based on MAP
- Fuel control is better when based on MAF

Table 9-2.Comparison of MAP and MAF

Load sensing	Year/ engine	Characteristics
MAP	**'82–84 L83, '90–92 L98, '93 LT1, all ZR-1**	MAP reacts faster to rpm/load changes than MAF, but needs reprogramming for most engine mods—camshaft, compression ratio, exhaust back pressure, ignition.
MAF	**'85–89 L98, LS1, LS6**	MAF controls fuel injection and timing. MAF handles most mods by direct measurement of intake air mass.
MAP + MAF	**'94-01 LT1**	MAF is primarily fuel injection. MAP is for spark timing and backup for fuel injection.

6. Bolt-Ons and Modifications

The efforts of tuners are interesting and exciting. They also provide valuable insight into how the experts think about the problems and how they approach solving them. These insights, combined with your knowledge of the system basics, should help you decide which modifications hold promise for your application.

"IF SOME IS GOOD, MORE IS BETTER, AND TOO MUCH IS JUST RIGHT."
Tuner's creed.

Some of these modifications are well-known and well-tested. I can give you a fairly accurate idea of the results, if any, that you might expect. For many others, the suppliers' results are much less certain. I've described these modifications and the theories by which they are supposed to work. I'll give you the plusses and minuses as I see them.

6.1 Increasing the Air Flow

To increase the W.O.T. airflow, reduce air restriction, supercharge, and the open exhaust:

- install a low restriction air filter
- change to a larger throttle body
- change to a larger Mass Airflow Sensor
- polish the intake manifolds (and exhaust manifolds)

Air Filter

K&N Engineering offers widely-used, low-resistance air filters. They reduce the restriction of the original air ducts and filter, thus increasing the intake air. K&N claims horsepower increases of *up to* 15. They are available for all **'82** and later Corvettes. Follow the maintenance instructions.

- **L83, '90-91 L98, '92-93 LT1:** The speed density system may need remapping to avoid lumpy idle and poor tip-in.
- **'94-96 LT1, LT4, LS1, LS6, '85-89 L98:** The MAF compensates for increased intake air, no problems.

Fig. 9-17. Replaceable air-filters are an inexpensive way to improve your system.

K&N's low-resistance air filter results in:

Plus: It increases the air flow up to 20%, it's reusable for the life of the car, and it's inexpensive. It's also easy to install (usually 5-20 minutes installation), and is certified street-legal.

Minus: The filter requires cleaning and oiling at every 15,000-30,000 miles, (a cleaning kit is about $10). Expect increased intake noise.

Fig. 9-18. LS1, LS6: High-flow washable air filter is designed for easy installation. This one is from Lingenfelter.

A high-flow washable air filter results in:

Plus: It reportedly increases rear-wheel horsepower, and is easy to install.

Minus: It's expensive, about $270, it's not certified street-legal, and it also requires a cleaning kit.

For those of you considering operating with the air cleaner removed, consider these warnings from the Owner's Manual:

NOTICE:

If the air cleaner is off, a backfire can cause a damaging engine fire. And, dirt can easily get into your engine, which will damage it. Always have the air cleaner in place when you're driving.

Throttle Body

Larger throttle bodies are popular. They flow more air, for improved performance at W.O.T., but be sure you know the effect of too much intake air on idling. **L98, LT1, LT4:** If you want to replace the stock twin-48 mm throttle body, use the twin-*52* mm. For Small Block Chevrolet (SBC), you will get all the intake air you can use at W.O.T., and still give you driveability and throttle response. A 58 mm throttle body would affect A/T shifting because it operates with different throttle positions. Save the 58 mm for extensive mods and for Big Block Chevrolet (BBC).

LS1: A single-74 mm throttle body will probably handle all the intake air the engine needs, for a while.

L83, '85-89 L98: The throttle bodies use an *adjustable* Throttle Position Sensor, 0.5v. at closed throttle. All **'90-96** use a fixed TPS, which automatically "zeros out" in the computer. These match any replacement TPS, although an adjustable one can substitute for a fixed one by setting it for start.

LT1, L98: A 52 mm throttle body can replace the stock 48 mm one. You can get a street legal replacement from TPIS, E.O. # 235, or Lingenfelter, E.O. #333:

The only negative is that a replacement throttle body flows about 5% more air with AirFoil than without, above 4500 rpm.

Fig. 9-19. L98, LT1: The TPIS patented Throttle Body Airfoil installs in about 10 minutes. About $60.

Mass Airflow Sensor

The Mass Air Flow (MAF) sensor is not easily serviceable, and cannot be electronically modified. It may look as if it could be serviceable, but be aware of hidden screws and solder joints. Also, be aware of the delicate wires.

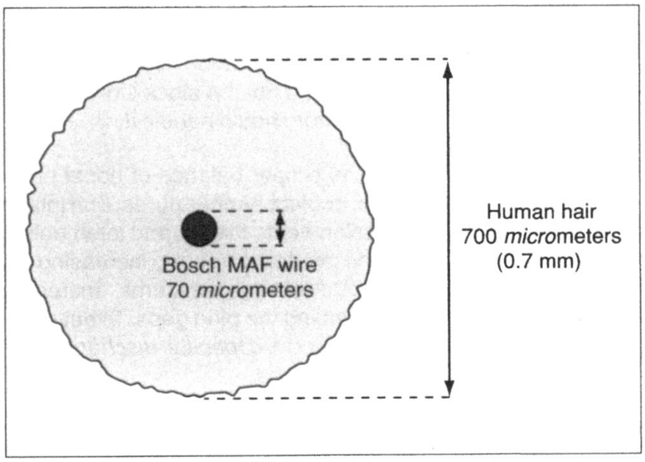

Fig. 9-20. Remember, the Bosch MAF hot wires are 1/10th the size of a human hair! This should discourage you from removing the screens.

If you do remove the MAF, check the arrow on the case to be sure that you are replacing it in the proper direction. It fits backward, but you'll have a lumpy idle.

Manifold

Intake and exhaust manifolds are part of flowing more air through the engine. But remember what I said in Chap. 2 about tuning the length of the intake runners according to where on the rpm curve you want the most ram effect.

- For higher rpm engine performance,and top speed, use short manifold runners.
- For low-mid rpm engine performance, launch and acceleration, use long manifold runners.

Fig. 9-23. This TPIS plenum chamber feeds *short* runners for the *same*-side cylinders. It's about $900, plus a Fuel Rail kit. The manifold is *lower*, if lowering the hood is of interest. It is reported to work well with the factory computer chip.

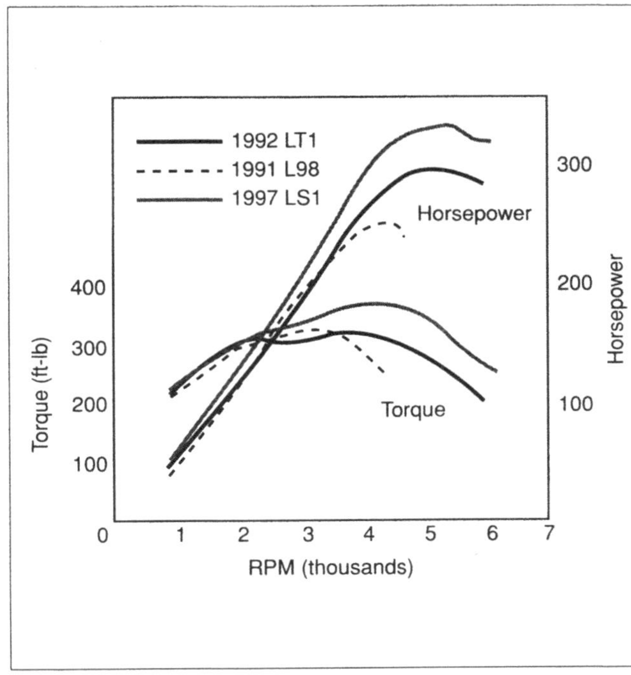

Fig. 9-21. **LT1:** The relatively flat torque curve, compared to the **L98**, shows that the **LT1** gives up a bit of torque at about 3,000 rpm. It gets a much flatter torque curve, and more torque above 4,000, and below 2000.
LS1: The power curve shows an increase over the **LT1**. And a similar torque curve also shows gains, mostly at the higher rpms.

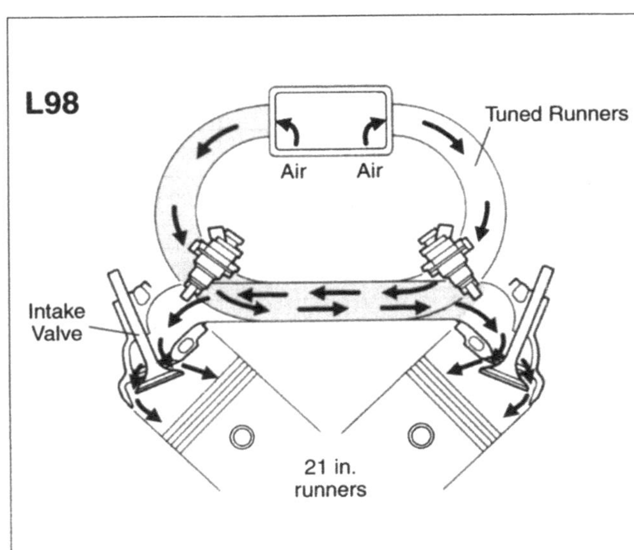

Fig. 9-22. **L98:** The stock TPI is 21 inches (530mm). Corvette's OEM *long*-runner resonance with valve timing is the basis for the *opposite*-side runners for port injection *mid-rpm* performance.

9

Fig. 9-24. This 383 SuperRam from Lingenfelter is for the **LT1** and **L98**. It is 383 cu. in, has an increased stroke, ported cylinder heads, 383 hp@6100 rpm, and 383 lb.ft.@4900 torque, with a peak at higher rpms. E.O.# 333

Supercharging

You want more air intake? *Push* it in. This is big time, a big increase for big bucks. A turbocharger is a form of supercharger, so I'll use the "supercharger" term in general, while distinguishing the mechanical superchargers from turbochargers according to current usage.

Specific systems may vary but these general numbers will give you a basis for comparison.

Table 9-3.Comparison of superchargers and turbochargers

	Mechanical Supercharger	Turbocharger
Drive	Belt and internal gearing	Exhaust turbine
Maximum rpm	60,000	120,000
Net horsepower @ rpm	50 @ 60,000	100 @ 120,000
Gross horse-power required for 100 hp gain @ 10 psi boost	150	100
Sense of acceleration	Begins at low-to-mid speeds	Stronger at mid-to-high speeds

Your first decision, when considering supercharging is:

- Do you want torque increases at *low-mid* rpms? If so, then go for a *mechanical* supercharger.
- Do you want horsepower increases at *mid-high* rpms? If so, then go for an *exhaust-driven turbocharger.*

Your engine needs extra fuel to match the extra air. When you supercharge a 5.7L Corvette to about a 7 psi boost, you're delivering air as if for a *8.7L displacement*, an increase of about 50%. (See Section 5.4 of this chapter for ways to add fuel.) In fact, if fuel flow fails to provide sufficient pressure, then a lean-burn will quickly cause detonation and "lunch" the engine. For outputs up to 300-400 hp, the stock Corvette fuel pump and fuel pressure regulator should handle it.

Supercharging requires the proper balance of boost pressure, spark timing, plug gap, coolant temperatures, and intake air temperatures. Compression heats the air, and intercooling can be used to reduce the air temperature, increasing its mass. When supercharged cylinder-pressures increase, sparks have a harder time jumping the plug gaps. To get more voltage at the gaps, you may need a *capacitor-discharge* ignition, and some heavy-duty plug wires.

Vortech recommends the following:

- Use premium fuel (there's a limit to KS spark retard).
- Always use an oil filter and keep renewal air filters (K&N) clean and oiled.
- Use a premium SH rated oil and OEM style oil filter, and change it every 3,000 miles. Since **'94**, Corvette factory installs synthetic oil in the crankcase.
- Never use platinum plugs unless OEM specified (superchargers normally reduce plug life, so use cheaper plugs and change them often).
- Change plugs every 15,000-20,000 miles. Check that the plug wires are within factory-specified resistance. Check the distributor cap and rotor every 15,000 miles. Do not use plug wires or a cap/rotor for more than 45.000 miles.
- Avoid high rpm operation until the engine/supercharger reaches full operating temperature.
- Avoid an aftermarket chip unless you know it is supercharger-compatible and that it will not increase detonation.

You may want to apply these Vortech instructions to any supercharger installation and operation. Remember, you are asking your engine to put out a real increase in its original output.

Representative Supercharger Installations

PROCHARGER (Accessible Technologies, Inc.)

Procharger offers an air-to-air intercooler. With the intercooler, the reduced intake-air temperature improves the air density. That's more pounds per kilograms, and more power. Compression raises the supercharger's outlet temperatures. Without intercooling, higher air temperatures reduce intake air quantity, and require spark retard.

- **L98:** Using standard gas, Procharger claims that the horsepower increase is 25-30%, and with the optional intercooler the increase is 60-70%.
- **LT1 '94-96:** There is a stage 1 boost of 7 psi (14 in.Hg.). Procharger claims outputs of around 420 hp. A higher boost of 9-15 psi (18-30 in.Hg.) produces 450-525 hp, but isn't recommended for long term use.

An optional Crane HI6 ignition system allows use of the factory plug-gap. With an OEM ignition, you must reduce the plug-gap.

Procharger's centrifugal supercharger uses an overall gear ratio of 7.3:1 to operate at 36,500 rpm/5,000 engine rpm. Corvette's P600B (for billet impeller) limit is 60,000 rpm.

The Procharger centrifugal supercharger's results are:

Plus: Stage 1 uses stock injectors and factory spark timing. The intercooler acts as a passive wastegate, flattening the boost curve at higher rpms. Lower intake air temperature reduces the tendency to knock, and permits running on pump gas. This supercharger is street-legal.

Minus: Installation requires moving the generator. Stage 2 requires larger injectors, high-octane race gas, or ignition with built-in retard. It also requires a high-volume fuel pump.

A Procharger kit with an intercooler costs about $5,000, without intercooler it goes for about $4,400. Superchargers alone cost about $2,000-3,000.

As the supercharger adds boost, a Fuel Monitoring Unit (also known as rising-rate Fuel Pressure Regulator) can increase the absolute fuel pressure by raising manifold pressure. It costs about $125.

VORTECH

Vortech superchargers are available directly from Vortech, and are used by other **LT1** tuner/suppliers, including Lingenfelter and Doug Rippie Motors, Tuned Port Injection Specialties (TPIS). Vortech superchargers result in:

Plus: They have a fast response at low-mid rpms.

Minus: They require spark retard and added fuel, during boost. The **LT1** is limited to a 8 psi (16 in.Hg.) boost. They may also require a high-delivery fuel pump, and there is some gear whine.

Fig. 9-25. LT1, '92-93 LT4: Vortech's VS1 kit costs about $5020 and includes all parts, including the necessary higher-flow injectors. The **'94-96** kit costs about $4900 and uses *stock* injectors. Installation time is about 16 hours. Up to 452 HP, and 470 lb.ft Torque. E.O.# D-213

Fig. 9-26. LT1, LT4: Vortech's supercharger installed. It has 10 psi (20 in.Hg.) boost superchargers, is street legal, and can maintain the original Corvette warranty. Other models are available with a higher boost, but they are for racing, and aren't street legal.

Fig. 9-27. Lingenfelter's **LS1** twin turbo with air-to-air intercoolers. This package shows the cutaway, in limbo. Installed, it's just about out of sight. The Stage I installed price is about $23,000. The Stage II version of this engine, rated at 650 hp@5800 rpm / 600 lb.ft. torque@5000, turned 0-100 mph in 6.7 seconds, and measured 226 mph. Stage II costs about twice Stage I.

In summary, about superchargers, forced-air induction raises the intake temperature. Some supercharger systems benefit from intercooling to reduce the intake air temperature. With a centrifugal blower, an intercooler can reduce the air-charge temperature by about 100° F (55° C), with a pressure drop of about 1 psi (2 in. Hg.).

9

You can see that increasing airflow has many dimensions, and your considerations are interrelated. How much horsepower you can achieve is related to several variables:

- Throttle body size
- Intake manifolds size
- How about smoothing the intake manifolds? Porting?
- Supercharger? Turbocharger?
- Headers and exhaust mods

6.2 Other Mods

Increasing the Fuel Injected

To achieve significant fuel-delivery increases without extremely high fuel pressures, you may also want to install larger injectors. At near-stock fuel pressures, you can expect poor fuel atomization. At idle, the minimum injector pulse-times may still add too much fuel, causing a lumpy idle. But a combination of larger injectors and boosted fuel pressure can produce mid-range and W.O.T. fuel delivery gains.

> ***CAUTION —***
> *If you install injectors other than ones with Corvette-part #s, be sure that the injector impedance matches the original Corvette injectors.*

This is no simple fix. Increasing the fuel pressure or installing larger injectors will increase fuel delivery throughout the engine-operating range, even though it is really only needed at mid or high rpm. So with modifications of this type, you must make a corresponding change to maintain the correct air-fuel mixtures at the low and middle rpm ranges. You may do one of three things: alter the inputs to the computer, reprogram the computer, or alter the computer's outputs.

Tuned Port Induction Systems, Inc. told me that they spend a lot of time on the phone trying to help customers who have been talked into larger injectors.

High-Flow Fuel Pump

If you're moving considerably more fuel (and I mean more than just a little bit more) through the injectors, you probably need to move more fuel from the tank to the fuel rails. Considering that inadequate fuel delivered to the fuel rails can mean lean mixtures and a "lunched" engine, most tuners opt for a higher-flow pump.

The Bosch high-flow fuel pump can increase delivery for up to 500 horsepower. It fits into an in-tank unit to eliminate high-end fuel starvation. It is CARB street legal and is not recommended for use with nitrous oxide. Some tuners report replacing the stock pump with a **ZR-1** dual fuel pump and controls.

Fuel Pressure Regulator

Increased pressure increases the amount of fuel injected per millisecond of pulse time. Remember, the oxygen sensor signals tend to cancel out this effect, except in W.O.T. (above 80% TPS). You may want to replace the stock Fuel Pressure Regulator with an adjustable unit. And, obviously, to set an adjustable FPR, you need a gauge.

Installing a new Fuel Pressure Regulator results in:

Plus: You will get more accurate control of the fuel pressures, which is especially important with the **LT1** and **LT4**. It tends to improve fuel atomization. And better atomization equals better burning, which means more power, and usually, lower emissions.

Minus: It can reduce performance and increase emissions because of overly rich mixtures. Excess pressure can lock the injectors closed or open. Either way, that's bad news.

Knowing what's right takes a bit of tuning. And what's right varies with different Corvette engines. Lingenfelter tends to stick with stock fuel pressures. Know what you're doing, or you'll get more pain than gain.

Table 9-4.GM's recommended range of fuel pressures

Engine	psi	kPa[1]	Notes
L83	9–13	62–90	TBI
'85 L98	34–40	235–275	MFI
'86–87 L98	37–43	255–300	MFI
'86–91 L98	41–47	285–325	MFI
LT1, LT4	41	285	SFI ('94-01)
LS1[2]	55–61	380–420	SFI
ZR-1	48–55	330–380	SFI

[1] Most kPa values are rounded to the nearest 5. Few gauges can read closer than this.

[2] **LS1** pressure is not affected by manifold pressure, so it remains the same whether the engine is on or off.

(See also Fig. 7-31.)

Tuned Port Induction Specialties also recommends:

- **L98** 49-52 psi (340-360 kPa)
- **LT1** 39 psi (270 kPa), it is more sensitive to fuel pressure variations than other engines

With disagreements about pressure recommendations, I suggest trying different pressures for your kind of driving.
In this case, the previously mentioned tuner's creed does *not* apply to fuel pressures. More is not always better, and too much may be just that, too much.

Chip Modules and Chips

You may want to replace the OEM chip, or change its input/output signals. The factory chip is designed to satisfy the whole range of conditions that Corvettes may encounter. It is calibrated for the government requirements for emissions and fuel economy. And it must remain untampered with for 6 months after a new-car delivery, if you want to retain your factory warranty. (See Chap. 3 and 4).

You may be operating at more specific air temperatures, and barometric air pressures. And you may decide that you want more low-end performance, or more high-end performance. Changing the chip to change the injection and spark-timing maps is usually simple to do, but the results are unpredictable. Results can vary according to the specific Corvette model, and even between two different "identical" vehicles.

- **'82-93** You probably don't want to replace the original microprocessor. Look for a chip-module to be connected between the engine-control module and its harness. You can modify the output of the factory installed computer.
- **'94-01** *EEPROM chips must be reprogrammed in the existing computer*, erasing all data and replacing it electronically. EEPROM chips are not removable. This is normally done at your friendly Chevrolet dealer, but can also be done by a few tuners.

For the first EEPROMs, in **'94-95**, reprogramming is relatively simple, and can be done by connecting the Tech 1 or 2 (or comparable scan tool) to the cigarette lighter and the Diagnostic Link Connector (DLC). (These systems are not true OBD-II, but a 16-pin DLC.) Follow the instructions for the scan tool. For the true OBD-II cars, **'96-01**, reprogramming may be more complex. Follow the instructions for your scan tool.

Hypertech Power Module

LT1, LT4: Hypertech, Inc. offers a Corvette power module for plugging into the original engine computer. There are different modules for 49-state and California cars, for manual transmission and automatic, and for specific rear-axle ratios, because the original engine calibrations are different. (Also, check NY and MA Corvettes, as some follow CA specifications)

Stand by for cautions:

- Check for detonation. Advance the spark timing and retard up to 2° if you hear detonation.
- Premium fuel (92-93 pump octane) may be required.
- Low-temperature thermostats may be required for competition chips, to lower the possibility of detonation, and to run with cooler, denser intake air.
- Look out for injector pulse-widths of more than 80% duty-cycle. There are two possible results of a control failure. Either the injectors remain fully open, as if there were a 100% duty cycle, creating an excessively rich A/F mixture. Or the injectors remain half-locked "stuck in the middle", creating a real danger of a lean mixture.

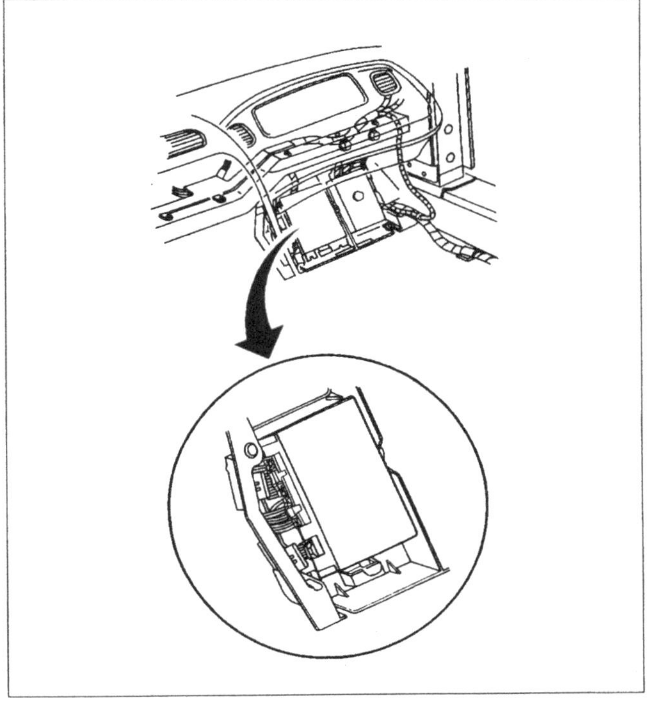

Fig. 9-28. EEPROM reprogramming may require more than plugging the scan tool into the Diagnostic Link Connector. Remove the passenger-side floor panel and install two star connector cables.

A word of warning for the chip-changers: In an SAE paper, the EPA reported on tests of several Aftermarket Programmable Read-Only-Memory (APROM) chips in performance cars. All increased emissions:

- HC up to 27%
- CO up to 59%
- NO_X up to 53%

And fuel economy went down. Performance chip modules may (or may not) improve your performance, but they can definitely be a threat to air quality. To a Roadside Remote Sensor, *they show up in less than 1 second.* Be wary of so-called "street-legal" chips, unless they include a CARB Executive Order number.

9

Fig. 9-29. **LT1, LT4:** Hypertech's Power Programmer Plus module installs between the computer and the harness connector. It modifies air-fuel ratios and spark timing. It is street legal. E.O.# 230.

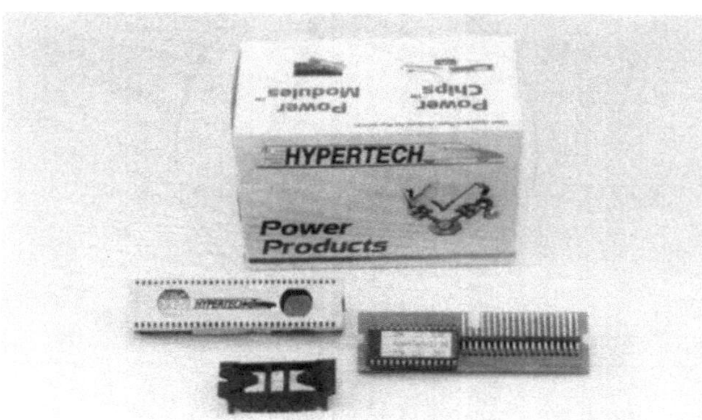

Fig. 9-30. Choose your power chip. You cannot tell a chip by its cover.

Hypertech offers two kinds of chips.

- The "Street Runner" delivers maximum power at full throttle, with no alterations at part throttle or idling. It is street legal. CARB E.O. #D-260-8. Use with the stock thermostat.
- The "ThermoMaster" controls all modes for maximum performance, when used with the Hypertech Powerstat, it lowers coolant temperature. It also controls other temperature-sensitive functions, including Closed Loop operation, Torque Converter Lock-up, and the fan. CARB E.O. # D-260-8.

Intake Air Temperature Extension

You can move the Intake Air Temperature (IAT) sensor so that it can read temperatures accurately, resulting in more accurate inputs to the engine computer, and improving power.

LT1, ZR-1, L98: Corvette's OEM IAT sensor is located under the intake manifold, near the rear of the engine. As a result, the IAT sensor sends excess temperature signals, particularly in warm weather driving, and temperature is an important factor in air density. High IAT retards spark timing and tends to produce lean mixtures. For speed/density operation, the Control Module calculates air mass, using pressure and temperature signals. Improper (too high) IAT signals reduce air mass and reduce fuel to match.

ZR-1 Power Plate

Fig. 9-31. **ZR-1:** The Power Plate from Lingenfelter reduces intake air temperature, thus increasing power.

The Power Plate for the **ZR-1**, from Lingenfelter, is a highly-resistant *thermal barrier*, which reduces heat transfer to the intake plenum and engine runners. The Power Plate also blocks coolant flow to the intake plenum and runners. It costs about $150.

7. Ignition System Mods

Some mods require attention to the ignition system because the they affect the air/fuel mixture at the plug gap, as well as affecting combustion pressures. These variations may require different spark timing or spark quality. Generally, aftermarket systems can deliver a higher voltage and more energy than OEM systems, at rpms above 5,000. When you increase the intake air with a supercharger, or other intake or exhaust modifications, you increase the compression pressures, requiring a hotter spark to fire.

Ignition Control

Start with a Multiple Spark Discharge (MSD)—Capactive Discharge (CD) module. Autotronics Controls Corporation makes all sorts of ignition goodies for the aftermarket. The CD unit draws battery power, steps it up to 300-450 volts, and stores the raised-voltage energy in a capacitor. When the spark signal arrives, the capacitor's energy is delivered to the coil, where it is transformed into the several-thousand volts needed to fire the plug.

CD ignition delivers a reliable spark, tending to pay off for serious engine mods, such as higher compression ratios, richer air/fuel mixtures, higher cylinder pressures resulting from supercharging, and nitrous oxide injection.

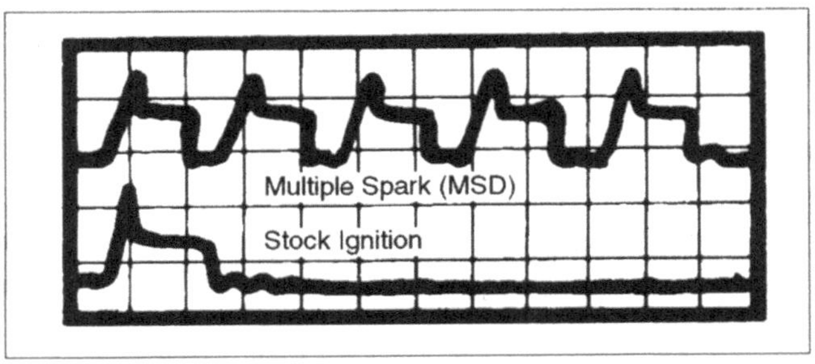

Fig. 9-32. An MSD capacitor discharge system produces sparks at over 20° of crankshaft rotation. At a low rpm, that means multiple sparks. O.E.# D-40.

The multiple sparks that are caused by an MSD capicitor discharge system show up more at lower revs, so you're likely to see improved results in starting, reduced plug fouling, better fuel economy, and lower idle emissions.

L98: You can get a CD ignition control module, coil, harness, and optional rev controller (rpm limiter). The module provides an improved capability to fire wet or fouled plugs. This is recommended for competition and all-out racing. With the ignition coil and wiring harness it connects directly to **L98** Corvette systems.

Fig. 9-33. An MSD Adjustable Timing Control puts timing control on your dashboard. Advance it towards the ping point. Retard timing for bad gas. E. O.# D-40-17

Fig. 9-34. **LS1:** Some tuners remove the engine cover, for racing. They claim that it results in cooler, faster-running electronics for a few tenths off the strip time. Incidentally, the **LS1** Camaro/Firebirds come from the factory without covers.

8. Complete Engines/Complete Cars

Some suppliers offer complete engines, some with E.O.#s, as well as components.

- Lingenfelter (LPE) offers just about any complete engine, **LS1**, **LT1**, **LT4**, and **LT5**. Many are street legal. Bring money, from about $9,000-30,000, if installed by an LPE dealer.
- GM Performance Parts, Chevrolet dealers offer a complete RAM JET 350 motor out of the crate and into your '55-65, or whatever. It is less than $5,000.
- Doug Rippie offers modified Corvettes which have passed certification tests. They are also available from some Chevrolet dealers under factory warranty.
- Arizona Speed and Marine offers engines as well.

9. Questionable Tricks

Modifications that attempt to trick the computer, by manipulating its input signals, may produce some performance gains under full-throttle acceleration. These modifications may also result in less-than-perfect air-fuel mixtures under most operating conditions, especially during transients, such as when you floor the throttle.

Steve Dinan, President of Dinan Engineering, one of the best-known aftermarket BMW tuning firms, told me that, "power gain is elusive. Somebody can show you tests with more power at one point on the curve, but you don't know that you are losing somewhere else. You might gain peak power, but driveability glitches can develop, such as poor idle, loss of low-end torque, and high HC (hydrocarbon exhaust emissions). Some of the most involved work we do is trying to restore driveability to a (car) that has been fooled with for a little gain at one point of the curve."

Fig. 9-35. Here is Lingenfelter's complete **LT1** engine package installed. The automatic transmission version retains the 5.7L displacement, delivering 390 hp @ 5500 rpm, 425 lb-ft @ 4000 rpm. The manual transmission version is larger, at 6.4L (383 cu.in.), it is rated at 440 hp @ 6000 rpm, and 435 lb-ft. torque @ 4500 rpm.

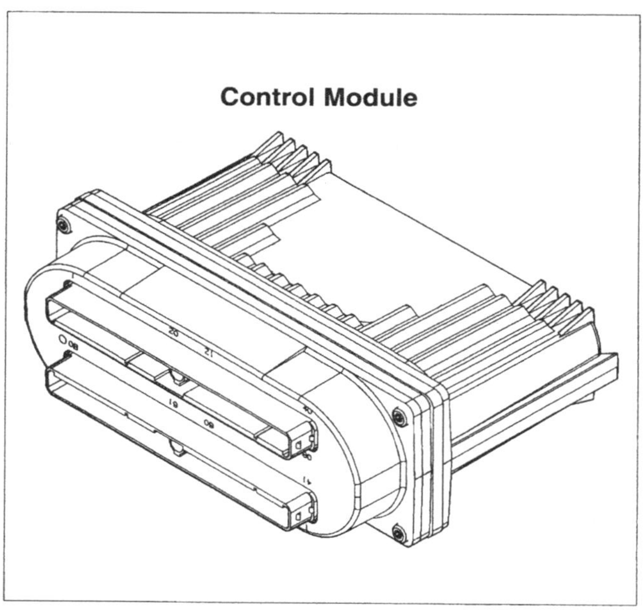

Fig. 9-36. Modifying the computer can improve performance in some rpms, but it may cause a drop in performance at other rpms. If you decide to go inside, know what you are doing, and be prepared to lose any factory warranty.

9.1 Fool the Coolant Temperature Sensor?

In Chapter 4, I showed the variable resistance of the coolant-temperature sensor as an input to the computer. At lower temperatures, the higher voltage-drop tells the computer to call for more fuel. Some will tell you that adding resistance to this circuit will trick the computer into thinking that the engine is cold, so it adds enrichment. Of course, cold enrichment and Closed Loop don't go together, so consider the computer confused. Even if it worked in Open Loop, it would be a bad trick. The extra fuel could cause wear of the rings and cylinder walls.

Here is the problem. Some systems operate based on the last "normal" temperature signal. Computers are programmed to operate within a certain band of voltage-drop values from the coolant-temperature sensor circuit. Instead of enriching the mixture, a radical change in resistance might cause the computer to revert to Limp Home mode, lighting your Check Engine light. In **'94-01** (OBD-II) engines, you can change that "might cause" to "*will* cause." Suffice it to say, I haven't seen any professional Corvette tuners trying this old trick.

9.2 Lower-Temperature Thermostat?

The factory-filled coolant is a glycol and water mix, and it is recommended year-round to raise the boiling temperature up to 250° F (120° C). Is cooler always better? Most will answer, "Yes," but I question that.

Many tuners recommend low-temperature thermostats. It is true that cooler intake air equals more pounds, and a little more power, but only by a small amount. Most late model Corvettes bring in intake air from ahead of the radiator, which is cooler than the engine compartment. A lower ECT, 175° F (80° C), can reduce engine heat-soak.

Jason Haines, the Project Engineer for Lingenfelter Performance, advised me about reduced engine coolant temperatures:

- **L98:** These are very sensitive, because of the long intake runners through the block.
- **LT1:** You can pick up 5-10 horsepower at the rear wheels.
- **LS1:** These are the least sensitive, because the nylon intake runners run cooler. There is minimal improvement.
- **ZR-1:** These are very sensitive to coolant and engine temperatures. An aftermarket cooler thermostat must avoid exposing the OEM radiator to the full flow of the water pump.

Jason also says that, whatever theory tells us, every test they run shows that cooler coolant temperatures shave a little time off of performance runs, but they also show an effect on emissions. (See Fig 9-37.)

LT1, LT4: Expect to see differences between nominal thermostat ratings and the dashboard instrument readings. This is a result of the cooling system design and of the thermostat location. Under normal operation and normal outside air temperatures, (up to 80° F (27° C), coolant temperatures will be indicated as:

- 193-197° F dashboard = 180° thermostat
- 173-177° F dashboard = 160° thermostat

You may see higher coolant temperatures on your gauge while your engine is running cooler.

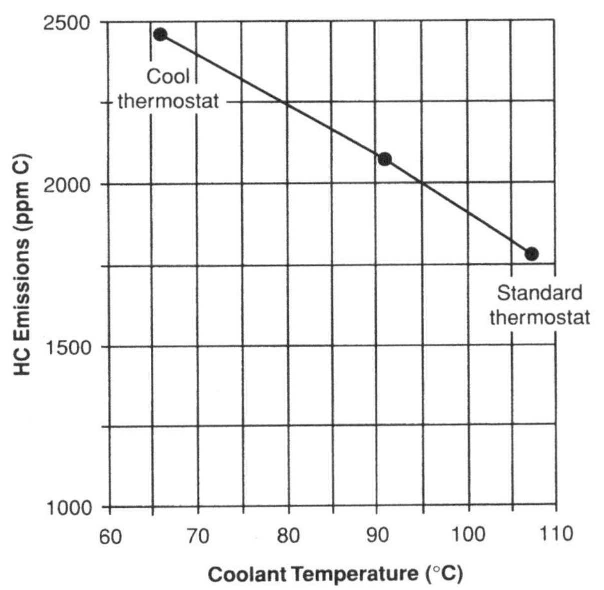

Fig. 9-37. Lower coolant temperatures are reported to increase HC emissions by as much as 30%, at 160°, as indicated in an SAE #950163 paper.

L98, LT1, LT4: Replacing the OEM 225° F (108° C) thermostat, with a 160° F (75° C) one, resulted in significant changes in emissions:

- an *increase* in HC, about 30%, because the aluminum pistons are slightly looser in the iron bore (**LS1**, aluminum block)
- a *decrease* in NO_X emissions, about 13%, because it has a slightly slower burn, and reduced peak pressures

There is one possible gain from a cooler thermostat. Lowering the head temperature may reduce the probability of detonation. That means that spark timing can advance, for increased power. In competition, a cooler 'stat seems to pay off at the strip, if you can cool the manifolds just before the run. Packing ice around the manifold seems to work. Cooler thermostats are certainly cheap, at $10-15. But what do they really give you for everyday street running? And what do they really cost? Except for racing, this is not one of your better ideas. I'd pass on cool thermostats, except at the strip or autocrossing.

9.3 Disconnect the Fuel-Pressure-Regulator Vacuum Line?

Disconnecting the Fuel-Pressure-Regulator vacuum line is one of the worst ideas. You're trying to improve performance by defeating the line that modifies the fuel pressure, according to manifold pressure. As you know, the **'85-96** Corvette system is designed to operate with constant *relative* fuel pressure. At W.O.T., the regulator will give you maximum fuel pressure, even with the vacuum line connected. At less than W.O.T., the injector pulse-times are calibrated for properly-reduced fuel pressures. Disconnecting the vacuum line causes rich running *only at part throttle.* What a scheme to foul plugs and burn up catalytic converters! And with no W.O.T. power gain. Fuhggedabouddit!

'97-98 LS1: The Fuel Pressure Regulator relates to the BARO, not MAP, using a small pressure line ahead of the MAF. The **'99-01** Fuel-Pressure Regulator is located in the fuel tank, and is part of the Fuel Pump Module. In the **LS1**, changing manifold pressures are compensated for in the computer, by changing pulse times.

9.4 Remove EGR (Exhaust Gas Recirculation)?

9

If you're tempted to block off the EGR flow, hold it. I know, it hurts to think about recirculating dirty old exhaust gases into the nice clean fresh air intake. When EGR first reared its ugly head in 1974, aftermarket blocking-off of the EGR was proba-

Fig. 9-38. Disabling the EGR valve is a no-no. Even if you don't care about emissions, the engine calibrations depend on EGR flow to reduce knock, and fuel flow.

bly the only thing that kept some of those engines running. But today, in these fuel-injected engines, I don't know anyone who's blocking off EGR. People who know don't do it for two reasons:

- The engine control already shuts off EGR under high-performance conditions, at idle, and during cold operation.
- The rate of fuel injection, and the spark advance are calculated for EGR under part-throttle conditions. If you block off the EGR, you stand a good chance of experiencing knock and wasting fuel. You might even lose power output, and you are dirtying the air!

9.5 Remove the Air Filter?

Remove a dirty air filter and replace it with a clean replacement filter, but do replace it. As they say in those ads, "Don't leave home without it!"

You might think of trying to run your engine without the air filter, for performance improvement. But it's not a good idea.

- The air cleaner keeps dirt out of engine. Aside from what dirt does to your oil, think what it does for the fine tolerances in the bearings and shafts. You know that, of course, so you inspect it and replace it as specified.
- The air filter is a flame arrester. Bet you didn't know that? If the engine backfires without a filter, you could be burned, perhaps by an underhood fire.
- Dirt on the sensing wires of the MAF can send lean signals, causing backfire on acceleration. (See Chap. 10 "Diagnosis".)

The bottom line: Small gain, could be a big pain.

9.6 Remove the Catalytic Converter or O_2S?

Early converters, '75 to early '80s, can be a drag with all that backpressure. Look for aftermarket converters that are designed for maximum flow. Random Technology converters use heavier coatings for the catalytic material, for greater efficiency, less backpressure, and greater cost. Later-model converters operate at a much lower resistance. Replacing the later converters is a much better idea than removing them, where the gain would be much less.

Removing the oxygen sensor is also not a great idea. At the track, I've watched the pros tuning the mixture for the race, using the oxygen sensor. They told me that the converter backpressure is important for better low-speed torque, for the launch. Before you pull the converter, remember that the whole system is tuned for that backpressure.

L83, L98, '90-92 ZR-1, '90-93 LT1: Replace unheated O_2S with *heated* O_2S. This requires rewiring the heating circuit.

9.7 Modify MAF sensor?

The MAF sensor seems like an easy target for improvement. Many tuners remove one or both screens, and some grind off the heat sink. But now you've exposed those fine wires, which are less than 1/10mm in diameter. Late model Bosch MAF sensors are made without a heat sink.

Fig. 9-39. '85-89 L98: Many tuners recommend removal of the screens from the Bosch MAF. TPIS and Lingenfelter also recommend removal of the heat sink, but this is *not* street legal.

You can install a new MAF sensor, which can result in:

Plus: Low to mid rpm HP/TORQUE gains are about 2-4%, there is no heat sink, it protects the wires, and it's street legal.

Minus: You could get inaccuracies, particularly during transients/acceleration because of interference with the flow through the central tube, causing leaner A/F mixtures. You could also get dirt on the hot wires, which could damage them. If a wire breaks, you'll pay a bunch, it could be $250. You may end up paying a big price for a small low to mid rpm gain, while getting small losses at the high end.

9.8 Fuel Additives

Higher fuel octane ratings reduce detonation, allowing spark advance to increase power. Refineries juggle fuel compositions to maintain the normal octane ratings. Some emission-control additives increase octane, which are simply aromatics. Octane ratings are not considered part of emission control.

Manganese additives, such as Methylcyclopentadienyl Manganese Tricarbonyl (MMT), have been illegal in California since 1977. You may still find this additive in states that are not as concerned as California about low emissions. Ask your service station operator.

If you burn fuel with MMT, your car may experience:
- reduced spark-plug life
- reduced emission-control system performance
- a lighted MIL (Malfunction Indicator Light)

10. Conclusion

In this chapter, I have presented a broad spectrum of ideas on the high-performance application of Corvette fuel-injection systems and on modifications. The subjects have ranged from inexpensive street-car modifications to expensive custom electronic controls made for racing.

The fundamental messages of the chapter are these:

- In the quest for more power, the stock fuel-injection system is not necessarily the weak link. Increasing intake air is your first bet, and the stock engine management system will handle significant increases in intake air without any changes.
- There are no "demon tweaks", and no magic chip that can miraculously unlock vast amounts of horsepower. Corvette engineers know that "Performance sells Corvettes." They may need to make some compromises, but they are not dummies. The need for fuel-injection system modifications depends directly on the needs of the engine. It all comes down to whether the needs of the engine exceed the capabilities of the injection system.
- There is no substitute for knowledge—knowledge of the engine's requirements, knowledge of the fuel-injection system limitations, and an understanding of the real effects of modifications. Thorough understanding of the detailed functions of each Corvette fuel-injection system should be the first, and most important step, in figuring out what adjustments and modifications will pay off for your car and your individual needs.

In this chapter, you have also seen:

- Corvette support of enthusiasts, shops, and racing endeavors
- legal issues involving warranties, tampering, and legislation
- your decisions, street or track?
- high-performance basics
- bolt-ons and mods
- questionable tricks

11. Representative Suppliers

The following list of representative suppliers will give you a starting point for investigation of possible modifications to your system. This is not a list of recommendations, but, at the time of writing, these companies have verified their street-legal registration, as applicable.

LINGENFELTER PERFORMANCE ENGINEERING
E.O. # D-333
1557 Winchester Rd, Decatur IN, 46733
PH: 219/724-2552
FAX: 219/724-8761
*Cars, engines **LT1**, **LT4**, **L98**, **LS1**, **ZR-1** packages, components*

DOUG RIPPIE MOTORSPORTS
E.O. # D-323
14070 23rd Av N, Plymouth MN, 55447
PH: 612/477-9272
FAX: 612/477-9277
*Complete cars, "Race-car performance for the streets", **LT1**, **ZR-1***

ACCEL (DANA ENGINE CONTROLS)
10601 Memphis Av. #12, Cleveland OH 44144
PH: 216/688-8300
Engine management systems, camshafts, throttle bodies, Ignition parts, complete conversion kits from carburetor to MFI, SFI, TBI

ACCESSIBLE TECHNOLOGIES-PROCHARGER
E.O.# D-365
14807 W. 114th St, Leneka KS 66215
PH: 913/338-2886
FAX: 913/338-2179
***LT1**, **L98** Supercharger/intercooler*

ARIZONA SPEED AND MARINE
6313 West Commonwealth Place, Chandler, AZ 85226
PH: 480/753-0208
*Engines, fuel injected, supercharged (based on **LT4**, **LS1**)*

AUTOTRONICS CONTROLS CORP.
E.O. # D-40
1490 Henry Brennan Dr., El Paso TX 79936
PH: 915/857-5200
MSD Ignition

GM PERFORMANCE PARTS
Call for catalog and nearest dealer
PH: 800-577-6888
Ram-Jet crate engines, parts

K&N ENGINEERING, INC
E.O. # 269
P. O. Box 1329, Riverside CA 92502
PH: 800/858-333; (CA: 909/684-9762)
Filtercharger intake air filters, renewable and replaceable

HYPERTECH, INC.
E.O. # 260
1910 Thomas Rd. Memphis TN 38134-6315
PH: 901/382-8888
FAX: 901/373-5290
Power chip, power programmer

(PROCHARGER—See ACCESSIBLE TECHNOLOGIES, INC., above)

TUNED PORT INDUCTION SPECIALTIES (TPIS)
E.O. # D-235
4255 County Rd 10 E., Chaska, MN 55318

PH: 612/448-6021
FAX: 612/448-7230
Packages, PROMs, throttle bodies, MINIRAM intake manifolds

VORTECH ENGINEERING
E.O.# D-213; D-365
1650 Pacific Ave, Channel Islands,CA 93033
PH:805/529-9330
FAX: 805/529-7831
Centrifugal superchargers, some street are legal

CORVETTE CENTER
1995 Lafayette St., Santa Clara, CA 95050
PH: 408/980-0866

Chapter 10

Diagnosis

Contents

IN THIS CHAPTER, LOOK FOR:

- control Module Diagnosis, major differences, OBD-I and OBD-II
- diagnostic procedures as they change—affected by the diagnostic capabilities of changing computers
- differences between Data Link Connectors (DLC)
- tables of Symptoms and Diagnostic Trouble Codes (DTC)

1. INTRODUCTION

So far, I've described Corvette's electronic engine control as it has increased in capability over the model years. All electronic fuel-injected Corvettes (**'82-01**) include some On-Board Diagnostics (OBD).

- OBD-I applies to **'82-93** Corvettes, although they didn't call it that. The name didn't come until 1988, when on-board diagnostics became a California requirement. In cars from the other 49-states, it is still often identified as plain old OBD. I'll use the term "OBD-I" to avoid confusion with OBD-II.
- OBD-II began in **'94**, becoming fully functional by **'96**. The second generation provides even greater usefulness. OBD-II requires a different approach to diagnostics, guiding you directly to the problem. You might call the **'94-95** an OBD-I with an OBD-II connector and with 86 different DTCs. *Note*: OBD-II in Canada is called "Enhanced Diagnostics."

Diagnosis requires an understanding of the system's logic:

- how the systems work
- how different systems affect each other
- finding the "root cause"

> Fix the problem, not the solution.

For the Corvettes in this book, it no longer works to "throw parts at the system." Unfortunately, we still see a lot of that, particularly from people who do not understand how the systems work, and how different systems affect each other. One shop owner actually argued that throwing parts is cheaper than buying diagnostic tools and paying techs who know how to run them! I would, personally, stay out of his shop. Corvette manuals warn: "Lack of basic knowledge of this powertrain when performing diagnostic procedures could result in incorrect diagnosis or damage to the powertrain components."

Failure of one of the components in these modern engine management systems may not be the fault of the component itself. You may replace the failed part, only to suffer the same failure again. Something else is causing the component to fail.

> We demonstrated the idea of *root cause* in one technician-training video I made. We showed a failure of a stuck EGR valve that was *not* solved by replacing the valve. The symptom was that a new valve would stick again in a month or two. The root cause was diagnosed as a crimped vacuum hose which controlled the valve. When we replaced the vacuum hose along with the EGR valve, we fixed the problem.

You'll want to make basic checks to be sure that you're not troubleshooting the engine management system when that may not be the problem. Replacing the computer will not solve problems caused by such engine mechanical conditions as low compression, vacuum leaks, exhaust system restrictions, or a worn ignition system wiring and parts.

Advances in Diagnostics

Three factors are behind the advances which have been made in diagnostics: 1) Customers are demanding better performance and reliability. 2) Demands for clean air are tightening the limits. 3) Computer capabilities continue to increase to satisfy the demands.

> Reductions in tailpipe and EVAP emissions from OBD-I to OBD-II include:
> - HC - over 80%
> - CO - over 80 %
> - NO_X - over 30%

OBD-II computers can perform many emission tests themselves, even as the car is being driven. Usually, these tests are so unobtrusive that you don't even feel them as you drive. GM chose Corvette to be one of two transition cars to "test the waters." In 1994, Corvette began partial OBD-II set-up to meet federal requirements for full implementation in 1996.

Table 10-1.Comparison of OBD-I and OBD-II

OBD-I	OBD-II
Oxygen sensor	Oxygen sensor (enhanced)
EGR system	EGR system (enhanced)
Fuel system	Fuel system (enhanced)
Electronic input components	Electronic input components
	Electronic output components
	Catalyst efficiency
	Catalyst heating
	Engine misfire
	EVAP system (leak check/function)
	Secondary Air system
Diagnostic information	Diagnostic information
Fault codes	Fault codes
	Engine parameter data
	"Freeze Frame" engine parameters
	Standardization

'82-93, OBD-I (and the **'94-95** transition years) monitors the systems and sensor signals to make sure that they stay within defined ranges.

'96-01 OBD-II monitors the components and systems for variations that could increase emissions above one and a half times the original test standards. As you drive, the computer checks on more components and systems. Your diagnostics are pinpointed.

Table 10-3.DVOM and SES readings for the '82-93, Field Service Mode

Open Loop	Rapid DVOM needle swings, about 2.5 a second or rapid SES flashes
Closed Loop	Slow DVOM needle swings, about 1 per second or slow SES flashes
Lean exhaust	DVOM low, less than 0.45 volt or SES off most of the time
Rich exhaust	DVOM high, more than 0.45 volt SES on most of the time

OBD-II means that you'll need a major change in your diagnostics. It depends, in part, on warm-up cycles and drive cycles. And it may take more care, but you can get a lot more information, with the many Diagnostic Trouble Codes (DTC). This means that you will depend less on the *symptoms* and more on the *root* problems.

Table 10-2.OBD-I and OBD-II parameters

	'82–93 (OBD-I)	'94–95 (transition years) '96-01 (OBD-II)
Diagnostic Trouble Codes	14 to 34 two-digit codes	86 to 131 five-digit codes freeze frame data
Read codes accessible through DTC connector	"Check Engine" blinks DVOM needle swings Scan tool (1)	Scan tool (2)
Data Link Connector	12-pin, pin M is serial data link	16-pin, pin 2 is J1850 bus
Diagnostic Mode **Scan tool function** ***Do not drive***	Ground Diagnostic Terminal A to B, ignition on, engine *off*	Diagnostic Executive management
Field Service Mode	Ground Diagnostic Terminal A to B, engine *running*	Diagnostic Executive management
Road Test Mode (3)	Scan tool function (3)	Real-time onboard diagnostics

(1) **'82-93** The Oxygen Sensor signals show up with the DVOM and Service Engine Soon (SES) light.
(2) **'94-01** The Malfunction Indicator Light (MIL) is specifically related to signal emission control problems. The MIL cannot read letters or 5-digit codes. Without a proper scan tool, "you got nutin', Charlie."
(3) During the Road Test:
- *deceleration*, expect programmed leaning, with the SES on, you'll get shorter times, and DVOM voltage will be lower
- *acceleration*, expect programmed enrichment, with the SES on, you'll get longer times, and DVOM voltage will be higher
- *idle below 1200 rpm*, expect the SES to be on longer, and DVOM voltage to be higher

This only applies to OBD-1, **'82-93**. In contrast, OBD-II shows none of these SES/MIL signals, but stores codes to indicate conditions.

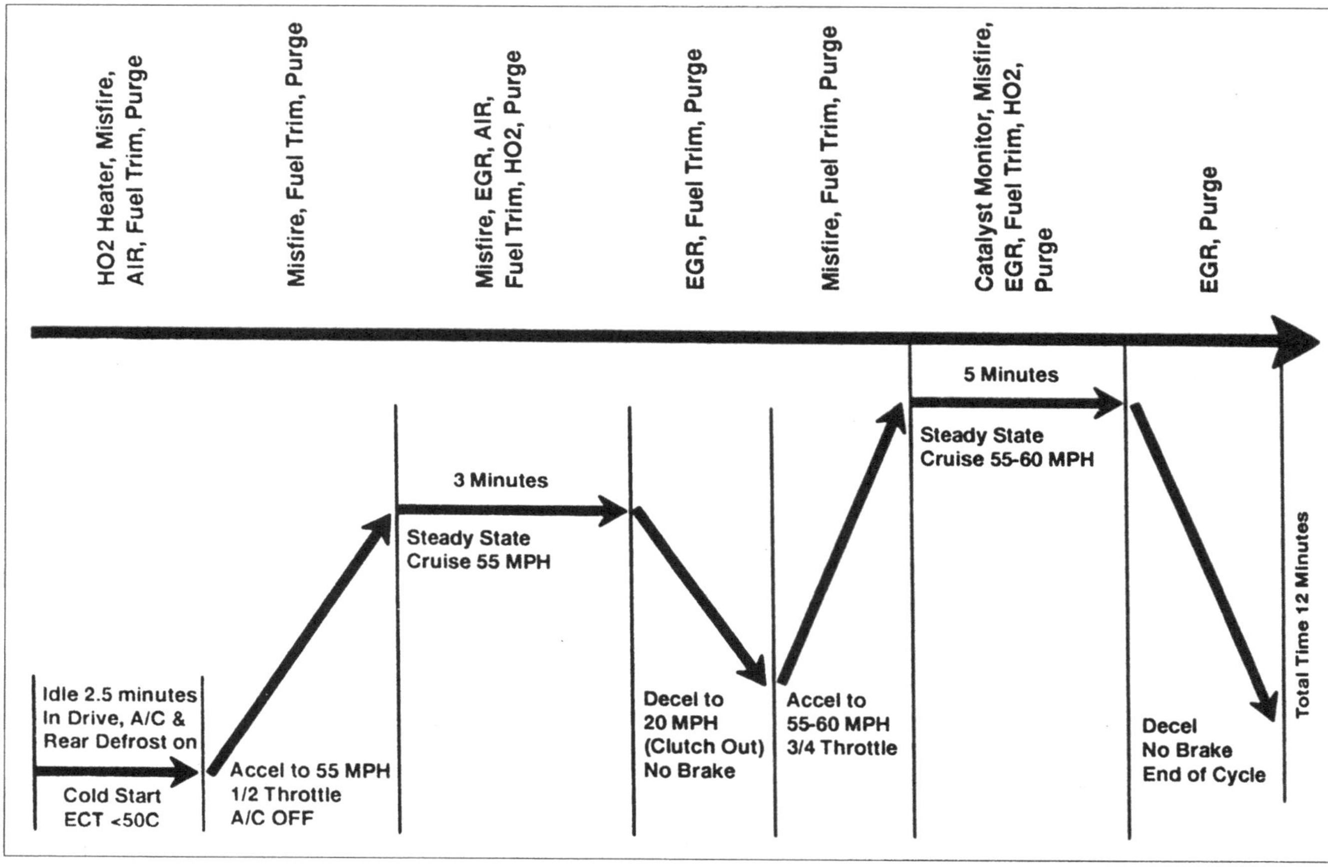

Fig. 10-1. The OBD-II's typical drive cycle is designed to exercise all of the different modes of OBD-II. *Note*: This drive cycle is Corvette-specific, and should not to be confused with the drive the cycle specified for other cars, or the IM240 version.

The **'94-01** OBD-II greatly extends the capabilities of the computer, providing sharper focus for a faster diagnosis.

- The PCM can momentarily (for 1/50th second) cause a deviation and observe the results. For example, the PCM increases the injector pulse-time and checks to make sure that the HO_2S indicates an increased voltage (a rich mixture). This deviation will show up with a scan tool, but it is too brief for you to feel otherwise.
- Some scan tools, with bi-directional control, can introduce modified sensor-input signals to observe PCM outputs. For example, it can momentarily close the drive IAC valve and check for reduced idle rpm.

2. OBD-I Typical Diagnostics

I'll discuss the typical OBD strategy-based diagnostics, using the **'91** VIN 8 Tuned Port engine as an example. Other model years, including the **ZR-1**, will be similar if not exactly the same. Keep in mind the differences between OBD-I and OBD-II, as in Table 10-1 (page 211). Following this section on OBD, I'll show the differences for OBD-II.

2.1 Diagnostic Circuit Check

For any driveability complaint, start with a Diagnostic Circuit Check. *Note*: the Malfunction Indicator Light (MIL) is the OBD-I version of the Service Engine Soon (SES) Light.

Here's a story from an OBD-II training session: The owner of a new OBD-II car (not a Corvette) complained to the dealer that the Malfunction Indicator Light was continuously lit. It turned out that the filler cap was repeatedly coming loose. So, since filler caps are an important part of emission control, the manufacturer took the car back, as it was still under warranty.

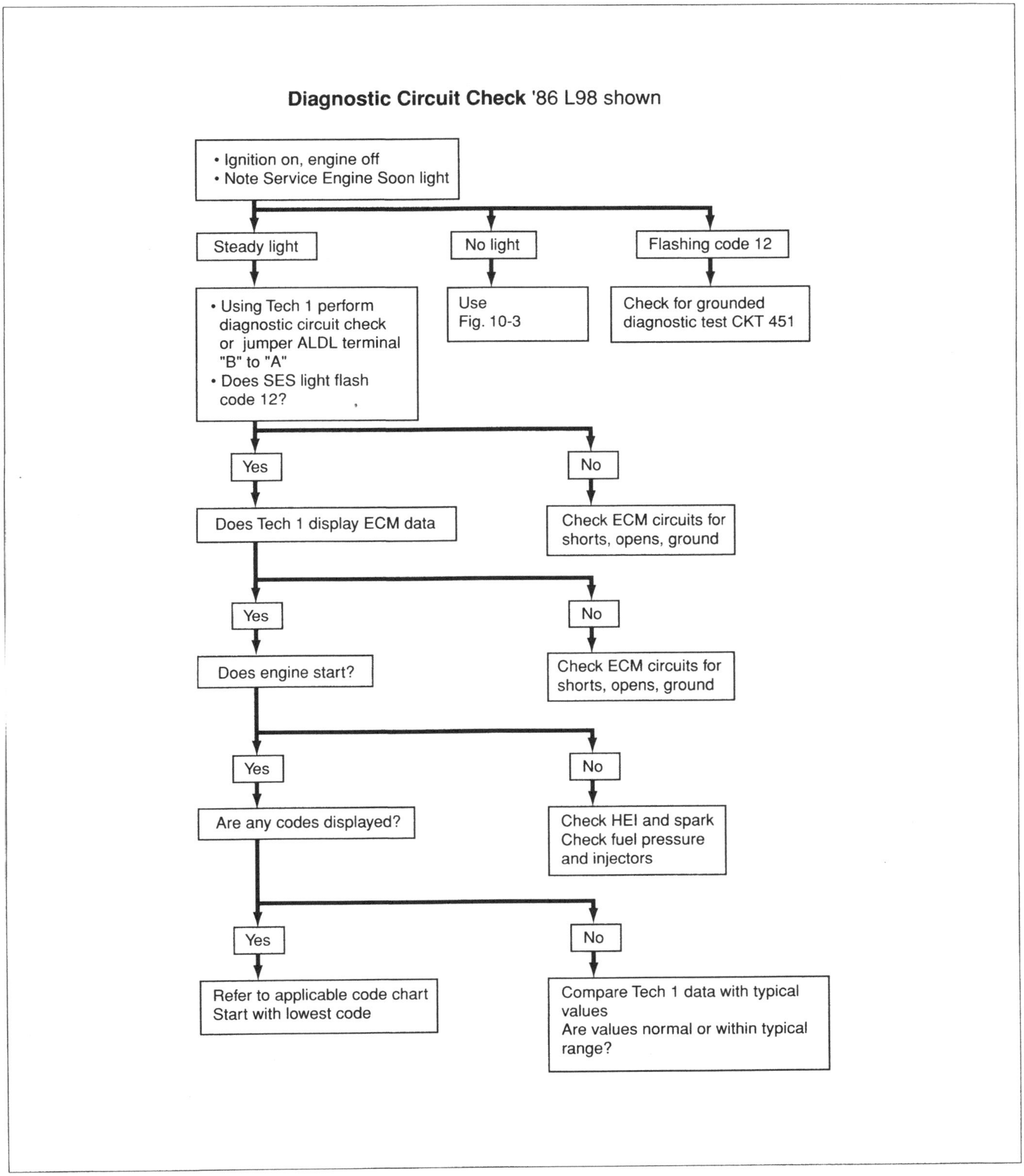

Fig. 10-2. Here is an example of a scan Diagnostic Circuit Check. It guides you through each step in diagnosing any complaint. Other advanced scan tools may be substituted for a Tech 1 tool.

10

Table 10-4.Typical Scan Tool Data Values
with a warm idling engine, in Closed Loop, with the accessories off ('91 L98 shown)

Scan Position	Units Displayed	Typical Data Values
Engine speed	RPM	±100 rpm from desired
Desired idle	RPM	ECM commanded (based on temp.)
Coolant Temp.	C°/F°	85°C–108°C (185°F–226°F)
Intake Air Temp.	C°/F°	10°C–90°C (50°F–194°F) (depends on underhood temp.)
MAP	KPa/Volts	29–48 kPa / 1–2 volts (varies with altitude)
Throttle Position	Volts	0.3–0.9
Throttle angle	Percentage	0% at idle
Oxygen Sensor	Millivolts	1–1000 mv (varies)
Open/Closed Loop	Open/closed	Closed Loop (may go open with extended idle)
Block Learn Cell (STFT)	Cell number	0 or 4 (depends on air flow and rpm)
Block Learn Enable (STFT)	No/yes	Yes
Fuel Integrator (LTFT)	Counts	Varies
Block Learn (STFT)	Counts	108–150 (84-125)
Spark Advance	Degrees	Varies
INJ Pulse Width	M/Sec	1–4 (varies)
Knock Retard	Degrees of retard	0
Knock Signal	Yes/no	No (unless detonation is detected)
Exhaust Recirc.	Percentage	0% at idle
Purge Duty Cycle	Percentage	0% at idle
Idle Air Control	Counts (steps)	5–50
Learn IAC	Counts (steps)	5–50
MPH/KPH	MPH/KPH	0
Torque Converter Clutch (A/T)	Off/on	Off (unless the TCC is commanded on)
Battery Voltage	Volts	13.5–14.5
Fuel Pump Volts	Volts	13.5–14.5
Air Control Sol.	Normal/divert	Normal
Air Switch Sol.	Port/converter	Converter
A/C Request	Yes/no	No (unless the A/C is requested)
A/C Clutch	On/off	Off (unless the A/C is commanded on)
Fan 1 Req. Fan 2 Req.	Yes/no	No (unless the fan is requested)
Fan 1	On/off	On at 108°C (226°F), off at 105°C (221°F)
Fan 2	On/off	On at 113°C (235°F), off at 108°C (226°F)
Park/Neutral	P-N-- and R-DL	Park/Neutral (P-N--)

Table 10-4. Typical Scan Tool Data Values with a warm idling engine, in Closed Loop, with the accessories off ('91 L98 shown)

Shift Light	Off/on	Off
1–4 Shift Active	No/yes	No
4th Gear Switch	No/yes	No
Oil Temperature	C°/F°	Varies with engine operating temperature
PASS-key	Enabled/disabled	Enabled
PROM ID	0–9999	PROM ID number (varies)
Time From Start	Hours/minutes/seconds	Depends on time elapsed from start

2.2 Strategy-based Diagnostics—'82-93

(See Fig. 10-3.)

1. Verify the complaint
   Try to duplicate the problem. Check the conditions: engine temperature, fuel, and weather.
2. Preliminary Checks
   Make a visual inspection, check the service history, listen for unusual sounds, check for strange odors, and check DTC's, but *do not erase* them.
3. Check Service Information
   Check any newsletters, bulletins, hotlines.
4. System Checks
   Verify the proper operation of the system. Have an organized approach to diagnostics and failure-mode testing.

5*a*. Hard code(s) are set for an existing fault (not intermittent). They will guide you to the fault. Perform Diagnostic Procedures.

b. No codes
   Based on your understanding of the system's operation, use the symptom as a clue for diagnosis.

c. No symptom that matches
   Use your knowledge of sensors, actuators, computers, strategies, and the schematics (Chapter 12) to make a hypothesis.

d. Intermittent
   Try to re-create the situation and use scan tool with data capture—Tech 1, Tech 2, Fluke 87.

e. System is operating as designed (see sidebars on right)

6. If all else fails re-examine the complaint, using these same procedures.

Knowing how the car should work is becoming more and more important as engine control becomes more sophisticated. Among many things, you have to know that, even with your right foot buried in the floor, and the Traction Control switched on, the **LS1**'s Throttle Actuator Control will cut back on the throttle if the drive wheels are spinning.

You need know your system well enough to recognize the difference between normal and abnormal operation. Here's another example of why this is important: One of the cars for which I produced training videos has an advanced computer-controlled heating system. In auto mode, the computer is designed to keep the blower off until the ECT signal reaches 100° F (40° C). That prevents cold air delivery when the driver wants heat, for one or two minutes, depending on the pre-start temperature. My car has that feature, so that when the blower comes on, I get warm air instead of cold ankles.

But, as I worked on the training for that system, we heard that angry customers were complaining to the dealers' service departments, "Something's wrong with my blower switch! The blower doesn't work after I start the engine, even with the switch on auto." Of course, the customers had not read the Owners Manual, but neither had some of the techs. The techs who did not have the training accepted the complaint of "something's wrong," and some *rewired the system* to bypass the signal from the CTS. This pemitted blower operation as soon as the engine started, which is normal for some cars, but not for the **LS1**.

We revised the training to explain the difference between blower operations. And the dealers' service departments showed the customers the normal operation, along with its benefits.

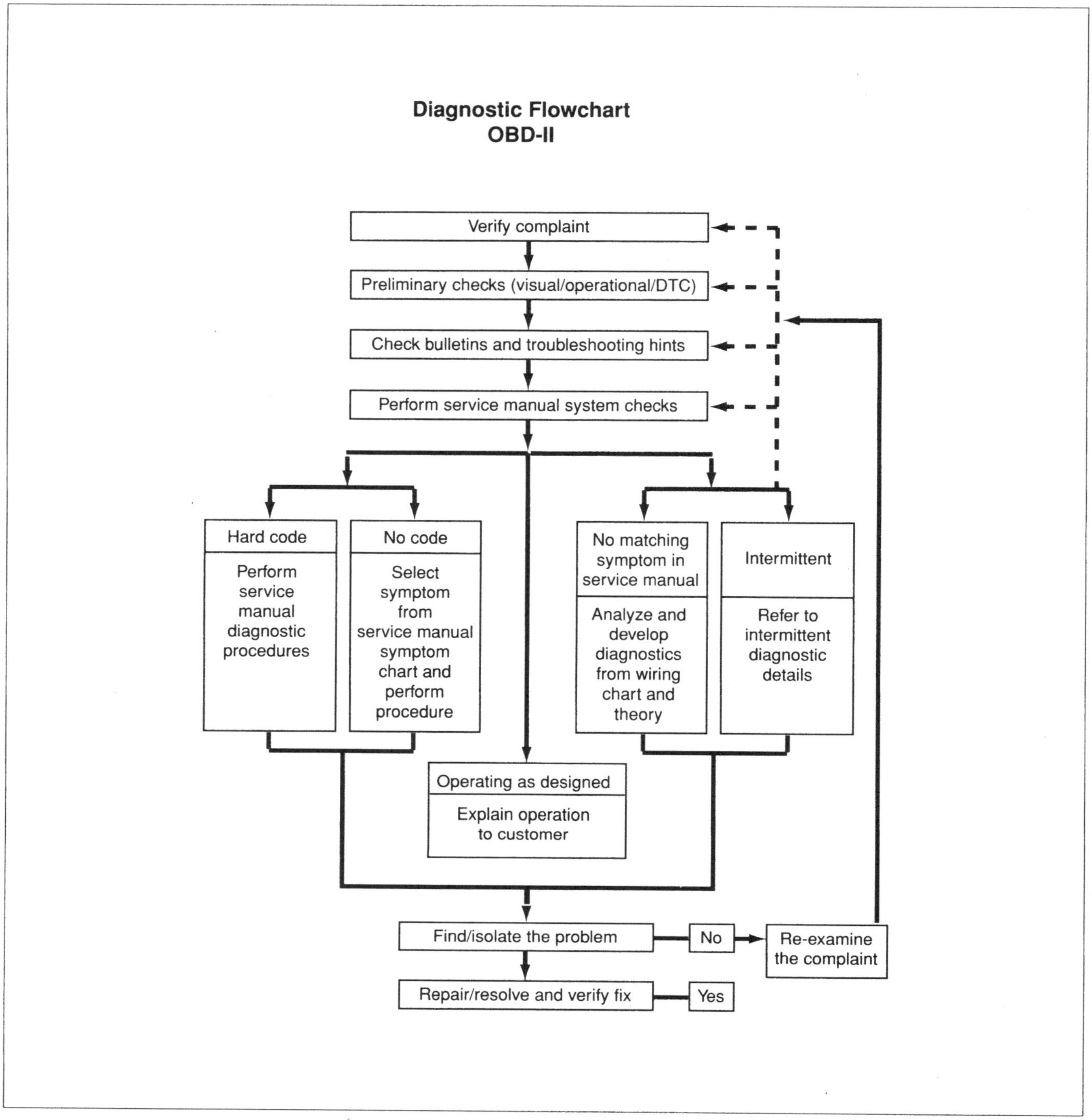

Fig. 10-3. L98, LT1, LT4, ZR-1: Here is an example of how strategy-based diagnostics guides you. You may have a hard code. You may have symptoms. You may have both, or neither. Or you may have an intermittent problem, not now present.

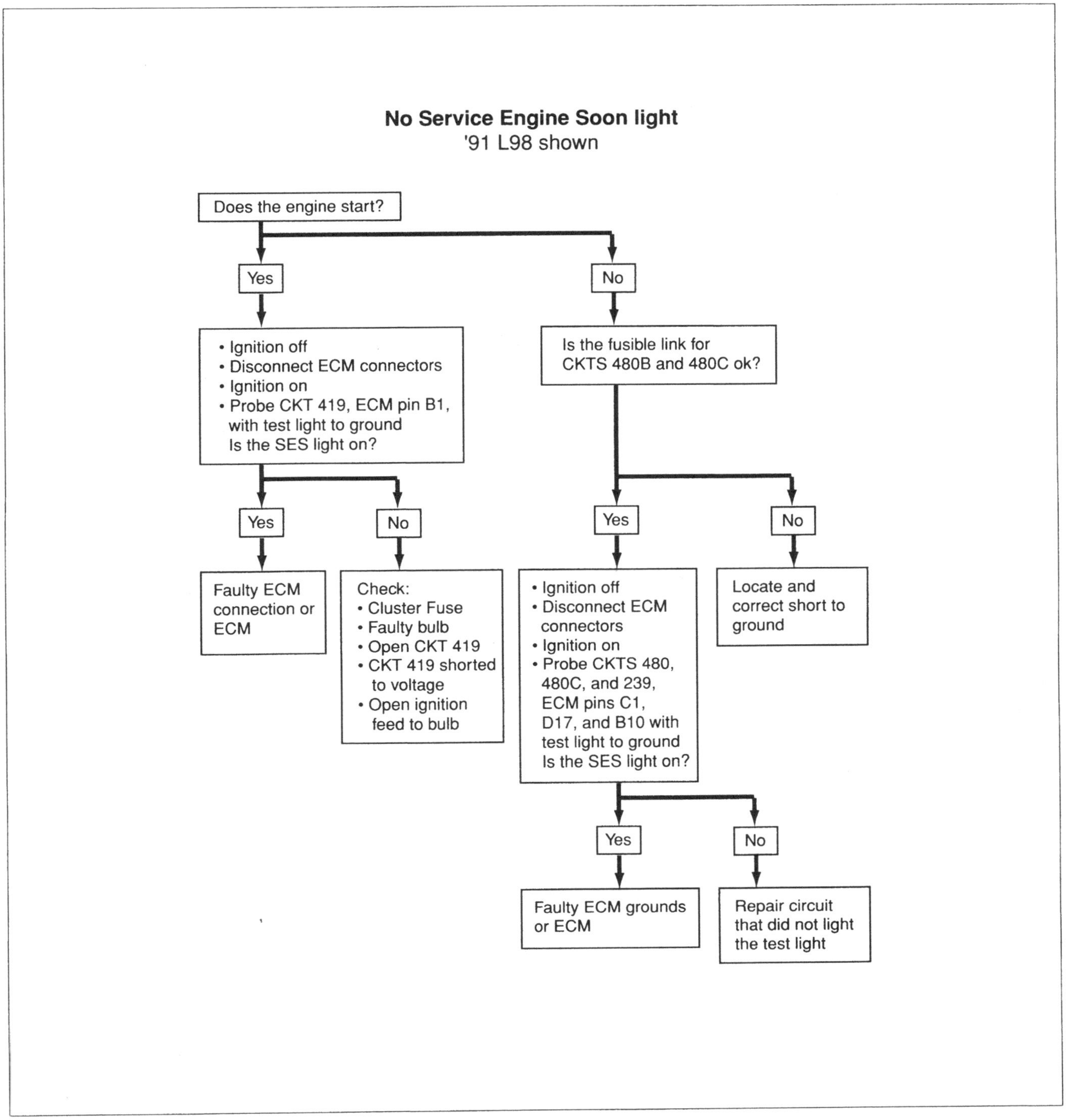

Fig. 10-4. Use this chart if the SES/MIL indicator does not illuminate during a Diagnostic Circuit Check.

Table 10-5.Scan tool data list— Idling values shown ('91 L98 shown)

Data	Value
Block Learn (LTFT)	115–141 counts[1] ('**94-01** between –10 and +10)
Open/Closed Loop	Check Closed Loop[1]
Oxygen Sensors	10–1000 mv[1]
ECT	185–226°F (85–108°C)
Engine rpm	varies ±100
EGR	0
Purge duty cycle	0
IAC counts	5-50
IAT	50–194°F (10–90°C)
Ignition	12.5–14.5 v
Injector PWM average	1–4 ms
MAP	20–48 kPa; 1.0–2.0 v[2]

[1] Check "Closed Loop". Idle exhaust may be cool enough to allow **'82-93** unheated Oxygen Sensors to cool so that Closed Loop switches over to Open Loop.

[2] Varies with altitude

Table 10-6.Emission Failure Diagnosis

Problem	Cause	Symptom/s
High HC	**EVAP** canister purging all the time	Rough idle in Open Loop, especially on restart
	EVAP canister not purging	Raw fuel smell
High HC and CO	**PCV** valve clogged	Oil leaks, high oil consumption
	PCV hose deteriorated	Rough run or unstable idle because of vacuum leak
	AIR to exhaust ports at all times	Poor fuel economy (false oxygen sensor signals) or backfire on deceleration
	AIR inoperative	Possible MIL/SES/Check Engine indicator
High HC, CO, and NO_X	**Catalytic Converter** plugged or restricted	No start, low power, poor fuel economy, or exhaust noise
High NO_X	**EGR** stuck open	Stalling
	EGR inoperative	Detonation

Counts/Block Learn Values

If the Block Learn values are:

Rich—less than 90% (115 Block Learn): look for leaking injectors, which could damage the rings and cylinder walls

Lean—more than 110% (141 Block Learn): look for clogged fuel injectors or low fuel pressure

2.3 Emission Failure Diagnosis

Experience shows that well-designed engines maintain their engine-out emissions up to 50,000 miles, or more, depending on the driver's habits and/or the use of synthetic oils. An increase in tailpipe-out emissions is the results of a catalytic converter thermal deterioration.

3. OBD-II Typical Diagnostics

1994-95 can be called the years of transition for OBD-II. The Corvette is a transition car, from the limited use in 1994, called "Pull Ahead" or partial-function OBD-II, to being used across the board in 1996. The benefits of OBD-II are:

- Many more components and systems can be tested.
- Tests of the components and systems are based on their effect on emissions rather than on mechanical limits.
- A DTC may be triggered before driveability symptoms appear to the driver
- Most, though not all, DTCs will illuminate the MIL.

Once you realize how OBD-II works, you understand why it requires a major change in diagnostics.

- You save time with OEM systems.
- When you start to modify aftermarket manifolding, intake systems, exhaust systems, or even axle ratios and tire sizes, you must consider the precise monitoring of OBD-II and the potential for setting codes.

3.1 Partial OBD-II

'94-95 LT1 includes some OBD-I and some OBD-II:

- It has a 16-pin connector, as opposed to an 8-pin one. Use special adapter to access data with a 12-pin scan tool
- The quantity of trouble codes is more than OBD-I, but less than OBD-II.
- Trouble codes are read through a scan tool.
- For Field Service Mode, the ground terminal is #6. *(Don't ground terminal #6 on a '96-01 OBD-II!)*

3.2 Diagnostic Investigation, OBD-II

(See Fig. 10-5.)

3.3 Driving Cycles and Warm-up Cycles

The OBD-II Self-Diagnosis System monitors eight kinds of engine-management data. It tests the powertrain control, during warm-up, in order to involve conditions of normal-use driving and IM240 testing. Some tests involve more than one warm-up because that's one of the bases for setting certain DTCs. Of course, it also tests during acceleration and deceleration, all adding up to a good trip. You'll want to test after any repairs, even a battery replacement. And here's another tip: Prolonged idle, as in basic tailpipe sensing, has no meaning with OBD-II systems.

Check back with Fig. 10-1 (page 212). OBD-II "drive cycles" and "warm-up" may change according to model year, so check out what applies to your Corvette.

Some OBD-II data is continuously monitored, while other data is monitored only once or twice per driving cycle, or pattern of operation.

> New terms for the OBD-II:
> "good trip"
> "warm-up"

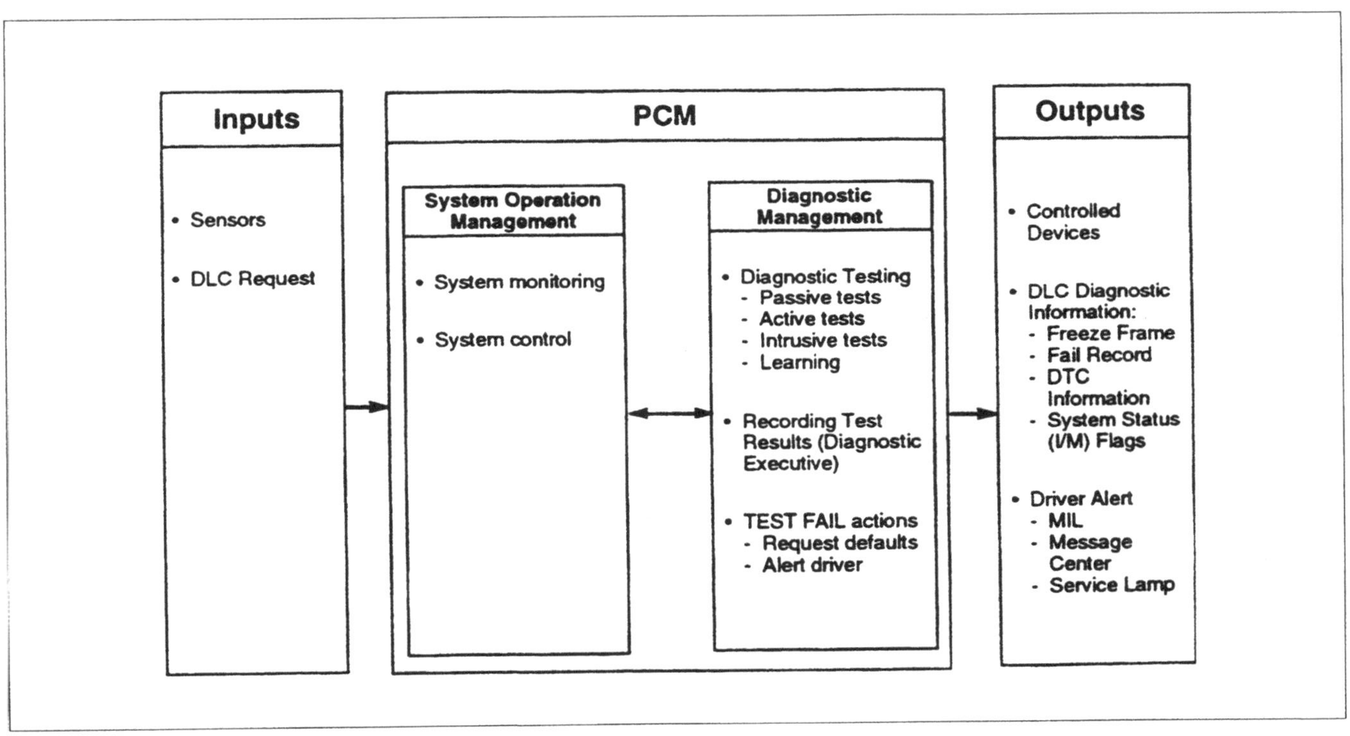

Fig. 10-5. The Power Control Module (PCM) includes System Operation Management, Diagnostic Management, and many more outputs.

At least once per trip the following are checked:

- catalyst efficiency (conversion efficiency)
- catalyst heating (time to operating temperature)
- EVAP (leak detection)
- HO_2S (output voltage, response rate)
- HO_2S heater
- EGR flow rate

The following are continuously monitored:

- misfire (%)
- the fuel system (delivery control)
- emission-related comprehensive components

CAUTION —
Even a few misfires can dump raw fuel into the converter, and possibly cause an overheat.

Don't disconnect the battery before reading this!
Be careful about that old saw about disconnecting the battery as the first step before diagnosis or repairs. Well, that's good advice, but first, save those memories by putting a power source across the line. Corvettes wire the lighter to the battery (other cars wire it through the ignition switch), so that you can plug in a temporary power source.

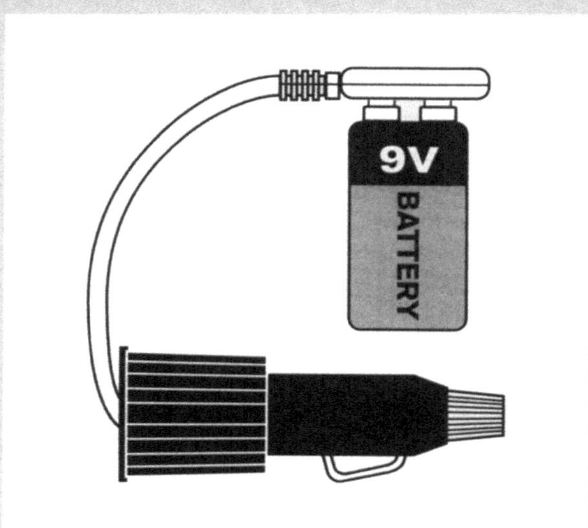

Fig. 10-6. '82-93: A 9v. dry cell, plugged into the cigarette lighter will maintain your system's power.
'94-01 models needs 11v.; any less will set several DTCs. To get the necessary 11v, rig up a set of eight dry cells (total 12v.) across B+ and ground.

3.4 Visual Inspection

The following are examples of the kinds of checks that you'll want to make for a thorough diagnostic check.

Table 10-7. Visual inspection and Non-emission Failures

Diagnostic Step		Yes	No
1	Review the IM test report for specific failure information. Did the vehicle fail the IM240 test for a non-emissions-related problem?	Vehicles will fail the IM240 test if the brakes are unable to maintain vehicle speed within required limits, during deceleration. Vehicles stalling more than three times will be recorded as an IM240 failure.	Go to step 2.
2	Did the vehicle fail the EVAP purge or pressure test?	Test EVAP.	Go to step 3.
3	Review the diagnostic worksheet for driveability and transmission symptoms. Record any additional symptoms .		

Table 10-8. Diagnostic Trouble Code check

Diagnostic Step		Yes	No
1	Perform an OBD system check. Were any DTCs found?	Note the DTCs, then go to step 2	Go to step 2
2	Bring the engine to normal operating temperature, and raise the engine speed to 2,000 rpm. Is there a misfire or is exhaust smoke present?	Note the symptoms then go to step 3	Go to step 3
3	Reduce the engine speed to idle. Wait 30 seconds. Is there a misfire or is exhaust smoke present?	Note the symptoms then go to step 4	Go to step 4
4	Note the PROM ID or EEPROM calibration, then check the service history.	—	—

'94-01 LT1, LT4: Using a scan tool, select "Eng DTC", "DTC Info", and "Last Test". If no codes are set, then select "Eng DTC", "DTC Info", and "History." Now "Capture" this info.

'97-01 LS1: Check the column lock items: Fuel Disable BCM , BCM Com, and Fuel Disable ABS.

- Be sure that the HO_2S (both front and rear) are sending the proper signals: HO_2S1-front should be Closed Loop 0-1v., while the HO_2S2-rear should be almost steady.
- Check the scan tool data.
- With OBD-II, notice that the DTC check comes early.

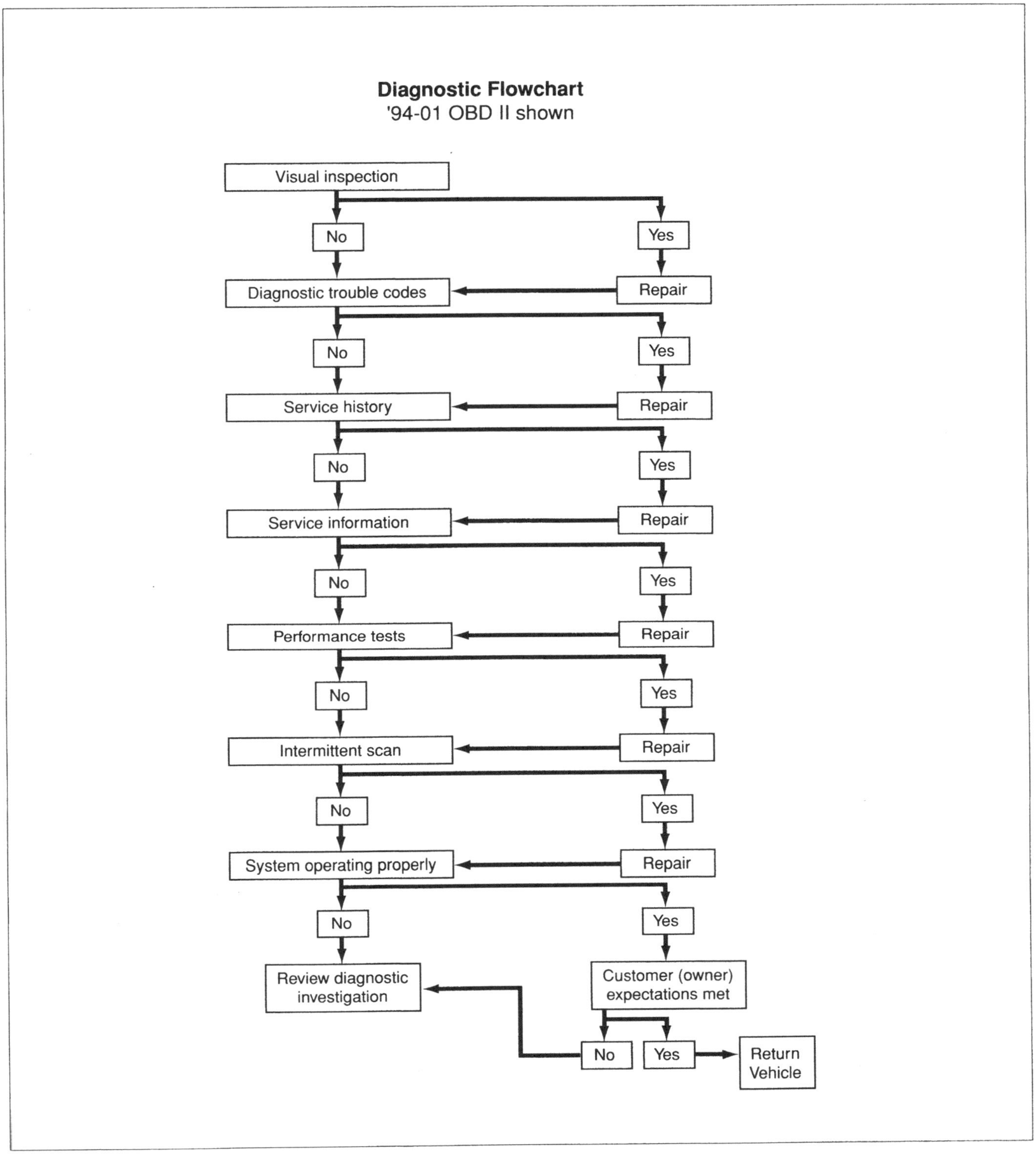

Fig. 10-7. Diagnostic investigation is based on the greatly increased number and detail of DTCs found in OBD-II Corvettes. Notice the consideration of DTCs is the first step after visual inspection, compared to OBD-I, where it is step #5.

Diagnostic Worksheet

Driveability

Symptom—Engine

❍ Hard starting or will not start (but cranks OK)

❍ Idle is rough ❍ Idle is too high ❍ Idle is too low

❍ Hesitates or stumbles

❍ Stalls

❍ Poor power/performance

❍ Ping or knocks

❍ Service Engine Soon light comes on

❍ Sulphur/rotten egg odor

❍ Backfires (popping noise)

❍ Underhood ❍ Tailpipe

❍ Increased oil consumption

❍ Runs on after key is turned off

❍ Speed fluctuates without moving accelerator

❍ Poor city gas mileage

❍ Poor highway gas mileage

How often does it occur?

❍ Always ❍ Sometimes ❍ Rarely

❍ Ever since vehicle was new ❍ Problem just started

When does it occur?

When engine temperature is:

❍ Cold ❍ Hot

❍ All the time ❍ During warmup

Weather conditions:

❍ Cold days ❍ Hot days ❍ Humid/rainy days

Driving conditions:

❍ Accelerating

❍ Light ❍ Medium ❍ Hard

❍ Decelerating

❍ When shifting

❍ Cruising steadily at________mph

❍ A/C-Defroster

❍ On ❍ Off

❍ In neutral ❍ In drive ❍ At idle

What type of fuel?

❍ Regular ❍ Premium

What brand?______________________________

Additional comments____

____

____

Fig. 10-8. Try to duplicate a problem. Look for the conditions that go with it. The diagnostic worksheet can help. The "What?" "Where?" "When?" and "How much?" can help you separate the likely from the unlikely.

3.5 Scan Tool Data List—OBD-II

'94-01: At warm idle, check the scan tool data against these typical data values. Set the car in park or neutral, keep the parking brake on, and the wheels chocked.

Table 10-9.'96-01 Scan tool data list—typical idling values ('96 LT1 shown)

Data	Value
BARO	65–104 kPa; 3.5–4.5 v[a]
CKP engine speed	±75 rpm from engine speed
CKP lo-res angle	0° (varies ± 3°)
ECT	185–225°F (85–105°C)
Engine load	3% idle; 9% 2500 rpm
Engine rpm	±100
Fuel trim cell	16
HO_2S	10–1000 mv, varying
IAC counts	10–50
IAT	50–194° F (10–90° C)
Ignition	12.5–14.5 v
Injector PWM average	1–4 ms
Loop status	Closed
MAF	5–9 g/s idle; 20–26 g/s @2500 rpm[a]
MAP	20–48 kPa; 1.0–2.0 v[a]
TP angle	0%
TP sensor	0.55–0.90 v
Warmup cycles	0–40
[a] Varies with altitude	

Table 10-10.Additional typical values for the '98 LS1

Data	Value
APP sensor #1	0.25–1.1 v
APP sensor #2	3.9–4.5 v
APP sensor #3	3.2–4.5 v
Air/fuel ratio	14.0
Engine load	2% idle; 9% @2500 rpm
TP desired angle	3–10%
TP indicated angle	3–9%
TP sensor #1	0.25–1.5 v
TP sensor #2	3.0–1.5 v
Warmup cycles	0–255

3.6 Service History & Service Information

Table 10-11.Service history

	Action	Yes	No
1	Review the service history, if available. Were any services performed or required that may affect emissions?	Note and go to step 2.	Go to step 2.
2	Does the vehicle have a history of failing components that may be related to the emission problem?	Note and go to step 3.	Go to step 3.
3	Perform adjustments as necessary, then check the service information chart.	—	—

10

Table 10-12.Service information

	Action	Yes	No
1	Search, by VIN and fuel system, for service and diagnosis information. Were bulletins or other information found?	Go to step 2	Go to the perform-ance test.
2	Review bulletins or other information for applicability to current diagnosis. Note any relevant information, and then check the performance test.	—	—

Corvette owners often keep complete service records that may show a history. Shops may cross-check.

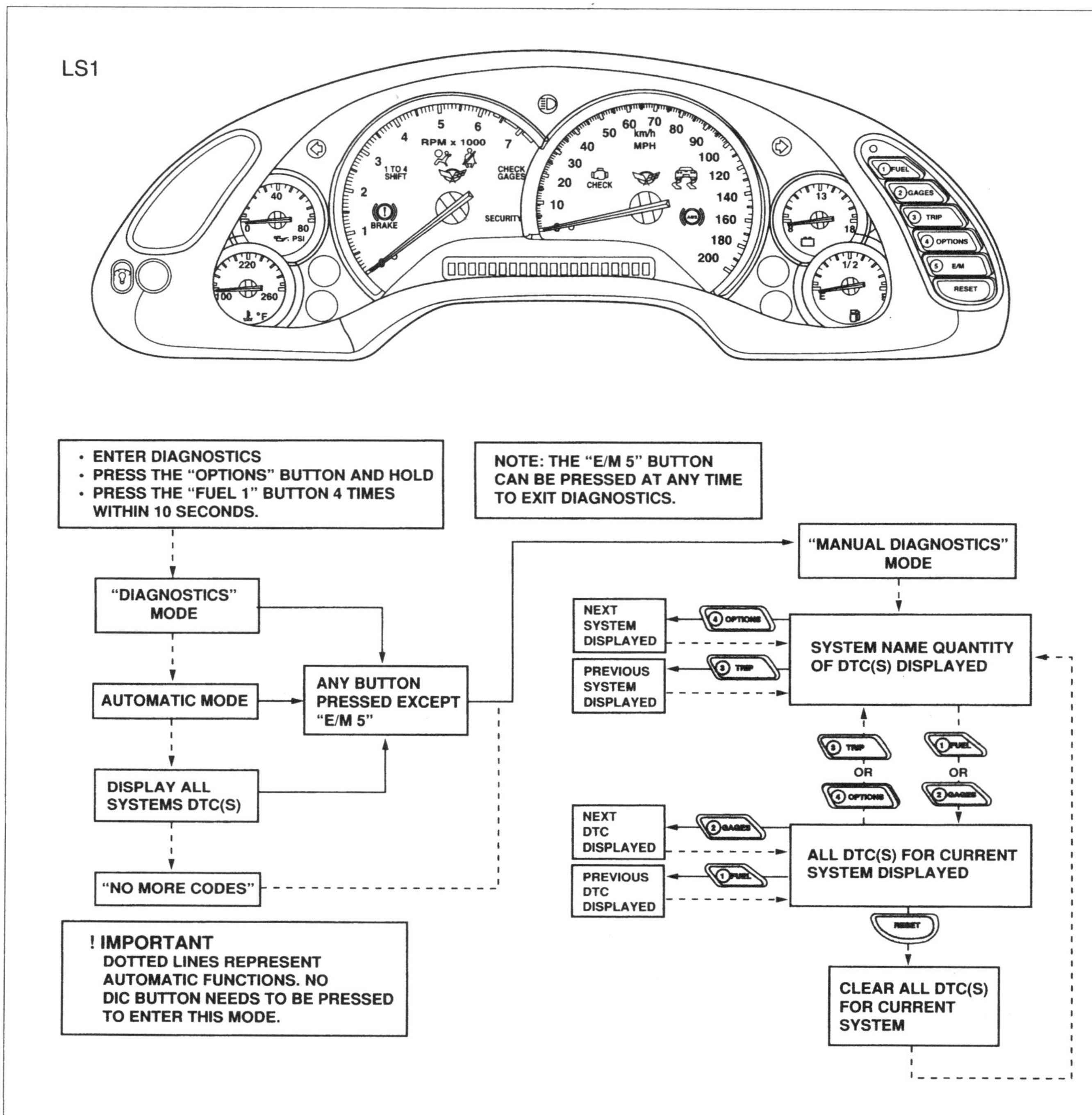

Fig. 10-9. Did you know that you can check for any DTCs without a scan tool? That's one of the things that makes Corvettes special. Check the 20-character vacuum fluorescent display, just above the steering column. You'll find both current and history codes, including PCM and Body Control Module (BCM). The **ZR-1** LCD display also shows DTCs as well as information from the computer-based diagnostic system.

Table 10-13.Phase information

Mode	HC reading[1]	CO reading[1]	NO_x reading[1]	Possible causes
4 or 9	High	High	High–Low	Leaky injector, fuel shutoff
1 or 5 and 5 or 9	Low High	Low High	High–Normal High–Normal	Fuel pressure: fuel pressure regulator, pump
7 and 8, and 2, 3, or 5	Normal Normal	Normal Normal	High Normal	Partially blocked passage to EGR valve
1 and 4. or 2 and 6	Normal High or Low	Normal High or Low	Normal–Low Normal–Low	Lazy O_2 sensor
3 and 6	Normal	Normal	Normal–High	Crimped EGR vacuum hose
1, or 7	High Low	High Low	Normal Normal	Dirty system, improper catalyst temperature, or a leaky injector
All	Normal–High	High	Normal–Low	Faulty catalyst, or a rich air-fuel ratio
All	High	High	High	Faulty catalyst, or tampering

[1] Relative to mode average
Check manual transmission for proper clutch operation and Computer Aided Gear Shift (CAGS) lockout ('**90-01**).
Check automatic transmission for proper shift speeds and Torque Converter Clutch actuation and release.

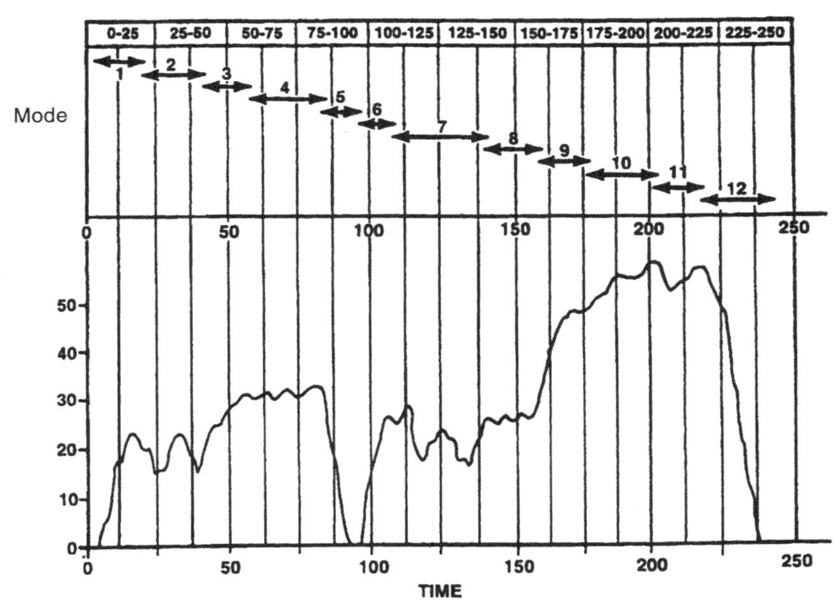

Fig. 10-10. IM 240 emission data is collected during acceleration, deceleration, and transmission shifts. Misfire or detonation may show up during loaded operation. A/T problems, such as an early or delayed shift, can affect engine rpm, and would probably contribute to emissions problems.

3.7 Performance Tests

The performance test is based on the emissions measured during each separate phase of the IM240 test, rather than the collective total for the test, as it would be in Acceleration Simulation Mode. The IM240 test will probably disappear in 20 years, with OBD being used in most cars. The IM240 test is divided into 12 phases, at varying speeds:

1. first acceleration
2. rolling deceleration and acceleration
3. rolling light acceleration
4. part throttle cruise
5. first deceleration
6. second acceleration
7. rolling deceleration and acceleration
8. part-throttle cruise
9. rolling moderate acceleration
10. rolling high-speed acceleration
11. high-speed deceleration and acceleration
12. final deceleration

3.8 Intermittent Scan

See Fig. 10-11.

3.9 System Operating Properly

After the repairs, validate the system to verify that the problem has been corrected. Perform system checks and/or road tests, preferably under conditions noted. Possible problems include:

- IM240 test results are reported in grams/mile (g/mi), in other words, *absolute* mass emissions.
- Exhaust Gas analyzers, such as BAR 90, report ppm or percent, in other words, *relative* mass emissions. BAR 97 is dynamometer-based and reports g/mi., and can be compared to IM240.

4. Symptoms

Preliminary Checks

- PCM and MIL are operating properly
- there are no DTCs
- ECT is below 266° F (130° C)
- scan tool data is within normal operating range

Visual/Physical Check

- PCM grounds are clean and tight
- MAF is installed with arrows pointing toward engine
- no air leaks at the throttle-body, MAF, or intake manifold

Intermittents

- may or may not turn on the MIL, or store a DTC
- often caused by faulty electrical connections or wiring
- conduct a road test with the scan tool—Snapshot feature, Freeze Frame, Failure Records buffer
- intermittent MIL with no stored DTC indicates that the coil(s) are shorted to ground

Once you've made the general preliminary and visual/physical (and intermittant, if necessary) checks, you should then proceed to the checking the specific symptoms.

4.1 Hard Start

Definition: The engine cranks but does not start for a long time. It may eventually run or it may start but immediately dies.

Preliminary Checks

- Check that the computer grounds are clean, tight, and in their proper locations. Refer to the wiring diagrams in Chapter 12.
- Search for bulletins.

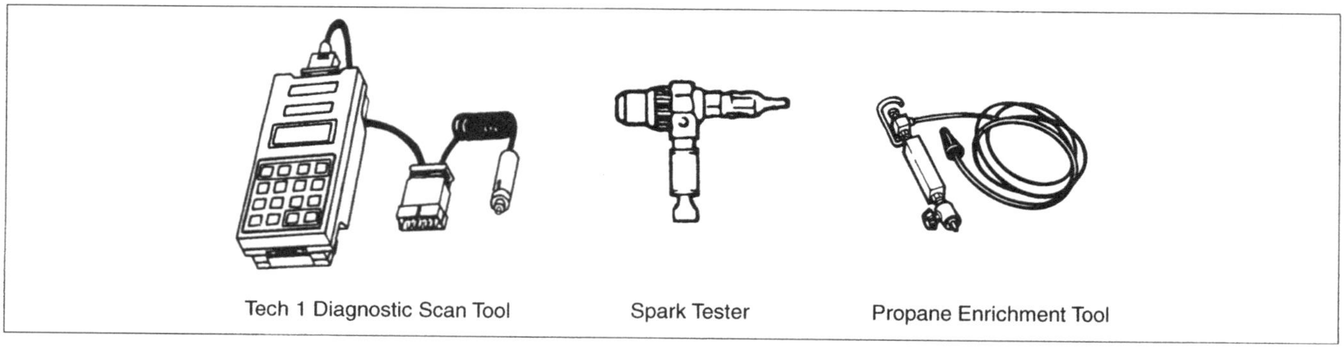

Fig. 10-11. Special tools may be required for an intermittent scan, and to validate emission-related analysis. The most important tools are: the scan tool and exhaust-gas analyzer (not shown). In some states, EVAP testing is limited to filler-cap leakage.

Sensor/System

- Check for proper operation of the column lock system.
- Check for a shifted value in the Engine Coolant Temperature (ECT) sensor. Connect a scan tool. Compare the ECT against the Intake Air Temperature (IAT) on a cold engine. The sensor values should be within 5° F (3° C) of each other. If the ECT sensor is out of range of the IAT sensor, check the resistance of the ECT sensor, and replace it if it is not within the specification. If it is within the specification, repair the signal circuit for high resistance.
- Check for proper installation of the Mass Air Flow (MAF) sensor. An incorrectly installed MAF sensor may cause a hard start. The embossed arrows on the sensor indicate the direction of the intake airflow, and must point toward the engine.

Fuel System

- Check for proper fuel pump relay operation. The fuel pump should come on for about two seconds when you turn on the ignition.
- Check for a faulty in-tank fuel pump check-valve. A faulty valve will allow the fuel in the lines to drain back to the tank, after the engine stops.
- Check that both injector fuses are intact. An open injector fuse will cause the four injectors and four ignition coils to fail. Replace the fuse if necessary. Also, inspect the injector circuits and coil circuits for an intermittent short to ground.
- Check for low fuel pressure.
- Check for a restricted fuel filter.
- Check for contaminated fuel.

Ignition System

- Check for proper ignition voltage output (using a spark tester).
- Remove the spark plugs and check for: wet plugs, cracks, wear, improper gap, burned electrodes, or heavy deposits. If the plugs are fouled with gas or oil, determine the cause of fouling before replacing.
- Check for bare or shorted ignition wires.
- Check for loose ignition coil grounds—the **LS1** has eight, and **ZR-1** has four.
- Check for the proper spark-plug heat range.

Engine Mechanical

- Check for excessive oil in the combustion chamber (leaking valve seals).
- Check for low cylinder compression.
- Check for excessive carbon buildup in the combustion chambers. Clean the chambers using a top engine cleaner. Follow the instructions on the can.
- Check for incorrect basic engine parts: cylinder heads, camshaft, pistons, and related parts.

Diagnostic Aids

- Check the EGR valve; a stuck-open valve can cause a low A/F ratio.
- Check the MAF for malfunction. In the **'94 LT1, LT4, LS1**: disconnect the MAF sensor—the PCM should default to MAP. If the engine starts, replace the MAF.

4.2 Surges/Chuggles

Definition: An engine power variation under steady throttle or cruise. It feels like the vehicle speeds up and slows down with no change in the accelerator pedal position.

Preliminary

- Search for bulletins.
- Check that the computer grounds are clean, tight, and in their proper locations. Refer to the wiring diagrams in Chapter 12.
- Be sure that you understand the operation of the Torque Converter Clutch and the air-conditioning compressor, as explained in the owner's manual.

Sensor/System

- Check for a quick response of the Heated Oxygen Sensors (HO_2S) in different throttle positions. If they do not respond quickly, check the sensors for silicon or other contaminants, from fuel or the use of an improper RTV sealant. Check the sensors for a white, powdery coating on them, and produce a high, but false, signal voltage (indicating rich exhaust).
- Check for proper connection of the Mass Air Flow (MAF) sensor. Repair or replace the terminals if necessary.

Fuel System

- Check for low fuel pressure.
- Check for a restricted fuel filter.
- Check for contaminated fuel.
- Check that each injector harness is connected to the correct injector/cylinder, according to the firing order.
- Check for conditions that could cause the engine to run rich or lean (Long Term Fuel Trim near –13% or +23%, respectively).

Ignition System

- Check for damaged or deteriorated components. To help detect these components, wet down the secondary ignition system with water from a spray bottle, while watching and listening for arcing or misfiring.
- Check for proper ignition voltage output (using a spark tester).
- Remove the spark plugs and check for: wet plugs,

cracks, wear, improper gap, burned electrodes, or heavy deposits. An improper plug gap will cause a driveability problem. If the plugs are fouled with gas or oil, determine the cause of fouling before replacing.

- Check for loose ignition coil grounds—the **LS1** has eight, and the **ZR-1** has four.
- Check for proper spark-plug heat range.

Engine Mechanical

- (**LS1** only) ensure that the Engine Coolant Temperature (ECT) is under 266°F (130°C). Above this, the computer operates in Engine Coolant Overtemperature–Fuel Disabled mode, which will turn off the fuel to four cylinders at a time, in order to keep the engine temperature from reaching damaging levels. The system will perceive this as a lack of power, miss, or rough idle.

Additional

- Check for vacuum hose splits or kinks, and also for proper connections and routing, as shown on the Vehicle Emission Control Information (VECI) label.
- Check for proper Torque Converter Clutch (TCC) operation. If it applies too soon it can cause spark knock.

4.3 Lack of Power, Sluggishness, Sponginess

Definition: The engine delivers less than the expected power and produces little or no increase in speed, when the accelerator pedal is pushed down part way.

Preliminary

- Search for bulletins.
- Check that the computer grounds are clean, tight, and in their proper locations. Refer to the wiring diagrams in Chapter 12.
- Remove the air filter element and check for dirt or restriction.

Fuel System

- Check that both injector fuses are intact. An open injector fuse will cause four injectors and four ignition coils to fail. Replace the fuse if necessary, and inspect the injector circuits and coil circuits for an intermittent short to ground.
- Check for low fuel pressure.
- Check for a restricted fuel filter.
- Check for contaminated fuel.
- Check for conditions that could cause the engine to run rich or lean (Long Term Fuel Trim near –13% or +23%, respectively).

Sensor/System

- Check for proper operation of the column lock system
- Check for excessive spark retard in the Knock Sensor (KS) system, using a scan tool.

Ignition System

- Check that both injector fuses are intact. An open injector fuse will cause four injectors and four ignition coils to fail. Replace the fuse if necessary. Inspect the injector circuits and coil circuits for an intermittent short to ground.
- Check for damaged or deteriorated components. To help detect these components, wet down the secondary ignition system with water from a spray bottle, while watching and listening for arcing or misfiring.
- Check for proper ignition voltage output (using a spark tester).
- Remove the spark plugs and check for: wet plugs, cracks, wear, improper gap, burned electrodes, or heavy deposits. An improper plug gap will cause a driveability problem. If the plugs are fouled with gas or oil, determine the cause of fouling before replacing.
- Check for loose ignition coil grounds—the **LS1** has eight, and the **ZR-1** has four.
- Check for proper spark-plug heat range.

Engine Mechanical

- Check for excessive oil in the combustion chamber (leaking valve seals).
- Check for low cylinder compression.
- Check for incorrect basic engine parts: cylinder heads, camshaft, pistons, and related parts.
- (**LS1** only) ensure that the Engine Coolant Temperature (ECT) is under 266°F (130°C). Above this, the computer operates in Engine Coolant Overtemperature–Fuel Disabled mode, which will turn off the fuel to four cylinders at a time, in order to keep the engine temperature from reaching damaging levels. The system will perceive this as a lack of power, miss, or rough idle.

Additional

- Check for restrictions in the exhaust system: damaged or collapsed pipes, a collapsed inner wall of the exhaust manifold, heat distress or internal failure of the mufflers, and plugged Three-Way Catalyst converters (compare back pressure on each side of the engine after removing the AIR check valves near the exhaust manifolds).
- Check for proper Torque Converter Clutch (TCC) operation

4.4 Detonation/Spark Knock

Definition: A mild to severe ping, usually worse under acceleration. The engine makes sharp metallic knocks that change with throttle opening.

Preliminary

- Search for bulletins.
- Check that the computer grounds are clean, tight, and in their proper locations. Refer to the wiring diagrams in Chapter 12.
- If the scan tool readings are normal and there are no engine mechanical faults, fill the fuel tank with a premium gasoline (minimum octane rating of 92) and reevaluate the vehicle's performance.

Fuel System

- Check for low fuel pressure.
- Check for a restricted fuel filter.
- Check for contaminated fuel.
- Check for conditions that could cause an engine to run lean (indicated by a Long Term Fuel Trim near 23%).

Ignition System

- Check for the proper spark-plug heat range.

Engine Cooling System

- Check for obvious overheating problems: low engine coolant, restricted air or coolant flow for the radiator, or inoperative electric cooling-fan circuits.

Engine Mechanical

- Check for excessive oil in the combustion chamber (indicating leaky valve seals).
- Check for low cylinder compression.
- Check for excessive carbon buildup in the combustion chambers. Clean the chambers using a top engine cleaner. Follow the instructions on the can.
- Check for incorrect basic engine parts: cylinder heads, camshaft, pistons, and related parts.

Additional

- Check for proper operation of the Park/Neutral Position (PNP) switch.
- Check for proper Torque Converter Clutch (TCC) operation. If the TCC is applied too soon you can get spark knock.

4.5 Hesitation, Sag, Stumble

Definition: A momentary lack of response as the accelerator is pushed down. This can occur at any speed and is usually more pronounced when first trying to make the vehicle move, as from a stop. It may cause the engine to stall if severe enough.

Preliminary

- Search for bulletins.
- Check that the computer grounds are clean, tight, and in their proper locations. Refer to the wiring diagrams in Chapter 12.

Sensor/System

- Check for proper operation of the Manifold Absolute Pressure (MAP) sensor.
- Check for proper Throttle Position (TP) sensor voltage.

Fuel System

- Check for low fuel pressure.
- Check for a restricted fuel filter.
- Check that both injector fuses are intact. An open injector fuse will cause four injectors and four ignition coils to fail. Replace the fuse if necessary, and inspect the injector circuits and coil circuits for an intermittent short to ground.
- Check for conditions that can cause an engine to run rich or lean (indicated by a Long Term Fuel Trim near –13% or +23%, respectively).

Ignition System

- Check for damaged or deteriorated components. To help detect these components, wet down the secondary ignition system with water from a spray bottle while watching and listening for arcing or misfiring.
- Check for proper ignition voltage output (using a spark tester).
- Remove the spark plugs and check for: wet plugs, cracks, wear, improper gap, burned electrodes, or heavy deposits. An improper plug gap will cause a driveability problem. If the plugs are fouled with gas or oil, determine the cause of fouling before replacing.
- Check for loose ignition coil grounds—the **LS1** has 8, and the **ZR-1** has 4.
- Check for proper spark-plug heat range.
- **LT1** and **LT4** only: Check for proper distributor vent system operation.

Engine Cooling System

- Check the engine thermostat for proper operation and heat range.

Additional

- Check for proper generator output voltage. Repair the charging system if output is not between 9 and 16 volts.

4.6 Cuts Out or Misses

Definition: Steady pulsation or jerking that follows engine speed, usually more pronounced as the engine load increases. This condition is not normally felt above 1,500 rpm or 30 mph (48 km/h). Also, the exhaust has a steady spitting sound at idle or low speed.

Preliminary

- Search for bulletins.
- Check that the computer grounds are clean, tight, and in their proper locations. Refer to the wiring diagrams in Chapter 12.
- Remove the air filter element and check for dirt or restriction.

Fuel System

- Check for low fuel pressure.
- Check for a restricted fuel filter.
- Check for conditions that could cause an engine to run rich or lean (indicated by a Long Term Fuel Trim near −13% or +23%, respectively).

Sensor/System

- Check for excessive spark retard in the Knock Sensor (KS) system, using a scan tool.

Ignition System

- Check for damaged or deteriorated components. To help detect these components, wet down the secondary ignition system with water from a spray bottle while watching and listening for arcing or misfiring.
- Check for proper ignition voltage output (using a spark tester).
- Remove the spark plugs and check for: wet plugs, cracks, wear, improper gap, burned electrodes, or heavy deposits. An improper plug gap will cause a driveability problem. If the plugs are fouled with gas or oil, determine the cause of fouling before replacing.
- Check for ignition wires which are cross-firing, arcing to ground, or improperly routed.
- Check for ignition coils which are cracked or showing carbon tracking.

Engine Mechanical

- Check for proper compression.
- Check for stuck or leaky valves.
- Check for worn camshaft lobes.
- Check for proper valve timing.
- Check for bent pushrods.
- Check for worn rocker arms.
- Check for broken valve springs.
- Check for excessive oil in the combustion chamber (indicating leaky valve seals).
- Check for low cylinder compression.
- Check for incorrect basic engine parts: cylinder heads, camshaft, pistons, and related parts.

Additional

- Check for restrictions in the exhaust system: damaged or collapsed pipes, a collapsed inner wall of an exhaust manifold, heat distress or internal failure of the mufflers, or plugged Three-Way Catalyst (TWC) converters (compare the back pressure on each side of the engine, after removing the AIR check valves near the exhaust manifolds).
- Check for electromagnetic interference (EMI) on the reference circuit. This can cause engine miss. A sudden increase in rpm with little change in actual engine rpm indicates the EMI is present. Check the high-voltage components (near the ignition control circuits) and the routing of the secondary ignition wires.
- Check the intake and exhaust manifold passages for casting flash.

4.7 Poor Fuel Economy

Definition: Fuel economy, as measured by a road test, is noticeably lower than expected, and noticeably lower than on the same vehicle in previous road tests.

Preliminary

- Search for bulletins.
- Check that the computer grounds are clean, tight, and in their proper locations. Refer to the wiring diagrams in Chapter 12.
- Check the owner's driving habits: ensure that the air conditioning or defroster is not on continuously, the tires are at the correct pressure, no excessively heavy loads are being carried, and the acceleration rate is not too much too often.
- Remove the air filter element and check for dirt or restriction.

Fuel System

- Check for low fuel pressure.
- Check for a restricted fuel filter.
- Check for contaminated fuel.
- Check that each injector harness is connected to the correct injector/cylinder, according to the firing order.
- Check for injectors that are shorted internally. Compare the injector resistances, which should be within 1 ohm of each other.
- Check for foreign material which may have accumulated in the throttle bore, or for coking on the throttle valve or throttle shaft. Also inspect for throttle-body tampering.
- Check for conditions that could cause an engine to run rich or lean (indicated by a Long Term Fuel Trim near –13% or +23%, respectively).

Sensor/System

- Check for air leaks in the air intake system and crankcase.
- Check for proper operation of the crankcase ventilation valve. Place a finger over the inlet hole, in the valve end, several times. The valve should snap back. If it doesn't, replace the valve.
- Check for proper calibration of the speedometer.
- Check for excessive spark retard in the Knock Sensor (KS) system, using a scan tool.

Ignition System

- Check for proper ignition voltage output, using a spark tester.
- Remove the spark plugs and check for: wet plugs, cracks, wear, improper gap, burned electrodes, or heavy deposits. An improper plug gap will cause a driveability problem. If the plugs are fouled with gas or oil, determine the cause of fouling before replacing.
- Check for ignition wires which are arcing to ground or improperly routed.
- Check for damaged or deteriorated components. To help detect these components, wet down the secondary ignition system with water from a spray bottle while watching and listening for arcing or misfiring.
- Check for loose ignition coil grounds—the **LS1** has 8, and the **ZR-1** has 4.

Engine Cooling System

- Check for low engine coolant .
- Check the engine thermostat for proper operation and heat range.

Engine Mechanical

- Check for proper compression.
- Check for stuck or leaky valves.
- Check for worn camshaft lobes.
- Check for proper valve timing.
- Check for bent pushrods.
- Check for worn rocker arms.
- Check for broken valve springs.
- Check for excessive oil in the combustion chamber (indicated by leaky valve seals).
- Check for low cylinder compression.
- Check for incorrect basic engine parts: cylinder heads, camshaft, pistons, and related parts.

Additional

- Check for vacuum hose splits or kinks, and for proper connections and routing, as shown on the Vehicle Emission Control Information (VECI) label.
- Check for proper Torque Converter Clutch (TCC) operation. The scan tool should indicate a drop in rpm when the system commands the TCC on.
- Check for restrictions in the exhaust system: damaged or collapsed pipes, a collapsed inner wall of an exhaust manifold, heat distress or internal failure of the mufflers, or plugged Three-Way Catalyst (TWC) converters (compare the back pressure on each side of the engine, after removing the AIR check valves near the exhaust manifolds).
- Check for electromagnetic interference (EMI) on the reference circuit. This can cause engine miss. A sudden increase in rpm with little change in actual engine rpm indicates that EMI is present. Check the high-voltage components (near the ignition control circuits) and the routing of the secondary ignition wires.
- Check for proper operation of the Park/Neutral Position (PNP) switch circuit.
- Check for faulty motor mounts.
- Check the intake and exhaust manifold passages for casting flash.

4.8 Rough, Unstable, Incorrect Idle, Stalling

Definition: The engine runs unevenly at idle. If severe, the engine or vehicle may shake. Engine idle speed may vary in rpm. Either condition may be severe enough to stall the engine.

Preliminary

- Search for bulletins.
- Check that the computer grounds are clean, tight, and in their proper locations. Refer to the wiring diagrams in Chapter 12.
- Remove the air filter element and check for dirt or restriction.

Fuel System

- Check for proper injector balance.
- Check for low fuel pressure.
- Check for a restricted fuel filter.
- Check for contaminated fuel.
- Check that each injector harness is connected to the correct injector/cylinder, according to the firing order.
- Check for injectors which are shorted internally. Compare the injector resistances, which should be within 1 ohm of each other.
- Check for conditions that could cause an engine to run rich or lean (indicated by a Long Term Fuel Trim near –13% or +23%, respectively).

Sensor/System

- Check for proper operation of the crankcase ventilation valve. Place a finger over the inlet hole in the valve end several times. The valve should snap back. If it doesn't, replace the valve.
- Check for excessive spark retard in the Knock Sensor (KS) system, using a scan tool.

Ignition System

- Check for proper ignition voltage output using a spark tester.
- Remove the spark plugs and check for: wet plugs, cracks, wear, improper gap, burned electrodes, or heavy deposits. An improper plug gap will cause a driveability problem. If the plugs are fouled with gas or oil, determine the cause of fouling before replacing.
- Check for ignition wires which are cross-firing, arcing to ground, or improperly routed.
- Check for damaged or deteriorated components. To help detect these components, wet down the secondary ignition system with water from a spray bottle while watching and listening for arcing or misfiring.
- Check for loose ignition coil grounds.

Engine Mechanical

- Check for proper compression.
- Check for stuck or leaky valves.
- Check for worn camshaft lobes.
- Check for proper valve timing.
- Check for bent pushrods.
- Check for worn rocker arms.
- Check for broken valve springs.
- Check for excessive oil in the combustion chamber (indicating leaky valve seals).
- Check for low cylinder compression.
- Check for incorrect basic engine parts: cylinder heads, camshaft, pistons, and related parts.

Additional

- Check for restrictions in the exhaust system: damaged or collapsed pipes, a collapsed inner wall of an exhaust manifold, heat distress or internal failure of the mufflers, or plugged Three-Way Catalyst (TWC) converters (compare the back pressure on each side of the engine, after removing the AIR check valves near the exhaust manifolds).
- Check for electromagnetic interference (EMI) on the reference circuit. This can cause engine miss. A sudden increase in rpm with little change in actual engine rpm indicates that EMI is present. Check the high-voltage components (near the ignition control circuits) and the routing of secondary ignition wires.
- Check for proper operation of the Park/Neutral Position (PNP) switch circuit.
- Check for faulty motor mounts.
- Check the intake and exhaust manifold passages for casting flash.

4.9 Dieseling, Run-On

Definition: The engine continues to run after the key is turned off, and runs roughly. (If the engine runs smoothly, check the ignition switch and the ignition switch adjustment.)

Preliminary

- Search for bulletins.
- Check that the computer grounds are clean, tight, and in their proper locations. Refer to the wiring diagrams in Chapter 12.

Fuel System

- Check for injector leakage

4.10 Backfire

Definition: Fuel ignites in the intake manifold or exhaust system, making a loud popping noise.

Preliminary

- Search for bulletins.
- Check that the computer grounds are clean, tight, and in their proper locations. Refer to the wiring diagrams in Chapter 12.

Fuel System

- Check for low fuel pressure.
- Check for a restricted fuel filter.
- Check for contaminated fuel.
- Check that each injector harness is connected to the correct injector/cylinder, according to the firing order.

Sensor/System

- Check for proper operation of the Secondary Air Injection (AIR) system.
- Check for air leaks in the air intake system and crankcase.
- Check for proper operation of the crankcase ventilation valve. Place a finger over the inlet hole in the valve end several times . The valve should snap back. If it doesn't, replace the valve.
- Check for proper calibration of the speedometer.
- Check for excessive spark retard in the Knock Sensor (KS) system, using a scan tool.

Ignition System

- Check for proper ignition voltage output using a spark tester.
- Check for an intermittent malfunction in the ignition control circuit. Use the scan tool's snapshot feature to help locate an intermittent ignition failure.
- Remove the spark plugs and check for: wet plugs, cracks, wear, improper gap, burned electrodes, or heavy deposits. An improper plug gap will cause a driveability problem. If the plugs are fouled with gas or oil, determine the cause of fouling before replacing.
- Check for ignition wires which are cross-firing or arcing to ground
- Check for ignition coils which are arcing to ground.
- Check for damaged or deteriorated components. To help detect these components, wet down the secondary ignition system with water from a spray bottle while watching and listening for arcing or misfiring.
- Check for loose ignition coil grounds—the **LS1** has 8, and the **ZR-1** has 4.

Engine Cooling System

- Check for low engine coolant.
- Check the engine thermostat for proper operation and heat range.

Engine Mechanical

- Check for proper compression.
- Check for stuck or leaky valves.
- Check for worn camshaft lobes.
- Check for proper valve timing.
- Check for bent pushrods.
- Check for worn rocker arms.
- Check for broken valve springs.
- Check for excessive oil in the combustion chamber (indicating leaky valve seals).
- Check for low cylinder compression.
- Check for incorrect basic engine parts: cylinder heads, camshaft, pistons, and related parts.

Additional

- Check for vacuum hose splits or kinks, and for proper connections and routing, as shown on the Vehicle Emission Control Information (VECI) label.
- Check for proper Torque Converter Clutch (TCC) operation. The scan tool should indicate a drop in rpm when the TCC is commanded on.
- Check for restrictions in the exhaust system: damaged or collapsed pipes, a collapsed inner wall of an exhaust manifold, heat distress or internal failure of the mufflers, or plugged Three-Way Catalyst (TWC) converters (compare the back pressure on each side of the engine, after removing the AIR check valves near the exhaust manifolds).
- Check for electromagnetic interference (EMI) on the reference circuit. This can cause engine miss. A sudden increase in rpm with little change in actual engine rpm indicates that EMI is present. Check the high-voltage components (near the ignition control circuits) and the routing of the secondary ignition wires.
- Check for proper operation of the Park/Neutral Position (PNP) switch circuit.
- Check for faulty motor mounts.
- Check the intake and exhaust manifold passages for casting flash.

4.11 Restricted Exhaust System, '97-01

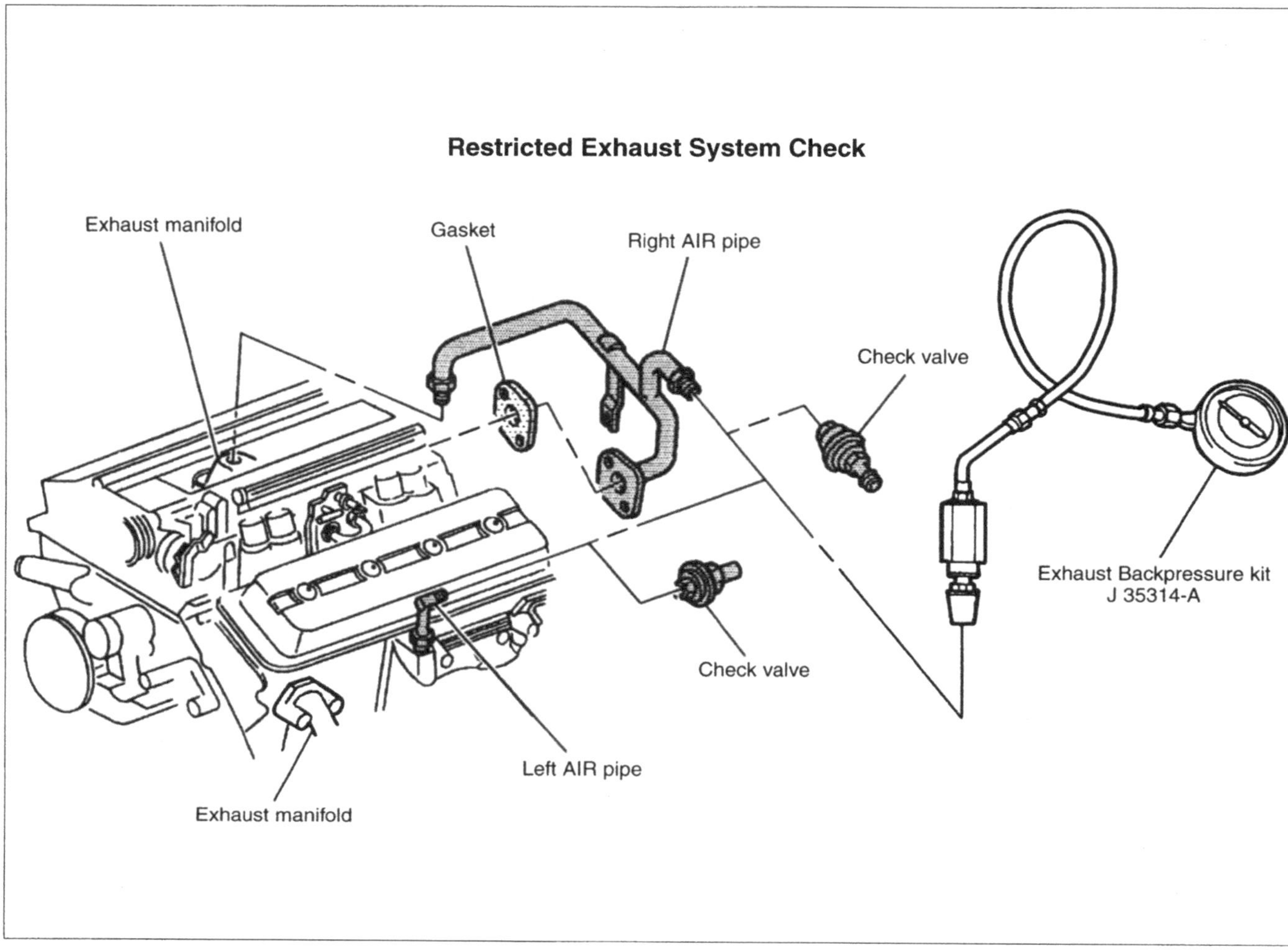

Fig. 10-12. LT1 shown.

5. Diagnostic Trouble Codes (DTC)

All information applies to *domestic* Corvette engines, no transmission codes are listed here.

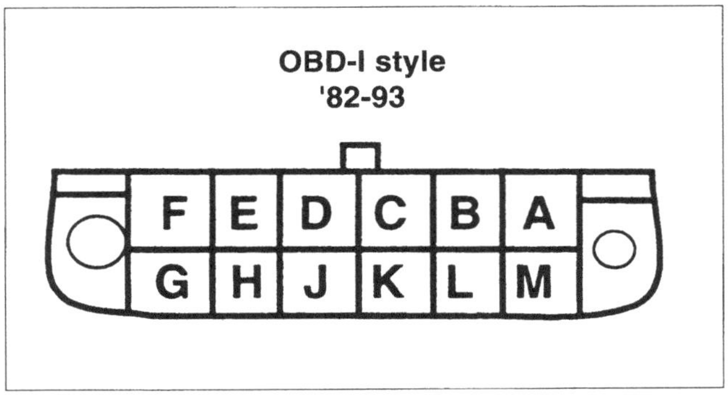

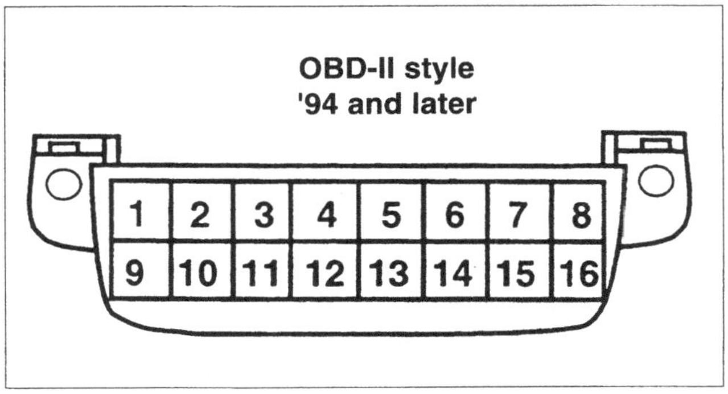

Fig. 10-13. This style Diagnostic Link Connector provided 12 channels. The **'94-95 LT1** has the OBD-II style but has 16 channels. It is the transition car, and is not fully compliant. The **'96 LT1**, **LT4**, **LS1** are fully compliant.

12 pin identification (typical)

PIN A-DLC Ground
PIN B-Diagnostic Request (ECM)
PIN C-AIR solenoid
PIN D-Malfunction Indicator Lamp
PIN E-Serial Data Link (160 baud)
PIN F-Torque Converter Clutch Solenoid
PIN G-Fuel Pump
PIN H-Anti-lock Brake System
PIN J-Entertainment and Comfort
PIN K-SIR Diagnostic Request
PIN L-Serial Data Loop (8192 baud)
PIN M-Serial Data Link (8192 baud)

16 pin identification (typical)

PIN 1-Secondary UART 8192 Baud Serial Data (CKT 800), Secondary Class B (CKT 710) or 160 Baud Serial Data (1995 only) (CKT 461)
PIN 2-J 1850 Bus + L Line on 2-Wire Systems, or Single Wire (Class 2) (CKT 1807)
PIN 3-Ride Control Diagnostic Enable (CKT 1826)
PIN 4-Chassis Ground Pin (CKT 150)
PIN 5-Signal Ground Pin (CKT 451)
PIN 6-PCM/VCM Diagnostic Enable (CKT 448)
PIN 7-K line for International Standards Organization (ISO) Application (Catera ECM/TCM Data)
PIN 8-Keyless Entry Enable (CKT 1455) or MRD Theft Diagnostic Enable (CKT 477)
PIN 9- Primary UART (CKT 1061)
PIN 10-J 1850 Bus-Line for J 1850-2 Wire Applications
PIN 11-Electronic Variable Orifice (EVO) Steering (CKT 1294) or Magnetic Steering Variable Assist (MSVA)
PIN 12-ABS Diagnostic (CKT 799) or COM Diagnostic Enable (CKT 555)
PIN 13-SIR Diagnostic Enable (CKT 326)
PIN 14-E & C Bus (CKT 835)
PIN 15-L Line for International Standards Organization (ISO) Application
PIN 16-Battery Power from Vehicle Unswitched (4 AMP MAX)

L83 and L98

Table 10-14. Diagnostic Trouble Codes for the L83 and L98

L83	L98	Problem
13	13	Left O_2 sensor open
14	14	ECT high temperature, above 130˚ C
15	15	ECT low temperature, below –32˚ C, or open
21	21	TPS voltage high, idle above 0.9 v
22	22	TPS voltage low, idle 0.3–0.9 v
—	23	IAT (MAT) low temperature, below 35˚ C, or open
24	24	VSS circuit
—	25	IAT high temperature, greater than 150˚ C for 12 seconds
—	32	EGR circuit
33	33	MAF '85–89 (or MAP '90–91) circuit—voltage high
34	34	MAF '85–89 (or MAP '90–91) circuit—voltage low
—	36	MAF burnoff circuit, '85–88
—	41	Cylinder Select error
42	42	EST circuit
43	43	ESC circuit
44	44	O_2 sensor lean
45	45	O_2 sensor rich
—	46	Personal Automotive Security System (or Vehicle Anti-Theft System) circuit, '87-and-later
51	51	MEM-CAL (PCM) error
—	52	EOT low temperature; fuel CAL-PAK missing, '85
—	53	System over-voltage
—	54	Fuel pump low voltage
55	55	Replace ECM
—	62	EOT high temperature

ZR-1

Table 10-15. Diagnostic Trouble Codes for the ZR-1 '90-95

LT5	Problem
13	Left O_2 sensor circuit open
14	ECT high temperature
15	ECT low temperature, below –32˚ C
16	DIS (EI) fault line circuit
21	TPS voltage low
22	TPS voltage high
23	IAT low temperature
24	VSS circuit
25	IAT high temperature
31	Camshaft Position Sensor (CMP) circuit
33	MAP voltage high
34	MAP voltage low
35	MAP low vacuum
36	DIS (EI) missing or extra EST signal
41	Cylinder Select error
42	EST circuit
43	ESC circuit
44	Left O_2 sensor circuit—lean exhaust
45	Left O_2 sensor circuit—rich exhaust
46	Personal Automotive Security System circuit
51	MEM-CAL (PROM) error
52	EOT low temperature, below –38˚ C
53	System voltage
54	Fuel-pump circuit low voltage
55	Fuel lean monitor
56	Vacuum sensor circuit high or low
61	Secondary port throttle system
62	EOT high temperature
63	Right O_2 sensor circuit open
64	Right O_2 sensor circuit—lean exhaust
65	Right O_2 sensor circuit—rich exhaust
66	Engine power switch

5. *Diagnostic Trouble Codes (DTC)*

'92-93 LT1

Table 10-16. Diagnostic Trouble Codes for the '92–93 LT1

'92–93 LT1	Problem
13	Left O_2 sensor circuit open
14	ECT high temperature, above 130˚C
15	ECT low temperature, below –32˚ C, or open
16	DI low-resolution pulse
21	TP sensor voltage high, idle above 0.9 v
22	TP sensor voltage low, idle 0.3–0.9 v
23	IAT (MAT) low temperature, below 35˚ C, or open
24	VSS circuit
25	IAT high temperature, greater than 150˚ C for 12 seconds
26	Quad driver #1 circuit
27	Quad driver #2 circuit
28	Quad driver #3 circuit
32	EGR circuit
33	MAP circuit, voltage high
34	MAP circuit, voltage low
36	DI faulty hi-res pulse or extra lo-res pulse
41	IC circuit open or shorted
42	IC circuit grounded
43	ESC circuit
44	Left O_2 sensor circuit—lean exhaust
45	Left O_2 sensor circuit—rich exhaust
46	Personal Automotive Security System (or Vehicle Anti-Theft System) circuit
52	EOT low temperature; fuel CAL-PAK missing, '85
53	System voltage
55	Fuel lean monitor
62	EOT high temperature
63	Right O_2 sensor circuit open
64	Right O_2 sensor circuit—lean exhaust
65	Right O_2 sensor circuit—rich exhaust
66	Air conditioner pressure sensor circuit open or shorted
67	Air conditioner pressure sensor or clutch problem
68	Air conditioner relay circuit shorted
69	Air conditioner clutch circuit
72	Gear selector switch circuit

Table 10-16. Diagnostic Trouble Codes for the '92–93 LT1

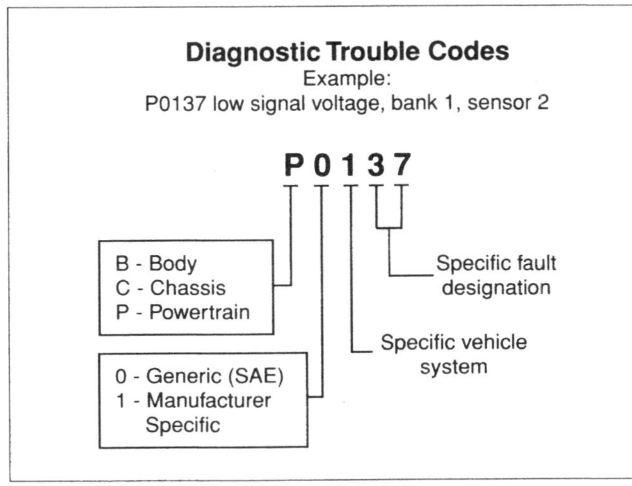

Fig. 10-14. New 5-digit codes began, with implementation in the **'94 LT1**, but not in the **ZR-1**. Full implementation occured by **'96 LT1, LT4**.

Table 10-17. Diagnostic Trouble Codes

Character	Indication
First letter	Location of problem (Body, Chassis, or Powertrain)
Second digit	0 = SAE generic code, all OBD-II cars
	1 = manufacturer specific, Corvette
Third digit	Specific vehicle system
	1 = fuel and air metering systems
	2 = fuel and air metering systems (fuel delivery)
	3 = ignition system, misfire
	4 = auxiliary emission controls
	5 = vehicle speed, idle-speed control systems
	6 = computer output controls
	7 = transmission
Fourth and fifth digits	Specific fault designation (similar to previous two-digit codes)

'94-95 LT1

Table 10-18. Diagnostic Trouble Codes for the '94–95 LT1

'94–95 LT1	Problem
Fuel and air metering systems	
13	Left O_2 sensor circuit open
14	ECT high temperature, above 140° C
15	ECT low temperature, below –40° C, or open
16	DI low-resolution pulse
18	Injector circuits
21	TP sensor voltage high, idle above .9 v
22	TP sensor voltage low, idle .3–.9 v
23	IAT (MAT) low temperature, below 35° C, or open
24	VSS circuit
25	IAT high temperature, greater than 150° C for 12 seconds
32	EGR circuit
33	MAF '85–89 (or MAP '90–91) circuit, voltage high
34	MAF '85–89 (or MAP '90–91) circuit, voltage low
41	Cylinder Select error
42	EST circuit
43	ESC circuit
44	O_2 sensor circuit—lean exhaust
45	O_2 sensor circuit—rich exhaust
46	Personal Automotive Security System (PASS or VATS) circuit, '87-and-later
51	MEM-CAL (PCM) error
52	EOT low temperature; fuel CAL-PAK missing, '85
53	System over-voltage
54	Fuel pump low voltage
55	Replace ECM
62	EOT high temperature
	Bank 2
63	Right front O_2 sensor circuit open
64	Right front O_2 sensor circuit—lean exhaust
65	Right front O_2 sensor circuit—rich exhaust
66	Air conditioner pressure sensor circuit open or shorted
67	Air conditioner pressure sensor or clutch problem
68	Air conditioner relay circuit shorted
69	Air conditioner clutch circuit
70	Air conditioner clutch relay driver circuit
72	VSS loss
73	Pressure control solenoid circuit, current error
74	TCS circuit low
75	Transmission system voltage low
77	Fan control PCM terminal A11
78	Fan control PCM terminal A10
79	Transmission fluid overtemperature
80	'94 only transmission component slipping
81	Transmission 2–3 shift solenoid circuit
82	Transmission 1–2 shift solenoid circuit
83	TCC PWM solenoid circuit fault
84	Manual transmission 2–3 blockout relay circuit *or* Automatic transmission 3–2 solenoid control circuit
85	TCC stuck on
90	TCC solenoid circuit
91	Manual transmission 1–4 upshift light
97	VSS output circuit
99	Tach output circuit
P01XX: Fuel and air metering systems	
P0116	ECT circuit range (performance problem)
P0117	ECT low voltage
P0118	ECT high voltage
	Bank 1
P0131	HO_2S1 low voltage
P0132	HO_2S1 high voltage
P0133	HO_2S1 slow response
P0134	HO_2S1 insufficient activity
P0135	HO_2S1 heater circuit
P0151	HO_2S1 low voltage
P0152	HO_2S1 high voltage
P0153	HO_2S1 slow response

Table 10-18. Diagnostic Trouble Codes for the '94–95 LT1

P0154	HO_2S1 insufficient activity
P0155	HO_2S1 heater circuit
P0158	HO_2S2 high voltage
P0160	HO_2S2 insufficient activity
P0161	HO_2S2 heater circuit
	Bank 1
P0171	Fuel trim lean
P0172	Fuel trim rich
	Bank 2
P0174	Fuel trim lean
P0175	Fuel trim rich
P04XX: Auxiliary emission controls	
P0420	TWC low efficiency, bank 2
P11XX: GM-specific fuel and air metering systems	
P1114	ECT intermittent low voltage
P1115	ECT intermittent high voltage
P1121	TP intermittent high voltage
P1122	TP intermittent low voltage
P1133	HO_2S1 insufficient switching
P1153	HO_2S2 insufficient switching
P1158	HO_2S2 shifted rich

'96 LT1 and LT4

Table 10-19. Diagnostic Trouble Codes for the '96 LT1 and LT4

'96 LT1; LT4	Problem
P01XX: Fuel and air metering systems	
P0101	MAF system performance
P0102	MAF low frequency
P0103	MAF high frequency
P0107	MAF low voltage
P0108	MAF high voltage
P0112	IAT low voltage
P0113	IAT high voltage
P0117	ECT low voltage
P0118	ECT high voltage
P0121	TP sensor circuit insufficient activity
P0122	TP sensor low voltage
P0123	TP sensor high voltage
P0125	ECT excessive time to Closed Loop
	Bank 1
P0131	HO_2S1 low voltage
P0132	HO_2S1 high voltage
P0133	HO_2S1 slow response
P0134	HO_2S1 insufficient activity
P0135	HO_2S1 heater circuit
P0137	HO_2S2 low voltage
P0138	HO_2S2 high voltage
P0140	HO_2S2 insufficient activity
P0141	HO_2S2 heater circuit
	Bank 2
P0151	HO_2S1 low voltage
P0152	HO_2S1 high voltage
P0153	HO_2S1 slow response
P0154	HO_2S1 insufficient activity
P0155	HO_2S1 heater circuit
P0157	HO_2S2 low voltage
P0158	HO_2S2 high voltage

10

Table 10-19. Diagnostic Trouble Codes for the '96 LT1 and LT4

P0160	HO_2S2 insufficient activity
P0161	HO_2S2 heater circuit
	Bank 1
P0171	Fuel trim lean
P0172	Fuel trim rich
	Bank 2
P0174	Fuel trim lean
P0175	Fuel trim rich
P02XX: Fuel injector system	
P0200	Injector control circuit
P03XX: Ignition system or misfire	
P0300	Engine misfire detection
P0323	DI low-/hi-res circuit intermittent
P0325	KS system
P0327	KS circuit bank 1
P0332	KS circuit bank 2
P0335	CKP sensor
P0336	CKP sensor circuit performance
P0372	DI hi-res circuit
P04XX: Auxiliary emission controls	
P0400	EGR system
P0403	EGR solenoid control circuit
P0410	AIR system
P0412	AIR solenoid relay control
P0420	TWC low efficiency, bank 1
P0430	TWC low efficiency, bank 2
P0441	EVAP, no flow during purge
P0443	EVAP purge solenoid control
P05XX: Vehicle speed and idle control	
P0500	VSS circuit (manual transmission)
P0506	Idle speed low, IAC responding
P0507	Idle speed high, IAC responding
P0530	Air conditioner refrigerant pressure
P0562	System voltage low
P0563	System voltage high

Table 10-19. Diagnostic Trouble Codes for the '96 LT1 and LT4

P06XX: Computer and output circuits	
P0601	PCM memory
P0602	PCM not programmed
P11XX: GM-specific fuel and air metering systems	
P1107	MAP intermittent low voltage
P1111	IAT intermittent high voltage
P1112	IAT intermittent low voltage
P1114	ECT intermittent low voltage
P1115	ECT intermittent high voltage
P1121	TP intermittent high voltage
P1122	TP intermittent low voltage
P1133	HO_2S1 insufficient switching
P1134	HO_2S1 transition time ratio
P1153	HO_2S1 insufficient switching
P1154	HO_2S2 transition time ratio
P1171	Fuel system lean during acceleration
P1187	EOT sensor low voltage
P1188	EOT sensor high voltage
P12XX: GM-specific fuel injector system	
P1222	Injector control circuit intermittent
P13XX: GM-specific ignition system or misfire	
P1351	IC circuit high voltage
P1351	IC circuit low voltage
P1371	DI low-res circuit
P1380	[Anti-Lock Braking System] rough road, data [unstable]
P1381	Misfire detected—no EBCM/PCM serial data
P14XX: GM-specific auxiliary emission controls	
P1415	AIR system bank 1
P1416	AIR system bank 2
P1441	EVAP flow during non-purge
P15XX: GM-specific vehicle speed and idle control	
P1508	IAC low rpm
P1509	IAC high rpm
P1539	Air conditioner clutch status, high voltage

Table 10-19. Diagnostic Trouble Codes for the '96 LT1 and LT4

P1545	Air conditioner clutch relay control
P1546	Air conditioner clutch status, low voltage
P1572	ASR/TCS active, low voltage too long
P16XX: GM-specific computer and output circuits	
P1641	Fuel control relay #1
P1642	Fuel control relays #2 and #3
P1643	Engine speed output
P1652	VSS output
P1657	Manual transmission 1–4 upshift solenoid circuit
P1661	MIL control circuit
P1664	Manual transmission 1–4 upshift light circuit

LS1 '97-01

DTCs are identified as emission related or non-emission-related, and classified as to storage and illumination of MIL/SES.

Emissions related
Type A:
- lights the Malfunction Indicator Light (MIL) after 1-trip failure
- stores history codes and failure records
- stores freeze frames

Beginning '98, most type A codes redesignated type B

Type B:
- lights the MIL if the fault is present after 2 drive trips
- stores freeze frames

Non-emissions related
Type C:
- lights the Service Engine Soon (SES) indicator after 1-trip failure (the MIL does not light)
- stores history codes

Type D:
- stores history code and failure record after 1 failure
- does not light MIL or SES

GM is phasing out type D DTCs

Table 10-20. Diagnostic Trouble Codes for the LS1

LS1	Problem	'97	'98–01
P01XX: Fuel and air metering systems			
P0101	MAF system performance	B	B
P0102	MAF low frequency	A	B
P0103	MAF high frequency	A	B
P0107	MAP low voltage	A	B
P0108	MAP high voltage	A	B
P0112	IAT low voltage	B	B
P0113	IAT high voltage	B	B
P0117	ECT low voltage	A	B
P0118	ECT high voltage	A	B
P0120	APP sensor #1 circuit ('97 only; see P1275 for '98–99)	C	—
P0121	APP sensor #1 circuit performance ('97 only; see P1276 for '98–99)	C	—
P0125	ECT excess time to Closed Loop control	B	B
	Bank 1		
P0131	HO_2S1 low voltage	A	B
P0132	HO_2S1 high voltage	A	B
P0133	HO_2S1 slow response	B	B
P0134	HO_2S1 insufficient activity	A	B
P0135	HO_2S1 heater circuit	B	B
P0137	HO_2S2 low voltage	B	B
P0138	HO_2S2 high voltage	B	B
P0139	HO_2S2 slow response	B	B ('00-01)
P0140	HO_2S2 insufficient activity	B	B
P0141	HO_2S2 heater circuit	B	B
	Bank 2		
P0151	HO_2S1 low voltage	A	B
P0152	HO_2S1 high voltage	A	B
P0153	HO_2S1 slow response	B	B
P0154	HO_2S1 insufficient activity	A	B
P0155	HO_2S1 heater circuit	B	B

10

Table 10-20. Diagnostic Trouble Codes for the LS1

P0157	HO_2S2 low voltage	B	B
P0158	HO_2S2 high voltage	B	B
P0159	HO_2S2 slow response	B	—
P0160	HO_2S2 insufficient activity	B	B
P0161	HO_2S2 heater circuit	B	B
	Bank 1		
P0171	Fuel trim lean	B	B
P0172	Fuel trim rich	B	B
	Bank 2		
P0174	Fuel trim lean	B	B
P0175	Fuel trim rich	B	B
P02XX: Fuel injector system, ['98–99]			
P0220	APP sensor #2 circuit ('97 only; see P1280 for '98–99)	C	—
P0221	APP sensor #2 circuit performance ('97 only; see P1281 for '98–99)	C	—
P0225	APP sensor #3 circuit ('97 only; see P1285 for '98–99)	C	—
P0226	APP sensor #3 circuit performance ('97 only; see P1286 for '98–99)	C	—
P0230	Fuel pump control circuit	A	B
P03XX: Ignition system or misfire			
P0300	Engine misfire detected	B	B
P0325	KS system	B	B
P0327	KS circuit front	B	B
P0332	KS circuit rear	B	B
P0335	CKP circuit	A	B
P0335	CKP performance	A	B
P0341	CMP circuit performance	B	B
P0342	CMP low voltage	B	B
P0343	CMP high voltage	B	B
P0351	IC #1 circuit	A	B
P0352	IC #2 circuit	A	B
P0353	IC #3 circuit	A	B
P0354	IC #4 circuit	A	B
P0355	IC #5 circuit	A	B
P0356	IC #6 circuit	A	B
P0357	IC #7 circuit	A	B
P0358	IC #8 circuit	A	B
P04XX: Auxiliary emission controls			
P0410	AIR system	B	B
P0412	AIR solenoid relay control	B	B
P0418	AIR pump relay control	B	B
P0420	TWC low efficiency, bank 1	A	A
P0430	TWC low efficiency, bank 2	A	A
P0441	EVAP no flow during purge	B	B
P0443	EVAP purge solenoid control	B	B
P0461	Fuel level sensor #1 performance	C	C
P0462	Fuel level sensor #1 low voltage	C	C
P0463	Fuel level sensor #1 high voltage	C	C
P0480	Fuel control relay #1	B	B
P0481	Fuel control relays #2 and #3	B	B
P05XX: Vehicle speed and idle control			
P0500	VSS circuit (manual transmission)	B	—
P0506	Idle speed low	B	—
P0507	Idle speed high	B	—
P0522	Engine oil pressure circuit low voltage ('98-01)	—	C
P0523	Engine oil pressure circuit high voltage ('98-01)	—	C
P0530	Air conditioner refrigerant pressure	C	C
P0562	System voltage low	C	C
P0563	System voltage high	C	C
P0567	Cruise resume circuit	C	C
P0568	Cruise set circuit	C	C
P0571	Cruise brake switch circuit (manual transmission)	C	C
P06XX: Computer and output circuits			
P0601	PCM memory	A	A
P0602	PCM not programmed	A	A
P0604	PCM RAM performance ('98–01)	—	A
P0606	PCM internal communication interrupted	A	A
P0608	VSS output check	C	C

Table 10-20. Diagnostic Trouble Codes for the LS1

P0650	MIL control circuit—no MIL	B	B
P0654	Engine speed output check	C	C
P07XX: Transmission			
P0704	Clutch switch circuit (manual transmission)	B	B
P0705	Transmission range switch circuit (automatic transmission)	C	C
P0706	Transmission range switch performance (automatic transmission)	C	C
P08XX: Transmission			
P0801	Reverse inhibit control circuit (manual transmission)	C	C
P0803	1–4 upshift solenoid control (manual transmission)	B	B
P0804	1–4 upshift light control (manual transmission)	C	C
P11XX: GM-specific fuel and air metering systems			
P1111	IAT intermittent high voltage	C	C
P1112	IAT intermittent low voltage	C	C
P1114	ECT intermittent low voltage	C	C
P1115	ECT intermittent high voltage	C	C
P1120	TP sensor #1 circuit	A	A
P1125	APP system	A	A
P1133	HO_2S1 insufficient switching	B	B
P1134	HO_2S1 transition time	B	B
P1153	HO_2S2 insufficient switching	B	B
P1154	HO_2S2 transition time	B	B
P1187	Engine oil pressure circuit low voltage ('97 only)	C	—
P1188	Engine oil pressure circuit high voltage ('97 only)	C	—
P12XX: GM-specific fuel injector system			
P1220	TP sensor #1 circuit	A	A
P1221	TP sensors #1 and #2 performance	A	A
P1258	Engine coolant overtemperature—fuel disabled	A	A
P1275	APP sensor #1 circuit	C	C
P1276	APP sensor #1 circuit performance	C	C
P1280	APP sensor #2 circuit	C	C

Table 10-20. Diagnostic Trouble Codes for the LS1

P1281	APP sensor #2 circuit performance	C	C
P1285	APP sensor #3 circuit	C	C
P1286	APP sensor #3 circuit performance	C	C
P13XX: GM-specific ignition system or misfire			
P1380	EB(T)CM DTC detected—rough road, data unusable	C	C
P1381	Misfire detected—no EB(T)CM serial data	C	C
P14XX: GM-specific auxiliary emission controls			
P1415	AIR system bank 1	B	B
P1416	AIR system bank 2	B	B
P1431	Fuel level sensor 2 circuit performance	C	C
P1432	Fuel level sensor 2 circuit low voltage	C	C
P1433	Fuel level sensor 2 circuit high voltage	C	C
P1441	EVAP flow during non-purge	B	B
P15XX: GM-specific vehicle speed and idle control			
P1514	TAC system MAF performance	A	A
P1515	Command/actual TP performance (PCM module)	A	A
P1516	Command/actual TP performance (TAC module)	A	A
P1517	TAC module processor	A	A
P1518	PCM-to-TAC-module serial data circuit	A	A
P1539	Air conditioner clutch status circuit high voltage	C	C
P1545	Air conditioner clutch relay control circuit	C	C
P1546	Air conditioner clutch status circuit low voltage	C	C
P1571	ASR desired torque	C	C
P1574	Stoplight control circuit	C	C
P1575	Extended-travel brake-switch circuit high voltage	C	C
P16XX: GM-specific computer and output circuits			
P1624	PCM RAM performance ('97 only)	A	—
P1626	Theft deterrent system, fuel enable circuit	C	C

Table 10-20. Diagnostic Trouble Codes for the LS1

P1630	Theft deterrent system, PCM in learn mode	C	C
P1631	Theft deterrent system, password incorrect	C	C
P1635	5 v reference #1 circuit	B	B
P1639	5 v reference #2 circuit	B	B
P1644	Delivered torque output circuit	C	C
P1652	Powertrain-induced chassis pitch output	C	C

In this chapter, you have seen:

- Control Module Diagnostics, the major differences, OBD-I and OBD-II
- diagnostic procedures as they change, and how they are affected by the diagnostic capabilities of changing computers
- the differences between Data Link Connectors
- symptoms and Diagnostic Trouble Codes

IN THE NEXT CHAPTER:

Now that you've been pointed in the right direction by diagnosis, Chapter 11 shows you typical service procedures, augmented by the schematics in Chapter 12.

Chapter 11

Servicing

Contents

In this chapter, look for ways to:

- check specifics on the fuel systems, ignition systems and idle-air systems
- run specific tests of components indicated by DTCs/symptoms
- list possible causes of failure
- observe and measure details of sensors

1. INTRODUCTION

In the previous chapter, you saw how to use tools such as scan tools to look for trouble codes, and to run Strategy-Based diagnostics of the engine-control system.

Tools

> ***CAUTION —***
> *In general, GM discourages backprobing. Specifically, GM does not recommend back-probing wherever moisture can enter the cable. This includes back-probing underhood; we're talking **ZR-1, LT1, LT4, LS1**.*

If you must backprobe, push in a straight pin through the back of a connector, probing for a contact that allows making an electrical test (voltage/ground/resistance) on that circuit. Some people use a paper clip, since it's easier to attach the meter's alligator clips.

- Be careful when backprobing connectors because it can destroy the waterproofing of a connector. Moisture entering the opening can corrode the contacts inside the connector. If in doubt, don't backprobe, especially underhood.
- When back probing connectors, insert the pin from the harness side.
- Never force a probe into a female connector. On the male-pin side, probe the pin. Be sure you do not short between the pins.
- Connect or disconnect multiple connectors and test leads only with the ignition off. Disconnect test leads

Fig. 11-2. LT1, LT4, ZR-1, LS1 The underhood location of the engine computer makes backprobing a no-no, as there is too much of a chance of moisture entering the cables/connectors.

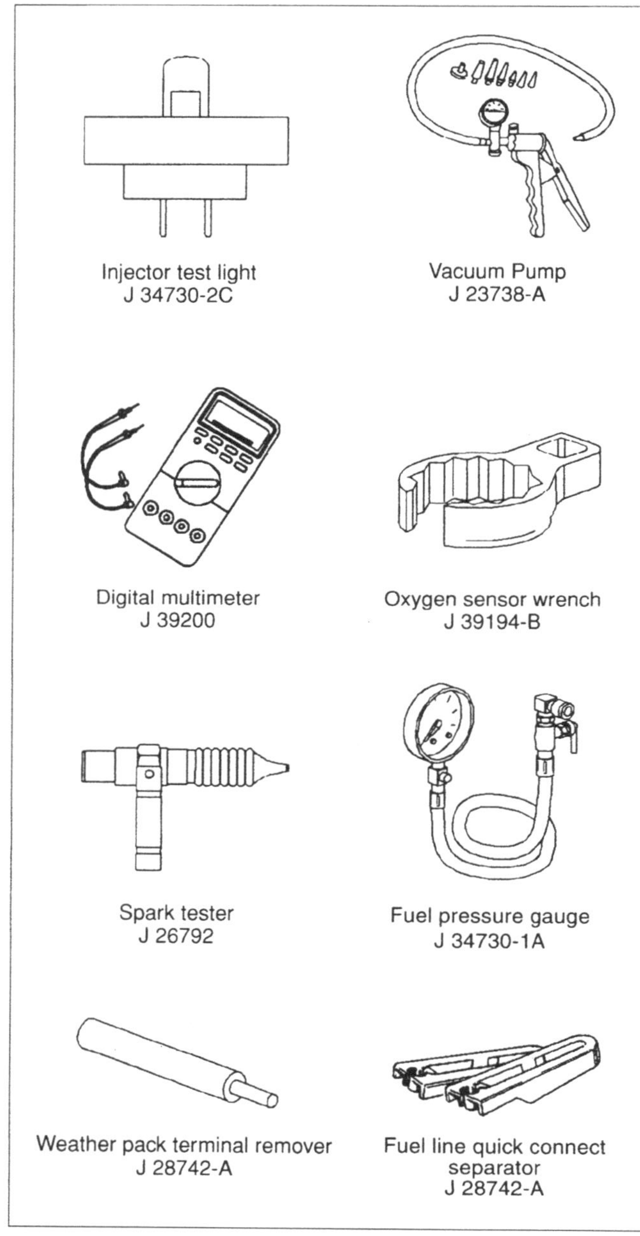

Fig. 11-1. Shops may want all of these tools and more. Owners on a budget may choose more carefully.

before switching multimeter functions or measurement ranges.

- Use an LED (light emitting diode) test lamp, not an incandescent bulb.

> Use a high quality Digital Multi-Meter (DMM) with a high input impedance (at least 10 megohms). Do not use an analog (swing-needle) meter to check circuit resistance or continuity on electronic (solid state) components. When I say DMM, this includes the earlier Digital Volt/Ohm Meter (DVOM).

2. Servicing Electrical Problems

You may want to skip section 2 if you're good at checking by codes and symptoms. Most of these procedures are useful for troubleshooting Corvettes with fewer codes (**'82-93**).

As you may know, four things are required for current to flow in most automotive circuits: a voltage source, wires or connections to transport the voltage, a device that uses the electricity, and a connection to ground.

For trouble-free operation of the engine control systems, the ground connections, including the negative battery cables and the body ground straps, must remain clean and free from corrosion. You can find most problems using only a DVOM/DMM to check for voltage, resistance, breaks in the wiring (infinite resistance/no continuity), or a path to ground that completes the circuit. Most modern cars provide, and actually *need*, multiple grounds. And of course you remember that Corvette body panels are plastic which is non-conducting.

You can locate electrical faults through a process of elimination. When you troubleshoot a complex circuit, separate it into smaller parts.

2.1 Testing for Voltage and Ground

The most useful and fundamental electrical troubleshooting technique is checking for voltage and ground. Scan tool displays can show direct circuit information, but here's how to check with simple tools.

I'll use the Corvette brake switch to show simple troubleshooting tests. When you step on the brake, the normally-closed switch *opens* to remove B+ from this circuit to the computer. The Powertrain Control Module (**'94-01**), or Transmission Control Module (TCM), uses this signal to de-energize the following three circuits, during braking:

- the Torque Converter Clutch Circuit, to prevent stalling
- the Cruise Control Circuit, to release power
- 4th-gear shifts to 3rd, to provide engine braking

Ther is also a separate stop-light switch that *closes* to apply power to the stoplamps.

Check for voltage

Connect the test light wire to a known good ground. Use the pointed end of the light to probe the positive (+) connector.

- light *on* at battery connection, power to the switch
- light *off* at switch connection to solenoid (load), switch is *open*

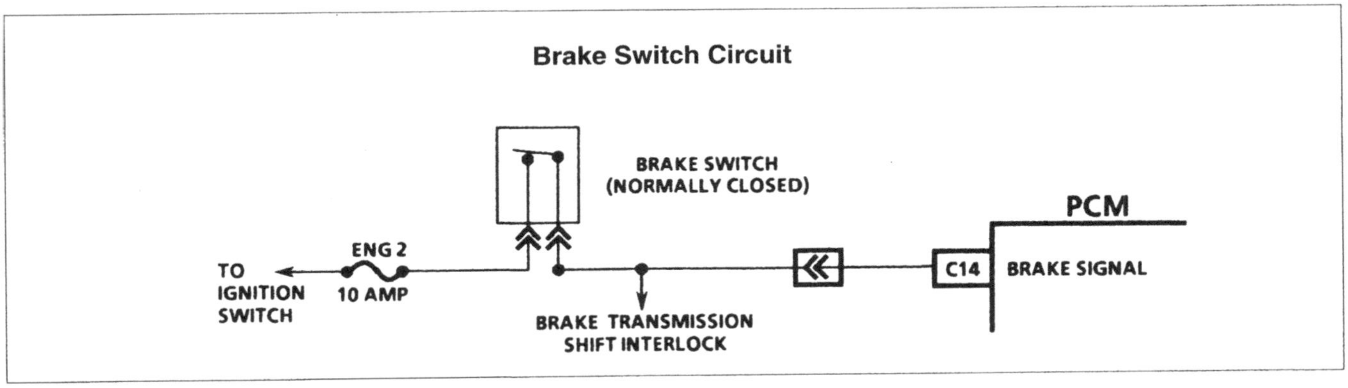

Fig. 11-3. This brake switch (not the stop *lamp* switch) should open when you step on the brake. If it is stuck closed, check for voltage at the ignition-feed circuit of the brake switch. Then apply the brake, to check switch operation. (**'94 LT1** A/T shown.)

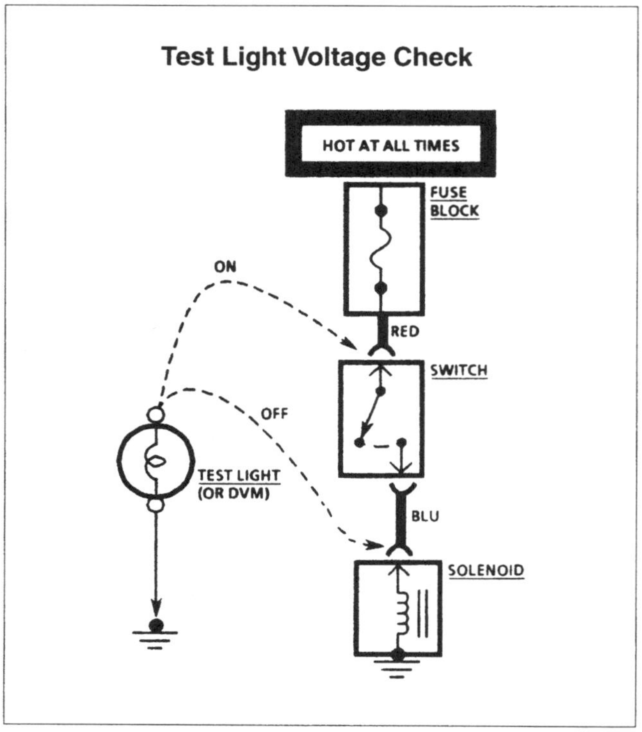

Fig. 11-4. A test-light check for battery voltage (B+), as in a Normally Open (N.O.) switch, for example. With a test light, you know you've got voltage. To know *how much*, you need a DMM.

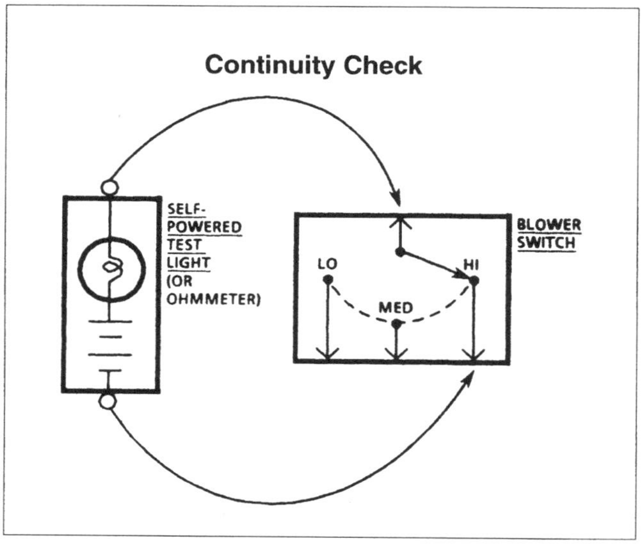

Fig. 11-5. A continuity check through a switch, with a self-powered test light or DMM.

Check for Continuity

Use self-powered test light or DMM, switched to read ohms, to check for continuity. Remove the fuse from the circuit, and connect the test light/meter to the input to the switch. Probe the connector to the output of the switch. With the switch *on*, the test light should light up, indicating continuity. The meter can show the actual resistance of the circuit.

A test light only determines *if* voltage or ground is present. It does not determine *how much* voltage or how good the path to ground is. If the voltage reading is important, use a digital voltmeter. To check the condition of the ground connection, check the voltage drop on the suspected connection.

Check for Voltage Drop

The wires, connectors, and switches that carry current are designed with low resistance so that current will flow with a minimum loss of voltage. This is particularly important for the low currents carried by the sensors. A voltage drop is caused by higher than normal resistance in a circuit. This additional resistance decreases the flow of current. A voltage drop can be noticed by problems ranging from dim headlights to rough running. Some common sources of voltage drops are: faulty wires or switches, dirty or corroded connections or contacts, and loose or corroded ground wires and ground connections.

A voltage drop can only be checked when current is running through the circuit. Since these measurements are usually small, use a DMM to ensure accurate readings. If a voltage drop is suspected, turn the circuit on and measure the voltage at the circuit load.

- A voltage-drop test is generally more accurate than a simple resistance check because the resistances involved are often too small to measure with most ohmmeters.
- The maximum voltage drop, as recommended by the Society of Automotive Engineers (SAE), is:
  0v. for small-wire connections
  0.1v. for high-current connections
  0.2v. for high-current cables
  0.3v. for switch or solenoid contacts.
  On longer wires or cables, the drop may be slightly higher. In any case, a voltage drop of more than 1.0v. indicates a problem.

When you're checking voltage, remember that you have two different conditions:

- With the key on and the engine off the battery voltage is about 12.6 volts.
- With the engine running, the charging voltage is about 14.5 volts.

You may want to measure your *actual* voltage at the battery with the ignition on (engine off), and then with the engine running to get your exact measurements. How are your battery/generator voltages today?

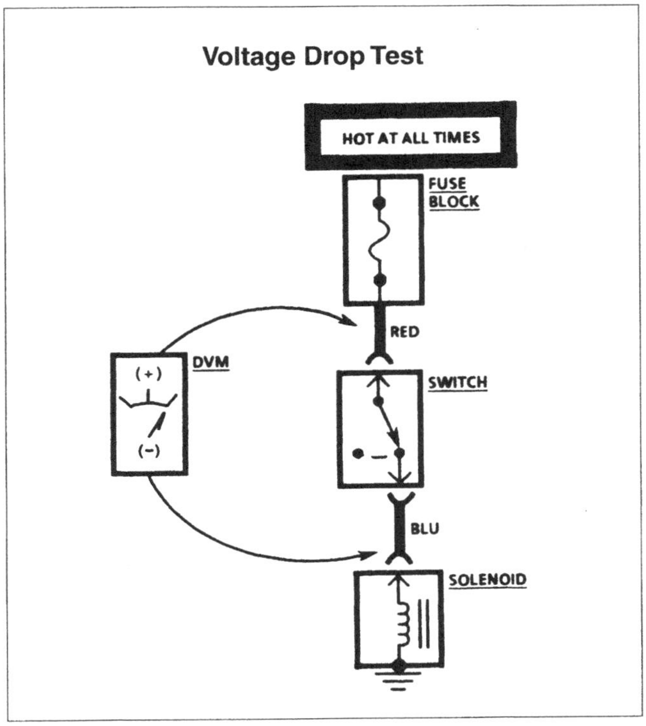

Fig. 11-6. To check for voltage drop along a wire, or through a connection or switch, use a DVM or DMM.

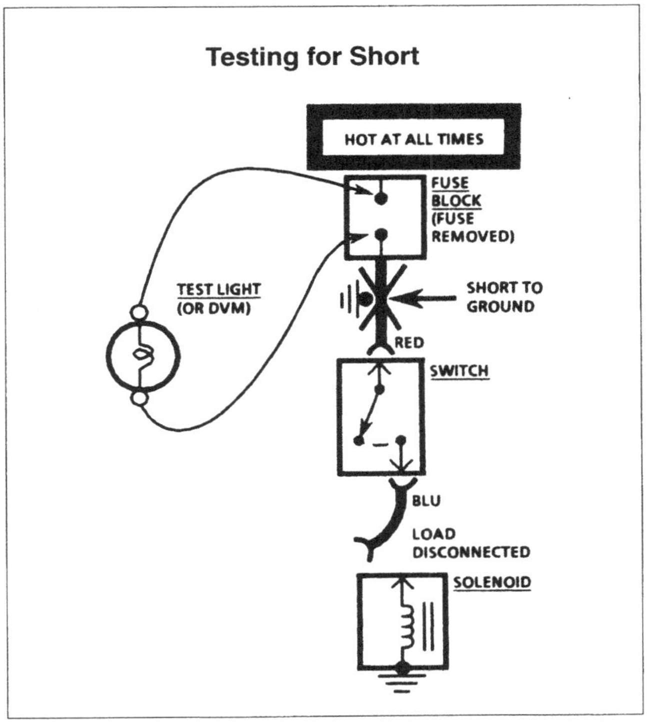

Fig. 11-7. You can check for a possible short to ground with a test light or a self-powered DVM or DMM.

To check for voltage drop, set the meter to the correct scale. Connect the negative (–) test lead to the negative side of the switch or circuit being tested. Touch the positive (+) test lead to the positive wire or connector. Read the voltage drop on the meter.

Check for Short to Ground

A short circuit is exactly what the name implies. The circuit takes a shorter path than it was designed to take. The most common short that causes problems is a short to ground where the insulation on a positive (+) wire wears away and the metal wire is exposed. If the exposed wire touches anything conductive, a fuse will blow and the circuit may be damaged.

To check for a short to ground, remove the blown fuse and disconnect the load. Connect the test light or voltmeter across the fuse terminals. (Be sure the fuse is powered). Beginning near the fuse block, wiggle the harness while watching the DMM. Do it again about 6 inches further. When the test light glows, or DMM registers, you've found a short to ground nearby. (See Fig. 11-7.)

When checking continuity, keep the ignition off. On circuits which are powered at all times, disconnect the battery, but if you're going to do this, remember to check for codes first. Using the wiring diagrams in Chapter 12, you can test a circuit for faulty connections, wires, switches, relays, and engine sensors by checking for continuity.

Short circuits are often difficult to locate. So check your schematic diagrams in Chapter 12. Use a logical approach to short circuits, based on the current path. And notice the symbols for connectors, which give you access to the circuits.

One you've determined that there is a short to ground, you can locate it using an ohmmeter. First, disconnect the harness connector from the circuit load, then connect one test lead to the load-side terminal (the terminal leading to the circuit) and connect the other test lead to a ground. (See Fig. 11-8.) You'll find the connectors' circuits in Chapter 12's schematics.

Working from the wire harness nearest to the fuse/relay panel, move or wiggle the wires, while observing the test light or the meter. Continue to move down the harness until you see the test light blink or a reading on the meter. This is the location of the short. Visually inspect the wire harness at this point for any faults. If no faults are visible, carefully slice open the harness cover or the wire insulation for further inspection. Solder and shrink-wrap the wire to repair any faults found.

You can also check for a short to the control module by disconnecting both the component and control module harness connectors, and then probing two of the terminals, say the

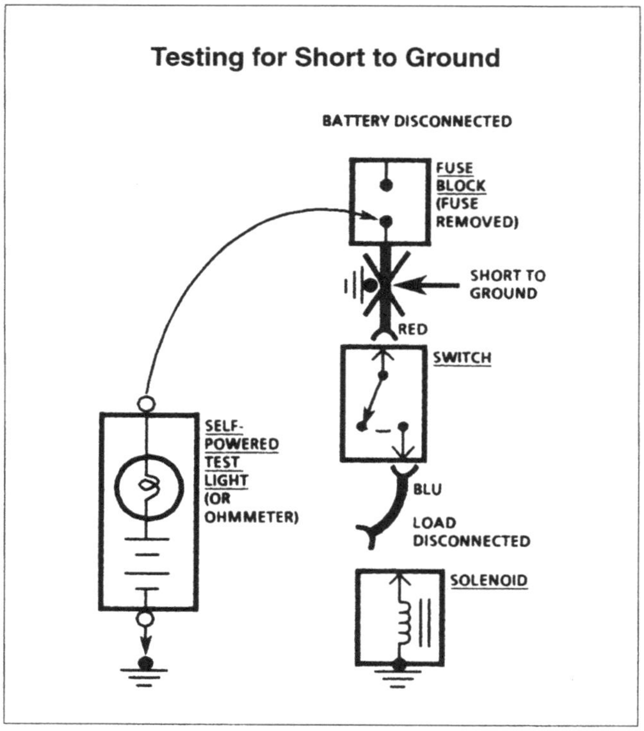

Fig. 11-8. You can locate an indicated short to ground with an ohmmeter, a voltmeter or a test light.

MAF and MAF RTN. If there is continuity, then there is a short somewhere in the harness.

3. Servicing the Fuel System

You can't pull codes which tell you directly when the fuel system is not operating properly. The engine does not have a sensor for low fuel flow or fuel pressure. But you can infer such problems from oxygen sensor indications. You can often consider improper fuel delivery as a possible cause of a driving complaint.

If you've come to this section from the Diagnostic Routines, you're specifically looking for a problem that would cause incorrect fuel flow through the injectors. In your fuel delivery checks, look for answers to three questions:

1. Is the pressure regulator controlling the amount of pressure?
2. Is the fuel pump creating *enough* pressure?
3. Are the injectors operating correctly?

If you suspect fuel system problems, first remove the filler cap and listen to make sure the pump runs when you turn on the ignition. Remember, to run the pump (which should start after the 2-second key on/engine off mode), you'll need to have someone cycle the ignition, or cycle the pump with a scan tool. To hear better, some techs use a stethoscope against the tank. A screwdriver can be a crude substitute:

If you don't hear the pump(s), check the system relay, the fuel pump relay, and the Vehicle Anti-Theft System (VATS) (**'86-01**). You'll be looking for power going to the relay, a ground coming from the relay, and for the relay to shwitch on when power is applied.

Also make sure that the control module is receiving the signal, indicating that the engine is turning over. This signal comes from different locations, depending on the engine type:

- **L83, L98:** HEI distributor
- **ZR-1:** CranKshaft Position (CKP) sensor
- **LT1, LT4, LS1:** CaMshaft Position (CMP) sensor.

If the pump runs, continue to check basic system pressures–Fuel Pressure Regulator (FPR), and operation of the fuel injectors.

> Remember that Corvette port-injection systems operate at high pressures, and can be as high as 60 psi (400 kPa).
> The fuel lines are usually pressurized, even when the engine is not running. Opening a line under pressure can spray gasoline. Believe me, that is a severe fire hazard. And a hazard to your eyes.
> *Caution*: Do not smoke or cause sparks. Have an approved fire extinguisher handy.

3.1 Pre-checks

Start by checking over the complete fuel-delivery system. You're looking for leakage in fuel lines or at fuel-line connections, loose wiring connectors, cracks, pinching, kinking, corrosion, or grounding abrasion. Check around all of the components, including the fuel tank(s), filter, pump(s), Injectors and fuel rail, and pressure regulator. *Note*: In the **'99-01** the pressure regulator is inside the tank

Check the battery voltage and the fuses controlling the fuel-delivery system. Check the fuel level. (I know, that's an obvious one, but it's the simple things that pull a "gotcha.")

Excess fuel pressure may enrich the mixture, while insufficient pressure may lean the mixture. Why? Because injection pulse-times are calculated on the basis of specified fuel pressures. You'll be checking four pressures, during the following:

- engine off, to check fuel-pressure regulation
- engine running, to check the pressure-regulator response to changing manifold pressure (except **LS1**)
- rest (or residual) pressure, to check leakage in the system
- drop-off during acceleration

Opening Fuel Lines—Relieving Fuel Pressure

Relieve the fuel pressure before servicing the system, to reduce the risk of fire and personal injury. Remember, I'm talking pressures as high as 60 psi (400 kPa), not the low-psi pressure used in carburetors. There are two ways to relieve fuel pressure.

The *book* way:

1. Turn the ignition off.
2. Remove the pump fuse. This is better than disconnecting the battery cable, which can lose memories.
3. Loosen the filler cap.
4. **LS1**—Remove the left fuel-rail cover.
5. Wrap a shop towel around the fuel pressure connection while you connect fuel-pressure gauge.
6. Install the bleed hose of the gauge into an approved container—metal or plastic better is than glass.
7. Open the valve on the gauge to bleed the pressure. If your gauge doesn't have a fitting, slightly loosen the connection and catch the fuel in a metal or plastic container.
8. Drain any fuel in the gauge into the container.
9. Keep the fuel-wetted towel in a safe container. You'll want to use it to wrap the fuel-pressure regulator or any fittings before disconnecting. Even though you have relieved the pressure, small amounts of unpressurized fuel remain in the lines.

Or the *simple* way:

1. Remove the pump fuse.
2. Start and run the engine until it stops, which it does after using up all the residual fuel.
3. Now you can wrap a shop towel around the fuel pressure connection and connect the fuel-pressure gauge with little or no fuel pressure leakout.
4. See #8 and #9 above and replace the pump fuse.

Warning: In my early work on servicing Bosch fuel injection, the "head guru" gave me a demonstration. He said, "Before loosening any fuel-system fitting, remove the gas-tank cap." Why? "Although most people don't realize it, gasoline vapor-pressure builds in fuel tanks. The tank pressure is controlled by the valve in the cap, at about 1 psi (7 kPa)."

If you open a fuel-return line with the gas-tank cap in place, the small pressure can force fuel out of the line when you least expect it, not a pleasant surprise! Next time you refill in a quiet place, listen as you unscrew the filler cap. You should hear the faint hissing sound, relieving the pressure.

3.2 Check the Engine-Off Fuel Pressure

Attach a fuel pressure gauge as described above.

1. Build engine-off fuel pressure by turning the key on several times. Each Key-on runs the pump for about two seconds, to build pressure. Or you can simply run the engine for a minute, then turn it off. Or you can command the pump to turn on with most scan tools. When the pump stops the last time, you may see the pressure vary slightly, then hold steady.
2. Check the engine-off fuel pressure according to the specifications. For GM recommended pressures, see Chap. 7 Fuel Systems.
3. Check *residual* pressure for about 1 minute. The system should hold at least the lowest recommended pressures.

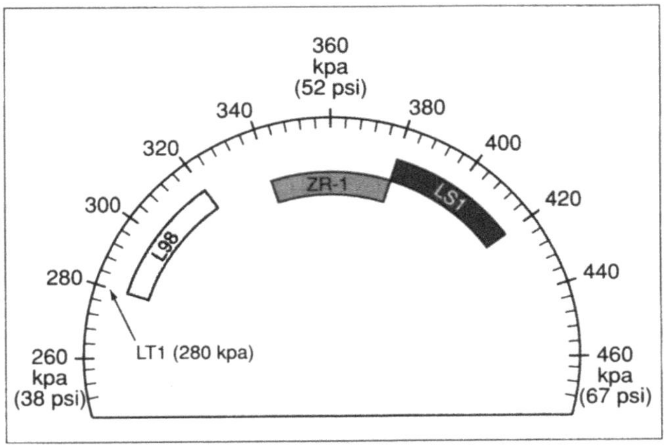

Fig. 11-9. Fuel pressures vary with engine type, but the principles are the same.

Causes of *low* engine-off fuel pressure may be:

- a damaged fuel filter, though this rarely occurs in the in-tank filter
- leakage in the fuel-pressure regulator valve or valve seat
- low voltage in the pump relay and circuits
- a malfunctioning fuel-pump check valve
- leaking injectors, check the spark plugs for saturation or soot-fouling

Causes of *high* engine-off fuel pressure include:

- a clogged return line
- a faulty fuel-pressure regulator

11

3.3 Check the Engine-Running Fuel Pressure

A symptom of low engine-running fuel pressure is a lean-fuel ratio. Some limited correction may occur in the engine computer, as the Oxygen Sensor's lean signals cause an increase in injector pulse-times, but that can result in reduced performance, hesitation, or misfire.

L98, LT1, LT4, ZR-1:

- 1. Check the engine-off pressure.
- 2. Start the engine.
- 3. AC (Air Conditioning) should be off.
- 4. Check the engine-running fuel pressure. It should drop about 3-10 psi (21-70 kPa) from the engine-off pressure.

Causes of low engine-running fuel pressure include:

- a faulty or stuck-open fuel-pressure regulator
- a clogged filter (rarely occurs in the in-tank filter)
- **'94** a clogged inlet filter in FPR. For this GM recommends removing it, as it is generally unnecessary. (It was designed to trap particles introduced during engine assembly).

Check during acceleration or cornering:

Your Corvette may test ok in the shop, but with high-performance driving, you may experience surging, loss of power or misfires during acceleration and cornering. Any of these can

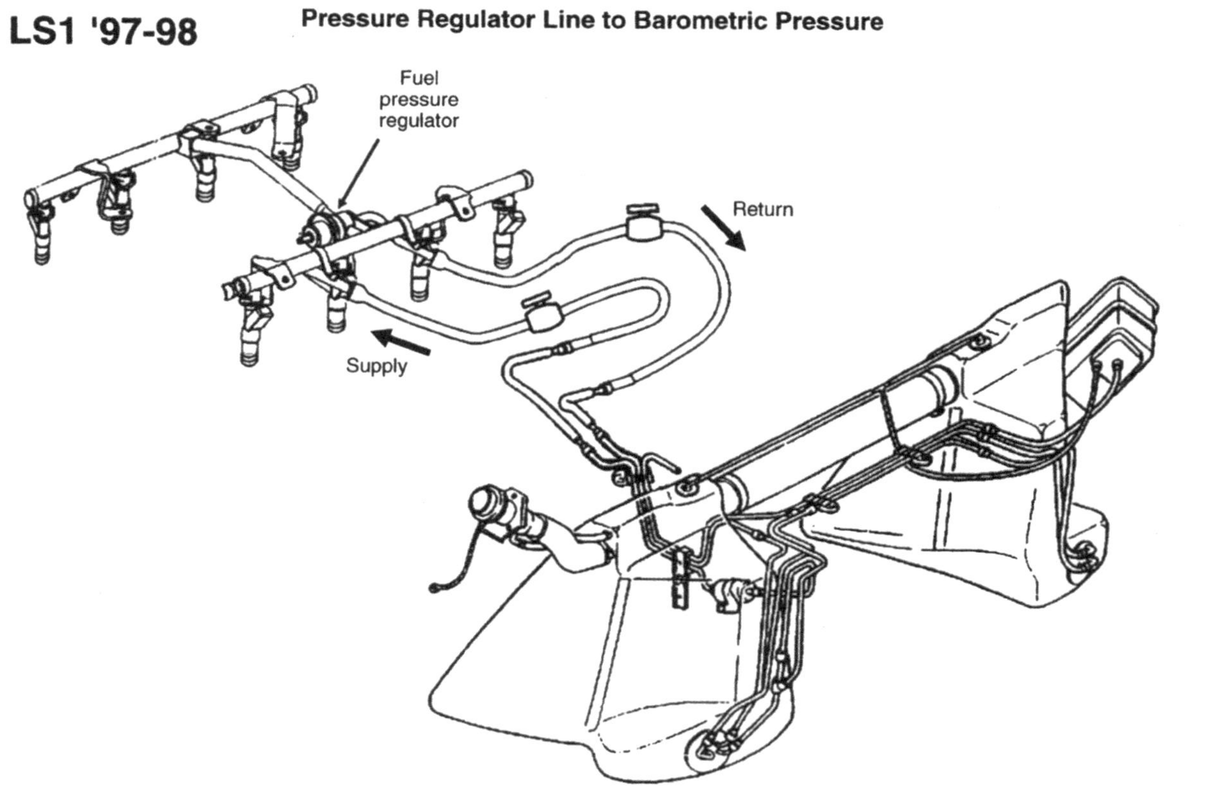

Fig. 11-10. Fuel pressure regulator line. **LS1** '97-98 shown

L98, LT1, LT4, ZR-1 Remember, with engine running at idle, fuel-pressure regulator reduces fuel pressure as shown on your gauge. This compensates for reduced manifold pressure, maintaining constant *relative* fuel pressure at the injectors. This is required for fuel metering control by computer.

L83 TBI, LS1 fuel pressures should be same, Engine OFF, Engine RUN.

- **L83** Fuel Pressure Regulator not referenced to BARO
- **LS1, '97-98** Fuel Pressure Regulator compensates for BARO, not MAP.
- **LS1, '99+** Fuel Pressure Regulator in fuel tank, not referenced to BARO or to MAP. Returnless fuel system. Engine-running pressure should be the same as engine-off.

be caused by lean mixtures resulting from low fuel pressures. Use a scan tool to diagnose this on the road, or actually drive the car and read the trouble code. For example: P0131means that the $H0_2S1$ voltage is low and exhaust is lean, or 44 means that there is a problem with the O_2 Sensor Circuit, and exhaust is lean.

> If you don't have a scan tool, you may be tempted to try an on-the-road test by taping your fuel-pressure gauge to the windshield. But high-G driving while watching a gauge filled with gasoline under pressure is not for me!

A rich-running engine is an indication of high fuel pressure. Some internal correction is also possible in this case, as rich oxygen sensor signals will cause decreased injector pulse times. This may cause: reduced mpg, surging, stalling, misfire, high emissions, possible black smoke, plug fouling.

Causes of high fuel pressure, engine running:

- a restricted return line (this can lead to high current draw and may be the root cause of a pump failure)
- the fuel-pressure regulator stuck closed
- **L98, LT1, LT4, ZR-1:** a loss of the vacuum to the fuel-pressure regulator, check the hose
- a catalytic converter failure

3.4 Check Fuel Delivery, LS1 only

1. Disconnect the fuel-feed pipe to the engine from the right fuel-sender (at the tank).
2. Insert a hose in an unbreakable container, which can hold at least one quart.
3. Run the pump (by repeatedly turning the ignition on, or using a scan tool) for about 8 seconds.
4. Measure the volume expelled into the container. It should be about 330 mL (11 oz.).

3.5 Checking Injectors

Injector Operation

> *CAUTION —*
> *Do not apply voltage to the fuel injectors in an attempt to test them. Remember, the injectors are powered whenever the ignition is on, and they are computer-grounded to deliver fuel.*

You can usually check injector operation, or engine idling, by the vibration, which indicates that they are opening and closing. Feel by your fingertip. If they're too hot to touch, use a mechanic's stethoscope or a screwdriver. Listen for vibration, a sharp metallic buzzing or clicking sound. Look out for a voltage jolt from the plug wires. You don't want to take the 40 kV jolt from **LS1**'s ignition.

- A dull sound suggests a clogged injector.
- No vibration indicates a bad injector or harness.
- A different pitch of vibration in one, compared to the others, may indicate a bad injector or injector clogging.

Try interchanging the injectors (MFI on the same circuit, and same side). If the same injector is still faulty, replace the injector. If the injector now works, check the wiring.

3.6 Clean and Inspect the Fuel Rail & Injectors

Before removal, clean the fuel rail assembly with a spray-type engine cleaner. Do not soak the rails in liquid cleaning solvent.

Fuel Rail Removal

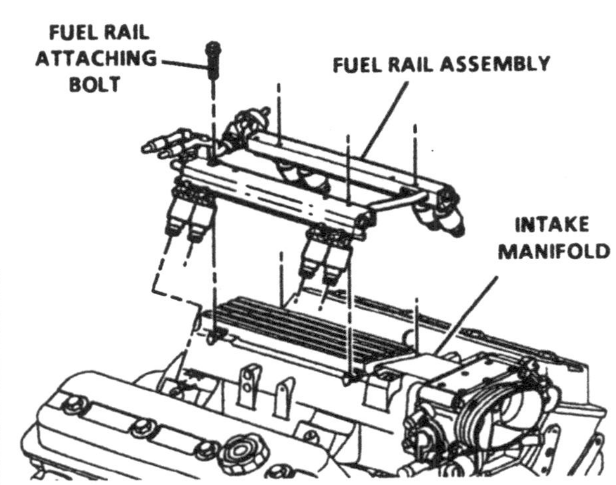

Fig. 11-11. Carefully remove the fuel rail assembly. Support the assembly to prevent damage to the injectors. Plug the holes in the heads and cap the fittings to keep out dirt. **LT1** shown.

1. Relieve the fuel system pressure.
2. **LS1:** Remove the fuel rail covers.
3. Slide the rubber dust covers from the quick-connect fittings, on the engine fuel pipes.
4. Loosen any dirt within the fittings by twisting the female connectors back and forth for a quarter turn.
5. Blow any dirt from the fittings with compressed air.
6. Push/pull inward to release the locking tabs, so that you can pull the connections apart. This may require a special tool which fits into the female connector.

11

7. Wipe off the male connector and inspect both ends for burrs and dirt.
8. Remove the vacuum line from the pressure regulator.
9. Remove the electrical connectors, and label for proper replacement.
10. Remove the fuel-rail bolts and lift out the fuel rail.

NOTE: the **ZR-1** requires removing intake plenum assembly.

Injector O-rings

On each injector, remove both O-rings. Notice that they are the same size, but that they are made with different materials, and are identified by different colors. The lower ring is brown, and the upper one is black. Discard both O-ring seals but save the O-ring backup, which is the nylon collar that helps to position the lower O-ring on the injector. If you leave it out, you can expect vacuum leaks and driveability complaints.

- Lubricate and install a new lower O-ring (brown) on the spray tip of each injector.
- Lubricate and install a new upper O-ring (black) on the rail tip of each injector.

To re-install:

1. Replace the fuel rail and tighten the bolts to 10 Nm (89 lb.in.)
2. Redo the electrical connections. Double-check to be sure that the proper connector is on the proper injector. In the **ZR-1**, **'94-96 LT1**, **LT4**, and **LS1** proper sequential firing order is a must. After the fittings are connected, pull on both sides to check security.
3. Replace the vacuum line to the pressure regulator.
4. Slide the rubber dust covers over the quick-connect fittings on the engine fuel pipes.
5. **LS1:** Replace the fuel rail covers.

4. Air Flow (Load) Measurement

4.1 Mass Air Flow Sensor '85-89 L98, '94-96 LT1, LT4, LS1

- The **'85-89 L98** has a Bosch MAF sensor. (See Fig. 11-13.)
- The **'94-96 LT1**, **LT4**, **LS1,** and **LS6** have an AC MAF sensor. (See Figs. 11-14 and 11-15.)

Check for MAF sensor trouble codes, for power, ground and signal. *NOTE*: In the **L83**, **'90-91 L98**, **'92-93 LT1**, and **ZR-1** there is no MAF sensor.

MAF sensors are not serviceable. If you replace them, *check the arrow* for direction of flow. The MAF will fit in either direction, but if the arrow points forward, the engine will run poorly.

Fig. 11-12. Be sure that the direction-of-flow arrow points into engine when replacing the MAF sensor.

4.2 Manifold Absolute Pressure (MAP) Sensor

Check for MAP sensor trouble codes, and for open, short and signal. Also check for a plugged or leaky vacuum line. MAP sensors are also not serviceable. Replace as necessary.

Tip from the field: This is probably the only place where "try a known good part" may diagnose a bad MAP sensor. They're almost interchangeable. If the MAP seems to test OK but the problem persists, try a replacement.

REMEMBER: In the **'94-96 LT1**, **LT4**, and **LS1** the MAP is a backup. The MAF is the basic engine-load sensor for these engines.

5. Servicing the Ignition System

Ignition system troubles usually relate to incorrect timing or misfires.

LT1, LT4, ZR-1, LS1: Timing is not adjustable. This kind of timing is known as "net build". Timing modifications are managed by the computer, based on inputs from one or more of the sensors.

L83, L98: Timing is adjustable, using a timing light and timing marks. (See Fig. 11-17.)

5.1 Pre-checks

Begin with a visual inspection. Check that all plug wires are properly routed and secure. Be sure that the plug wires are not lying directly together, but are separate and secure in their clamps. **LS1:** Remove the fuel-rail covers.

Check condition of the spark plugs. If you have a wet plug, check for a leaking injector. Hook up spark tester. See if you can check on an engine analyzer/scope.

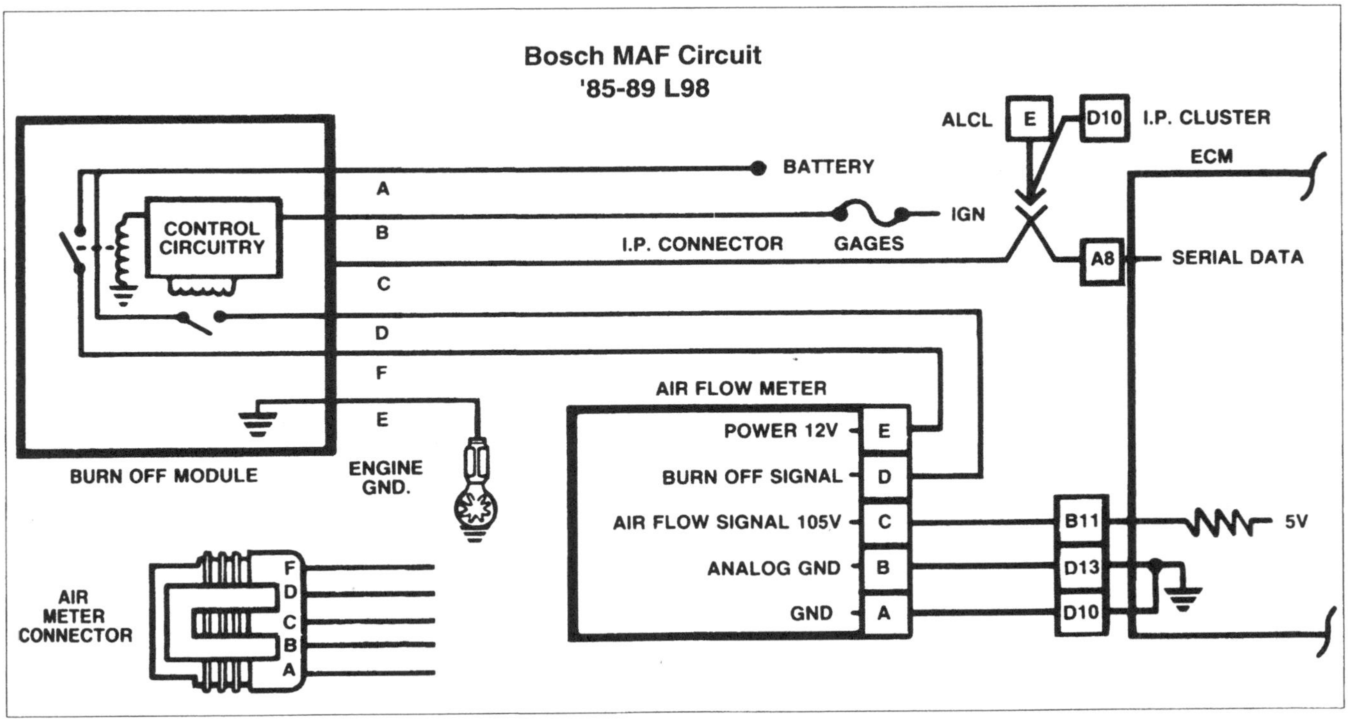

Fig. 11-13. The **L98 '85-89:** uses a Bosch MAF circuit.

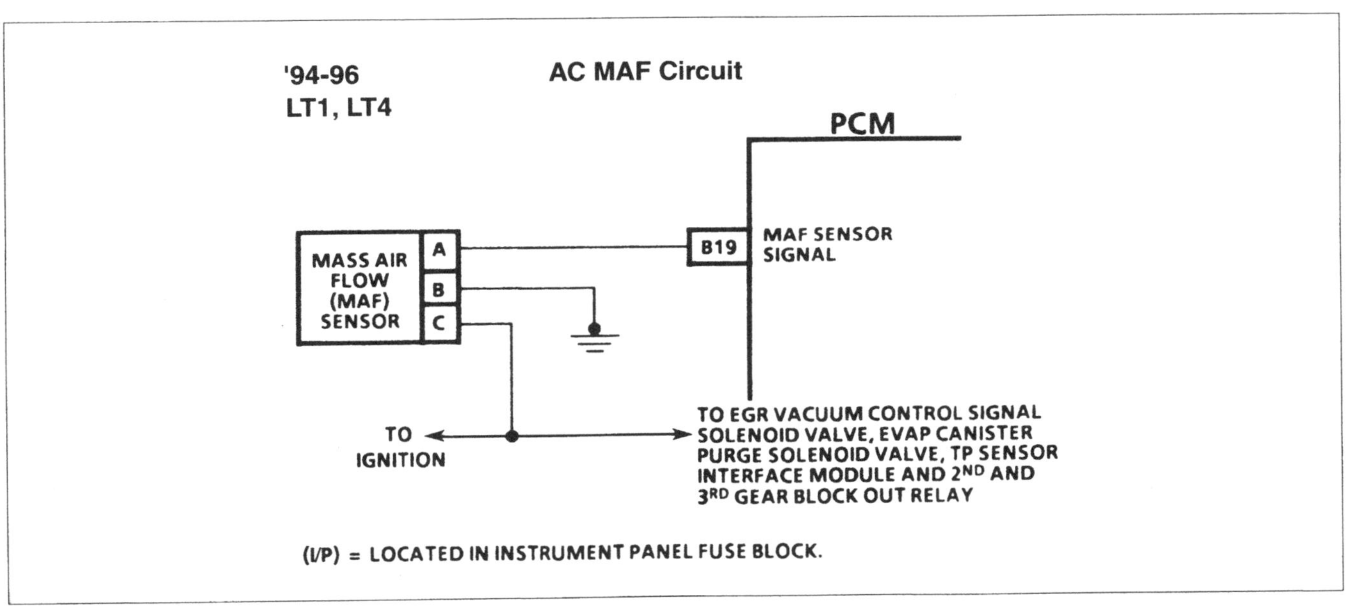

Fig. 11-14. The **'94-96 LT1** and **LT4** use an AC MAF circuit.

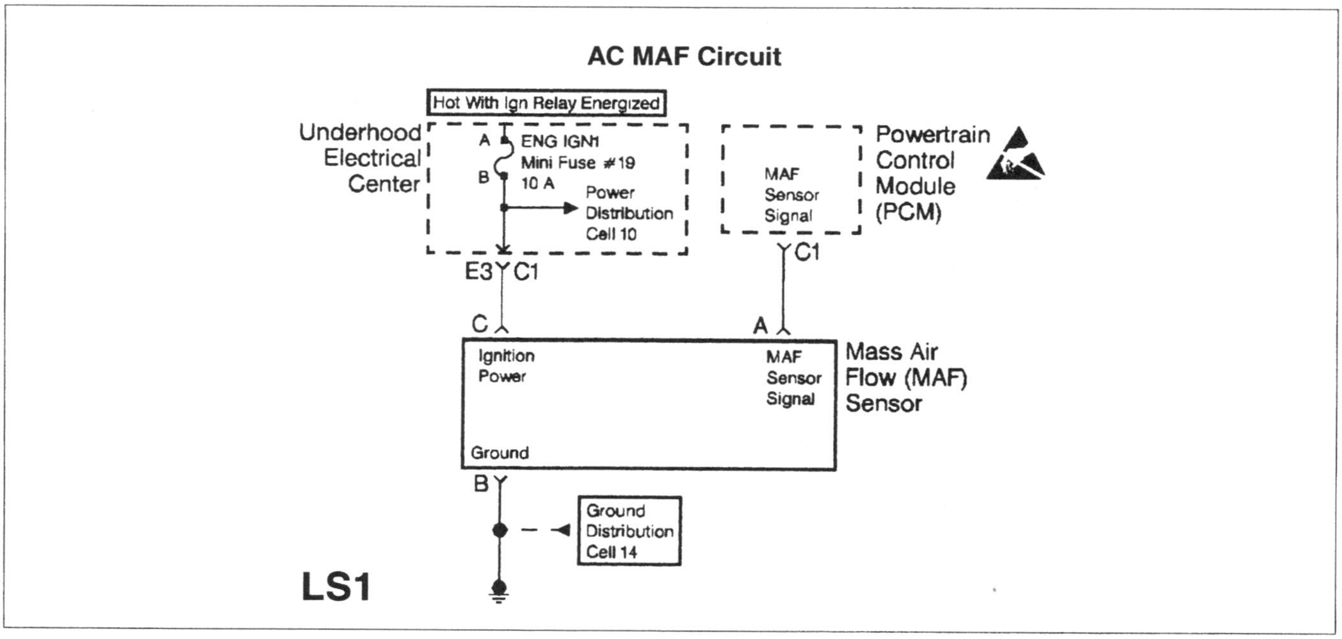

Fig. 11-15. The **LS1** and **LS6** use an AC MAF circuit.

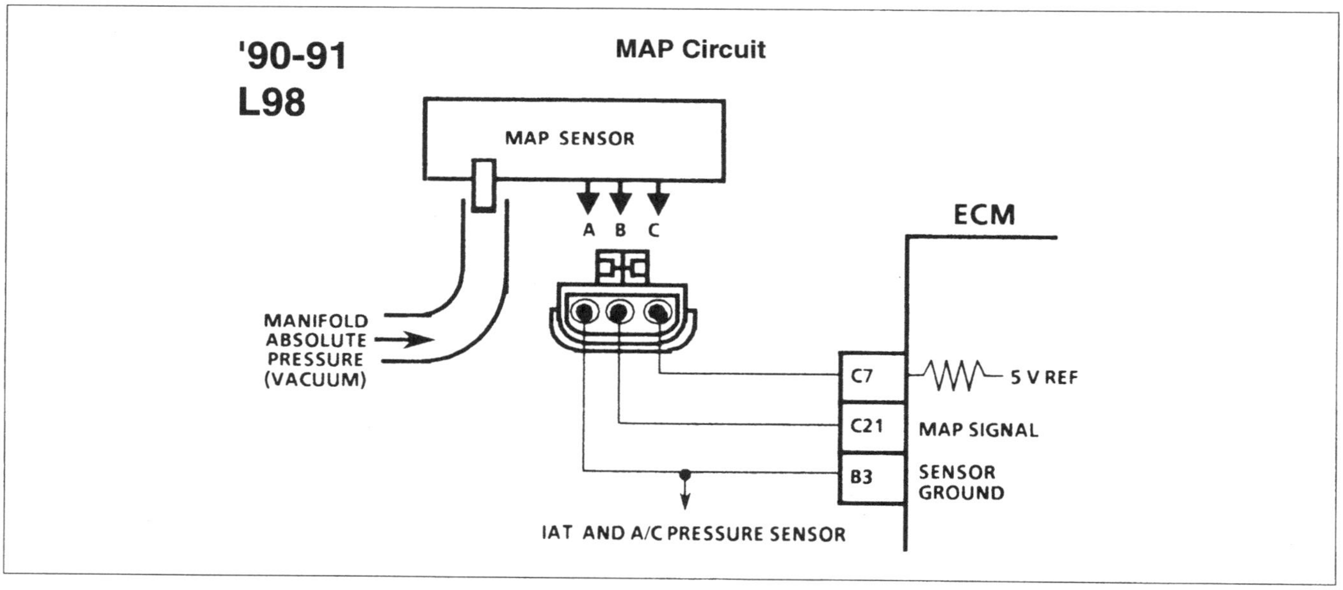

Fig. 11-16. This is the **'90-91 L98** MAP sensor. The **'92-93 LT1**, **LT4**, **ZR-1**, **LS1**, and **L83** are similar.

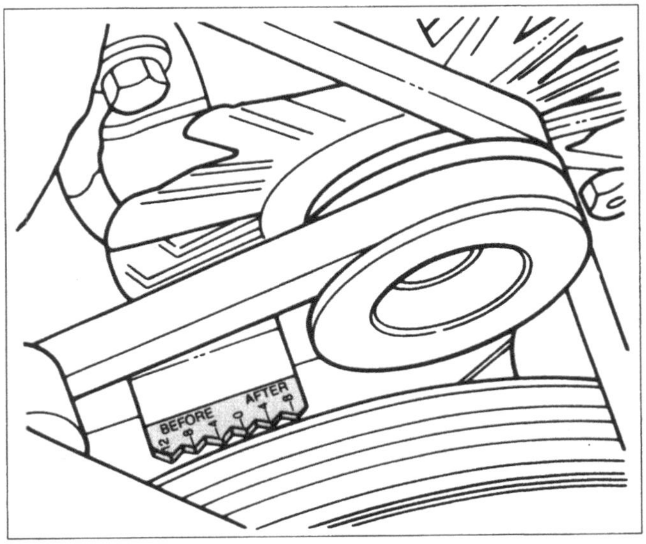

Fig. 11-17. L83, L98: Aim a timing light at the timing marks. The timing is adjustable at the distributor. Follow the instructions on the underhood Vehicle Emissions Control Information label.

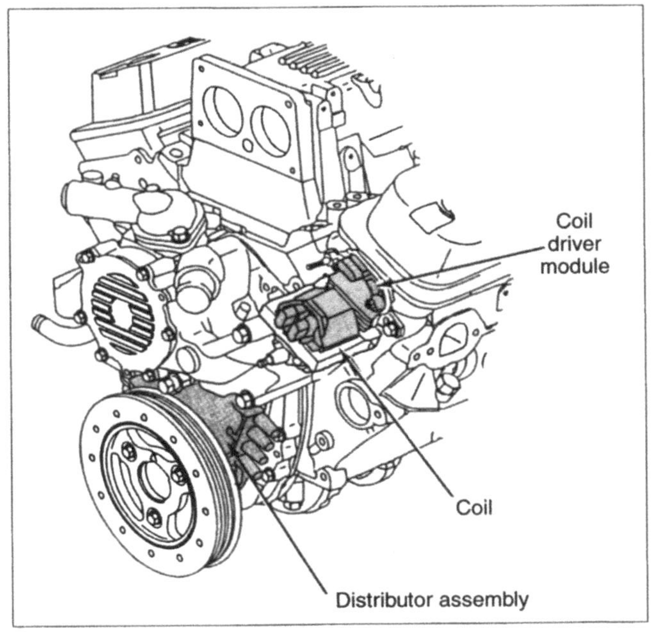

Fig. 11-19. Look closely, or you may decide that the **LT1** and **LT4** have no distributor.

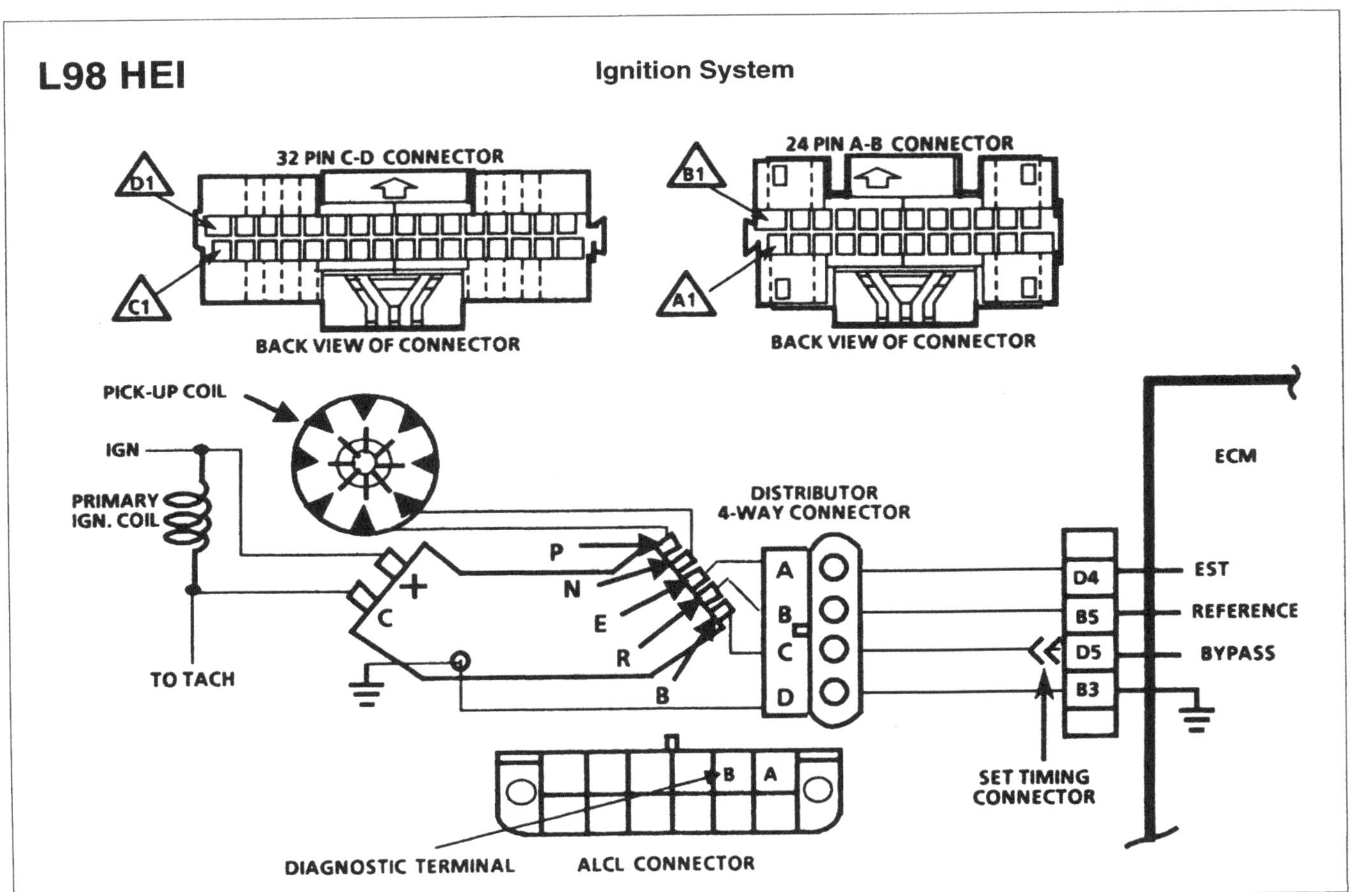

Fig. 11-18. The **L98** HEI Schematic. The **L83** is similar

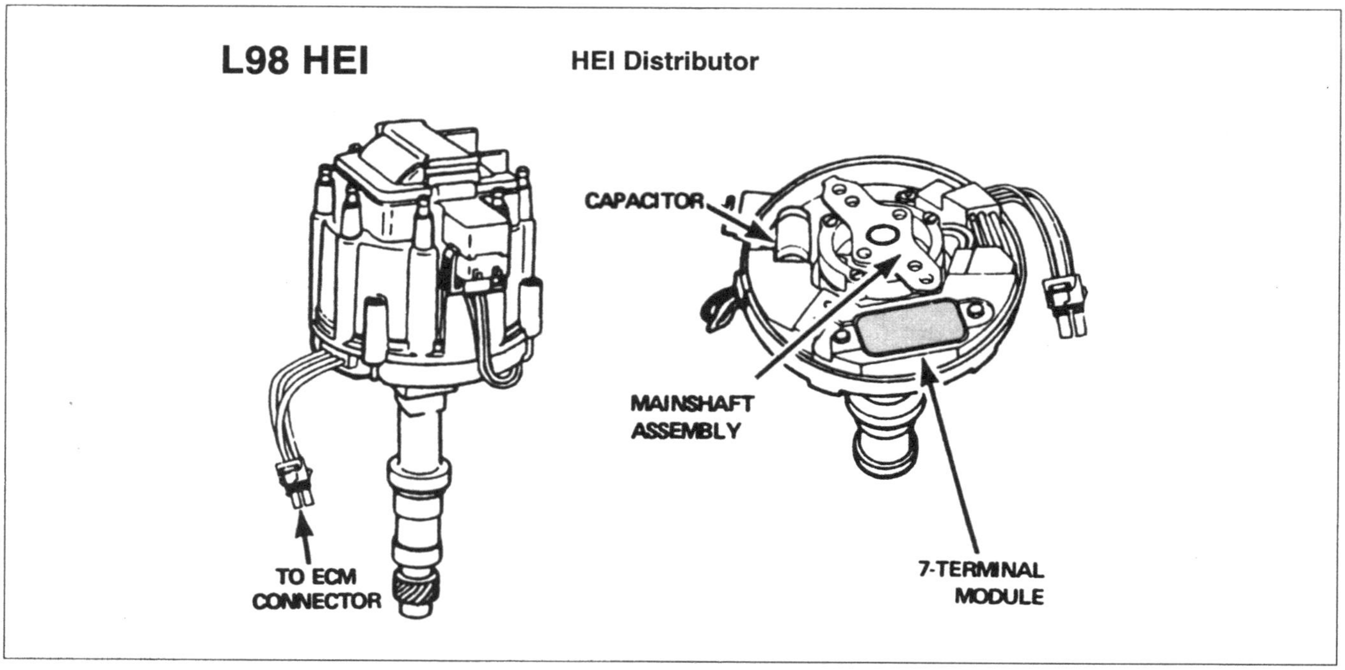

Fig. 11-20. A seven-terminal module controls the EST, including bypass timing for starting, and ESC. The **L83** is similar.

> *NOTE—*
> A spark tester requires more than 25 kV to fire. Remember, the ignition system may provide enough voltage to run the engine but not enough to fire plugs under heavy load.

With all accessories off, the engine at normal temperature, and with the transmission in neutral (or park), check the wiring harnesses and connectors for insulation damage, loose connections, or burned or broken connectors. Also check that the generator indicator shows *OK*.

5.2 Misfires

Look for disconnected plug wires, carbon tracking on the coils and plugs, cracked insulators, rubbing plug wires, and arcing to the block.

Spark Plugs

Plugs are a lot more reliable than once upon a time. I remember when we changed plugs (or at least, cleaned and re-gapped) every 5,000 miles. If we got 10,000, that was some kind of miracle. Modern plugs can run 100,000 miles. But even the best plugs can develop problems, including:

- carbon fouling, from slow running or short trips, a faulty ignition, or a worn engine, from a rich A/F mixture
- excessive gap, from high speeds or plug overheating, often from lean A/F mixtures
- a broken insulator, from careless wrenching or high-speed operation

Plug-wire material is very soft, and is designed to carry a higher voltage and withstand heat. Make sure to route the wires correctly, to avoid chaffing or cutting. When you pull the wire from a plug, twist the boot about one-half a turn. We call it a *wire*, but it's really a circuit of soft carbon. *Lead* (pronounced "leed") may be a better term than wire.

Engine Running

Listen for the snapping sound of a spark jumping to the block (not good). Or check using a scope. Check at various engine speeds including at idle and under load.

One way to test the ignition system under "load":

1. Disable fuel delivery by pulling the pump fuse.
2. Hold the accelerator at W.O.T.
3. Crank the engine, but for no more than 15 sec.

On an ignition scope, look for coil output well above 40 kV.

Distributor Ignition and Optispark—LT1, LT4

As the camshaft turns, reference signals toggle between zero and five volts. Lack of pulsing at the computer can be caused by: open, short to voltage, short to ground, or a defective sensor inside the distributor

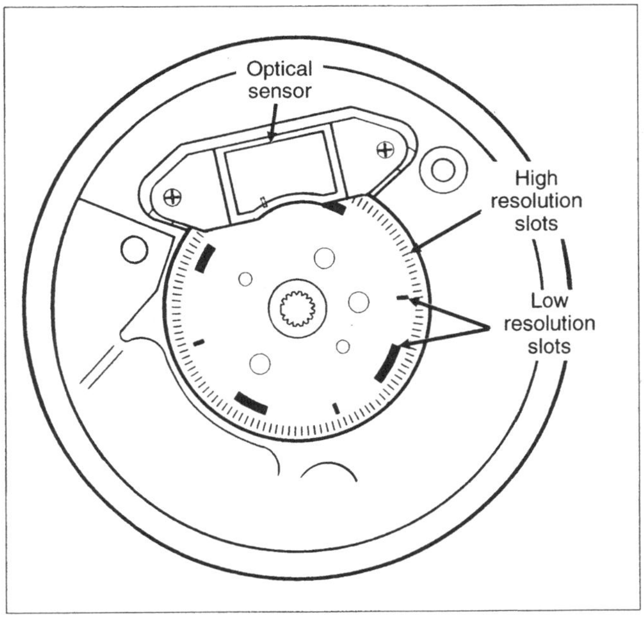

Fig. 11-21. The **LT1** and **LT4** have two timing inputs per camshaft revolution (two crankshaft revolutions); a 720X high-resolution, and a 16X low-resolution signal.

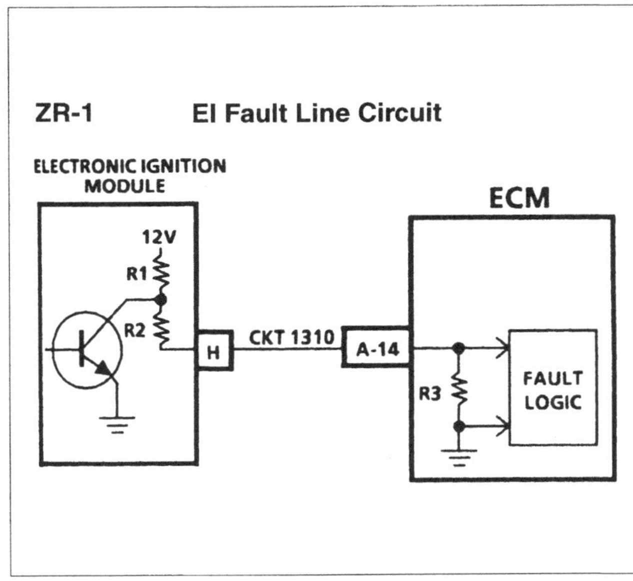

Fig. 11-23. ZR-1: The EI Fault line signals the computer when the ignition-control module detects a fault line with the computer inputs. The normal EI signal, with the ignition on, is about 7 v., which is 50% of an engine running-B+ (14 v.).

Electronic Ignition—ZR-1, 4-Coil

Monitoring the voltage across R3, the computer may read logic at 0 (low) on the EI Fault line. If so the problem could be: an IC fault input on Circuit #423, an EI fault input that #1310 is open, or that an EI fault line is shorted to ground. The computer may store a DTC 16 or 36 error and turn on the MIL.

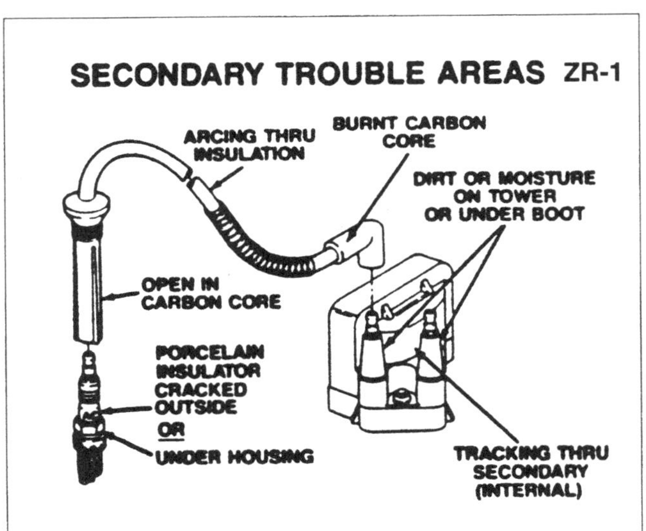

Fig. 11-22. ZR-1: For intermittent miss or miss under load, check the secondary trouble areas. Leads must be at least one inch from the signal wires (EST, REF, etc.)

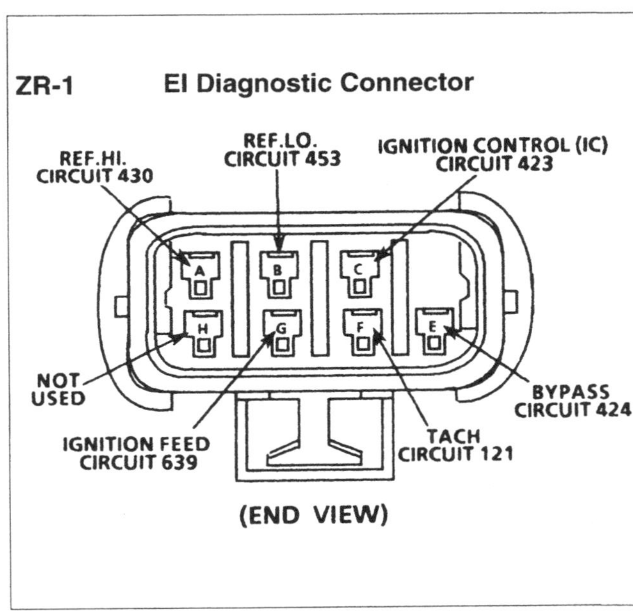

Fig. 11-24. ZR-1: Use the EI diagnostic connector, near the brake booster, to check six of the circuits. This saves removing the intake plenum, to reach the Ignition-Control Module. Remember, bypass advance is determined purely by the engine's speed.

11

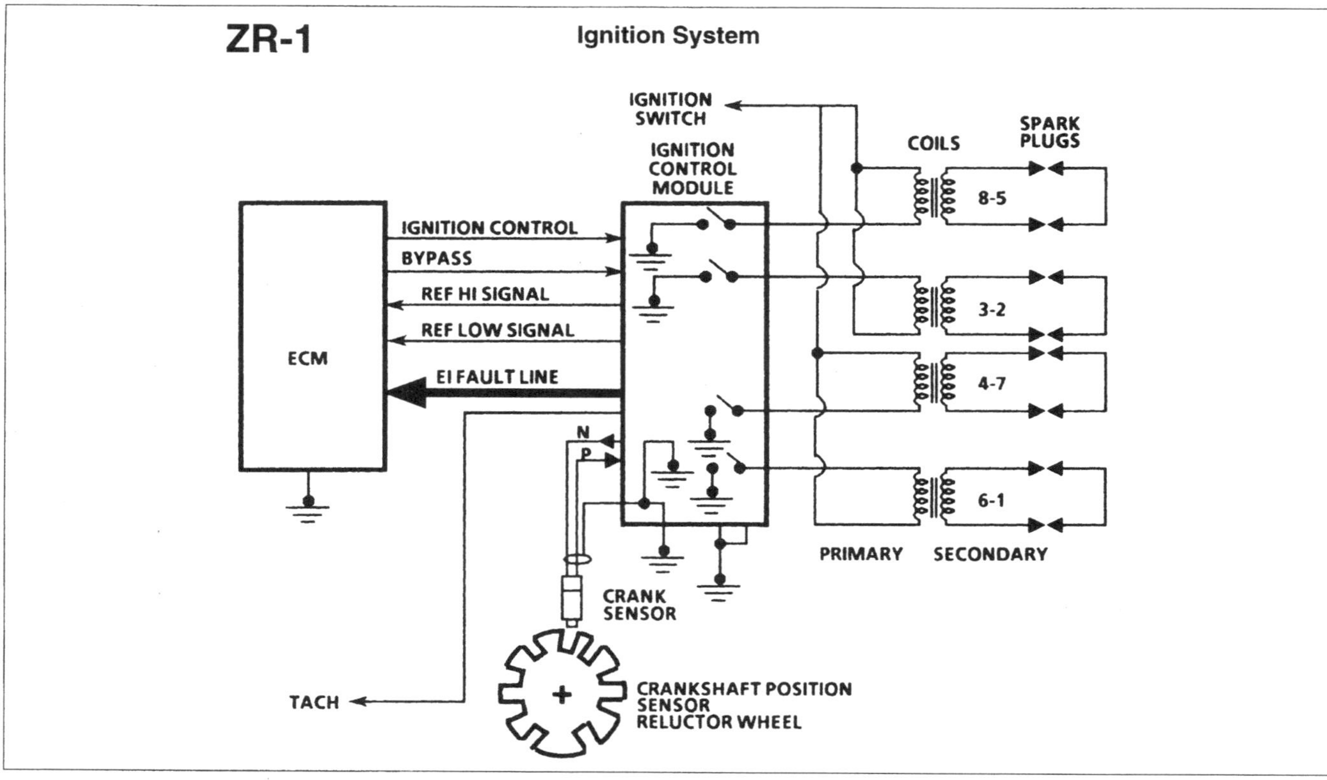

Fig. 11-25. ZR-1: Check the Ignition Control Module separately from the ECM. The Crankshaft Position Sensor provides Distributor Reference signals.

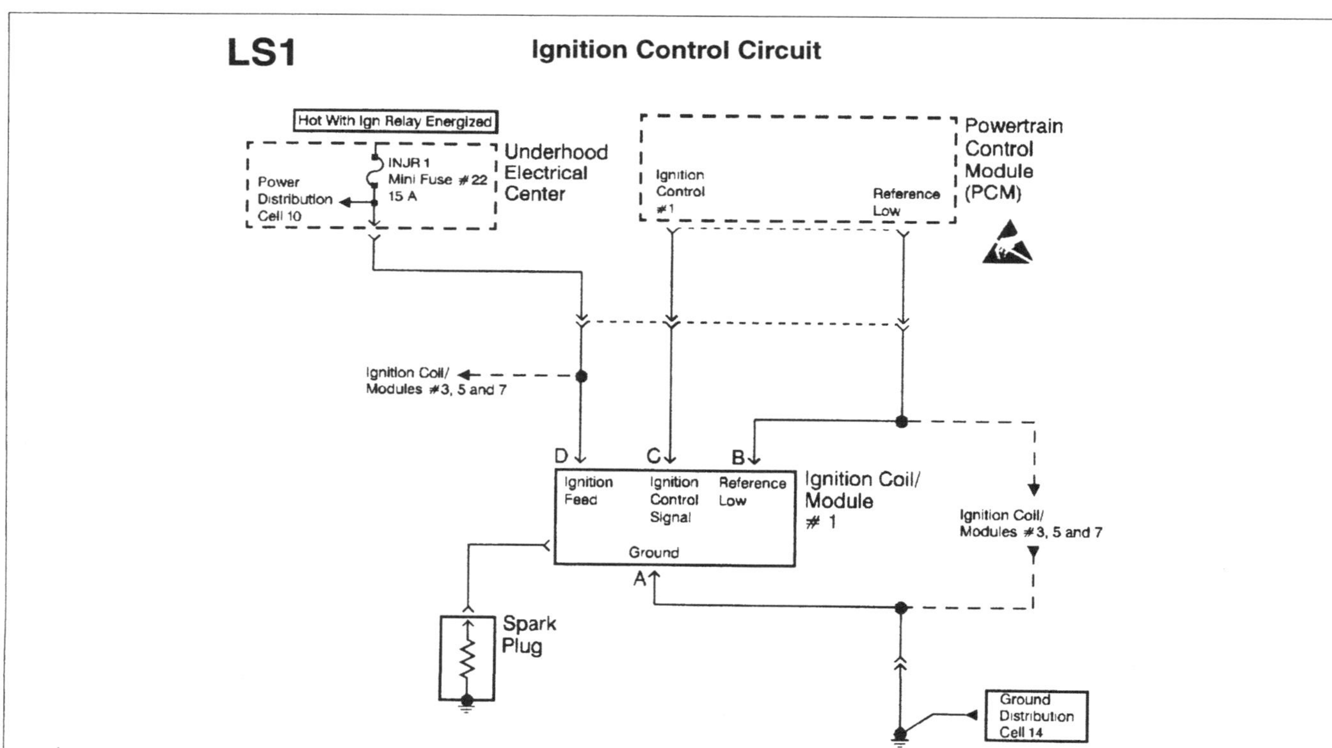

Fig. 11-26. LS1: This is the Ignition Control Circuit #1 (of 8) There is one for each cylinder. The computer turns on the Malfunction Indicator Light if the diagnostic fails.

Electronic Ignition—LS1 8-Coil

Causes of an intermittent problem include: poor connections, corrosion, a mis-routed harness, or a damaged wire or insulation. Remember:

- Avoid contact with the high-voltage secondary wires. The over-40 kv can do more than sting. Never pierce or puncture a secondary wire or boot for testing purposes. Today, it may look OK. Tomorrow, trouble.
- The Crankshaft Position (CKP) sensor is critical part. If damaged, you're engine won't start. A sensor touching the rotating interrupter ring means damage. Bent interrupter ring blades means a destroyed sensor.

5.3 HEI Ignition Timing—L83, L98

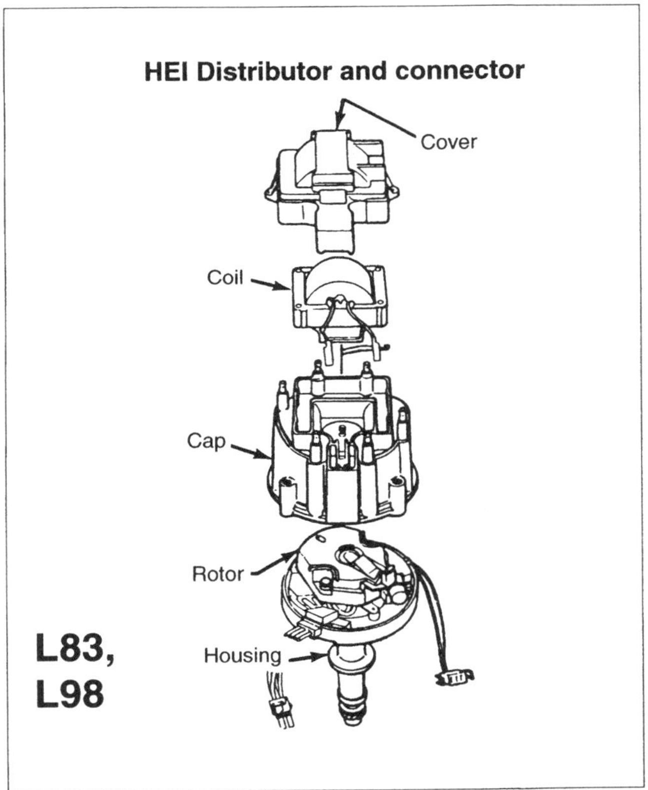

Fig. 11-27. L83, L98: The distributor is adjustable, for timing.

In the **L83** and **L98**, base timing is adjustable at the distributor. All other timing/dwell is controlled by the computer, based on signals from several sensors.

If the ignition timing is off, you can have all sorts of problems, including: sluggish operation, pinging, overheating, and high HC and NO_X emissions. No wonder proper timing is so important.

Set Ignition Timing—L83, L98

To set the ignition timing, disconnect the timing connector. Connect a timing-light plug-lead to the #1 plug. Use a jumper between the plug and wire, or an inductive pickup. Start the engine and let it idle. Aim the timing light at the timing mark. Even with the engine running, the mark should look static when you shine the timing light on it. (See Fig. 11-17.)

To change the timing, loosen the clamp bolt at the base of the distributor. Rotate the distributor slightly until you see the correct timing. Tighten the bolt.

A Code 42, EST, will be stored as a result of setting the timing. You should clear the memory of the code, when you are finished.

> ***NOTE—***
> The Knock Sensor (KS) circuit relies on correct base timing. If the system starts off wrong, it will never get to the proper adjustment for pinging.

6. Checking Idle RPM

Many factors determine idle rpm, as discussed in Chap. 6. If idle rpm is not correct, check for:

- a malfunctioning Park/Neutral switch; if it does not signal the computer about a Drive/Reverse position, idle rpm will not be controlled properly
- leaky injectors; vacuum or crankcase leaks lead to lumpy idle
- a sticking throttle shaft or TPS, which may prevent the computer from controlling idle properly
- faulty battery cables, which may cause computer to compensate erratically
- compressor loads, via the A/C signal
- malfunctioning coolant fans; they should be off because fan current draw will cause the computer to increase the idle rpm, to increase charging.
- **LS1:** Idle Learn Procedure (see section 6.3)

6.1 Setting Idle Speed—'82-96

11

If the IAC is disconnected and reconnected with the engine running, idle rpm may be wrong. You can reset it:

L83: (A/T) With a warm engine, engage Drive and hold idle

L98:

1. Depress the accelerator pedal slightly.
2. Start the engine and run it for 5 seconds.
3. Turn the ignition off for 10 seconds.

When you restart the idle rpm should be correct.

Idle-Learn Procedure

LT1, LT4, ZR-1: The computer learns the correct position for start and idle. The Idle Air Control *count* is stored in RAM in the Control Module. If the battery is disconnected, or if you disconnect and reconnect the IAC with the engine running, the computer must re-learn the IAC starting count. Do the 1-2-3 as in the **L98** (previous page).

What happens if the IAC valve sticks?

- stuck open = fast idle, A/T creep at stoplight
- stuck closed = slow idle, A/T stalls at stoplight
- stuck in the middle = rough idle, idle rpm does not respond to engine changes

CAUTION —
Do NOT apply voltage across the IAC terminals. If you do, you will permanently damage the IAC motor windings.

L83: The distance between the tip of the pintle and the motor should be no more than 32 mm (1.25 in). (See Fig. 11-28.)

L98, LT1, LT4: The distance between the tip of the pintle and the motor should be no more than 28 mm (1.125 in). (See Fig. 11-29.)

If the distance is greater than specified, try the following:

- If it is a new IAC valve with a collar, press in firmly on pintle. Wiggling it slightly may also help.
- If it is a new IAC valve without a collar, compress the spring and turn the pintle clockwise a little.
- Do not push or pull on an IAC valve that has been in service. If you force the pintle, you may damage the threads of the worm drive.

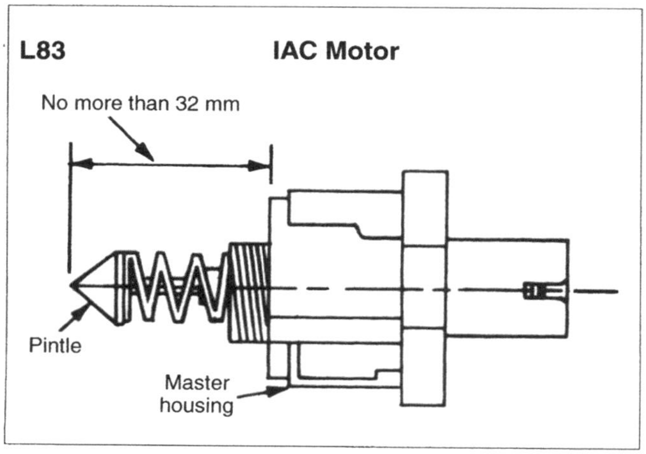

Fig. 11-28. L83: If you install a new IAC unit, be sure the distance from the tip of the pintle to the front of the motor is no more than 32 mm (1.25 in.). If greater, expect damage to the throttle body housing or the IAC motor.

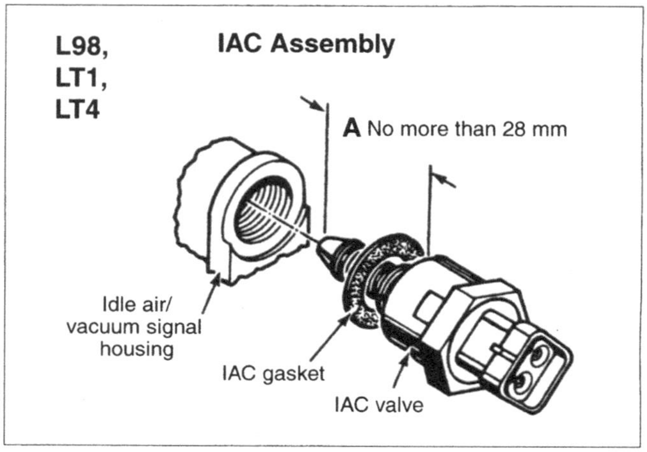

Fig. 11-29. L98, LT1, LT4: Be sure distance A is no more than 28mm (1.125 in.).

'96 LT1, LT4: If the computer detects an engine rpm outside the range of the IAC's control, under the specified conditions, it will perform an active test. The computer commands the IAC valve to move while monitoring the MAF. An active IAC test lasts for at least 5 seconds, and occurs while the car is in motion. The conditions for such a test include: a vehicle speed between 20-77 mph, an MAF flow between 13-35 gm/s, and a steady TPS and rpm.

Idle RPM Low

'96 LT1, LT4: If idle rpm is lower than specified for current conditions,the computer calls for a diagnostic, in which it:

1. commands the IAC valve to open
2. monitors the MAF signals, which should increase
   - if the MAF changes flow by 6 g/sec, the computer sets code P0507, indicating that the IAC is responding, but the MIL does not light
   - if the MAF does not change, the computer sets code P1508, indicating that the IAC did not operate

The MIL lights only if the diagnostic fails on the second consecutive drive trip. The control module turns the MIL off if the diagnostic does not fail after three consecutive trips.

Idle RPM High

'96 LT1, LT4: If idle rpm is higher than specified for current conditions, the computer calls for a diagnostic, in which it:

1. commands the IAC valve to close
2. monitors the MAF sensor, which should decrease
   - if the MAF changes flow by 6 g/sec, the computer sets code P0507, indicating that the IAC is responding, but the MIL does not light
   - if the MAF does not change, the computer sets code P1509, indicating that the IAC did not operate

Again,the MIL illuminates only if the diagnostic fails on the second consecutive drive trip. And the control module turns off the MIL if the diagnostic does not fail after three consecutive drive trips.

Throttle Body Carbon Buildup

L98, LT1, LT4, ZR-1: Carbon buildup in the throttle body bores, behind the throttle valves, may produce one or more of the following symptoms:

- a high idle speed
- high accelerator pedal effort
- engine stalling at idle or during deceleration
- a rough or low idle
- a code 34 on restart

Servicing Carbon Buildup:

1. Measure the minimum idle speed.
2. If less than specified: disconnect the negative battery terminal, after recording any DTCs
3. Disconnect the MAF duct, at the throttle body.
4. Cover the TPS and the IAC valve, to protect them from overspray.
5. Plug the IAC air inlet passage with clean rags.
6. Using an appropriate solvent, clean the bores.

6.2 Checking Throttle Actuator Control (TAC)

Scan Tool Procedure (applies to OBD-II **LS1, '96 LT1, LT4**):

1. Check for trouble codes.
2. Using a scan tool, with the engine running, command the engine up to 1500 rpm, then down to 500 rpm, do this a couple of times. Check that the engine speed is within 100 rpm of the commanded speed. In other words, when set at 1500 rpm the actual engine speed should be within the 1400-1600 rpm range.
3. If higher, check for: vacuum leaks or a faulty PCV valve.
4. If lower, check for: excessive throttle-body deposits, or a parasitic load on the engine, such as a transmission problem

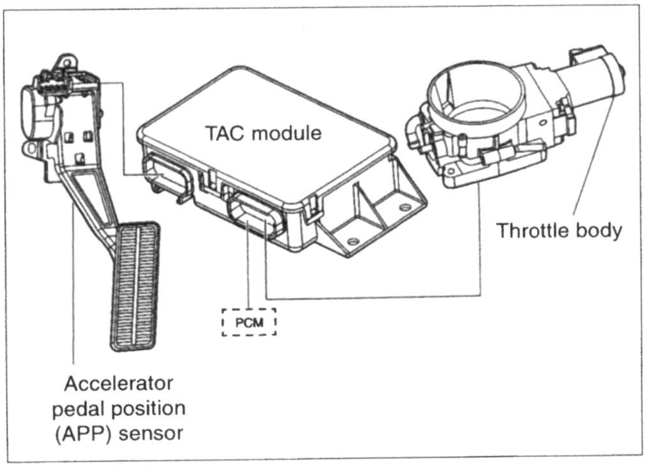

Fig. 11-30. LS1: The TAC controls idle rpm with the main throttle. There is no Idle Air Control (IAC) with this engine.

6.3 LS1 Idle Learn Procedure

The computer's learned-idle is lost any time the computer loses power, as with a disconnected battery. The result: engine idle is unstable. Here's how you help the computer relearn the minimum throttle angle:

Automatic Transmission LS1

1. Turn the key off, reconnect the battery, turn the AC control off, set the parking brake, and block the drive wheels.
2. Start the engine, shift to *drive*, let the engine idle for five minutes.
3. Turn the AC control on (make sure the AC button light is on). Idle for another five minutes.
4. Leave AC on, shift into to *park*, and idle for yet another five minutes.
5. Turn the AC off, and idle for a final five minutes.
6. Finally, turn the engine off for at least 30 seconds.

Manual Transmission LS1

1. Turn the key off, reconnect the battery, turn the AC control off, set the parking brake, and block the drive wheels.
2. With the transmission in *neutral*, start the engine.
3. Turn the AC control on (make sure the AC button light is on), and idle for five minutes.
4. Turn the AC off, and idle for another five minutes.
5. Turn the engine off for at least 30 seconds.

7. Checking the Temperature Sensors

11

7.1 Reference Voltage

The computer provides a Reference Voltage (VRef) of 5 v.

7.2 Resistance-Engine Off

Remember, some Corvette ECT sensors operate with a two-step signal, shunting the voltage at about 50° C. If you monitor the circuit with a voltmeter, your voltage reading could be misleading.

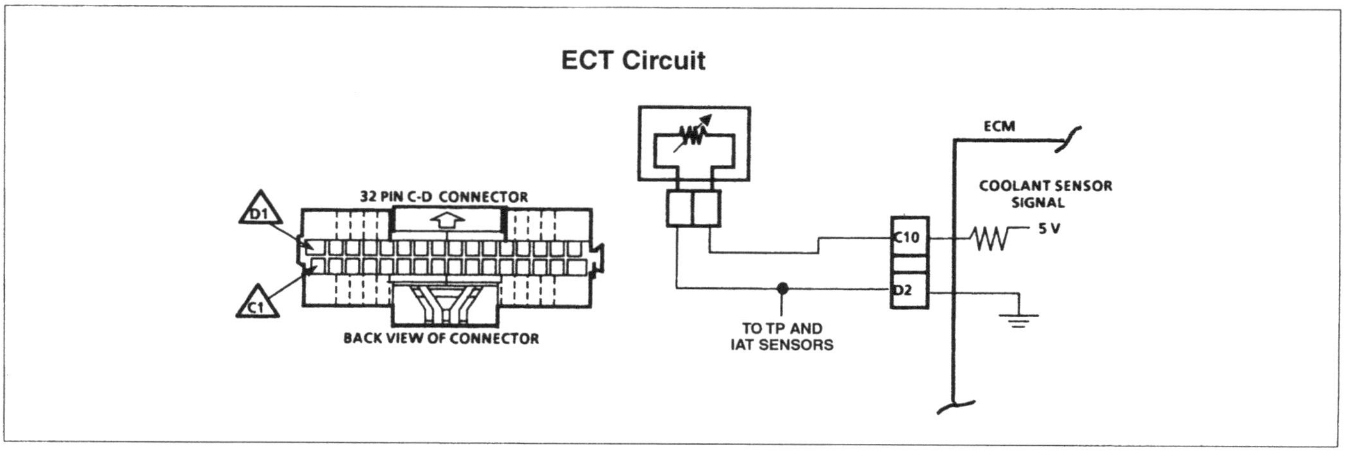

Fig. 11-31. Use the Temperature Sensor circuit to check VRef, either at the sensor, or at the 32-pin C-D connector. Measure the voltage between the VRef and the ground, with the key turned on. *Note*: the ECT is shown; the IAT and MAT are similar, but have different wiring colors.

Check resistance at the Control Module, at an ambient temperature:

- For an open circuit, read the VRef signal at the computer.
- For a short circuit, read zero volts at the computer.

NOTE—
A scan tool can read sensor temperatures in degrees Celsius and Fahrenheit.

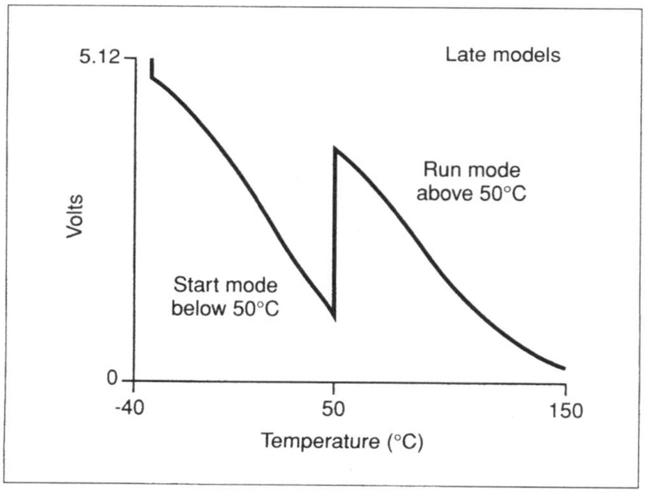

Fig. 11-32. Check resistance across the ECT sensor, noting temperature and observing resistance.

8. CONCLUSION

You have now seen how to service the engine-management's electrical system, fuel system, air-intake, ignition, idle air-control, and temperature sensors.

Now you are ready to apply all that you've learned in the first 11 chapters to understanding and using the wiring diagrams in Chapter 12.

Chapter 12

Schematics

Contents

Front-mounted EBTCM/BPMV Power, Ground, DLC, MIL

Figure 1 of 30

LS1 1998-2001 VIN G

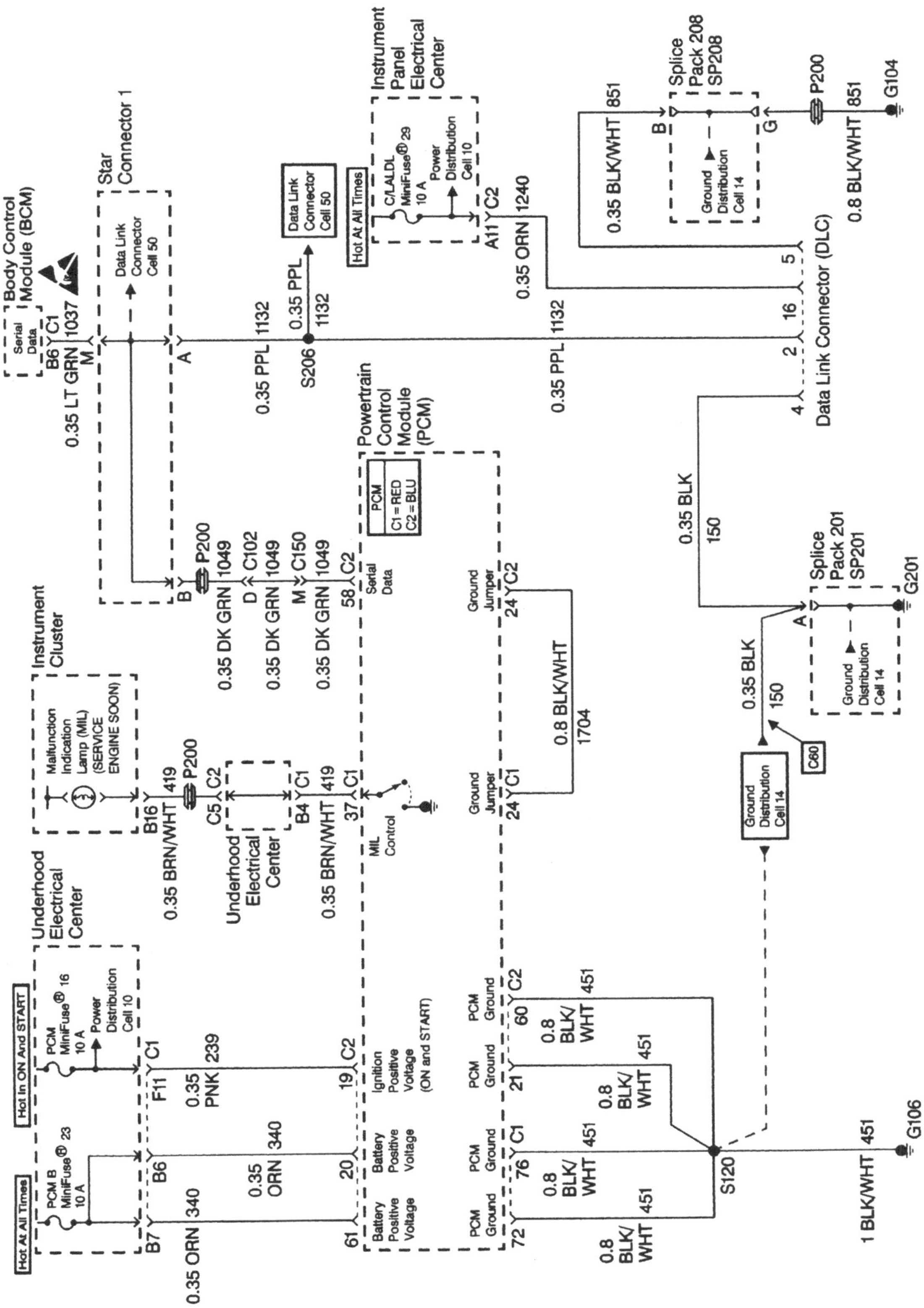

LS1 1998-2001 VIN G

Rear-mounted EBTCM/BPMV Power, Ground, DLC, MIL

Figure 2 of 30

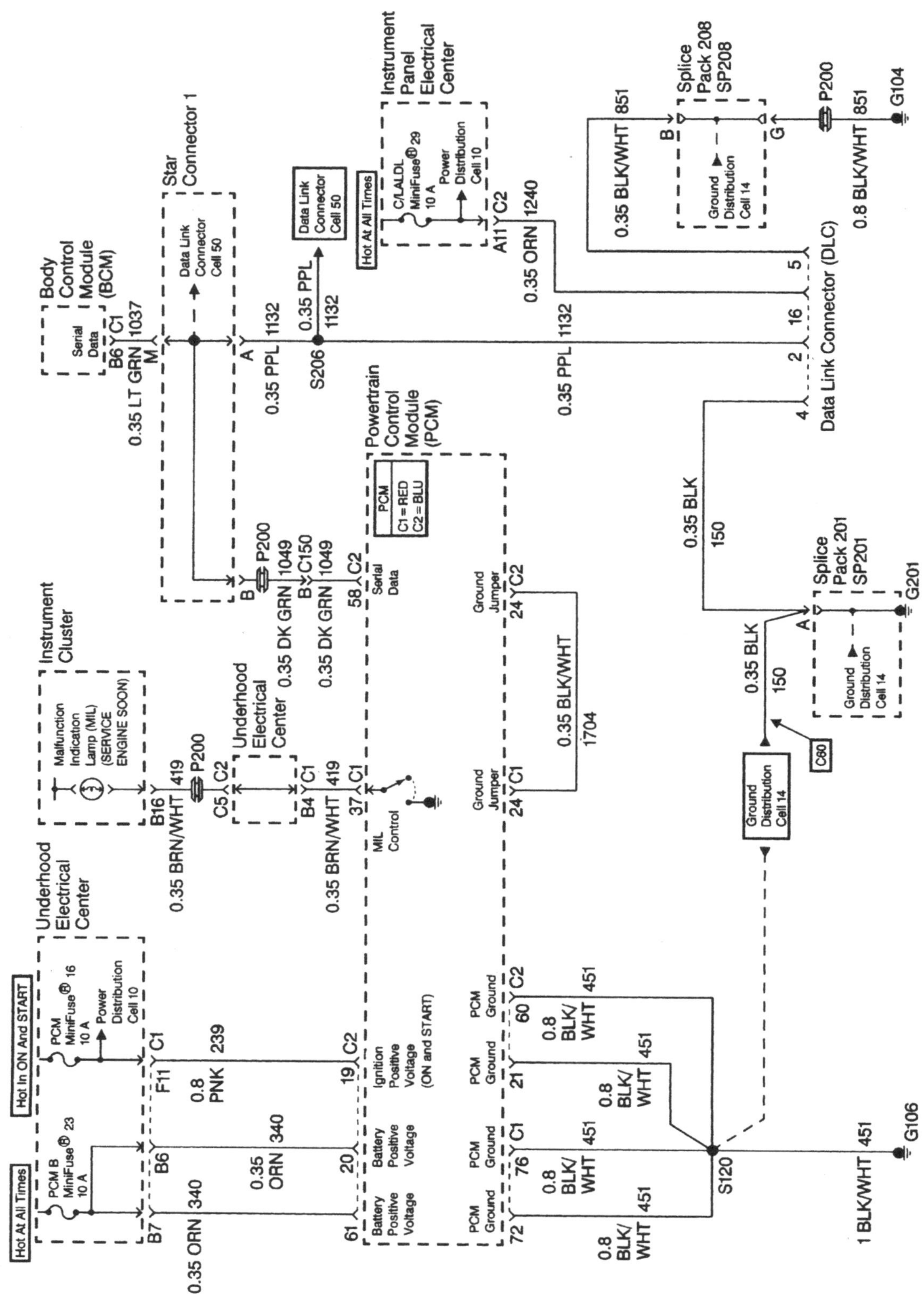

All EBTCM/BPMV
Ignition Coil/Module Cyl. #1, #3

Figure 3 of 30

LS1 1998-2001 VIN G

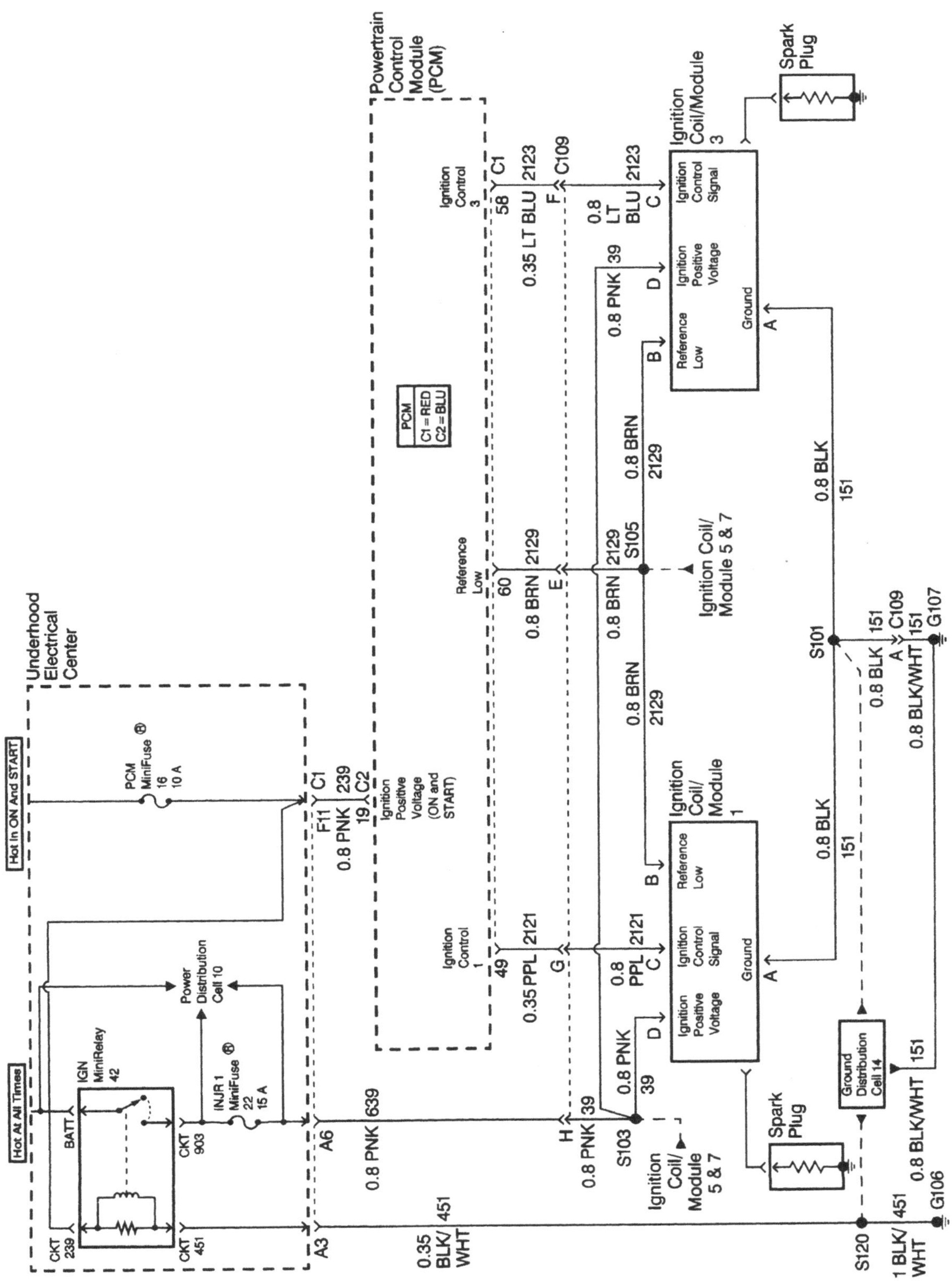

12

LS1 1998-2001 VIN G

All EBTCM/BPMV
Ignition Coil/Module Cyl. #5, #7

Figure 4 of 30

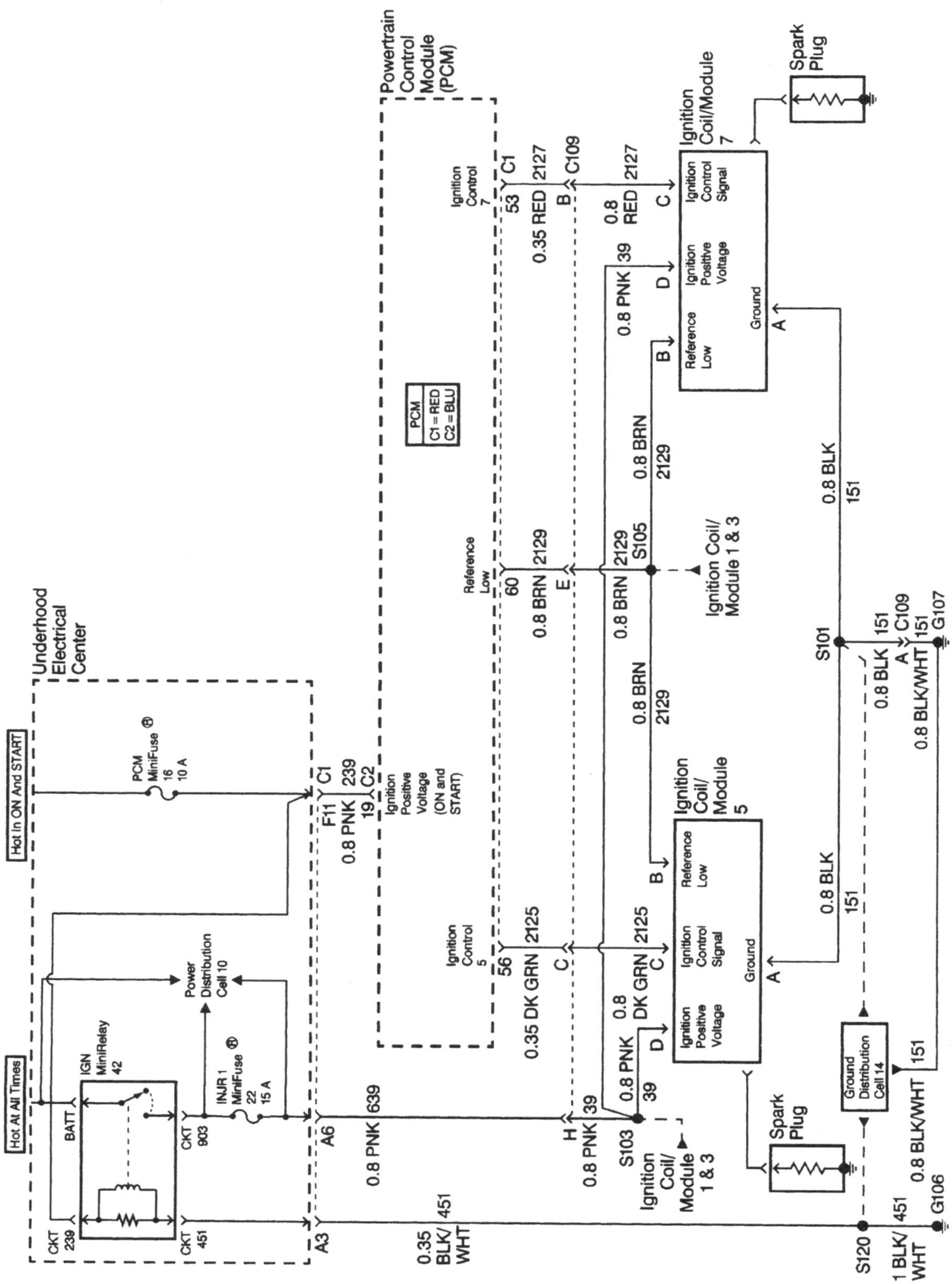

All EBTCM/BPMV
Ignition Coil/Module Cyl. #2, #4

Figure 5 of 30

LS1 1998-2001 VIN G

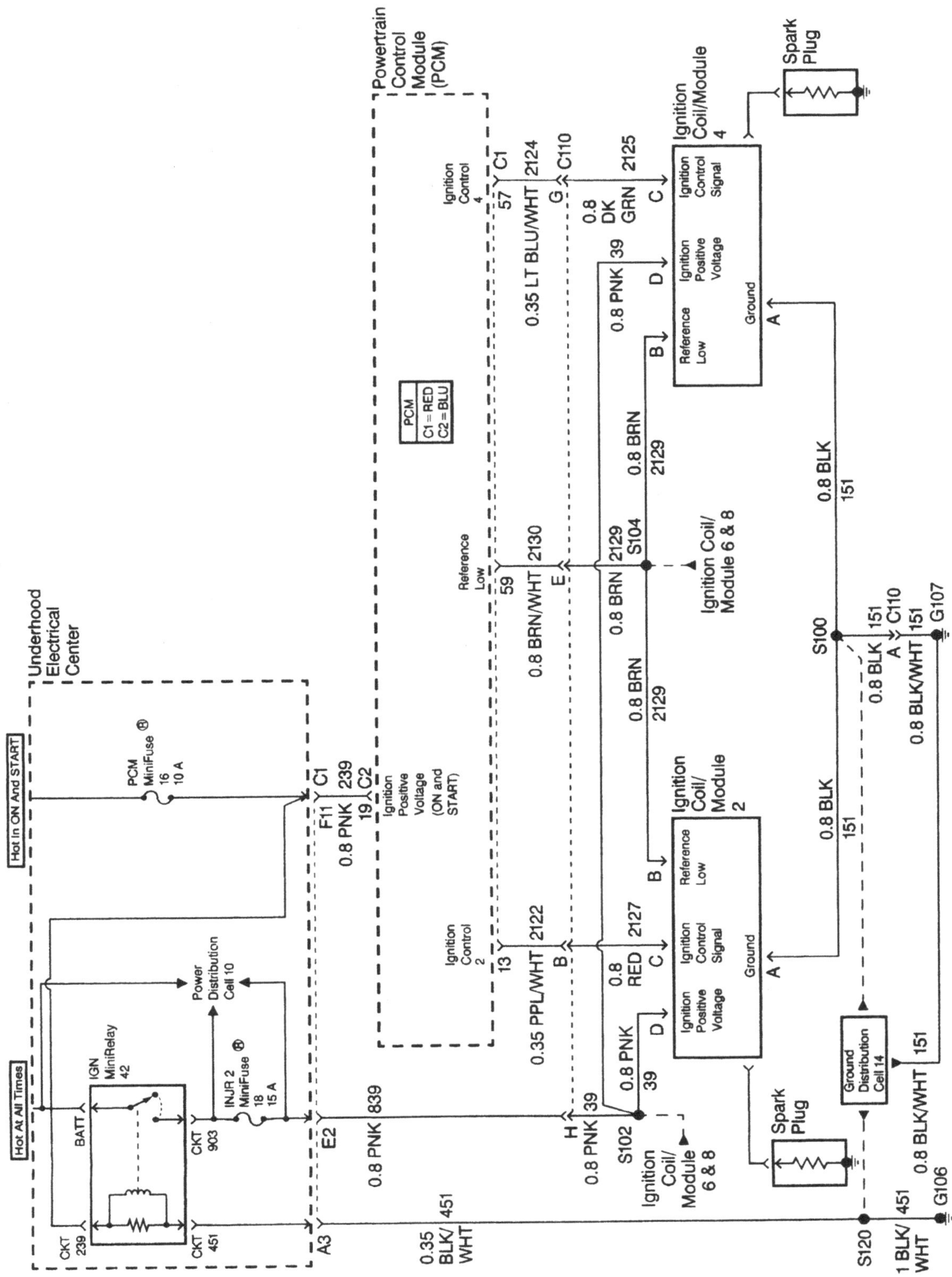

12

LS1 1998-2001 VIN G

All EBTCM/BPMV
Ignition Coil/Module Cyl. #6, #8

Figure 6 of 30

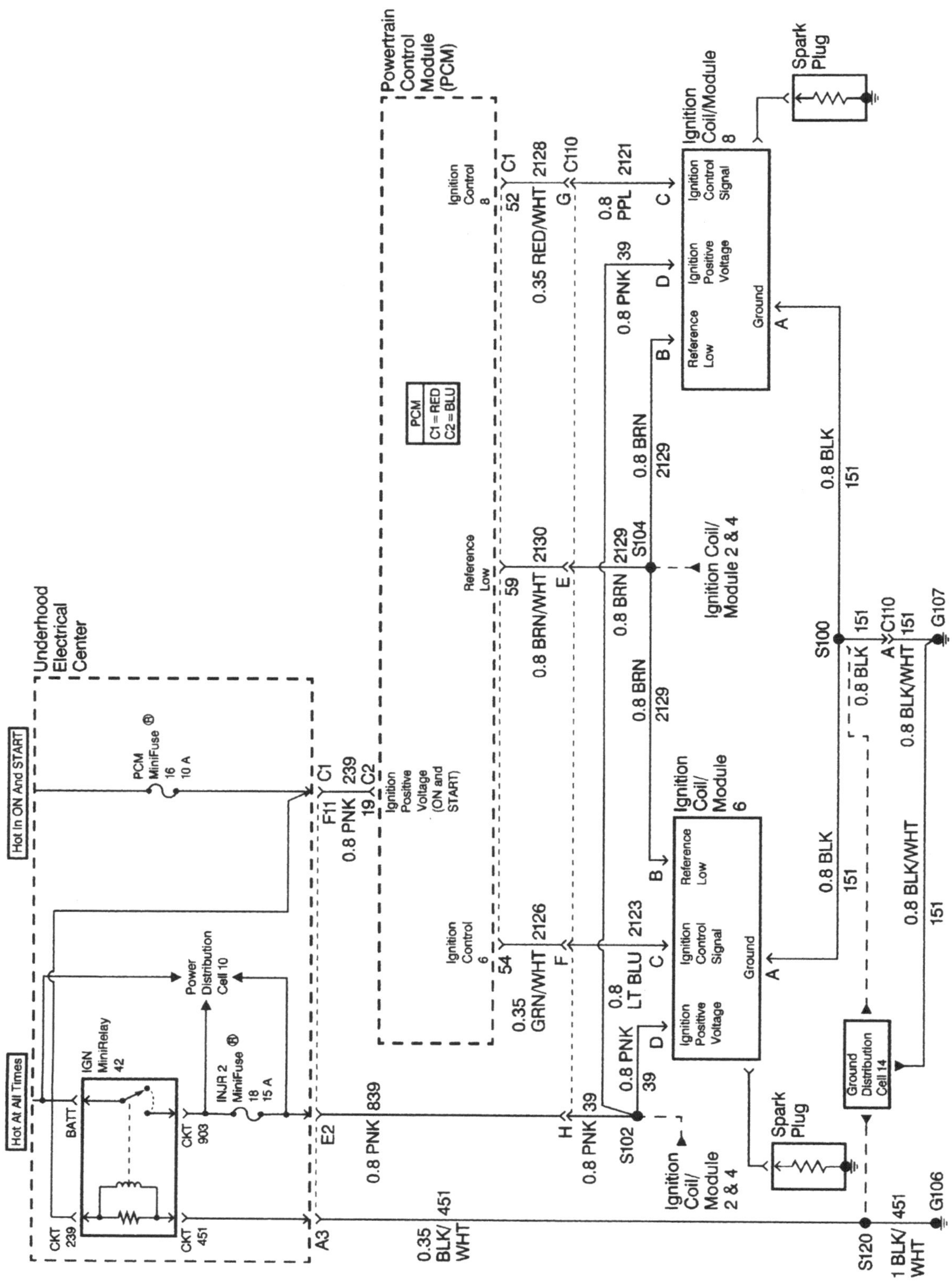

Front-mounted EBTCM/BPMV CKP, CMP, KS, Traction Control

Figure 7 of 30

LS1 1998-2001 VIN G

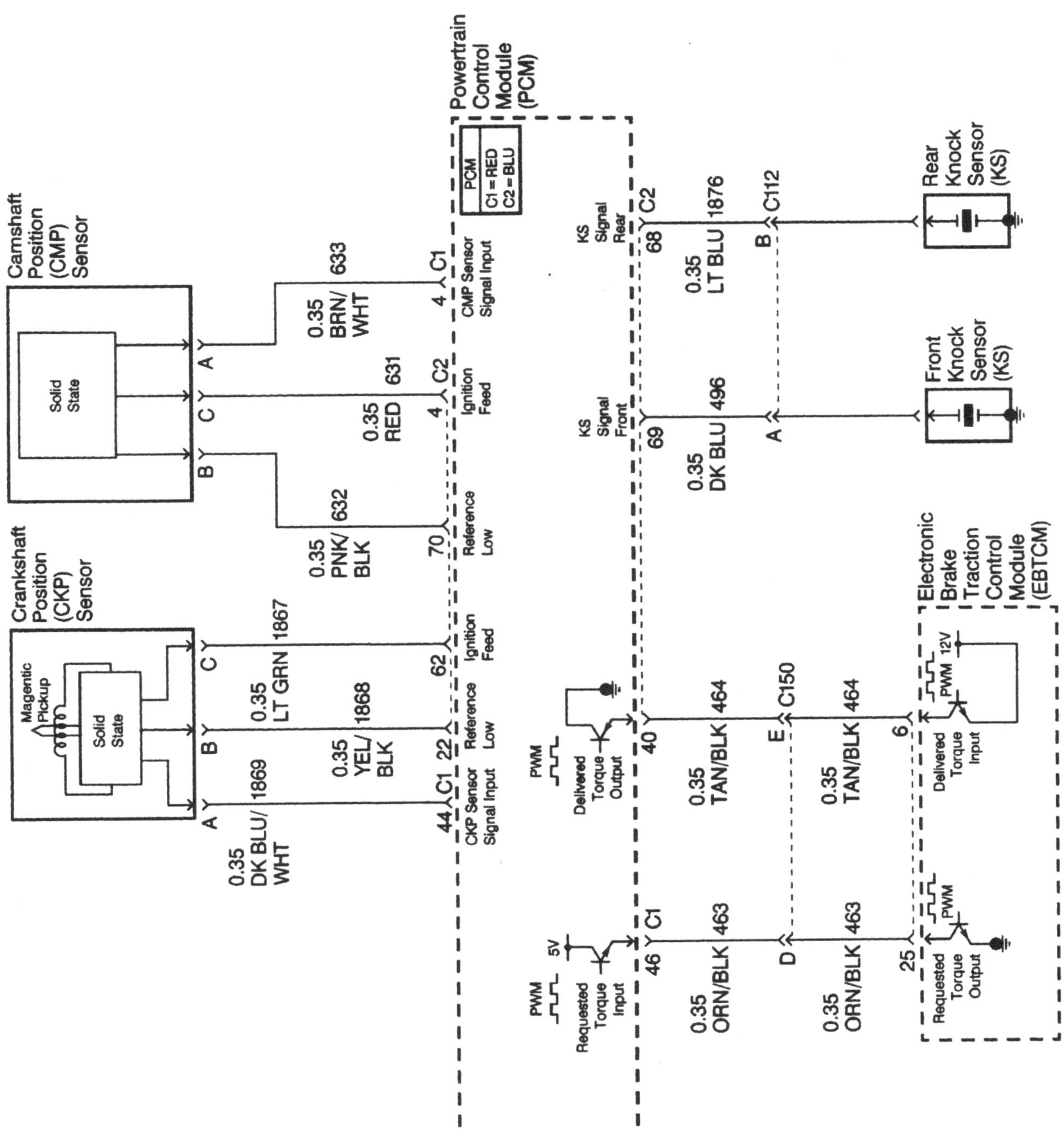

LS1 1998-2001 VIN G

Rear-mounted EBTCM/BPMV
CKP, CMP, KS, Traction Control

Figure 8 of 30

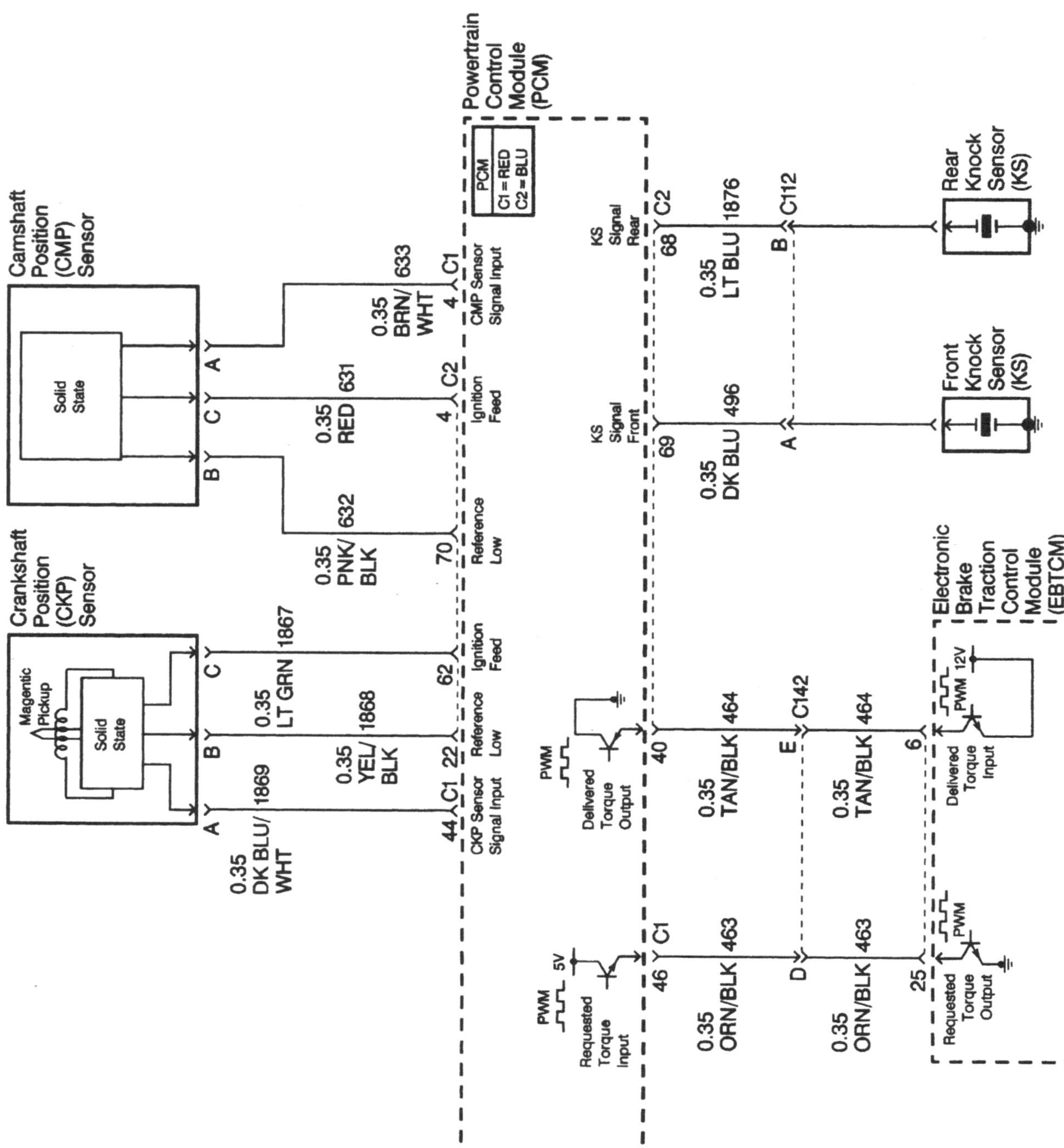

All EBTCM/BPMV
Fuel Pump and Fuel Level Sensors

Figure 9 of 30

LS1 1998-2001
VIN G

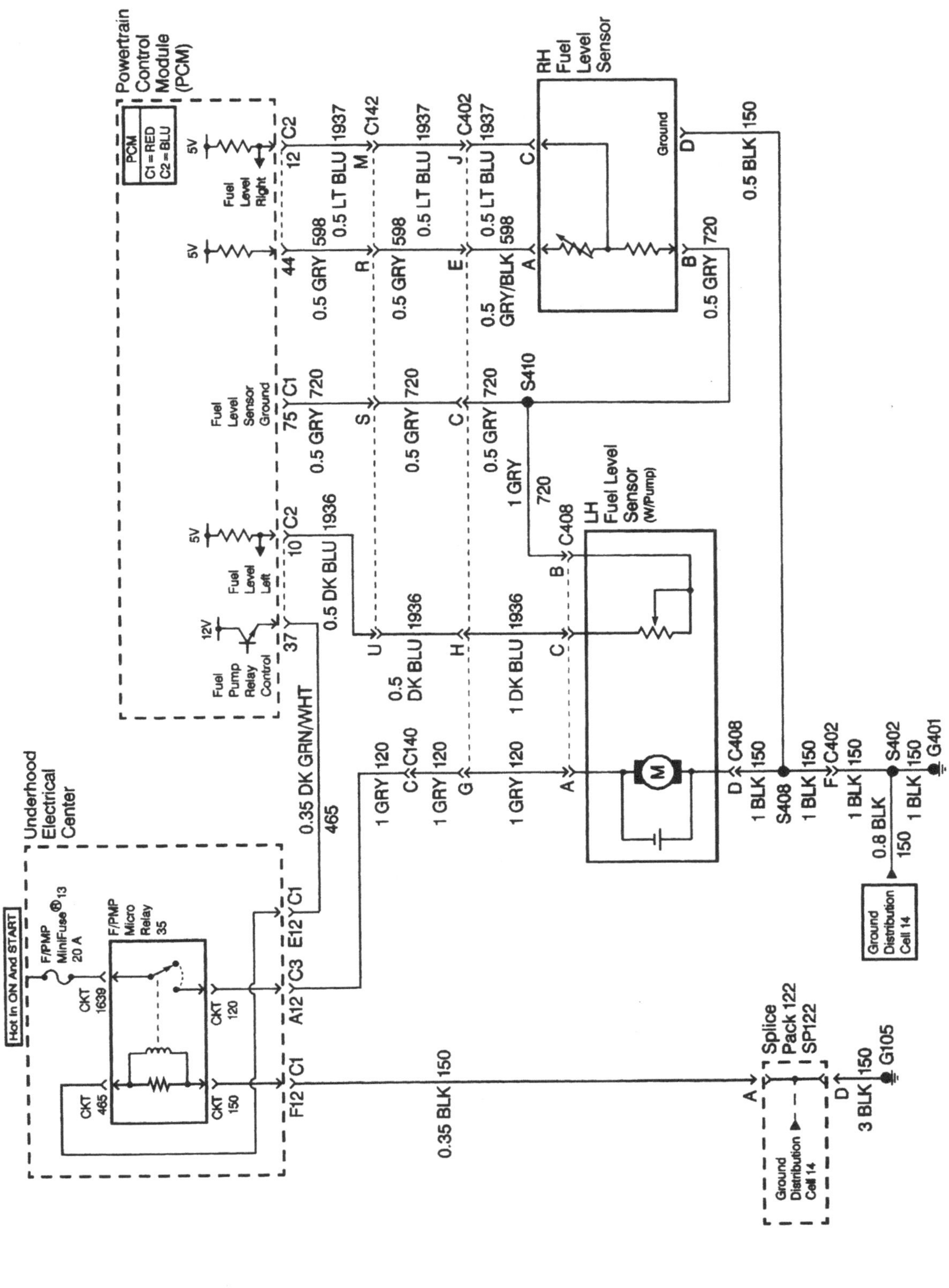

12

LS1 1998-2001 VIN G

All EBTCM/BPMV Injectors #2, #4, #6, #8

Figure 10 of 30

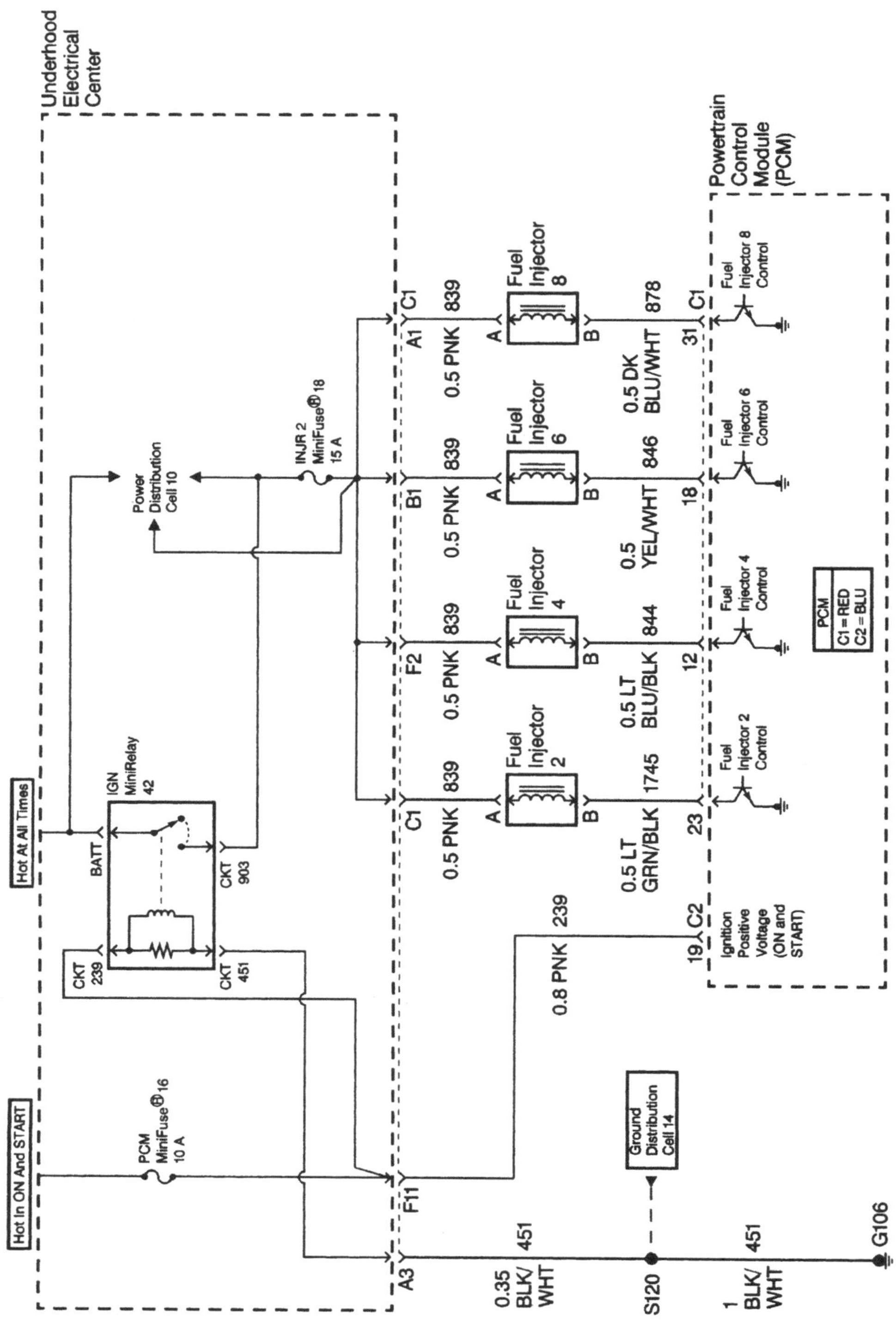

All EBTCM/BPMV
Injectors #1, #3, #5, #7

Figure 11 of 30

LS1 1998-2001 VIN G

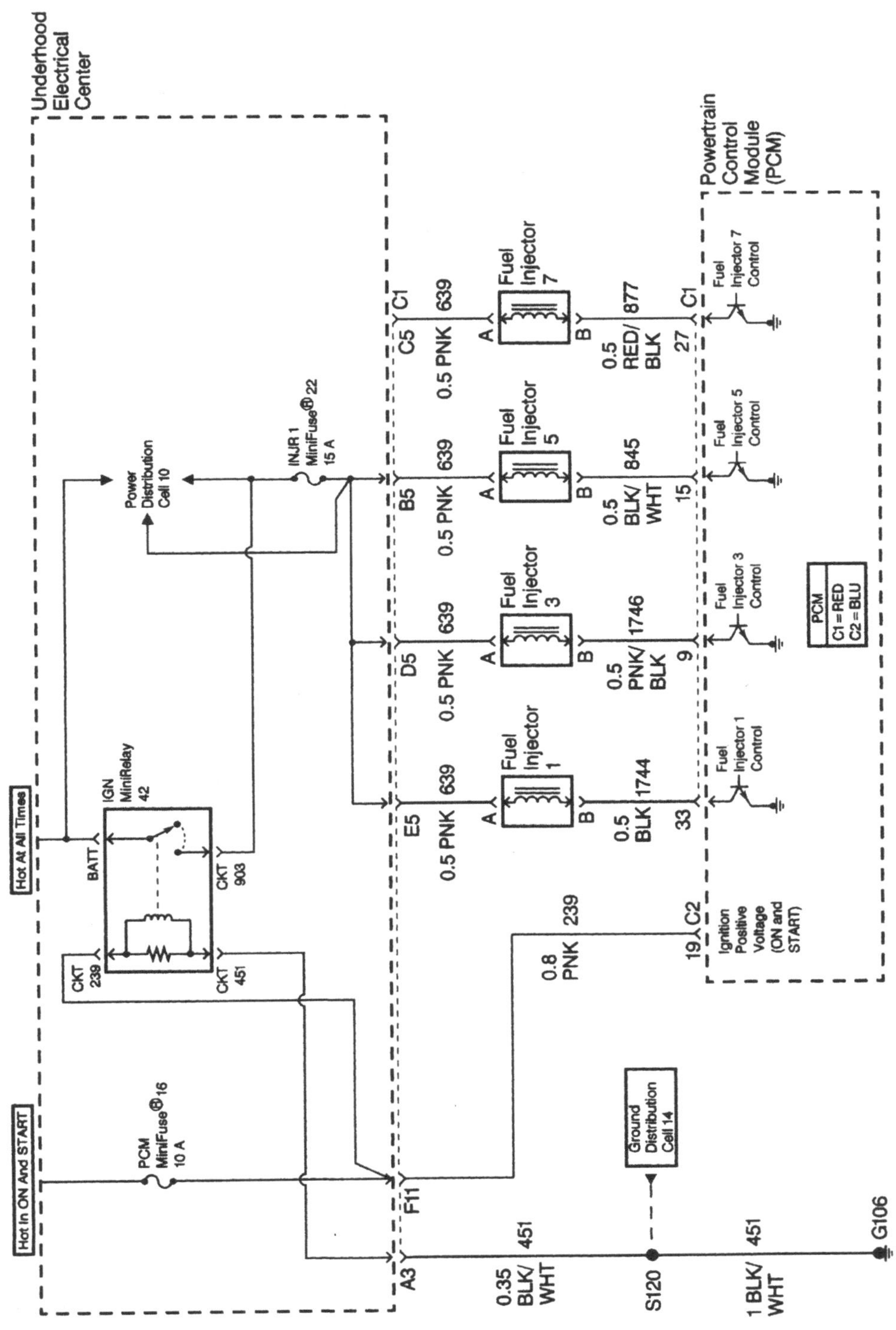

12

LS1 1998-2001 VIN G

All EBTCM/BPMV Cruise Control

Figure 12 of 30

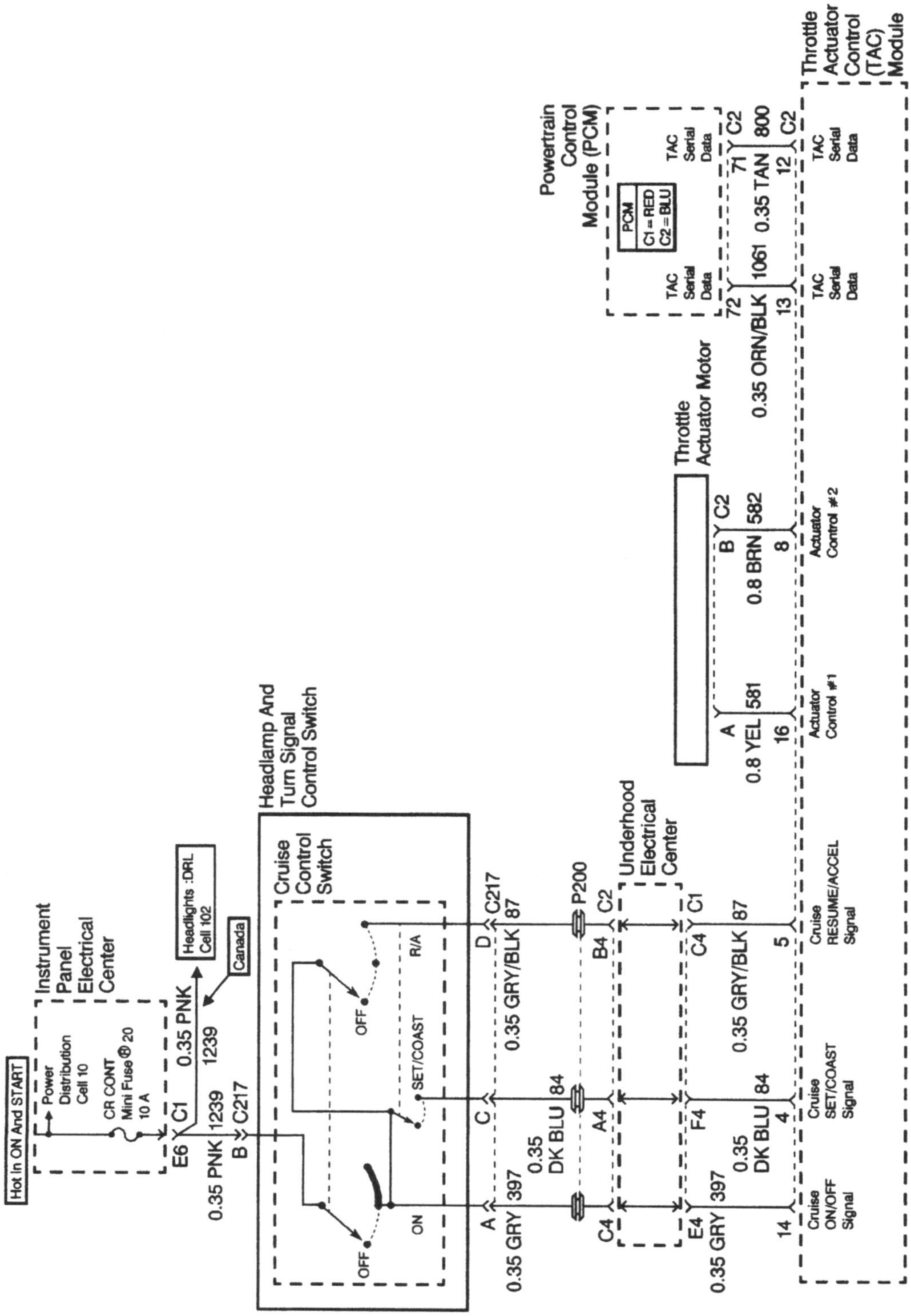

All EBTCM/BPMV
Stop Lamp Switch
Figure 13 of 30

LS1 1998-2001
VIN G

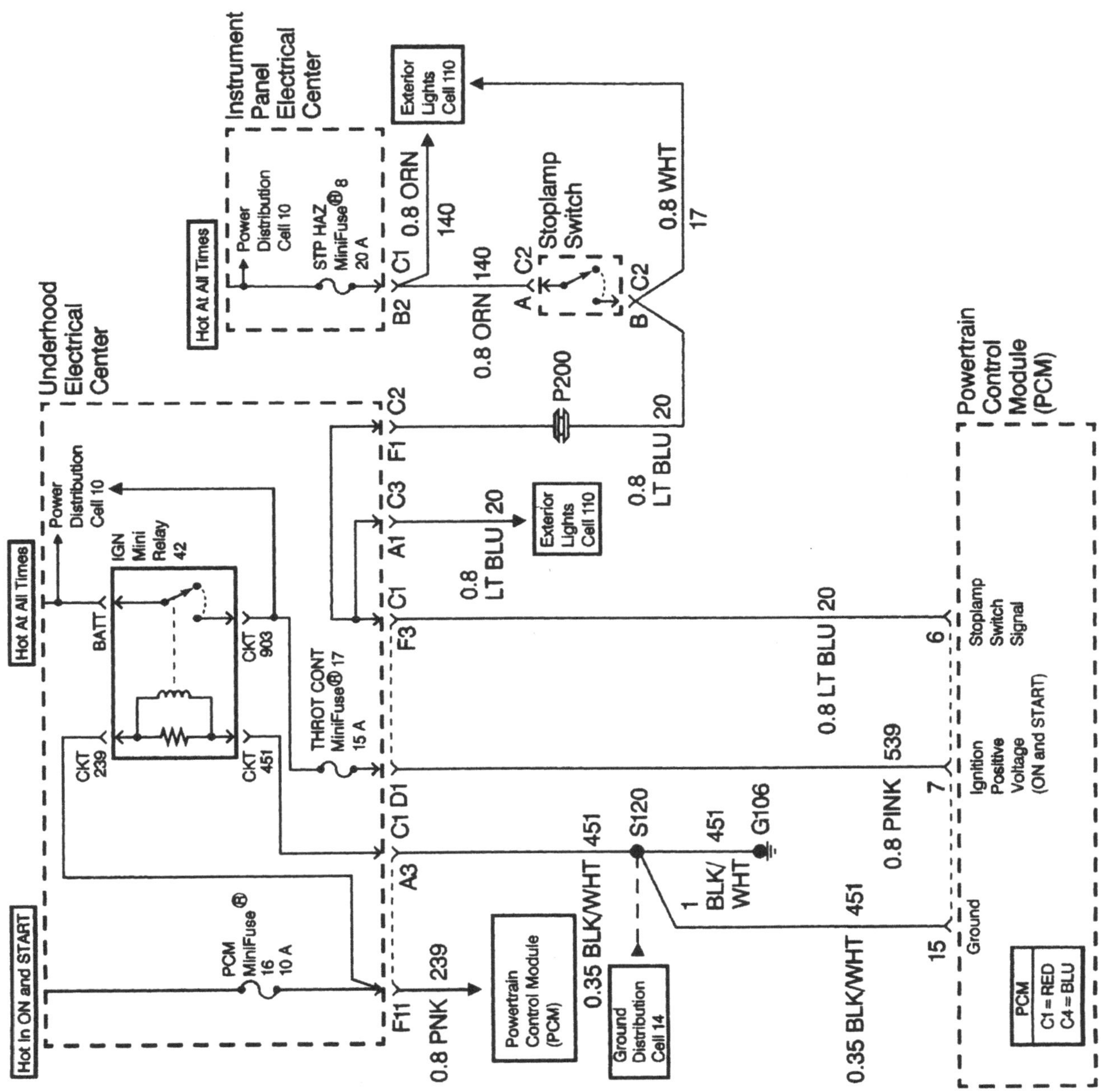

12

LS1 1998-2001 VIN G

All EBTCM/BPMV TP and APP Sensors

Figure 14 of 30

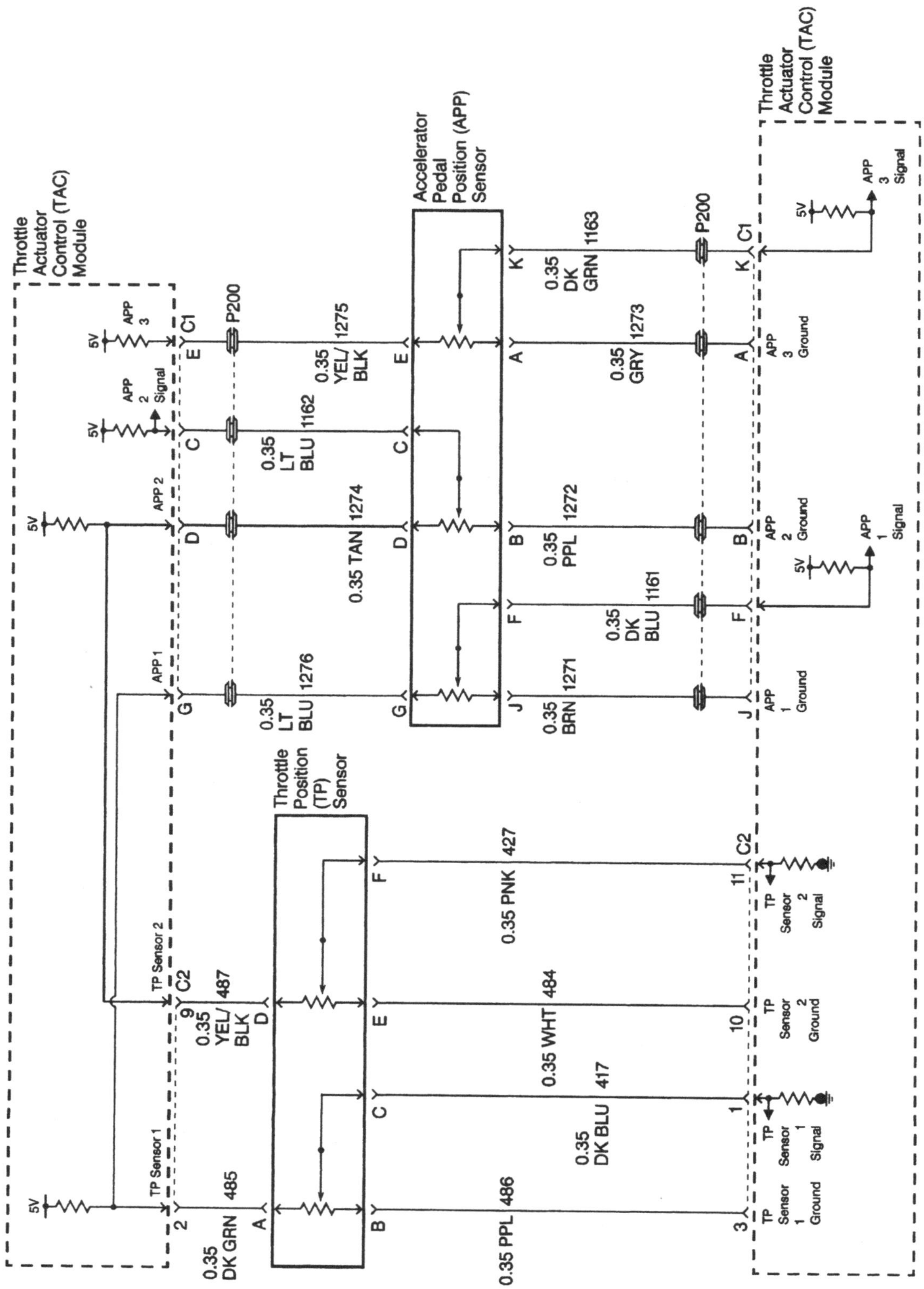

Front-mounted EBTCM/BPMV
Engine Data Sensors

Figure 15 of 30

LS1 1998-2001 VIN G

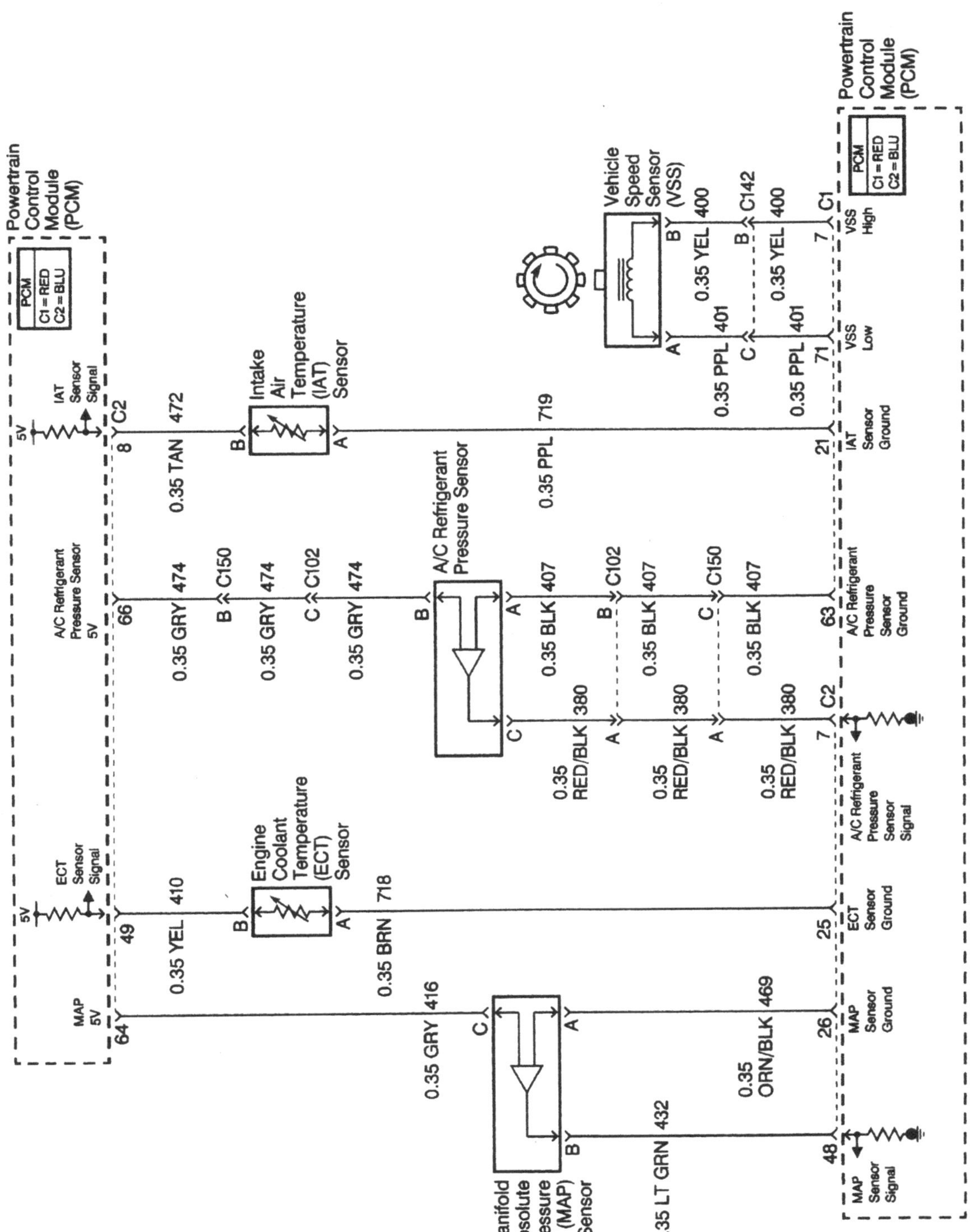

12

LS1 1998-2001 VIN G

Rear-mounted EBTCM/BPMV Engine Data Sensors

Figure 16 of 30

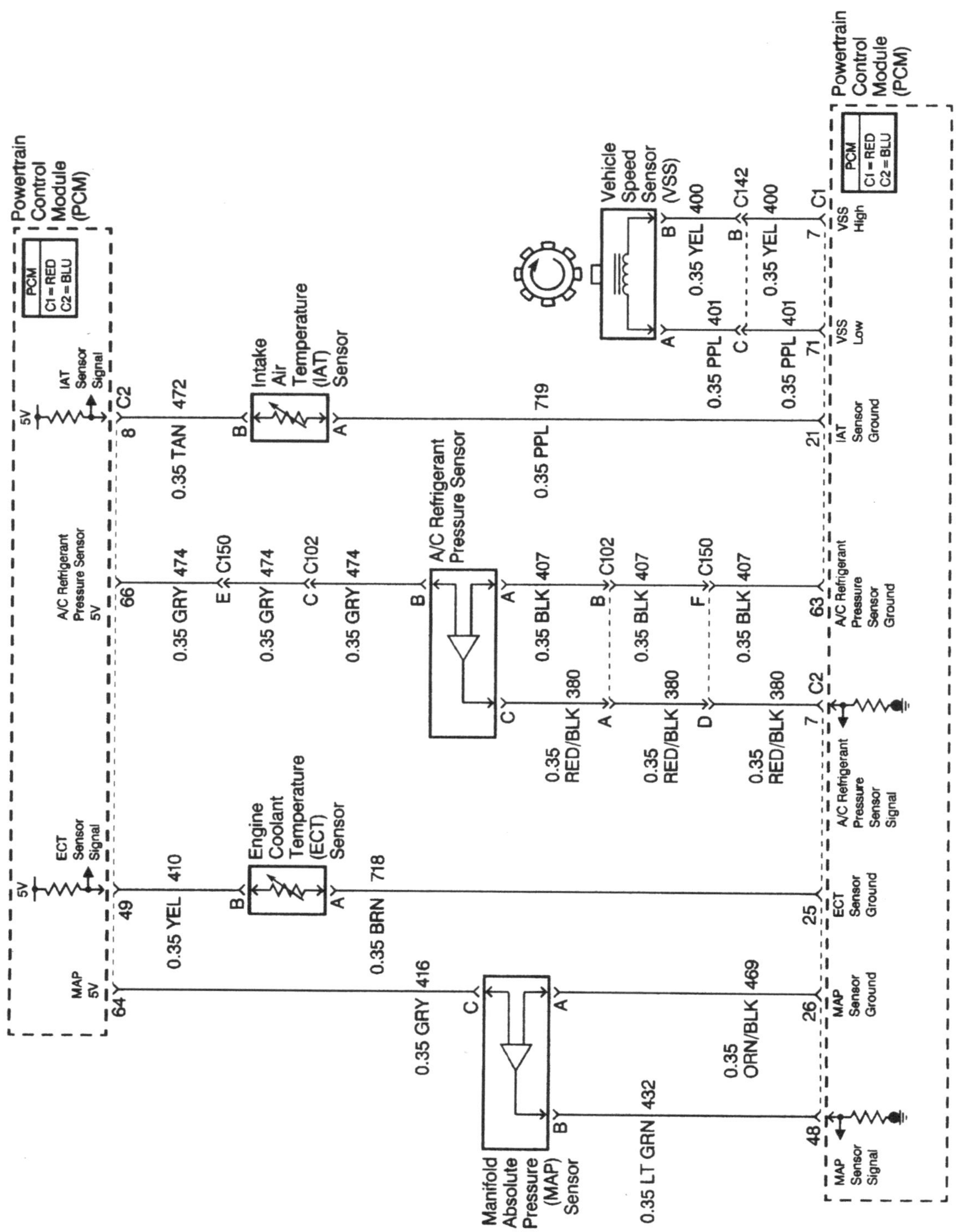

All EBTCM/BPMV
HO_2S
Figure 17 of 30

LS1 1998-2001 VIN G

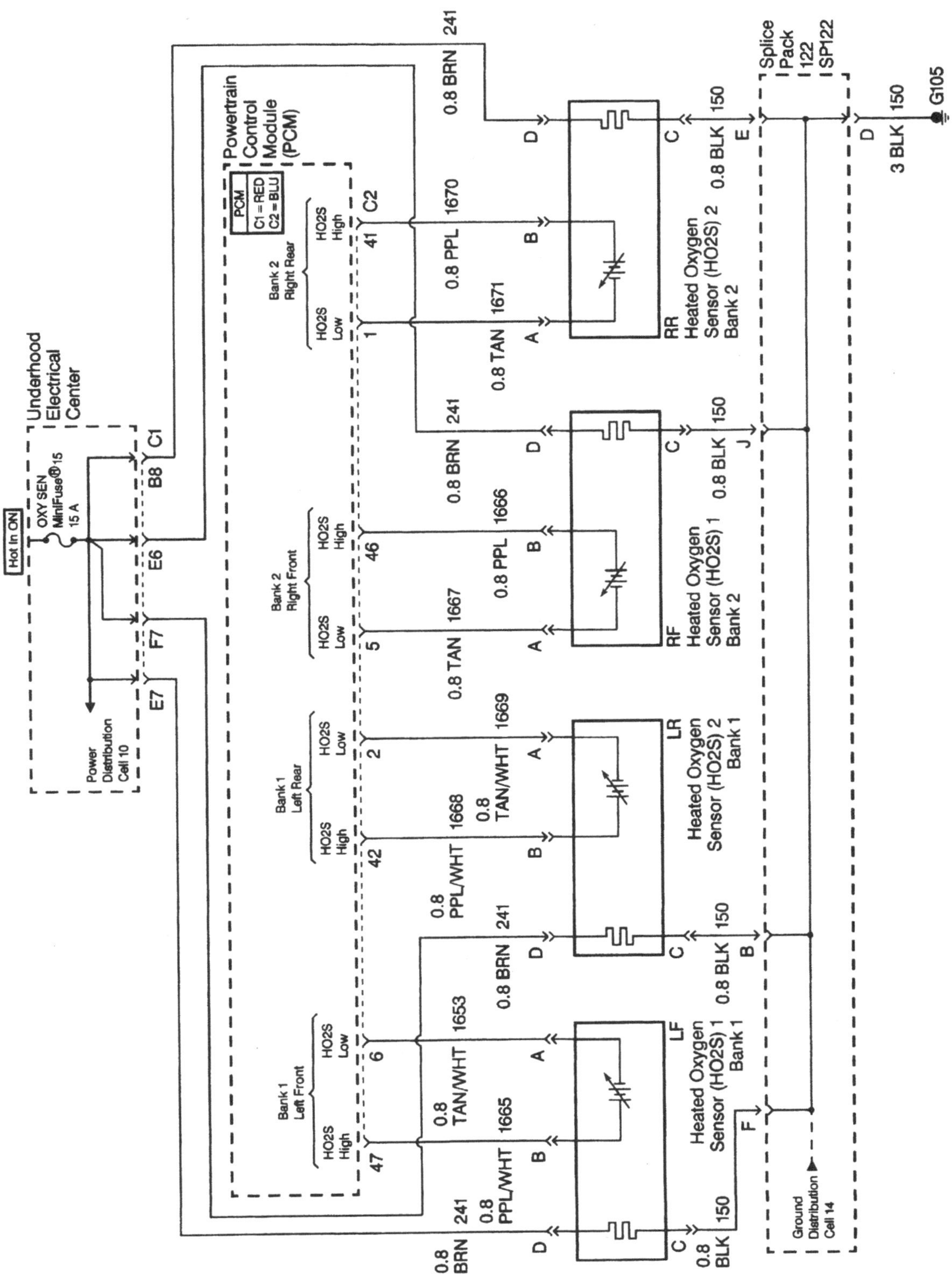

12

LS1 1998-2001 VIN G

All EBTCM/BPMV EVAP and AIR

Figure 18 of 30

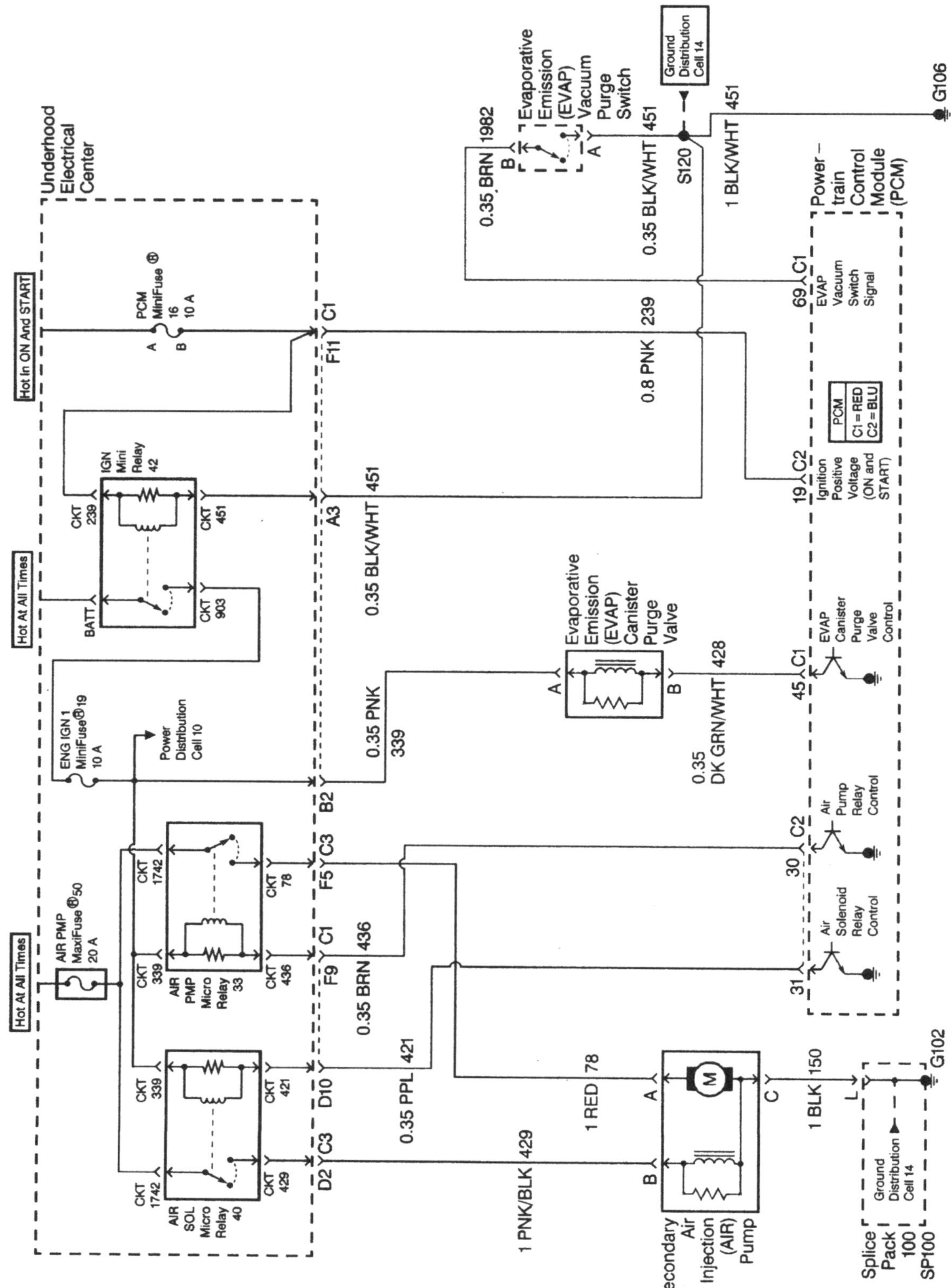

Front-mounted EBTCM/BPMV Man/Trans

Figure 19 of 30

LS1 1998-2001 VIN G

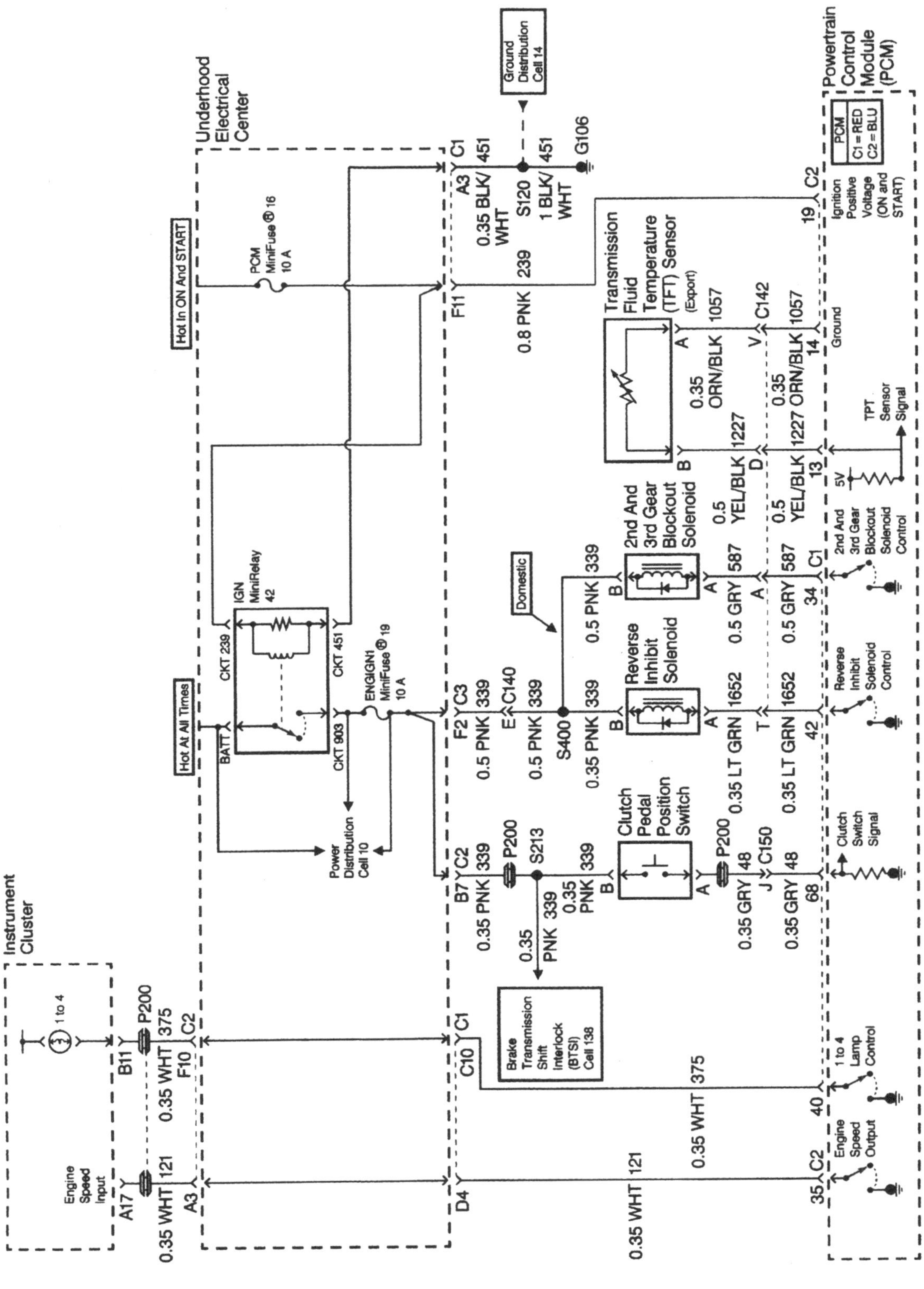

LS1 1998-2001 VIN G

Rear-mounted EBTCM/BPMV Man/Trans

Figure 20 of 30

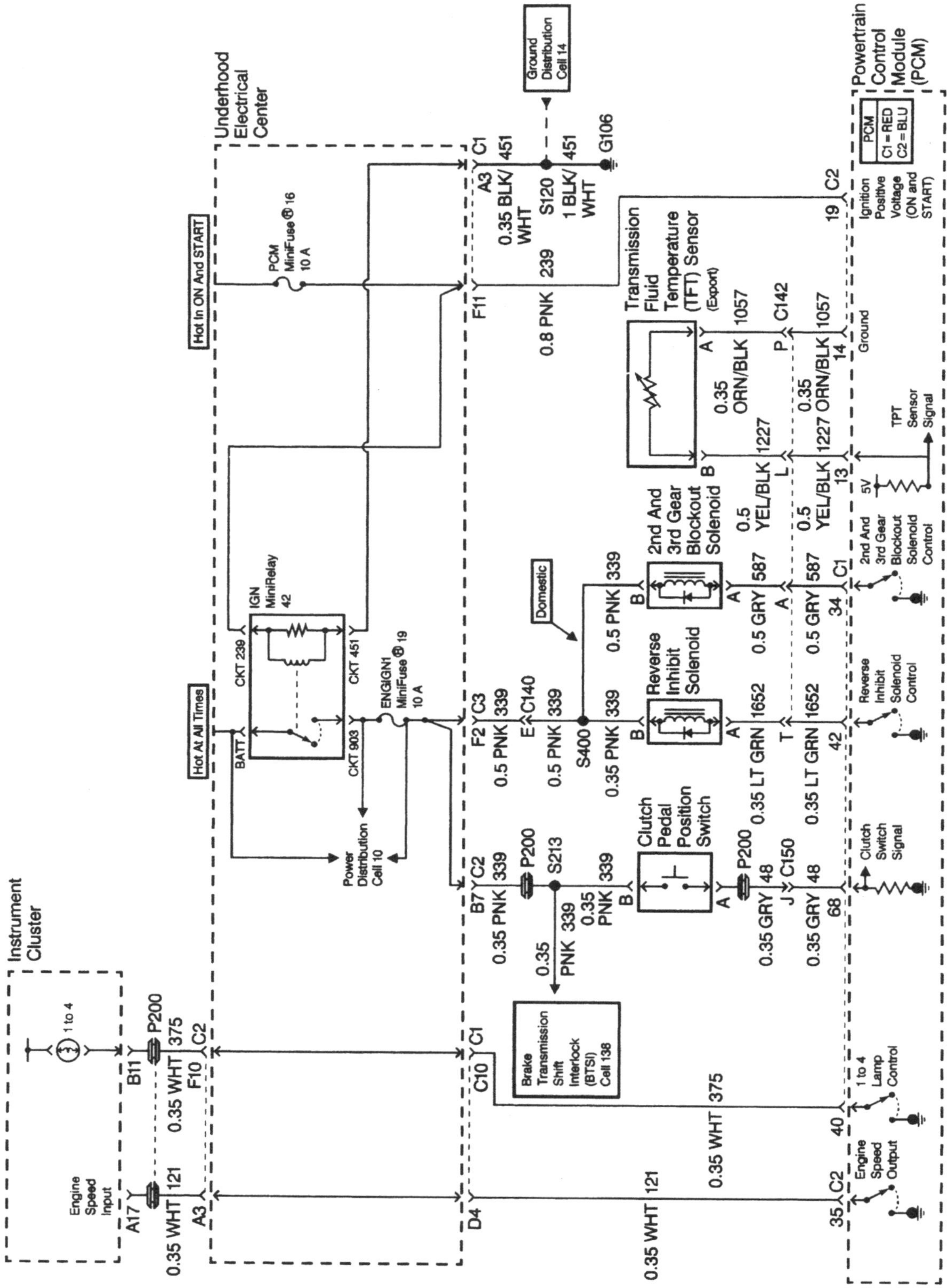

Front-mounted EBTCM/BPMV
Auto/Trans

Figure 21 of 30

LS1 1998-2001
VIN G

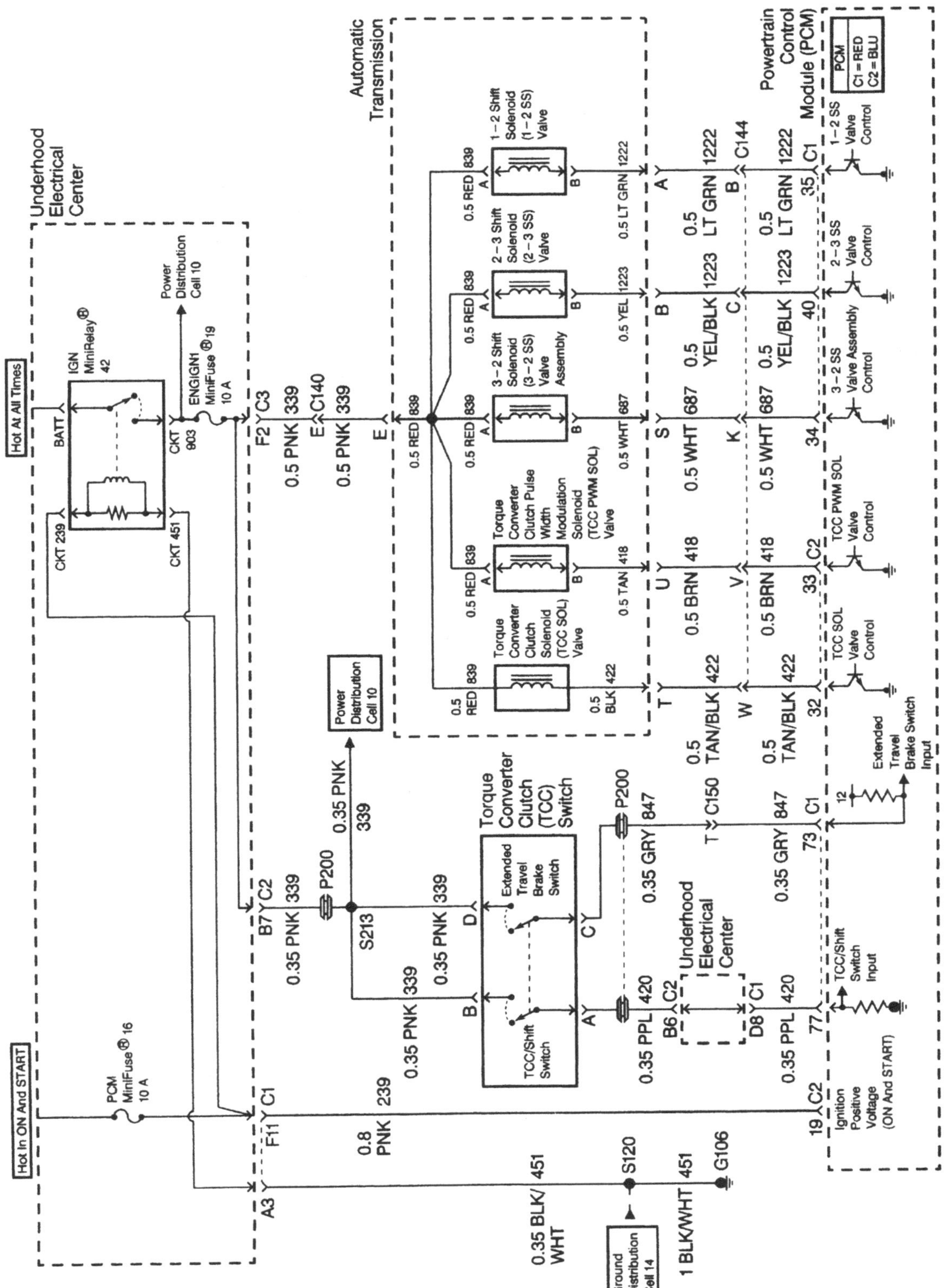

12

LS1 1998-2001 VIN G

Rear-mounted EBTCM/BPMV Auto/Trans

Figure 22 of 30

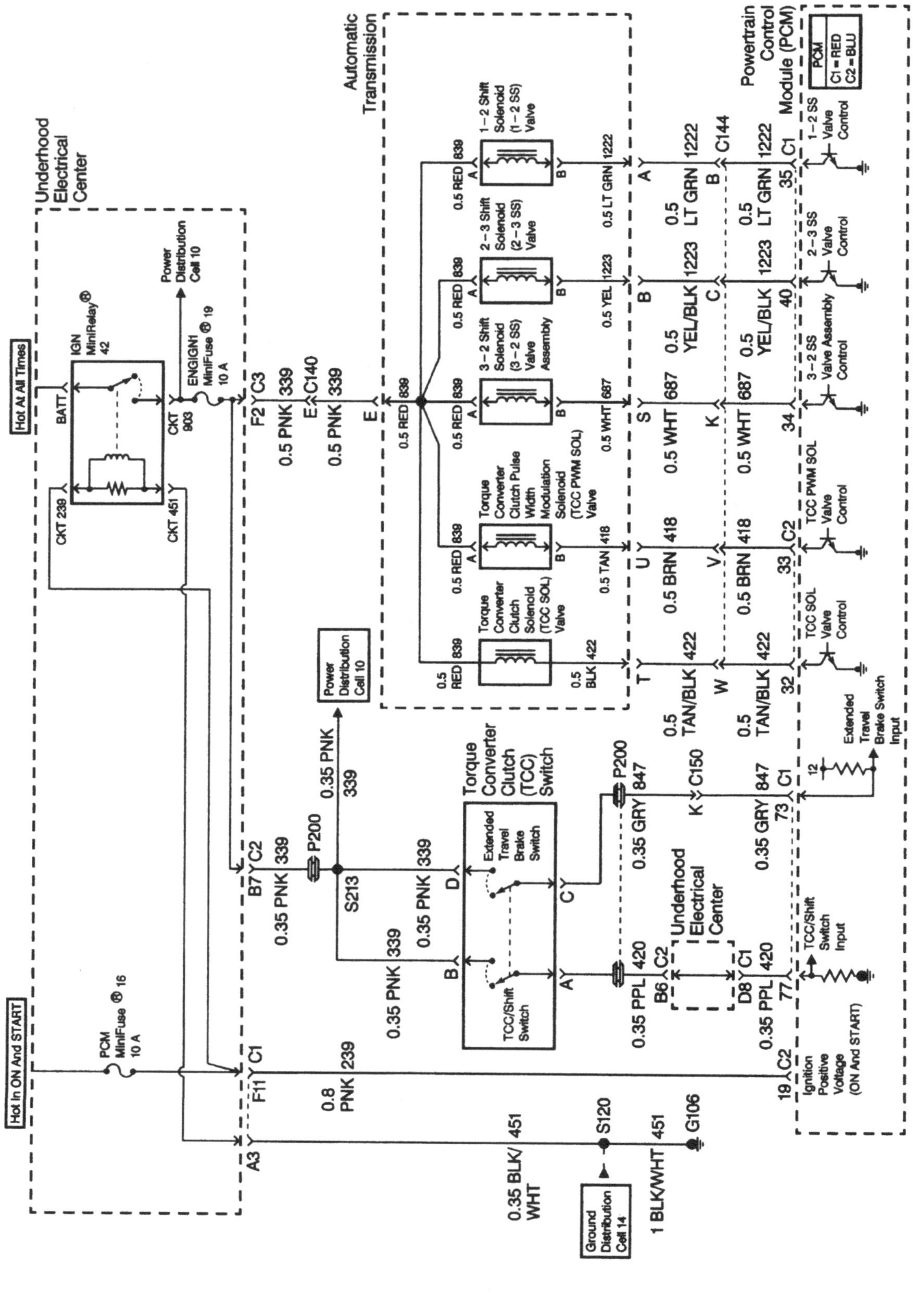

Front-mounted EBTCM/BPMV Auto/Trans and VSS Output

Figure 23 of 30

LS1 1998-2001 VIN G

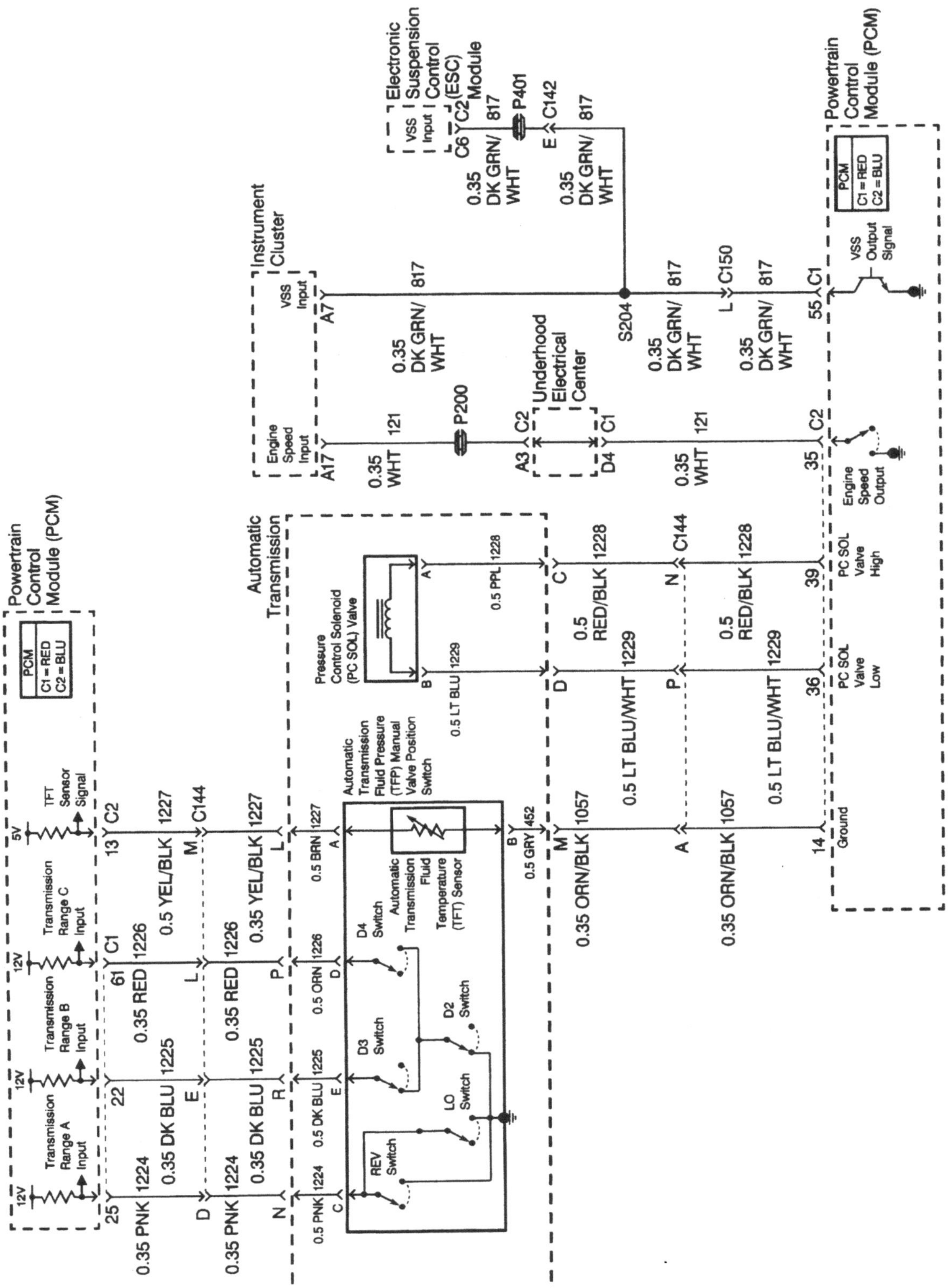

12

LS1 1998-2001 VIN G

Rear-mounted EBTCM/BPMV Auto/Trans and VSS Output

Figure 24 of 30

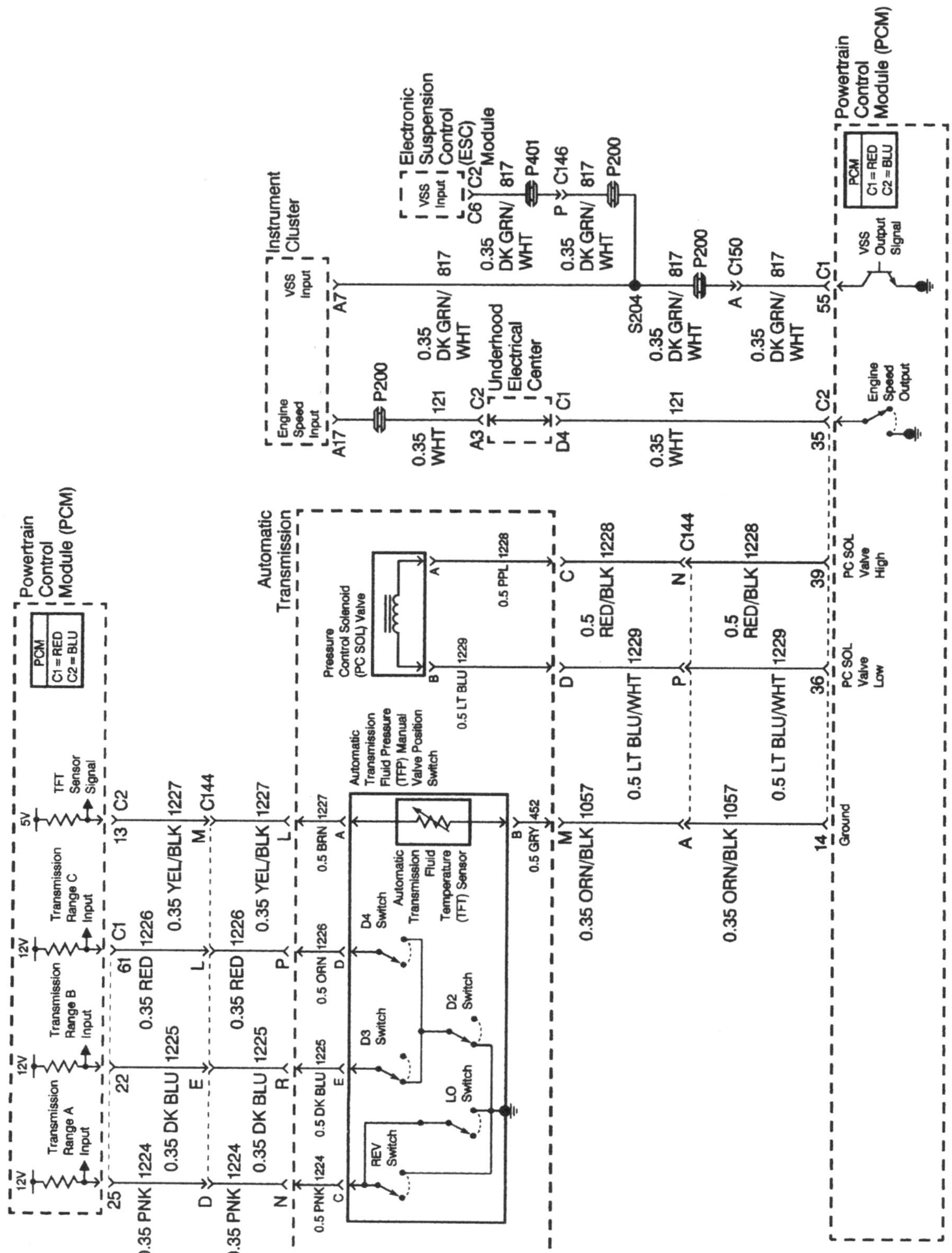

Front-mounted EBTCM/BPMV
PNP Switch

Figure 25 of 30

LS1 1998-2001 VIN G

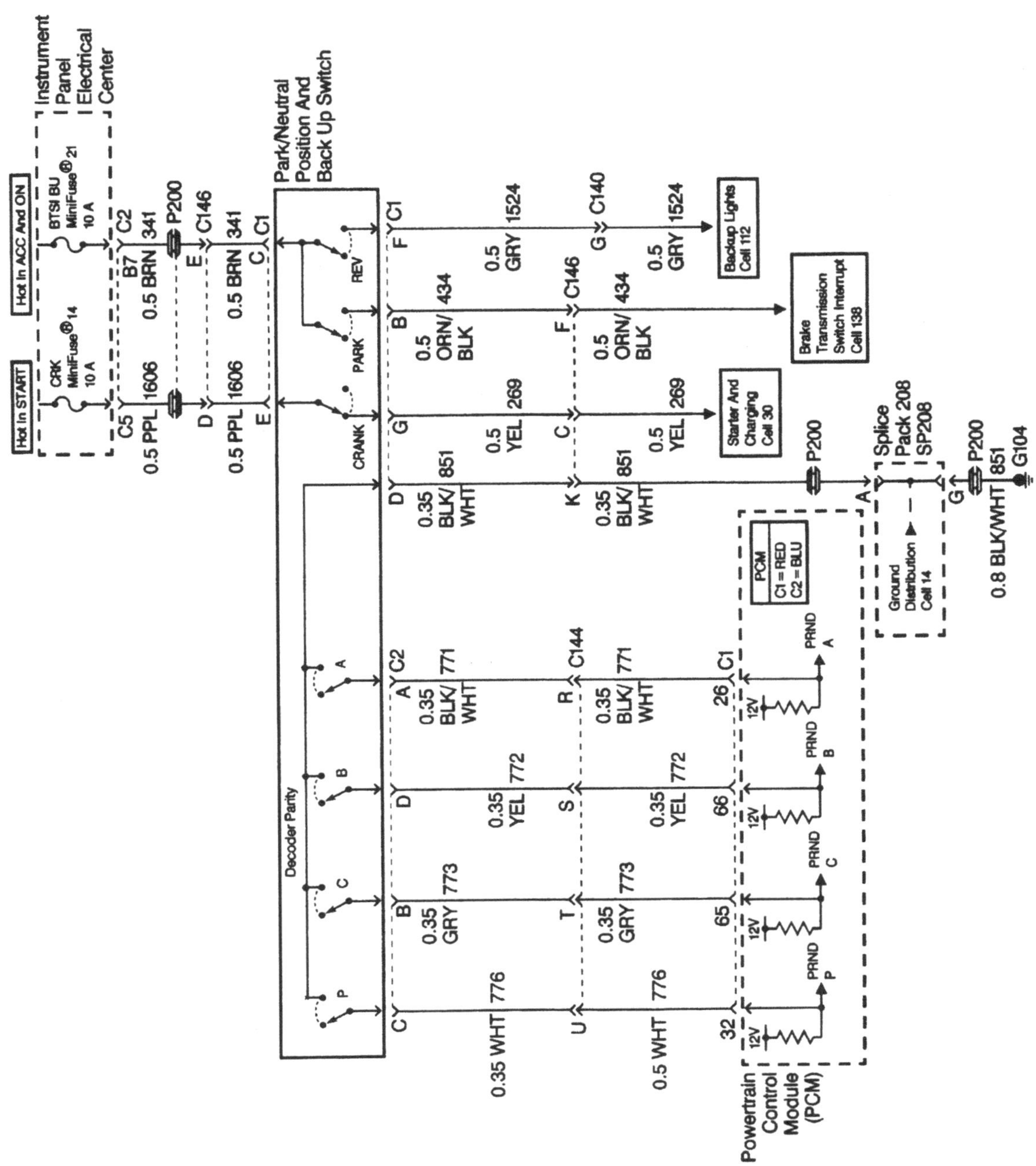

12

LS1 1998-2001 VIN G

Rear-mounted EBTCM/BPMV PNP Switch

Figure 26 of 30

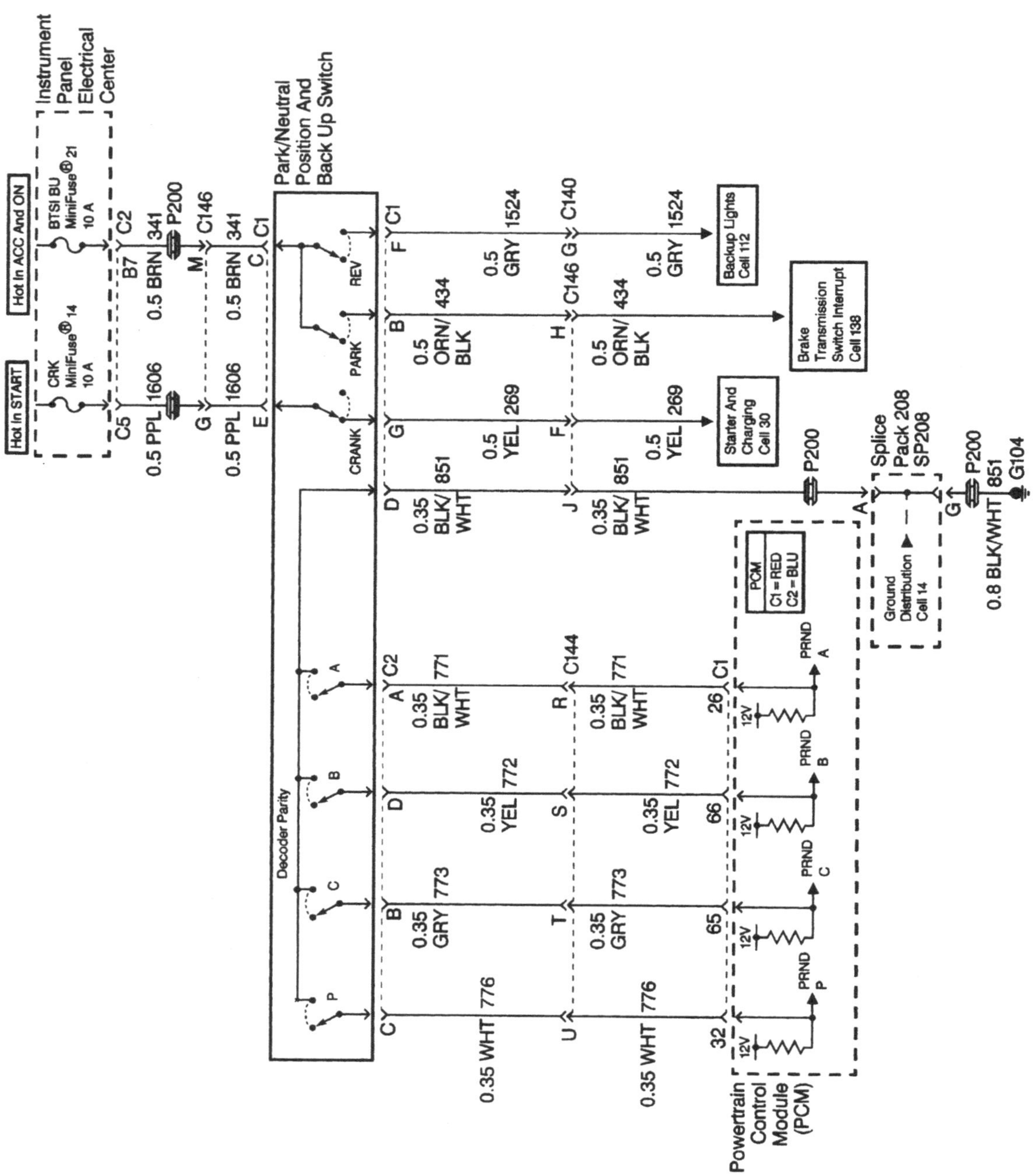

All EBTCM/BPMV Miscellaneous

Figure 27 of 30

LS1 1998-2001 VIN G

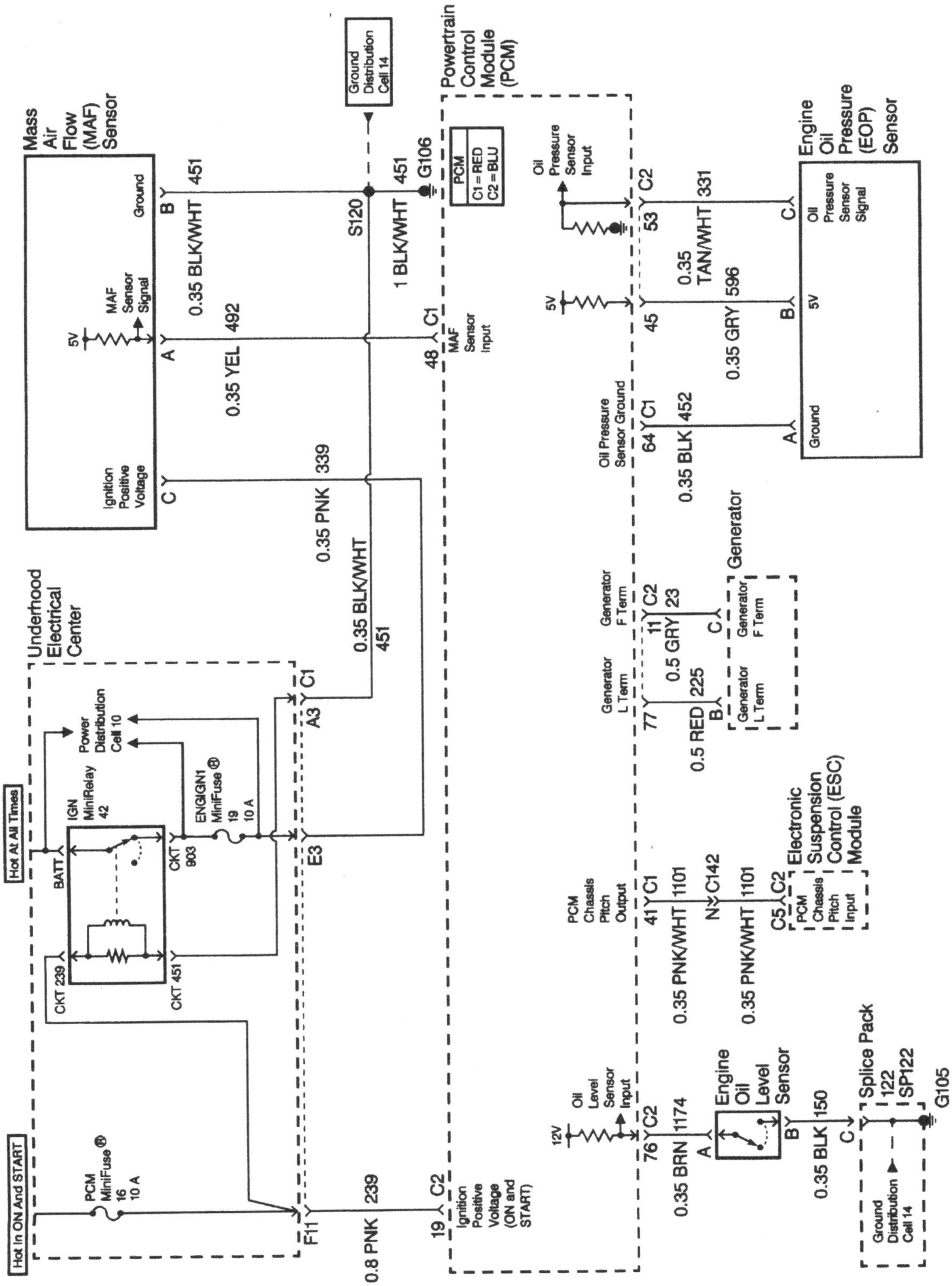

12

LS1 1998-2001 VIN G

Front-mounted EBTCM/BPMV A/C Compressor Cont

Figure 28 of 30

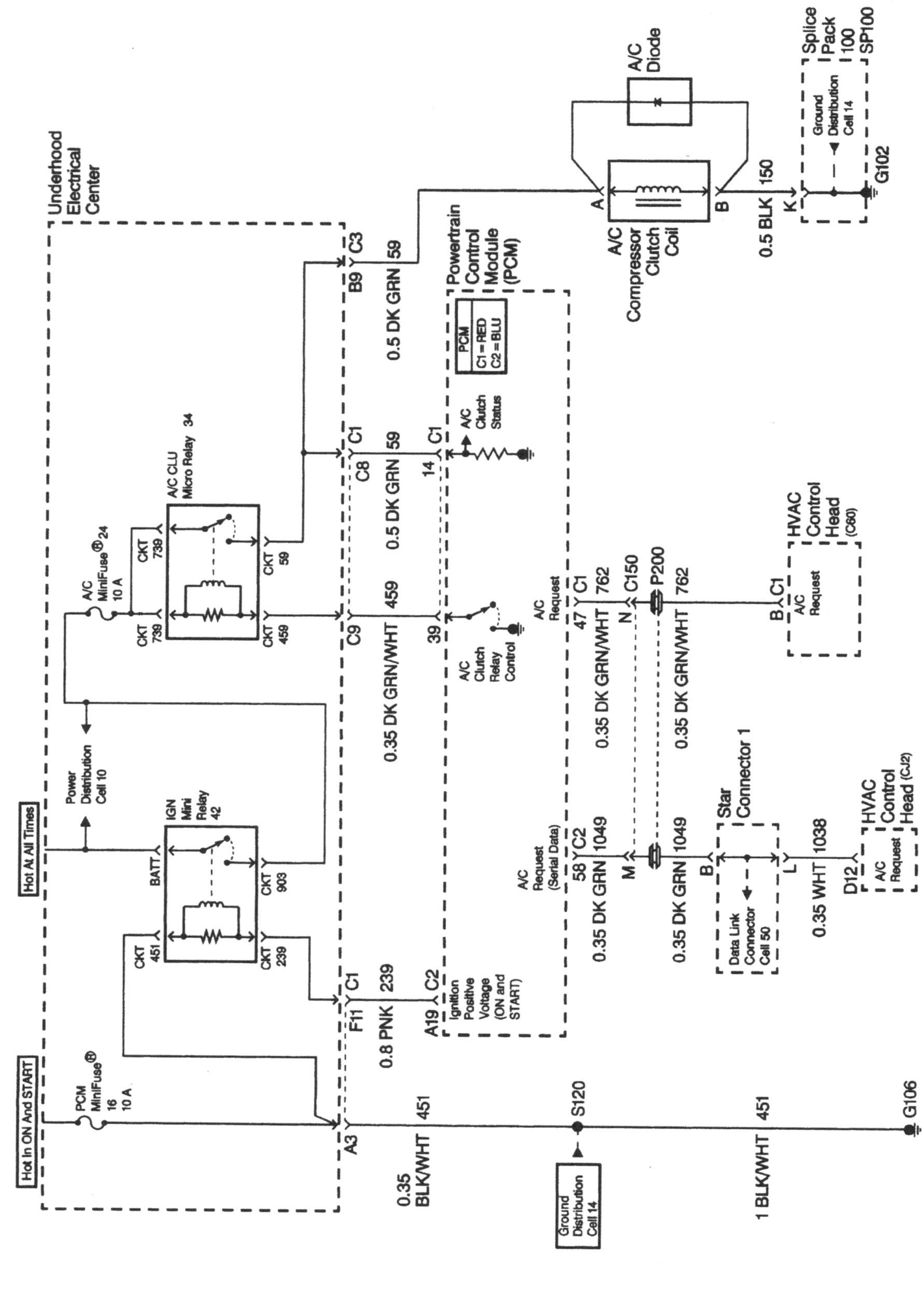

Rear-mounted EBTCM/BPMV
A/C Compressor Cont

Figure 29 of 30

LS1 1998-2001
VIN G

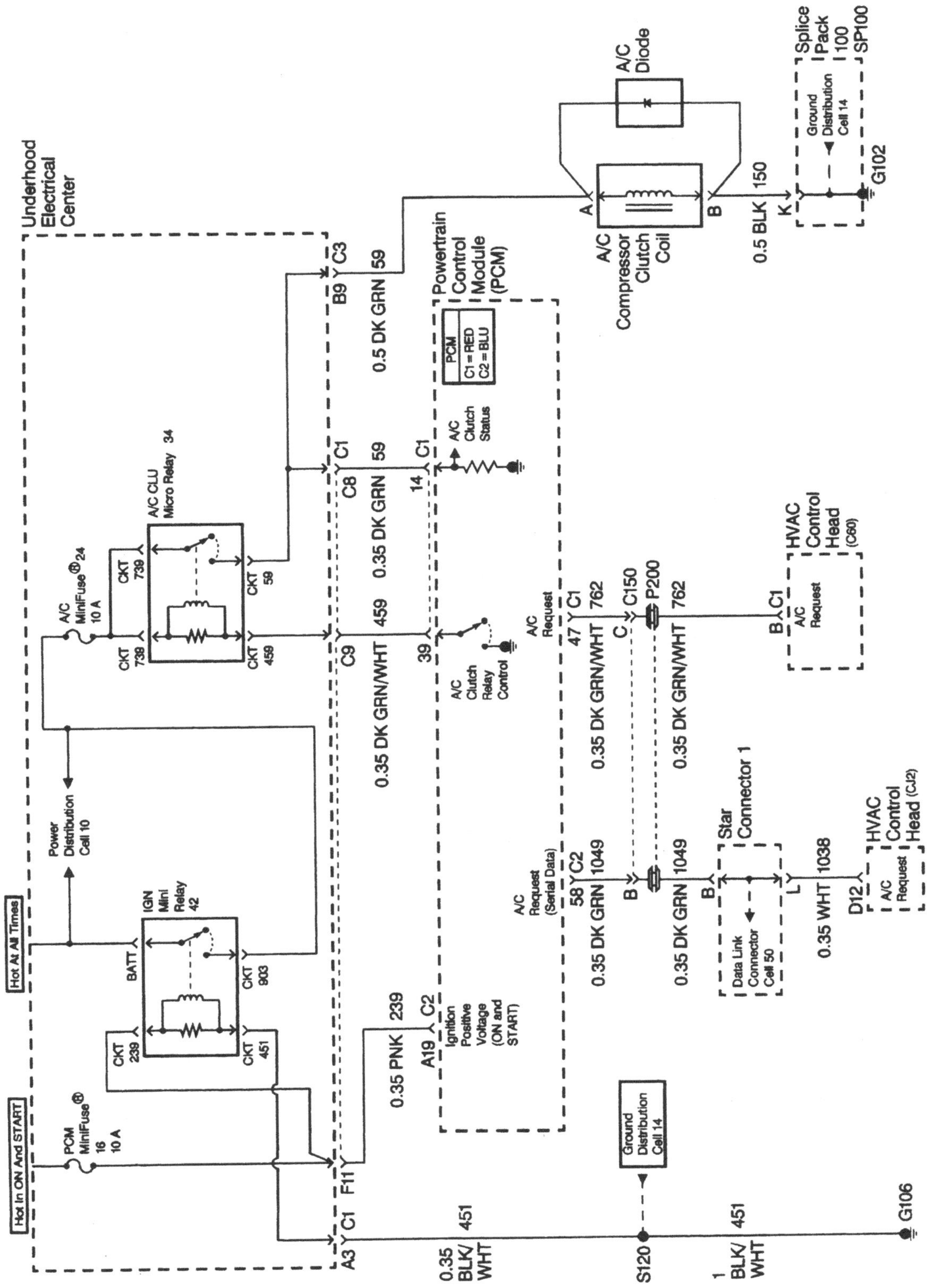

12

LS1 1998-2001 VIN G

All EBTCM/BPMV Cooling Fans

Figure 30 of 30

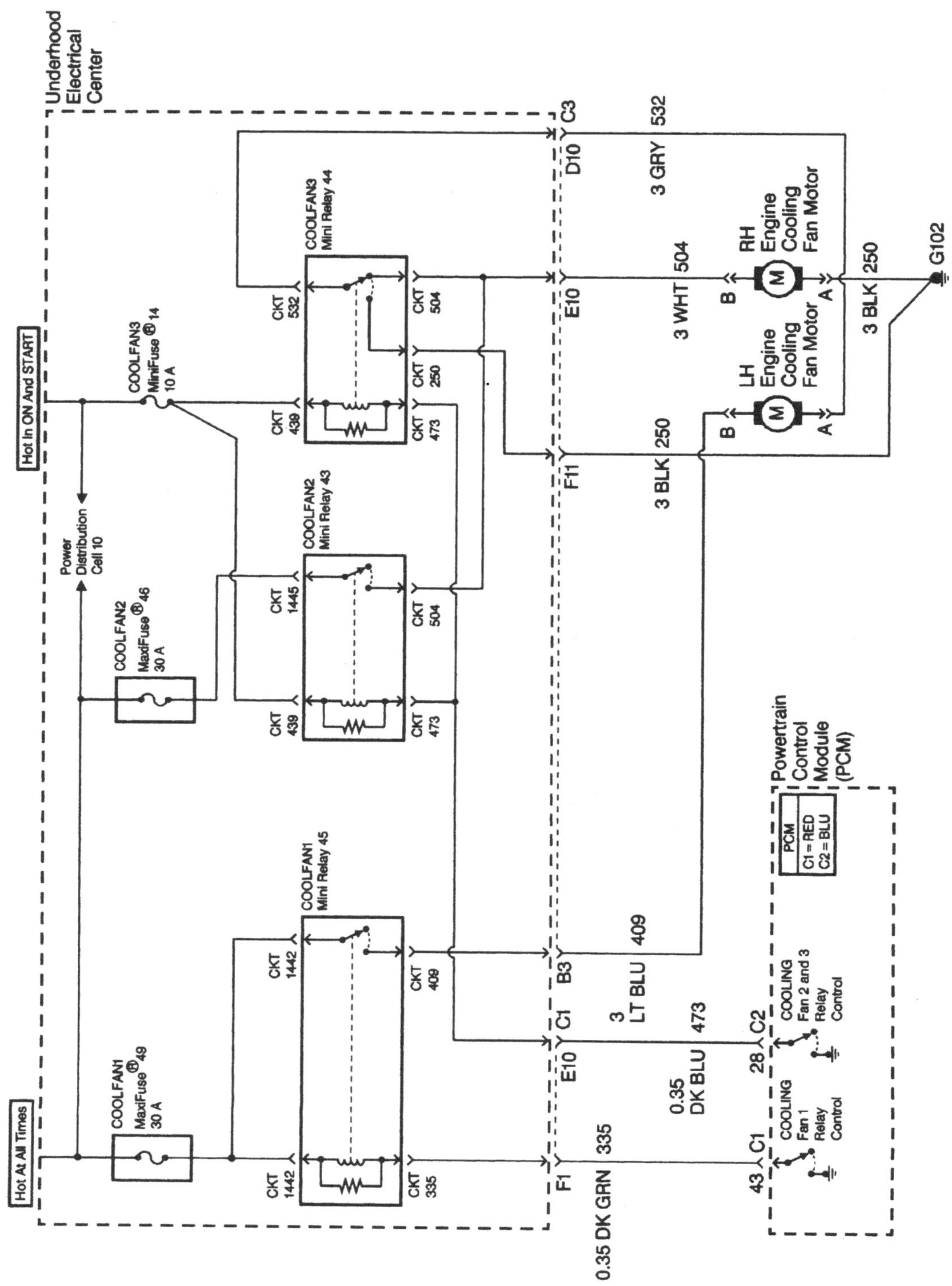

PCM C1 Red

Figure 1 of 15

LS1 1998-2001 VIN G

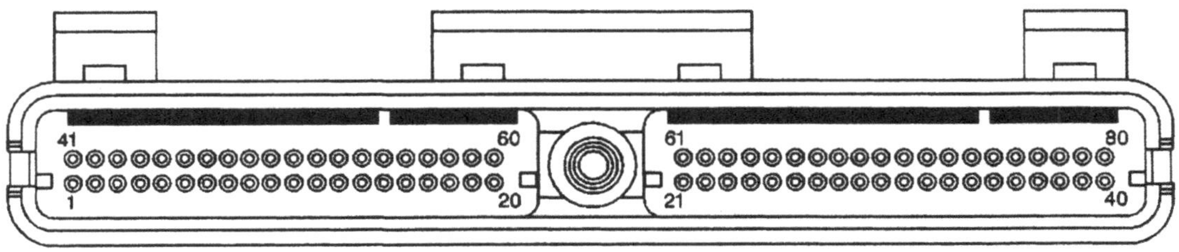

Connector Part Information		• PCM Connector C1 (Red) • 80 Pin Connector	
Pin	**Wire Color**	**Circuit No.**	**Function**
1–3	—	—	Not Used
4	BRN/WHT	633	Camshaft Position Sensor Signal
5–6	—	—	Not Used
7	YEL	400	VSS Signal High
8	—	—	Not Used
9	PNK/BLK	1746	Injector #3 Control
10–11	—	—	Not Used
12	LT BLU/BLK	844	Injector #4 Control
13	PPL/WHT	2122	Injector #2 Control
14	DK GRN	59	A/C Status
15	BLK/WHT	845	Injector #5 Control
16–17	—	—	Not Used
18	YEL/BLK	846	Injector #6 Control
19–20	—	—	Not Used
21	PPL	719	IAT Sensor Ground
22	DK BLU	1225	Transmission Range Signal B
23	LT GRN/BLK	1745	Injector #2 Control
24	BLK/WHT	1704	Sensor Ground Jumper
25	PNK	1224	Transmission Range Signal A
26	BLK/WHT	771	PRND A
27	RED/BLK	877	Injector #7 Control
28–29	—	—	Not Used
30	DK/BLU	449	Real Time Damping (If Equipped)
31	DK BLU/WHT	878	Injector #8 Control
32	WHT	776	PRND P
33	BLK	1744	Injector #1 Control
34	GRY or WHT	587 or 687	2nd and 3rd gear Block out Solenoid (M/T) or 3–2 Shift Solenoid (A/T)
35	LT GRN	1222	Shift Solenoid A
36	—	—	Not Used
37	BRN/WHT	419	MIL Control

LS1 1998-2001 VIN G

PCM C1 Red (continued)

Figure 2 of 15

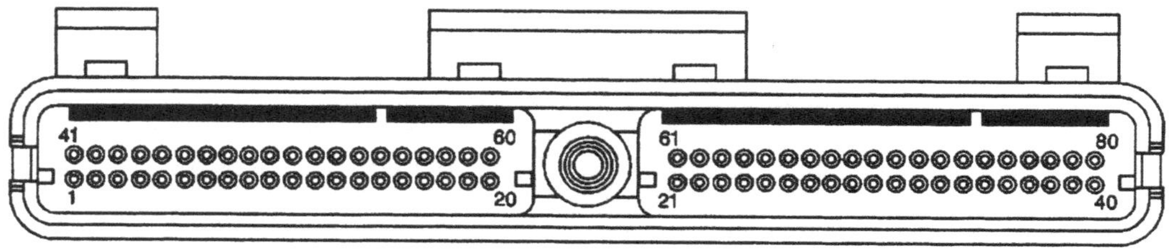

Connector Part Information	• PCM Connector C1 (Red) • 80 Pin Connector		
Pin	**Wire Color**	**Circuit No.**	**Function**
38	—	—	Not Used
39	DK GRN/WHT	459	A/C Clutch Relay Control
40	WHT or YEL/BLK	375 or 1223	1 to 4 Upshift Lamp Control (M/T) or Shift Solenoid B (A/T)
41	PNK/WHT	1101	PCM Chassis Pitch Output Circuit
42	LT GRN	1652	Reverse Inhibit Solenoid Control
43	DK/GRN	335	Cooling Fan Relay #1 Control
44	DK BLU/WHT	1869	Crankshaft Position Sensor Signal
45	DK GRN/WHT	428	EVAP Canister Purge Valve Control
46	ORN/BLK	463	Desired Torque
47	DK GRN/WHT	762	A/C Request Signal (C60)
48	YEL	492	MAF Signal
49	PPL	2121	Ignition #1 Control
50–51	—	—	Not Used
52	RED/WHT	2128	Ignition #8 Control
53	RED	2127	Ignition #7 Control
54	DK GRN/WHT	2126	Ignition #6 Control
55	DK GRN/WHT	817	VSS Output Signal
56	DK GRN	2125	Ignition #5 Control
57	LT BLU/WHT	2124	Ignition #4 Control
58	LT BLU	2123	Ignition #3 Control
59	BRN/WHT	2130	Ignition Control Reference Low Bank 2
60	BRN	2129	Ignition Control Reference Low Bank 1
61	RED	1226	Transmission Range Signal C
62	—	—	Not Used
63	BLK	407	A/C Refrigerant Pressure Sensor Ground
64	BLK	452	Engine Oil Pressure Sensor Ground
65	GRY	773	PRND C
66	YEL	772	PRND B
67	—	—	Not Used
68	GRY	48	Clutch Switch Signal
69	BRN	1982	EVAP Purge Vacuum Switch Signal

PCM C1 Red (continued) and C2 Blue

Figure 3 of 15

LS1 1998-2001 VIN G

PCM Connector C1 (Red) (cont'd)

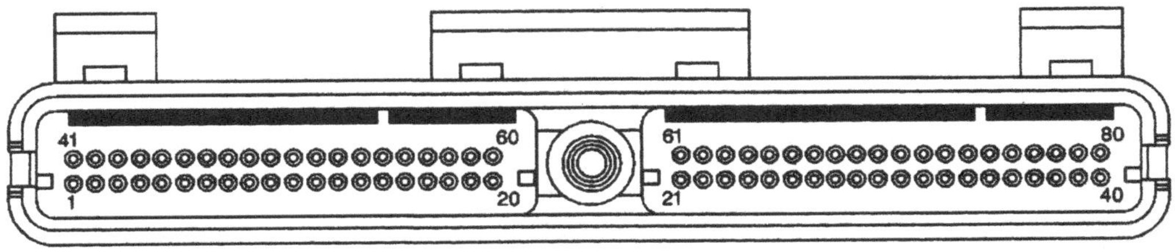

Connector Part Information		• PCM Connector C1 (Red) • 80 Pin Connector	
Pin	Wire Color	Circuit No.	Function
70	—	—	Not Used
71	PPL	401	VSS Signal Low
72	BLK/WHT	451C	PCM Ground
73	GRY	847	Extended Travel Brake Switch Signal
74	—	—	Not Used
75	GRY	720	Left and Right Fuel Tank Level Sensor Ground
76	BLK/WHT	451H	PCM Ground
77	PPL	420	Cruise/TCC Brake Switch Signal
78–80	—	—	Not Used

PCM Connector C2 (Blue)

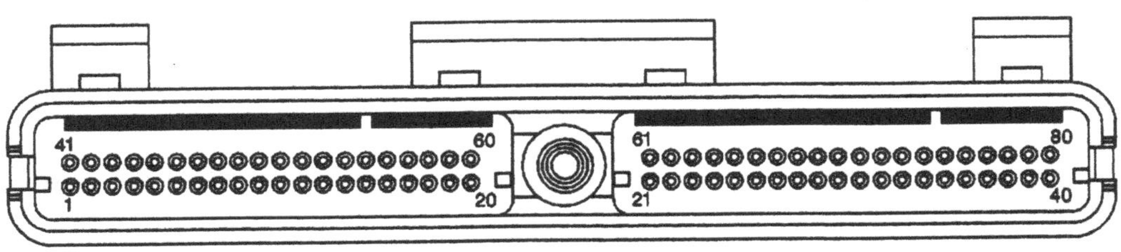

Connector Part Information		• PCM Connector C2 (Blue) • 80 Pin Connector	
Pin	Wire Color	Circuit No.	Function
1	TAN	1671	HO2S Signal Low Bank 2 Sensor 2
2	TAN/WHT	1669	HO2S Signal Low Bank 1 Sensor 2
3	—	—	Not Used
4	RED	631	Camshaft Position Sensor B+ Supply
5	TAN	1667	HO2S Signal Low Bank 2 Sensor 1
6	TAN/WHT	1653	HO2S Signal Low Bank 1 Sensor 1
7	RED/BLK	380	A/C Refrigerant Pressure Sensor Signal

LS1 1998-2001 VIN G

PCM C2 Blue (continued)

Figure 4 of 15

PCM Connector C2 (Blue) (cont'd)

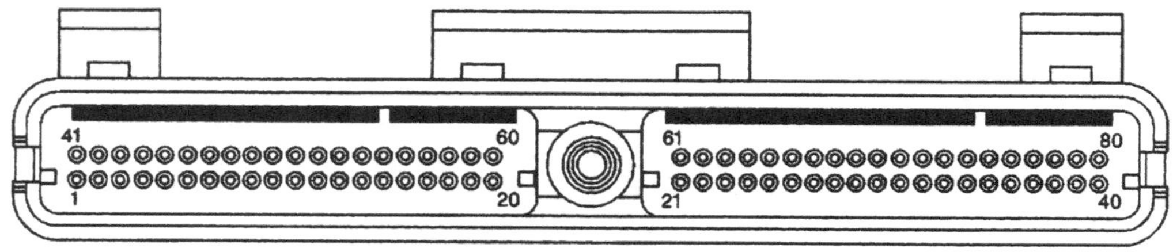

Connector Part Information		• PCM Connector C2 (Blue) • 80 Pin Connector	
Pin	**Wire Color**	**Circuit No.**	**Function**
8	TAN	472	IAT Sensor Signal
9	—	—	Not Used
10	DK BLU	1936	Fuel Level Sensor (Left Tank) Signal
11	GRY	23	Generator F Terminal
12	LT BLU	1937	Fuel Level Sensor (Right Tank) Signal
13	YEL/BLK	1227	Transmission Fluid Temperature Sensor Signal
14	ORN/BLK	1057	Transmission Fluid Temperature Sensor Ground
15–18	—	—	Not Used
19	PNK	239	PCM Ignition 1 Feed
20	ORN	340B	PCM Battery Feed
21	BLK/WHT	451F	PCM Ground
22	YEL/BLK	1868	Crankshaft Position Sensor Ground
23	—	—	Not Used
24	BLK/WHT	1704	Sensor Ground Jumper
25	BRN	718	ECT Sensor Ground
26	ORN/BLK	469	MAP Sensor Ground
27	—	—	Not Used
28	DK BLU	473	Engine Cooling Fan Relay #2 and #3 Control
29	—	—	Not Used
30	BRN	436	AIR Pump Relay Control
31	PPL	421	AIR Solenoid Relay Control
32	TAN/BLK	422	TCC Enable Solenoid
33	BRN	418	TCC Control (PWM) Solenoid Control
34	—	—	Not Used
35	WHT	121	Engine Speed (Tach) Output Signal
36	LT BLU/WHT	1229	Transmission Fluid Pressure Control Solenoid Control Low
37	DK GRN/WHT	465	Fuel Pump Relay Control
38	—	—	Not Used
39	RED/BLK	1228	Transmission Fluid Pressure Control Solenoid Control High
40	TAN/BLK	464	Torque Delivered
41	PPL	1670	HO2S Signal High Bank 2 Sensor 2
42	PPL/WHT	1668	HO2S Signal High Bank 1 Sensor 2

PCM C2 Blue (continued)

Figure 5 of 15

LS1 1998-2001 VIN G

PCM Connector C2 (Blue) (cont'd)

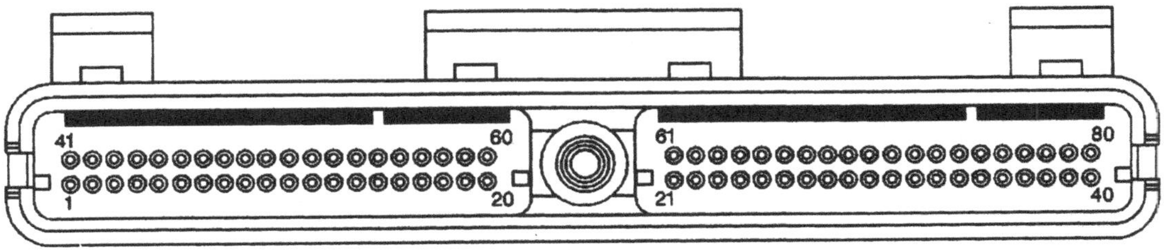

Connector Part Information		• PCM Connector C2 (Blue) • 80 Pin Connector	
Pin	Wire Color	Circuit No.	Function
43	—	—	Not Used
44	GRY	598	Fuel Level Sensors 5.0 Volt Feed
45	GRY	596	Engine Oil Pressure Sensor 5.0 Volt Reference
46	PPL	1666	HO2S Signal High Bank 2 Sensor 1
47	PPL/WHT	1665	HO2S Signal High Bank 1 Sensor 1
48	LT GRN	432	MAP Sensor Signal
49	YEL	410	ECT Sensor Signal
50–52	—	—	Not Used
53	TAN/WHT	331	Engine Oil Pressure Sensor Signal
54–57	—	—	Not Used
58	DK GRN	1049	Serial Data Class 2
59	—	—	Not Used
60	BLK/WHT	451E	PCM Ground
61	ORN	340C	PCM Battery Feed
62	LT GRN	1867	Crankshaft Position Sensor B+ Supply
63	—	—	Not Used
64	GRY	416	MAP Sensor 5.0 Volt Reference
65	—	—	Not Used
66	GRY	474	A/C Refrigerant Pressure Sensor 5.0 Volt Reference
67	—	—	Not Used
68	LT BLU	1876	Knock Sensor (KS) Rear
69	DK BLU	496	Knock Sensor (KS) Front
70	PNK/BLK	632	Camshaft Position Sensor Ground
71	TAN	800	TAC Module Serial Data
72	ORN/BLK	1061	TAC Module Serial Data
73–75	—	—	Not Used
76	BRN	1174	Oil Level Sensor Signal
77	RED	225	Generator L Terminal
78–80	—	—	Not Used

LS1 1998-2001 VIN G

TAC Module Connector C1

Figure 6 of 15

TAC Module Connector C1

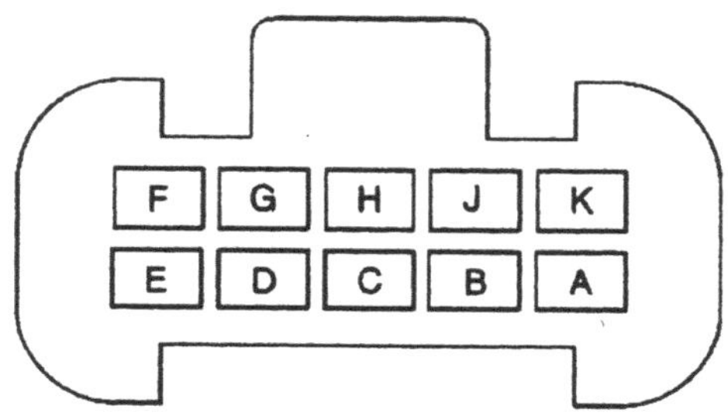

Connector Part Information		• TAC Module Connector C1 • 10 Pin Connector	
Pin	Wire Color	Circuit No.	Function
A	GRY	1273	APP #3 Ground
B	PPL	1272	APP #2 Ground
C	LT BLU	1162	APP #2 Signal
D	TAN	1274	APP #2 5.0 Volt Reference
E	YEL/BLK	1275	APP #3 5.0 Volt Reference
F	DK BLU	1161	APP #1 Signal
G	LT BLU	1276	APP #1 5.0 Volt Reference
H	—	—	Not Used
J	BRN	1271	APP #1 Ground
K	DK GRN	1163	APP #3 Signal

TAC Module Connector C2

Figure 7 of 15

TAC Module Connector C2

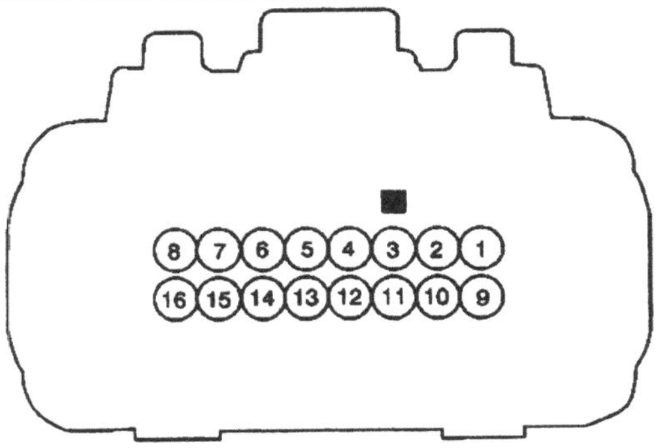

Connector Part Information		• TAC Module Connector C2 • 16 Pin Connector	
Pin	**Wire Color**	**Circuit No.**	**Function**
1	DK BLU	417	TP Sensor #1 Signal
2	DK GRN	485	TP Sensor #1 5.0 Volt Reference
3	PPL	486	TP Sensor #1 Ground
4	DK BLU	84	Cruise Set/Coast Signal
5	GRY/BLK	87	Cruise Resume/Accel Signal
6	LT BLU	20	Stop Lamp Switch Signal
7	PNK	539	Battery Positive Voltage Supply
8	BRN	582	Throttle Actuator Control Motor #2
9	YEL/BLK	487	TP Sensor #2 5.0 Volt Reference
10	WHT	484	TP Sensor #2 Ground
11	PNK	427	TP Sensor #2 Signal
12	TAN	800	Serial Data
13	ORN/BLK	1061	Serial Data
14	GRY	397	Cruise On/Off Signal
15	BLK/WHT	451	Ground
16	YEL	581	Throttle Actuator Control Motor #1

LS1 1998-2001 VIN G

CMP, CKP, KS, Left Fuel Tank
Figure 8 of 15

Engine Controls Connector End Views

Camshaft Position (CMP) Sensor

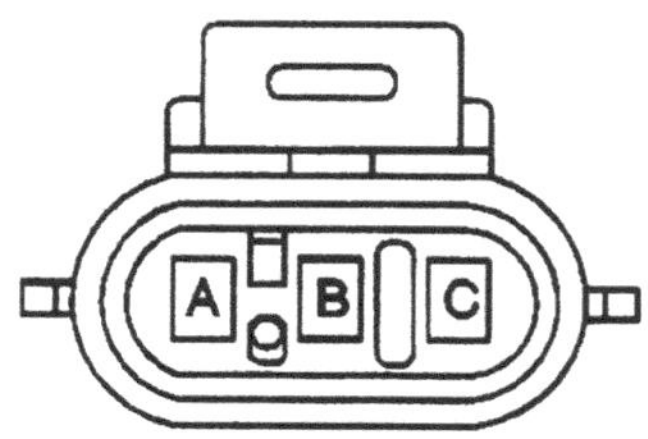

Connector Part Information	• 1205-9595 • 3 Way F Metri-Pack 150 Series (BLK)		
Pin	**Wire Color**	**Circuit No.**	**Function**
A	BRN/WHT	633	Camshaft Position (CMP) Sensor - Signal
B	PNK/BLK	632	Reference Low
C	RED	631	Ignition Feed

Crankshaft Position (CKP) Sensor

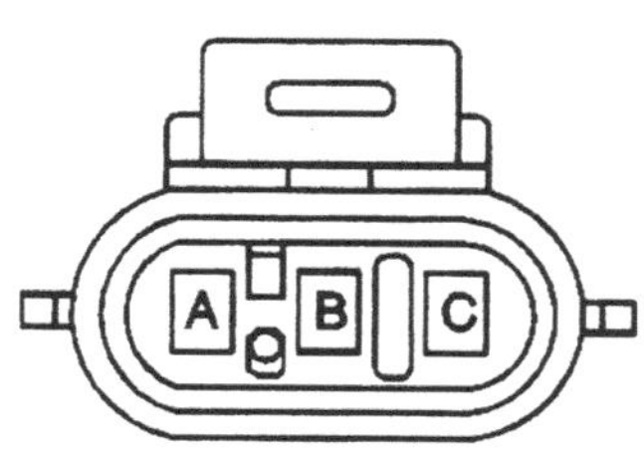

Connector Part Information	• 1205-9595 • 3 Way Metri-Pack 150 Series (BLK)		
Pin	**Wire Color**	**Circuit No.**	**Function**
A	DK BLU/WHT	1869	Crankshaft Position (CKP) Sensor - Signal
B	YEL/BLK	1868	Reference Low
C	LT GRN	1867	Ignition Feed

Knock Sensors (KS)

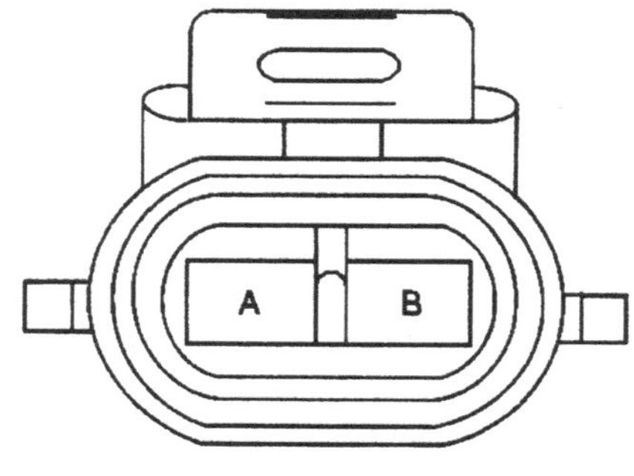

Connector Part Information	• 1205-2641 • 2 Way F Metri-Pack 150 Series Sealed (BLK)		
Pin	**Wire Color**	**Circuit No.**	**Function**
A	DK BLU	496	Knock Sensor (KS) Signal - Front
B	LT BLU	1876	Knock Sensor (KS) Signal - Rear

Left Fuel Tank

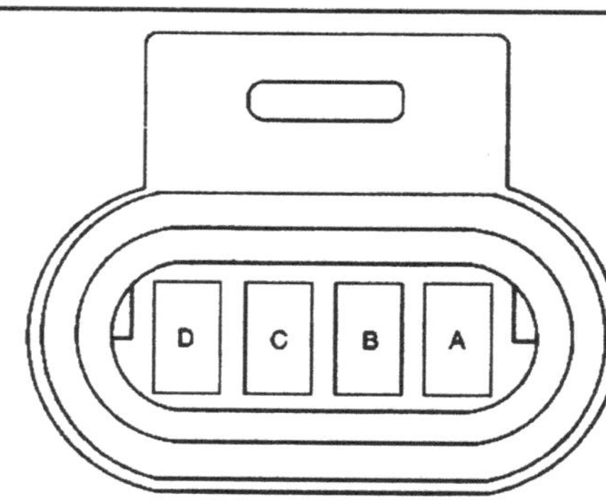

Connector Part Information	• 1216-2188 • 4 Way F Metri-Pack 1500.2 Series Pull To Seat		
Pin	**Wire Color**	**Circuit No.**	**Function**
A	GRY	120	Left Fuel Pump Motor - Feed
B	GRY	720	Sensor Return
C	DK BLU	1936	Left Fuel Level Sensor - Signal
D	BLK	150	Ground

Right Fuel Tank, Fuel Injector #1, #3, #5

Figure 9 of 15

LS1 1998-2001 VIN G

Right Fuel Tank

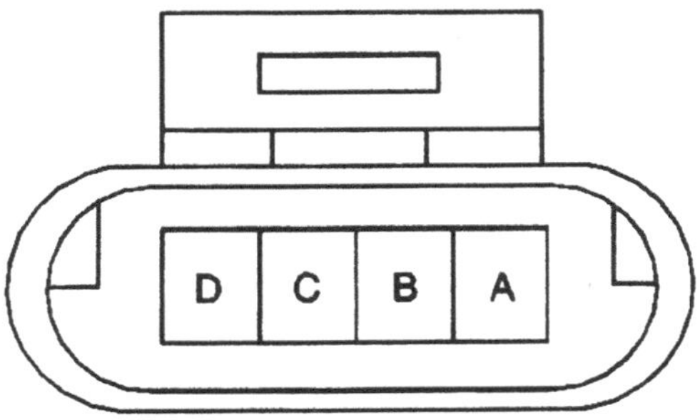

Connector Part Information	• 1216-2189 • 4 Way F Metri-Pack 150.2 Series Pull To Seat (BLK)		
Pin	**Wire Color**	**Circuit No.**	**Function**
A	GRY/BLK	598	5.0V Reference
B	GRY	720	Sensor Return
C	LT BLU	1937	Right Fuel Level Sensor - Signal
D	BLK	150	Ground

Fuel Injector #1

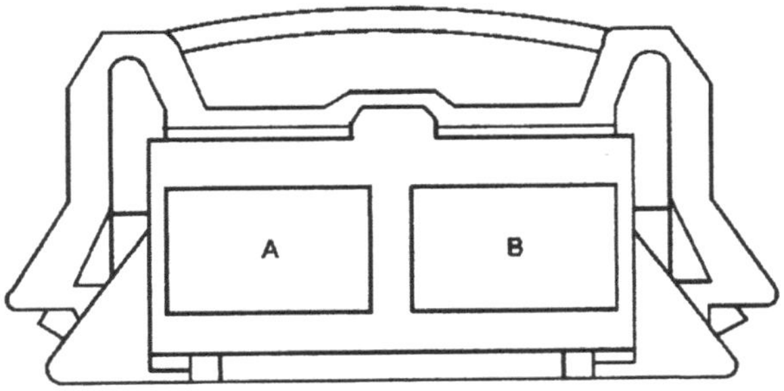

Connector Part Information	• 1212-9140 • 2 Way F		
Pin	**Wire Color**	**Circuit No.**	**Function**
A	PNK	639	Ignition (ON and START)
B	BLK	1744	Fuel Injector #1 Control

Fuel Injector #3

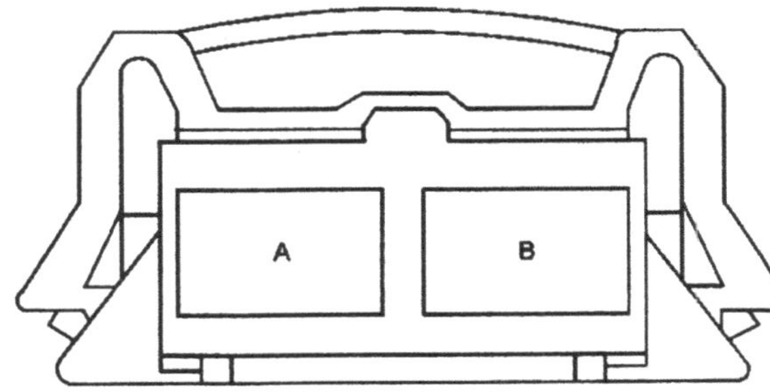

Connector Part Information	• 1212-9140 • 2 Way F		
Pin	**Wire Color**	**Circuit No.**	**Function**
A	PNK	639	Ignition (ON and START)
B	PNK/BLK	1746	Fuel Injector #3 Control

Fuel Injector #5

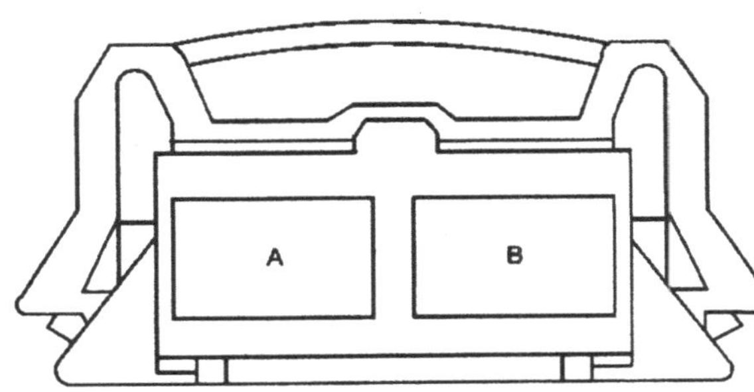

Connector Part Information	• 1212-9140 • 2 Way F		
Pin	**Wire Color**	**Circuit No.**	**Function**
A	PNK	639	Ignition (ON and START)
B	BLK/WHT	845	Fuel Injector #5 Control

LS1 1998-2001 VIN G

Fuel Injector # 8, TPS, IAT, MAP

Figure 10 of 15

Fuel Injector #8

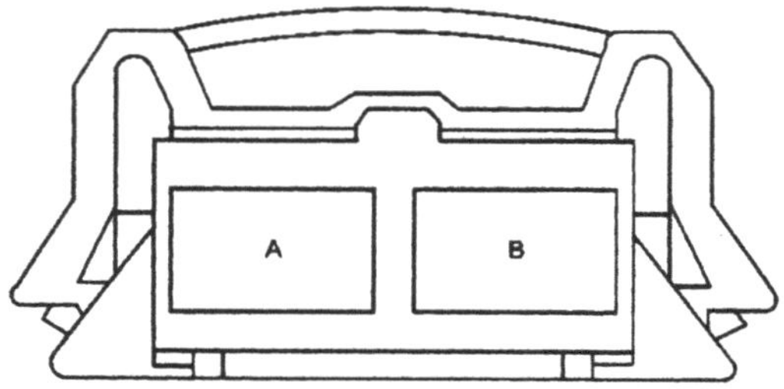

Connector Part Information	• 1212-9140 • 2 Way F		
Pin	**Wire Color**	**Circuit No.**	**Function**
A	PNK	839	Ignition (ON and START)
B	DK BLU/WHT	878	Fuel Injector #8 Control

Intake Air Temperature (IAT) Sensor

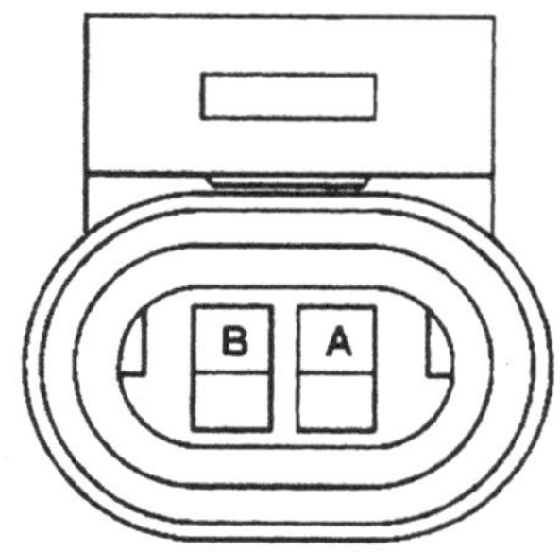

Connector Part Information	1216-2199		
Pin	**Wire Color**	**Circuit No.**	**Function**
A	PPL	719	Intake Air Temperature (IAT) Sensor - Ground
B	TAN	472	Intake Air Temperature (IAT) Sensor - Signal

Throttle Positon (TP) Sensor

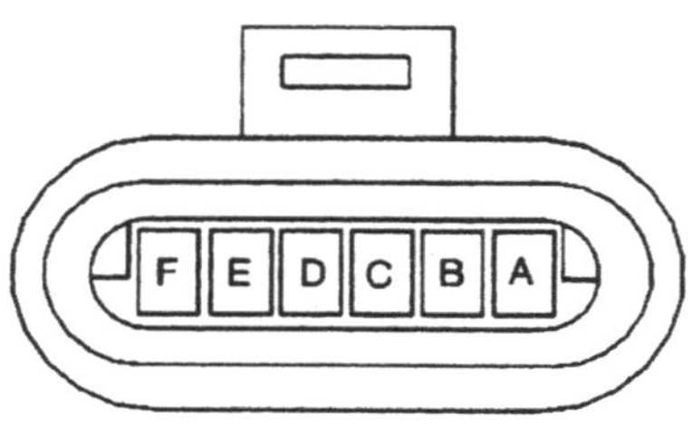

Connector Part Information	• 1216-2261 • 6 Way F Metri-Pack 150.2 Series Pull To Seat (BLK)		
Pin	**Wire Color**	**Circuit No.**	**Function**
A	DK GRN	485	Throttle Position (TP) Sensor #1 - 5V Reference
B	PPL	486	Throttle Position (TP) Sensor #1 - Ground
C	DK BLU	417	Throttle Position (TP) Sensor #1 - Signal
D	YEL/BLK	487	Throttle Position (TP) Sensor #2 - 5V Reference
E	WHT	484	Throttle Position (TP) Sensor #2 - Ground
F	PNK	427	Throttle Position (TP) Sensor #2 - Signal

Manifold Absolute Pressure (MAP) Sensor

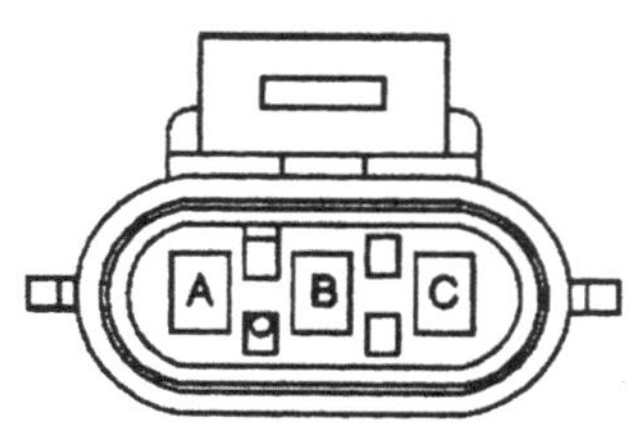

m

Connector Part Information	• 1212-9946 • 3 Way F Metri Pack 150 Series (GRY)		
Pin	**Wire Color**	**Circuit No.**	**Function**
A	ORN/BLK	469	Manifold Absolute Pressure (MAP) Sensor - Ground
B	LT GRN	432	Manifold Absolute Pressure (MAP) Sensor - Signal
C	GRY	416	Manifold Absolute Pressure (MAP) Sensor - 5.0V Reference

A/C Refrig. Pressure Sensor, EVAP Canister Purge Valve, AIR pump, LF HO_2S

Figure 11 of 15

LS1 1998-2001 VIN G

A/C Refrigerant Pressure Sensor

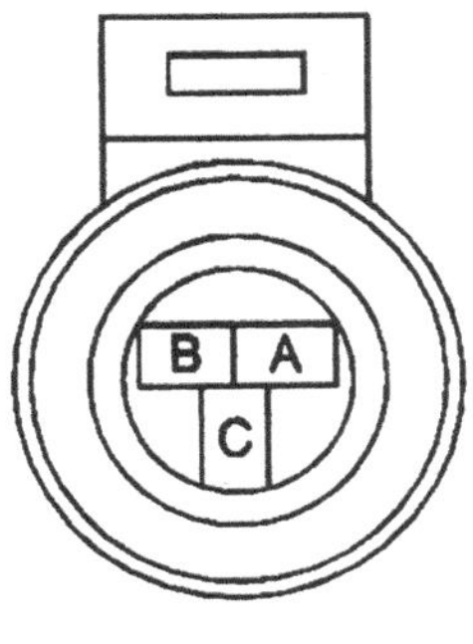

Connector Part Information	1211-0192		
Pin	**Wire Color**	**Circuit No.**	**Function**
A	BLK	407	A/C Refrigerant Pressure Sensor - Return
B	GRY	474	A/C Refrigerant Pressure Sensor -5.0V Reference
C	RED/BLK	380	A/C Refrigerant Pressure Sensor - Signal

Secondary Air Injection (AIR) Pump

Connector Part Information	• 1211-0293 • 3 Way F Metri-Pack 150 Series (BLK)		
Pin	**Wire Color**	**Circuit No.**	**Function**
A	RED	78	Secondary Air Injection Pump Motor - Feed
B	PNK/BLK	429	Secondary Air Injection Solenoid - Feed
C	BLK	150	Ground

Evaporative Emission (EVAP) Canister Purge Valve

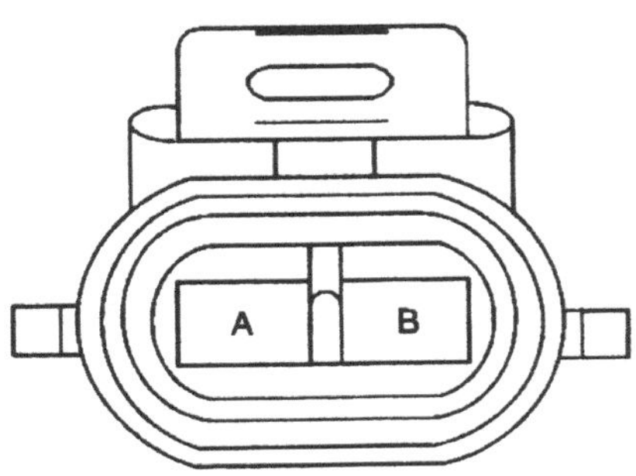

Connector Part Information	• 1205-2643 • 2 Way Metri-Pack 150 Series Sealed (RED)		
Pin	**Wire Color**	**Circuit No.**	**Function**
A	PNK	339	Ignition (Hot W/IGN Relay Energized)
B	DK GRN/WHT	428	EVAP Canister Purge Valve Control

LF Heated Oxygen Sensor (HO2S)

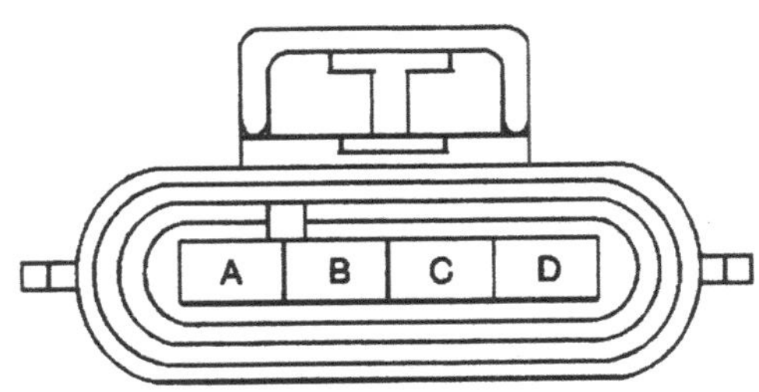

Connector Part Information	• 1218-6694 • 4 Way Metri-Pack 150 Series (BLK)		
Pin	**Wire Color**	**Circuit No.**	**Function**
A	TAN/WHT	1653	LF Heated Oxygen Sensor (HO2S) - Low
B	PPL/WHT	1665	LF Heated Oxygen Sensor (HO2S) - High
C	BLK	150	Ground
D	BRN	241	Ignition (ON)

LS1 1998-2001 VIN G

LR HO_2S, RF HO_2S, RR HO_2S, Clutch Switch

Figure 12 of 15

LR Heated Oxygen Sensor (HO2S)

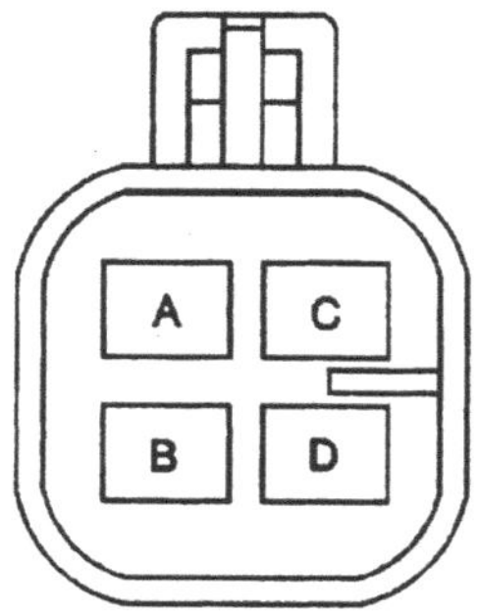

Connector Part Information	• 1216-0825 • 4 Way Metri-Pack 150 Series (BLK)		
Pin	**Wire Color**	**Circuit No.**	**Function**
A	TAN/WHT	1669	LR Heated Oxygen Sensor (HO2S) - Low
B	PPL/WHT	1668	LR Heated Oxygen Sensor (HO2S) - High
C	BLK	150	Ground
D	BRN	241	Ignition (ON)

RR Heated Oxygen Sensor (HO2S)

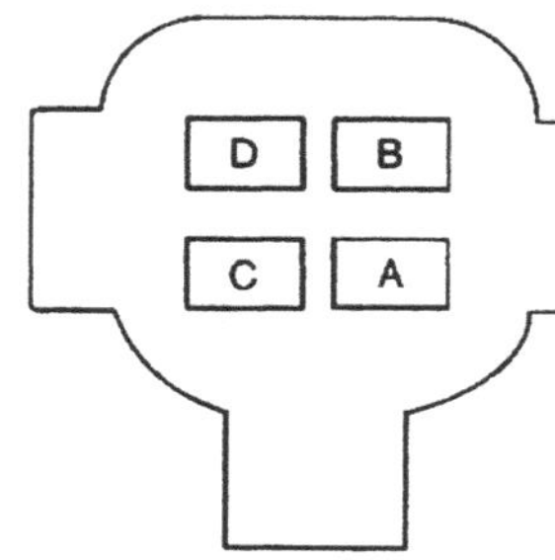

Connector Part Information	• 1216-0825 • 4 Way Metri-Pack 150 Series (BLK)		
Pin	**Wire Color**	**Circuit No.**	**Function**
A	TAN	1671	RR Heated Oxygen Sensor (HO2S) - Low
B	PPL	1670	RR Heated Oxygen Sensor (HO2S) - High
C	BLK	150	Ground
D	BRN	241	Ignition (ON)

RF Heated Oxygen Sensor (HO2S)

Connector Part Information	• 1218-6694 • 4 Way Metri-Pack 150 Series (BLK)		
Pin	**Wire Color**	**Circuit No.**	**Function**
A	TAN	1667	Heated Oxygen Sensor (HO2S) - Low
B	PPL	1666	Heated Oxygen Sensor (HO2S) - High
C	BLK	150	Ground
D	BRN	241	Ignition (ON)

Clutch Switch

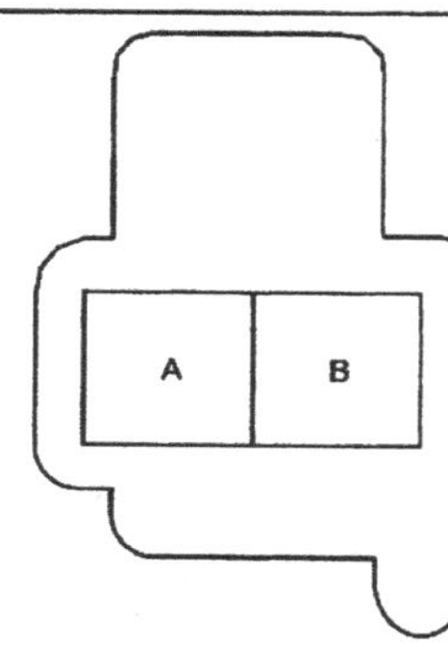

Connector Part Information	• 1204-1433 • 2 Way		
Pin	**Wire Color**	**Circuit No.**	**Function**
A	GRY	48	Clutch Switch Signal
B	PNK	339	Ignition Feed

2nd and 3rd Blockout Solenoid, Reverse Inhibit Solenoid, MAF, Engine Oil Level Sensor

Figure 13 of 15

LS1 1998-2001 VIN G

2nd and 3rd Gear Blockout Solenoid

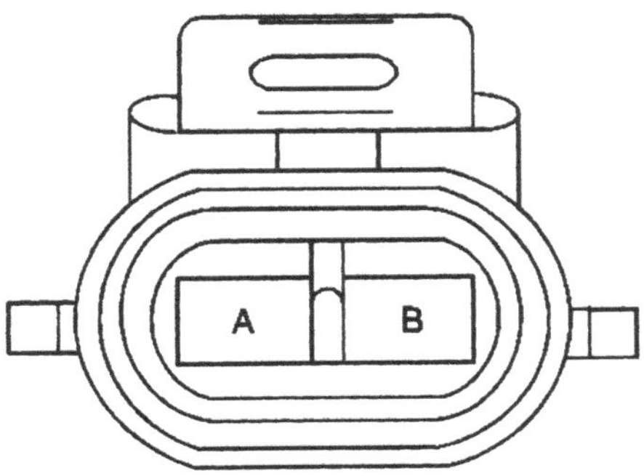

Connector Part Information	• 1205-2646 • 2 Way F Metri-Pack 150 Series Sealed (WHT)		
Pin	**Wire Color**	**Circuit No.**	**Function**
A	GRY	587	2nd and 3rd Gear Blockout Solenoid - Output
B	PNK	339	Ignition Feed

Reverse Inhibit Solenoid

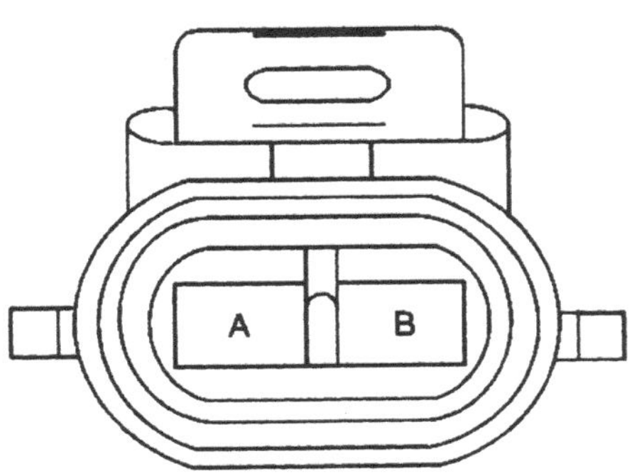

Connector Part Information	• 1205-2646 • 2 Way F Metri-Pack 150 Series Sealed		
Pin	**Wire Color**	**Circuit No.**	**Function**
A	LT GRN	1652	Reverse Inhibit Solenoid - Output
B	PNK	339	Ignition Feed

Mass Air Flow (MAF) Sensor

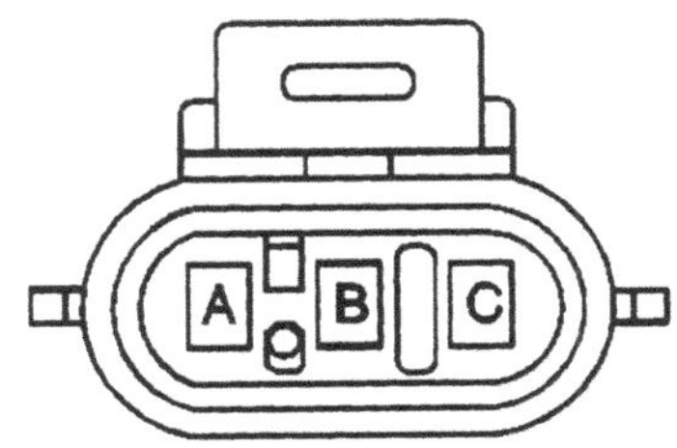

Connector Part Information	• 1205-9595 • 3 Way F Metri-Pack 150 Series (BLK)		
Pin	**Wire Color**	**Circuit No.**	**Function**
A	YEL	492	Mass Air Flow (MAF) Sensor - Signal
B	BLK/WHT	451	Ground
C	PNK	339	Ignition Feed

Engine Oil Level Sensor

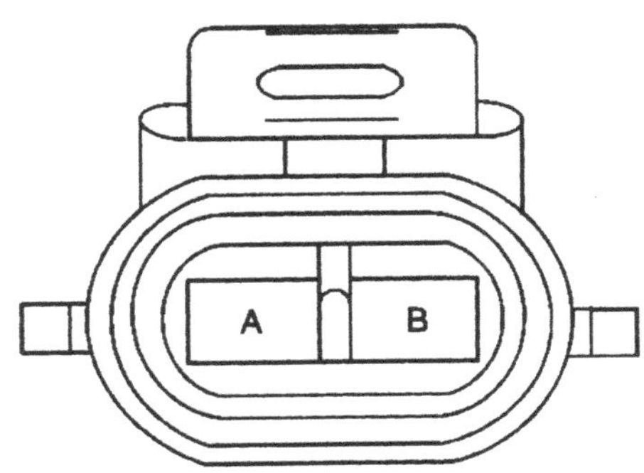

Connector Part Information	• 1205-2641 • 2 Way F Metri-Pack 150 Series Sealed		
Pin	**Wire Color**	**Circuit No.**	**Function**
A	BRN	1174	Engine Oil Level Sensor - Signal
B	BLK	150	Ground

LS1 1998-2001 VIN G

Engine Oil Pressure Sensor, ECT Sensor, A/T Park/Neutral/Back-Up Switch C-1

Figure 14 of 15

Engine Oil Pressure Sensor

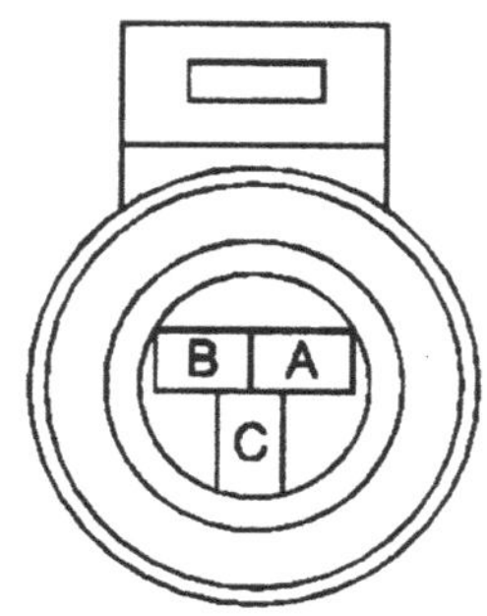

Connector Part Information		• 1211-0192 • 3 Way F Metri-Pack 150 Series	
Pin	**Wire Color**	**Circuit No.**	**Function**
A	BLK	452	Engine Oil Pressure (EOP) Sensor - Return
B	GRY	596	5.0V Reference
C	TAN/WHT	331	Engine Oil Pressure (EOP) Sensor - Signal

Engine Coolant Temperature Sensor

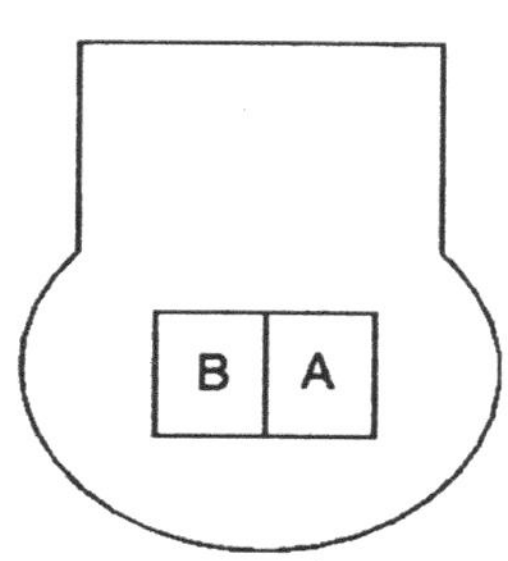

Connector Part Information		• 1216-2195 • 2 Way F Metri-Pack 150.2 Series (BLK)	
Pin	**Wire Color**	**Circuit No.**	**Function**
A	BRN	718	Engine Coolant Temperature (ECT) Sensor - Ground
B	YEL	410	Engine Coolant Temperature (ECT) Sensor - Signal

Park/Neutral Position and Back Up Switch – C1

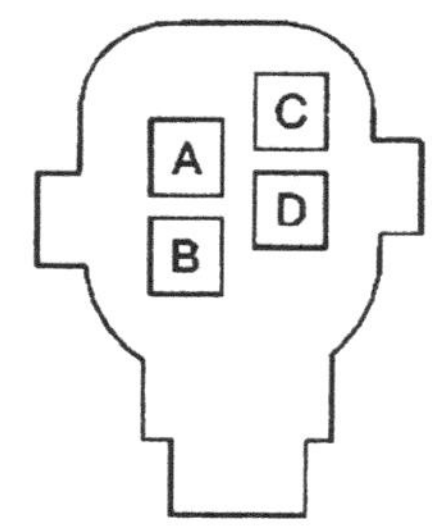

Connector Part Information		• 1212-9804 • 4 Way F Metri-Pack 150 Series (LT GRY)	
Pin	**Wire Color**	**Circuit No.**	**Function**
A	BLK/WHT	771	PRND - A
B	GRY	773	PRND - C
C	WHT	776	PRND - P
D	YEL	772	PRND - B

A/T Park/Neutral Switch C-2, EVAP vacuum purge switch

Figure 15 of 15

Park/Neutral Position Switch – C2

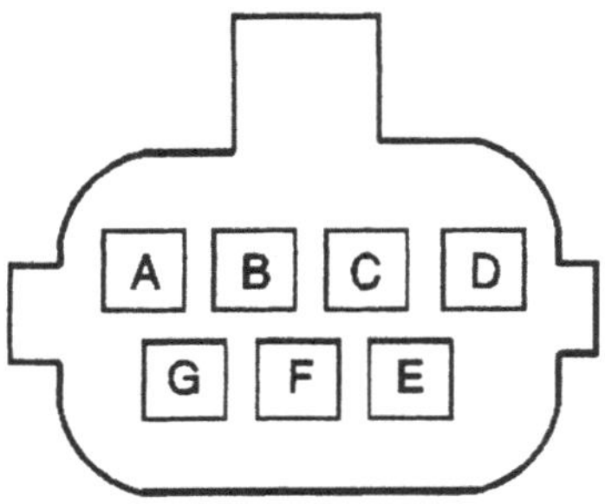

Connector Part Information	• 1212-9841 • 7 Way F Metri-Pack Mixed Series (GRY)		
Pin	**Wire Color**	**Circuit No.**	**Function**
A	–	–	Not Used
B	ORN/BLK	434	Trans Mounted Neutral Safety Switch Signal
C	BRN	341	Ignition Feed
D	BLK/WHT	851	Ground
E	PPL	1606	Ignition Feed
F	GRY	1524	Back - Up Lamp Feed - Monitored
G	YEL	269	Theft Deterrent Relay Feed - Coil

Evaporative Emission (EVAP) Vacuum Purge Switch

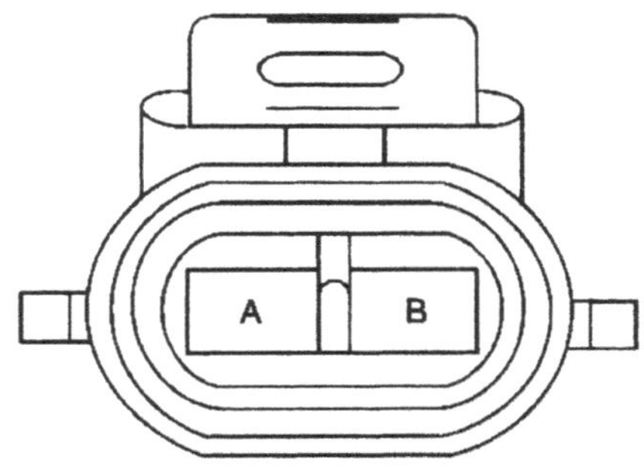

Connector Part Information	• 1205-2641 • 2Way F Metri-Pack 150 Series (BLK)		
Pin	**Wire Color**	**Circuit No.**	**Function**
A	BLK/WHT	451	Ground
B	BRN	1982	EVAP Vacuum Purge Switch - Signal

LS1 1997 VIN G

Power, Ground, DLC, MIL

Figure 1 of 16

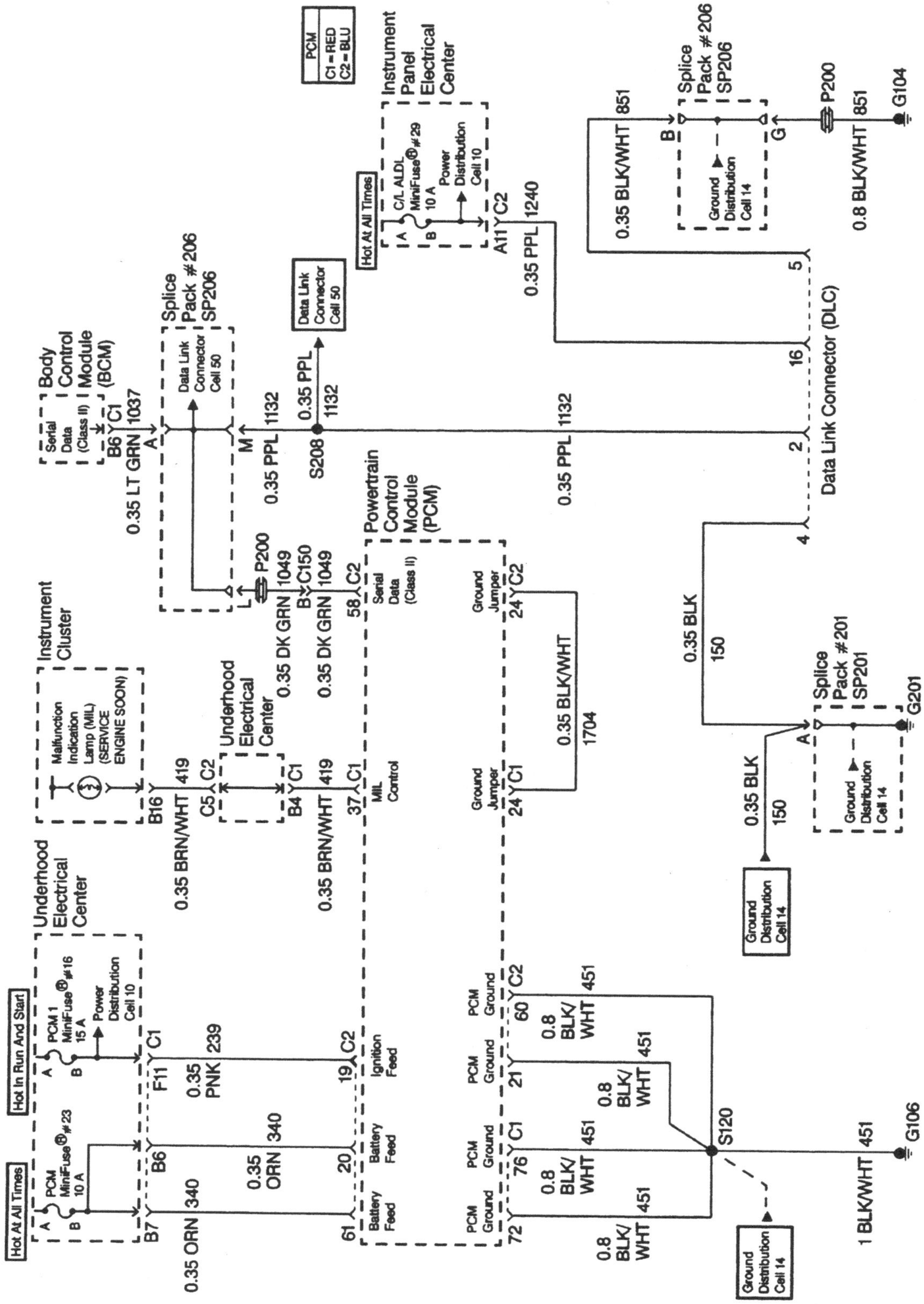

Ignition, Bank 1

Figure 2 of 16

LS1 1997
VIN G

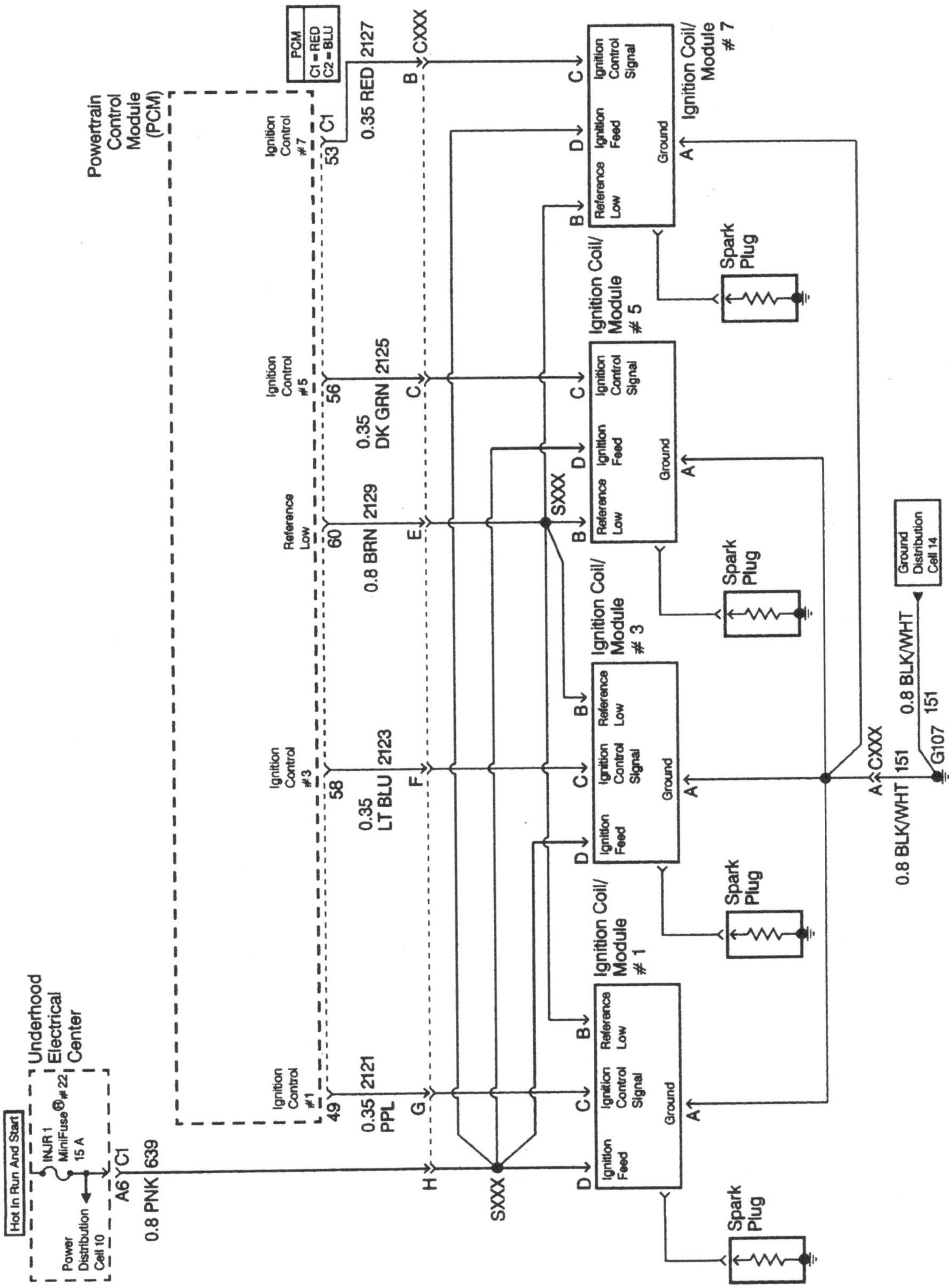

12

LS1 1997
VIN G

Ignition, Bank 2

Figure 3 of 16

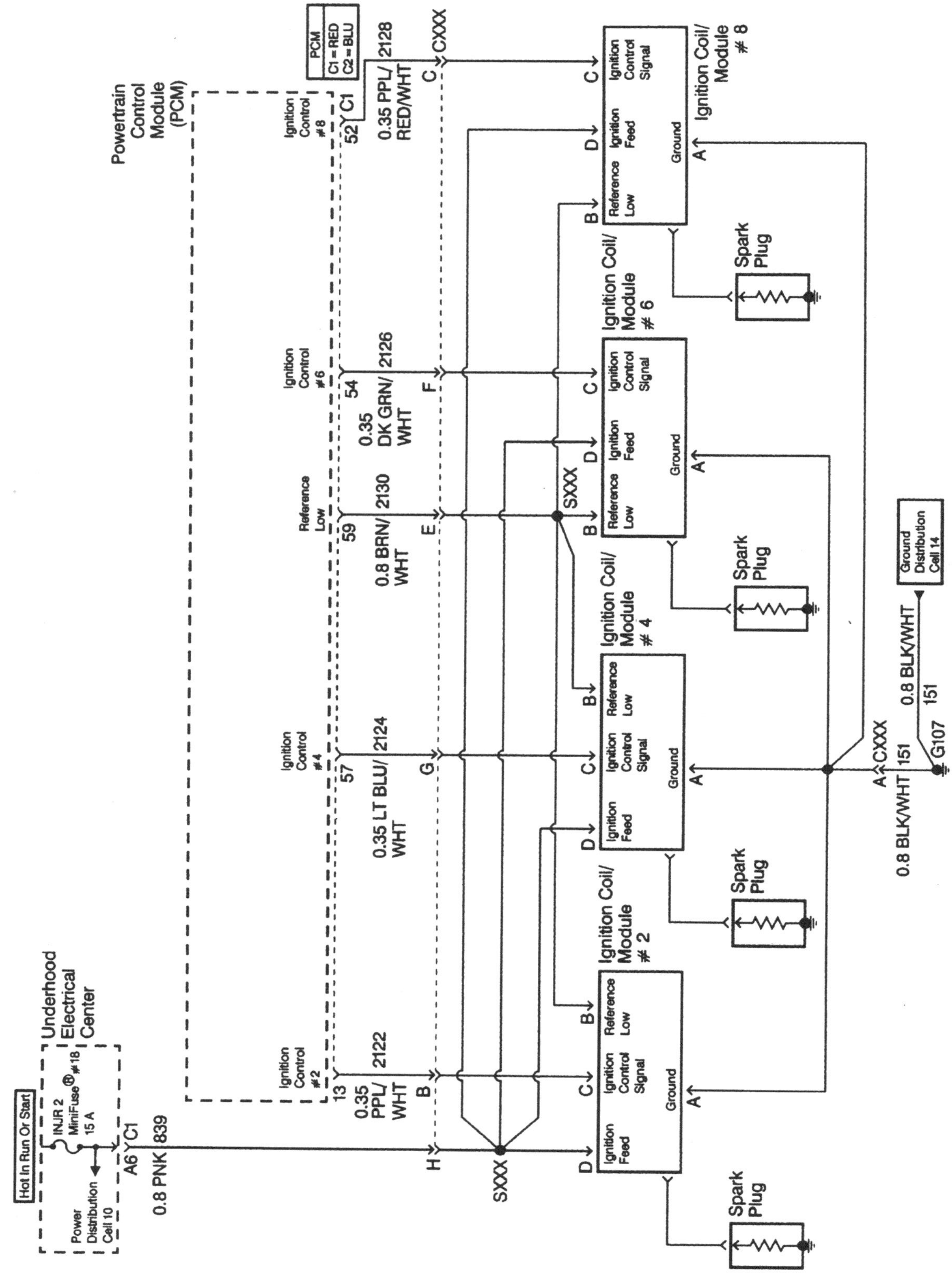

CKP, CMP, KS, Traction Control

LS1 1997
VIN G

Figure 4 of 16

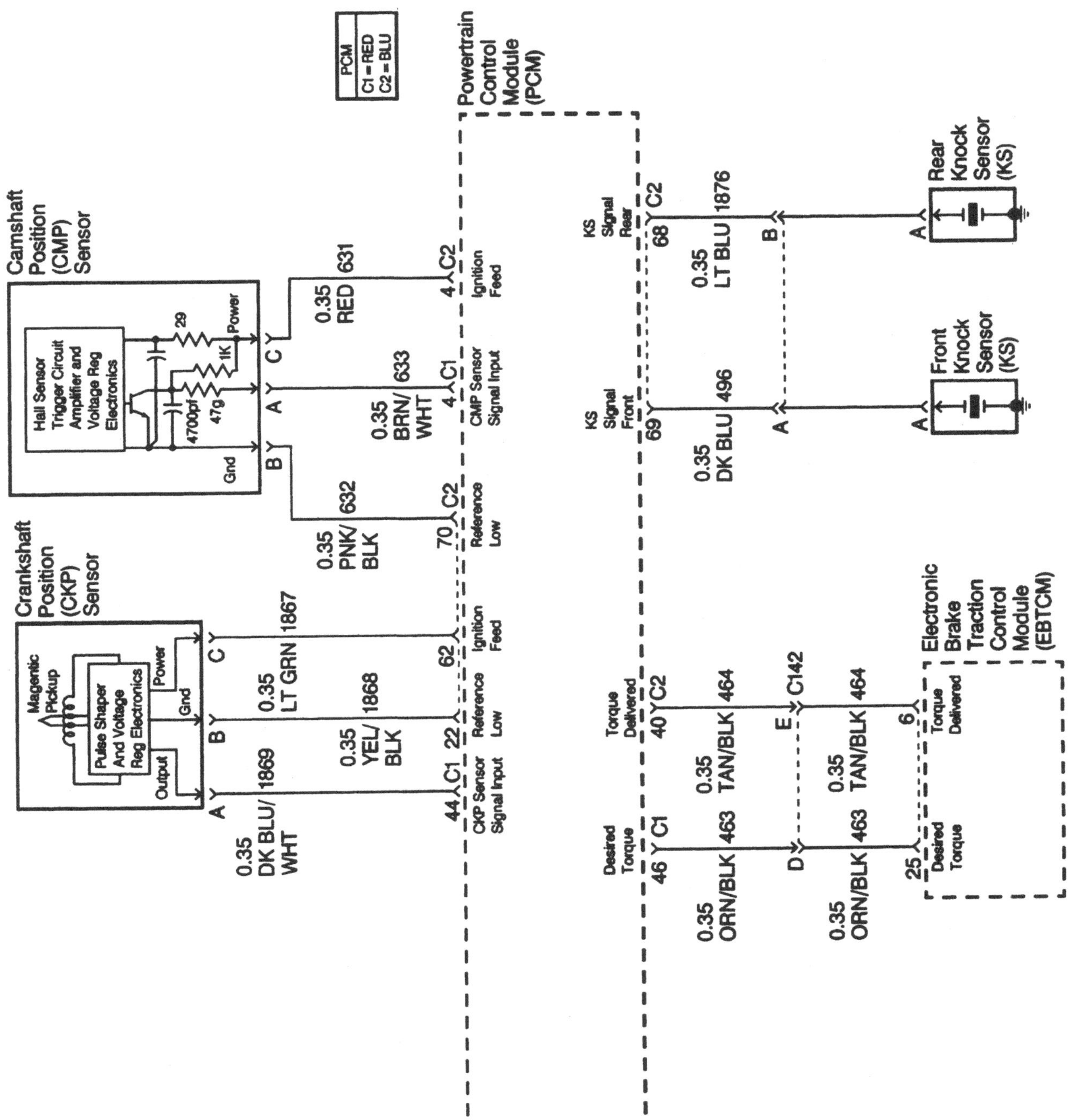

12

LS1 1997 VIN G

Fuel Pump and Fuel Level Sensors

Figure 5 of 16

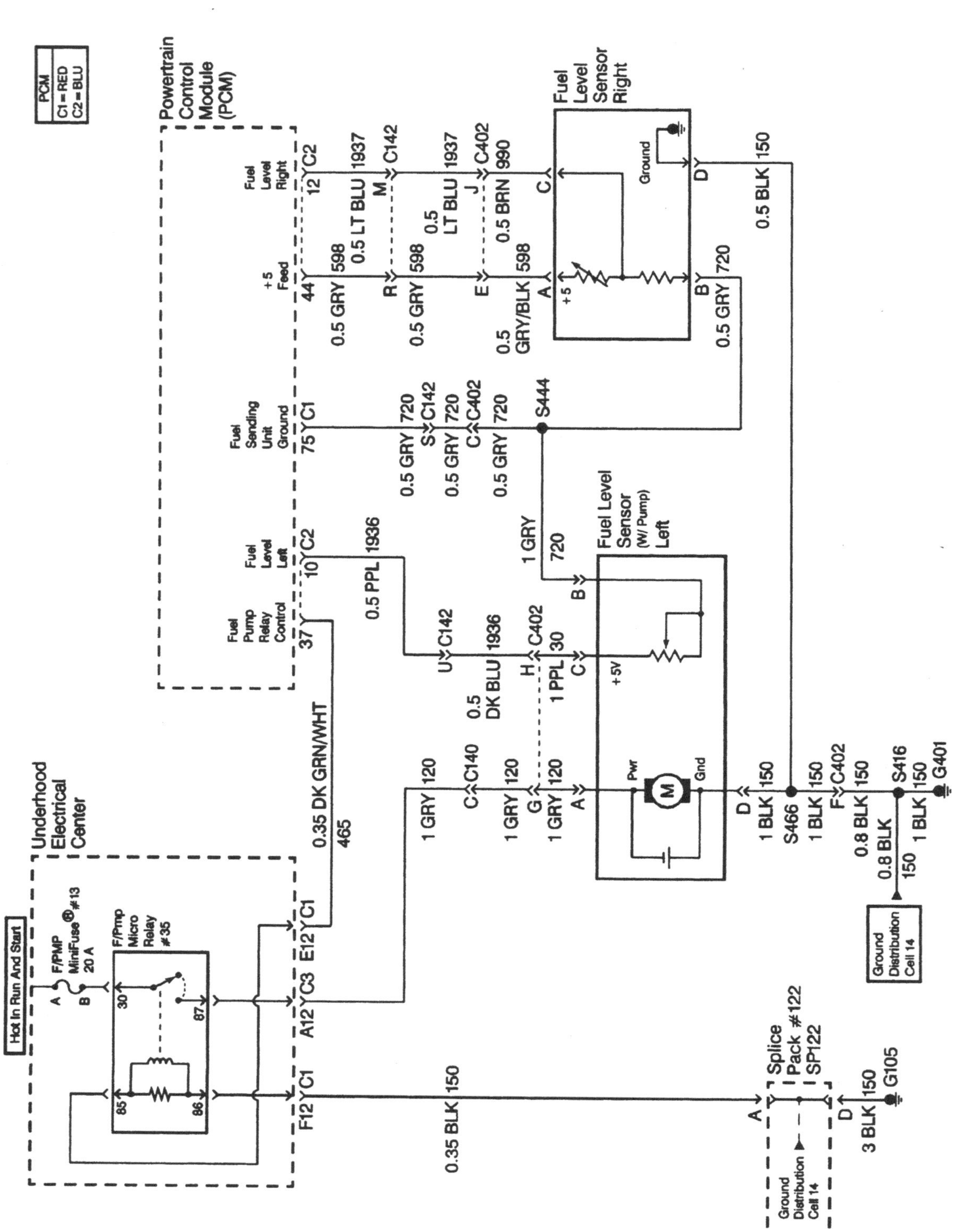

Injectors

Figure 6 of 16

LS1 1997
VIN G

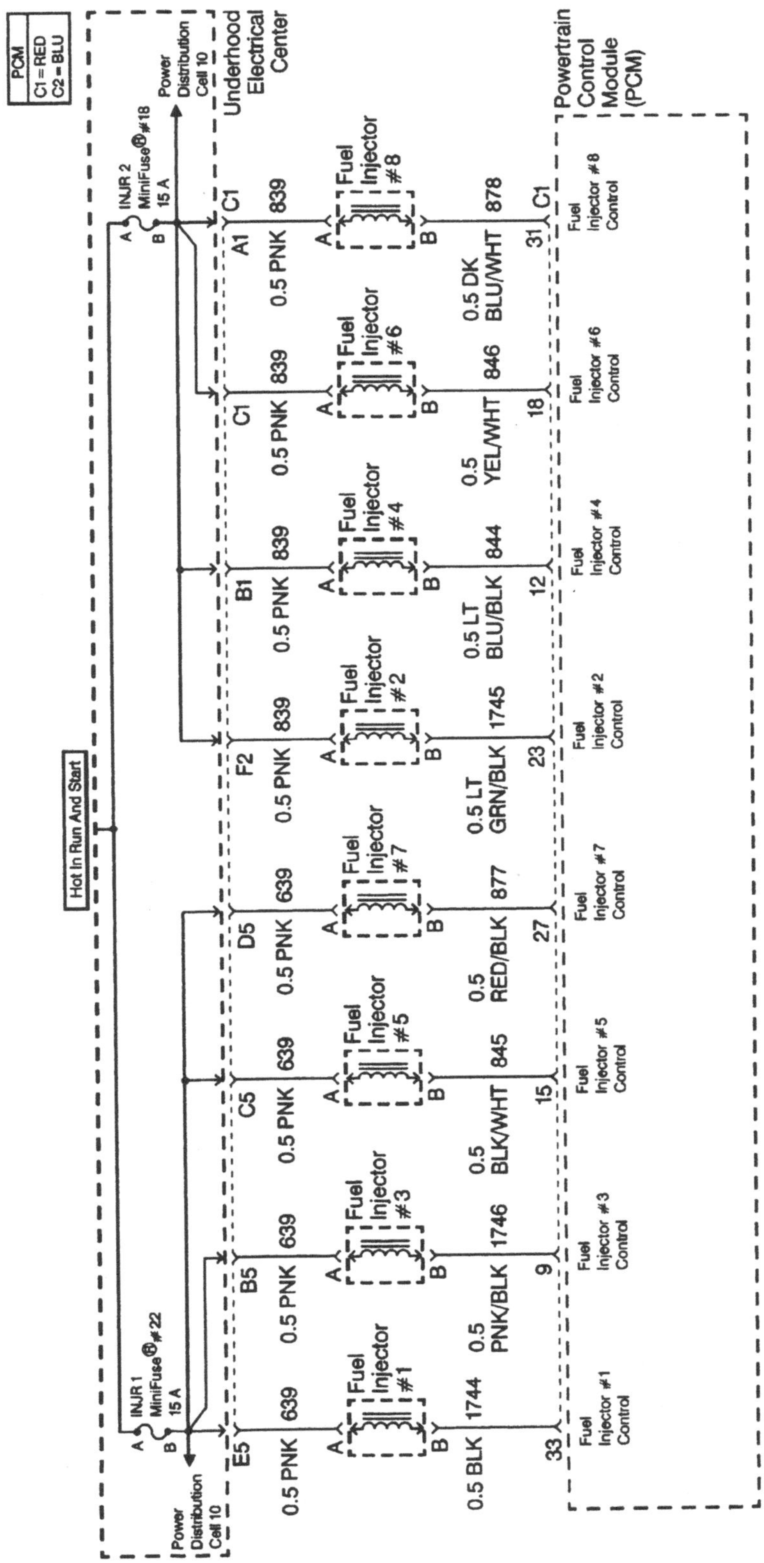

LS1 1997 VIN G

Cruise Control and Stop Lamp

Figure 7 of 16

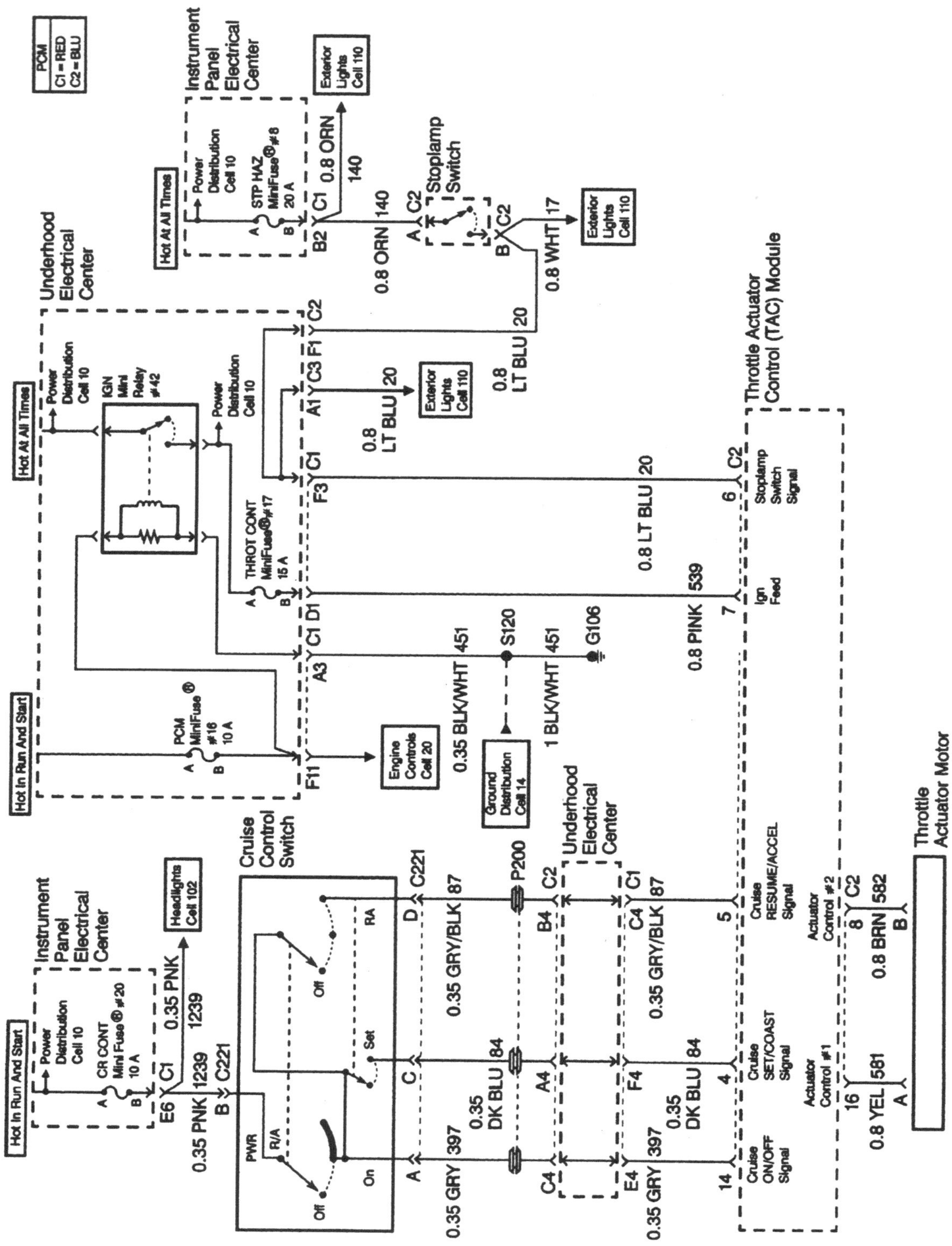

TP and APP PSensors

Figure 8 of 16

LS1 1997
VIN G

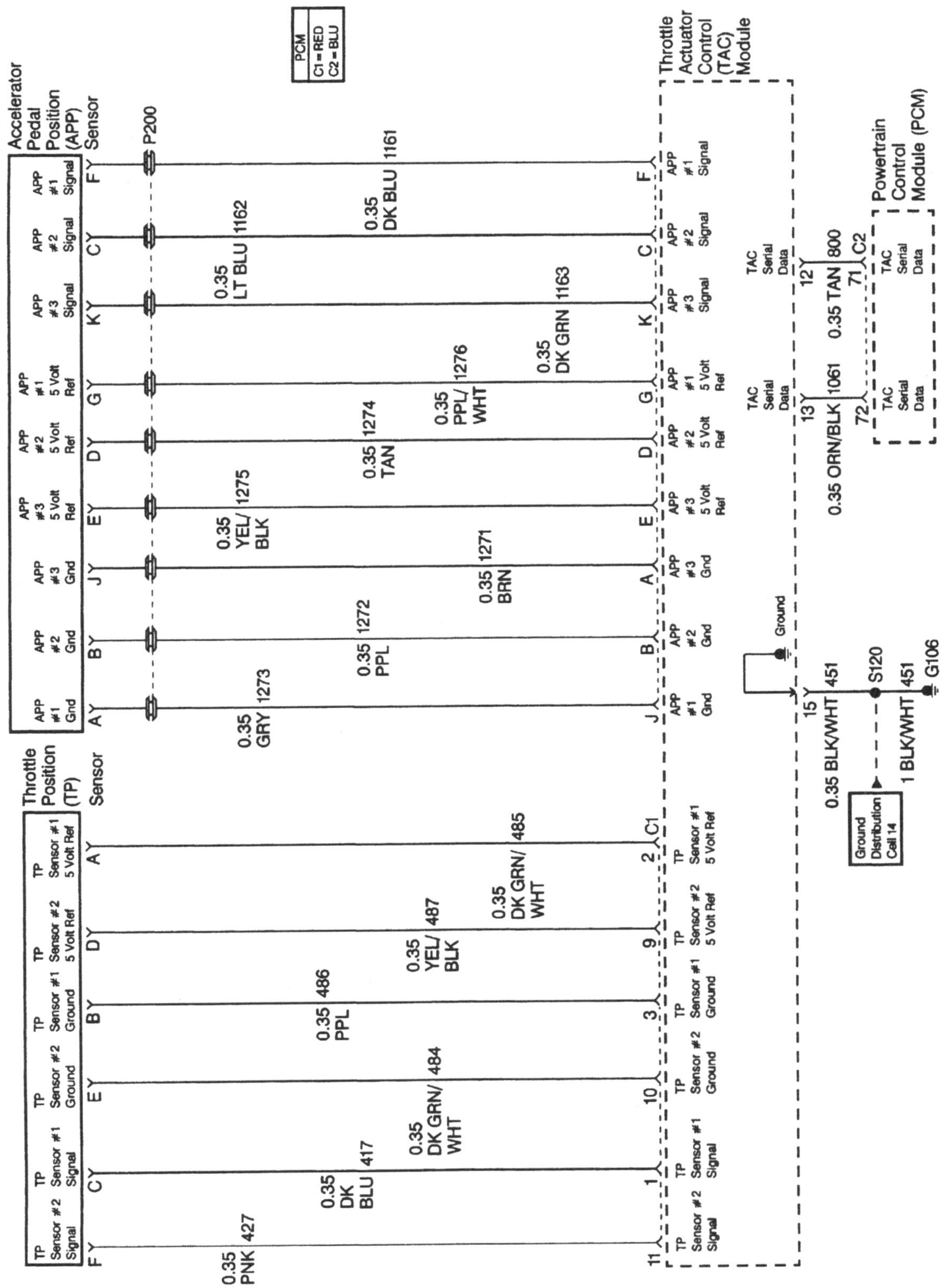

LS1 1997
VIN G

Engine Data Sensors

Figure 9 of 16

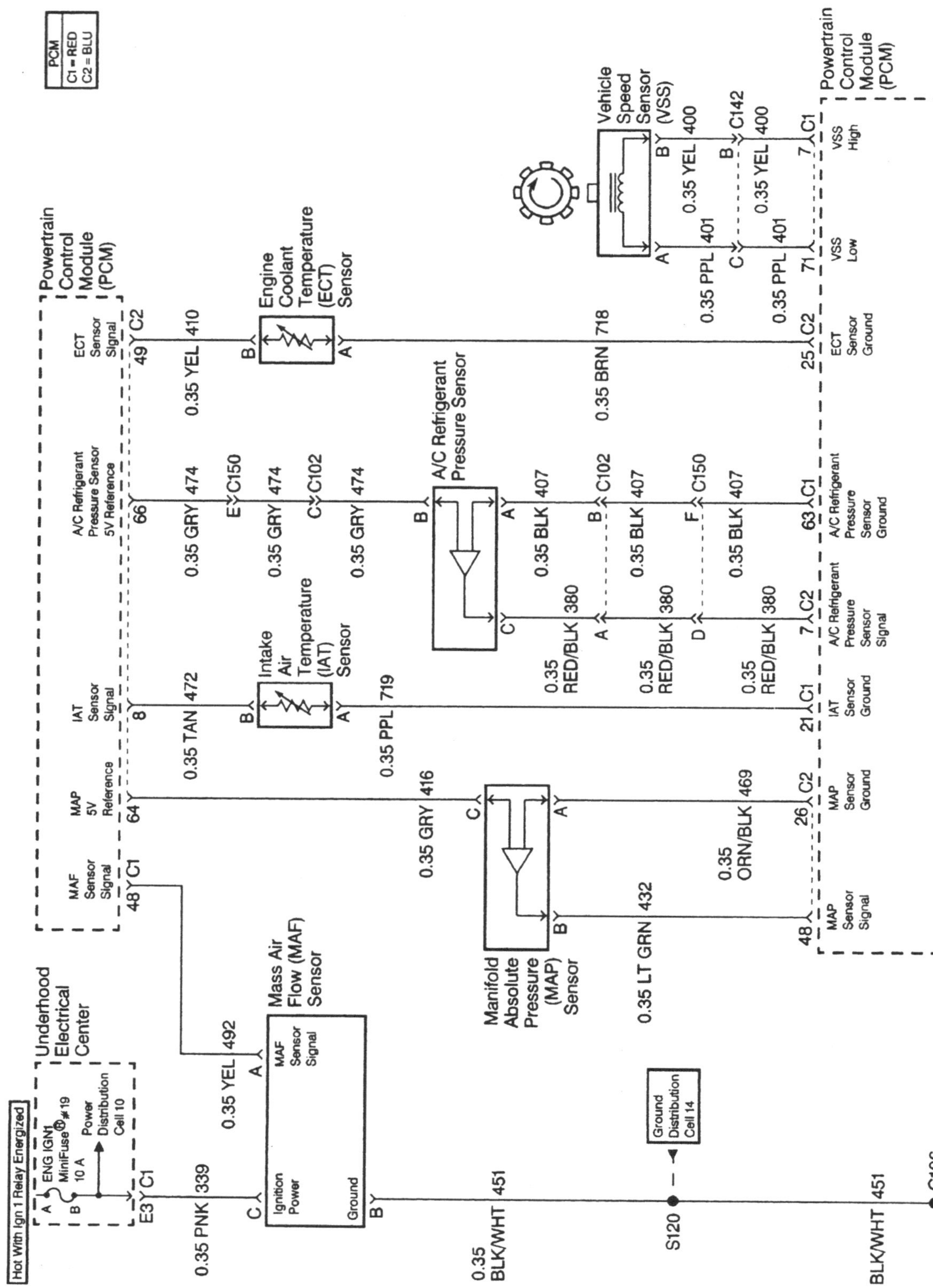

Engine Control (HO_2S)

LS1 1997
VIN G

Figure 10 of 16

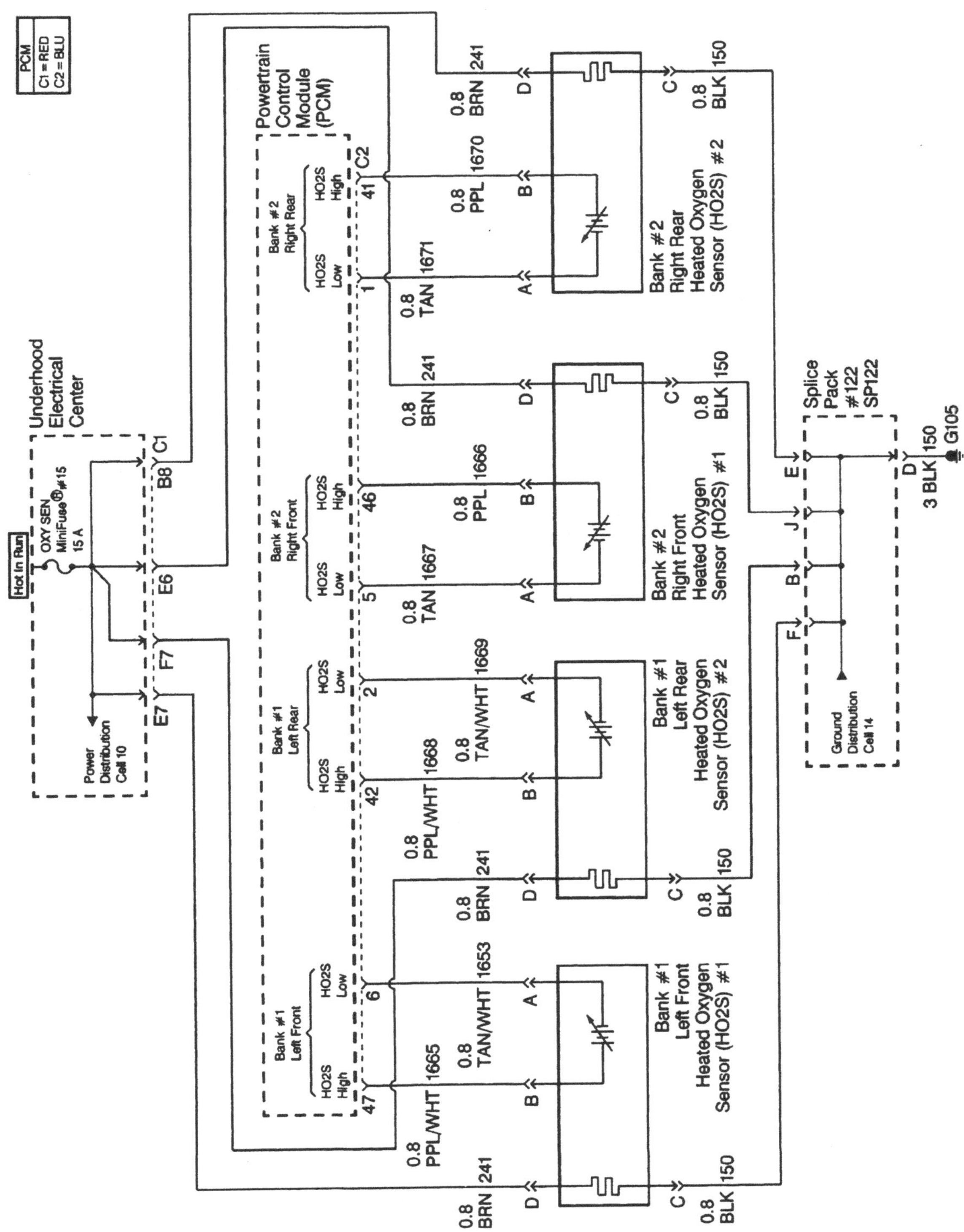

LS1 1997 VIN G

EVAP and AIR

Figure 11 of 16

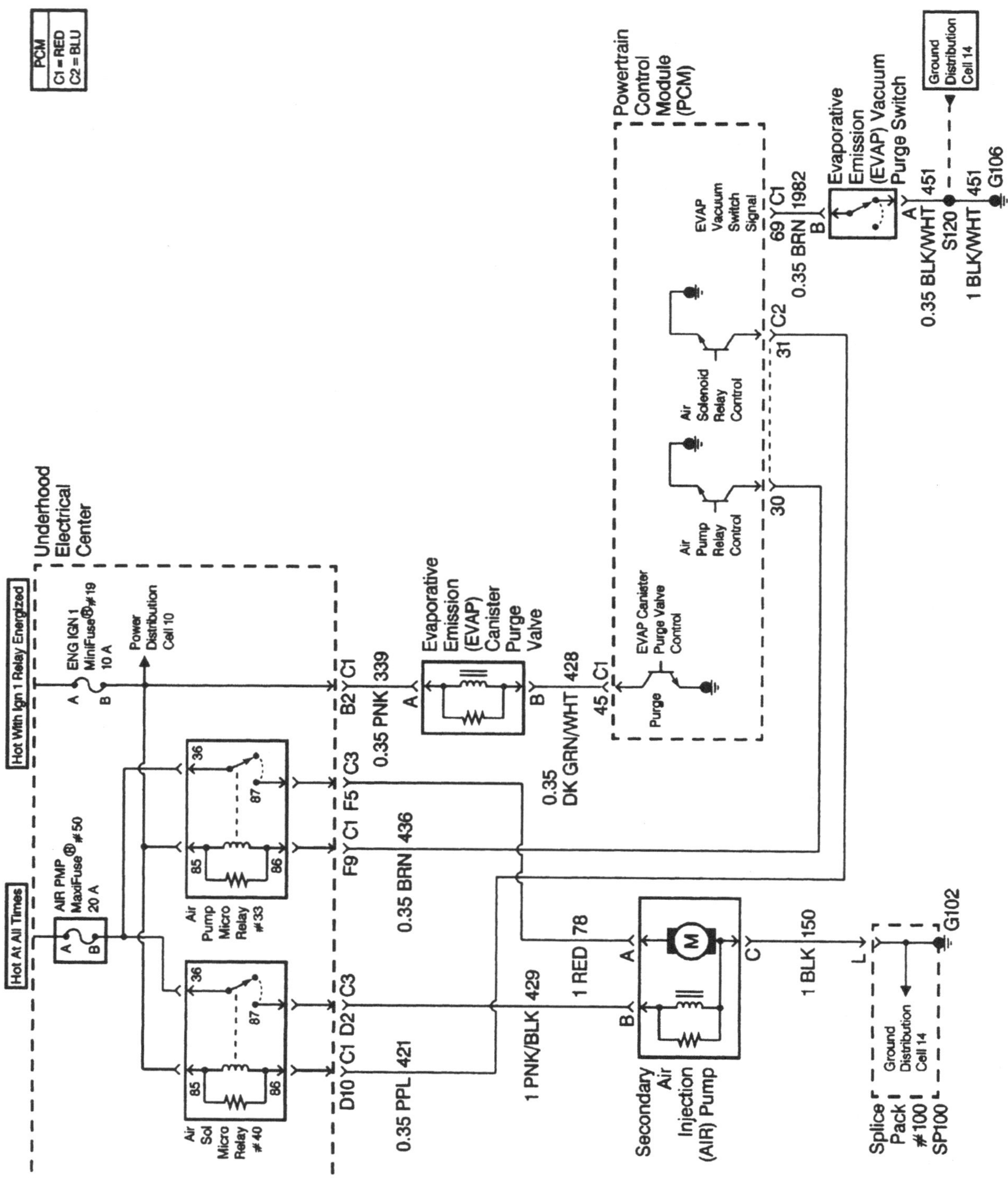

M/T

Figure 12 of 16

LS1 1997 VIN G

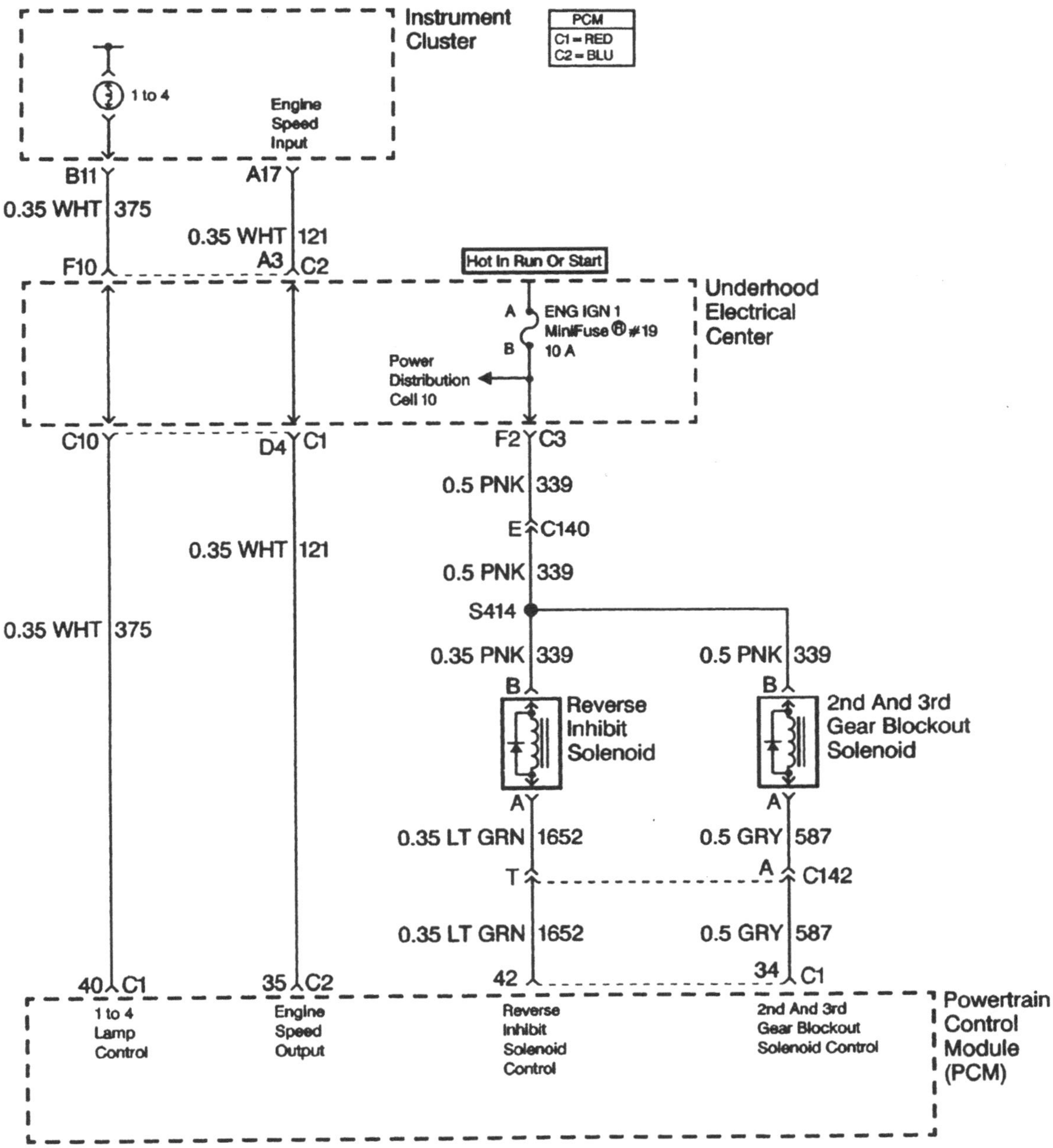

LS1 1997 VIN G

A/T

Figure 13 of 16

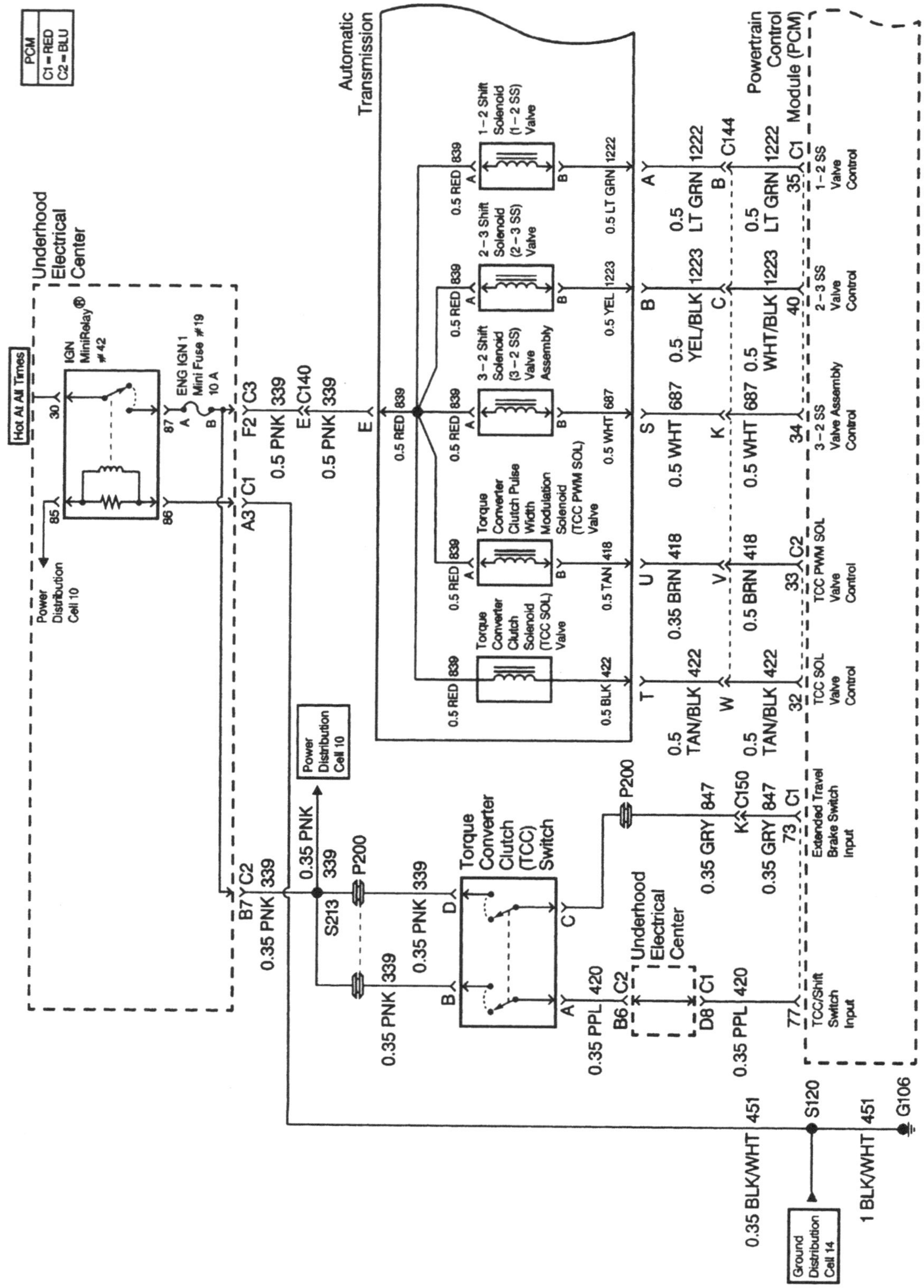

A/T and VSS

LS1 1997
VIN G

Figure 14 of 16

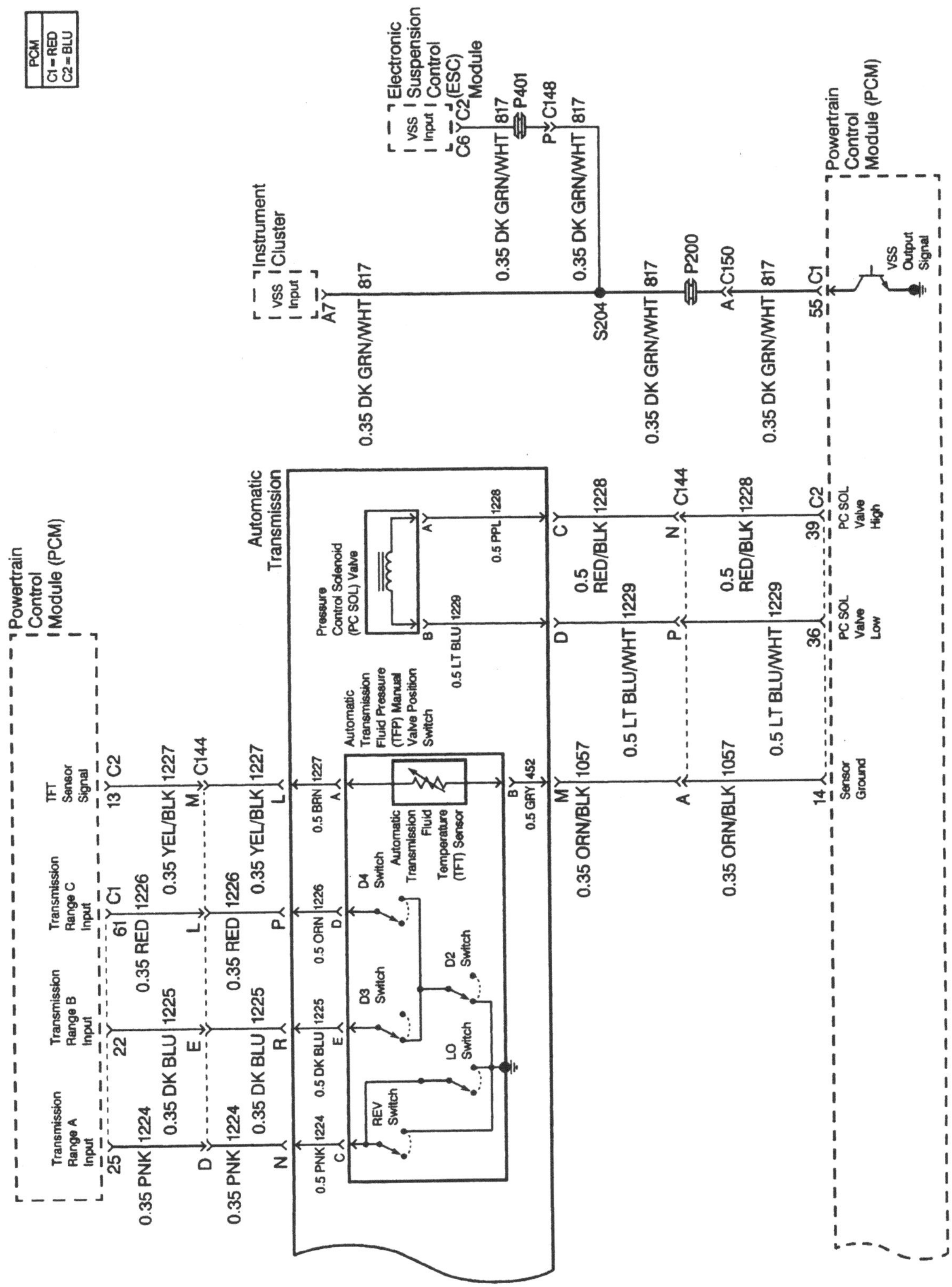

12

LS1 1997
VIN G

PNP Switch

Figure 15 of 16

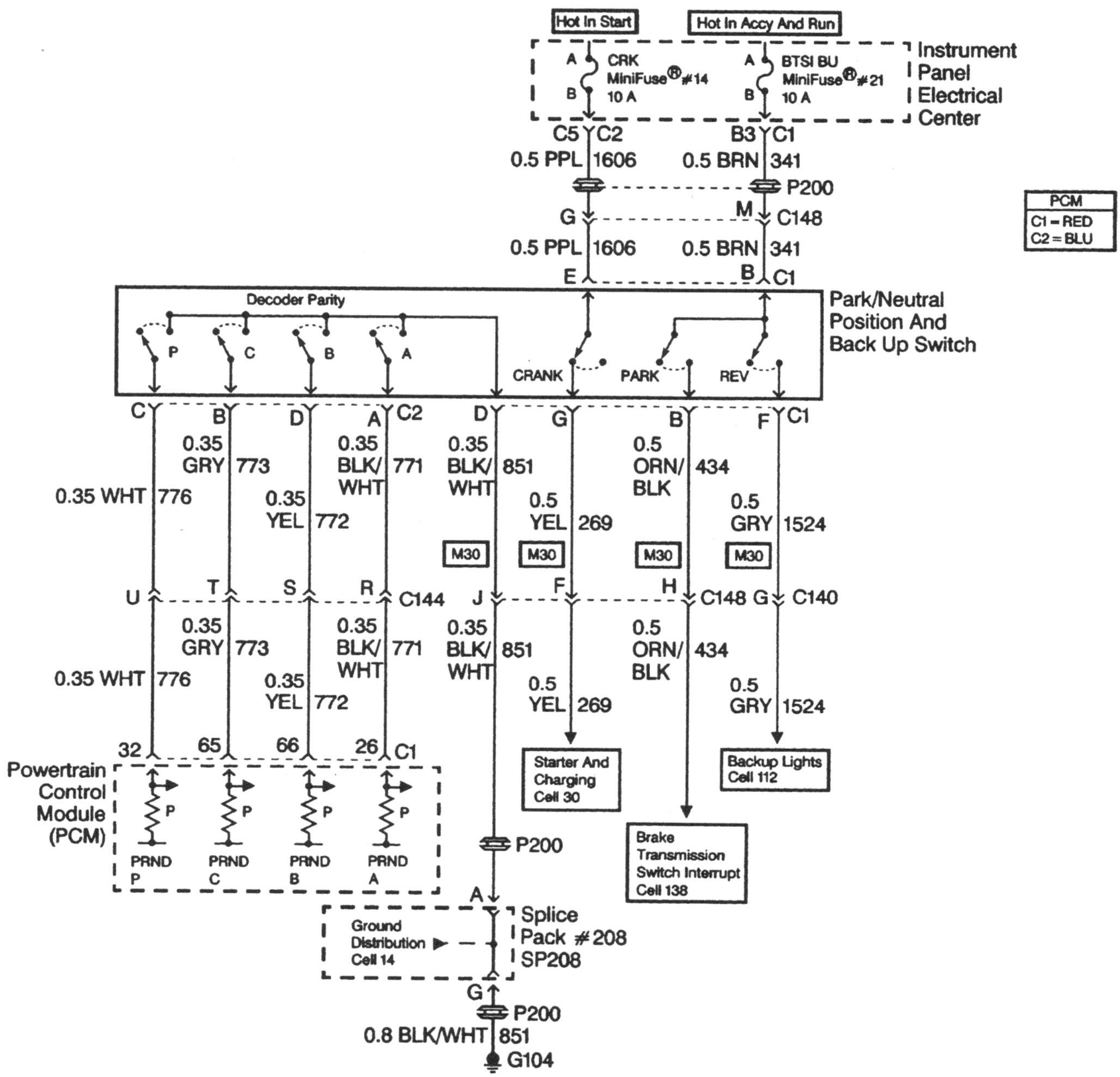

Miscellaneous

Figure 16 of 16

LS1 1997 VIN G

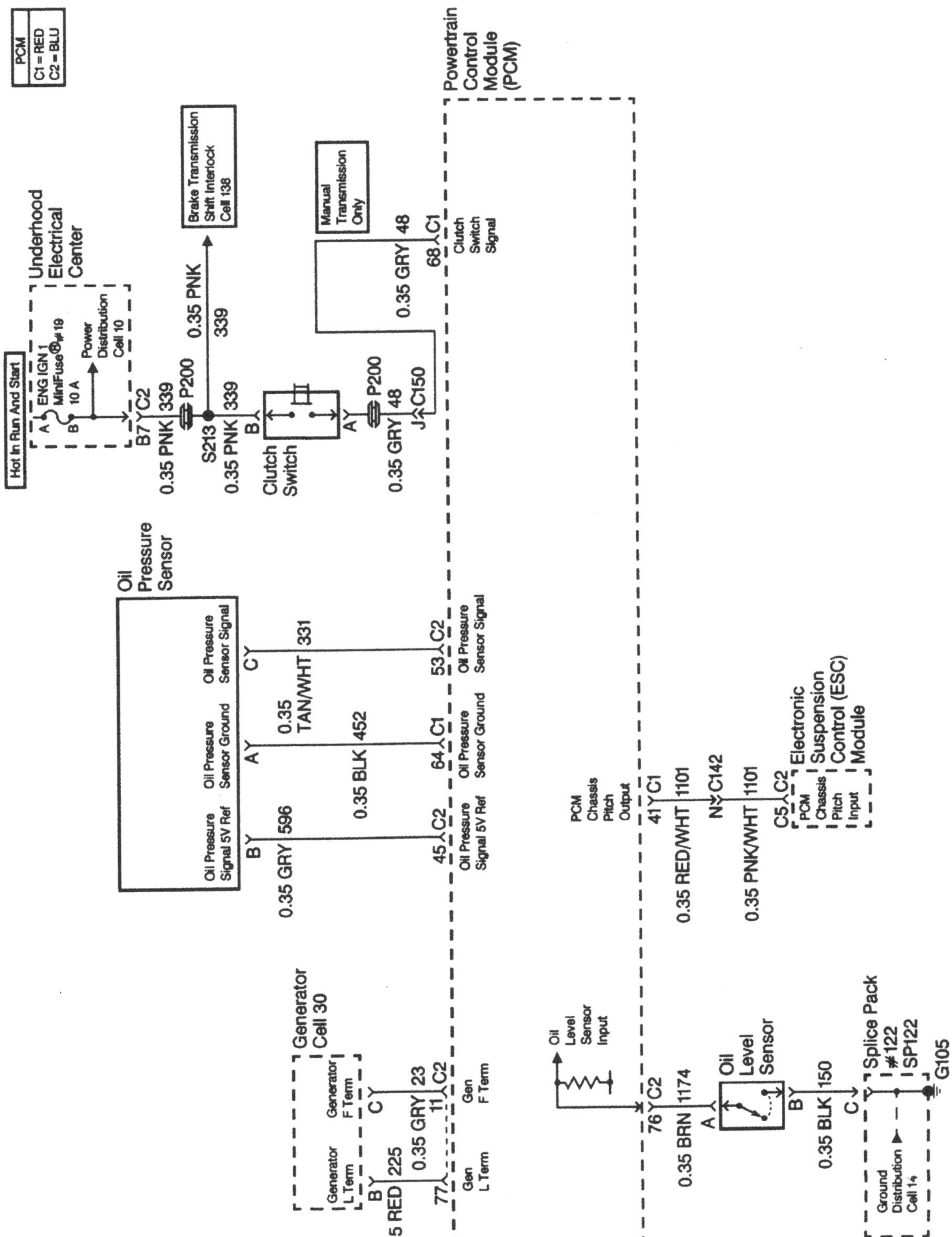

12

LT1/LT4 1996 VIN P/G

Power, Ground, and MIL

Figure 1 of 16

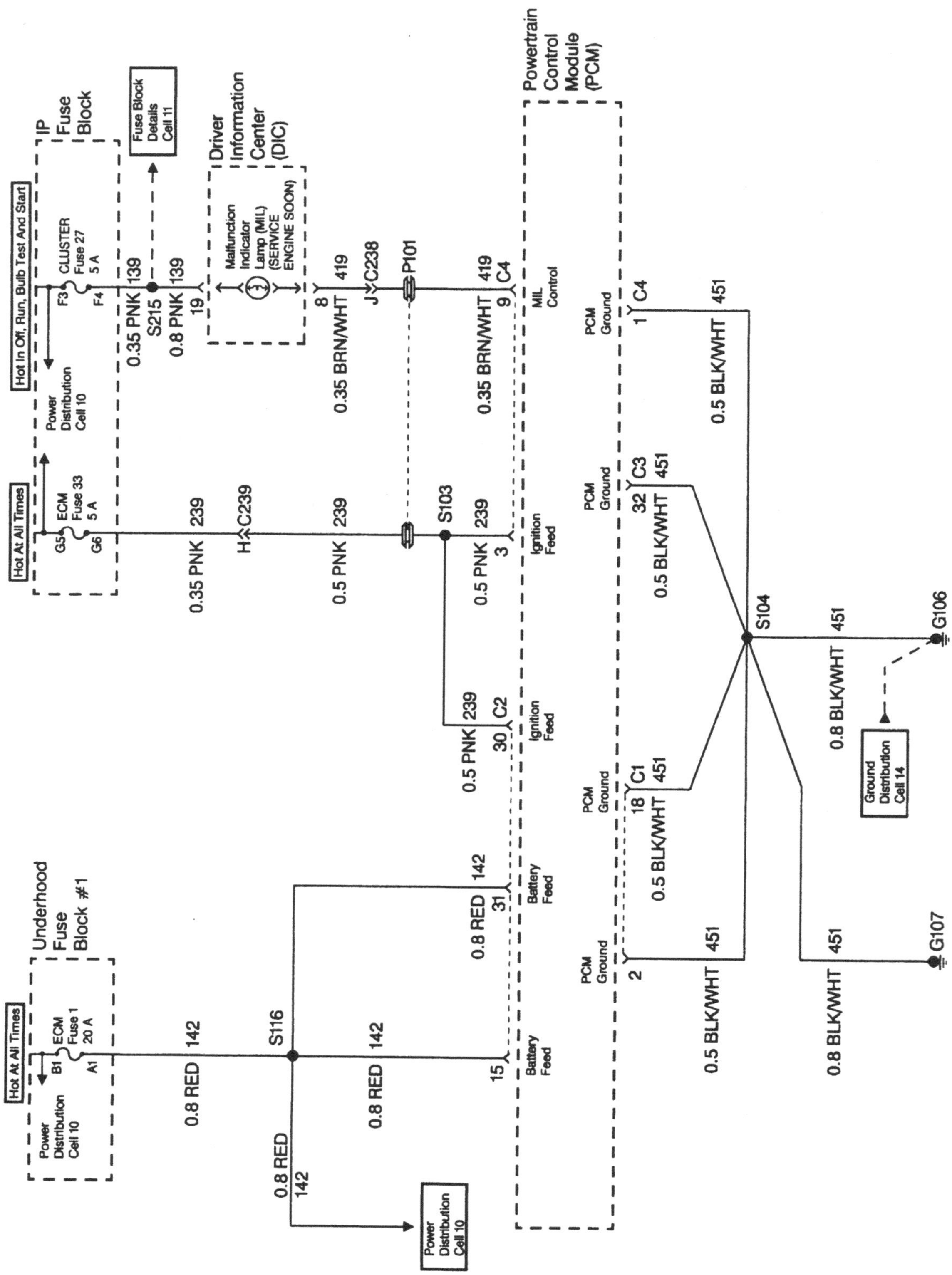

DLC

LT1/LT4 1996
VIN P/G

Figure 2 of 16

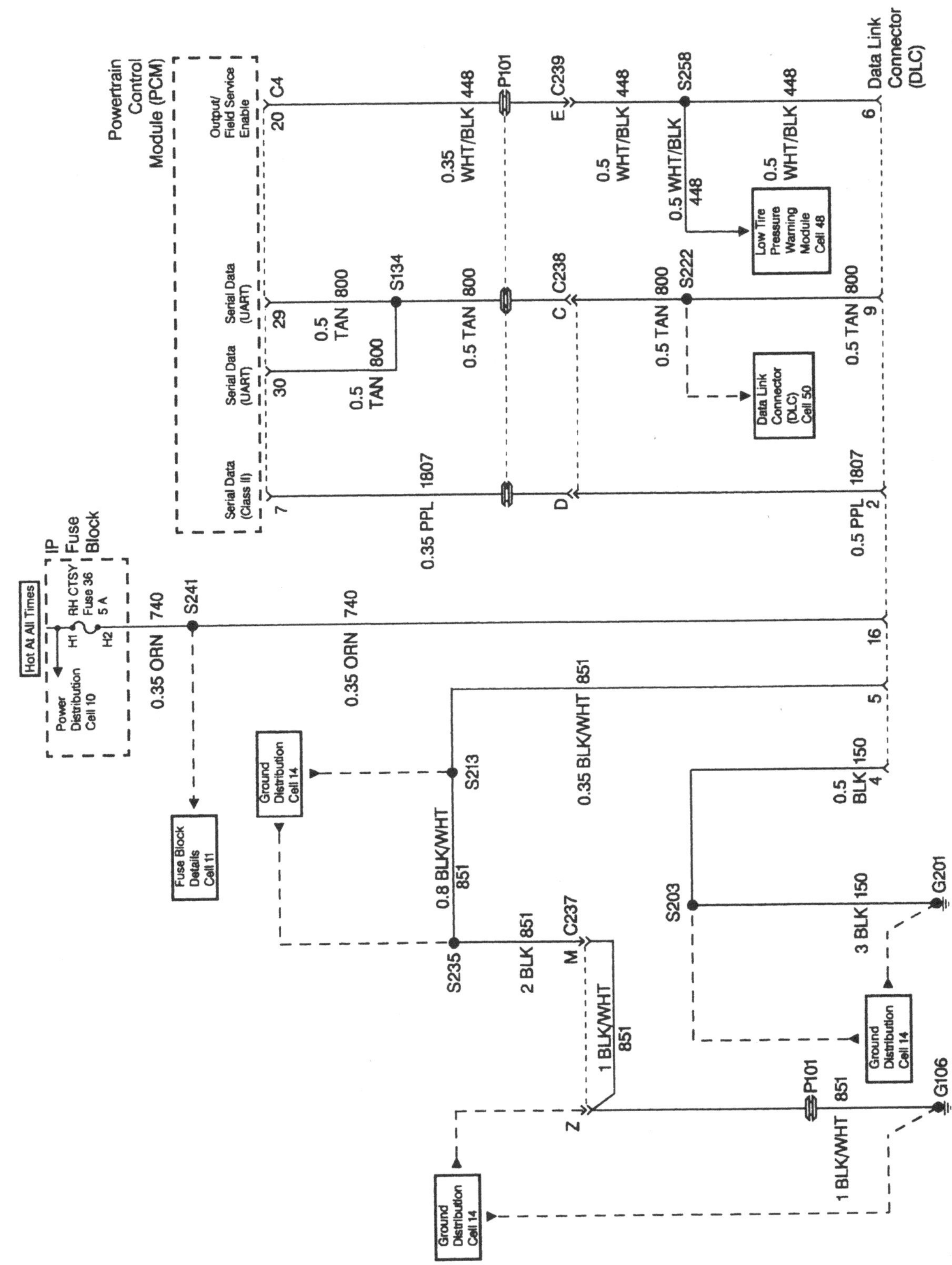

LT1/LT4 1996 VIN P/G

Distributor Ignition

Figure 3 of 16

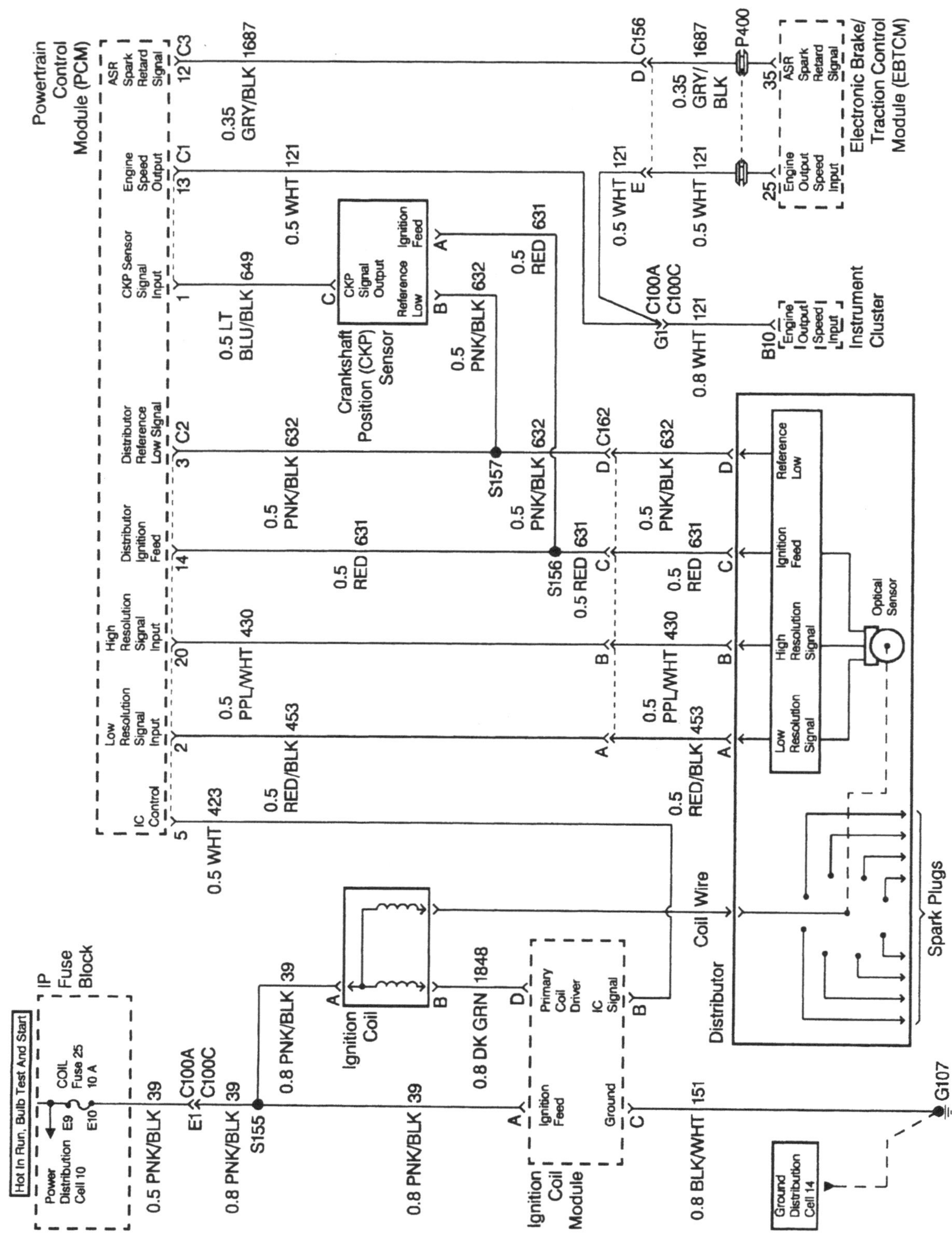

Fuel Pump

LT1/LT4 1996
VIN P/G

Figure 4 of 16

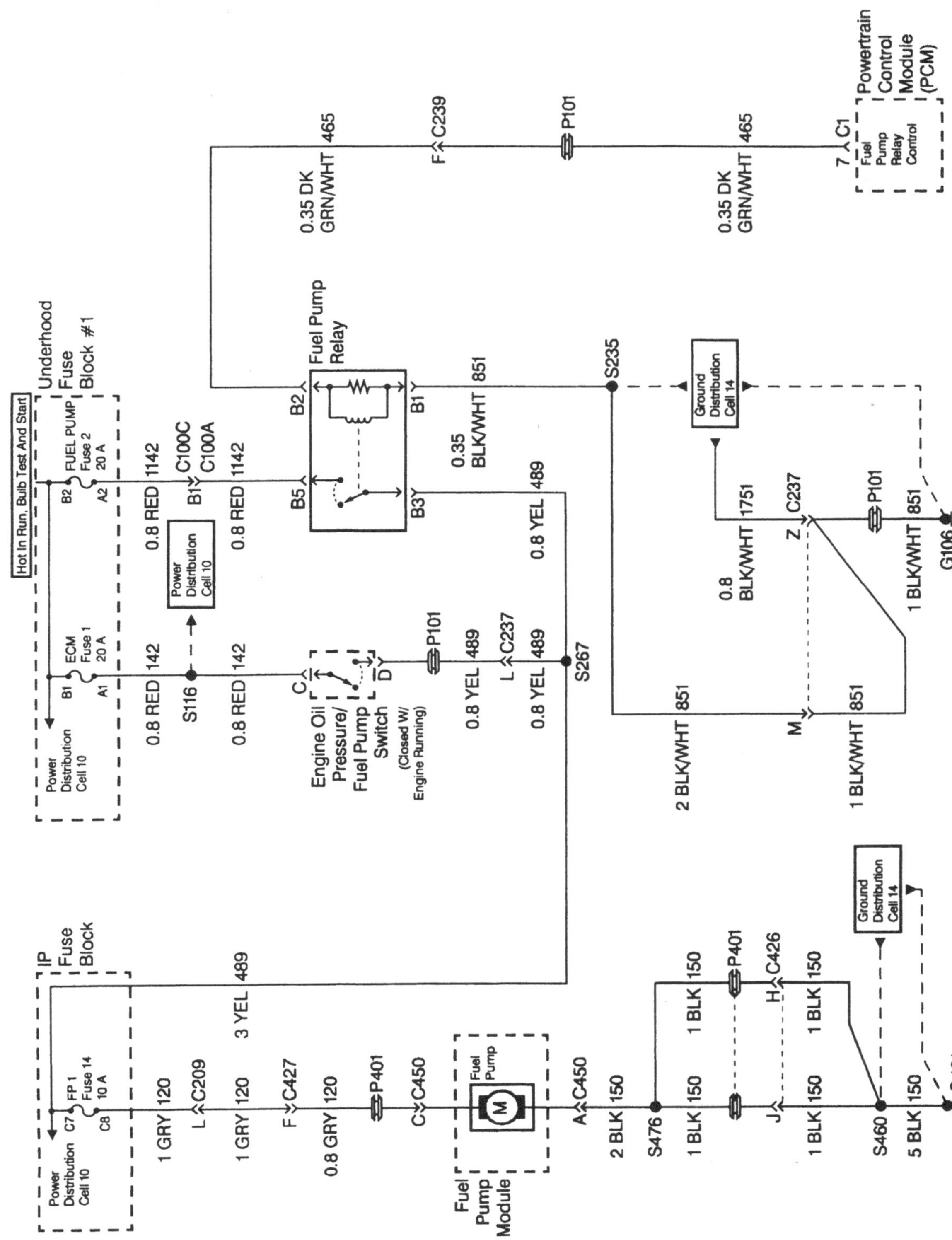

12

LT1/LT4 1996 VIN P/G

Injectors

Figure 5 of 16

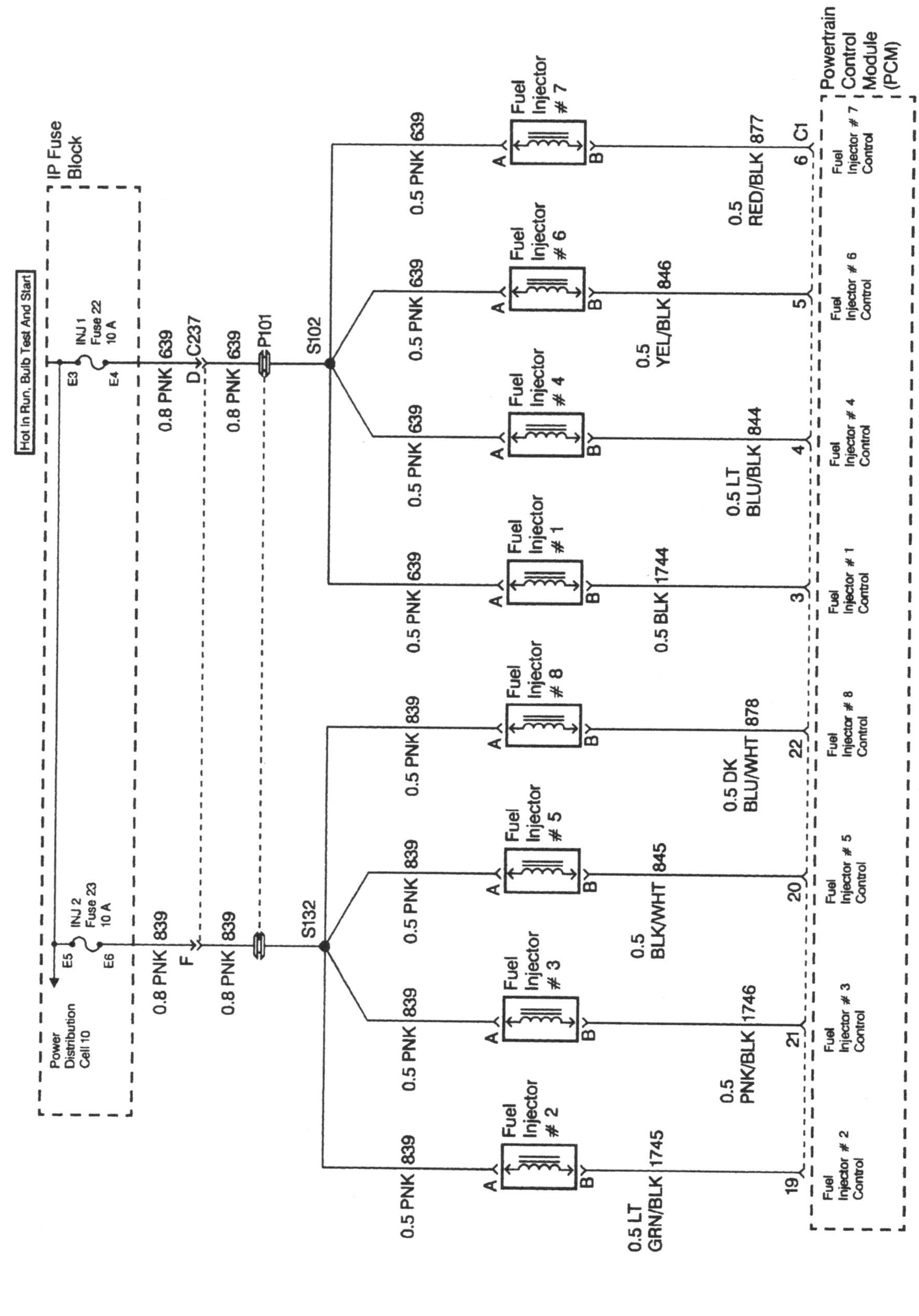

Engine Data Sensors

Figure 6 of 16

LT1/LT4 1996
VIN P/G

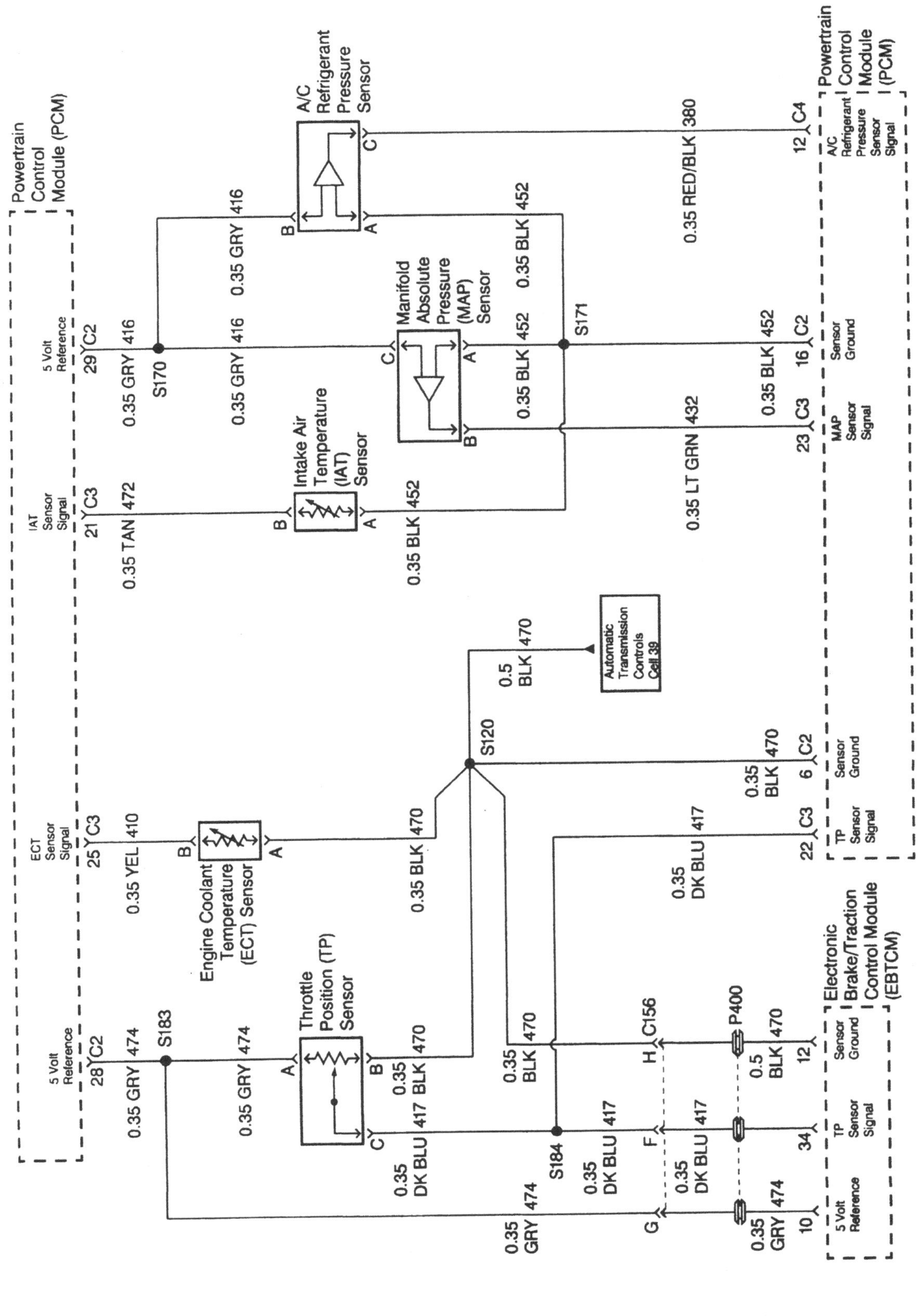

12

LT1/LT4 1996 VIN P/G

HO_2S

Figure 7 of 16

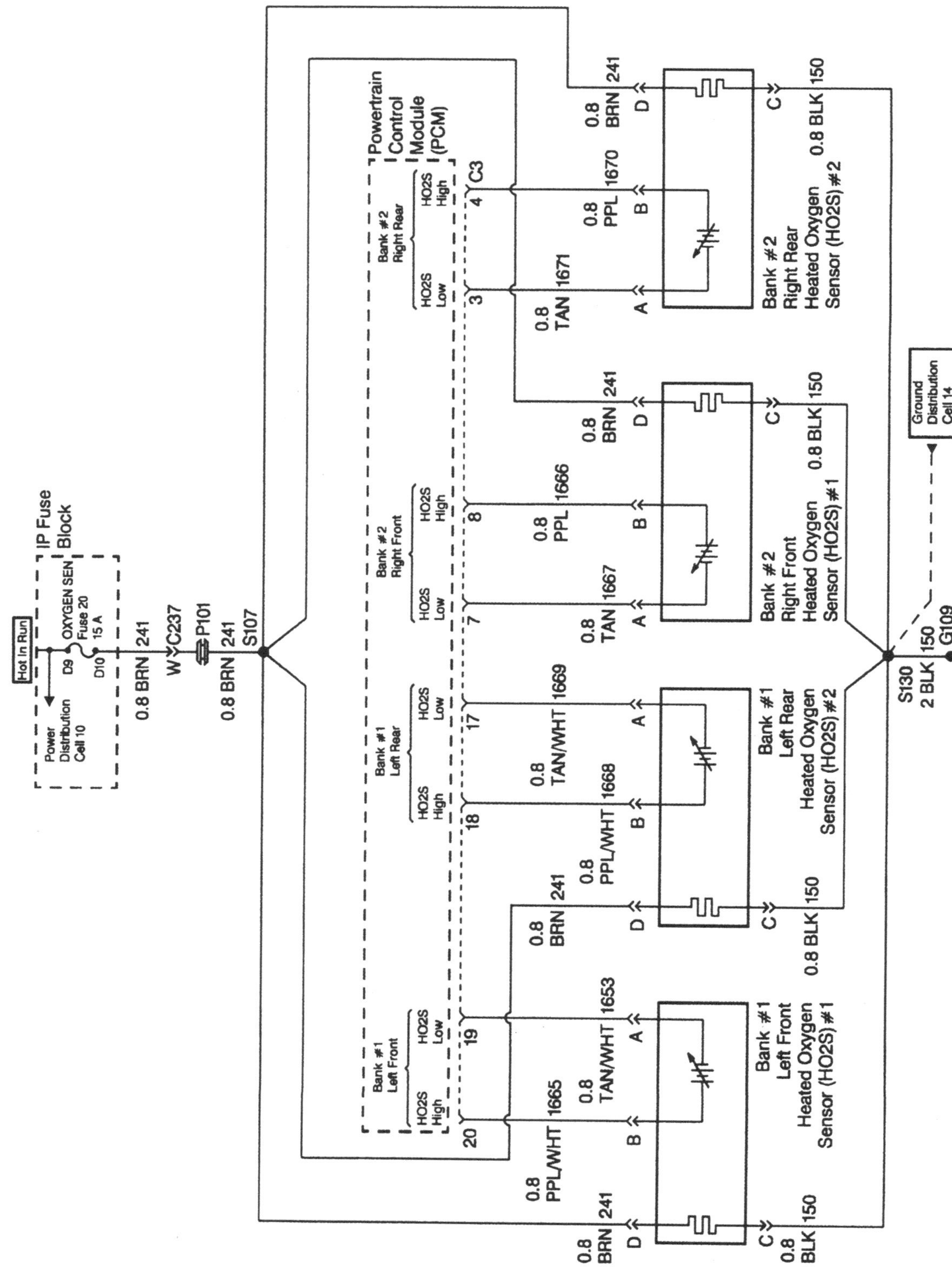

Cruise Control

LT1/LT4 1996
VIN P/G

Figure 8 of 16

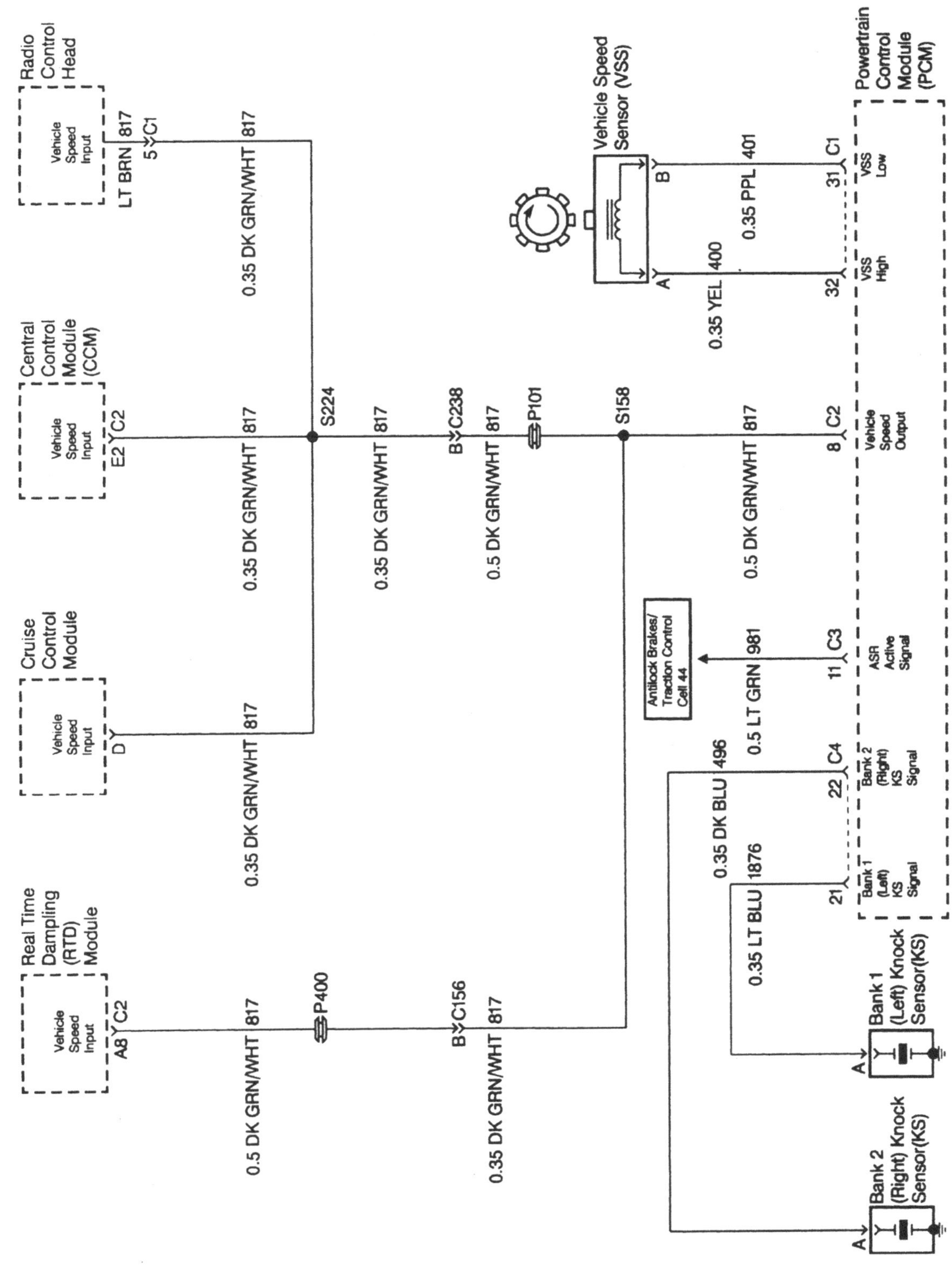

12

LT1/LT4 1996 VIN P/G

MAF and EVAP; EGR-LT1 only

Figure 9 of 16

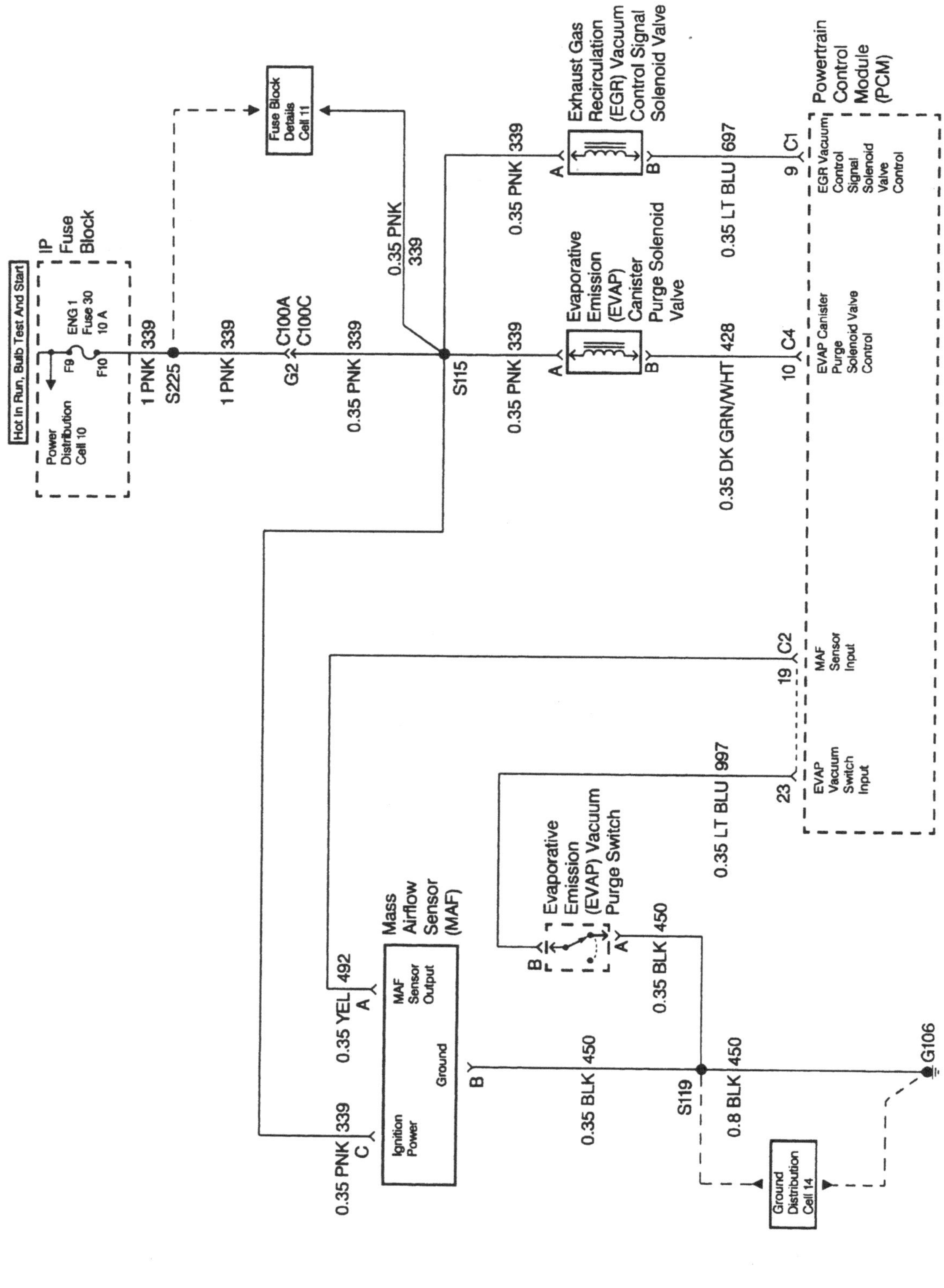

AIR and IAC

Figure 10 of 16

LT1/LT4 1996 VIN P/G

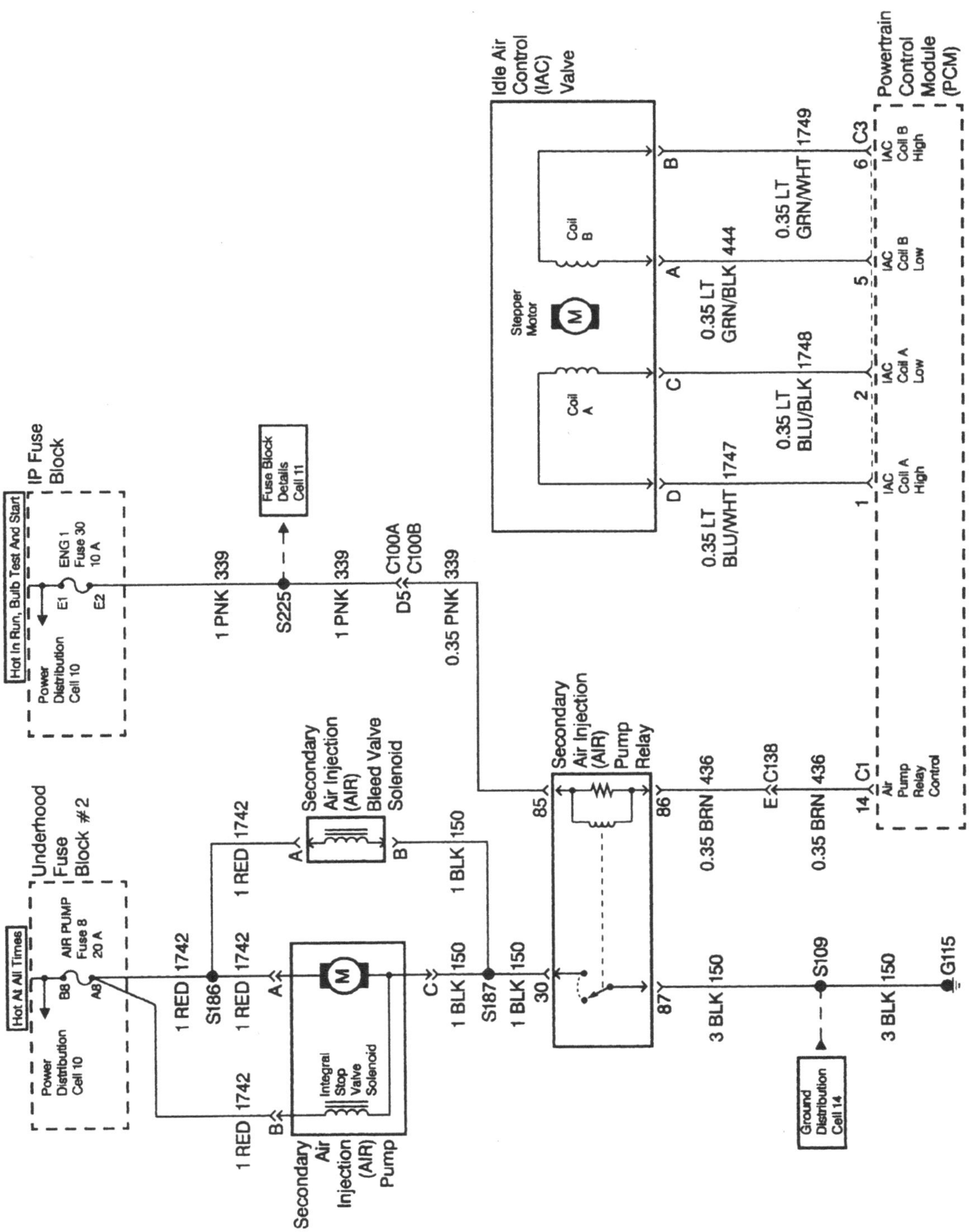

12

LT1/LT4 1996 VIN P/G

1-4 Upshift, M/T (2-3 Blockout)

Figure 11 of 16

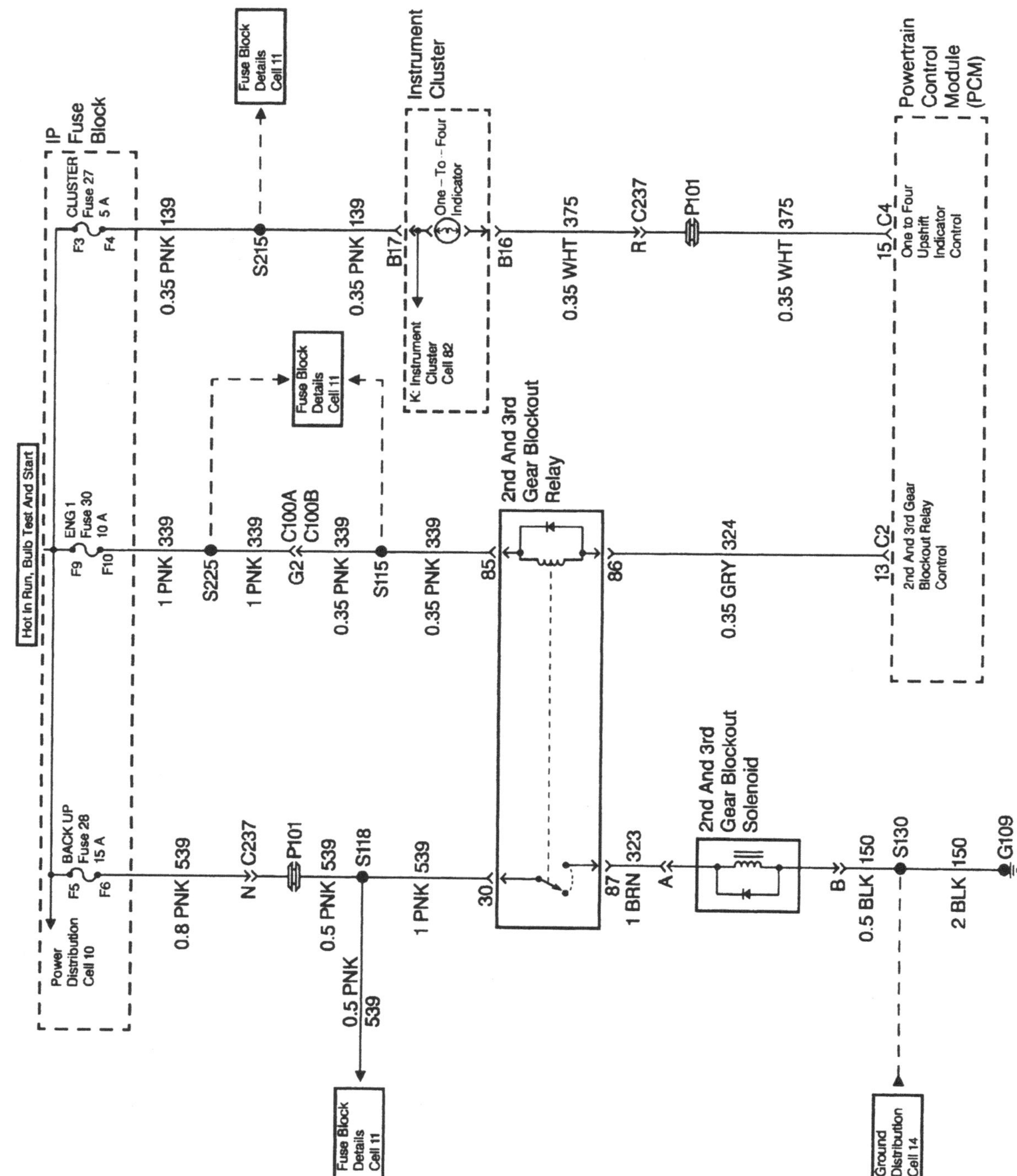

A/T

Figure 12 of 16

LT1/LT4 1996 VIN P/G

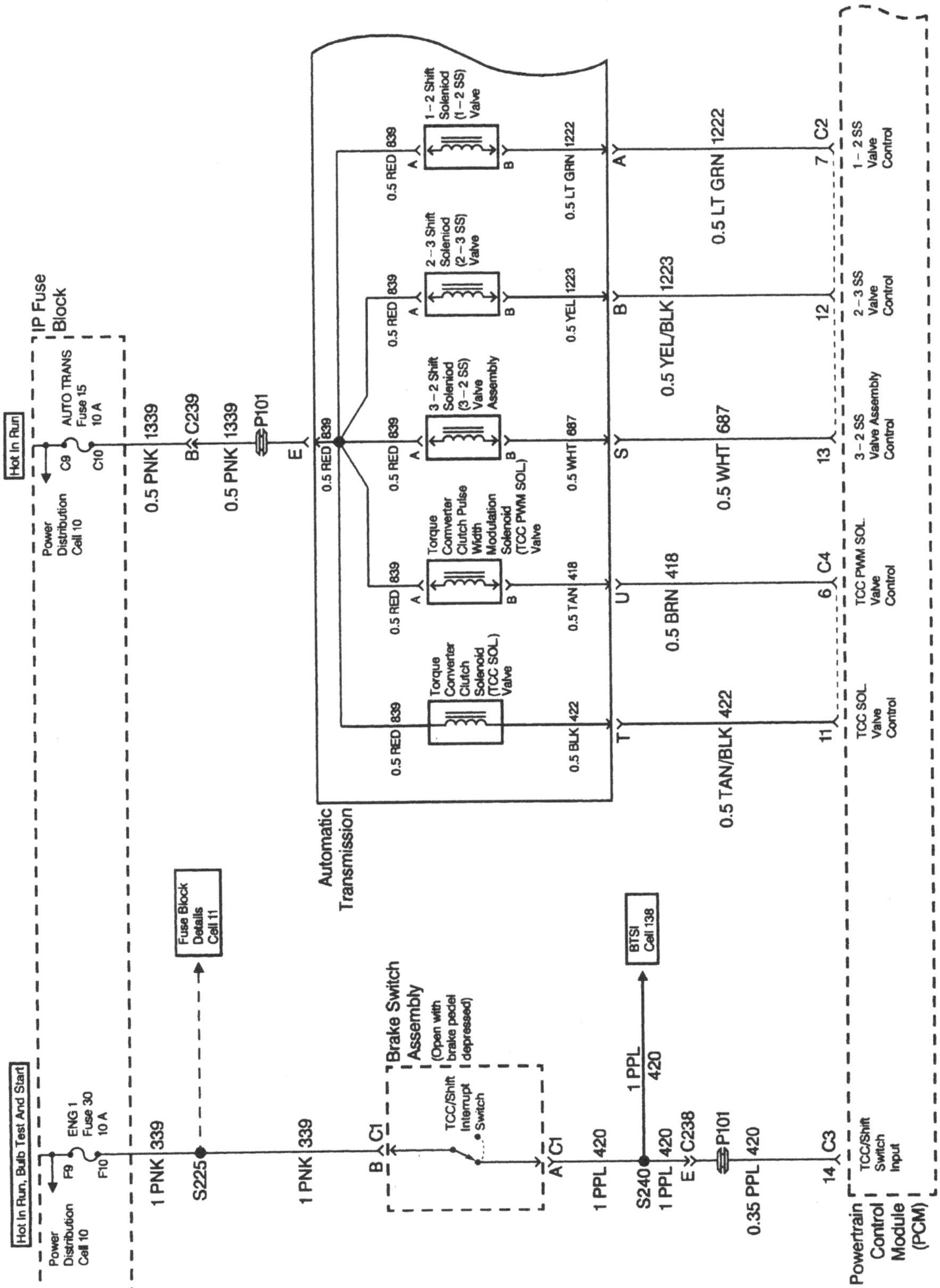

12

LT1/LT4 1996 VIN P/G

A/T

Figure 13 of 16

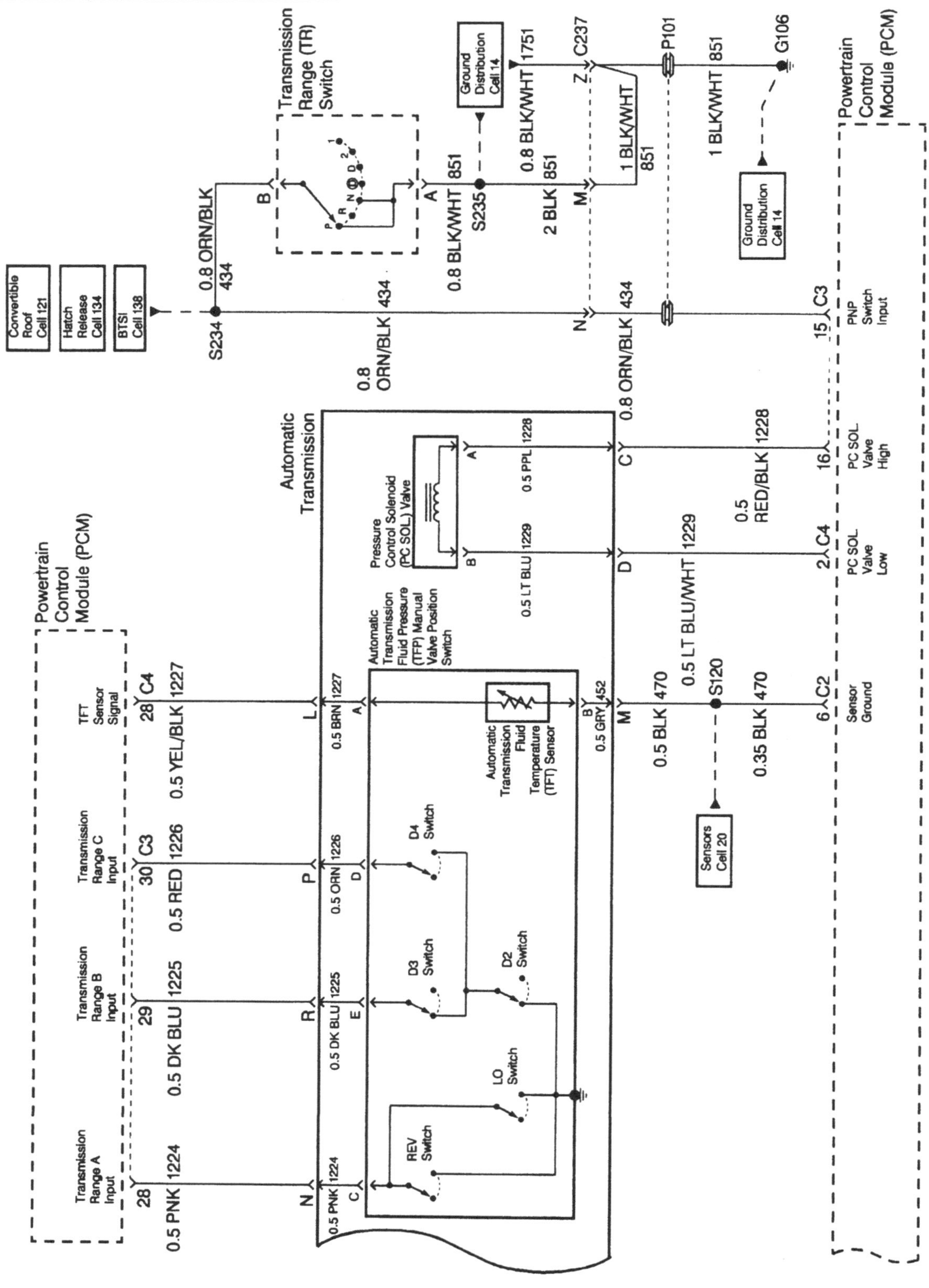

A-C Compressor Clutch (C60—Manual)

LT1/LT4 1996 VIN P/G

Figure 14 of 16

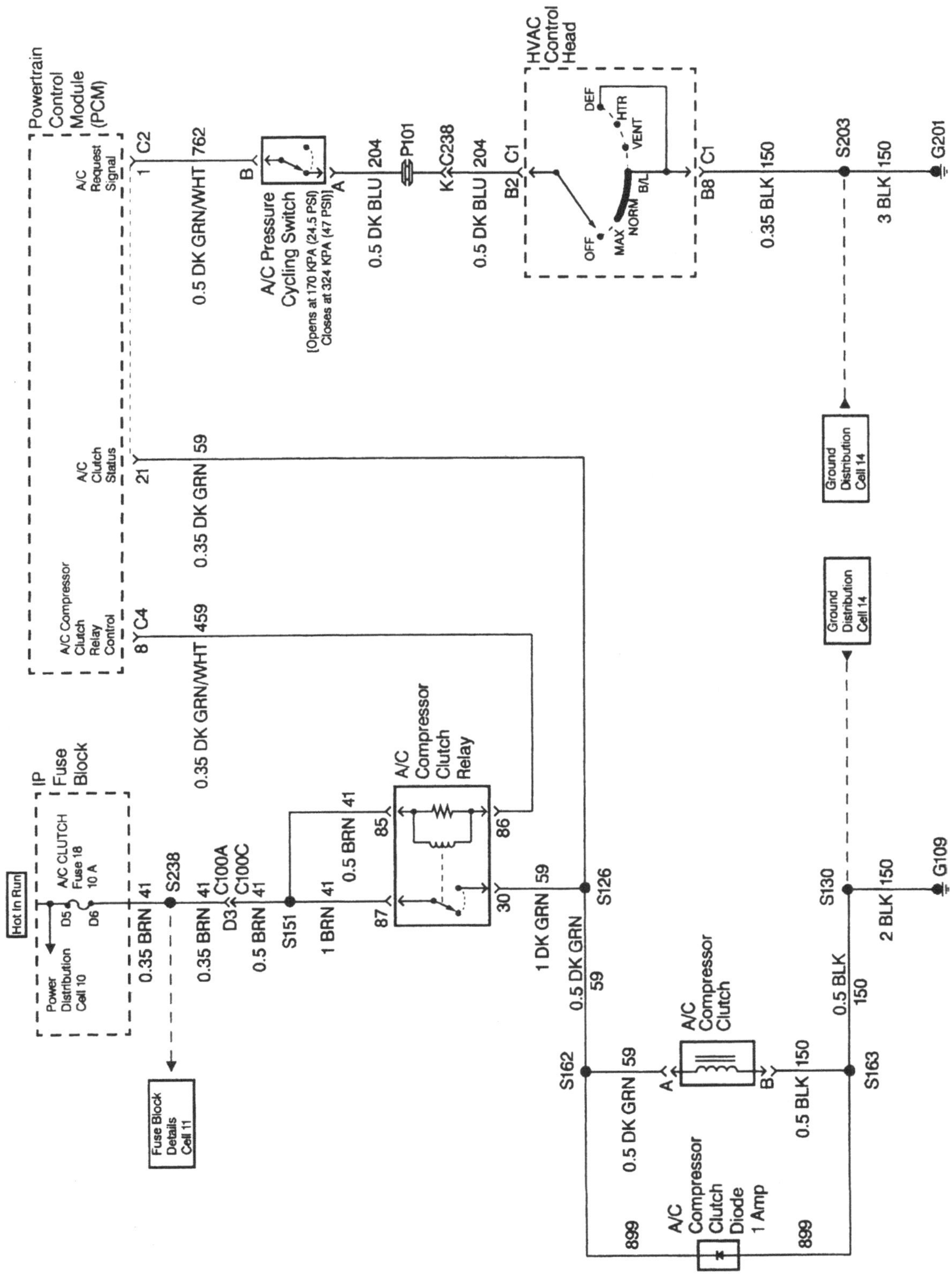

LT1/LT4 1996
VIN P/G

A-C Compressor Clutch (C68—Automatic)

Figure 15 of 16

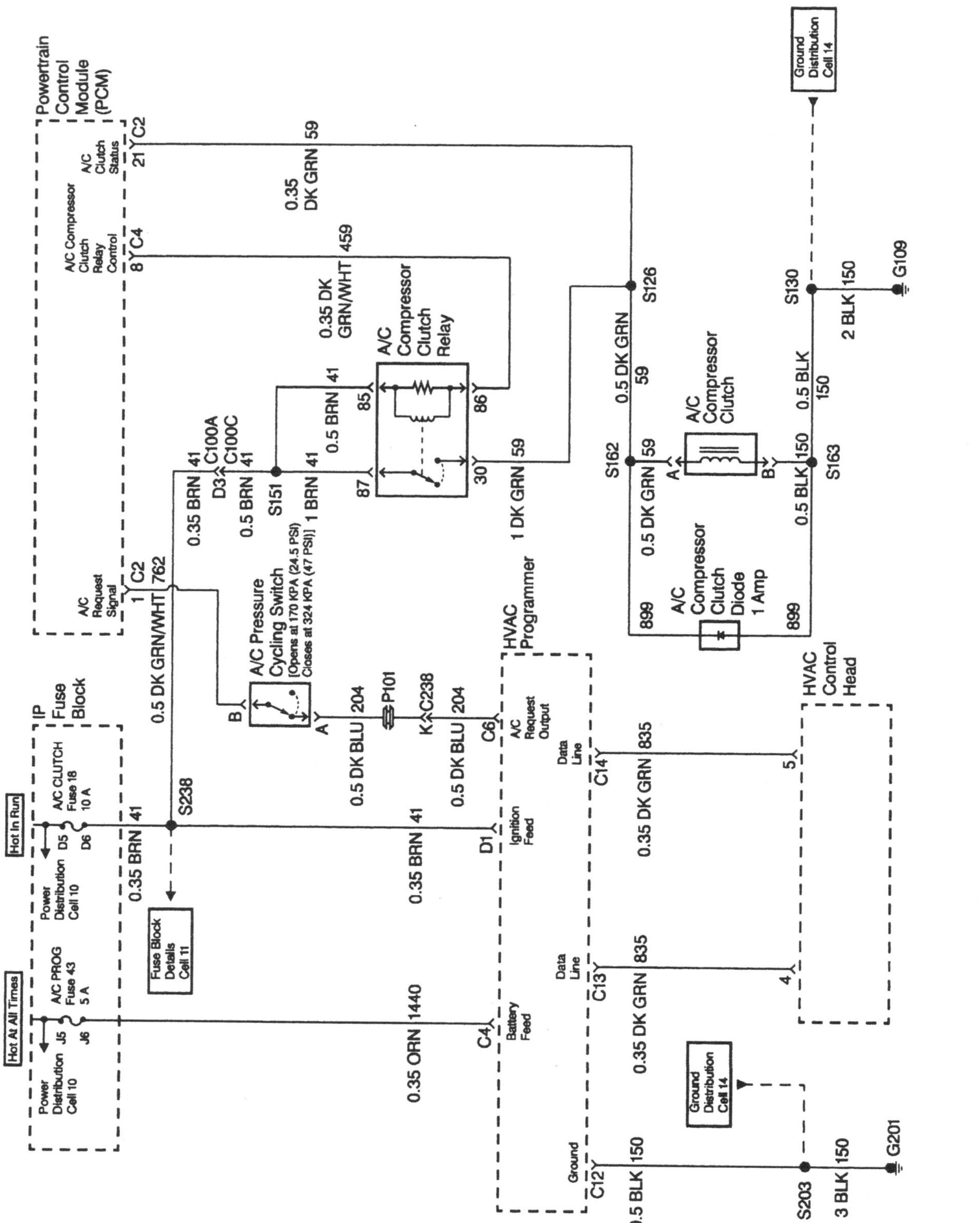

Fan Controls

LT1/LT4 1996
VIN P/G

Figure 16 of 16

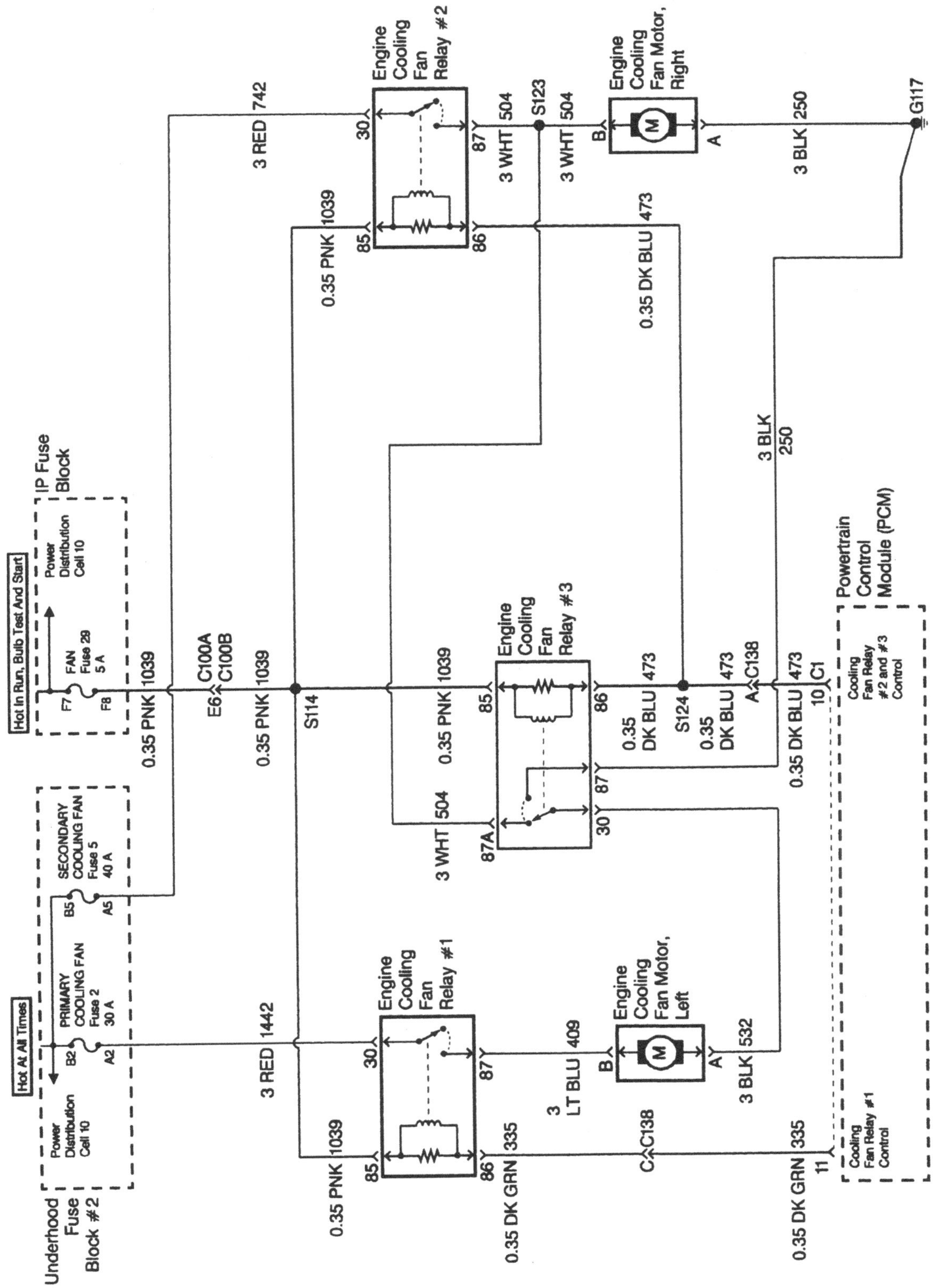

LT1 1994-95 SFI VIN P

Engine Data Sensors

Figure 1 of 7

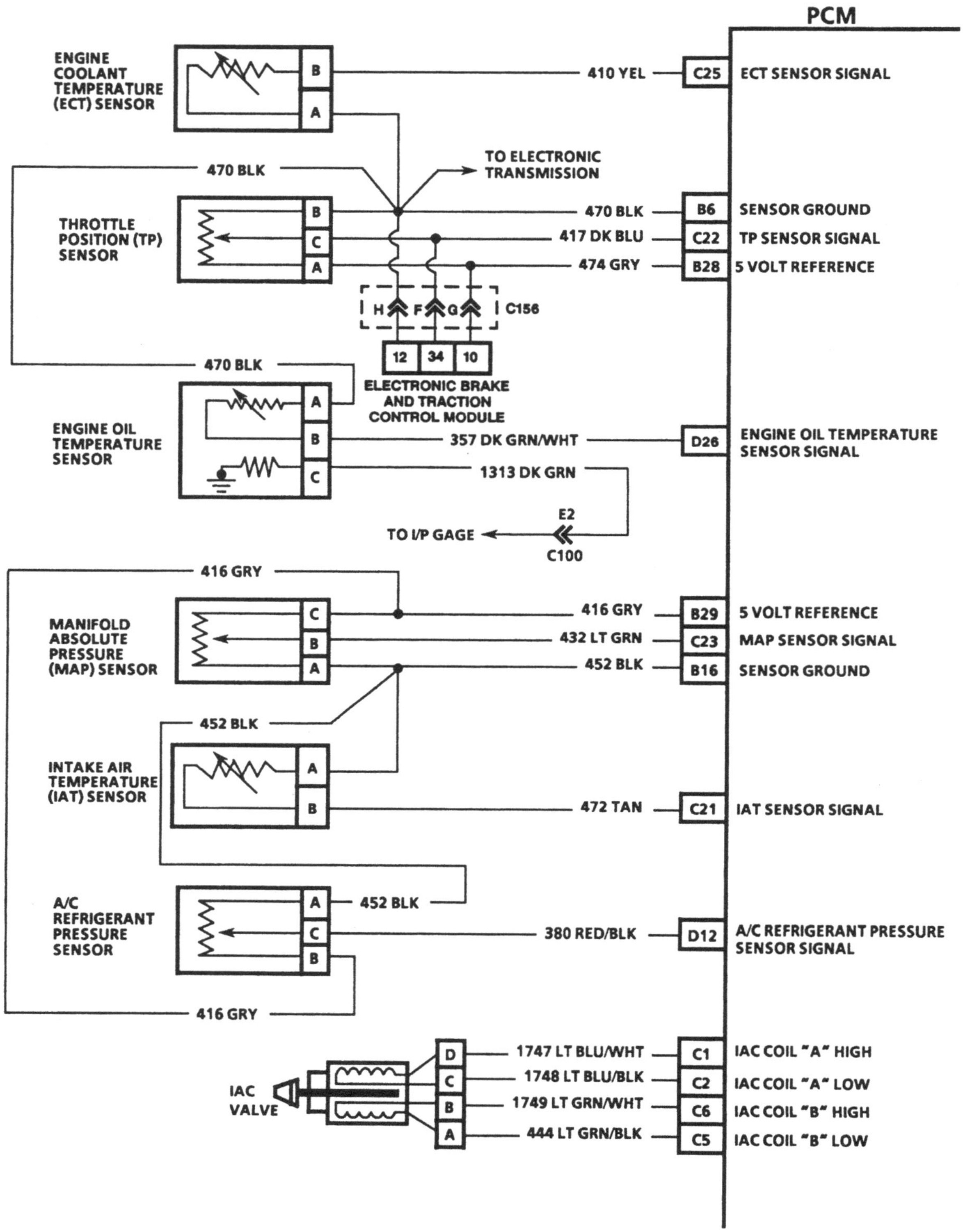

HO_2S, VSS, KS

Figure 2 of 7

LT1 1994-95 SFI VIN P

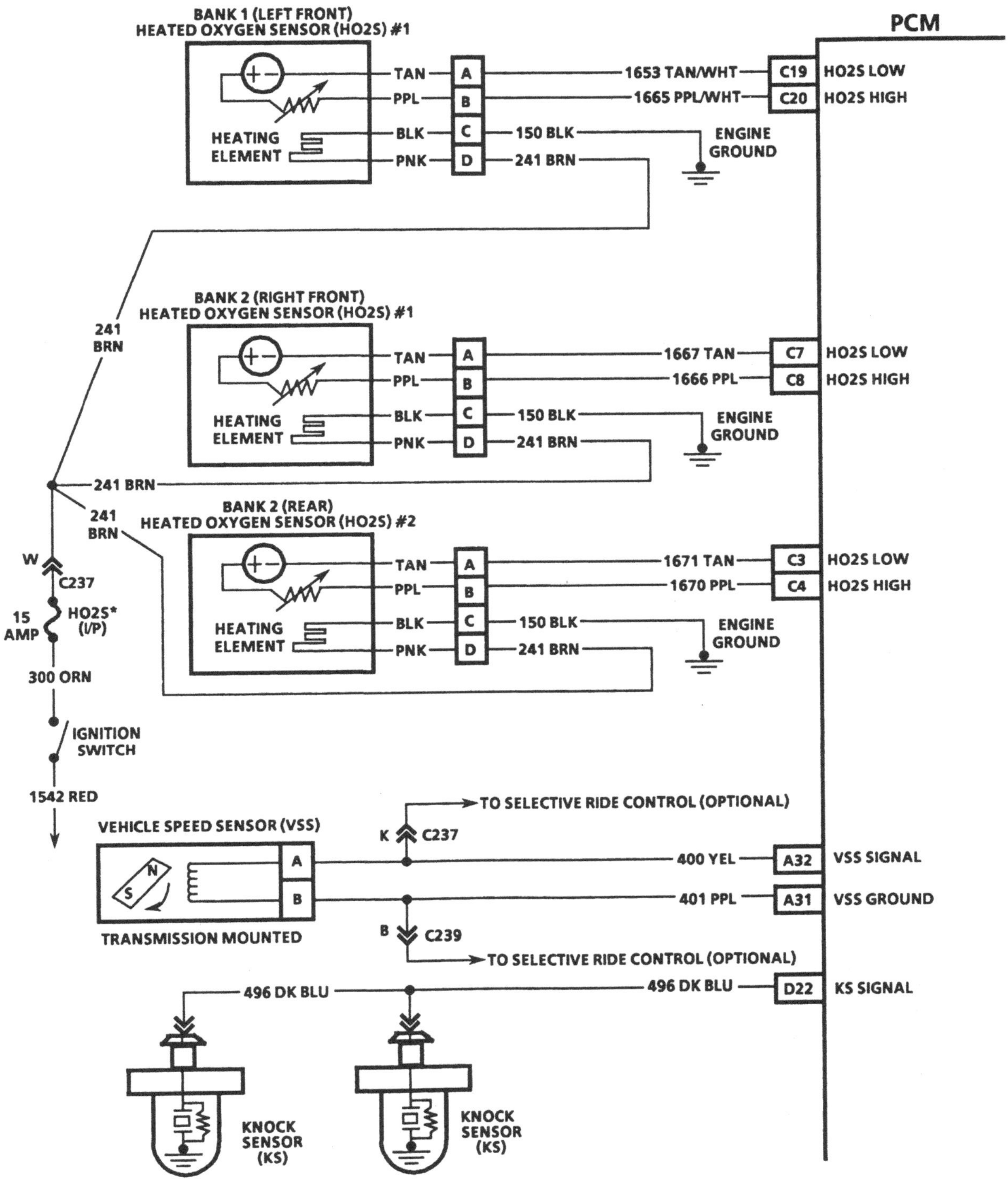

* LABELED " OXYGEN SEN" IN I/P FUSE PANEL

LT1 1994-95 SFI VIN P

AIR, EVAP, EGR, MAF, 1-4, PNP

Figure 3 of 7

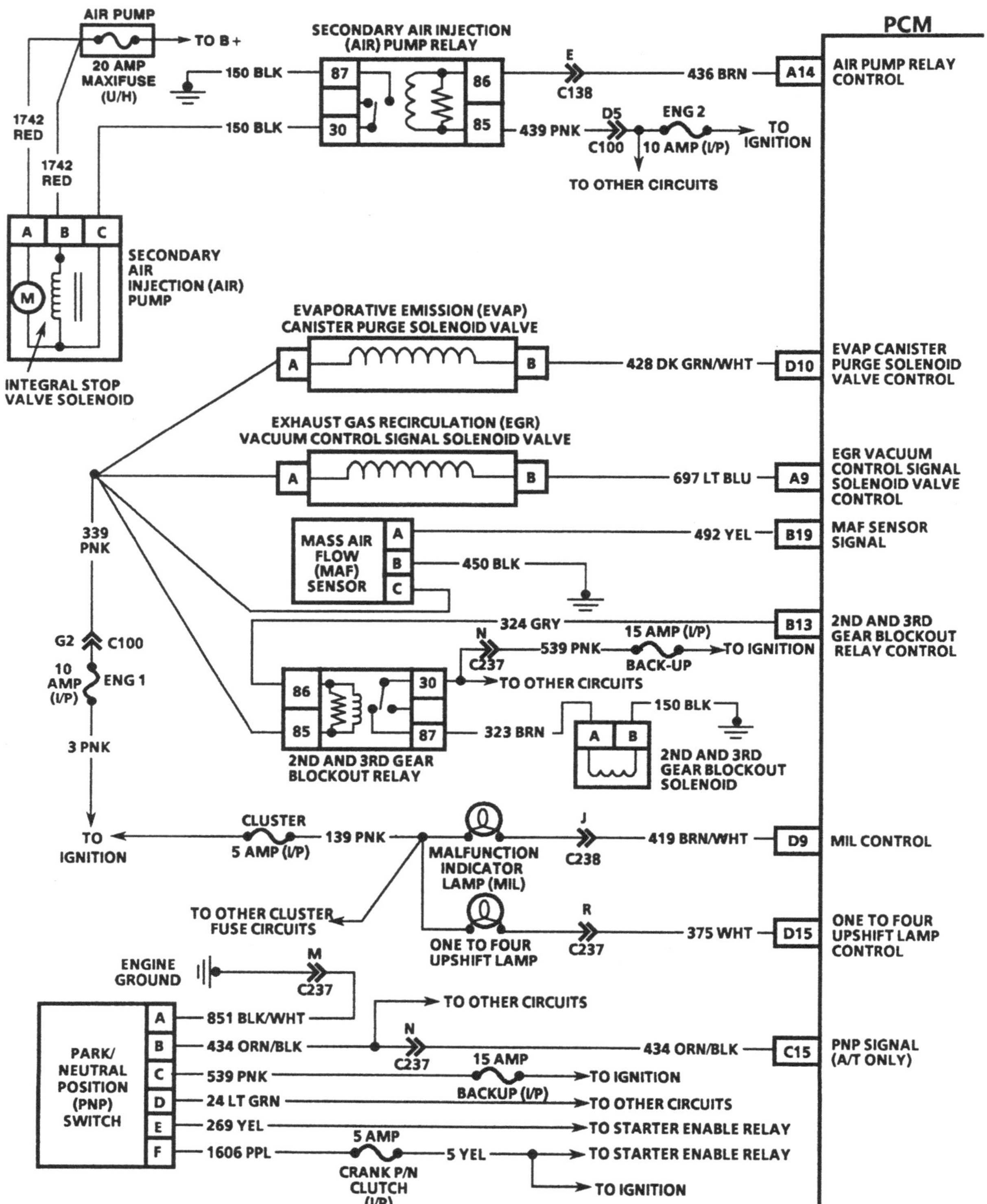

(I/P) = LOCATED IN INSTRUMENT PANEL FUSE BLOCK
(U/H) = LOCATED IN UNDERHOOD ELECTRICAL CENTER

Injectors, Distributor

Figure 4 of 7

LT1 1994-95 SFI VIN P

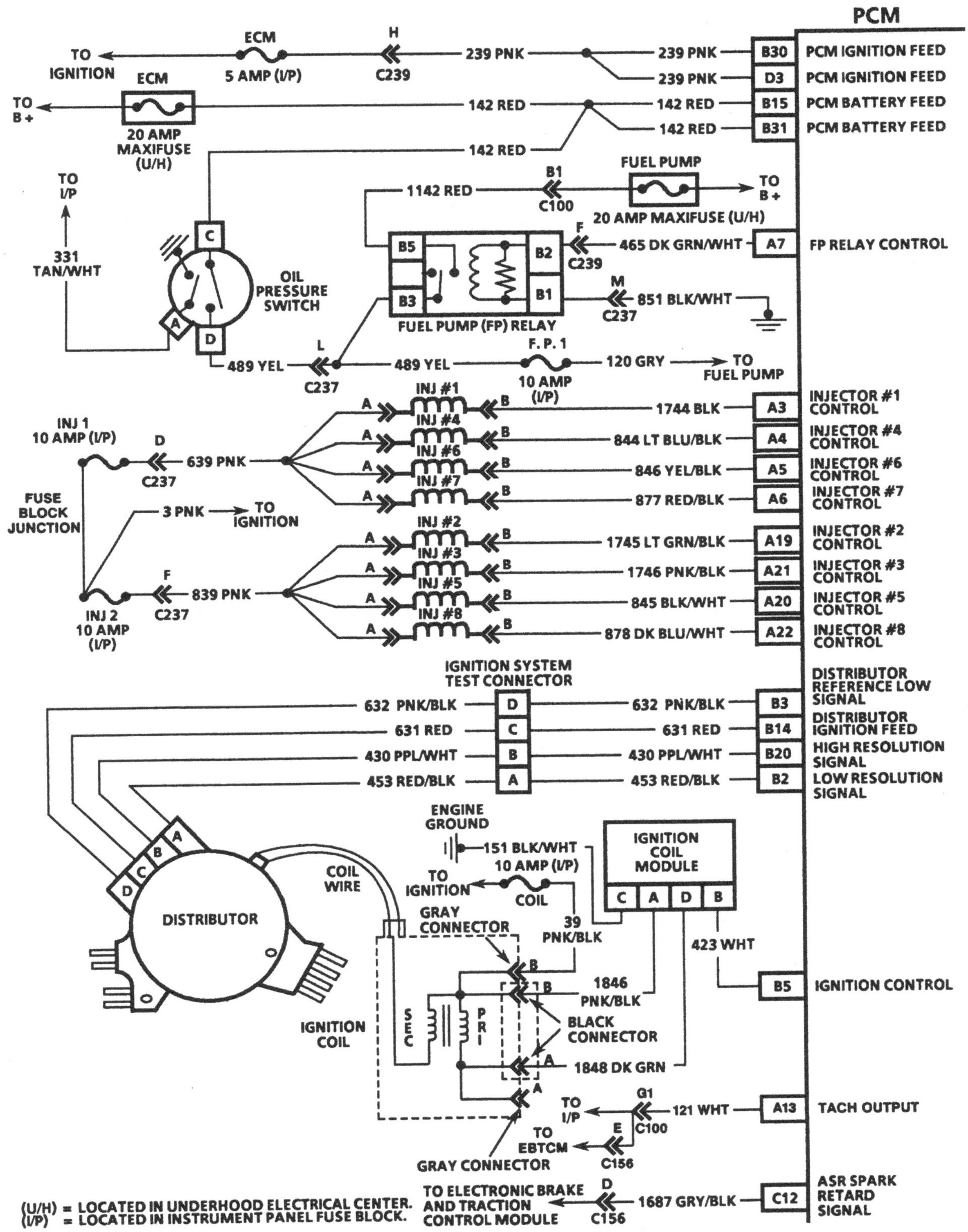

(U/H) = LOCATED IN UNDERHOOD ELECTRICAL CENTER.
(I/P) = LOCATED IN INSTRUMENT PANEL FUSE BLOCK.

LT1 1994-95 SFI VIN P

Engine Cooling Fans, HVAC

Figure 5 of 7

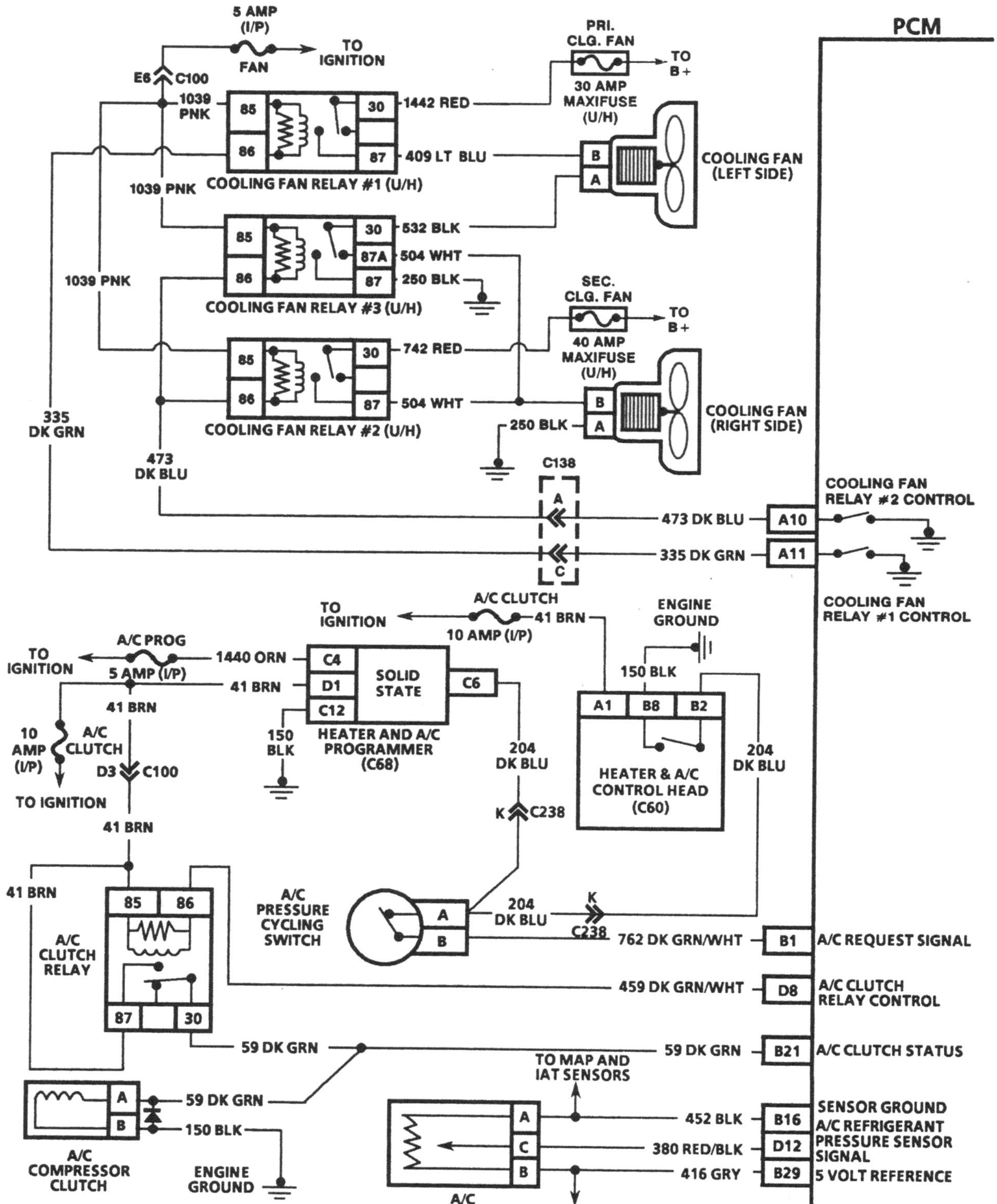

(U/H) = LOCATED IN UNDERHOOD ELECTRICAL CENTER.
(I/P) = LOCATED IN INSTRUMENT PANEL FUSE BLOCK.

DLC, misc.

Figure 6 of 7

LT1 1994-95 SFI VIN P

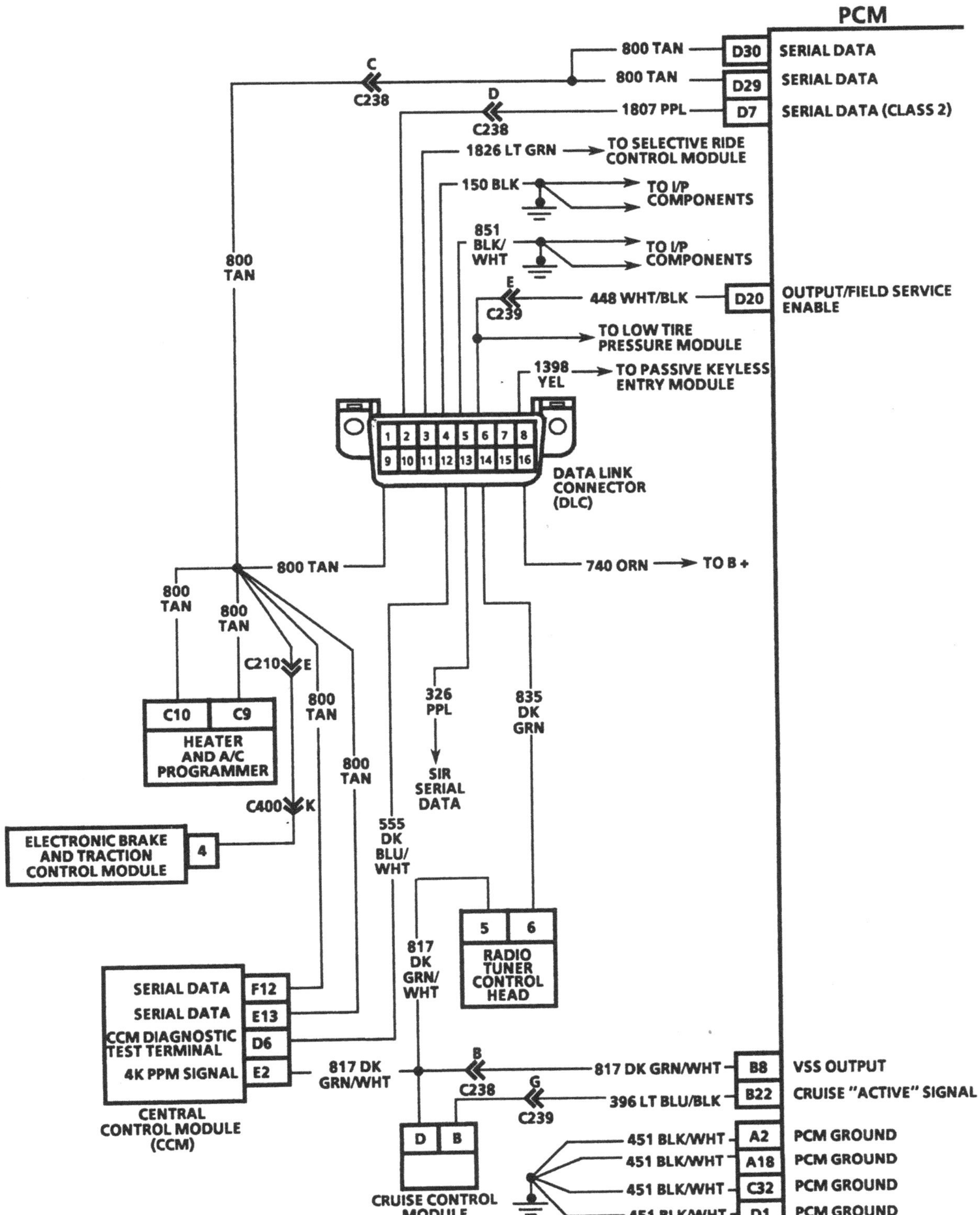

12

LT1 1994-95 SFI VIN P

A/T

Figure 7 of 7

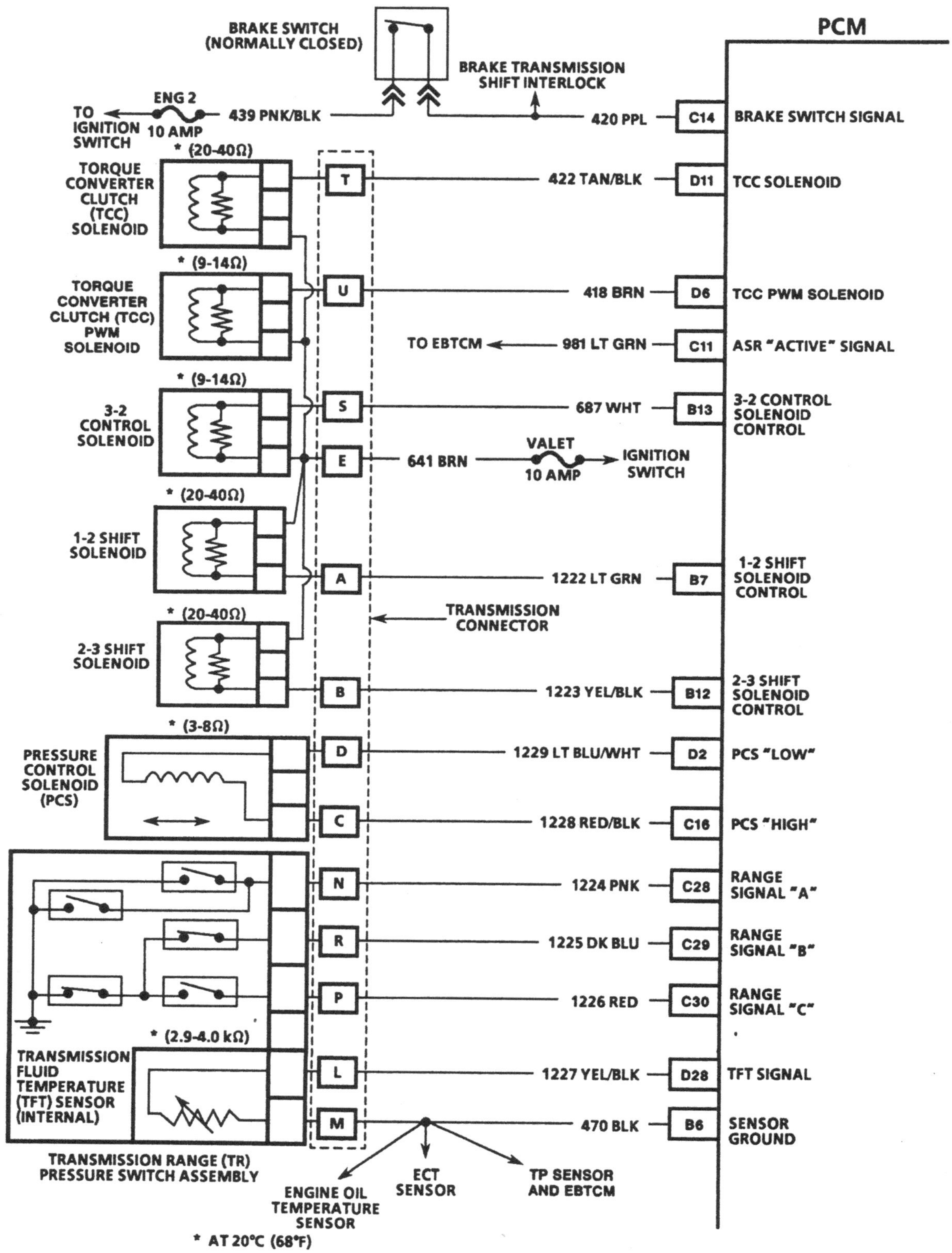

Engine Data Sensors

Figure 1 of 7

LT1 1992-93
MFI VIN P

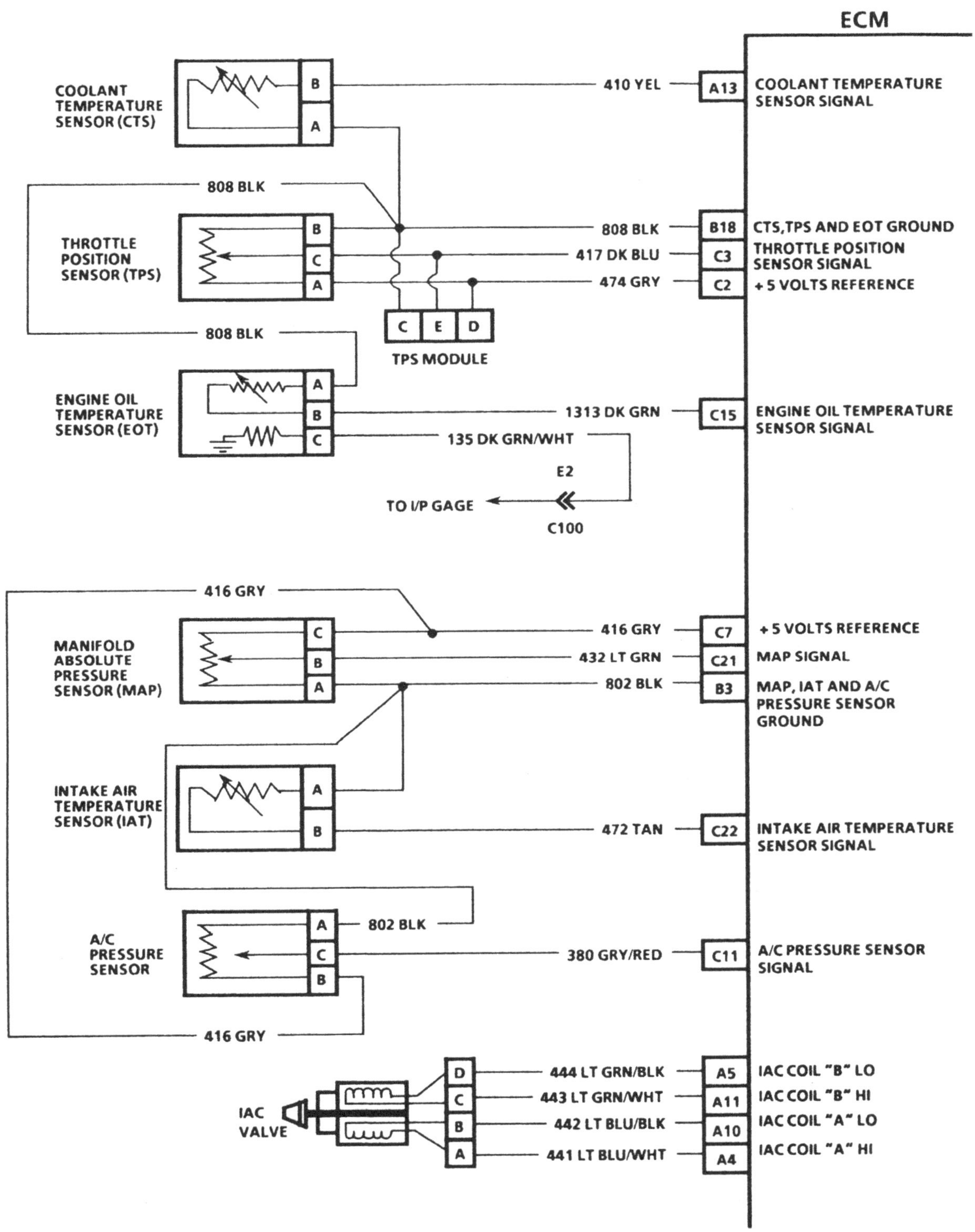

LT1 1992-93 MFI VIN P

A/T, M/T, EVAP, EGR

Figure 2 of 7

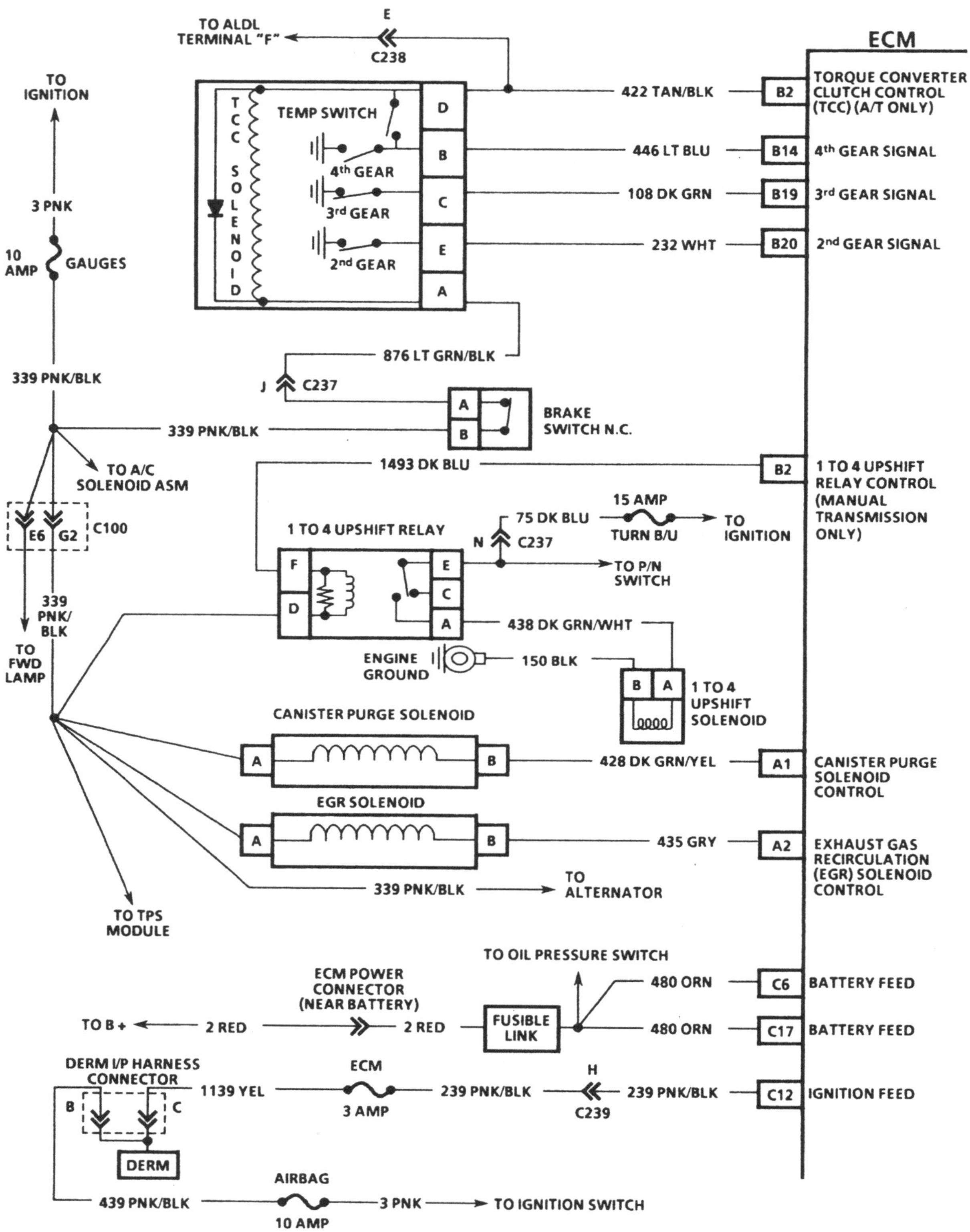

O_2S, PNR, VSS, KS

Figure 3 of 7

LT1 1992-93 MFI VIN P

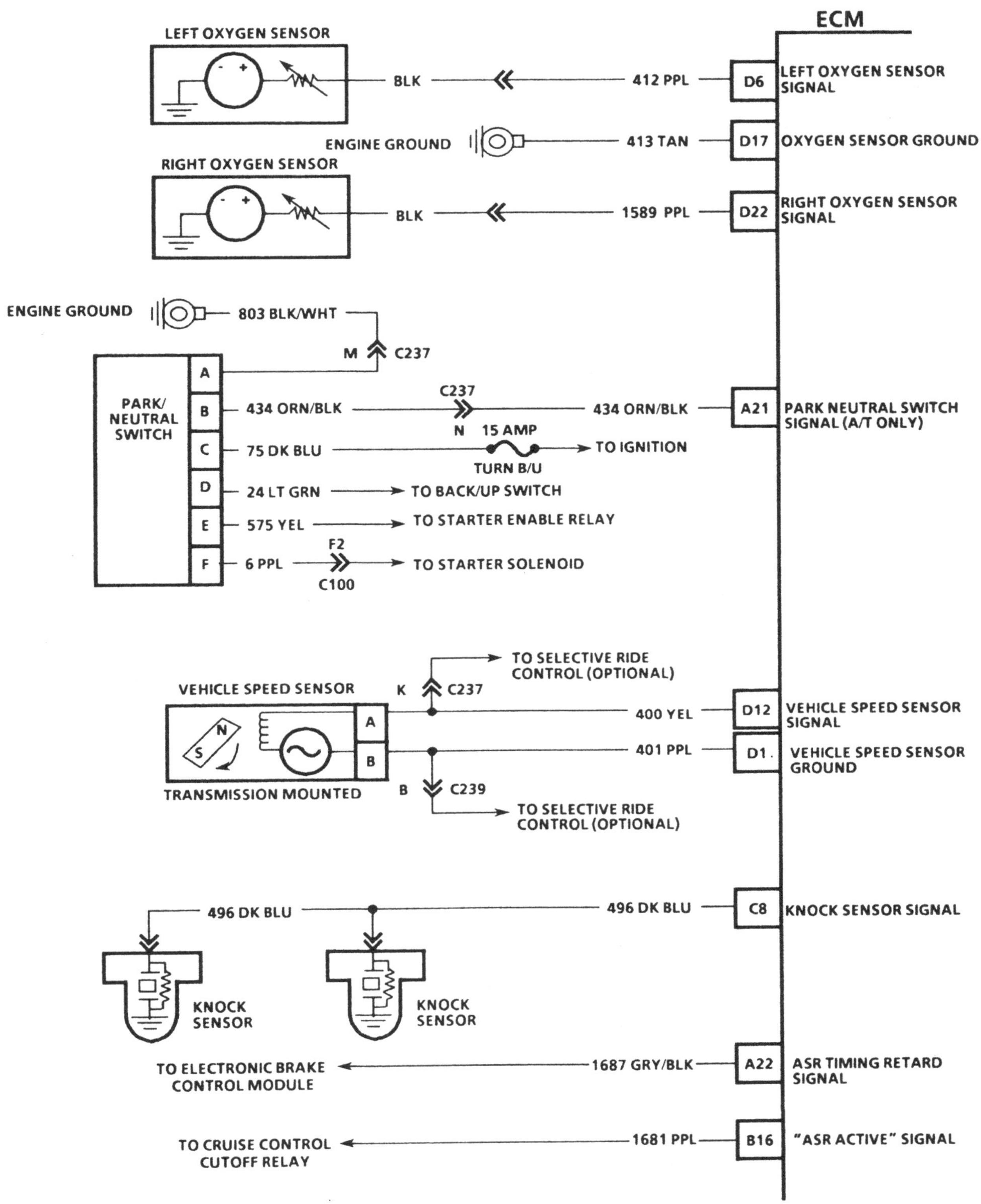

LT1 1992-93 MFI VIN P

Injectors, Fuel Pump Relay

Figure 4 of 7

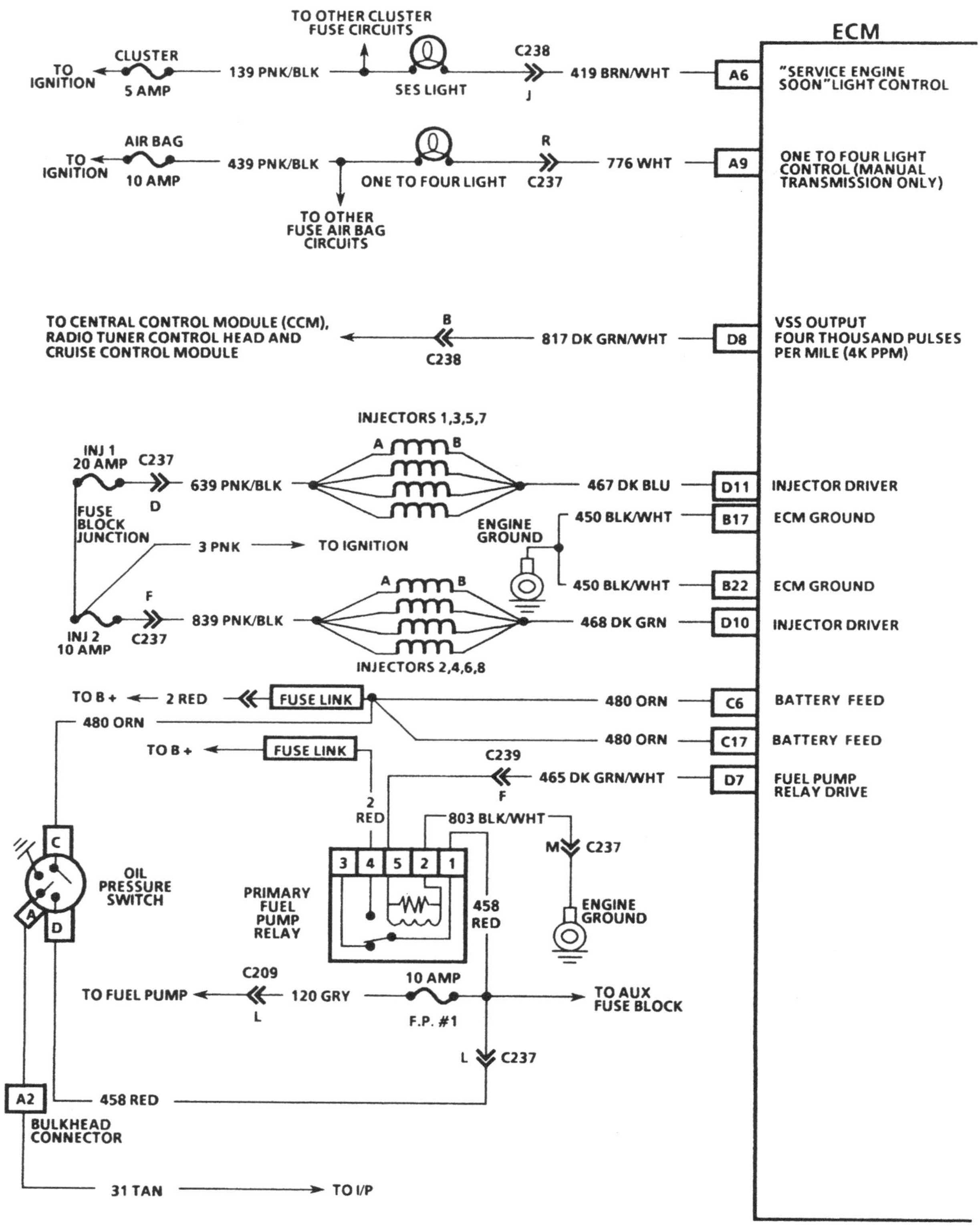

Engine Cooling Fans, HVAC

LT1 1992-93 MFI VIN P

Figure 5 of 7

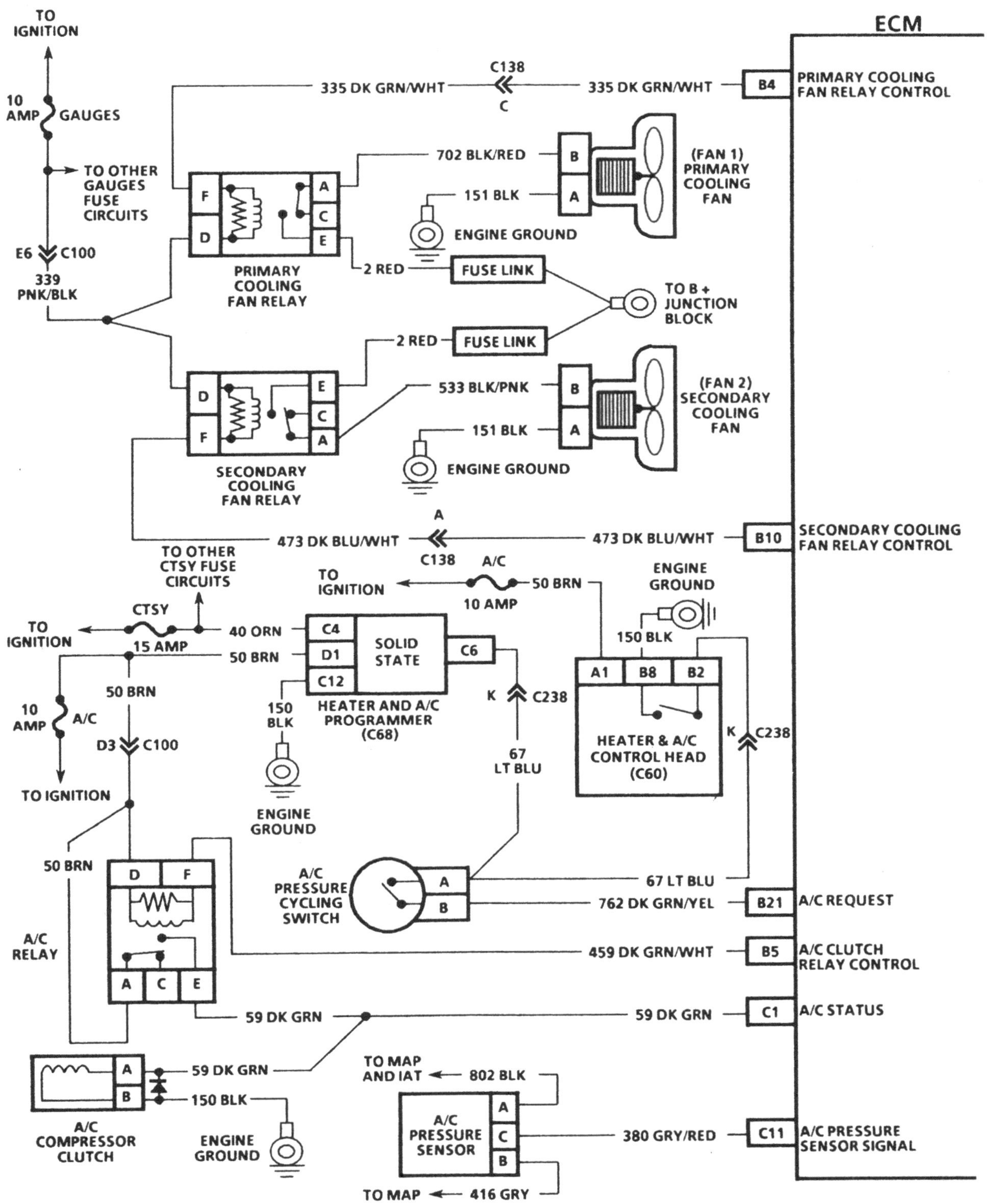

12

LT1 1992-93
MFI VIN P

Distributor, AIR

Figure 6 of 7

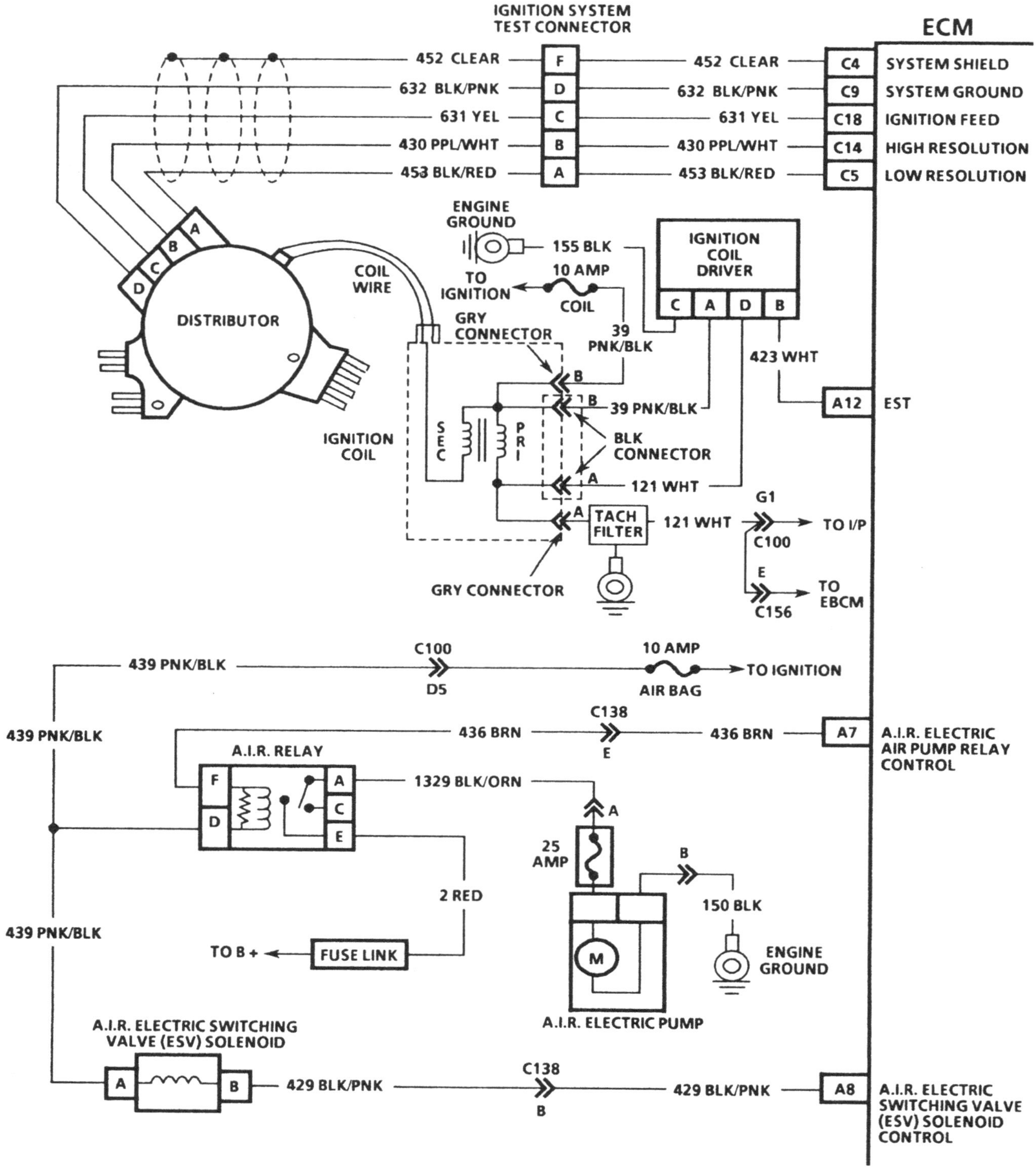

DLC, misc.

Figure 7 of 7

LT1 1992-93 MFI VIN P

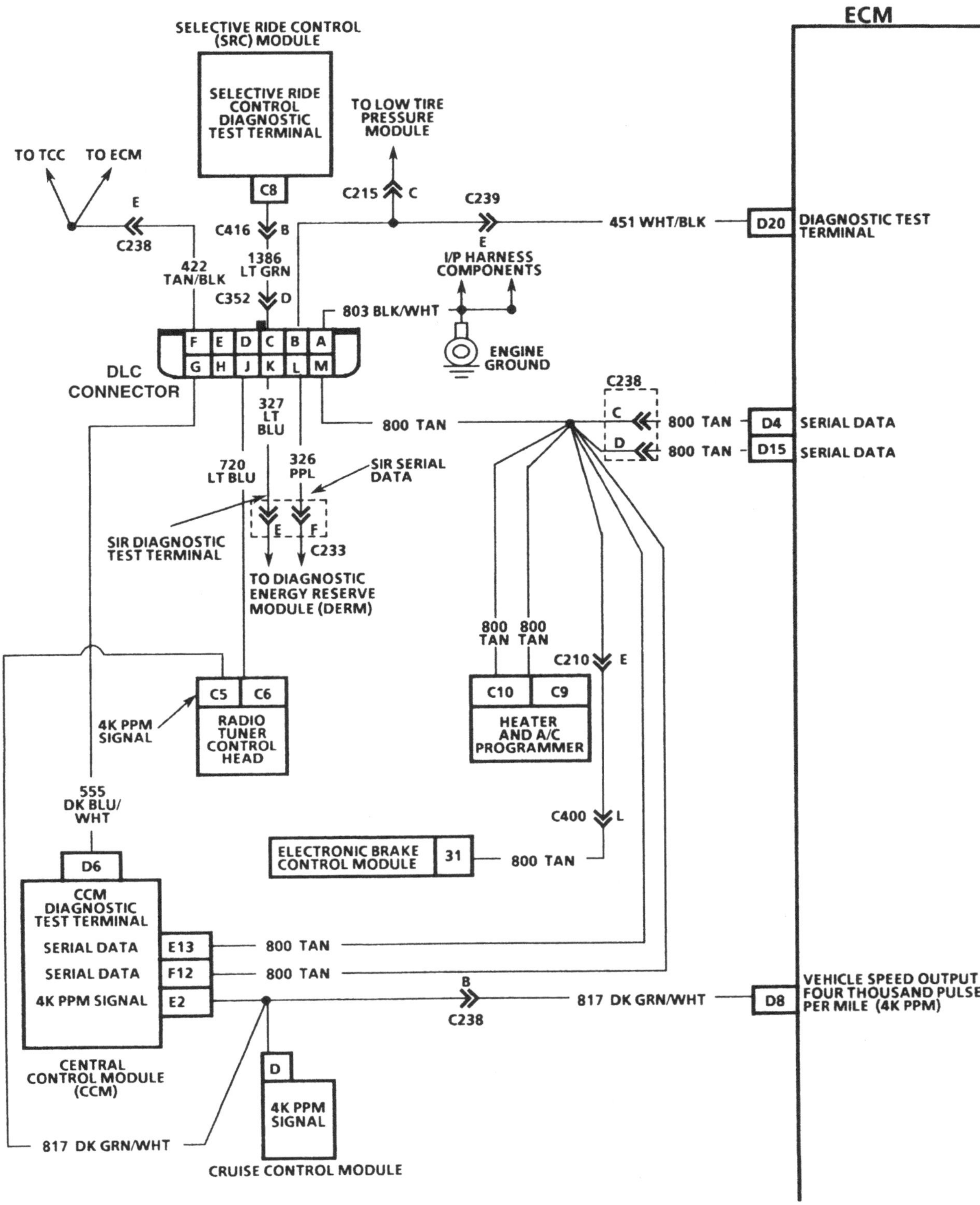

12

Engine Data Sensors, Ignition

L98 1986 MFI 8

Figure 1 of 4

For additional schematics and technical notes on the L98 engine covering years 1985-1991, please visit the following Bentley Publishers web address:

http://www.BentleyPublishers.com/errata/

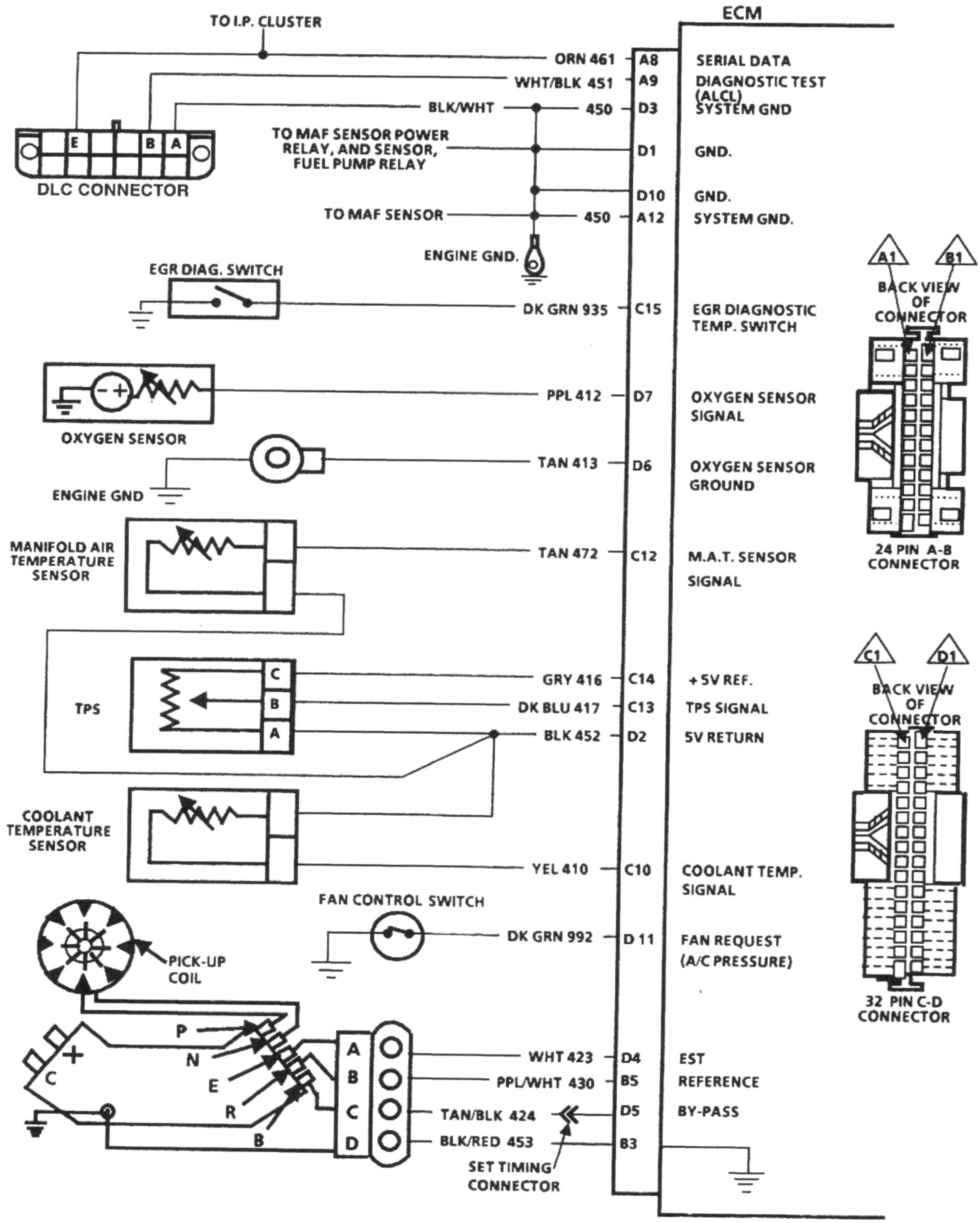

IAC, KS, ESC, Injectors

L98 1986 MFI 8

Figure 2 of 4

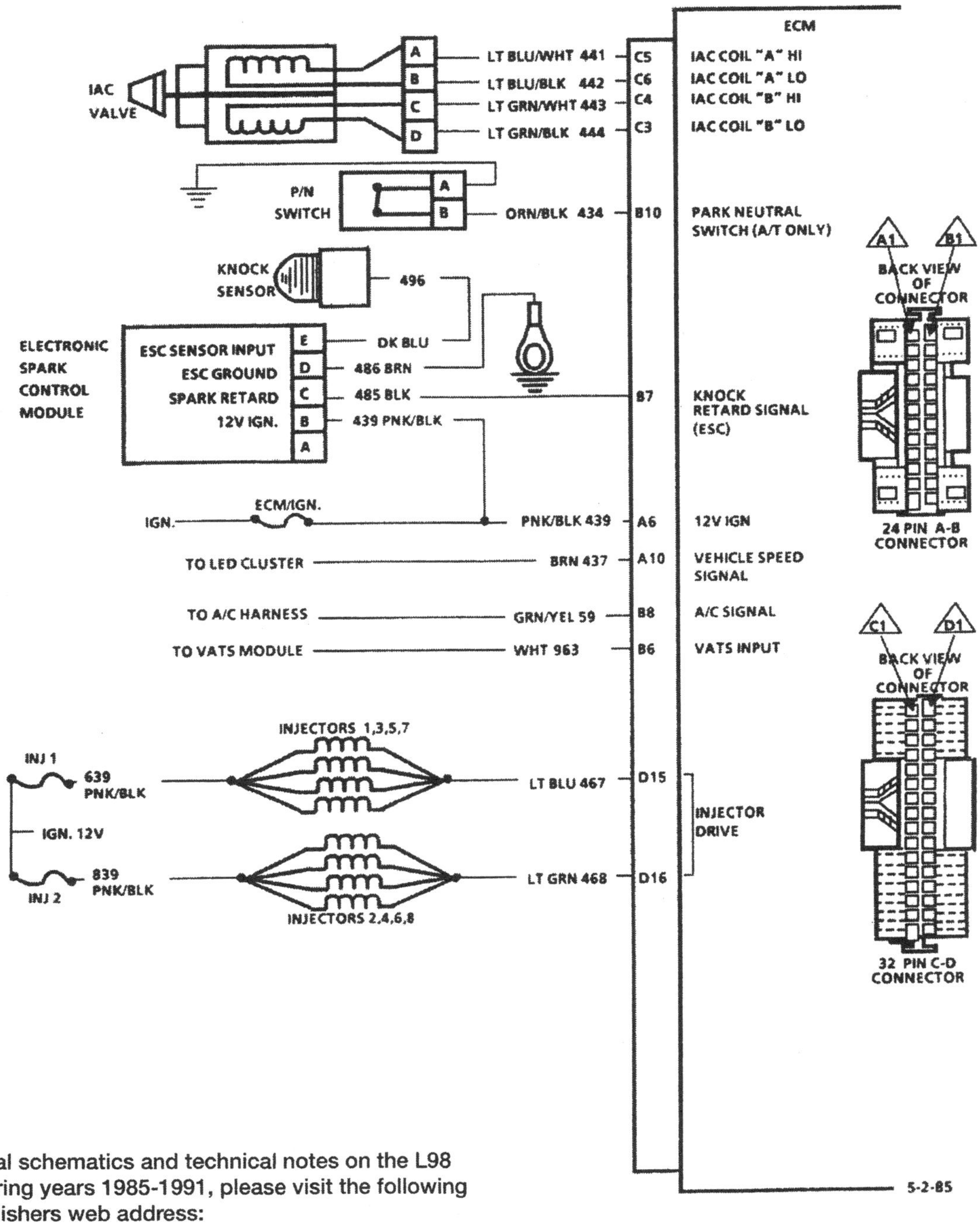

For additional schematics and technical notes on the L98 engine covering years 1985-1991, please visit the following Bentley Publishers web address:

`http://www.BentleyPublishers.com/errata/`

L98 1986 MFI 8

Fuel Pump Relay, MAF

Figure 3 of 4

For additional schematics and technical notes on the L98 engine covering years 1985-1991, please visit the following Bentley Publishers web address:

`http://www.BentleyPublishers.com/errata/`

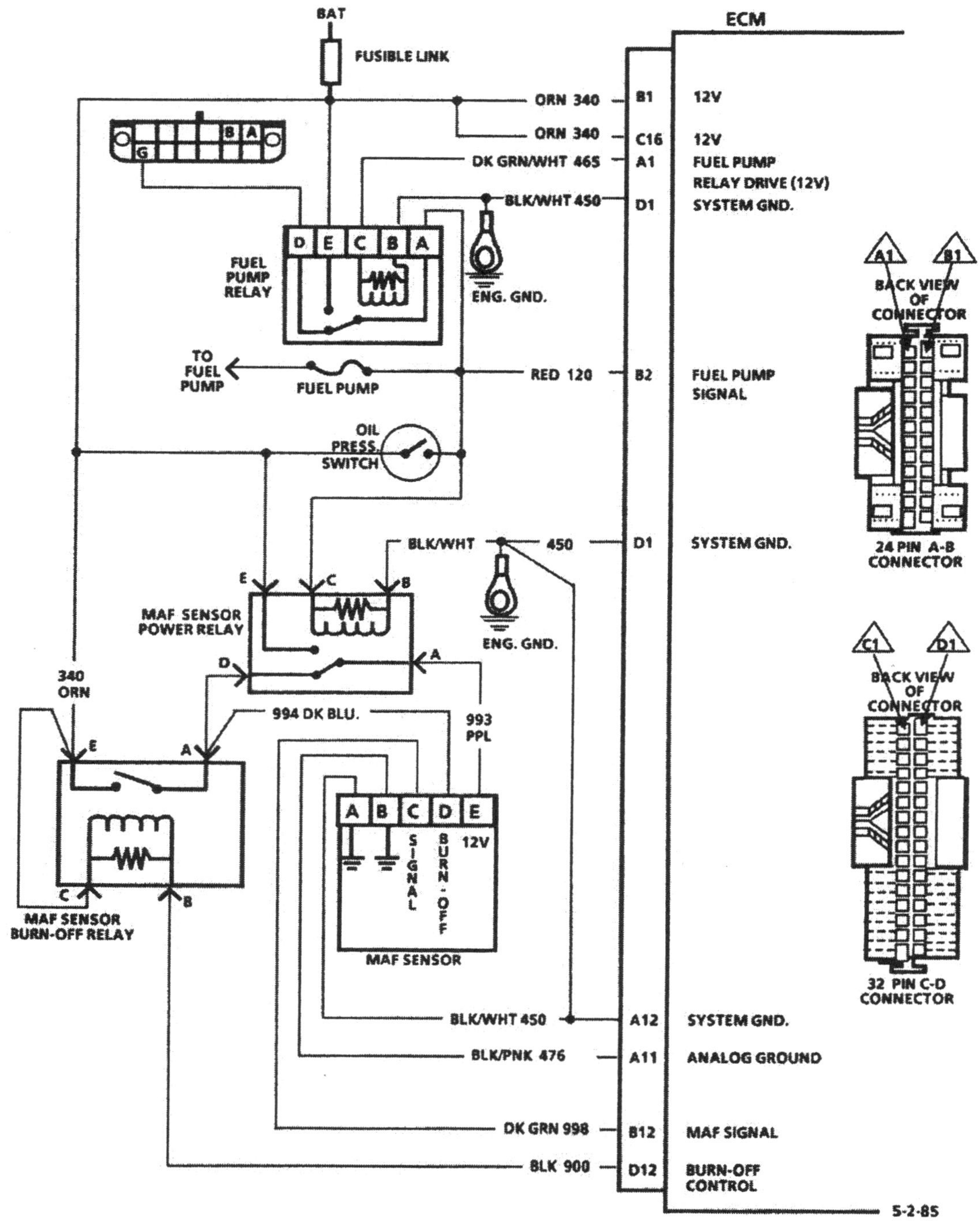

Figure 4 of 4

L98 1986 MFI 8

For additional schematics and technical notes on the L98 engine covering years 1985-1991, please visit the following Bentley Publishers web address:

`http://www.BentleyPublishers.com/errata/`

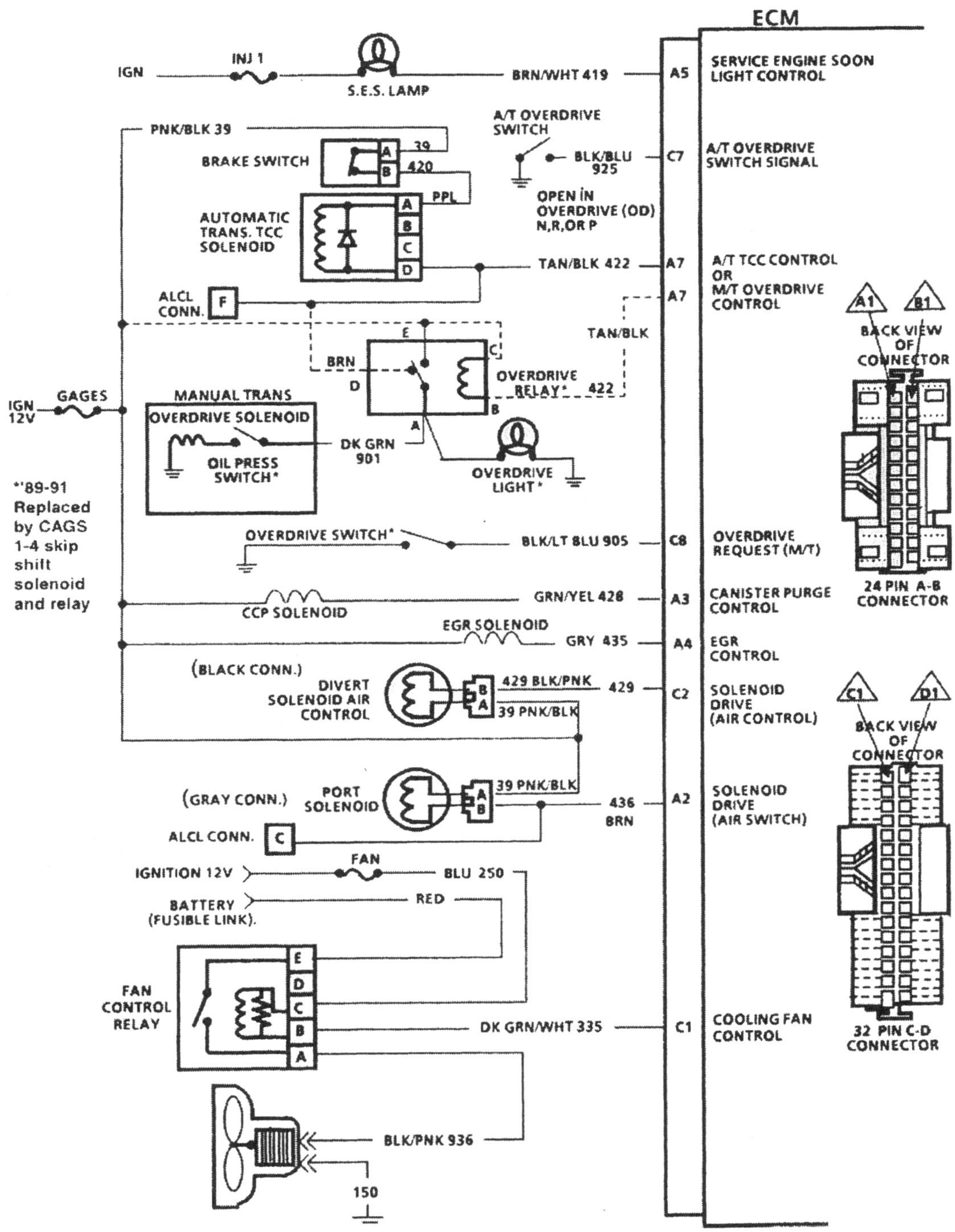

12

L83 1982, 84 TBI 8

White Connector

Figure 1 of 2

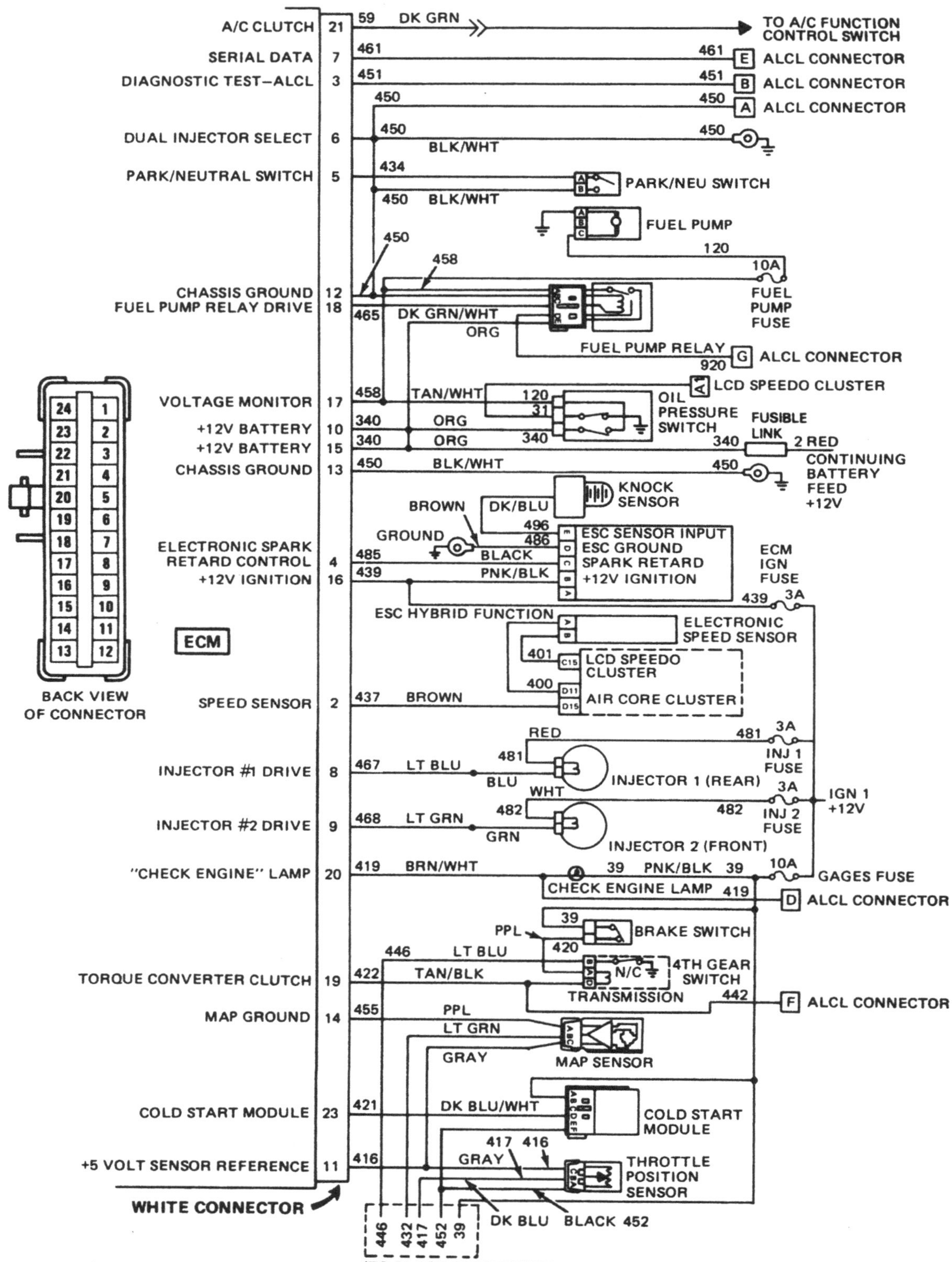

Black Connector

Figure 2 of 2

L83 1982, 84 TBI 8

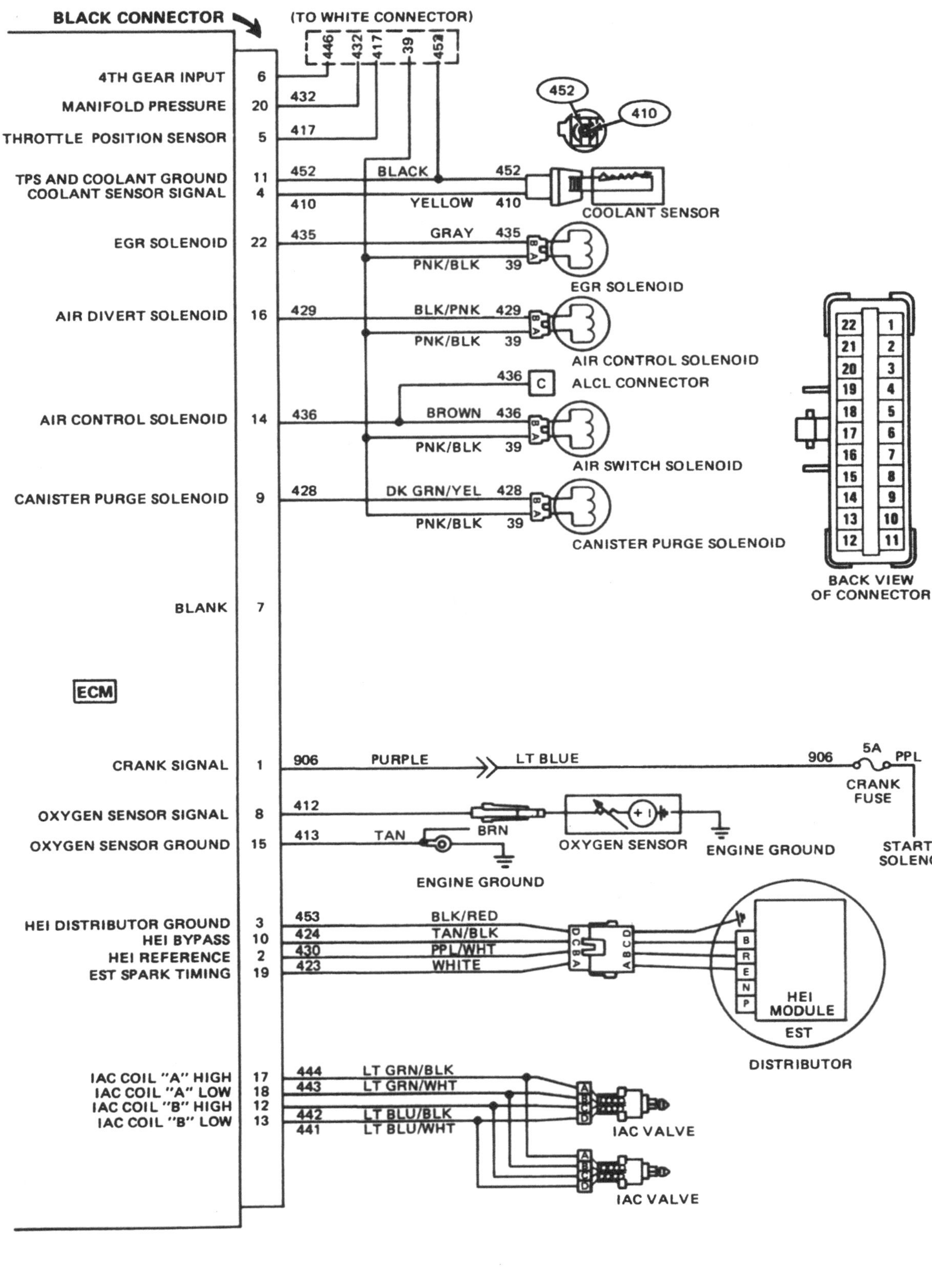

12

LT5 1993-95 MFI VIN J

Engine Data Sensors, IAC

Figure 1 of 8

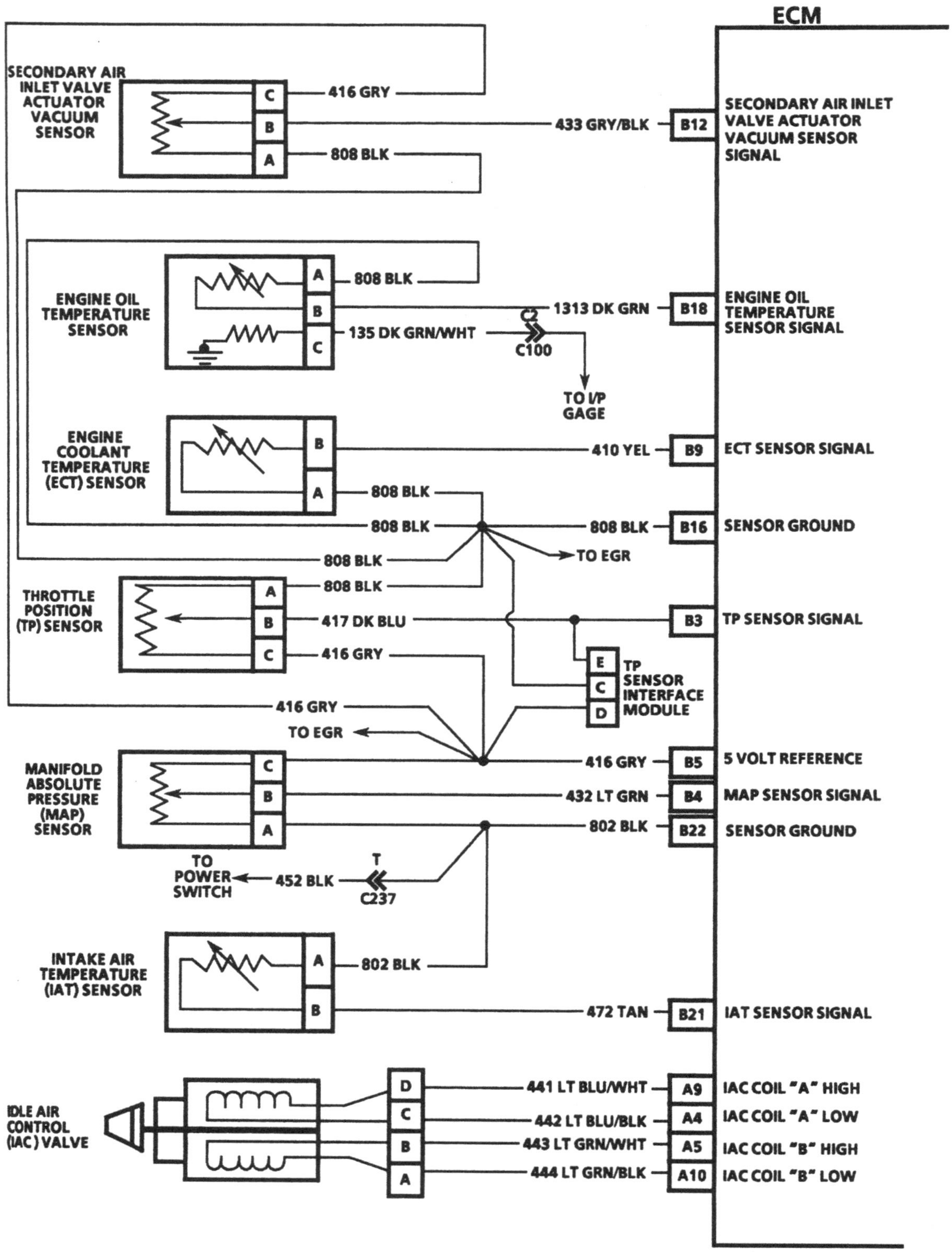

Injectors, Secondary SFI

Figure 2 of 8

LT5 1993-95 MFI VIN J

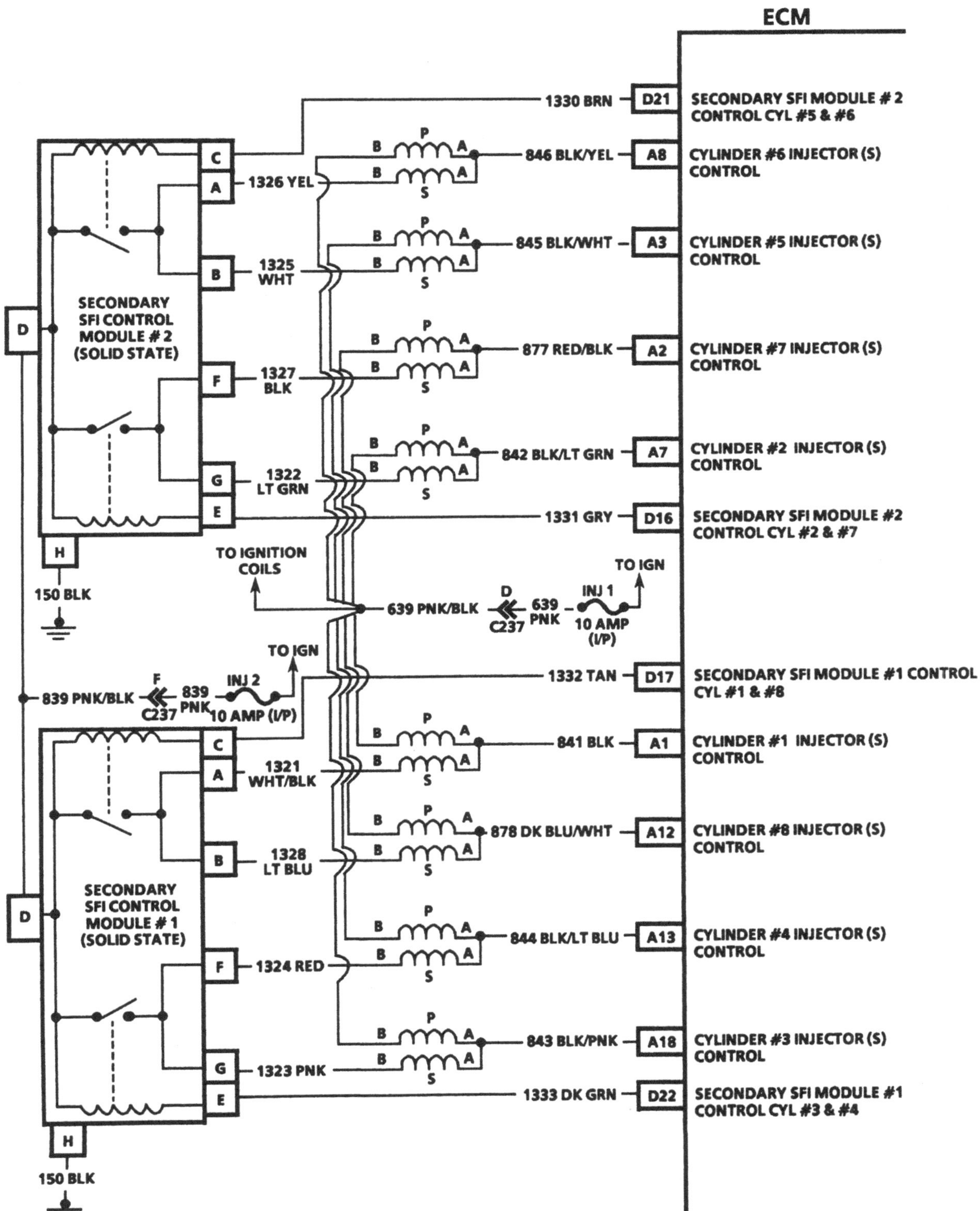

(I/P) = LOCATED IN INSTRUMENT PANEL FUSE BLOCK

12

LT5 1993-95 MFI VIN J

CMP, VSS, KS, Fuel Pumps

Figure 3 of 8

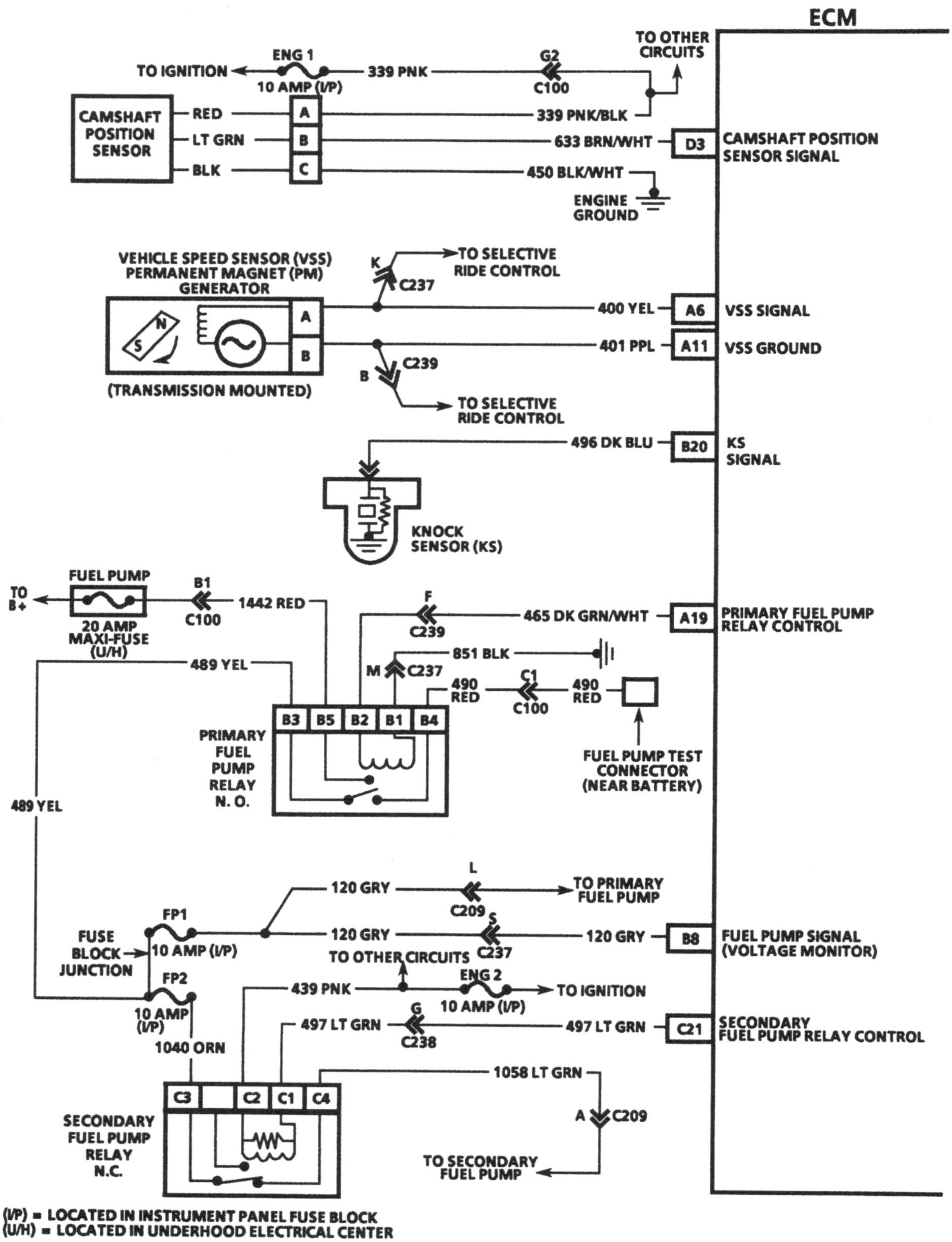

Ignition

LT5 1993-95 MFI VIN J

Figure 4 of 8

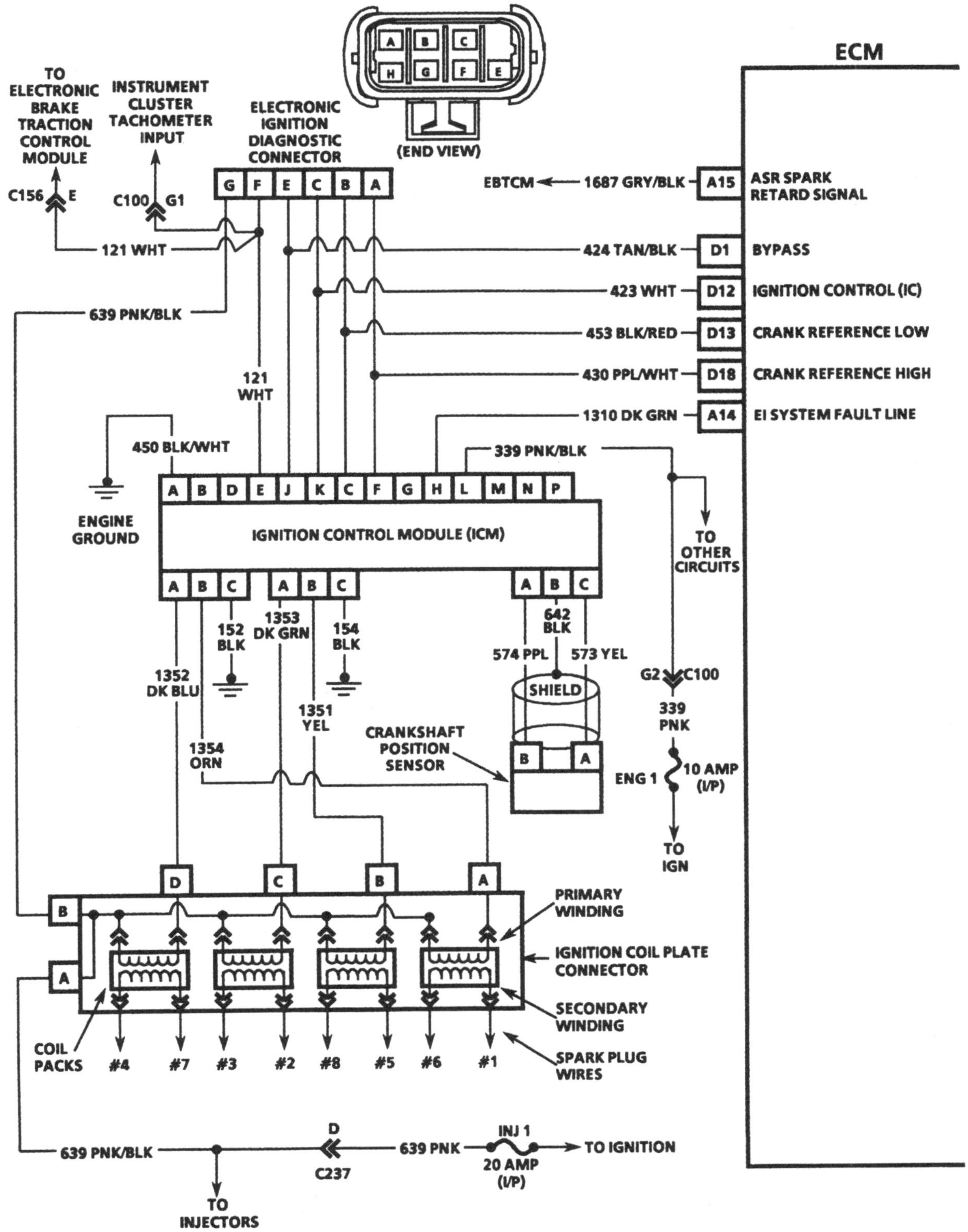

(I/P) = LOCATED IN INSTRUMENT PANEL FUSE BLOCK

12

LT5 1993-95 MFI VIN J

HO$_2$S, EGR, ENgine Power Mode

Figure 5 of 8

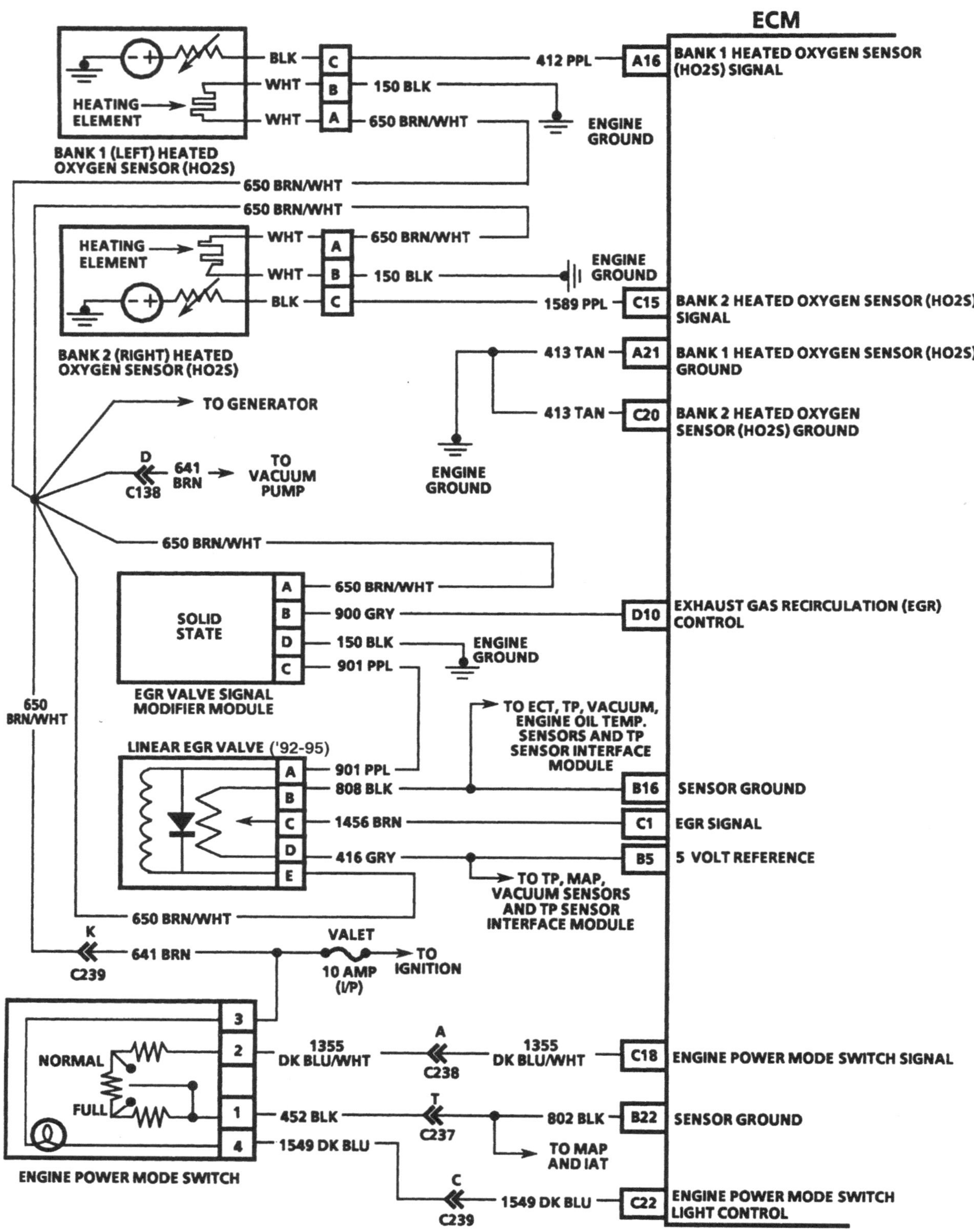

(I/P) = LOCATED IN INSTRUMENT PANEL FUSE BLOCK

HVAC, AIR

Figure 6 of 8

LT5 1993-95 MFI VIN J

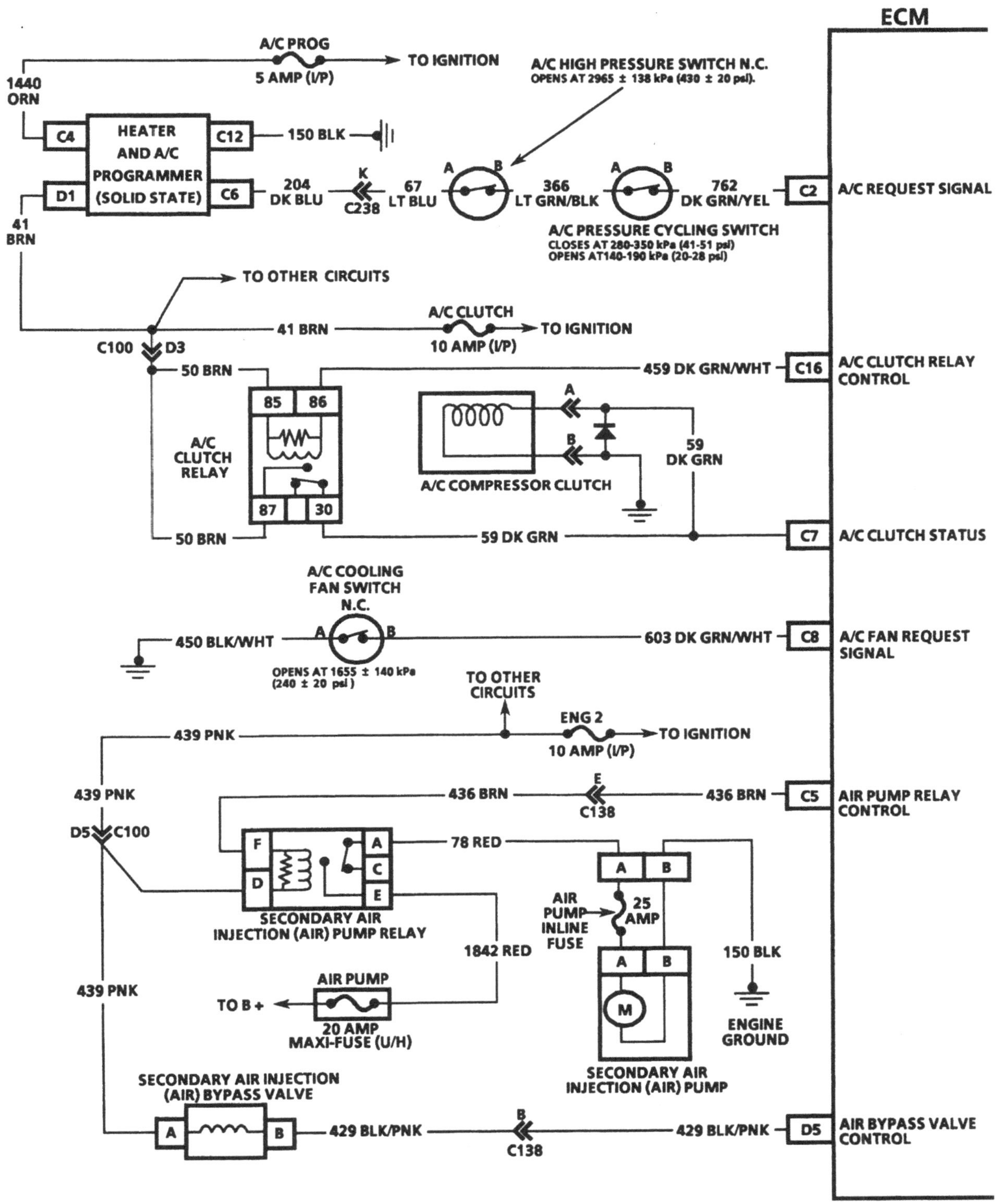

(I/P) = LOCATED IN INSTRUMENT PANEL FUSE BLOCK.
(U/H) = LOCATED IN UNDERHOOD ELECTRICAL CENTER.

LT5 1993-95 MFI VIN J

1-4, MIL, Cooling Fan Relays, AIR, EVAP

Figure 7 of 8

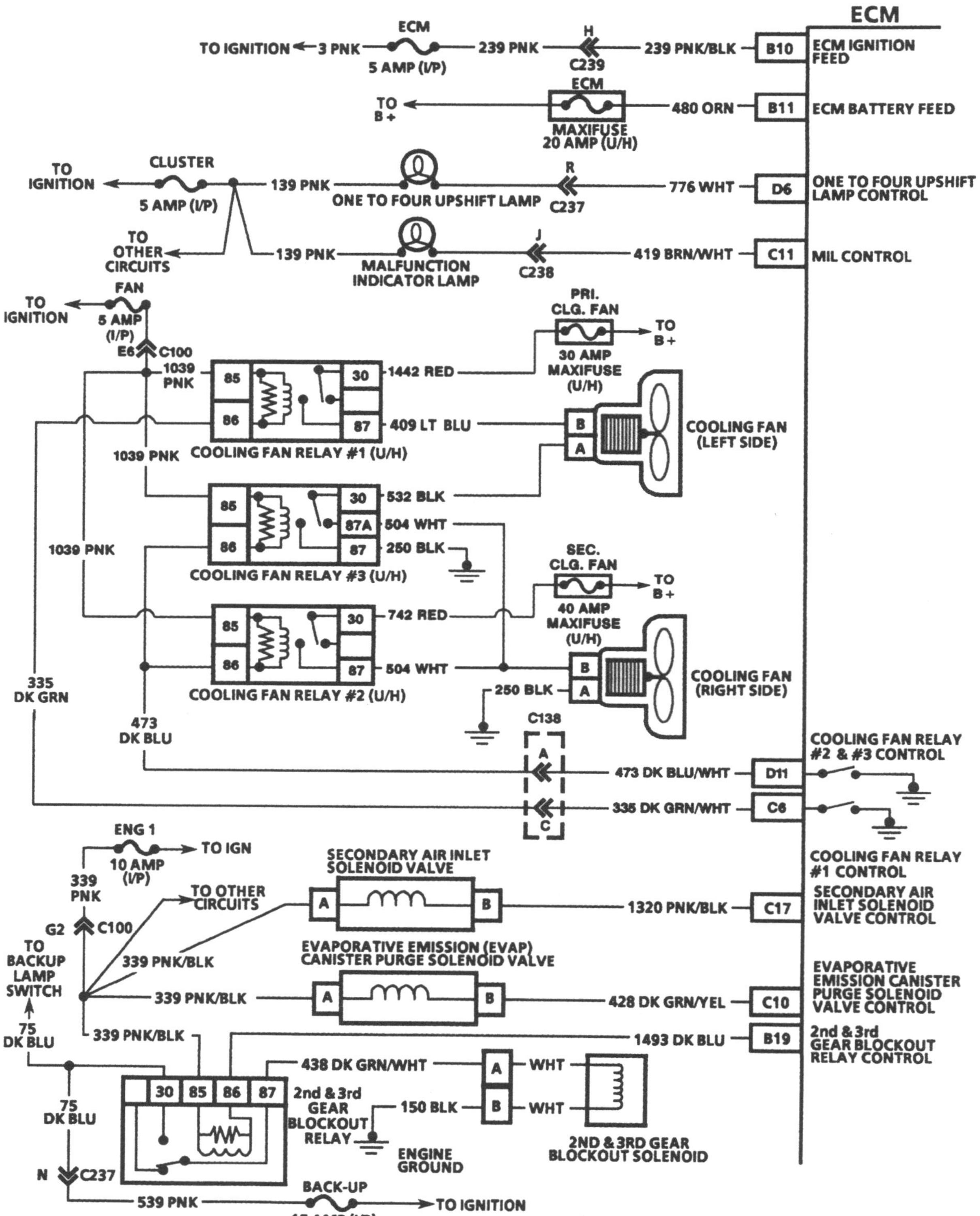

(U/H) - LOCATED IN UNDERHOOD ELECTRICAL CENTER
(I/P) - LOCATED IN INSTRUMENT PANEL FUSE BLOCK

DLC, misc.

Figure 8 of 8

LT5 1993-95 MFI VIN J

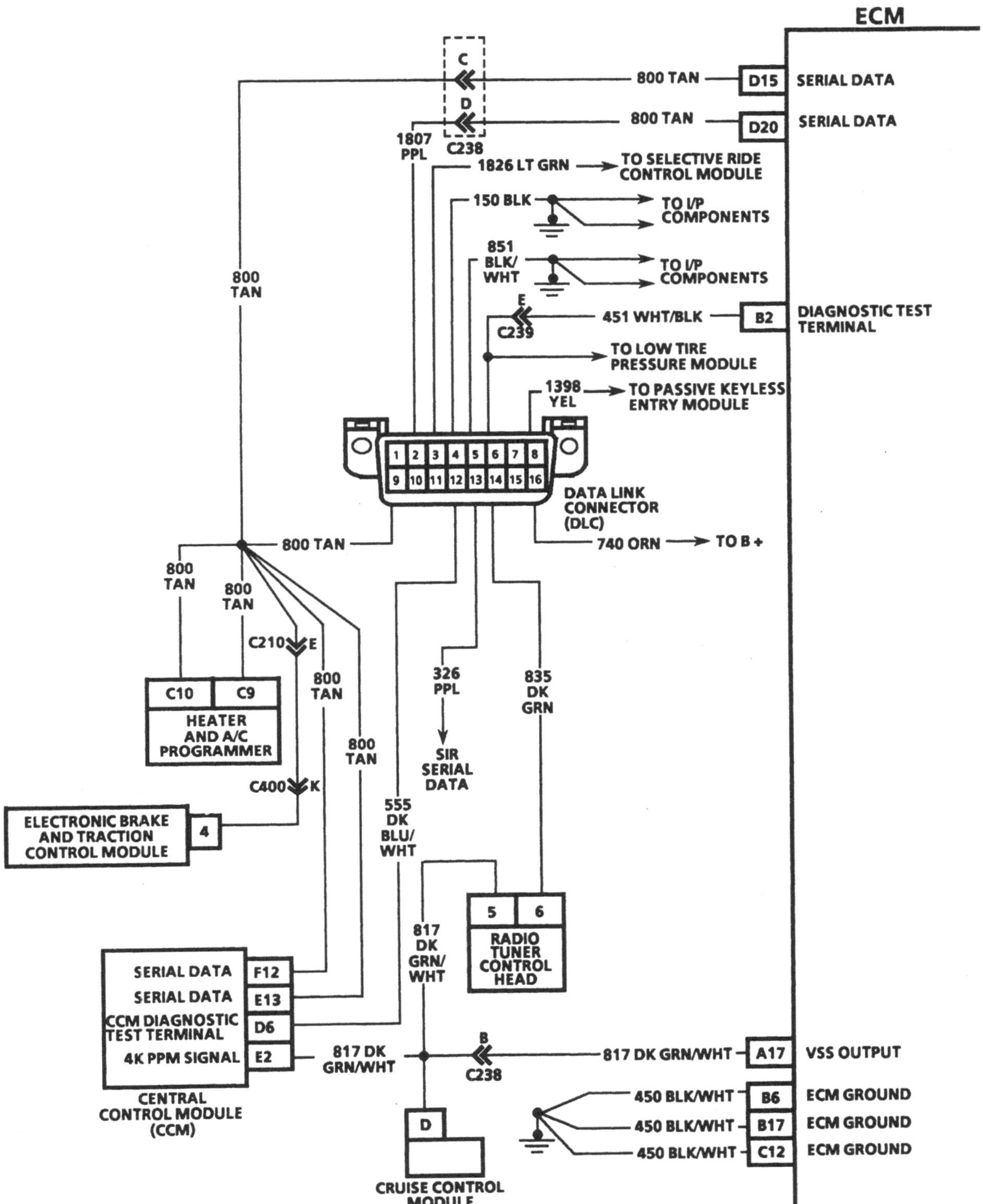

12

LT5 1990-92 MFI VIN J

Engine Data Sensors, IAC

Figure 1 of 8

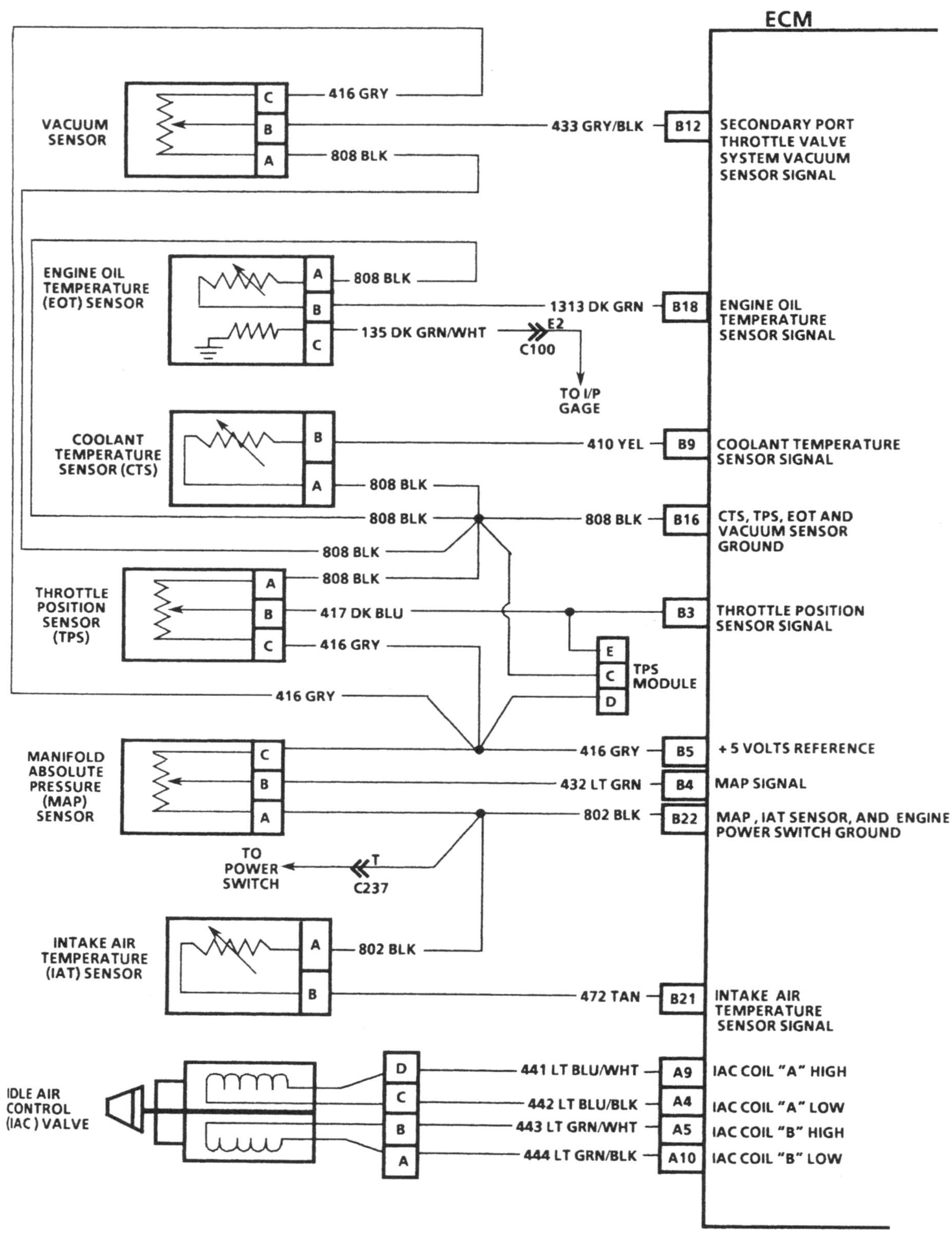

Injectors

LT5 1990-92 MFI VIN J

Figure 2 of 8

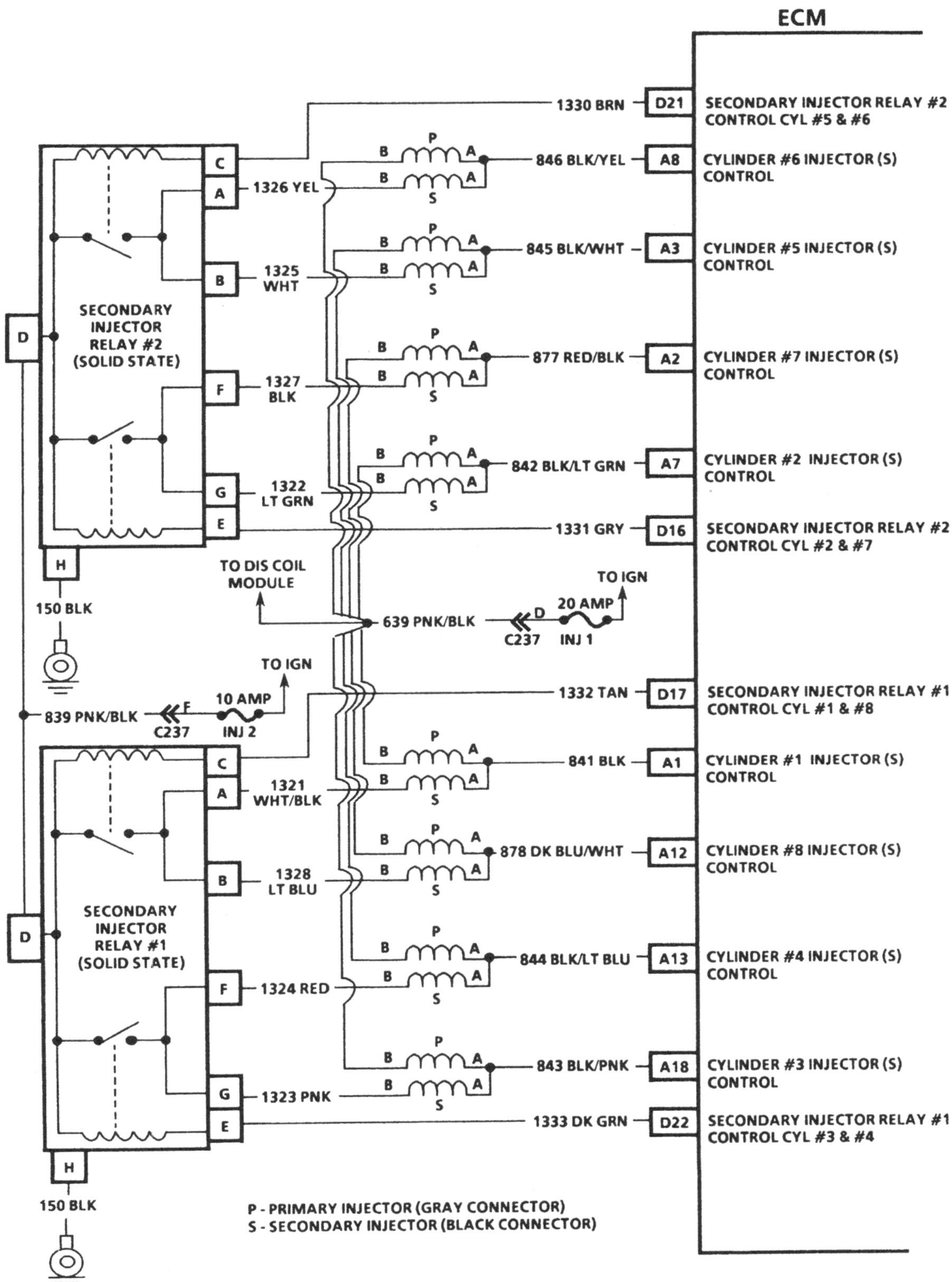

P - PRIMARY INJECTOR (GRAY CONNECTOR)
S - SECONDARY INJECTOR (BLACK CONNECTOR)

LT5 1990-92 MFI VIN J

CMP, Engine Power Mode, Fuel Pumps

Figure 3 of 8

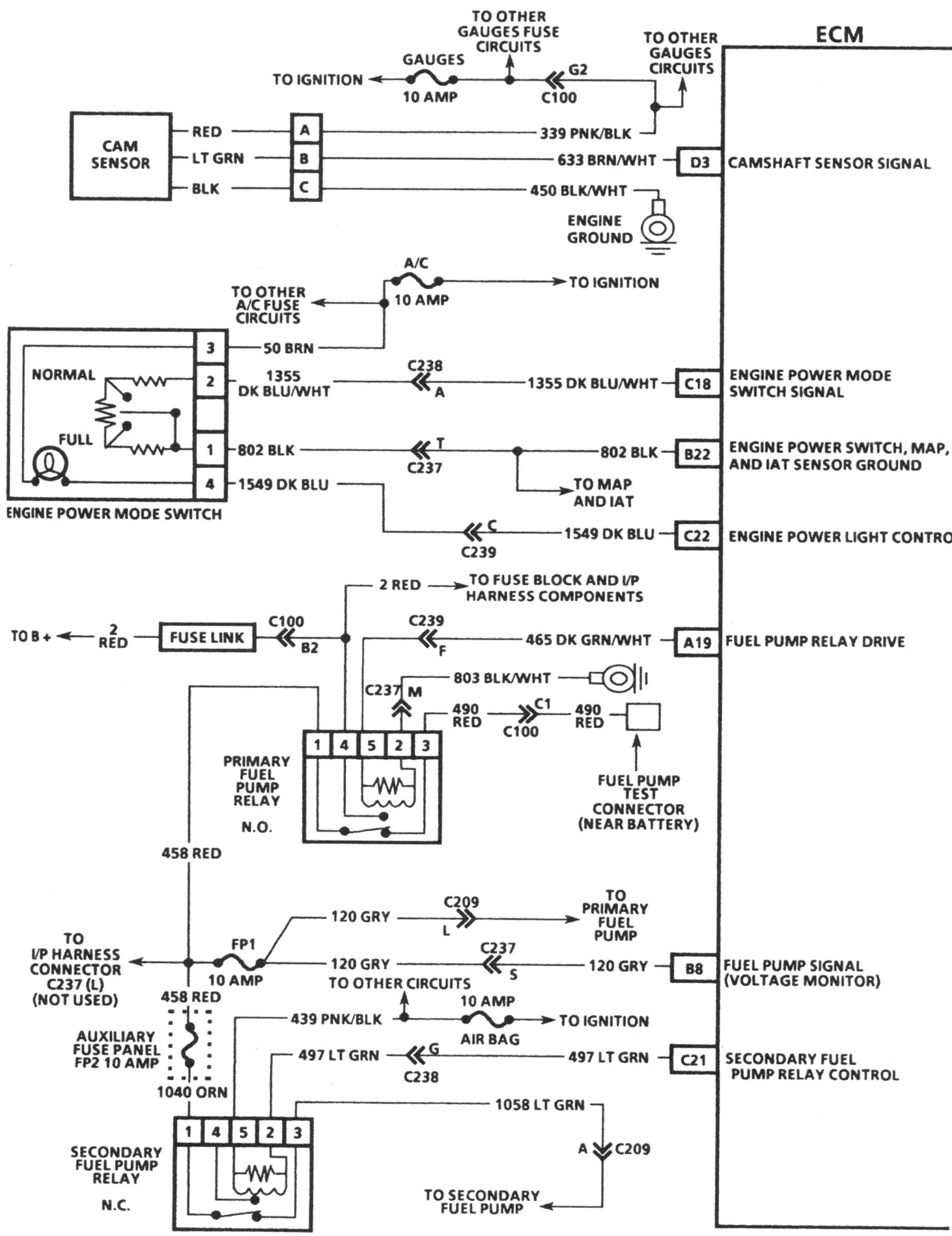

Ignition

Figure 4 of 8

LT5 1990-92
MFI VIN J

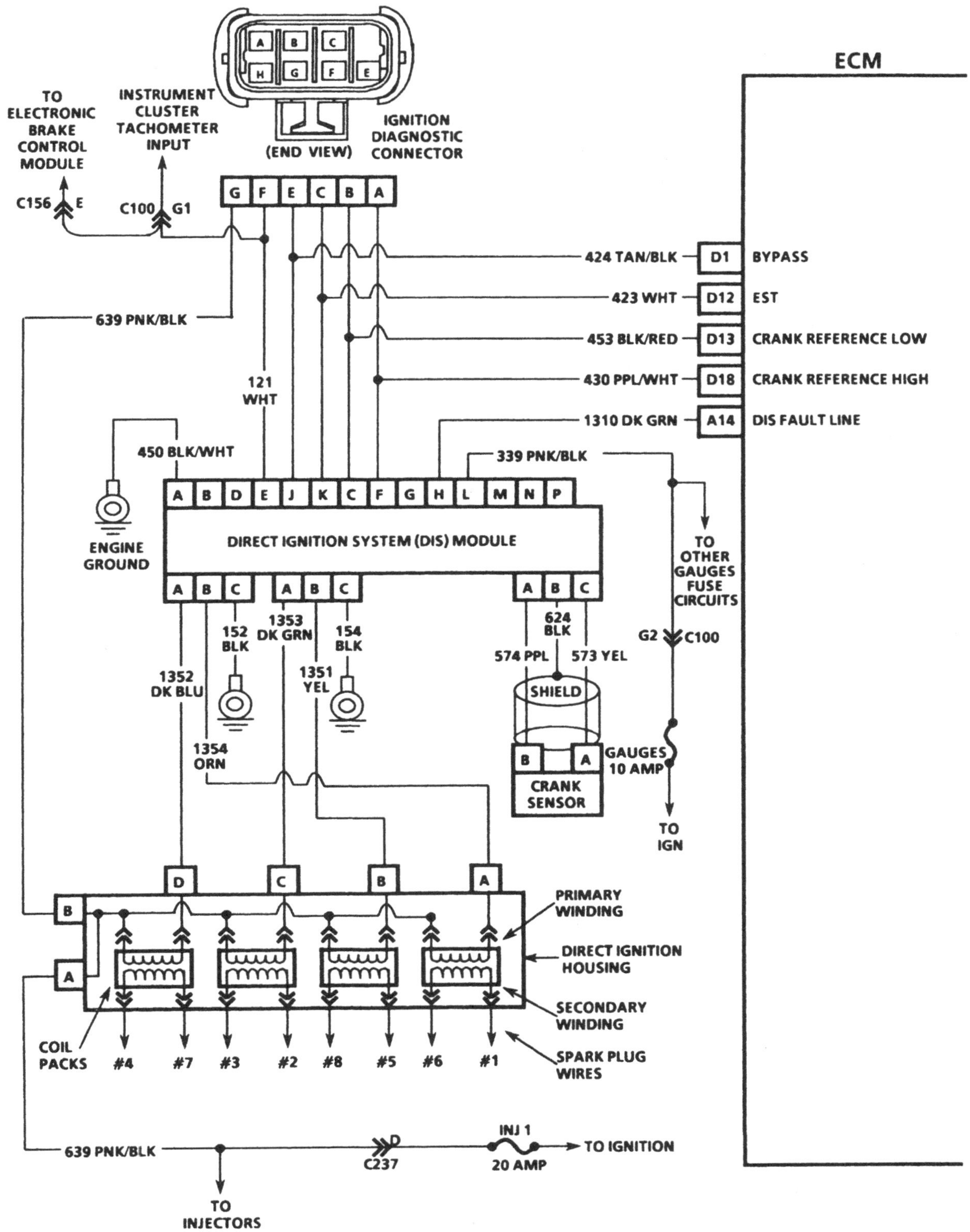

12

LT5 1990-92
MFI VIN J

O_2S, Knock Sensor, Vehicle Speed Sensor

Figure 5 of 8

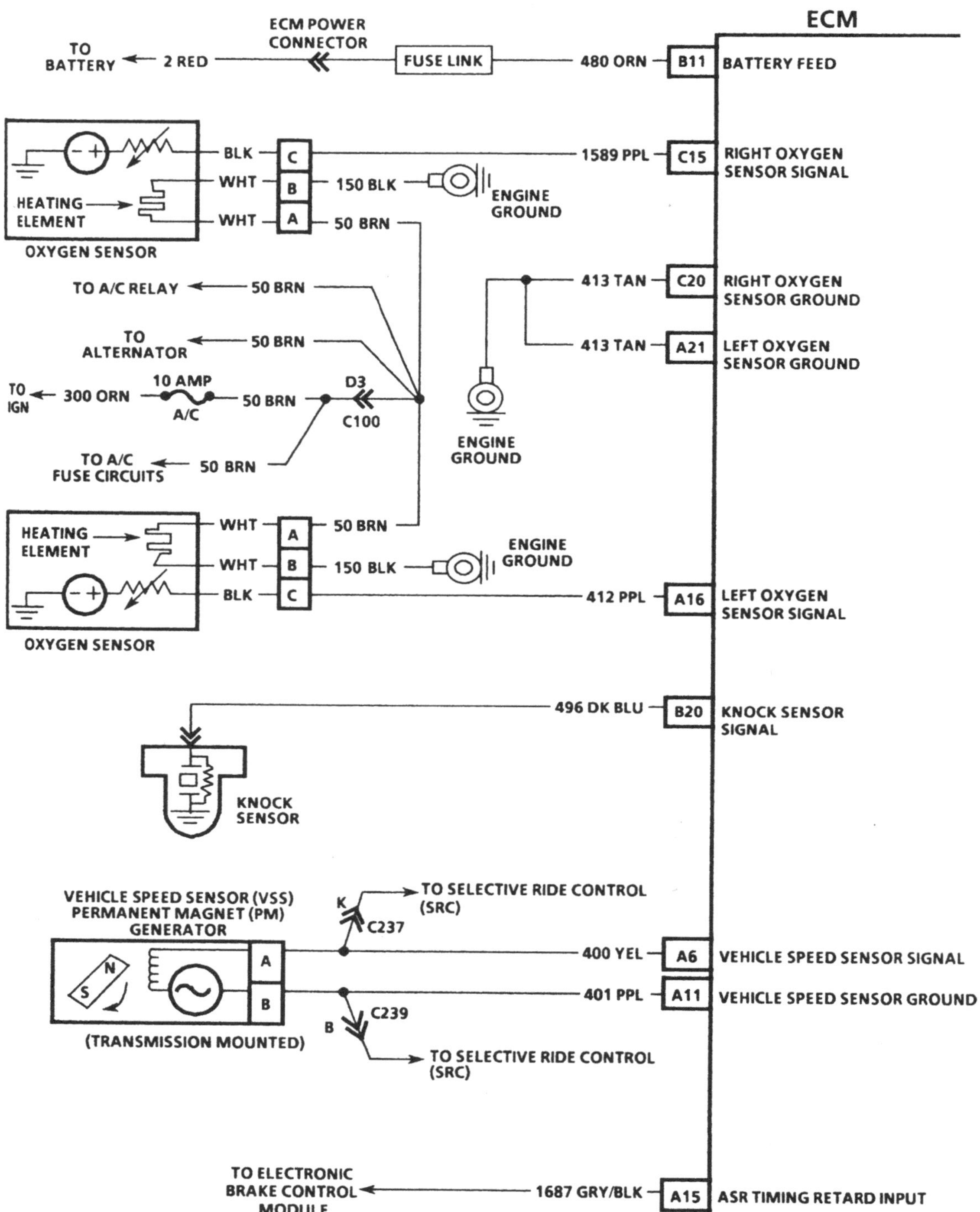

HVAC, AIR

LT5 1990-92 MFI VIN J

Figure 6 of 8

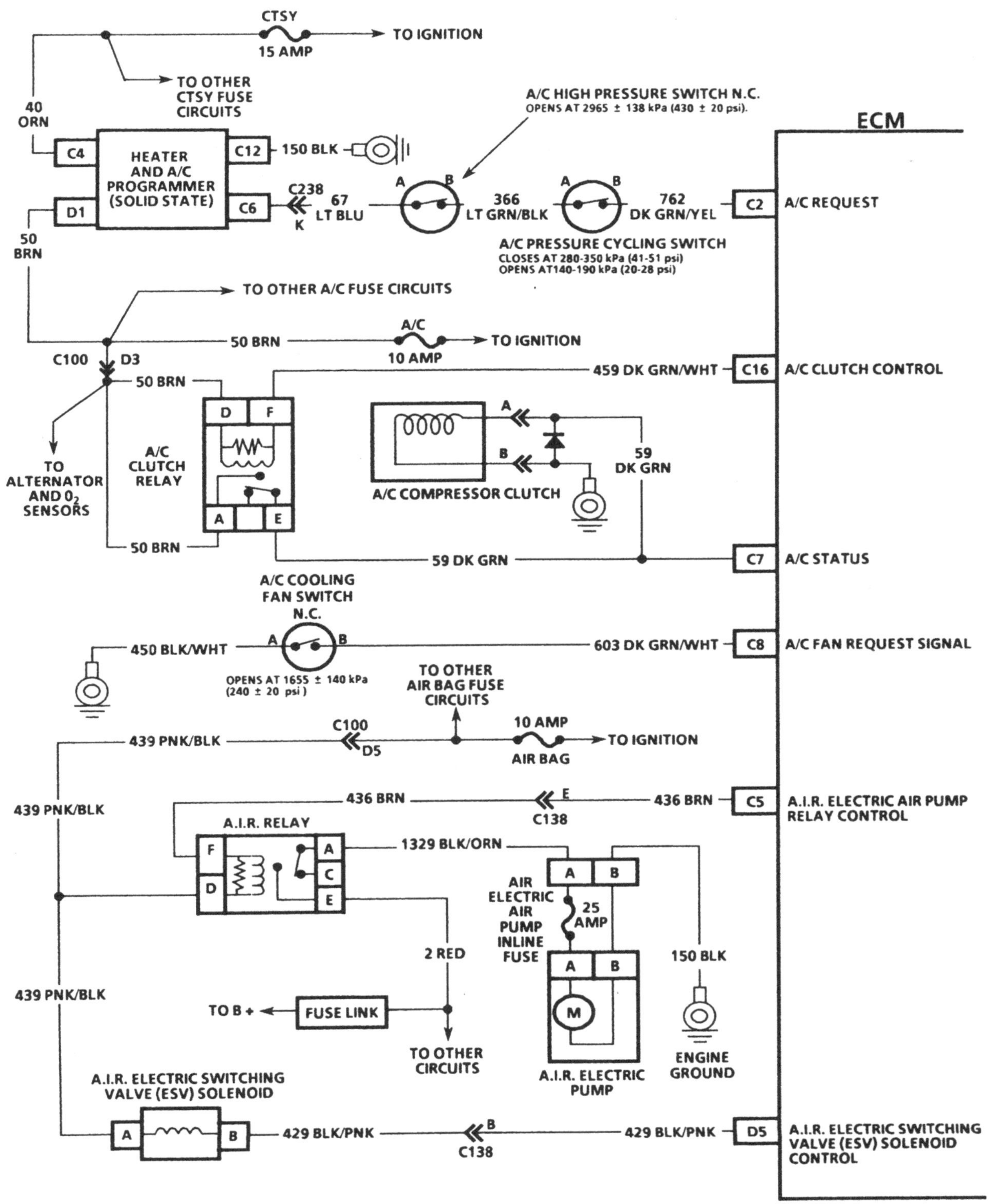

12

LT5 1990-92 MFI VIN J

1-4, Cooling Fan Relays

Figure 7 of 8

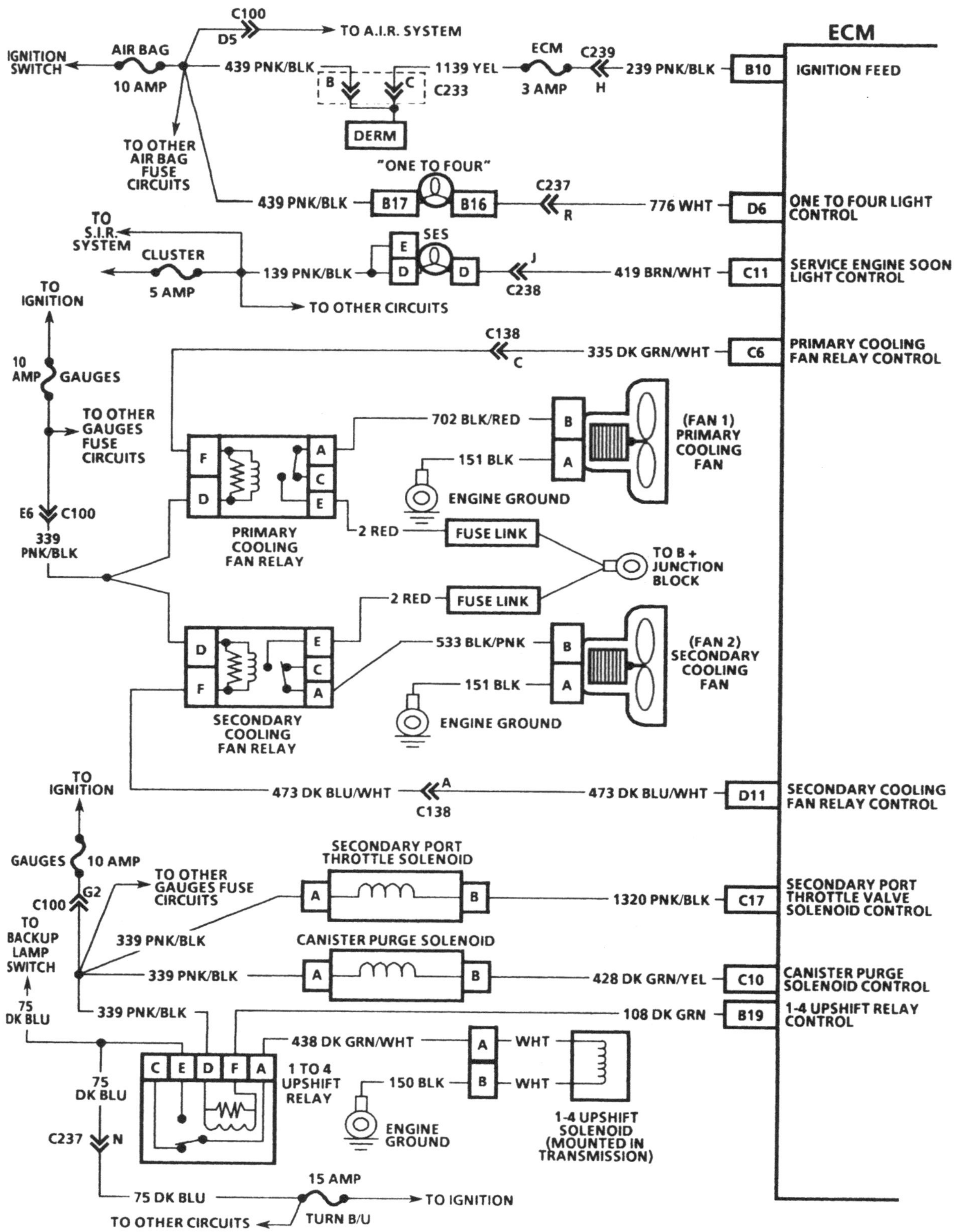

DLC, misc.

LT5 1990-92 MFI VIN J

Figure 8 of 8

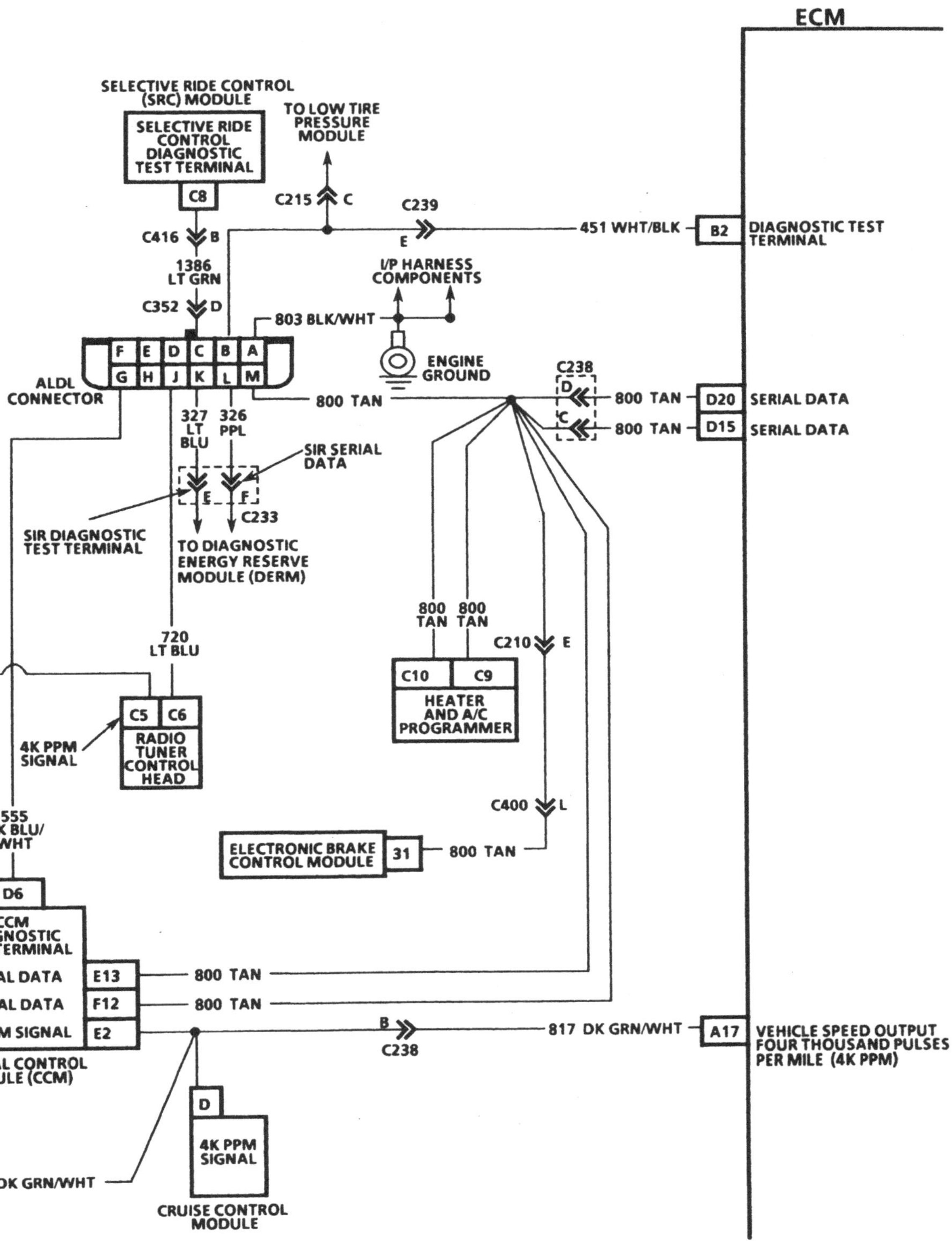

12

Corvette Model Specifications

Model Year/ Generation	Engine	HP/RPM	Fuel Injection Control[a]	Production Quantity[b]	Production of Manual Transmission (MT) vs. Automatic Transmission (AT)
2001 C5	LS1 LS6[c]	345@5600 385 hp@6000	SFI-MAF SFI-MAF	NA NA	NA 100% MT 6-speed
2000 C5	LS1	345@5600	SFI-MAF	NA	NA
1999 C5	LS1	345@5600	SFI-MAF	33,372	NA
1998 C5	LS1	345@5600	SFI-MAF	19,236	78% AT, 22% MT ($815 option)
1997 C5	LS1	345@5600	SFI-MAF	9,752	71% AT, 22% MT ($815 option)
1996 C4	LT1 LT4	300@5000 330@5800	SFI-MAF SFI-MAF	15,536—70% 6,359—30%	100% AT 100% MT (6-speed)
1995 C4 **ZR-1 C4**	LT1 LT5	300@5000 405@5800	SFI-MAF SFI-MAP	20,294—98% 448—2%	79% AT, 21% MT (6-speed) 100% 6-speed
1994 C4 **ZR-1 C4**	LT1 LT5	300@5000 405@5800	SFI-MAF SFI-MAP	22,882—98% 448—2%	76% AT, 24% MT (6-speed) 100% 6-speed
1993 C4 **ZR-1 C4**	LT1 LT5	300@5000 405@5800	MFI-MAP SFI-MAP	21,135—98% 455—2%	77% AT, 24% MT (6-speed) 100% 6-speed
1992 C4 **ZR-1 C4**	LT1 LT5	300@5000 375@5800	MFI-MAP SFI-MAP	19,977—98% 502—2%	78% AT, 23% MT (6-speed) 100% 6-speed
1991 C4 **ZR-1 C4**	L98 LT5	245@4000 375@5800	MFI-MAP SFI-MAP	18.595—90% 2,044—10%	81% AT, 19% MT (6-speed) 100% 6-speed
1990 C4 **ZR-1 C4**	L98 LT5	245@4000 375@5800	MFI-MAP SFI-MAP	20,597—85% 3,049—15%	79% AT, 19% MT (6-speed) 100% 6-speed
1989 C4	L89	245@4000	MFI-MAF	26,412	84% AT, 16% MT (6-speed)
1988 C4	L98	240@4000	MFI-MAF	22,798	84% AT, 16% MT (4-speed+3)
1987 C4	L98	240@4000	MFI-MAF	30,632	86% AT, 14% MT (4-speed+3)
1986 C4	L98	230@4000[d]	MFI-MAF	35,109	81% AT, 16% MT (4-speed+3)
1985 C4	L98	230@4000	MFI-MAF	39,729	76% AT, 24% MT (4-speed+3)
1984 C4[e]	L83	205@4000	TBI-MAP	51,547	87% AT, 13% MT (4-speed+3)
1982 C3	L83	200@4000	TBI-MAP	25,407	100% AT

a SFI–Sequential Fuel Injection
MAF-Mass Air Flow
MAP-Manifold Absolute Pressure
MFI–Multiport Fuel Injection
TBI-Throttle Body Injection

b Figures derived from automotive registrations in the US

c With Z06 option (hardtop)

d 235 HP with aluminum heads

e There were no Corvettes designated as Model Year 1983. Model Year 1984 sold beginning March 1983

Index

A

B

C

WARNING—

- Automotive service and repair is serious business. You must be alert, use common sense, and exercise good judgement to prevent personal injury and complete the work safely.
- Before beginning any work on your vehicle, thoroughly read the safety notice, disclaimer, and all the Cautions and Warnings listed near the front of this manual.
- Always read the complete procedure before you begin the work. Pay special attention to any Cautions and Warnings that accompany that procedure, or other information on a specific topic.

13

D

E

F

WARNING—

- Automotive service and repair is serious business. You must be alert, use common sense, and exercise good judgement to prevent personal injury and complete the work safely.
- Before beginning any work on your vehicle, thoroughly read the safety notice, disclaimer, and all the Cautions and Warnings listed near the front of this manual.
- Always read the complete procedure before you begin the work. Pay special attention to any Cautions and Warnings that accompany that procedure, or other information on a specific topic.

WARNING—

- Automotive service and repair is serious business. You must be alert, use common sense, and exercise good judgement to prevent personal injury and complete the work safely.
- Before beginning any work on your vehicle, thoroughly read the safety notice, disclaimer, and all the Cautions and Warnings listed near the front of this manual.
- Always read the complete procedure before you begin the work. Pay special attention to any Cautions and Warnings that accompany that procedure, or other information on a specific topic.

M

N

O

P

Q

R

S

WARNING—

- Automotive service and repair is serious business. You must be alert, use common sense, and exercise good judgement to prevent personal injury and complete the work safely.
- Before beginning any work on your vehicle, thoroughly read the safety notice, disclaimer, and all the Cautions and Warnings listed near the front of this manual.
- Always read the complete procedure before you begin the work. Pay special attention to any Cautions and Warnings that accompany that procedure, or other information on a specific topic.

Art Credits

The following individuals and companies have graciously provided the art indicated:

Chapter 1

General Motors: Figures 11-14
Charles Probst: Figures 4-10, 15-18
General Motors (photo provided by Dr. David E. Cole): Figure 1
Bentley: Figure 3

Chapter 2

General Motors: Figures 9, 16, 18-21, 28, 29, 38
Charles Probst: Figures 2-8, 10, 15, 17, 22, 27, 30-4, 35
Bentley: Figure s 11, 12, 26
Ford Motor Company: Figures 1, 25
Robert Bosch Corporation: Figures 13, 14, 24
Lingenfelter: Figure 37

Chapter 3

General Motors: Figures 4, 10, 20, 22-27, 31
Charles Probst: Figures 2, 3, 5, 9, 11, 12, 14, 19, 21, 28-30, 33
Bentley: Figure s 8, 13, 15, 16, 32
Ford Motor Company: Figures 1, 17
Robert Bosch Corporation: Figures 7, 18

Chapter 4

General Motors: Figures 1-4, 6-10, 12-15, 17, 18, 20, 21, 23, 24, 27, 29, 30, 33, 35, 40, 42, 44, 46-54, 56-62, 64, 65, 67, 68, 70-73, 75-81
Charles Probst: Figures 5, 17, 19, 22, 26, 28, 31, 32, 41, 45, 55, 63, 74
Ford Motor Company: Figure 69
Robert Bosch Corporation: Figure 43

Chapter 5

General Motors: Figures 2-7, 9-12, 15, 18-21, 24, 25
Charles Probst:: Figures 1, 8, 14, 16, 17 22, 23

Chapter 6

General Motors: Figures 1-4, 6-10, 13-27, 29-33, 35-38, 40-51, 54, 55, 57-65, 66 bottom, 67-69
Charles Probst: Figures 5, 11, 12, 28, 39, 52, 53, 56, 66 top

Chapter 7

General Motors: Figures 1-4, 5 top, 6-19, 22-28, 29 bottom, 30, 32-36, 37 bottom, 38, 39
Charles Probst: Figures 5 bottom, 20, 21, 29 top, 31, 37 top and middle

Chapter 8

General Motors: Figures 1, 2, 5-13
Charles Probst: Figures 3, 4

Chapter 9

General Motors: Figures 4-6, 12, 15, 17, 21, 22, 28, 36, 37
Charles Probst: Figures 2, 9, 13, 16, 20, 38
Lingenfelter Performance Engineering: Figures 1, 3, 8, 10, 18, 24, 27, 31, 35, 39
Robert Bosch Corporation: Figure 14
Tuned Port Induction Specialties: Figures 19, 23
Hypertech, Inc: Figures 29, 30
MSD: Figures 32, 33
Vortech Engineering: Figures 25, 26
Ford Motor Company: Figure 11
Specialty Equipment Manufacturers Association: Figure 7

Chapter 10

Bentley: Figure 6
General Motors: Figures 1-5, 7-14

Chapter 11

General Motors: Figures 1, 3-8, 10, 11, 13-33
Charles Probst: Figures 2, 9, 12

Chapter 12

General Motors: All figures

Acknowledgements

It is an honor, albeit a sad one, for me to include a few acknowledgements on behalf of Chuck Probst. Although the writing and editorial work for this book had been completed before Chuck's sudden death in September, he had not yet compiled the list of people he wanted to thank for their help with this project. I have a few notes from Chuck on this subject, as well as the memory of my conversations with him as we pulled this book together. However, since Chuck's friendships in the automotive world were numerous, I have inevitably left out many important people who should have been thanked. If we owe you an acknowledgement, please accept my apologies and contact me so that you can be included in future printings of this book.

I know Chuck wanted to thank Karl Ludvigsen for providing details on the development of Rochester fuel injection in his book *Corvette: America's Star-Spangled Sports Car – The Complete History.* Chuck was also grateful to Max Krewson for allowing him to take many of the photographs for this book at the Corvette Center in Santa Clara, California. John Lingenfelter and his staff were extremely helpful by providing many photographs that were indispensable in Chuck's research. Chuck did include in his notes a special thanks to Buzz Marston and the Bay Area "Team ZR-1" car club. Buzz is the proud owner of a '95 ZR-1 which is shown in this book. Additionally, thanks go to the many people at GM who made information available to Chuck throughout his meticulous research on the topic of Corvette fuel injection. Chuck was quite grateful to Dave McLellan for taking the time to write the foreword to this book.

I'd like to include my thanks here to Chuck's wife Sally and his son Richard for helping with the final stages of getting this book published. I'm certain Chuck would have wanted to thank them too - their graceful stepping in to the publication process and generosity with spirit and time kept Chuck's book on schedule despite our sudden loss.

Janet E. Barnes
Director of Publishing
February, 2001